AAH 4141

Water Treatment
Principles and Design

Water Treatment Principles and Design

JAMES M. MONTGOMERY, CONSULTING ENGINEERS, INC.

A Wiley-Interscience Publication

JOHN WILEY & SONS

New York · Chichester · Brisbane · Toronto · Singapore

Library of Congress Cataloging in Publication Data:

Main entry under title:
Water treatment principles and design.
    "A Wiley-Interscience publication."
    Includes bibliographies and index.
    1. Water—Purification.   2. Water treatment plants—
Design and construction.   I. James M. Montgomery,
Consulting Engineers.
TD430.W375   1985   628.1′62   85-5344
ISBN 0-471-04384-2

Printed in the United States of America

10 9 8

Foreword

This book is dedicated to the memory of James McKee Montgomery (1896–1969), a man who devoted his engineering career to mainly one area of engineering—water treatment. He founded James M. Montgomery, Consulting Engineers, Inc. in 1945, and his influence on the philosophy of our company has continued over the years of its growth. Mr. Montgomery was a gentleman, admired and respected by all who worked closely with him. He was our Socrates, a man to whom we could go with our engineering problems. He would lead us to solutions through discussions on water chemistry and all the associated aspects of electrical, mechanical, and civil engineering.

Mr. Montgomery was a practitioner of the art, science, and engineering of the treatment of water. He was a chemical engineer but was also registered as a professional engineer in the fields of electrical, mechanical, and civil engineering. However, first and foremost, he was a designer and innovator of design concepts.

One of the major reasons for writing this book and dedicating it to his memory on the 40th Anniversary of his founding of our company is that he published very little in his own name, gave few speeches, and did not like to attend conventions and seminars, yet he influenced and encouraged so many engineers to write and present papers and to attend professional meetings. He shared his knowledge in a different way and it was limited mainly to personal contact. Because of this, many of us who were his students and others in our company who are involved with the "Montgomery" spirit wish to share with our colleagues in a broader sense the knowledge we have gained through actual design experience.

This book is devoted to the principles of water treatment and the design of water treatment facilities. It is intended for the practitioner, the on-the-line person who actually is responsible for putting something on paper so that it can be constructed and operated. The end product has to be a potable water that people can drink and utilize in a safe manner. The book also should be of value to the student as a reference on some process or design problem in which he or she may be interested. It covers both theory and practice to the best of our knowledge, but with the constraint of wanting to put the information in one reasonable-size book.

Relative to this last point, it took a large amount of restraint and discipline not to write a book on each of the subjects covered. We hope we have succeeded in being concise, yet covering the entire subject in an understandable and adequate manner. Since the book has been prepared over a period of several years, another difficulty has been to include the very latest developments in this multidisciplinary field. We have done our best not to inadvertently omit any critical advancement.

The entire field of water chemistry and water treatment processes is dynamic. Concepts, knowledge, and solutions change with time. The true student adapts to these changes and continually updates his or her knowledge. It is hoped that this book does its part in that continual updating process.

WILLIAM J. CARROLL,
PRESIDENT AND CHIEF EXECUTIVE OFFICER
JAMES M. MONTGOMERY,
CONSULTING ENGINEERS, INC.
Pasadena, California
June 1985

Acknowledgments

The preparation of this book has involved the services of many of the staff members in the various offices of the company who participated directly or indirectly, but all meaningfully. The project was supervised by an advisory committee, which included William W. Aultman, Dr. Michael C. Kavanaugh, Dr. Susumu Kawamura, Dr. Carol Tate, and Dr. R. Rhodes Trussell. Day-to-day direction was handled initially by Dr. Kavanaugh and later by Mr. Aultman. Coordination of the work and final in-house editing was performed by Jeanne-Marie Hand.

Those members of the firm who contributed either as authors or reviewers of the various chapters include:

Madan L. Arora
William W. Aultman
William P. Ball
Mark A. Biggers
A. Eugene Bowers
Scott D. Boyce
William J. Carroll
Junn-Ling Chao
Ben H. Chen
Bruce M. Chow

James R. Corning
John DeBoice
Ann E. Farrell
Kenneth G. Ferguson
L. Gerald Firth
Katherine P. Fox
Harold T. Glaser
Frank A. Grant
Jeanne-Marie Hand
David S. Harrison

Lewis K. Hashimoto
Richard S. Holmgren, Jr.
E. Barcus Jernigan
Michael C. Kavanaugh
Susumu Kawamura
Peter H. Kreft
John S. Lang
Lawrence Y. C. Leong
William T. McGivney
Paul F. Meyerhofer
John G. Moutes
Philip A. Naecker
Richard D. Pomeroy
Brian L. Ramaley
Kenneth D. Reich
Lawrence L. Russell

Norbert Schneider
Robert C. A. Siemak
J. Edward Singley
Donald J. Spiegel
Brian G. Stone
Carol H. Tate
Rudy J. Tekippe
Murray A. Todd
Murli Tolaney
Gordon P. Treweek
Albert R. Trussell
R. Rhodes Trussell
Robert B. Uhler
Mark D. Umphres
John J. Vasconcelos
Joseph A. Wojslaw

In addition, valuable suggestions regarding portions of the text were made by Dr. Betty H. Olson of the University of California, Irvine, and Dr. John C. Crittenden of the Michigan Technological University. Dr. Philip C. Singer of the University of North Carolina reviewed and commented on the book for the publisher.

vii

Contents

Water Treatment
Principles and Design

—1—

Introduction

HISTORICAL BACKGROUND

The future of water treatment was viewed as a relatively static undertaking in a well-known book on water treatment plant design prepared by several professional engineering societies in 1969. "In the immediate future, drastic changes in the design of water treatment plants are unlikely. New plants are expected to resemble those in present use (ASCE, 1969)." A few years later in 1974, organic chemicals suspected of being carcinogens were found in drinking water, and the Safe Drinking Water Act was passed by the U.S. Congress. Events such as these have caused water treatment professionals to undertake a thorough reevaluation of water treatment practices. Although the basic facilities of water treatment plants remain essentially the same, the processes required to provide the quality of water desired are significantly more sophisticated.

Applied and basic research conducted over the past 15 years is leading to substantial if not dramatic changes in water treatment practices. Treatment plant designers now must combine their concerns for the elimination of contaminants causing acute health affects (primarily human pathogens) and the production of an aesthetically pleasing water with a new objective of producing a water that minimizes long-term or chronic health effects.

One of the major factors governing the development of human settlements has been the preoccupation with securing and maintaining an adequate supply of water. Water quantity concerns dominated the earliest developmental phases. Population increases, however, exerted more pressure on limited high-quality surface sources and contaminated water sources with human wastes, which led to deteriorating water quality. Thus the water quality of a source could no longer be overlooked in water supply development.

A classic example of this progression is the history of water supply development for the City of New York (Wiedner, 1974). In the early eighteenth century, the only source of acceptable water on Manhattan Island was shallow groundwater wells. As those wells became contaminated due to waste disposal practices, alternative sources were evaluated, leading to the development of surface sources such as the Croton supply which were used without treatment. Subsequently longer water transport systems from the Catskills were developed as water supply demands increased. More recently, however, the discovery that even protected watersheds are subject to contamination has forced the city to investigate treatment for its supplies. Other major U.S. cities such as Los Angeles, San Francisco, Boston, and Seattle have passed through similar

1

stages of water supply development. Thus, the inevitable consequence of population growth and economic development is the need to design water treatment facilities to provide a water of acceptable quality from contaminated surface sources.

Water treatment can be defined as the manipulation of a water source to achieve a water quality that meets specified goals or standards set by the community through its regulatory agencies. The evolution of water treatment practices is a rich history of empirical developments. As outlined by M. N. Baker in *The Quest for Pure Water* (Baker, 1949), the earliest water treatment techniques were primarily conducted in batch operations in the households utilizing the water. From the sixteenth century onward, however, it became increasingly clear that some form of treatment of large quantities of water was essential to maintaining water supply in large human settlements.

The challenge to water treatment professionals in the nineteenth century was the elimination of waterborne diseases. In developed countries during this period great successes in water treatment were achieved as shown by the virtual elimination of the most deadly waterborne diseases including typhoid, cholera, and amebiasis. Since the early twentieth century, however, public health concerns have shifted from acute illnesses to the chronic health effects of trace quantities of organic, inorganic and microbiological contaminants.

This shift is reflected in the increasing number of regulated contaminants. Approximately 35 water quality parameters are included in the current U.S. Environmental Protection Agency's (U.S. EPA) primary and secondary drinking water standards. By the 1990s, it is anticipated that up to 50 additional parameters, principally organic chemicals of synthetic origin, may be incorporated into the U.S. EPA's drinking water standards.

In the United States, and other developed nations throughout the world, future demands on water treatment professionals will be extensive. The global water supply is constant while demand continues to increase. In addition use of surface supplies for dilution of wastewaters continues. In the United States, the fraction of treated wastewaters in surface supplies is relatively low. However, in some areas, especially under drought conditions, this fraction can increase significantly. In more arid nations, the proportion of wastewater in surface sources can be as high as 100%. Despite the dramatic improvements in wastewater treatment since

the 1960s, further degradation of surface water quality from wastewaters can be anticipated although at a much slower rate than in the past.

An additional challenge is the growing awareness of increased contamination of groundwater resources. Previously, major groundwater problems resulted from agricultural activities; contaminants were predominantly nitrate and other inorganic contaminants. More recently, however, widespread contamination of usable groundwater aquifers by synthetic organic chemicals has been observed, raising concerns that a major water resource may be irreparably degraded.

Because of these constraints on available water supplies, the need for health and water treatment professionals is as great now or greater than it was in the early part of this century. Continued economic development and population growth is closely tied to long-range water transportation schemes and the development of improved technologies to provide water of acceptable quality from contaminated sources.

Yet, in many areas of the world, waterborne diseases are still the principal water quality concerns. Massive investments of resources to provide the world's population with a satisfactory water supply are needed. The governments of the world have recognized this major challenge in developing countries through the United Nations' declaration of the 1980s as the Decade of Water and Sanitation. As we approach the mid 1980s, it appears unlikely that the laudatory and ambitious goal of providing the world population with potable water within this decade will be achieved. Nonetheless this will remain a key objective for water treatment professionals in the coming decades.

PRINCIPLES AND DESIGN OF WATER TREATMENT FACILITIES

Although many would relate water treatment to the production of potable or drinking water, treatment encompasses a much wider range of problems and ultimate uses. Water treatment includes home treatment units, community treatment plants, and facilities for industrial water treatment with highly variable water quality requirements dependent on the specific industrial type. Water treatment may also be an important component in agricultural use when crops have specific water quality requirements and when wastewater is utilized for irrigation.

Perhaps the most critical determinants in the selection of water treatment processes are the nature of the water source and the intended use of the treated water. The two principal sources are groundwater or surface water. Depending on the hydrogeology of a basin, the levels of human activity in the vicinity of the source, and other factors, a wide range of water qualities can be encountered. One major distinction is based on the level of dissolved salts or total dissolved solids (TDS) present in the water source. Fresh waters are considered to be those sources with TDS less than 1000 mg/L. Brackish sources, between 1000 and 10,000 mg/L, can be used under special circumstances for specific uses with adequate treatment. Finally, the most abundant source, the ocean, contains approximately 35,000 mg/L dissolved salts and consequently requires demineralization prior to use. Each of the predominant types of surface sources, such as natural or man-made lakes and rivers, require an alternative water treatment strategy.

Solution of a water treatment problem generally depends on five major steps:

1. Characterization of the source and definition of the treated water quality goals or standards.
2. Predesign including process selection.
3. Detailed design of the selected alternative.
4. Construction.
5. Operation and maintenance of the completed facility.

These five steps require input from a wide range of disciplines including engineering, chemistry, microbiology, geology, architecture, and financial analysis. Each plays an important role at various stages in the process. The predominant role, however, rests with professional engineers who carry the responsibility for the success of the water treatment undertaking.

SCOPE OF BOOK

This book is a compilation of over 40 years of experience by James M. Montgomery, Consulting Engineers, Inc., in all phases of water treatment. Each of the key phases in the successful completion of a water treatment project has been addressed in detail to provide the reader with the analytical and experiential tools needed to determine the most cost-effective solution to a water treatment problem regardless of source quality and finished water criteria or standards.

The book is intended for both students and practicing engineers. Source characterization and process analysis is the predominant theme of the first part of the book. These chapters present the fundamental physical, chemical, and microbiological principles that are the basis of the treatment techniques utilized for removal of contaminants. Where possible, analytical models of treatment processes have been discussed. Often, these models are useful in process selection and definition of design criteria. In the selection of optimum design criteria this is particularly important when previous experience is lacking. The second part of the book addresses detailed design and operational issues. Design options and alternative design approaches are presented for each of the major treatment processes used in water treatment. The approach is sufficiently general to address water treatment for potable supplies, water reuse applications, and industrial water treatment problems.

REFERENCES

ASCE, AWWA, CSSE *Water Treatment Plant Design,* American Water Works Association, Inc., New York (1969).

Baker, M. N., *The Quest for Pure Water,* 2nd ed., American Water Works Association, Inc., New York (1981).

Wiedner, C. H., *Water For a City,* Rutgers University Press, New Brunswick, N.J. (1974).

—2—

Physical and Chemical Quality

Water is the most abundant compound on the surface of the earth (Eisenberg and Kauzman, 1969). Without it, life as we know it would cease to exist. The structure of water, while inherently simple, leads to unique physical–chemical properties. These properties have practical significance for water supply, water quality, and water treatment engineers. Examples include water's capacity to dissolve numerous materials, its effectiveness as a heat exchange fluid, its high density and pumping energy requirements, and the inverse effect of temperature on its viscosity. In dissolving or suspending materials, water gains chemical characteristics of biological, health, and aesthetic importance. The type, magnitude, and interactions of these materials determine properties of water such as its potability, corrosivity, and taste and odor. Treatment affects these properties by forcing limits on levels of specific constituents. As subsequent chapters will show, the technology exists to remove essentially all of the dissolved and suspended components of water. Normally, however, the extent of treatment applied is determined by the existing characteristics of a potential water supply and limits specified by its designated end use.

The purpose of this chapter is to develop the background information behind the limits that typically apply to drinking water. Initial sections briefly review the physical structure and properties of water. Next, chemical characteristics are discussed within the broad categories of inorganic minerals, particles, organics, and radionuclides.

PHYSICAL STRUCTURE AND PROPERTIES OF WATER

Within the following two sections the structural components of water, its composition, heat of formation, dimensions, polarity, and structure are defined in order to explain properties of water pertinent to water engineering. These properties include water's density, melting point, vaporization, freezing and boiling points, heat capacity, heat of fusion, viscosity, surface tension, dipole moment, dielectric constant, and conductivity.

Structure

Composition. The water molecule is composed of two hydrogen atoms and one oxygen atom. Accord-

4

ingly, its chemical formula is H_2O and its molecular weight is 18 (1 mole, or 6.023×10^{23} molecules, weighs 18 g).

Water is actually a mixture of several species differing in molecular weight, since there are three known isotopes of hydrogen [^1H, ^2H or D (deuterium), and ^3H or T (tritium)] and six of oxygen (^{14}O, ^{15}O, ^{16}O, ^{17}O, ^{18}O, and ^{19}O). The ^{14}O, ^{15}O, and ^{19}O isotopes are radioactive, short-lived, and do not occur significantly in natural waters. Tritium is radioactive, but its half-life of 12.5 yr is sufficient for it to be observed in natural waters. With the three hydrogen and the three stable oxygen isotopes, there are 18 possible isotopic combinations for water. The precise isotopic content of natural water depends on the origin of the sample, but typical abundances of combinations other than $H_2^{16}O$ are: $H_2^{18}O$ (0.20 percent), $H_2^{17}O$ (0.04 percent), and HDO (0.03 percent) (Eisenberg and Kauzman, 1969). In subsequent discussions, the term *water* refers to the natural mixture of its various isotopic forms.

Heat of Formation. The standard heat of formation ($\Delta H_f°$ at 25°C and 1 atm) for water is -68.317 kcal/mole (Barrow, 1966). On forming water from the elements, heat is released; the reaction is exothermic. The product (water) contains less energy than the reactants (hydrogen and oxygen). The O–H bond energy is 109.7 kcal/mole at 0 °K (Eisenberg and Kauzman, 1969).

Dimensions. The dimensions of the water molecule are shown schematically in Figure 2-1. The H_2O molecule forms an isosceles triangle, with oxygen at the vertex, the HOH bond angle of 104.523° and the O–H bond distance of 0.95718×10^{-8} cm, 0.95718 angstrom (Å) (Eisenberg and Kauzman, 1969). Due to the nature of the chemical bonds, which contain $2sp^3$ hybridized orbitals from the oxygen, the structure of water is tetrahedral. The oxygen nucleus is in the center, hydrogens bound to the oxygen are at two vertices, and unshared electron pairs from the oxygen are at the remaining two vertices (Brown, 1963).

Polarity. The asymmetric water molecule contains an unequal distribution of electrons. The oxygen, which is highly electronegative, exerts a stronger pull on the shared electrons than the hydrogen; also, the oxygen contains two unshared electron pairs. The net result is a slight separation of charges or dipole, with the slightly negative charge (δ^-) on

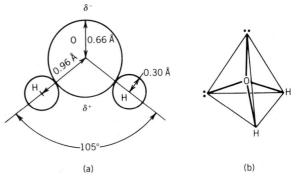

FIGURE 2-1. Shape of water molecule. (a) Dimensions. (b) Tetrahedral arrangement of electronic structure.

the oxygen end and the slightly positive charge (δ^+) on the hydrogen end. Attractive forces exist between one polar molecule and another such that the water molecules tend to orient themselves with the hydrogen end of one directed toward the oxygen end of another.

The attractive interaction between a hydrogen atom of one water molecule and the unshared electrons of the oxygen atom in another water molecule is called a hydrogen bond, represented schematically in Figure 2-2. Estimates of hydrogen bond energy between molecules range from 1.3 to 4.5 kcal/mole (Eisenberg and Kauzman, 1969), approximately 1–4 percent of the O–H bond energy within a single molecule.

Structure. Water molecules attracted to each other form aggregates with distinct structural properties. The structure of ice is an example of the regular array of water molecules determined by hydrogen bonding, as shown in Figure 2-3. The regular three-dimensional arrangement of water molecules in ordinary hexagonal ice (ice I) shows each oxygen surrounded tetrahedrally by four hydrogen atoms, two

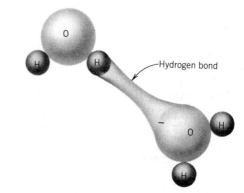

FIGURE 2-2. Hydrogen bonding between water molecules.

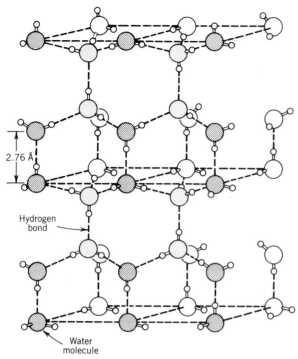

2.76 Å

Hydrogen
bond

Water
molecule

FIGURE 2-3. The structure of ice (Brown, 1963; Pauling, 1960).

of which are bonded to the oxygen by the normal O–H bond and two of which are held by hydrogen bonds. The oxygen-to-oxygen distance is 2.76 Å (Eisenberg and Kauzman, 1969). An important characteristic of this structure is the presence of vacant "shafts" running both parallel and perpendicular to the vertical axis. The open structure produced by the shafts accounts for the fact that ice floats on water.

Ordinary ice I is one of at least nine polymorphic forms of ice. Other forms, found at high pressures or low temperatures, have different crystal forms and densities, as described by Eisenberg and Kauzman (1969). Only ice I is of significance to sanitary engineering processes, which occur at ambient temperatures and pressures.

The structure of liquid water is less well understood than the structure of ice; liquid water models are based on the structure of ice as a starting point with the assumption that the greater energy of the liquid state produces a less regular array than the solid, but a more orderly arrangement than would exist in a liquid in the absence of hydrogen bonding. Classes of models, described by Eisenberg and Kauzman (1969), can be subdivided as follows:

1. SMALL AGGREGATE MODELS. Liquid is depicted as an equilibrium mixture of small aggregates

of water molecules, such as H_2O, $(H_2O)_2$, $(H_2O)_3$, $(H_2O)_4$, and $(H_2O)_8$. Once widely accepted, these models did not fit observed data, such as spectroscopic results, and are now of only historical interest.

2. MIXTURE AND INTERSTITIAL MODELS. Mixture models show liquid water composed of at least two different molecular species, a bulky species representing some type of structured units and a dense species, such as monomeric water molecules. Interstitial models are a class of mixture models wherein one of the species forms a hydrogen-bonded framework and the other species resides in the cavities that exist in this framework. Frank and Wen's "flickering cluster" model and Pauling's "clathrate cage" or "water hydrate" model are notable examples of mixture and interstitial models.

3. DISTORTED HYDROGEN BOND MOLECULES. Hydrogen bonds in liquid water are regraded as distorted to varying degrees, a phenomenon known to occur in ices II, III, V, and VI. Water molecules in the liquid are considered to be four-coordinated, but the networks of linked molecules are depicted as irregular and varied. The random network model proposes rings composed of four to seven (or more) water molecules. The distorted hydrogen bond models seem to be in accord with most of what is known about the structure of water from experiments.

Drost-Hansen (1967) categorizes theoretical models for the structure of liquid water as (1) uniformist, average models; (2) mixture models, broken down ice structures; (3) significant structure theory and polymer models; (4) cluster theories; and (5) clathrate-cage model. He suggests that structural units responsible for observed anomalies probably range from about 20 to 100 molecules at room temperature, and most evidence points to the correctness of mixture models. Furthermore, cage-like models may have an advantage over cluster models.

Properties

Hydrogen bonding accounts for many of the unique properties of water. Compared to other species of similar molecular weight, water has a higher melting and boiling point, making it a liquid rather than a gas at room temperature. Hydrogen bonding also explains its high heat capacity, density, viscosity, and surface tension. These properties and their significance to environmental engineers are summarized in Table 2-1 and described more fully below.

Since water is such a common substance, many units of measurement are based on its properties. For example, 0°C is defined as the freezing point of water and 100°C is its boiling point. A density of 1 g/mL and viscosity of 1 cP (centipoise) are based on water.

Density. The density of water depends on temperature (assuming atmospheric pressure). At 0°C ice has a density (ρ) of 0.91671 g/mL (Eisenberg and Kauzman, 1969), while melted water at 0°C has a density of 0.99867 g/mL. The molar volume of water (18.0 cm³/mole) at 0°C increases by 8.3 percent to 19.5 cm³/mole as liquid turns to ice. This is a result of the open structure of ice, due to the regular hydrogen bonding arrangement shown previously.

TABLE 2-1. Properties of Water and Their Significance to Environmental Engineering

Property	Value	Significance to Environmental Engineering
Molecular weight of H_2O	18	Different isotope ratios can be used to trace source of water
Heat of formation (ΔH_f°) at 25°C, 1 atm	−68,317 cal/mole	Water is stable relative to hydrogen and oxygen
Dipole moment	1.84 debyes	Water is good solvent for ionic species and accelerates weathering of rocks; natural waters contain dissolved ions
Dielectric constant, ε at 25°C	78.5	High constant makes water effective at retaining ions in solution
Conductivity, κ	70 μmho/m	Pure water is not a good conductor of electricity; dissolved ions in environmental samples increase conductivity
Viscosity, η, at 20°C	1.005 cP	Viscosity decrease at higher temperatures improves effectiveness of physicochemical treatment processes; cold-weather designs are more conservative
Surface tension at 20°C	72.75 dynes/cm	High value causes raindrops to assume near spherical shape and water to rise by capillary action in plants
Density Ice at 0°C Water at 0°C Water at 4°C	 0.91671 g/mL 0.99867 g/mL 1.00000 g/mL (62.4 lb/ft³)	Density value needed to calculate pumping energy requirements. Fact that ice expands 8.3% on freezing allows aquatic life to exist below frozen surface of lakes; also requires design to protect water in treatment plants from freezing and bursting pipes
Melting point	0°C (32°F)	Facilities exposed to cold weather should be designed to account for freezing; covered filters and alteration to water velocity are examples
Boiling point	100°C (212°F)	High value keeps water in liquid state at ambient temperatures. Increased pressure increases boiling point; high-temperature steam used in industrial applications for heat transfer
Heat of fusion at 0°C	80 cal/g	Water requires considerable heat to melt from ice to liquid
Heat of vaporization at 100°C	540 cal/g	Steam has high energy content and is effective heat transfer medium
Vapor pressure at 20°C	17.54 mm Hg	Liquid water in equilibrium with vapor, exerts slight vapor pressure under ambient conditions, and leads to evaporative losses from lakes, basins, other open surfaces
Heat Capacity Ice Water Vapor	 0.5 cal/g-°C 1 cal/g-°C (1 Btu/lb-°F) 0.5 cal/g-°C	High heat capacity makes water effective heat transfer medium for industrial processes; also precludes economical heating or cooling of water in treatment plants for domestic uses

The lighter ice floats on water. If water did not expand on freezing, the surface ice would not insulate lower water layers, causing them to freeze solid so aquatic life could not exist.

The densest point for water is at 3.98°C, where its density is 1.00000 g/mL. The density decreases slightly as temperature increases until at 100°C the density is 0.95835 g/mL. These small differences in liquid water's density with temperature do not affect engineering design.

Practical implications of water's density properties follow. By knowing the weight of water, 1.0 kg/L (8.34 lb/gal or 62.4 lb/ft³), engineers can calculate energy requirements for pumping. Knowing that ice expands on freezing is important in design; by properly protecting pipes in cold climates, engineers can prevent water from freezing and bursting the pipes. The fact that ice expands on freezing is also responsible for accelerated weathering of rocks, since water gets into crevices, freezes, cracks rocks, thaws, and carries away dissolved minerals.

Melting and Vaporization. Ice melts at 0°C to form liquid water. In melting ice, the molecules' energy of motion must be increased sufficiently to break a large number of the hydrogen bonds. Thus,

TABLE 2-2. Temperature Dependence of Water's Vapor Pressure

Temperature (°C)	Vapor Pressure (mm Hg)
0	4.58
10	9.21
20	17.54
30	31.82
40	55.32
50	92.51
60	149.38
70	233.7
80	355.1
90	525.8
100	760.0

a higher temperature is needed to melt water than other low-molecular-weight compounds.

Liquid water boils or is converted to gas at 100°C and 1 atmosphere (atm). The boiling point of water is abnormally high compared to the hydrides of other group VI elements (H_2S, H_2Se, H_2Te), as shown in Figure 2-4. This can be explained by the hydrogen bonds between water molecules, which require more applied energy to break and convert the liquid to gas. Similar anomalies are shown in Figure 2-4 for HF and NH_3, which are also hydrogen bonded, but not for CH_4, which is not hydrogen bonded.

At higher pressures, the boiling temperature increases while the opposite occurs at lower pressures. For example, at 220 atm water boils at 374°C. At 1 atm pressure and temperatures below the boiling point, an equilibrium is established between liquid water and water vapor. The gaseous water concentration, expressed in units of pressure, is called the vapor pressure. The vapor pressure of water (in mm Hg) ranges, as shown in Table 2-2 (Barrow, 1966), from 4.58 at 0°C to 760.0 at 100°C, where water boils. Vapor pressure explains why water will evaporate from a surface even when the temperature is below the boiling point. Solutes in water lower the vapor pressure (raise the boiling point), but in concentrations found in environmental samples, this effect is negligible.

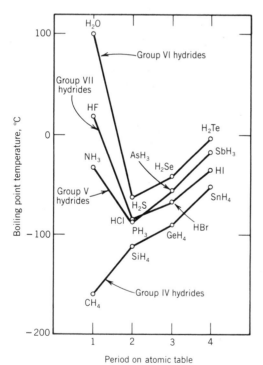

FIGURE 2-4. Anomalous boiling point behavior of water and other hydrogen-bonded hydrides (Brown, 1963).

Freezing Point Depression and Boiling Point Elevation. If water contains a solute, the liquid range is extended; the melting point is depressed and the boiling point is elevated. For 1 mole of nonvolatile solute in 1000 g of water, the freezing point depression (K_f) is 1.86°C and the boiling point elevation (K_b) is 0.512°C. It would take about 29.3 g (0.5 mole of NaCl, but 1.0 mole of ionic species) of NaCl in 1000 g of water to depress the freezing point by 1.86°C. While this principle has practical applications such as street or sidewalk deicing, it is of little significance to environmental engineers since solutes in water are present in the milligram per liter rather than the multigram per liter range.

Heat Capacity. The heat capacity at constant pressure (C_p) is the amount of heat energy required to raise the temperature of a given amount of a substance. The heat capacity of water depends on its state (Eisenberg and Kauzman, 1969), as shown in Figure 2-5. As ice, its heat capacity is about 9 cal/mole-°C (0.5 cal/g-°C) at the melting point. On fusion, it doubles to 18 cal/mole-°C (1 cal/g-°C). As a liquid, the heat capacity remains nearly constant but experiences a slight minimum around 35°C. On vaporization, the heat capacity falls back to about 9 cal/mole-°C.

Water is the standard by which other heat capacitances are calculated. Its heat capacity is commonly expressed as 1 cal/g-°C or 1 Btu/lb-°F. The specific heat per gram of other compounds is calculated by comparing them to water.

Heat of Fusion. The heat of fusion (ΔH_c) is the heat absorbed in converting a solid to a liquid at its melting point. For water, the heat of fusion is 1437

cal/mole (Brown, 1963) or 80 cal/g. Thus, converting ice to water at 0°C requires as much energy as heating liquid water from 0 to 80°C. Converting ice to water requires considerable energy to break some of the hydrogen bonds.

Heat of Vaporization. The heat of vaporization (ΔH_v) is the heat required to convert a substance to the gaseous state; its negative is the heat of condensation. The heat of vaporization depends on temperature. For water at 100°C, the heat of vaporization is 9717 cal/mole (Brown, 1963) or 540 cal/g. Thus, it takes over five times as much energy to convert liquid water at 100°C to a vapor (steam) as it does to heat liquid water from 0 to 100°C. For water to vaporize, all remaining hydrogen bonds must be broken, so the heat of vaporization is relatively high.

Viscosity. Viscosity is a measure of the resistance to flow of a bulk quantity of liquid. The coefficient of viscosity (η) of water is high compared to other substances of similar molecular weight due to intermolecular forces or hydrogen bonds. The coefficient of viscosity for water decreases with increasing temperature as the intermolecular forces decrease, as shown in Table 2-3 (Barrow, 1966).

The fact that viscosity decreases as temperature increases affects physiochemical water treatment processes. Coagulation, flocculation, sedimentation, and filtration are all more effective at higher temperatures. Plants designed for cold-weather conditions are necessarily more conservative in their design criteria. Pumping rates, such as for backwash, should be designed to account for varying viscosity throughout the year for surface water supplies.

TABLE 2-3. Temperature Dependence of Water's Viscosity

Temperature (°C)	Viscosity (cP)[a]
0	1.792
20	1.005
40	0.656
60	0.469
80	0.356
100	0.284

[a] 1 cP = 0.01 g/cm-sec.

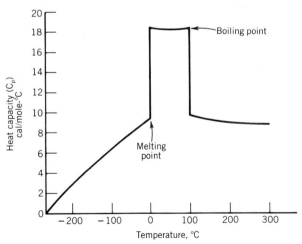

FIGURE 2-5. Heat capacity of water as function of temperature (Eisenberg and Kauzman, 1969).

Surface Tension. Surface tension refers to the forces that act to minimize the surface area of a given volume of liquid. It is responsible for the tendency of water droplets to assume a spherical shape when falling through the air. Intermolecular forces or hydrogen bonds are responsible for the high surface tension of water compared to other liquids of similar molecular weight. The surface tension of water, measured in dynes/cm, is 72.75 at 20°C, 66.18 at 60°C, and 58.85 at 100°C (Barrow, 1966).

Dipole Moment. The dipole moment of water is a consequence of the fact that the center of positive charge does not coincide with the center of negative charge; the molecule is polar. The dipole moment, measured in debyes (10^{-18} esu-cm), is a product of the magnitude of the charges and the distance separating them. For water, the dipole moment is 1.84 debye, which is higher than the majority of other polar compounds.

Although water is a covalent molecule, its high dipole moment makes it a good solvent for ionic species. The polar water molecules are attracted to the surface of an ionic crystal through charge-dipole forces. As ions leave the surface, they are completely surrounded by water molecules, as shown in Figure 2-6. *Note:* the oxygen end of the water molecule is attracted to positive ions and the hydrogen end to negative ions.

The ability of water to dissolve ions accounts for the presence of small quantities of inorganic constituents in natural waters, described in more detail later in this chapter. The behavior of ions in solution is a complex subject that has been described in detail by others (Stumm and Morgan, 1970; Snoeyink and Jenkins, 1980).

Dielectric Constant. The dielectric constant (ε) of water is a measure of its ability to maintain a separation of charges. Water's dielectric constant of 78.5 (at 25°C) is unusually high and ranges from 88.00 at 0°C to 55.44 at 100°C (Weast, 1964). Thus, it is difficult for ions once dissolved to combine again to form a solid, making water an effective solvent.

Conductivity. Pure water is not a good conductor of electricity. Ordinary distilled water in equilibrium with carbon dioxide of the air has a conductivity (κ) of about 70×10^{-6} $\Omega^{-1} \cdot m^{-1}$ (70 μmho/m). After 42 successive distillations in water under vacuum, a conductivity of 4.3×10^{-6} $\Omega^{-1} \cdot m^{-1}$ was obtained (Moore, 1972). Current technology permits production of large quantities of water with specific conductance of 5.5×10^{-6} $\Omega^{-1} \cdot m^{-1}$ for critical uses such as rinse water for the semiconductor industry. Conductivity increases as water dissolves ionic species.

QUALITY CHARACTERIZATION OF WATER

Engineers, health officials, scientists, water purveyors, and legislators function within the constraints of available or attainable water quality and the water's designated end use. Quality characterization and quantification are basic to this functioning. Initial planning of any project generally defines baseline and peak conditions for the constituents in question. Turbidity in a river planned for use as a raw water supply, for example, may typically measure 20–30 turbidity units (TU), while during spring runoff conditions, peaks up to 1000 TU might be observed. If the ultimate purpose of a project is design or modification of an existing design, then bench and/or pilot-scale studies may be utilized to gather data under simulated project conditions. Subsequent to construction of the selected design, further monitoring verifies design parameters. Routine monitoring for selected (indicator) constituents ensures operational control and/or compliance with appropriate regulations.

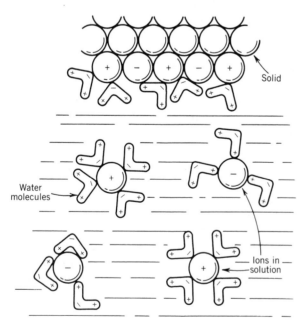

FIGURE 2-6. Hydration of ions in aqueous medium (Brown, 1963).

Water quality characteristics are often classified as physical, chemical (organic or inorganic), or biological and then further classified as health related or aesthetic. While for descriptive and analytical purposes these and further subclassifications are warranted, they do oversimplify natural aquatic systems. Many examples can be given:

- Biological and physical parameters merge as bacteria and/or viruses adsorb onto particulate material; possible treatment then combines chemical oxidation (disinfection) with the physical–chemical processes of flocculation and filtration.

- Geochemistry and biology interact to control nitrogen cycling within natural waters: the bacteria *Nitrosomonas* oxidizes ammonia to nitrate, and *Nitrobacter* oxidizes nitrite to nitrate (Lee and Hoadley, 1967).

- In waters of high organic content, organisms may produce amounts of organic carbon or ammonia sufficient to affect the water's pH and buffer capacity (Beck, Reuter, and Perdue, 1974; Lee and Hoadley, 1967).

- Chemical reactions are frequently biologically catalyzed, as is postulated to occur in the oxidation of pyritic compounds to produce "acid mine" drainage waters (Lee and Hoadley, 1967).

Despite such observed complexity, in the interests of clarity, the remainder of this chapter follows generally accepted water quality parameter classifications. Separate sections address inorganic minerals, particulates, organic chemicals, and radionuclides. Chemically related parameters, commonly utilized as composite water quality indicators, including hardness, total dissolved solids (TDS), turbidity, total organic carbon (TOC), and color are also discussed.

INORGANIC MINERALS

The predominant inorganic minerals in natural waters are calcium, magnesium, sodium, potassium, bicarbonate/carbonate, sulfate, and chloride. Typical freshwater concentrations of individual constituents range from 1.0 to 1000 ppm. Other ions, generally present in lesser amounts (0.01–10 ppm range), are primarily important biologically or because of industrial water quality constraints. These include nitrogen, phosphorus, iron, manganese, silica, and fluoride. Still others, normally designated minor or trace constituents, are found in the parts-per-billion range but may be important as micronutrients, as indicators of local mineral deposits or of agricultural or industrial pollution.

Figure 2-7 shows the range of concentrations found for individual constituents in natural waters. The data indicate the percent of time each constituent was found below a specified concentration. Potassium, for example, occurs over a range of 0.4–15 mg/L; however, 80 percent of the waters sampled would show potassium below 5 mg/L. Thus, if numerous streams and rivers around the United States were randomly sampled, approximately half of them would have TDS levels below 350 mg/L, bicarbonate below 200 mg/L, calcium below 50 mg/L, and so on.

Livingstone in 1963 and later Turekian (1971) utilized a different method to characterize the world's water quality. Figure 2-8 shows data that would theoretically result if the rivers on each continent were sampled and the concentration of each constituent were weighted to reflect the relative volume of discharge of its source. These data are compiled to generate Livingstone's classic "world average" river (Table 2-4).

As these data show, the bulk of water in the world tends to be of the calcium bicarbonate or "hard" type. At the other end of the scale are waters in which sodium and chloride assume greater relative importance. These are generally designated as sodium chloride or "soft" waters. Sodium chloride waters are more typical of coastal regions subject to the influence of marine aerosols, while cal-

TABLE 2-4. Concentration of Major Ions in a "World Average" River

Constituent	Concentration (mg/l)	Cations (meq/l)	Anions (meq/l)
Ca	15	0.750	—
Mg	4.1	0.342	—
Na	6.3	0.274	—
K	2.3	0.059	—
Fe	0.67	—	—
SiO_2	13.1	—	—
HCO_3	58.4	—	0.958
SO_4	11.2	—	0.233
Cl	7.8	—	0.220
NO_3	1	—	0.017
Sum	120	1.428	1.425

Source: Livingstone (1963).

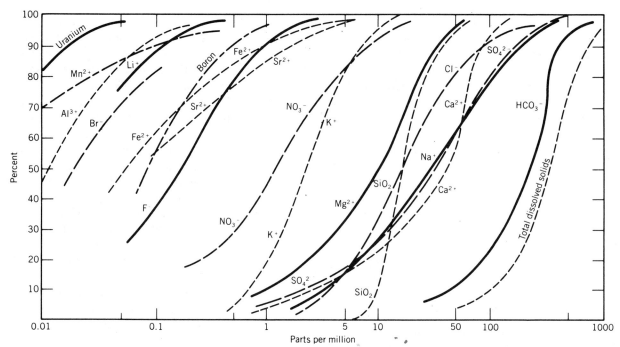

FIGURE 2-7. Cumulative curves showing the frequency distribution of various constituents in terrestrial water. Data are mostly from the United States from various sources (Davies and Wiest, 1966). (Reproduced with permission from John Wiley & Sons.)

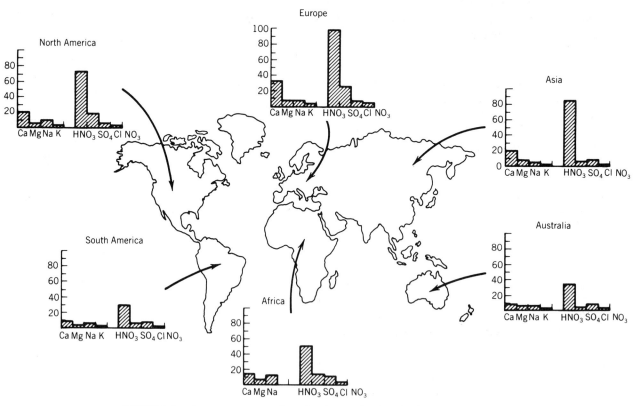

FIGURE 2-8. Average concentration of cations and anions in rivers worldwide.

cium bicarbonate waters are more typical of inland (continental) regions.

Sources of Inorganic Water Quality

The chemical composition of natural waters originates from reactions during rock weathering and soil and sediment leaching, is modified through biological metabolism, and may be intermittently concentrated or diluted via hydrologic transport processes. These major pathways typically are considered to be significant for most engineering applications and are the ones discussed below. A more macroscopic view of solute transport and transformation process is the ''solusphere'' shown in Figure 2-9.

Abiotic Reactions: Weathering and Leaching of Minerals. Weathering is an interaction of water and atmospheric gases with the surfaces of minerals, while soil and sediment leaching occurs as the exchange of cations between the negatively charged clay particles and the surrounding solution. Reactions that occur during weathering or leaching are dissolution, oxidation–reduction, ion exchange, and complexation.

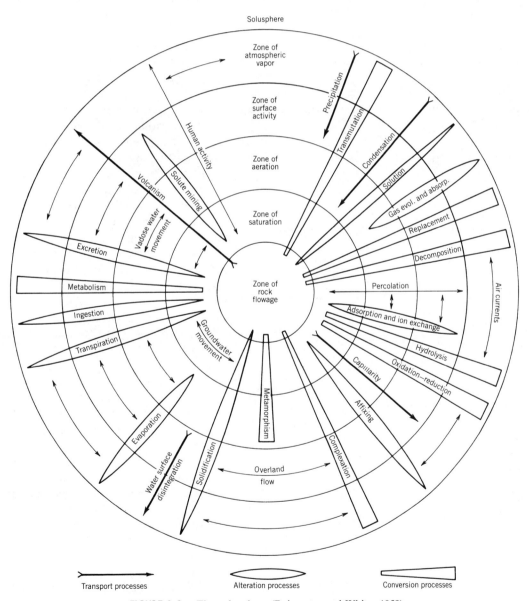

FIGURE 2-9. The solusphere (Rainwater and White, 1958).

In typical dissolution reactions, quartz breaks down to silicic acid [$SiO_2(s) + 2H_2O = H_4SiO_4$] and calcite to calcium and bicarbonate ions.

$$CaCO_3(s) + H_2O = Ca^{2+} + HCO_3^- + OH^-$$

Another type of dissolution reaction occurs as one mineral converts to another through the exchange of a solution ligand such as hydroxide with a mineral ligand like fluoride. In this way fluoroapatite converts to hydroxyapatite.

$$Ca_5(PO_4)_3F(s) + H_2O$$
$$= Ca_5(PO_4)_3(OH)(s) + F^- + H^+$$

Oxidation–reduction reactions involve changes of oxidation state, such as the reduction of ferric iron in low oxygen or pore waters ($Fe^{3+} + e^- = Fe^{2+}$), or the oxidation of pyrite to ferric hydroxide and sulfate as, for example, when anoxic sediments become exposed to air.

$$4FeS_2(s) + 16O_2 + 12H_2O$$
$$= 4Fe(OH)_3 + 12H^+ + 8SO_4^{2-}$$

Note that this latter reaction releases acid, which may then react with minerals in dissolution reactions.

Ion exchange is a reaction of clays or clay minerals with cations from solution. In soils, ion exchange is the primary reaction in soil leaching. It is also important in sediment pore water reactions. Soils and sediments are comprised of varying proportions of clay, silt, sand, and organic materials. The negatively charged clay particles are the principal ion exchange component. With their large surface-area to volume ratios, they are able to adsorb a significantly greater proportion of cations than is either silt or sand. As leachate passes through soil, and pore waters through sediment, they become selectively enriched or depleted of cations, depending on the relative degree of saturation of the particles contacted.

Ion exchange reactions may indirectly act to increase the rate or type of weathering reactions. For example, sodium in solution may exchange with the potassium in orthoclase to produce albite and free potassium ions (Stumm and Morgan, 1970):

$$KAlSi_3O_8(s) + Na^+ = K^+ + NaAlSi_3O_8$$

This new mineral is structurally weaker than its predecessor and is thus more susceptible to dissolution reaction.

In waters of high ionic concentration or high organic content, complexation or ion-pairing reactions may alter the activities of the reactants or products in the above reactions. For example, organic metal complexation can increase the effective solubility of a potential metal precipitate by sequestering the ion from its coprecipitants.

Biological Metabolism. Aquatic organisms can influence the concentration of compounds directly by metabolic uptake, transformation, storage, and release or indirectly by changing the concentration of solutes which are important in abiotic equilibria. Although little quantitative information is available, without the constant turnover of chemical compounds provided by aquatic fauna and flora, water chemistry reactions would proceed largely in one direction. While the activities of higher plants and animals may become significant when their numbers are large, bacteria and microscopic plants and animals ordinarily play dominant roles in biological transformation reactions. Of particular importance is the high surface-area to volume ratio of microorganisms and their biochemical versatility.

In waters where transport and physical–chemical reactions are minimal, that is, where no precipitation and little or no weathering or leaching occur, aquatic organisms may control the concentration of an element in solution. Direct biological control is most obvious for those elements that are major constituents of cellular materials, for example, carbon, nitrogen, and phosphorus. Also directly influenced by the aquatic biota are those elements that form principal components of the hard parts of many organisms, such as calcium in mollusks and silicon in diatoms. Figure 2-10 demonstrates observed fluctuations in silica concentrations in Lake Windemere corresponding to population dynamics of four diatom species.

Indirect influence of aquatic organisms on the chemistry of elements in natural waters is generally the result of assimilation or excretion of compounds that react chemically with the elements in water. For example, where large populations of organisms are present, biological uptake and release of CO_2 can establish the pH of the water. The pH in turn determines the distribution of carbonic species, phosphates, sulfides, iron and aluminum oxides, and organic acids as well as the saturation values of

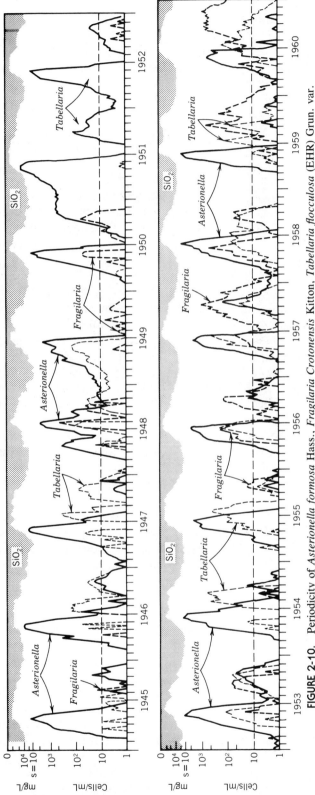

FIGURE 2-10. Periodicity of *Asterionella formosa* Hass., *Fragilaria Crotonensis* Kitton, *Tabellaria flocculosa* (EHR) Grun. var. *Asterionelloides* (Grun.) Knuds. and fluctuations in the concentration of dissolved silica in 0–5-m water column of the northern basin of Windermere from 1945–1960, inclusive (Lee and Hoadley, 1967).

15

precipitating species that enter into acid–base reactions. During photosynthesis, aquatic plants may remove sufficient CO_2 to raise the pH to the value at which the concentrations of carbonate and calcium exceed the solubility product for calcium carbonate. If nucleation occurs, calcium carbonate may precipitate. Figure 2-11 illustrates diurnal variations in pH, dissolved oxygen, and temperature attributed to biological activity within the Trinity River in North Central Texas.

In addition to altering the pH of natural waters, biological organisms may excrete compounds, particularly organic compounds, which can then enter into chemical reactions with other compounds in solution. For example, organic compounds excreted by aquatic organisms may result in the formation of complexes with transition metals in natural waters. Dead organisms and/or their waste products can comprise a significant fraction of detrital particles, which serve as sorption sites or flocculants in natural waters. Aquatic organisms can also significantly affect the oxidation state of some elements in natural waters. An example of this latter effect is the biologically catalyzed reduction of sulfate to sulfide in anoxic waters. The sulfide formed from this reaction may then react with ferric iron and reduce it to ferrous iron. The resulting ferrous iron and other transition metals may then precipitate with additional sulfide formed from additional biochemically reduced sulfate (Lee and Hoadley, 1967).

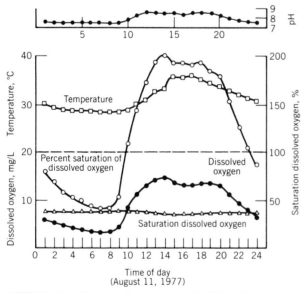

FIGURE 2-11. Water quality in daily cycle, Trinity River (James M. Montgomery, 1978).

Hydrologic Transport Processes. Further alteration of the chemical composition of a water system occurs through hydrologic transport processes. Atmospheric transport of ions within particulate aerosols, evaporation and precipitation, groundwater seepage, and surface runoff collectively comprise the major components of the hydrologic cycle. These processes impact water systems through dilution or concentration of the system's ionic content with attendant shifts in chemical equilibria. The degree of this impact varies geographically and according to the type of water system involved. Atmospheric transport of ions within particulate aerosols, for example, can be a major source of ions in freshwater systems near the ocean. As water evaporates from airborne salts, they are borne inland on wind to precipitate as dry fallout or to form the nuclei for rain and snow. Loose soils can be another source of aerosols, and in industrial or highly populated areas pollutant aerosols are important.

In general, transport processes affect small water systems more than large, and rivers more than lakes. To some extent lakes may become more concentrated through evaporation; however, temporal changes in dissolved substances within lakes are typically restricted to biologically significant constituents and attendant reactions.

Major Cations and Anions

The ability to model a water system and/or predict water quality changes depends on a knowledge of the relative importance of the aforementioned reactions and processes for the specific site and time frame of interest. The accuracy of this assessment in turn depends in part on a knowledge of the constituents comprising the system: their sources, typical concentrations, and reactivities. Some of this background material is presented below. More rigorous treatment can be found in other texts (Hem, 1971; Wetzel, 1975; McKee, 1963; Stumm and Morgan, 1970; Hutchinson, 1957).

Calcium. Calcium is the second major constituent after bicarbonate in most surface waters of the world and is generally among the top three or four ions in groundwaters. In arid regions or under conditions where hydrogen ion activity is increased above that found under atmospheric equilibrium conditions (e.g., a groundwater under reducing conditions), the calcium concentration may reach several hundred milligrams per liter (Hem, 1971). This would be an unusual situation, however, and waters with calcium levels between 40 and 100 mg/L are generally considered hard to very hard.

Weathering and soil ion exchange reactions are the main sources of calcium in natural waters. The predominance of calcium in the relatively easily weathered precipitate rock is responsible for its widespread distribution in surface and groundwaters. Interestingly, Livingstone's world average calcium concentration for river water (15 mg/L) is close to the theoretical solubility of calcium in water which is at equilibrium with the atmosphere (16 mg/L) (Hem, 1971).

Common mineral forms of calcium are calcite and aragonite ($CaCO_3$), gypsum ($CaSO_4 \cdot H_2O$), anhydrite ($CaSO_4$), and fluorite (CaF_2). In natural waters and adsorbed onto soil particles, calcium is generally present as the free ion Ca^{2+}. The ion pair $CaSO_4(aq)$ occurs in waters exposed to significant amounts of gypsum and in highly alkaline waters. The calcium hydroxide [$Ca(OH)_2$] or calcium carbonate ($CaCO_3$) ion pairs may also be present. Industry may contribute other ionic species such as the highly explosive calcium nitrate [$Ca(NO_3)_2$], used in manufacturing matches and explosives, or calcium chloride, derived from oil field brines or salt works (McKee and Wolf, 1963).

Calcium is required as a nutrient for higher plants and as a micronutrient for some, if not all, algae. Although calcium is a required mineral for humans and other animals, the levels found in water are not great enough to contribute significantly. On the other hand, calcium is a primary constituent of water hardness. As a composite indicator of the polyvalent cation content of water, hardness has been inversely correlated with the incidence of cardiovascular disease. Calcium is of importance to industry as a component of scale. The precipitation of $CaCO_3$ scale on cast iron and steel pipes helps inhibit corrosion, but the same precipitate in boilers and heat exchangers adversely affects heat transfer. Waters high in calcium have a nuisance value in washwaters as the alkaline environment causes precipitation of the calcium and soap.

Magnesium. Magnesium salts are more soluble than calcium; however, they are less abundant in rocks than calcium and therefore less available for weathering reactions. Concentrations above 10–20 mg/L in surface waters and above 30–40 mg/L in groundwaters are unusual. Magnesium solubility is difficult to describe. In part, this is because of the variety of forms of magnesium minerals, some of which may not participate in reversible reactions. In addition, magnesium precipitates as an impurity in limestones. The effect of the activity of the resulting impure solids on overall solubility relationships is not known.

In addition to weathering, soil and rock ion exchange reactions may affect local concentrations of magnesium. This is not, however, a very significant source of magnesium in most water supplies. The predominant form of magnesium in natural waters is the free ion Mg^{2+}. Magnesium sulfate and magnesium chloride are found in solution and may derive either from natural deposits or from industry.

Magnesium is an essential mineral for humans with an acceptable intake level listed as 3.6–4.2 mg/kg/day (NAS, 1977). Like calcium, the concentration of magnesium in drinking water is generally insignificant for meeting intake requirements. At high concentrations (400 mg/L for sensitive people, 1000 mg/L for normal population) magnesium salts may have a laxative effect. Calcium and magnesium together comprise most natural water hardness and may or may not contribute to the inverse relationship with cardiovascular disease. The EPA does not limit magnesium in water supplies although the World Health Organization suggests that the highest desirable level is 30 mg/L and the maximum permissible should be 50 mg/L. For irrigation purposes, magnesium is a necessary plant nutrient as well as a necessary soil "conditioner." Magnesium is a problem for industry in only a few instances. As a component of hardness, treatment for the more problematic calcium will generally also reduce magnesium concentrations to acceptable levels. Magnesium may be deleterious in washing, brewery operations, and film development. At high temperatures and pressures found in some boilers, magnesium silicates and phosphates may deposit as sludges; however, these may be controlled with dispersants (Snoeyink and Jenkins, 1980).

Sodium. Sodium compounds comprise almost 3 percent of the earth's crust. Most sodium-containing rock deposits are relatively susceptible to weathering reactions. Another reservoir for sodium is soil, which serves as a medium for sodium participation in ion exchange reactions. Near the coast, marine aerosols can also be large contributors of sodium. Industrial wastes often contain high concentrations of sodium since it is the common cation of industrial salts. Road salting, too, may contribute to local waters' sodium concentrations if runoff and leaching are not controlled. Finally, municipal water treatment chemicals may add sodium to the water supplies; chemicals utilized include sodium hypochlorite, sodium hydroxide, sodium carbonate, and sodium silicate.

Sodium ions and salts are highly soluble and tend to remain in solution, although significant evaporation in arid regions may create "evaporite" deposits of halite (NaCl) or sodium carbonate (Na_2CO_3) (Hem, 1971). Brines associated with such deposits can result in sodium concentrations greater than 100,000 mg/L. More normal sodium concentrations would be less than 100 mg/L. A 1975 survey of 630 U.S. public water supplies showed a range of 0.4–1900 mg/L, with 42 percent containing greater than 20 mg/L and 3 percent having greater than 200 mg/L (NAS, 1977).

In natural waters, sodium is generally present as the free ion. Several complexes and ion pairs may occur, however, including sodium carbonate ($NaCO_3^-$), sodium bicarbonate ($NaHCO_3$), sodium sulfate ($NaSO_4^-$), and sodium chloride (NaCl).

While sodium is required in limited amounts for most plant growth, high concentrations in the soil can be toxic and may, depending on the relative amounts of calcium and magnesium, decrease soil permeability. Sodium salts can be toxic to fish and wildlife at high levels and may affect osmoregulation in aquatic life. In industry, NaCl and Na_2SO_4 contribute to increased rates of corrosion (McKee and Wolf, 1963).

Some evidence links excessive sodium intake to hypertension; however, sodium in drinking water is generally a small dietary component (NAS, 1977). For the general population, the National Academy of Sciences (NAS) suggests a drinking water concentration below 100 mg/L. On the other hand, for that portion of the population on a diet restricted to less than 500 mg of sodium per day, the NAS suggests a drinking water concentration of less than 20 mg/L (NAS, 1977). More recently, the EPA's 1980 Amendments to the Interim Primary Drinking Water Regulations suggested an optimum level for all drinking water supplies as 20 mg/L (EPA, 1980).

Potassium. Potassium is also a common element of the earth's crust; however, it is generally incorporated into deposits that weather less readily than those containing sodium. Because of this and because potassium is preferentially reincorporated into minerals, its concentration in natural waters is generally much lower than sodium. Concentrations of potassium of more than a few tens of milligrams per liter are very unusual. Livingstone reports a world average river potassium concentration of 2.3 mg/L. In highly cultivated areas, runoff may contribute to temporarily high water concentrations since plants take up potassium in fairly significant percentages and their decay releases it.

Potassium occurs in nature only in ionic or molecular forms. It is similar in many properties to sodium and occasionally replaces it in industrial applications. Potassium is more expensive than sodium, however, and consequently is generally utilized only when sodium cannot be. Some common industrial potassium salts are potassium bicarbonate ($KHCO_3$), utilized in baking powders; potassium chlorate ($KClO_3$), utilized in matches and explosives; potassium ferricyanide [$K_3Fe(CN)_6$], utilized in blueprint textile processes and electroplating; and potassium thiocyanate (KSCN), utilized as an intensifier in film development and printing and dying of textiles. In addition, potassium fluoride (KF) is used as an insecticide and potassium permanganate ($KMnO_4$) is used in water treatment for taste and odor control (McKee and Wolfe, 1963).

Potassium is a necessary nutrient to humans, animals, and plants, although excessive quantities can be deleterious to any of them. A dose of 1–2 g is reported to be cathartic to humans. The low levels found in water are not likely to contribute significantly to such a dose, however, and there is no potassium drinking water standard.

Bicarbonate. The carbonate–bicarbonate system in natural water performs important functions in acid–base chemistry, buffer capacity, metal complexation, solids formation, and biological metabolism. Species comprising the carbonate systems include CO_2, H_2CO_3, HCO_3^-, CO_3^{2-}, OH^-, H^+, Ca^{2+}, and $CaCO_3$ (Lee and Hoadley, 1967). Within this framework, pH [$-\log(H^+)$] and bicarbonate (HCO_3^-) typically control reactions among the rest.

The dominant role of the carbonate system in acid–base chemistry of natural waters is well documented (Stumm and Morgan, 1970; Hem, 1971; Wetzel, 1975; Hutchinson, 1957; Snoeyink and Jenkins, 1980). Exceptions occur in waters with very high concentrations of dissolved organics (Beck, Reuter, and Perdue, 1974) or in high-sulfate groundwaters (Hem, 1971).

A series of reactions describes the carbonate system:

1. The protonation of aqueous CO_2:

$$CO_2(aq) + H_2O \rightleftharpoons H_2CO_3$$
$$K_m = 10^{-2.8} \text{ at } 25°C \quad (2\text{-}1)$$

2. The generation of bicarbonate ions:

$$H_2CO_3 \rightleftharpoons H^+ + HCO_3^-$$
$$K_1 = 10^{-3.5} \text{ at } 25°C \quad (2\text{-}2)$$

3. The generation of carbonate ions:

$$HCO_3^- \rightleftharpoons H^+ + CO_3^{2-}$$
$$K_2 = 10^{-10.3} \text{ at } 25°C \quad (2\text{-}3)$$

Because it is difficult to analytically differentiate between $CO_2(aq)$ and H_2CO_3, an arbitrary titration species, $H_2CO_3^*$, which represents the sum of the concentrations of $CO_2(aq)$ and H_2CO_3, is utilized in most calculations. Actually, because H_2CO_3 is much lower in concentration than $CO_2(aq)$, the concentration of $H_2CO_3^* \cong CO_2(aq)$. Thus, the generation of bicarbonate ions in Eq. 2-2 is more typically represented as follows:

$$H_2CO_3^* \rightleftharpoons H^+ + HCO_3^- \quad K = 10^{-6.35} \text{ at } 25°C$$

These equations can be combined to delineate the relative distribution of carbonate species with respect to pH (Figure 2-12). As shown, if pH and one other species are measured, then the concentration of the others are defined.

Carbonate and bicarbonate are the constituents generally responsible for the acid-neutralizing capacity of water. *Alkalinity* is defined as the capacity of a solution to neutralize added acid down to a specified pH value, which for most natural (carbonate) waters is a pH of 4.5–4.8. At pH 4.5, H_2CO_3 comprises the reactive carbonate form, although $CO_2(aq)$ is physically dominant.

Typically observed concentrations of bicarbonate are less than 10 mg/L in rainwater and less than 200 mg/L in surface streams. In some groundwaters where calcium and magnesium are in low concentration, bicarbonate concentrations may go as high or higher than 1000 mg/L (Hem, 1971). Important metal carbonate complexes and solids are formed with calcium, magnesium, and sodium, iron, copper, and zinc.

Iron. Iron is found in rocks, soils, and water in a variety of forms and oxidation states. Common mineral sources (deposits) of iron include ferric oxides and hydroxides such as hematite (Fe_2O_3) and ferric hydroxide [$Fe(OH)_3$]. This latter species gives rocks and soils their red and yellow colors. Sedimentary forms of iron include sulfides, such as pyrite and marcasite; two minerals with identical chemical composition (FeS_2) but different crystalline structures; carbonates such as siderite ($FeCO_3$); and mixed oxides such as magnetite (Fe_3O_4). The ferrous oxides and sulfides are the usual sources of dissolved iron in groundwaters. Weathering of iron silicates can produce dissolved iron in near surface waters; however, this is a relatively slow process.

In a manner analogous to the solubility controls exerted by pH over a number of elements, the oxidation–reduction, or "redox," potential of a water greatly affects the stability of several constituents including iron, carbon, oxygen, nitrogen, sulfur, and manganese. The redox potential of a water reflects its electron activity in the same way that pH reflects hydrogen ion activity. For example, when oxygen is dissolved in water, a potential ($E°$) of $+1.23$ V is generated:

$$O_2(g) + 4H^+ + 4e^- \rightleftharpoons 2H_2O \quad E° = +1.23 \text{ V}$$

The electrons taken up in the above reduction reaction derive from a coupled oxidation reaction such as

$$Fe(s) \rightleftharpoons Fe^{2+} + 2e^- \quad E° = +0.44 \text{ V}$$

Standard half-cell potentials for a number of such reactions may be found in a variety of texts (Stumm and Morgan, 1970; Snoeyink and Jenkins, 1980).

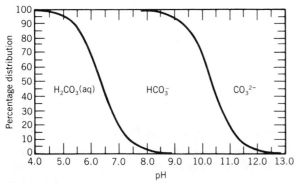

FIGURE 2-12. Percentages of total dissolved carbon dioxide species in solution as a function of pH, 25°C, pressure 1 atm (from Hem, 1971).

The redox potential of a given water is typically defined by one or more dominant redox reactions. This ambient or "cell" potential (denoted as Eh) in turn influences the speciation of other redox reactants. The utility of describing the stability of given reactants in terms of redox equilibria calculations depends on how rapidly the relevant equilibria are approached and to what extent they are affected by other types of reactions such as acid–base, complexation, biological, and so on. When these other types of reactions are important, composite equilibrium relationships must be considered.

Iron in water is appropriately described within the context of an Eh–pH diagram since the relevant redox and acid–base reactions tend to occur quite rapidly. As shown in Figure 2-13 for iron, an Eh–pH diagram presents the system redox potential along the y axis and the system pH along the x axis. The equilibrium form of iron which would be ex-

pected to occur at any specific Eh–pH combination is shown within a boundary defined by the Eh–pH limits of water. The system represented in Figure 2-13 has a total iron activity at 10^{-7} molar or 5.6 $\mu g/L$; other species for this figure are sulfur at 96 mg/L as SO_4^{2-}, CO_2 species at 1000 mg/L as HCO_3, temperature at 25°C, and pressure at 1 atm. The shaded areas in Figures 2-13 indicate solid forms of iron. The diagram shows that under reducing conditions (Eh below 0), over a wide range of pH, iron solubility is low and pyrite tends to precipitate. In addition, under oxidizing conditions (Eh above 0) above pH 5, ferric hydroxide [$Fe(OH)_3$] tends to precipitate. Between these two regions iron is relatively soluble. In particular, for the region corresponding to typical Eh–pH conditions of groundwater (pH 5–9 and Eh 0.20 to −0.10 V) ferrous iron (Fe^{2+}) is quite soluble.

Another characteristic of iron highlighted by the Eh–pH diagram is that metallic iron is unstable within the bounds of water stability. Thus, water tends to corrode iron ($Fe^0 \rightarrow Fe^{2+}$). If, however, the solution is in the region where $Fe(OH)_3$ is stable, the corroding iron may reprecipitate as $Fe(OH)_3$ and form a protective coat against further corrosion.

Significant deviations from Figure 2-13 occur primarily in the region of iron–sulfur reactions where biological activity can be an important factor or where species not shown in the diagram are at significant concentrations. Since iron does form complexes with dissolved organics and colloidal suspensions of ferric hydroxide are not uncommon in surface waters, this latter variable may frequently be significant.

In oxygenated surface waters (pH 5–8) typical concentrations of total iron are around 0.05–0.2 mg/L. In groundwater, the occurrence of iron at concentrations of 1.0–10 mg/L is common. Higher concentrations (up to 50 mg/L) are possible in low-bicarbonate, low-oxygen waters. If water under these latter conditions is pumped out of a well, for example, red-brown ferric hydroxide will begin to precipitate as soon as oxygen begins to mix with the water. This may result in subsequent staining of fixtures.

A flocculant suspension or colloidal particles of ferric hydroxide contribute to the solubility and availability of iron in surface waters. These may join with clay particles, organic colloidals, and other suspended solids material (including metals) and form precipitates. In addition, ferrous and fer-

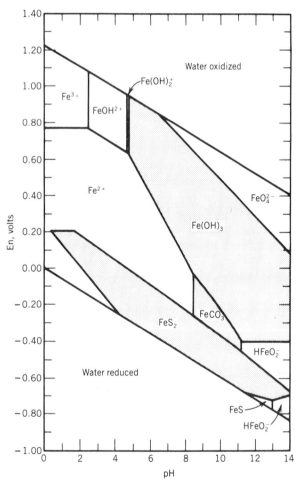

FIGURE 2-13. Forms of iron in water and respect to pH and redox potential (from Hem, 1971).

ric iron complex with organic molecules. Iron enrichment of surface waters in these ways is common when humic and tannic acids are present.

Iron bacteria can speed up thermodynamically slow reactions and/or promote endothermic processes. *Crenothrix* and *Gallionella* utilize ferrous iron as an energy source and precipitate ferric hydroxide.

$$4Fe(HCO_3)_2 + O_2 + 6H_2O \rightarrow$$
$$4Fe(OH)_3 + 4H_2CO_3 + 4CO_2$$

These bacteria are common in wells, treatment plants, and pipelines as long as oxygen is present and local concentrations of soluble iron exceed 0.2 mg/L. Ferrous iron will in fact precipitate spontaneously as ferric hydroxide where oxygen mixes into the water, so these bacteria must exist in a metastable region and both compete with oxygen as well as use it for precipitation of iron. The iron precipitated in these reactions may coat the bacterial colonies and nearby extrusions to a notable degree; ultimately, they may produce pipe clogging.

The EPA Secondary Drinking Water Regulations (EPA, 1979) limit iron to 0.3 mg/L based on aesthetic and taste considerations. Iron is a nutritional requirement, on the order of 1–2 mg/day, and most people ingest around 7–35 mg/day, so water does not comprise a significant source of total daily intake. Iron has a taste threshold reported around 0.1–0.2 mg/L of ferrous sulfate or ferrous chloride. Iron (and manganese) tend to precipitate as ferric and manganous hydroxides at high concentrations and may stain laundry and/or sink fixtures. The secondary limit for drinking water is also a good guide for industries where iron levels are typically recommended below 0.1–0.2 mg/L.

Manganese. Manganese is very abundant in rocks and soils, typically in the form of manganese oxides and hydroxides in association with other metallic cations. At low and neutral pH values, the predominant dissolved form of manganese is as the divalent cation Mn^{2+}. Concentrations on the order of 0.1–1 mg/L are common, although in low pH waters higher concentrations can occur.

Manganese often is present with iron in groundwaters and, like iron, may cause aesthetic problems such as laundry and fixture staining. At concentrations around 0.2–0.4 mg/L manganese may also impart an unpleasant taste to water and can foster the growth of microorganisms in reservoirs and distribution systems. It's staining and deposit formation activities also make it undesirable in many industrial waters. The U.S. EPA specifies a secondary maximum contaminant level (MCL) for manganese at 0.5 mg/L because of these aesthetic problems.

Manganese is essential for nutrition of both plants and animals with a typical daily intake (primarily from food sources) of 10 mg. For plants manganese acts as an enzyme activator and in animals it is important in growth and in nervous system functioning.

Manganese can be removed from water via oxidation. Typically aeration is not as efficient at removing manganese as it is for iron; however, stronger oxidants such as permanganate are very effective (see Chapter 15.)

Chloride. Chlorine is present in water supplies almost exclusively as the chloride ion, although hydrolysis products of chlorine (HOCl and OCl^-) exist temporarily where chlorine has been added as a disinfectant. In typical surface waters, the concentration of chloride ion is less than 10 mg/L; however, in areas subject to seawater intrusion or hot spring inflows or where evaporation greatly exceeds precipitation, concentrations can approach seawater levels.

The high concentration of chloride in seawater is thought to originate from fairly complete leaching of crustal rocks during early geologic time. The implication is that rocks at depth (not subject to such leaching) would show a much higher percentage of chlorine in their makeup. This has not, however, been verified. The concentration of chloride in inland waters is assumed to come from rainfall; again, however, no critical measures of inflow versus outflow have been made to verify this (Hem, 1971).

Chloride is among the most conservative of ions in solution. It participates in very few redox, solubility, complexation, or adsorption reactions and has little biochemical significance. Transport is almost completely via physical processes.

In drinking water, chloride concentrations above 250 mg/L cause an objectionable salty taste. The EPA has established a secondary contaminant level of 250 mg/L for chloride. Consumption of chloride in reasonable concentrations is not harmful to most people; however, if the chloride is present as sodium chloride, the sodium ion may be undesirable to persons requiring salt restriction. Chloride concentrations in drinking water normally contribute relatively little (0.3–2 percent) to total chloride intake (NAS, 1977). Chloride can be significantly corrosive to steel and aluminum at levels of 50 mg/L or even less. Chlorides are toxic to many plants at a variety of levels.

Sulfur. Sulfur occurs in natural waters as various species of sulfates (e.g., $CaSO_4$, $NaSO_4$, $MgSO_4$, etc.) and sulfides (e.g., H_2S, HS^-, $Na_2S_2O_3$). The primary sources of sulfates are evaporite rocks such as gypsum ($CaSO_4 \cdot H_2O$) and anhydrite ($CaSO_4$), sedimentary rock such as pyrite (FeS_2) rainfall, and bacterial metabolism. The concentration of sulfate in precipitation can be quite high (up to 10 mg/L) (Hem, 1971) even in areas where there is little contribution from fossil fuel combustion. The probable sources of such concentrations are sulfur from volcanoes and hot springs as well as dust and terrestrial biotic activity. Concentrations of sulfate in oxidized waters typically range between 5 and 30 mg/L.

The primary sources of sulfide in water are aquatic volcanic activity, organic decomposition, and bacterial activity. Levels of sulfides in natural waters are normally very low. H_2S released from organic decomposition is oxidized very rapidly under toxic conditions and readily reacts to form insoluble metal sulfides under reducing conditions. Polluted bodies of water or swampy areas of naturally high organic decomposition, or some anoxic groundwaters, can sustain levels of H_2S high enough to cause unpleasant odors.

The cycling of sulfur within natural waters relies heavily on bacterial activity. In fact, reduction of sulfate species proceeds extremely slowly without bacterial catalysis. Figure 2-14 shows the forms of sulfur that would be predicted to occur at equilibrium under various Eh–pH conditions and the domains of four major groups of bacteria without whose activity such equilibria would rarely be approached.

The major observed health effect of sulfate is its laxative action. However, such effects are typically not encountered until sulfate levels are well above 100 mg/L. Outside the United States, waters having sulfate levels greater than 1000 mg/L are utilized for drinking supplies without ill effects. Humans are able to acclimate to such high levels relatively rapidly, so the laxative effect of high sulfate is most pronounced on travelers. Reported levels of sulfate or sulfate salts causing laxative effects range from 300 mg/L for sodium sulfate to 390 mg/L for magne-

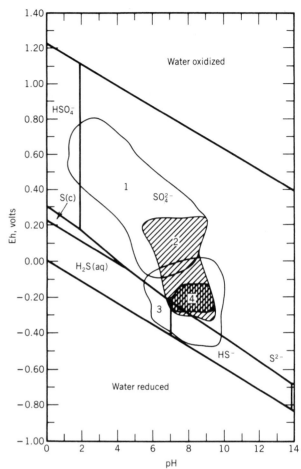

FIGURE 2-14. Stable forms of sulfur under various Eh–pH conditions and domains of bacteria catalyzing their occurrence. Group 1—Chemosynthetic (colorless) sulfur-oxidizing bacteria (aerobic) (some anaerobics use $NO_3 \rightarrow N_2$ instead of O_2): $H_2S + 2O_2 \rightarrow H_2SO_4 \rightarrow SO_4^{2-} + 2H^+$. Group 2—Photosynthetic sulfur-oxidizing purple bacteria (high pH): $3CO_2 + 3H_2S + H_2O \rightarrow 3CH_2O + 2S + H_2SO_4$. Group 3—Anaerobic sulfur-reducing bacteria: $H_2SO_4 + 2(CH_2O) \rightarrow 2CO_2 + 2H_2O + H_2S$. Group 4—Photosynthetic sulfur-oxidizing green sulfur bacteria: (a) $CO_2 + 2H_2S \rightarrow (CH_2O) + H_2O + 2S$; (b) $2CO_2 + 2H_2O + H_2S \rightarrow 2(CH_2O) + H_2SO_4$.

sium sulfate. The taste threshold for sulfate in water has been reported to lie between 300 and 400 mg/L, but some individuals are able to detect as little as 200 mg/L. The EPA secondary drinking water standard is set at 250 mg/L.

Sulfates are corrosive to concrete structures and pipes and asbestos–cement pipes, although at low levels (less than 350 mg/L) the rate of attack is typically very slow. Above 1000 mg/L, attack will be rapid. For irrigation waters with sulfate above 500 mg/L, plant life may be harmed. Generally, sulfate levels below 300 mg/L are preferred for irrigation.

Nitrate. Nitrate represents the most highly oxidized form of nitrogen whose oxidation states range from -3 to $+5$. Other forms of nitrogen in water include organic compounds such as urea (NH_2CONH_2), amino acids and their breakdown products, ammonia (NH_3), ammonium ion (NH_4^+), hydroxylamine (NH_2OH), nitrogen gas (N_2), and nitrite (NO_2^-). Ammonia, ammonium ion, and protein by-products are all reduced species; N_2 gas is in the zero oxidation state; nitrite is at $+3$ and nitrate at $+5$. Transformation from one state to another is closely tied to biological activity, the influx of domestic wastes, and the local use of nitrogen fertilizers.

Nitrate is a major nutrient for vegetation and is an essential nutrient for all living organisms. Certain species of bacteria in soil, the blue-green algae, and other aquatic microbes can fix atmospheric nitrogen and convert it to nitrate. Other organisms oxidize organic forms of nitrogen to nitrate; however, unless there are significant sources of organic nitrogen present, most nitrate formed by either of these two routes is rapidly taken up by plants. Thus, surface water concentrations of nitrate are seldom very high. Livingstone's world average river concentration is 1 mg/L. Groundwater levels can get much higher where a more limited vegetative sink for nitrate exists. In excessive amounts nitrate may produce infant methemoglobinemia, and a limit of 45 mg/L has been set to avert this condition (Environmental Protection Agency, 1980).

High nitrate concentrations are a problem for the brewing industry in that above 15–30 mg/L they may give a bad taste to beer and/or they may be partly reduced to nitrite during fermentation and thence poison the yeast. Excessive nitrates can reduce soil permeability; however, normally nitrates are of benefit as fertilizers (McKee and Wolf, 1963).

The role of nitrate in accelerated eutrophication is generally overshadowed by that of phosphate. The carbon–nitrogen–phosphorus ratio of plants is roughly 40 : 7 : 1 by weight. In most lakes, inorganic carbon exceeds nitrogen by an order of magnitude and exceeds phosphorous by 2–3 orders of magnitude. Similarly the nitrogen of most natural waters exceeds phosphorous by at least an order of magnitude. In general terms, then, normalized natural water supplies of carbon, nitrogen, and phosphorus can be estimated at 100 : 10 : 1, and phosphorus is the limiting factor. When sewage effluents with high proportions of phosphorus (e.g., C–N–P ratio of 6 : 4 : 1) load a natural water, however, the phosphorus limitation is overcome and nitrogen becomes the critical factor. Under these conditions blue green algae, able to fix atmospheric nitrogen, are likely to dominate.

Silica. Silica is present in almost all rocks, soils, and natural waters. The Si^{4+} ion fits closely within a matrix of four oxygen or hydroxide ions. The SiO_4^{4-} tetrahedron is the most fundamental form of silica found, and a variety of minerals are formed by joining these tetrahedra with divalent cations or additional oxygens. In water, silica exists as hydrated H_4SiO_4 or $Si(OH)_4$, although water analyses commonly represent dissolved silica as SiO_2. Actually the precise form of silica dissolved in water is not known: it does not behave as a charged ion such as SiO_3^{2-} or as a colloid. In addition, silica precipitates from solution as amorphous silica but may convert to more crystalized forms of lower solubility (Hem, 1971).

The range of concentrations of silica most commonly found in natural waters is between 1 and 30 mg/L. It has not yet been satisfactorily determined what controls the solubility of silica. Temperature is a critical factor; the solubility of quartz was found to be 6.0 mg/L at 25°C and 26 mg/L at 84°C (as SiO_2). Amorphous silica, on the other hand, has a solubility of 115 mg/L at 25°C and 370 mg/L at 100°C (Hem, 1971).

Silica is a biologically active material as it is assimilated by diatoms and precipitated with their carcasses. Surface waters of (productive) lakes and of seawater are frequently depleted in silica due to this activity. Definite seasonal cycles are observed on silica associated with diatom metabolism (see Figure 2-10).

Silicon is used in metallurgy, and silica is employed in making glass, enamels, abrasives, and so on. Sodium silicates have been used as coagulants in water treatment and as corrosion inhibitors on iron pipes. For most industrial applications the concentrations of silica in natural waters cause no problems. However, in high-pressure boilers silica can be disastrous as it may pass through the boilers in the steam and condense on heater tubes and turbine blades. Such silicon deposits may be hard and glassy and extremely difficult to remove (McKee and Wolf, 1963).

Fluoride. In natural waters fluoride is present primarily as the F^- ion or as a complex with aluminum, beryllium, or ferric iron. Fluoride ions have the same charge and nearly the same radius as hydroxide ions, and thus the ions may replace each other (Hem, 1971).

Although the amount of fluoride in crustal rocks is much greater than another member of the halogen group, chlorine, fluoride remains tied up in minerals to a much greater degree. Fluorite (CaF_2) is a common fluoride mineral. Apatite [$Ca_5F(PO_4)_3$] also commonly contains fluoride.

In waters with TDS < 1000, fluoride is typically <1 mg/L, although groundwaters are found with levels higher than 10 mg/L. Also, waters affected by volcanic activity may have higher levels. River waters are seldom more than a few tenths of a milligram per liter.

For more than 30 years fluoride has been added to public water supplies to reduce dental decay. Typically it is added as sodium fluoride, sodium silicofluoride, hydroflurosilicic acid, or ammonium silicofluoride at doses of 1 mg/L (NAS, 1977). Fluoride at higher levels may cause mottling of teeth (>4.0 mg/L) or crippling skeletal fluorosis (15–20 mg/L). Fluoride is fatal at around 4–5 g and toxic at around 250–450 mg.

Acceptable drinking water levels are described as a function of temperature. The EPA identifies a range of 1.4–2.4 mg/L while the World Health Organization recommends a range of 0.6–1.7 mg/L. The rationale behind the temperature range is that in warmer climates, people tend to drink more (McKee and Wolfe, 1963). For most industrial and agricultural purposes, the levels of fluoride in natural waters cause no deleterious effects (McKee and Wolfe, 1963).

Minor and Trace Inorganic Constituents

Constituents of natural waters found in the parts-per-billion to parts-per-trillion range may still be of significant health or water quality concern. This is particularly true for trace organics in water; and many inorganic constituents are also important. Table 2-5 summarizes information on the chemistry and water quality significance of several inorganic trace constituents.

Inorganic Water Quality Indicators

Several chemically related quality measures are utilized to indicate specific tendencies or properties of a water supply such as its polyvalent cation content (hardness), dissolved solids (TDS), corrosivity (TDS), electrical conductivity, hydrogen ion activity (pH), or value to irrigation (sodium adsorption ratio).

Hardness. Hardness is a composite measure of the polyvalent cation content of water and has been inversely correlated with the incidence of cardiovascular disease. Calcium and magnesium are often the primary components of water hardness (NAS, 1977). Hardness is an important parameter to industry as an indicator of potential (interfering) precipitation such as with carbonates in cooling towers or boilers, with soaps and dyes in cleaning and textile industries, and with emulsifiers in photographic development. Waters of 0–60 mg/L hardness (expressed as $CaCO_3$) are generally designated as "soft," of 61–120 as "moderately hard," of 121–180 as "hard," and of 180 or more as "very hard" (Todd, 1970).

Total Dissolved Solids. Total dissolved solids (TDS) is a measure of the total ions in solution. It is analyzed by filtering out a sample's suspended material, evaporating the filtrate, and weighing the remaining residue. Table 2-6 shows some recent estimates of the dissolved solids loading on streams around the world (Turekian, 1971). Local concentrations in arid regions or in waters subjected to pollution runoff can be much higher. Colorado River water, for example, after reaching southern California has a TDS content of 700–800 mg/L. Although high TDS waters have not been shown to be harmful, the EPA recommends a maximum of 500 mg/L in drinking water supplies. Above this level the water may have an increasingly salty taste.

Conductivity. A parameter related to TDS is electrical conductivity (EC) or specific conductance. EC is actually a measure (in micromhos/cm) of the ionic activity of a solution in terms of its capacity to transmit current. In dilute solutions, the two measures are reasonably comparable; that is, TDS = 0.5 (EC). However, as the solution becomes more concentrated (TDS > 1000 mg/L, EC > 2000 μmho/cm), the proximity of the solution ions to each other depresses their activity and consequently their ability to transmit current. The physical amount of dissolved solids, however, is not affected. Thus the ratio of TDS–EC increases and the relationship tends toward TDS = 0.9 × EC (Figure 2-15). Note,

TABLE 2-5. Minor and Trace Elements[a]

Constituent	Concentration in Natural Waters (μg/L)[b]	Chemistry	Significance in Water Supplies
Alkali metals			
Lithium	0.001–0.3		Potentially toxic to plants, but not at concentration likely to be encountered in irrigation waters
Rubidium	0.0015		
Cesium	0.05–0.02[c]		
Alkaline earths			
Beryllium	0.001–1	Found often as a constituent of beryl, an aluminum silicate. Typically occurs as a particulate, rather than dissolved	Highly toxic, but occurs at very low concentration
Strontium	0.6 Median river water 0.11 Median public waters	Can replace Ca and Mg in igneous rocks. Concentration in natural water is less than solubility	See strontium-90 in radionuclides section of this chapter
Barium	Median river 0.043 Median public water	Concentration controlled by $BaSO_4$ solubility $K_{sr} \sim 10^{-10}$	Ingestion of soluble barium salts can be fatal. Normal water concentrations have no effect
Other metallic elements			
Titanium	8.6 Median river <1.5 Median public waters	One of the 10 most abundant elements in the earth's crust, but quite insoluble. Highly resistant to corrosion	
Vanadium	<70	Occurs as anions and cations of V^{3+}, V^{4+}, and V^{5+}	May concentrate in vegetation
Chromium	5.8 Median river 0.43 Median public water	Cr^{3+} and Cr^{6+} stable in surface waters $Cr_2O_7^{2-}$ and CrO_4^{2-} stable in groundwaters	Industrial pollutant
Molybdenum	0.35 Median river 1.4 Median public water	Rare element. Concentrations in water typically reflect abundance in ambient minerals	Accumulated by vegetation. Forage crops may become toxic
Cobalt	ND–1.0	+2 and +3 valence states. Under normal redox +2 is found. Adsorption and complexation important determination of concentration	Essential in nutrition in small quantities
Nickel	10	Found generally at +2 oxidation state. Adsorbed by manganese + iron oxides	
Copper	10	Solubility limited to 64 μg/L at pH 7.0 and 6.4 at pH 8.0 by cupric oxide or hydroxy-carbonate minerals	Utilized in water treatment and metal fabrication. Used to inhibit algae growth in reservoirs. Essential for nutrition of flora and fauna

(*Table 2-5 continued on next page.*)

TABLE 2-5. (*Continued*)

Constituent	Concentration in Natural Waters (μg/L)[b]	Chemistry	Significance in Water Supplies
Silver	0.1–0.3	Solubility predicts concentration of 0.1–10 μg/L in dilute aerated water. Rare element probability often controlled by availability	Has been used as disinfectant
Gold	ND–trace	Very low solubility within Eh–pH region of natural waters. May form complexes with reduced sulfur spp.	
Zinc	10	Probability controlled by availability since natural water does not seem to contain levels dictated by hydroxide or carbonate solubility	Widely used in industry. Found in wastes, dissolved from galvanized pipes, cooling water treatment, etc.
Cadmium	ND–10	Relatively rare. Used in electroplating	Toxic. Presence may indicate industrial contamination
Mercury	ND–<10	Relatively rare in rocks or water	Highly toxic. Presence indicates pollution from mining, industry, or metallurgical works
Lead	1–10	Probable solubility limited by cerussite, a lead carbonate or $PbSO_4$, in oxidized systems, to ~2 μg/L	Older plumbing systems contain lead, which may dissolve at low pH. Effect of use as gas additive on rain + H_2O concentration not known
Nonmetals (occur as anions in natural waters)			
Arsenic	0–1000	Arsenate spp. $H_2AsO_4^-$ and $HASO_4^2$ equilibrium forms in natural waters. $HASO_2$(aq) may be found in reducing environment. Solution controlled by metal arsenates, e.g., Cu^{2+}. Adsorption also important.	Used in industry in some herbicides and pesticides. Lethal in animals above 20 mg/lb. Long-term ingestion of 0.21 mg/L reported to be poisonous
Selenium	0.2	Selenite form in natural waters SeO_3^{2-}. Elemental form found in mildly reducing waters. Low solubility	Taken up by vegetation
Bromine	20	Present in natural waters as Br^-. Present in rainwater	
Iodine	0.2–2	Cycling controlled through biochemical process. Its volatility allows significant atmospheric circulation. Not very abundant element	Essential nutrient in higher animals. Has been used to seed clouds

[a] Data from NAS (1977), Livingstone (1963), Turekian (1971), and Hem (1971).
[b] Values presented are approximate and represent one or more author's best estimate as to mean values for typical rivers and lakes.
[c] Values observed in six analyses of rivers in Japan.

TABLE 2-6. Estimates of the Total Runoff of the World and the Flux of Dissolved Solids

	Livingstone (1963)			Alekin and Brazhnikova (1961)		
	Water Flux (10^{15} l/yr)	Dissolved Solids		Water Flux (10^{15} l/yr)	Dissolved Solids	
		Concentration (mg/L)	Flux (10^{15} g/yr)		Concentration (mg/L, incl. organics)	Flux (10^{15} g/yr)
North America	4.55	142	0.646	6.43	89	0.572
Europe	2.50	182	0.455	3.00	101	0.303
Asia	11.05	142	1.570	12.25	111	1.360
Africa	5.90	121	0.715	6.05	96	0.581
Australia	0.32	59	0.019	0.61	176	0.107
South America	8.01	69	0.552	8.10	71	0.575
Total	32.33	120	3.957	36.45	88	3.498

SOURCE: Turekian (1971).

however, that the slope of the line in Figure 2-15 cannot be standardized for more than one water. The slope for any one sample will fall between 0.5 and 0.9, but for several samples having the same TDS, the slope will vary.

The importance of these two measures lies in their effect on the corrosivity of a water sample and in their effect on the solubility of slightly soluble compounds such as $CaCO_3$. In general, as TDS and EC increase, the corrosivity of the water increases (all other factors, such as dissolved oxygen content, remaining constant). In addition, with increasing TDS and EC, the activities of individual ions decrease and "ion pairing" occurs. Ion pairs are interactions between ions, weaker than intermolecular forces, but strong enough to depress other reactions such as the formation and precipitation of calcium carbonate. The net result is to increase the effective solubility of potential precipitates. (For example, water from the Colorado River contains sufficiently high sulfate concentrations that $CaSO_4$ ion pairs must be taken into account before accurately estimating the potential of $CaCO_3$ precipitation for scale to form and protect pipes.)

Sodium Adsorption Ratio. The sodium adsorption ratio (SAR) is a criterion for evaluating the sodium hazard in irrigation waters, determined by measuring the ratio of sodium to magnesium and calcium:

$$SAR = \frac{Na^+}{\sqrt{(Ca^{2+} + Mg^{2+})/2}} \text{ (in meq/L)}$$

With the proper amount of calcium and magnesium in the irrigation water, the irrigated soil will be granular in texture, easily worked, and permeable. With increasing proportions of sodium, the soil will tend to become less permeable and water logging may occur. Other more sophisticated SAR calculations are available in which the principle is the same but additional variables, such as alkalinity, are considered (EPA, 1977; Suarez, 1981).

PARTICLES IN WATER

Particles are defined as finely divided solids larger than molecules but generally not distinguishable by the unaided eye. The principal natural sources of particles in water are soil-weathering processes and

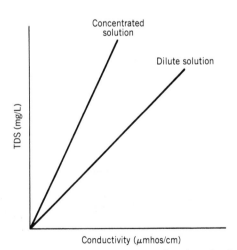

FIGURE 2-15. Relationship between TDS and conductivity varies according to solution concentration.

biological activity. Clays are the principal component of particles produced by weathering. Algae, bacteria, and other higher microorganisms are the predominant types of particles produced biologically. Some particles have both natural and anthropogenic sources, a notable example being asbestos fibers. Industrial and agricultural activities tend to augment these natural sources by increasing areas of runoff, through cultural eutrophication, or by direct pollution with industrial residues.

Particles have a range of impacts on water quality depending on their physical, chemical, and biological properties. Because of their fine size, particles exhibit a large surface area, which serves as a potential adsorption sink for toxic substances, such as heavy metals and chlorinated hydrocarbons. Ingestion of the particles may then cause acute or chronic toxic effects. The large particle surface area also causes strong scattering of incident light and leads to degradation of visibility in the water column. As a sink or source of various adsorbed ionic species, particles play an important role in biogeochemical cycles.

Methods of particulate measurement include turbidity readings and particle counts.

Turbidity

In the 15th edition of *Standard Methods* (1980), *turbidity* is defined as "an optical property of the sample causing light to be scattered and absorbed, not transmitted." Measurement thus requires a light source and a sensor of the scattered light beams.

While the suspended and colloidal material comprising turbidity are important for aesthetic reasons, turbidity itself is also a significant indirect health concern. Specific health-related characteristics of turbidity include the association of microorganisms with particulate material, with resulting interference with disinfection, and a distinct turbidity-related chlorine demand. In addition, chlorinated organic precursor materials have been related to aquatic particulate material. Depending on the source, turbidity can be the most variable of the water quality parameters of concern in drinking water sources. As such, it is often the key determinant in determining necessary elements of a treatment plant.

Figure 2-16 contrasts the relatively constant level of raw water turbidity, which occurred over a month's period in a raw water reservoir in Colorado, with the extreme fluctuations that occurred

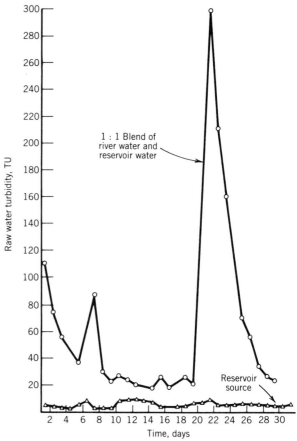

FIGURE 2-16. Levels of raw water turbidity (James M. Montgomery, 1981).

during the same period in a nearby river. The top line in Figure 2-16 represents a 1:1 blend of the reservoir water with the river water; the river water itself had turbidity values approaching 1000 TU.

Particle Counts

Particle concentration measurements provide additional information which supplements turbidity measurements. Particle counters have sensors available in different-size ranges, such as a 1.0–60-μm sensor or a 2.5–150-μm sensor. Particle counts are recorded in within about 12 subranges in the chosen micrometer sensor range. Particle count analyses reveal the distribution of particle sizes in a particular sample.

The design of water and wastewater treatment facilities depends in part on the characteristics of particles as discussed in Chapters 6, 7, and 8. Fine particles (below 10 μm) settle slowly, and efficient removal by sedimentation requires transformation into larger-size classes. Effective disinfection of water or wastewater demands removal of particles that could protect pathogens from attack. Particle removal is a prerequisite for proper operation of separation processes utilizing membranes (such as reverse osmosis, ultrafiltration), which clog rapidly if particle levels are high. Although regulations concerning particle concentrations do not exist in state or federal law, monitoring particle counts throughout a treatment process can aid in understanding the process.

ORGANICS

Organic compounds in water derive from the natural decomposition of plant and animal material; from industrial, urban, or agricultural pollution; and from the reaction of halides (most often chlorine) with natural organics during water treatment. Concentrations range from none in protected groundwaters to 10–30 mg/L in naturally productive or contaminated surface waters.

While the presence of organics as a composite parameter has been known and measured for a number of years, knowledge of the vast array of differing organic species that comprise this composite is relatively new. Only since the middle of this century has development of sophisticated methods and instrumentation allowed the necessary separation and measurement of their often parts-per-billion (μg/L) or parts-per-trillion (ng/L) concentrations.

Of the total aquatic organic material, one fraction is composed primarily of low-molecular-weight synthetic organics. An as yet undelineated fraction (more than 90 percent) comprises relatively refractory naturally occurring organics as well as some high-molecular-weight synthetic organics. Organics are discussed in detail in Chapter 16; some introductory material is given below.

Naturally Occurring Aquatic Organics

Naturally occurring organics represent the greatest proportion of aquatic organic material. The form of the organic material present in any one water source reflects the form of humification products in the soils which drain into the water. Humification of soil organic material originates with plant and animal degradation products. These condense and polymerize through fulvic and humic acids to kerogen and coal (Schnitzer and Khan, 1972; Jackson, 1975). The process is hypothesized to proceed chemically and/or microbially with functional groups and aliphatic side chains being split off, and with condensation polymerization of various reactive groups resulting in larger, more aromatic molecules. The process continues with a concurrent decrease in solubility until kerogen or humin is produced; the end products are nonsoluble in acid or alkali and resistant to degradation and reaction (Schnitzer and Khan, 1972; Jackson, 1975). This process is illustrated in Figure 2-17.

Synthetic Organics

Synthetic organics comprise an extremely diverse group of compounds. While generally found at very low concentrations in water, many are of significant health concern. Thus, a great deal of effort is currently being directed toward isolating, identifying,

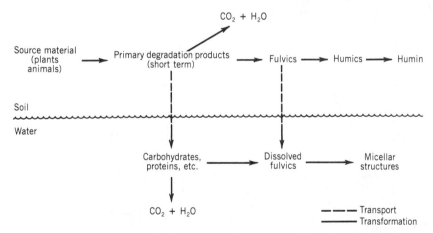

FIGURE 2-17. Hypothetical interactions of humic materials with soil and water.

and evaluating the health effects of synthetic organics. The NAS has compiled and analyzed available data on the sources, concentrations, and health effects of 129 synthetic organics found or likely to be found in drinking water supplies (NAS, 1977). Included in their study are pesticides, industrial contaminants, plasticizers, polychlorinated biphenyls (PCBs), polynuclear aromatics (PNAs), and halogenated methane and ethane derivatives. Enough information is available regarding the health effects of several pesticides and of halogenated methane derivatives (designated "total trihalomethanes") that they are included in EPA drinking water standards. (See Chapter 4.)

Organic Measures

Given the array of synthetic organics and the nonspecific nature of naturally occurring organics, it is not surprising that a variety of measures have been developed or adapted for the quantification of aquatic organic material. Two types of measures are in common use, nonspecific and specific. Nonspecific measures are intended to quantify part or all of the organic content of a water. They include color, ultraviolet (UV) absorbance, fluorescence, TOC, and total organic halogen (TOX). Specific measures, on the other hand, allow delineation and quantification of individual organic compounds from within the total present. These include gas chromatography, mass spectroscopy, and high-pressure liquid chromatography. The applicability of these measures is described in Chapter 16.

Color.   The measurement and control of color in water has previously been of interest to treatment plant operators for purely aesthetic reasons; however, more recently color has been used as a quantitative assessment of the organic materials causing it. The method of color quantification consists of comparing natural waters to a color scale with solutions made from serial dilutions of concentrations of a standard chloroplatinate solution. The chloroplatinate standard bears no relation to the color-producing substance in the water other than by its hue. The assumption behind the measure is that color-producing substances within water behave consistently in nature (i.e., light absorbance and scattering) such that their concentration should follow Beer's law. That the nature of the color molecule is not, in fact, consistent is evidenced by the variation in color value with pH, probably caused by a change in the ionization of the molecule with corresponding effects on bond lengths and configurations, and thus light absorption. Similar effects may be expected through the molecule's associating with dissolved and/or colloidal substances in the water. Iron in particular has been implicated as contributing a variable influence on color values of source waters (Lamar, 1968), and this influence in turn varies with pH (Ghassemi and Christman, 1968) and organic particle size (Lamar, 1968). Despite these deterrents, the color measure remains in use, primarily as an aesthetic check on finishing drinking waters.

Ultraviolet Absorbance.   Specific organic materials may show definitive UV absorbance bands reflecting their particular unsaturation pattern and/or aromatic components. Such configurations are susceptible to the short-wavelength/high-energy excitation of UV radiation, with increasingly shorter wavelengths required to excite more stable molecules. Thus, simple aliphatic molecules will not tend to absorb UV, whereas the complex multiaromatic, multiconjugated humic substances would be expected to absorb UV very strongly.

The UV absorption profile for colored water is smooth and nondescript, showing increasing absorbance with decreasing wavelength (Figure 2-18).

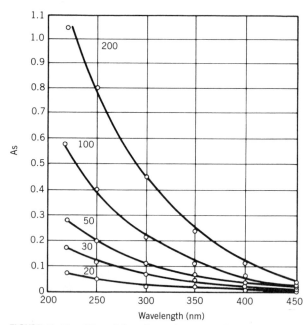

FIGURE 2-18.   Ultraviolet absorption as a function of color value. Each curve represents the results obtained with the sample diluted with pH 8.0 buffer to the color unit value indicated by the curves (from Black and Christman, 1963).

This may be attributed to the heterogeneity of the organic material, with no one configuration showing dominant absorption. Such an absorption curve may indicate scattering of light rather than absorbance (Black and Christman, 1963). While organics in water include colloidal particles (Beck, Reuter, and Perdue, 1974), the total organic matter consists of a range of particle sizes and true dissolved species (Ghassemi and Christman, 1968). Thus, both mechanisms, light absorbance and light scattering, may occur and the uniformity of the UV profile may result from the mixture of aliphatic chains and aromatic nuclei of the humic molecules and/or from the scattering effect of a range of particle sizes.

Fluorescence. Some organic molecules absorb UV energy and then release longer wavelengths of lower energy as they return to normal state (Sylvia, 1973). In a way similar to UV absorbance, then, such a measure (fluorescence) may reflect changes in concentration of organic materials in solution through decreased emission value. Fluorescence is considered a more sensitive measure than UV absorbance. While UV absorbance shows broad bands of absorbance reflecting bonding groups and particular molecular configurations (such as aromaticity), fluorescence is more characteristic of the specific molecule. Both the excitation and emission wavelengths can have multiple peaks characteristic of the compound. Thus, Sylvia's (1973) fluorescence scan (Figure 2-19a) of contaminated river water shows many individual peaks spread over the range of 200–700 nm wavelength, which are assumed to correspond to particular contaminants. In relatively uncontaminated source water, however, the fluorescence spectrum, like the UV profile, is unremarkable with only one or two excitation–emission wavelength couples showing significant peaks (Figure 2-19b). Again, this indicates a fairly heterogeneous mixture of material, but of generally similar structural components.

Total Organic Carbon. TOC analysis, as implied, quantifies the organic carbon contained in a sample. The TOC measurement does not give any indication as to what sorts of materials are present in the sample and thus, even more than UV or fluorescence or color, is strictly quantitative. On a carbon basis, it is correspondingly more exact than any of the latter methods.

TOC has been utilized in attempts to correlate chloroform in finished waters with source water or-

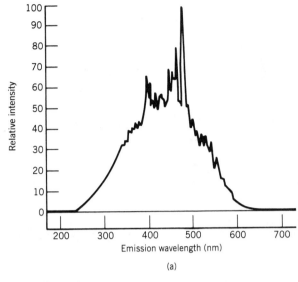

(a)

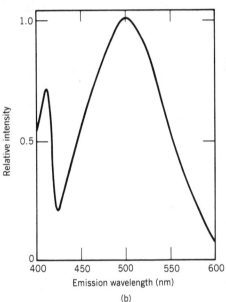

(b)

FIGURE 2-19. Fluorescence of aquatic organics. (a) Contaminated water fluorescence profile; excitation wavelength 340 nm (from Sylvia, 1973). (b) Uncontaminated raw water fluorescence profile (from Black and Christman, 1963).

ganic material (Rook, 1976; Rook, 1974; Symons et al., 1975; Stevens, Slocum, Seeger, and Robeck, 1975). This has been successful in two general situations: where the TOC represents a synthetic or extracted humic material and where the range of precursor and chloroform concentrations are great. Thus, Rook (1974) found a good linear relation between TOC and chloroform when his TOC was composed of fulvic acid extracted from peat and ranged in value from 10 to 250 mg/L (Figure 2-20).

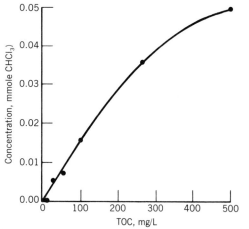

FIGURE 2-20. Chloroform concentration versus TOC (from Rook, 1976).

In an iron coagulation and resin adsorption experiment, where TOC ranged from 2.6 to 5.9 mg/L, such a relationship was not observed. Apparently, on a broad scale as in the former case, TOC may adequately reflect potential chloroform formation, while in a more narrow range, variation in the nature of the source material weakens the relationship. The EPA National Organics Reconnaissance

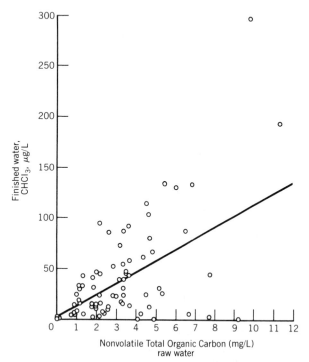

FIGURE 2-21. Concentrations of chloroform in finished water as a function of raw water nonvolatile TOC concentrations (data from Symons et al., 1975).

Survey (NORS) data (Symons et al., 1975) represents a wide range of values; TOC values from less than 0.05 to 19.2 mg/L span just about the entire range of values expected in natural waters, and chloroform ranges from 0.1 to 211 μg/L (Figure 2-21). The use of TOC to measure precursor in natural water samples, instead of extracted fulvic acid, probably accounts for the scatter in their data; however, the range of values observed allows for a statistically significant linear correlation.

RADIONUCLIDES

Radionuclides are radioactive atoms that break down to release energy (radioactivity). The energy is primarily released in one of three forms: (1) alpha radiation, consisting of large positively charged helium nuclei; (2) beta radiation, consisting of electrons or positrons; or (3) gamma radiation, consisting of electromagnetic wave-type energy similar to X-rays. Each of these forms has different health effects significance. Alpha particles travel at speeds as high as 10 million m/sec, and when ingested, these relatively massive particles can be very damaging. Beta particles travel at about the speed of light. Their smaller masses allow greater penetration but create less damage. Gamma radiation has tremendous penetrating power but has limited effect at low levels.

Radiation is generally reported in units of curies (Ci) or rems. One curie equals 3.7×10^{10} nuclear transformations per second. A common fraction is the picocurie (pCi), which equals 10^{-12} curies. A rem quantifies radiation in terms of its dose effect, such that equal doses expressed in rem produce the same biological effect regardless of the type of radiation involved. Numerically a rem is equal to the absorbed dose times a quality factor specific to the type of radiation which reflects its biological effects (Hem, 1971).

Radioactivity in water can be naturally occurring or man-made. Table 2-7 describes properties of some of the more common types of radiation from both sources. Naturally occurring radiation derives from elements in the earth's crust or from cosmic ray bombardment in the atmosphere. The two most generally abundant of these radionuclides are tritium (^{3}H) and potassium-40 (^{40}K).

Tritium is produced from cosmic ray reaction with atmospheric oxygen and nitrogen. It is then oxidized to tritiated water and incorporated into

TABLE 2-7. Radioactivity in Water

Nuclide	Half-life	Average Concentration in Natural Waters	Significance in Water
Naturally occurring			
Uranium-238 (α)[a]	4.5×10^9 yr	0.02–200 μg/L[b]	Probably produces greatest amount of radioactivity in waters (USGS). One area of midwest including parts of Iowa, Illinois, Wisconsin, and Missouri contains well waters having up to 25 pCi/L.
Radium-226 (α)	1.6×10^3 yr	0.01–0.1 pCi/L	Susceptible to removal via flocculation or softening
Radon-222 (α)	3.82 days	1 pCi/L surface >1000 pCi/L groundwater	Gas formed by decay of radium-226
Uranium-235 (α)	7×10^8 yr	0.02–200 μg/L[b]	
Radium-223 (α)	11.4 days		
Thorium-232 (α)	NA[c]	Extensively measured	
Radium-223 (β)[d]	5.8 yr	0.005–0.1 pCi/L	Decay gives rise to series of alpha-emitting daughters
Radium-224 (α)	3.6 days		
Radon-220 (α)	<1 min		
Tritium (^{3}H) (β)	12.3 yr	10–25 pCi/L	Occurs from cosmic ray bombardment → rainwater. Used for tagging water which is away from atmosphere. Increase with nuclear weapon tests
Carbon-14	5730 yr	6 pCi/g C[e]	Cosmic ray bombardment → atmospheric CO_2. Taken up in plant organic material. Used for dating purposes
Potassium-40 (β)	1.3×10^9 yr	<4 pCi/L	Occurs as constant percentage, 0.0118% of total potassium. U.S. adults ingest ~2300 pCi ^{40}K/day of which <0.70% is from water. Correlates with TDS in natural water
Rubidium-87 (β)	5×10^{11} yr		
Polonium-210 (α)	138 days		
Nuclear fission products (weapons testing)			
Strontium-90 (β)	30 yr	0.5 pCi/L	
Cesium-137 (β)	30 yr	0.1 pCi/L	
Iodine-131	8.1 days	NA	
Tritium (H^3)	12.5 yr	10–25 pCi/L	
Carbon-14	5730 yr	6 pCi/L	
Radiopharmaceuticals			
Iodine-125	60 days	Variable due to intermittent release into waters from medical facility effluents	Iodine isotopes are of concern due to potential effects on thyroid gland
Iodine-131	8.1 days		
Technetium-99m	6.0 hr		
Nuclear fuel			
Tritium (H^3)	12.5 yr	10–25 pCi/L	
Plutonium-239	24,360 yr		Released at ~1 mCi/yr from liquid storage sites
Iodine-129	1.7×10^7 yr	<0.01 pCi/L	

[a] Alpha-emitting nuclide. Radium-226 and the daughter products of radium-228 contribute the bulk of gross alpha activity in water. Drinking water estimated to contribute 0.01–0.05 pCi/day or 2% to total daily intake of alpha radiation.

[b] Represents the concentration of all uranium in water, primarily ^{238}U.

[c] Information not available.

[d] Beta emitting nuclide, consisting of electrons or positrons.

[e] That is, with 1 mg C/L would have ~0.006 pCi ^{14}C.

SOURCE: NAS (1977) and Hem, 1971.

rain or snow. ^{14}C is also formed by cosmic ray bombardment of the atmosphere. It is incorporated into atmospheric CO_2 and thence into organic material of growing plants. In fresh water, ^{14}C is generally present at approximately 0.006 pCi/mg carbon, and in oceans at approximately 0.1 pCi/mg carbon. ^{14}C is used to date wood or other carbonaceous materials that have been cut off from the atmosphere. In a similar way tritium is used to trace water. ^{40}K derives from the earth's crust and is present at a constant 0.0118 percent of the earth's total potassium. It is generally found in water at levels below 4 pCi/L. Adults consume approximately 2300 pCi ^{40}K/day, of which approximately 0.35–0.70 percent is contributed by drinking water (NAS, 1977).

Three high-atomic-weight, naturally occurring isotopes, uranium-238, thorium-232, and uranium-235 (along with their radioactive breakdown products, especially radium-226 and radium-228), also contribute significant but variable amounts of natural radioactivity to water. Other radionuclides of less quantitative significance may occasionally occur in water supplies, presumably due to local conditions. In Table 2-7 these are represented by rubidium-87 and polonium-210. These lesser isotopes are generally not well characterized in terms of distribution, average concentrations, or health effects.

Man-made or man-induced radionuclides are from three general sources: nuclear fission from weapons testing, radiopharmaceuticals, and nuclear fuel processing and use. Table 2-7 includes the major isotopes that derive from each of these sources. ^{90}Sr, ^{137}Cs, ^{131}I, ^{3}H, and ^{14}C are of most potential importance and have received most attention. Note that tritium and ^{14}C appear again on the man-made list. The contribution of these radioisotopes from weapons testing has been decreasing since 1963 when such testing stopped, while the net contribution to water supplies from the other two sources has remained constant or perhaps slightly increased. Although production and use of radionuclides in medicine and energy production has increased, source control has kept contributions to waters from these sources relatively constant.

Radionuclides in water are important primarily from their human health standpoint and to a lesser extent because of their use as tracers and age determinants. Although, as is shown in Table 2-8, the body dose that accrues from drinking water as compared to natural background radiation is quite small (estimated by the NAS at 0.24 percent), the EPA's

TABLE 2-8. Annual Radiation Dose by Source

Source	Annual Dose (mrem)
Cosmic rays	44
Terrestrial radiation	
External to the body	40
Internal to the body	18
Total natural sources	102
Hypothetical water	
Supply in the US	0.244

SOURCE: After NAS.

policy is to assume that there is potential for harm from any level of radiation dose.

Included in the primary drinking water regulations, therefore, are limits on the gross quantity of alpha radiation and on radium 226, radium 228, strontium 90, and tritium.

REFERENCES

Alekin, O. A., and Brazhnikova, L. V., "The Discharge of Soluble Matter from Dry Land of the Earth," *Gidrokhim. Materialy,* **32,** 12–24. (1961).

Barrow, G. M., *Physical Chemistry,* 2nd ed. McGraw-Hill, New York (1966).

Beck, K. C., Reuter, J. H., and Perdue, E. M., "Organic and Inorganic Geochemistry of Some Coastal Plain Rivers of the Southeastern United States," *Geochim. Cosmochim. Acta,* **38,** 341–364 (1974).

Black, A. P., and Christman, R. F., "Characteristics of Colored Surface Waters," *JAWWA,* **54,** 753 (1963).

Brown, T. L., *General Chemistry,* Charles E. Merrill, Columbus, OH (1963).

Davies, S. N., and DeWiest, R. J. M., *Hydrogeology,* Wiley, New York (1966).

Drost-Hansen, W. "The Structure of Water and Water-Solute Interactions," *Equilibrium Concepts in Natural Water Systems,* Advances in Chemistry Series, No. 67, American Chemical Society, Washington, DC (1967).

Eisenberg, D., and Kauzman, W., *The Structure and Properties of Water,* Oxford University Press, New York and London (1969).

Environmental Protection Agency, "National Secondary Drinking Water Regulations," *Fed. Reg.,* **44,** 153, 42195–42202 (July 19, 1979) (EPA, 1979a).

Environmental Protection Agency, "Process Design Manual for Land Treatment of Municipal Wastewater," EPA 625/1-77-008 (October 1977).

Environmental Protection Agency, "Interim Primary Drinking Water Regulations; Amendments," *Fed. Reg.,* **45**(168), 57332–57357 (August 27, 1980).

Ghassemi, M., and Christman, R. F., "Properties of the Yellow Organic Acids of Natural Waters," *Limnol. Oceanogr.* **13,** 583 (1968).

Hem, J. D., *Study and Interpretation of the Chemical Characteristics of Natural Water,* Geological Survey-Water Supply Paper 1473, U.S. Government Printing Office, Washington, DC (1971).

Hutchinson, G. E., *A Treatise on Limnology,* Wiley, New York (1957).

Jackson, T. A., "Humic Matter in Natural Waters and Sediments," *Soil Sci.* **119** (1975).

James M. Montgomery, Consulting Engineers, Inc., "Analysis of Water Quality Data From Continuous Automated Monitors at Beach Street 1976–77," Working Paper (August 1978).

James M. Montgomery, Consulting Engineers, Inc., "Ute Water Conservancy District, Western Engineers Pilot Studies for Ute Water Treatment Plant Expansion" (October 1981).

Lamar, W. L., "Evaluation of Organic Color and Iron in Natural Surface Waters," USGS Professional Paper 600-D (1968), pp D24–D29.

Lee, F. G., and Hoadley, A. W., "Biological Activity in Relation to the Chemical Equilibrium Composition of Natural Waters," *Equilibrium Concepts In Natural Water Systems,* American Chemical Society, Washington, DC (1967), Chap. 16.

Livingstone, D. A., *Chemical Composition of Rivers and Lakes, Data of Geochemistry,* 6th ed., Prof. Pop. U.S. Geological Survey, 440-G (1963), Chap. G.

McKee, J. E., and Wolf, H. W., *Water Quality Criteria,* 2nd ed., The Resources Agency of California State Water Quality Control Board, Publication No. 3-A (1963).

Moore, W. J., *Physical Chemistry,* 4th ed., Prentice-Hall, Englewood Cliffs, NJ (1972).

National Academy of Sciences, Safe Drinking Water Committee, *Drinking Water and Health,* National Academy of Sciences Printing and Publishing Office, Washington, DC (1977).

Pauling, L., *The Nature of the Chemical Bond,* Cornell University Press, Ithaca, NY (1960).

Rainwater, F. H., and White, W. F., "The Solusphere: Its Inferences and Study," *Geochemica et Cosmochimica Acta,* **14,** 244–249 (1958).

Rook, J. J. "Formation of Haloforms during Chlorination of Natural Water," *J. Water Treatment Exam.,* **23,** 234 (1974).

Rook, J. J., "Haloforms in Drinking Water," *JAWWA,* **68,** 168 (1976).

Schnitzer, M., and Khan, S. V., *Humic Substances in the Environment,* Marcel Dekker, New York (1972).

Snoeyink, V. L., and Jenkins, D., *Water Chemistry,* Wiley, New York (1980).

Standard Methods for the Examination of Water and Wastewater, 15th ed., American Public Health Association, American Water Works Association, Water Pollution Control Federation (1980).

Stevens, A. A., Slocum, C. J., Seegar, D. R., and Robeck, G. G., "Chlorination of Organics in Drinking Water," Proceedings of Conference on the Environmental Impact of Water Chlorination, Oak Ridge National Laboratory, Oak Ridge, Tennessee (October 22–24, 1975).

Stumm, W., and Morgan, J. J., *Aquatic Chemistry,* Wiley-Interscience, New York (1970).

Suarez, D. L., "An Alternative Method of Estimating the SAR of Soil or Drainage Water," *Soil Sci. Soc. Am.,* **45,** 469–475 (1981).

Sylvia, A. E., "Detection and Measurement of Microorganics in Drinking Water," *J. N. Engl. Water Works Association,* **87,** 2 (1973).

Symons, J. M., Bellar, T. A., Carswell, J. K., DeMarco, J., Kropp, K. L., Robeck, G. G., Seeger, D. R., Slocum, C. J., Smith, B. L., and Stevens, A. A., "National Organics Reconnaissance Survey for Halogenated Organics in Drinking Water," WSRL MDL and QAL, National Environmental Research Center, USEPA, Cincinnati, Ohio, *JAWWA,* **67,** 634 (1975).

Todd, D. K. (ed.), *The Water Encyclopedia,* Water Information Center, Inc., Port Washington, NY (1970).

Turekian, K. K., "Rivers, Tributaries and Estuaries," in D. W. Hood (ed.), *Impingement of Man on the Ocean,* Wiley, New York (1971).

Weast, R. C. (ed.), *Handbook of Chemistry and Physics,* 46th ed., The Chemical Rubber Company, Cleveland (1964).

Wetzel, R. G., *Limnology,* W. B. Saunders, Philadelphia (1975).

—3—

Microbiological Quality

INTRODUCTION

For the water engineer, microbiology is important for its effects on public health, physical and chemical water quality, and treatment plant operation. Waterborne microorganisms can be responsible for diverse public health problems including bacterial diseases such as cholera and gastroenteritis, viral infections such as hepatitis, amoebic dysentery or diarrhea originating from protozoa, and parasitic helminth (worm) infections such as tapeworm or roundworm. Around the world, the prevalence and severity of public health problems relate directly to an area's degree of development. In countries with low income and/or minimal education, a host of diseases associated with contaminated water may be observed, while in western Europe and the United States, local outbreaks of bacterial or viral gastroenteritis and giardiasis are the only common waterborne microbiological problems still occurring (Table 3-1). Within this context, developed countries have the resources to put more emphasis on microbiologically based problems of physical and chemical quality such as reservoir eutrophication and treatment plant operation problems such as filter clogging and taste and odor episodes.

The field of aquatic microbiology encompasses diverse organisms. Major groups of interest include bacteria, viruses, algae, fungi, protozoa, and helminths, or worms. Each of these groups is discussed in this chapter. By way of introduction, however, Table 3-2 summarizes the aspects of each group that are significant in terms of their ecology, public health impacts, and/or treatment effects. Table 3-2 shows the surface charge, shape, mode of existence, oxygen requirements, energy requirements, motility, and environmentally resistant stages of each group of microorganisms.

TABLE 3-1. Etiology of Waterborne Outbreaks of Disease in the United States, 1971–1981

Disease	Outbreaks	Illnesses
Gastroenteritis, unidentified etiology	192	39,845
Giardiasis	50	19,863
Shigellosis	25	5,448
Salmonellosis	8	1,150
Hepatitis A	16	463
Campylobacter diarrhea	4	3,902
Viral gastroenteritis	10	3,147
Vibrio cholerae	1	17
Rotavirus	1	1,761
	307	75,596

Source: After Craun et al. (1983).

TABLE 3-2. Physical and Biological Characteristics of Water Quality Importance

Organism	Size (Å)	Surface Charge	Shape	Status	Oxygen Requirements	Energy Requirements	Motility	Environmentally Resistant Stage
Viruses	10^2–10^3	Negative	Variable	Parasitic	NA[a]	NA[a]	Nonmotile	Viron
Bacteria	10^3–10^5	Negative	Rod, coccoid, spiral comma, pleomorphic	Free living, parasitic	Aerobic, anaerobic facultative	Chemolithotroph, chemoheterotroph, photolithotroph, photoheterotroph, organophototroph	Motile, nonmotile	Spores, cystlike
Blue-green algae	10^4	Negative	Coccoid, filamentous	Free living, symbiotic	Aerobic	Photoautotroph	Gliding	Cysts
Green algae	10^4–10^6	Negative	Colloid, pennatic	Free living, symbiotic	Aerobic	Photolithotroph	Motile, nonmotile	Spores, cysts
Protozoa	10^4–10^6	Negative	Variable	Free living, parasitic	Aerobic, anaerobic	Chemoorganotroph	Motile, nonmotile	Cysts
Fungi	10^4–10^6	Negative	Filamentous, coccoid	Free living, parasitic	Aerobic, anaerobic	Chemoorganotroph	Nonmotile	Spores
Helminths	10^4–10^9	Negative	Variable	Free living, parasitic	Aerobic	Chemoheterotroph	Motile	Eggs

[a] NA, not applicable.

Size is an important consideration in microorganism analysis and treatment. As is shown in Figure 3-1, for example, while most bacteria can be viewed with a light microscope, which has been in use since the late 1700s, viruses require an electron microscope to be seen. Until this technology was developed in the 1930s, therefore, the presence of viruses was only postulated. Figure 3-1 further illustrates the sizes of microorganisms within the context of higher organism sizes. The whole of the biological world is then shown encapsulated between the atomic world and the cosmic world.

Knowledge of the size of microorganisms to be removed in a water treatment plant can affect process selection. For example, if it is known that the protozoan *Giardia lamblia* is present in a raw water source, then coagulation and filtration operating parameters can be optimized to remove particles of size 5–25 μm, the range which encompasses *Giardia* cysts. The negative surface charge of all microorganisms can also impact treatment process design—specifically in the area of polymer selection. Cationic polymers would be expected to destabilize microorganisms better than would either anionic or nonionic polymers (Table 3-3).

TABLE 3-3. Efficacy of Polyelectrolytes (Coagulant Aids) in the Removal of Poliovirus Type 1 by Coagulation–Sedimentation[a]

Polyelectrolyte	Polyelectrolyte Concentration (ppm)	Virus Removal (%)
No polyelectrolyte		97.81
Cationic polyelectrolyte	0.25–1	99.72
Anionic polyelectrolyte	0.25	88.82
	1	95.83
Nonionic polyelectrolyte	0.25	92.55

[a] Coagulation with FeCl, at a concentration of 60 mg/L.
SOURCE: Bitton (1980).

The next four characteristics shown in Table 3-2, mode of existence, life style, oxygen requirements, energy requirements, and motility, are used to describe the ecology of microorganisms; that is, under what conditions they might be expected to be found, or under which adverse conditions they might be expected to be killed. Figure 3-2, for exam-

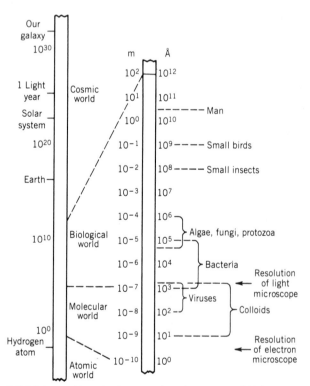

FIGURE 3-1. Size of microorganisms in context of high organisms sizes.

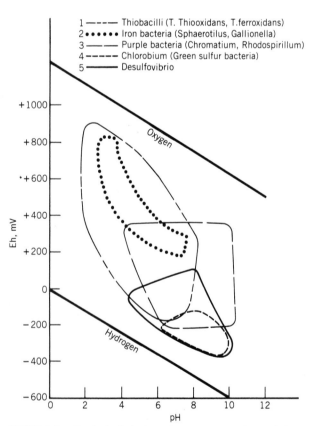

FIGURE 3-2. Ecological nitches of some groups of aquatic bacteria (Rheinheimer, 1980).

ple, illustrates the pH and Eh ranges under which various bacteria can grown and reproduce. Whereas most bacteria require a pH range of 4–9, and thrive between 6.5 and 8.5, a few, such as *Thiobacillus thiooxidans* and *Thiobacillus ferroxidans* can tolerate values down to pH 1. Similarly, the oxidation–reduction (redox) potential of a water selects for certain types of bacteria. Aerobic bacteria such as *T. thiooxidans* and *Sphaerotilus* require higher redox potentials than anaerobes such as *Desulfovibrio*.

Finally the environmentally resistant stages of various microorganisms are significant in terms of disinfection. Spores and cysts are much more resistant to disinfectants than are viable organisms. Thus if it is known that these forms are present in raw water, methods of control which are much more rigorous than those designed to remove coliforms, for example, must be used.

These and other characteristics of microorganisms which are significant for the water engineer are discussed below for each group.

BACTERIA

Bacteria are single-celled organisms ranging in size from 0.1 to 10 μm (Figure 3-1). Despite two orders of magnitude difference in size, bacteria are all of relatively simple structure. More specifically, bacteria contain only one membrane—the one bounding the cell itself. The bacterial interior contains only two major regions: the cytoplasm and the nucleus, both relatively uniform in appearance. Cells of this simple type are termed procaryotic and are also characteristic of the blue-green algae. All other animal cells, including those comprising algae, fungi, protozoa, plants, and animals, are termed eucaryotic and contain regions that are distinctive both morphologically and physiologically.

The simple appearance of bacteria is deceiving; physiologically they are more diverse than any other biological group. These two factors, structural uniformity and physiological diversity, make the classification of bacteria extremely difficult. Typically a bacterial genus is identified based on a combination of morphology and metabolic responses. For example, the genus *Escherichia* (including the species *E. coli*) is described as unicellular, nonphotosynthetic, nonsporing, straight, rod-shaped organisms, less than 2 μm wide, gram negative, aerobic, heterotrophic, acid and gas pro-

ducing from glucose and lactose within 48 hr, methyl red positive, Voges–Proskauer negative, and catalase positive, with four species differentiated on the basis of pigmentation, utilization of citrate, and production of H_2S. As might be suspected, the classification of bacteria is constantly changing; thus, *E. coli* were previously assigned to the genus *Bacterium* as *Bacterium coli* and to the genus *Bacillus* as *Bacillus coli*.

This method of bacterial classification is significant for the water engineer only as background information. For most applications, it is sufficient or even preferable to simply refer to the specific name of the organism of interest. Attempts to organize specific bacteria within groups as they relate to particular projects is usually not useful and is often confusing. On the other hand, certain investigations will require or benefit from a more thorough characterization of bacteria and their relationships to other species or genera; thus it is useful to be aware of the state of the art in this area. The best reference for an explanation and application of bacterial taxonomy is *The Microbial World* by Stanier, Douderoff, and Adelberg.

Types of Bacteria

Bacteria within water supplies are broadly classified into two groups: the autochthonous group (Greek autocthon "one sprung from the land itself"), which belong there, and the allocthonous group (Greek allos "other"), which arrived there via contamination, runoff, rainfall, and so on, and which typically will have a limited life span there. Bacteria that are of concern from the public health standpoint fall into this latter group; often their normal locale is warm blooded animal intestines and their presence in water is indicative of sewage or fecal contamination. Outside of the warm, nutritionally rich environment of a human or animal gut, they die off rather rapidly. Figure 3-3, for example, illustrates die-off rates for *E. coli* in a water supply reservoir at 20 feet depth. Table 3-4 lists bacteria of interest to the water engineer from a public health standpoint and describes their hosts, modes of transmission, disease symptoms, and worldwide occurrence.

Bacteria of interest from a water quality standpoint are typically autocthonous and include such diverse groups as the iron and sulfur bacteria, the methane bacteria, and the nitrate reducers. These are important to the water engineer in a variety of

TABLE 3-4. Bacteria of Public Health Significance in Water

Genus	Species	Host(s)	Disease	Transmission	Occurrence
Salmonella	Several hundred serotypes known to be pathogenic to man. Examples: *S. typhi, S. enteritides, S. typhimurium*	Found in human and animal guts, polluted water, contaminated foods	Typhoid fever from *S. typhi*, salmonellosis from other species; both cause acute diarrhea and cramping; typhoid fever can be fatal	Transmitted through water or food processed with contaminated water	Worldwide, typhoid fever common in areas of Far East, Middle East, Eastern Europe, Central and South America, Africa
Shigella	Four species cause disease in man: *S. sonnei, S. flexneri, S. boydii, S. dysenteriae*	Rarely found in hosts other than man	Shigellosis, or bacillary dysentery, causes fever and bloody diarrhea	Transmitted via contaminated food, polluted water, and person-to-person contact	Worldwide occurrence, endemic in tropical populations; in United States moderately endemic in lower economic areas
Leptospira	Common human isolates include *L. pomona, L. autumnalis, L. australis*	Found in infected humans and animals. Excreted via urine	Leptospirosis: acute infections of kidney, liver, central nervous system	Transmitted through skin abrasions or mucous membranes to blood stream; may come from contact with animal carriers and/or polluted water	Worldwide; occupational hazard to rice field workers, farmers, sewer workers, and miners
Pasturella	*P. tularensis*	Man and wild animals, especially rabbits	Tularemia: causes chills and fever, with an ulcer at site of infection, swollen lymph nodes	Handling of infected animals, arthropod bites (deer fly, wood tick), drinking contaminated water	North America, Europe, USSR, Japan
Vibrio	*V. cholerae*	Humans	Cholera: an acute intestinal disease causing vomiting, diarrhea, dehydration, can be fatal	Contaminated water Person-to-person contact	Endemic in parts of India, East Pakistan; occurs in Philippines, Indonesia, Thailand, Hong Kong, Korea
Entero-pathogenic *E. coli*	Various serotypes	Warm-blooded animals	Diarrhea, urinary infections	Sewage contaminated water or food; person to person via physical contact	Worldwide
Yersinia	*Y. enterocolitica, Y. pseudotuberculosis*	Birds and mammals	Diarrhea, fever, vomiting, anorexia, acute abdominal pain, abcesses, septicemia	Animal to man, person to person, contaminated water	Worldwide
Myco-bacterium	*M. tuberculosis, M. balnei, M. bovis*	Man, diseased cattle	Pulmonary or skin tuberculosis	Typically airborne, but sewage-contaminated water is demonstrated route	Worldwide

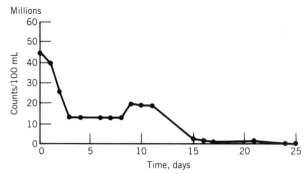

FIGURE 3-3. *E. coli* die-off in a water supply reservoir, depth 20 ft (data from Geldrich, 1980).

situations (e.g., pipe and well corrosion, metal leaching, acid mine drainage, sulfur odors). Many of these activities come together under the heading of distribution system problems. These are described and discussed in further detail in a subsequent section.

Bacteria of Public Health Significance in Water

Enteric bacteria enter water sources directly through human or animal contact or indirectly via sewage input or urban or rural runoff. Once in the water, various factors act to inhibit or enhance survival. In untreated waters, sunlight, thermal effects, sedimentation, and predation tend to reduce bacterial levels. Results shown in Table 3-5 indicate, for example, that increased temperatures cause more rapid inactivation, while Table 3-6 illustrates the effect of differing water sources under the same temperature and light conditions on bacterial die-off.

As is shown, filtered river water provided the best medium, probably because of the presence of organics and the lack of predators. At lower temperatures survival in both river water and tap water was greater, demonstrating the same trend shown in Table 3-5.

TABLE 3-5. Effect of Temperature on Bacterial Inactivation[a]

Temperature (°C)	Time for 100% Inactivation (days)
37	<1
22	<2
2	<17

[a] In polluted river; agent, *S. typhi;* initial concentration, 10^{-4} mL^{-1} (Berger, 1983).

TABLE 3-6. Effect of Differing Water Sources on Bacterial Inactivation[a]

Water	Temperature (·C)	Time for 99.9% Infectivity Loss (days)
River	23–25	4
Filtered river	23–25	21
Tap water	23–25	6
River	4–6	7
Filtered river	4–6	21
Tap water	4–6	21

[a] Agent, *Shigella flexneri;* initial concentration, 10^6–10^7/mL.
SOURCE: Berger (1983).

Of great importance in the area of indicator organisms is the relative survival of different types of bacteria under the same conditions. Table 3-7 illustrates a range of 2.4–27 hr for a 50 percent infectivity loss among different organisms in well water at 9.5–12.5°C.

It is significant that the survival time for coliforms is less than that for several other organisms monitored. As is discussed further in a later section, while the total coliform analysis has traditionally been used as an indicator of fecal contamination, the fact that coniforms may die off more quickly than other enteric bacteria, viruses, and microorganism cysts makes its comprehensive utility somewhat questionable.

Bacteria and Water Treatment

During the period 1971–1977, 67 percent of the largest waterborne disease outbreaks were due to source water contamination where treatment was either inadequate or nonexistent (Berger, 1983).

TABLE 3-7. Survival of Different Organisms under the Same Conditions

Agent	Water	Time for 99.9% Infectivity Temperature (°C)	Loss (days)
Coliforms (Aug.)	Well	9.5–12.5	17
Enterocoli (Aug.)		9.5–12.5	22
Shigella dysenteriae		9.5–12.5	22.4

SOURCE: Berger (1983).

Generally the more contaminated a raw water source, the more elaborate the treatment to control bacteria and other microorganisms must be. Studies conducted in 1935 by Streeter and his colleagues concluded that with filtration and chlorination, consistent finished water quality meeting a coliform standard of 1/100 mL required raw water levels of less than or equal to 5000/100 mL. For waters treated only with chlorination, a raw water density of less than or equal to 50/100 mL was necessary to consistently meet drinking water standards (Berger, 1983).

At the same time, consideration must be given to the specific types of organisms present. While coagulation and filtration processes are more a function of bacterial size and surface charge and would hence not be expected to react differently across genera, different groups of bacteria do react differently to disinfection.

Table 3-8 indicates typical bacterial reduction via coagulation and filtration. The range of 76–99 percent removal for coagulation and 50–99 5 percent for filtration reflect differences in process operations, not bacterial types. On the other hand, as Table 3-9 illustrates, in order to achieve 99–100 percent inactivation among different groups of bacteria via chlorine disinfection, significantly different concentration–time (*ct*) values for the same disinfectant must be utilized.

Autochthonous Bacteria: Distribution System Problems

Microorganisms that inhabit various niches in soil, sediments, ponds, and streams can also find their way into distribution systems and wells. Pipe joints,

encrustations, and deposits provide suitable habitats for a number of aerobic and anaerobic microorganisms. Problems deriving from these organisms include "red water," taste and odor, corrosion, and slime production. Table 3-10 lists some of these organisms and identifies their associated "nuisance" activities.

Several of these groups of bacteria can develop in water in which there is little or no organic matter, but where there are ferrous or manganous compounds, sulfides, or other incompletely oxidized mineral substance to serve as sources of energy. The iron bacteria oxidize ferrous iron to ferric iron and affect the precipitation of ferric hydrate. Bacteria in this category include members of the genera *Sphaerotilus, Leptothrix, Crenothrix,* and *Gallionella.* These bacteria are all aerobic and many are filamentous. They cause fouling of pipelines and well casings due to substantial extracellular buildup of rusty brown ferric hydrate. Such buildup can comprise up to 500 times the mass of the cells themselves (Starkey, 1945). The source of the ferrous iron is typically the water supply, although pipelines themselves can be the contributors. Control of iron bacteria can focus on the bacteria themselves or on controlling the input of iron to the system.

Sulfur-oxidizing bacteria such as *Beggiatoa* and *Thiobacillus* can be found in waters containing sulfide or elemental sulfur. Some filamentous types can cause fouling while others create corrosion problems. They are all characterized by an ability to oxidize sulfur or sulfide to sulfate. Filamentous forms may be encountered along with iron bacteria and contribute to formation of slimes. *Thiobacillus thioxidans* is a typical sulfur bacteria that oxidizes sulfur and sulfur compounds with concurrent pro-

TABLE 3-8.   Bacteria Reduction via Coagulation and Filtration

Agent	Type of Water	Coagulant Conditions		Filtration Condition	
		Coagulant	Dose	Filter Type	Removal (%)
Total coliform	River	$Al_2(SO_4)_3$	12.6		97
	River	$Al_2(SO_4)_3$	20		74
	River	$Al_2(SO_4)_3$	25		99.4
E. coli	River	$Al_2(SO_4)_3$	10.5		83
	Lake	$Al_2(SO_4)_3$	12.1		76
Total coliform				Slow sand	70–98
				Slow sand	96.3–99.5
				Slow sand	50

Source: Berger (1983).

TABLE 3-9. Bacterial Reduction by Disinfection

	Disinfectant Concentration (mg/L)	Contact Time (min)	Ct (mg-min/L)	pH	T (°C)	Percent Reduction
	OCl⁻ Disinfection					
	1.0	0.92	0.92	10	5	99
E. coli	0.3	10	3.0	10	20–25	100
	0.4	10	4.0	10	20–25	100
S. typhi	0.3	10	3.0	10.7	20–25	100
	HOCl Disinfection					
E. coli	0.1	0.4	0.4	6.0	5	99
L. pneumophila	3.3	<1	3.3	—	25	99.99
	1.0–1.5	19	19–24			99
Campylobacter jejuni	2.5	1	2.5			100
	0.63	30	19			100
	5.0	1	5.0			100

SOURCE: Berger (1983).

duction of very high acidity (see Figure 3-2). Under these conditions iron can be brought into solution.

The sulfate-reducing bacteria of the genera *Desulfovibrio* (also called *Sporovibrio*) are able to reduce sulfate to sulfide. They are strict anaerobes,

TABLE 3-10. Some Genera of Microorganisms Associated with Effects on the Quality of Water in Distribution Systems

Genera	Possible Effect
Actinomyces	Taste and odors
Arthrobacter	Color production (porphyrins): Slime formation
Bacillus	Nitrate reduction; corrosion potential indicator antagonist
Beggiatoa	"Red" water, sulfur oxidation
Crenothrix	"Red" water (iron bacteria)
Desulfovibrio	Corrosion: hydrogen sulfide production ("black" water)
Gallionella	"Red" water (iron bacteria)
Leptothrix	"Red" water (iron bacteria)
Methylmonas (Methanomonas)	Methane oxidation
Micrococcus	Nitrate reduction; corrosion; potential indicator antagonist
Nitrobacter	Nitrate production; possible corrosion and slime formation
Nitrosomonas	Nitrate production; possible corrosion and slime formation
Sphaerotilus	"Red" water
Streptomyces	Taste and odor
Thiobacillus	Corrosion, slime production

SOURCE: National Research Council (1982) and Larson (1966).

with sulfate replacing oxygen for the oxidation of organic matter to provide energy for growth. The sulfide produced by these bacteria can cause odor problems and suspended black particles and have a corrosive effect on steel and other metal. These bacteria can also enhance corrosion under anaerobic conditions by serving as cathodic depolarization agents. Under anaerobic conditions the normal series of reaction would comprise

$$Fe \rightarrow Fe^{2+} + 2e \quad \text{(anode)}$$

$$2H^+ + 2e \rightarrow H_2 \quad \text{(cathode)}$$

Typically oxygen present in the water would then react with the hydrogen to produce H_2O. Without O_2 *Desulfovibrio desulfuricans* can remove the hydrogen and oxidize it at the expense of sulfate:

$$4H_2 + SO_4 \rightarrow S^{2-} + 4H_2O \quad \text{(cathode)}$$

Pitting corrosion of steel pipe can occur in this manner, typically in pipe buried in anaerobic soils but also within anaerobic encrustations on interior pipe surfaces.

Nitrifying bacteria include groups that oxidize ammonia to nitrite, such as *Nitrosomonas*, and others that oxidize nitrite to nitrate, such as *Nitrobacter*. *Nitrobacter* is a soil bacteria that may or may not cause problems in distribution systems. *Nitrosomonas*, on the other hand, is known to develop as a slime on the walls of pipe in the presence of ammonia and dissolved oxygen. The slime can slough off and appear at hydrants or household taps. In addition, the organisms create a substantial chlorine demand, such that it may be difficult or impossible to achieve chlorine residuals at the far ends of distribution systems. Under anaerobic conditions, at the end of lines or localized areas of stagnant flow, putrification can result in taste and odor problems. While high levels of ammonia make the use of breakpoint chlorination impractical for *Nitrosomonas* control, the use of copper sulfate with combined chlorine can serve to keep the growth in check (Larson, 1966).

Other groups of bacteria can utilize nitrate in place of oxygen in the oxidation of organic substrates. Such denitrification reactions are carried out by species of *Bacillus, Micrococcus, Pseudomonas, Escherichia,* and *Achromobacter.* This activity can be described by the following reactions:

$$C_6H_{12}O_6 + 12NO_3^- \rightarrow 12NO_2^- + 6CO_2 + 6H_2O$$

Within this context, *Bacillus* and *Micrococcus* have been associated with corrosion problems due to local CO_2 production and decreased pH.

In a manner similar to *Nitrosomonas, Methylmonas* or *Methanomonas* is reported to create slime and associated problems in the presence of methane and dissolved oxygen. Other nuisance bacteria within distribution systems are associated with color production (*Arthrobacter*) and taste and odor problems (*Streptomyces* and *Actinomyces*).

VIRUSES

Viruses are structurally extremely simple organisms. All are parasites and are found in animals, plants, bacteria, fungi, and algae. Virus size relative to these other organisms was shown in Figure 3-1. The basic virus consists of a core of nucleic acid (either DNA or RNA) surrounded by a protein coat. Some have protective lipid envelopes. Despite this straightforward morphology, viruses are highly host specific; animal viruses do not infect bacteria and vice versa.

Because of their small size, viruses could not be detected until the advent of the electron microscope in 1931. Their presence prior to that time had, however, been postulated, and they were referred to as "filterable agents" since they would pass through standard filters used to retain bacteria.

Even with the advanced electron microscopy techniques available today, viruses are still difficult

to detect in environmental samples. Except in raw or partially treated wastewaters where their numbers may exceed 10,000 virus plaque-forming units (PFU) per liter, large volume sampling and subsequent concentration techniques are required. In drinking water supplies, where their numbers are typically on the order of 1–100 PFU/L, samples of 100–1000 L must be processed in order to achieve significant recovery.

Further sophistication is required to identify viruses collected. Given these problems, it is not surprising that definitive association of waterborne viruses with specific disease occurrences is rare. In fact, infectious hepatitis is the only epidemiologically established waterborne virus disease (Grabow, 1968). On the other hand, the circumstantial evidence linking viruses to various disease outbreaks is persuasive. As is discussed below, enteric viruses are regularly shed in animal and human feces; they are ubiquitous in untreated surface waters around the world, and although water treatment plants are typically effective in removing incoming viruses, inadequate treatment or temporary breaks in treatment effectiveness can allow some through.

Virus Types

Table 3-11 and Figure 3-4 describe some of the viruses that infect animals and list the diseases with which each is associated. Included within the animal group are the "enteric viruses," those able to enter the body via the oral route. Enteric viruses include adenoviruses, reoviruses, rotaviruses, en-

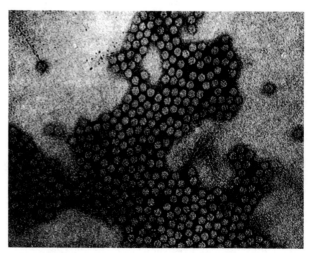

FIGURE 3-4. Electron micrograph of typical poliovirus particles. ×141,500 (Acton, 1974).

TABLE 3-11. Viruses that Infect Animals

Group	No. Types	Disease	Nucleic Acid	Size (nm)
Adenoviruses	37	Respiratory illness, conjunctivitis	DNA	70 × 90
Reoviruses	3	Respiratory disease, diarrhea, Colorado tick fever	RNA	75
Rotaviruses	>3	Infantile gastroenteritis	RNA	64–66
Enteroviruses				
Polioviruses	3	Poliomyelitis	RNA	28
Coxsackie viruses A	23	Aseptic meningitis	RNA	20–30
Coxsackie viruses B	6	Aseptic meningitis	RNA	20–30
Echoviruses	31	Aseptic meningitis, cold	RNA	20–30
Other enteroviruses	>4	AHC, encephalitis	—	—
Hepatitis A virus	1	Infectious hepatitis	Unclassified	27
Norwalk and related gastrointestinal viruses	>3	Gastroenteritis	—	—
Poxviruses	1	Smallpox	DNA	200–300
Herpesvirus	5	Fever blisters, cervical cancer, chicken pox shingles, mononucleosis	DNA	100 × 200
Mxyoviruses	±3	Influenza	RNA	80 × 120

SOURCE: Adapted from Bitton (1980), Berger (1983), and Acton (1974).

teroviruses, the hepatitus A virus, and the Norwalk and related gastrointestinal group. These are the viruses of most concern to the water engineer since they can be transmitted via drinking water. Enteric viruses multiply in the gut and are excreted in large numbers in feces. As is shown in Table 3-11, there are more than 100 types of enterics. Although they are the most likely to be transmitted via the water route, other disease-causing viruses could conceivably survive long enough to be transmitted in this fashion. Members of the myxovirus and paramyxovirus groups fall within this category, for example.

Viruses in Water

Viruses can enter the water via direct contamination from humans or animals or indirectly via sewage or urban and rural runoff. Various data illustrate that such contamination is ubiquitous. Table 3-12, for example, shows results of a survey of river waters and domestic water supplies in several developed countries where a universally high percentage of samples were positive for enteric viruses. Table 3-13 shows some of the numbers of viruses that have been found in surface waters in the United

States, Israel, Germany, and the United Kingdom. Finally, Table 3-14 presents some of data gathered on the survival time of specific virus groups in a variety of water environments.

Within the environment, a variety of factors act to inactivate viruses by damaging the lipid envelope (if present), the protein coat, or the DNA/RNA. In some cases, however, damage to the envelope or protein coat may still allow infection to occur. Conversely, loss of infectivity due to destruction of the DNA or RNA may still elicit immune responses from the host.

TABLE 3-12. Global Aspect of Virus Contamination of Water Supplies

Type of Water	Samples Positive for Enteroviruses (%)
River water (France)	21; 9
River water (USSR)	34
River water (Switzerland)	38; 63
River water (Illinois River)	27
Domestic water supply (Israel)	2
Domestic water supply (England)	56

SOURCE: Bitton (1980).

TABLE 3-13. Occurrence of Enteric Viruses in Natural Waters

Type of Water	Location	Ambient Concentration (per 100 mL)
River	River Wear, U.K.	0–2.5
River	River Wear, U.K.	0–2.5
Surface	German Democratic Republic	0–1.8[a]
River (bayou)	Houston, TX	Avg. 3.5
River	U.K.	0–57.4
Creeks and streams	Israel	0.6–15
Surface	Israel	8–19
River	River Trent, U.K.	ND–0.8
River	River Avon, U.K.	ND–54
Bayou	Brays Bayou, TS	0.29–22

[a] In virus-positive samples.

SOURCES: Berger (1983) and Leong (1983).

In water, important physical factors that may inactivate viruses include temperature, sunlight, and dessication. Important chemical factors include pH, heavy metals and oxidizing agents such as chlorine, ozone, bromine, and iodine. Algae, bacteria, and protozoa have also been implicated in viral inactivation. Figure 3-5 illustrates that poliovirus and Coxsackie virus are inactivated much more rapidly in nonsterile lake water than in lake water that has been previously sterilized. Within other environments, specific chemicals can destroy or alter specific viral components, rendering the viruses inactive or weakened. These include lipid solvents such as ether, protein degenerants such as phenol and low ionic strength solutions, and nucleic acid inactivators such as formaldehyde and ammonia.

Factors that tend to protect viruses in water (and enhance their transmissivity) are suspended solids and turbidity. As Table 3-15 illustrates, viruses adsorbed to solids remain nearly 100 percent viable. When subjected to inactivation agents such as chlorine, viruses adsorbed to solids tend to survive longer than their nonadsorbed counterparts. Such protection extends to other physical and chemical inhibitors. This likely accounts for the observed in-

creased survival time of viruses in sediments as compared to overlying waters. Table 3-16 shows that poliovirus 1, coxsackie virus B3, coxsackie virus A9, and echovirus 1 survive an average of 6 days in an estuarine water but an average of 14 days in the underlying sediments. Such sediment survival can constitute a "reservoir of virulence" in recreational lakes, rivers, seashores, and so on.

Viruses and Water Treatment

Under optimum conditions, virus removal via water treatment can be very good. Pilot-scale systems can demonstrate greater than 99.999 percent removal using coagulation, sedimentation, filtration, carbon adsorption, and chlorine disinfection (Guy et al., 1977). However, removals in the field are normally much lower than in controlled laboratory conditions.

Chemical coagulation acts to adsorb viruses so they may settle out during subsequent sedimentation and/or become entrained in filter media in plants without sedimentation. As was indicated earlier, viruses can survive substantial periods of time in this adsorbed state. As shown in Table 3-17, laboratory coagulation tests "removed" between 88 and 99.8 percent of incoming virus levels.

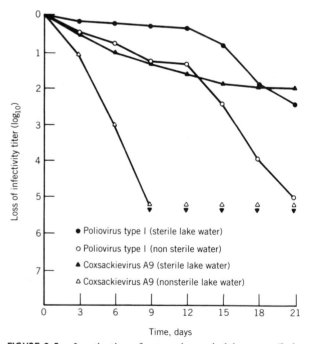

FIGURE 3-5. Inactivation of enteroviruses in lake water (Lake Wingra, WI) (Bitton, 1980).

TABLE 3-14. Survival of Viruses in Natural and Finished Waters

Agent	Initial Concentration (mL^{-1})	Type of Water	Temp. (°C)	Time for Infectivity Loss (days), T_{100}[a]
Poliovirus 2	10^4	River	16–20	29–35
Coxsackie virus B3	10^4	River	16–20	29–35
Echovirus 7	10^4	River	16–20	29–35
Coxsackie virus A4	0.32	River	4–8	>150
Coxsackie virus A4	32	River	20–22	>45
Echovirus 6, 11, 30, 33	$10^{5.1}$–$10^{8.1}$	Diluted river	8	>560
Echovirus 6, 11, 30, 33	$10^{5.1}$–$10^{8.1}$	Diluted river	20	70–322
Echovirus 7	$10^{2.3}$	Well	20	66
Echovirus 7	$10^{2.3}$	Well	10	113
Coxsackie virus B3	$10^{2.3}$	Well	20	66
	$10^{2.3}$	Well	10	1137
Echovirus 7	10^3	River	4–6	90
Vibrio cholerae Inaba	10^6	Spring	Room	<0.04
Vibrio cholerae Inaba	10^6	Tap (Calcutta)	Room	<0.75
Vibrio cholerae Inaba	10^6	River	Room	<0.75
V. cholerae	Unknown	Tank	Unknown	1–2
V. cholerae El Tor	Unknown	Tank	Unknown	<8
Yersinia enterocolitica	Unknown	Sterile	Unknown	157
Campylobacter jejuni	10^7–10^9	Stream (autoclaved)	25	4

[a] Time for 100% infectivity loss.
SOURCE: Berger (1983).

TABLE 3-15. Infectivity of Viruses in the Adsorbed State

Virus	Infectious Virus (%)
Virus adsorbed to bentonite clay[a]	
T2	82
T7	76
f2	3
Poliovirus 1	93
Virus adsorbed to magnetite[b]	
T2	75
MS2	100
Poliovirus 1	100
Natural solids from lakes[c]	
Encephalomyocarditis virus (EMC)	87–138

[a] Viruses adsorbed to bentonite in the presence of 0.01 M CaCl$_2$. Data adapted from Moore et al. (1975).
[b] Virus adsorbed to magnetite in the presence of 1610 ppm of CaCl$_2$. Data adapted from Bitton et al. (1976).
[c] Adapted from Schaub and Sagik (1975).
SOURCE: Bitton (1980).

TABLE 3-16. Survival of Selected Enteroviruses in Estuarine Water and Sediment

Virus	Maximum Length of Virus Survival (Days)	
	Water	Sediment
Poliovirus 1	10	14
Coxsackie virus B3	4	18
Coxsackie virus A9	2	4
Echovirus 1	7	>18

SOURCE: Bitton (1980).

TABLE 3-17.   Laboratory Tests of Virus Removal

Agentwater	Type of Water	Coagulant	Dose	Removals	
				Virus	Turbidity
Coxsackie A2	River	Alum	25	95	95
Coxsackie A2	River	$FeCl_3$	15	95	80–90
Poliovirus 1,2,3	River	$Fe(SO_4)_3$	40	99.8	—
Naturally occurring enterics	River	$Fe(SO_4)_3$	40	88.3	—
Poliovirus 1	—	Alum	10	86	96
Phage T4	—	Alum	25.7	98	99
Phage MS2	—	Alum	25.7	99.8	98

SOURCES: Berger (1983) and Bitton (1980).

TABLE 3-18.   Virus Reduction by Filtration

Agent	Type of Filter	Aerial Loading Rate	Initial Concentration	Percent Removal
Poliovirus	Slow sand	0.008 gpm/ft^2	10^6/L	98
	Slow sand	12 m/day	Unknown	98.25 (5°C)
	Slow sand	4.8 m/day	Unknown	99.999 (11°C)
	Sand	2 gpm/ft^2	10^7/L	10–58
	Sand	2–6 gpm/ft^2	10^7/L	98[a]
Coxsackie virus A5	Sand	0.2 gpm/ft^2	8/L	98
Coxsackie virus A5	Sand	2.0 gpm/ft^2	8/L	10
Poliovirus 1	Sand	Unknown	10^4/L	19–37.5[b]

[a] Alum pretreatment.
[b] Ferric pretreatment.
SOURCE: Berger (1983).

Removal of viruses in filters is highly variable and depends on the filter design and operation as well as on the type of pretreatment provided. Table 3-18, for example, illustrates the variability obtained under differing filter conditions. Table 3-19 illustrates that sand filtration alone, without prior

TABLE 3-19.   Removal of Poliovirus by Rapid Sand Filters[a]

Treatment	Percent Removal
A. Sand filtration	1–50
B. Sand filtration + coagulation with alum	
1. Without settling	90–99
2. With settling	>99.7

[a] Flowrate 2–6 gpm/ft^2.
SOURCE: Bitton (1980).

coagulation, removes only 1–50 percent of the incoming virus load. This is because the sand has no affinity for the virus, and the virus particles are small enough to pass through the filter pores. When the virus is adsorbed to alum or other polymers, entrainment within the filter can occur.

Water softening provides two potential mechanisms for virus removal, adsorption to precipitated $CaCO_3$ and/or $Mg(OH)_2$ and an increase in pH. As Table 3-20 illustrates, lime softening alone, with $CaCO_3$ precipitation and moderate pH values, does not result in substantial virus removals. $CaCO_3$ bears a slight negative charge and would not be expected to adsorb the negatively charged particles very well. With the lime soda ash process, however, the slightly positively charged $Mg(OH)_2$ precipitate adsorbs viruses into the precipitate, and the higher pH (generally >10.5) inactivates the virus through protein coat denaturization. Removals on

TABLE 3-20. Removal of Viruses by Water Softening Processes

	Initial Conditions Hardness as CaCO$_3$ (mg/L)		Final Conditions			
			Hardness as CaCO$_3$			Removal of Poliovirus (%)
	mgCl$_2$	Ca(HCO$_3$)$_2$	Mg	Ca	pH	
Lime softening		100	—	64	9.0	9
		300	—	218	8.1	70
Lime–soda ash process						
	67	133	50	76	10.8	99.90
	100	300	82	204	11.2	99.993

SOURCE: Bitton (1980).

the order of 99 percent can be expected with this latter process.

Virus removal via activated-carbon adsorption is relatively low. A study conducted with Trent River water in England showed a 21–86 percent removal range for attenuated poliovirus and phage T4. Such

TABLE 3-21. Relative Resistances of 20 Human Enteric Viruses to 0.5 mg/L Free Chlorine in Potomac Water (pH 7.8 and 2°C)

Comparison Based on First-Order Reaction		Comparison Based on Experimental	
Virus	Min[a]	Virus	Min[a]
Reo 1	2.7	Reo 1	2.7
Reo 3	<4.0	Reo 3	<4.0
Reo 2	4.2	Reo 2	4.2
Adeno 3	4.8	Adeno 3	<4.3
Cox A9	6.8	Cox A9	6.8
Echo 7	7.1	Echo 7	7.1
Cox B1	8.5	Cox B1	8.5
Echo 9	12.4	Echo 9	12.4
Adeno 7	12.5	Adeno 7	12.5
Echo 11	13.4	Echo 11	13.4
Adeno 12	13.5	Polio 1	16.2
Echo 12	14.5	Echo 29	20.0
Polio 1	16.2	Adeno 12	23.5
Cox B3	16.2	Echo 1	26.1
Polio 3	16.7	Polio 3	30.0
Echo 29	20.0	Cox B3	35.0
Echo 1	26.1	Cox B5	39.5
Cox A5	33.5	Polio 2	40.0
Cox B5	39.5	Cox A5	53.5
Polio 2	40.0	Echo 12	>60.0

[a] Minutes required to kill 99.99% of virus.
SOURCE: Adapted from Liu et al. (1971).

poor removal was attributed to competition between the virus and soluble organics in the water for adsorption sites.

Disinfection remains the most certain method for virus inactivation in water treatment (see Chapter 12). Relative effectiveness depends, however, on the organism itself, the type of disinfectant used, and the contact time between the viruses and the disinfectant. Table 3-21 illustrates the resistances of a variety of enteric viruses to 0.5 mg/L free chlorine in Potomac River water. As shown, from 2.7 to >60 min were required to inactivate 99.99 percent of the virus types. This is particularly significant when it is noted that bacterial indicators would be inactivated at the low end of the scale. Thus, as has been demonstrated in the field, coliform-free water does not necessarily indicate virus-free water.

ALGAE

Algae play important roles in reservoirs and lakes in the cycling of nutrients and other constituents as part of a food chain that progresses beyond algae, through zooplankton and fish, to birds, animals, and man and in establishing a water's taste, odor, and trophic status. A few are pathogenic to man, producing endotoxins that can cause gastroenteritis. Still others interfere with treatment plant operations, specifically with filter operations.

Discussion of algal roles in nutrient cycling and as the basis of food chains is outside the framework of this text. Complete descriptions of these roles can be found in limnology texts such as Hutchinson (1957) and Wetzel (1975). The following sections provide an introduction to algal ecology and nomenclature and discuss algae impacts on lake trophic status, taste, and odor, summarize what is currently known about algal endotoxins and indicate some of the problems algae may cause in treatment plant filter operations.

Algae Ecology and Nomenclature

Algae have limited powers of locomotion. Some use flagella or buoyancy mechanisms to achieve limited mobility, a few filamentous forms inhabit shorelines, and some others attach to bottom substrates. Most, however, are "free floating." Actually algae sink, since they are slightly more dense than water.

Certain groups of algae are often classified as other organisms. The blue-green algae, for example, are sometimes grouped with the bacteria because of their procaryotic structure. Some of the flagellated algae are grouped with the protozoa. All algae, however, are photoautotrophic, using photosynthesis as their primary mode of nutrition and as the basis for synthesis of new organic matter.

There are 10 basic algal groups, five of which contain members commonly associated in reservoir and/or lake studies. Table 3-22 indicates the 10

TABLE 3-22.   Major Algal Groupings and Common Members

Group Name	Common Members
Blue-greens (Cyanophyta)	*Anacystis (Microcystis)*
	Oscillatoria
	Aphanitomenon
	Anabaena
Green algae (Chlorophyta)	*Chlamydomonas*
	Sphaerocystis
	Volvox
	Scenedesmus
	Oocystis
	Selenastrum
Diatoms	*Asterionella melosira*
	Diatoma cyclotella
	Fragilaria navicula
	Synedra
Dinoflagellates	*Ceratium*
Yellow-greens (Xanthophyceae)	*Peridinium*
Golden browns (Chysophyceae)	*Dinobryon*
Cryptomonads	*Peridinium*
Euglenoide	—
Brown algae (Phalephyta)	—
Red algae (Rhodophyta)	—

SOURCE: Adapted from Wetzel (1975).

groups and lists various members of interest to the water quality engineer.

Algal Associations with Lake Trophic Status

The trophic level, or "fertility," of a body of water refers to the amounts of nutrients and organic matter being cycled through it. An oligotrophic lake is one with a low level of nutrients and organic matter. Often the water in such lakes appears very clear and free of plant life. Mesotrophic refers to a moderate amount of nutrient input with correspondingly moderate amounts of plant and animal life. A eutrophic lake is one through which large amounts of nutrients and organic matter are being cycled and which supports substantial plant life.

Progression of a lake from oligotrophic to eutrophic is a natural occurrence. Ultimately, the lake becomes clogged with organic sediments and plants and becomes dry land. With no external input, such a progression occurs over a very long time frame. Where excessive nutrient inputs are allowed, however, rapid eutrophication may occur. The term *cultural eutrophication* applies to this occurrence.

Algae speciation at these differing trophic levels is largely dictated by each strain's nutrient uptake capabilities and requirements. Table 3-23 indicates some of the major algae groups that might be expected to occur under a given trophic level and set of water characteristics. Seasonal variation in populations, micronutrient levels, and predator grazing intensity will create great shifts in these general trends, but they are useful as a baseline.

It is important to recognize that the algae are not responsible for a lake's trophic status. They are merely an indication of the degree of fertilization that has been occurring. Thus, applying copper sulfate to reservoirs to control algae blooms often treats the symptom rather than the disease. On the other hand, controlling inputs of nutrients such as nitrogen and phosphorous can act to slow the eutrophication process.

Various measures including analyses of nitrogen and phosphorous, algae activity or productivity, organic carbon, and so on, are used to gauge lake or reservoir trophic status. Table 3-24 summarizes the ranges of values likely to be cited in identifying such status. The first four measures provide an overall indication of algal activity. As was suggested in Table 3-23, identifying the dominant phytoplankton groups may provide an indication of trophic status.

TABLE 3-23. Characteristics of Common Major Algal Associations of the Phytoplankton in Relation to Increasing Lake Fertility

General Lake Trophy	Water Characteristics	Dominant Algae	Other Commonly Occurring Algae
Oligotrophic	Slightly acidic; very low salinity	Desmids *Staurodesmus*, *Staurastrum*	*Sphaerocystis, Gloeocystis, Rhizosolenia, Tabellaria*
Oligotrophic	Neutral to slightly alkaline; nutrient-poor lakes	Diatoms, especially *Cyclotella* and *Tabellaria*	Some *Asterionella* spp., some *Melosira* spp., *Dinobryon*
Oligotrophic	Neutral to slightly alkaline; nutrient-poor lakes or more productive lakes at seasons of nutrient reduction	Chrysophycean algae, especially *Dinobryon*, some *Mallomonas*	Other chrysophyceans, e.g., *Synura, Uroglena;* diatom *Tabellaria*
Oligotrophic	Neutral to slightly alkaline; nutrient-poor lakes	Chlorococcal *Oocystis* or chrysophycean *Botryoccocus*	Oligotrophic diatoms
Oligotrophic	Neutral to slightly alkaline; generally nutrient poor, common in shallow Arctic lakes	Dinoflagellates, some *Peridinium* and *Ceratium* spp.	Small chrysophytes, cryptophytes, and diatoms
Mesotrophic or entrophic	Neutral to slightly alkaline; annual dominants or in eutrophic lakes at certain seasons	Dinoflagellates, some *Peridinium* and *Ceratinum* spp.	*Glenodinium* and many other algae
Eutrophic	Usually alkaline lakes with nutrient enrichment	Diatoms much of year, especially *Asterionella*, spp. *Fragilaria crotonensis, Synedra, Stephanodiscus,* and *Melosira granulata*	Many other algae, especially greens and blue-greens during warmer periods of year; desmids if dissolved organic matter is fairly high
Eutrophic	Usually alkaline; nutrient enriched; common in warmer periods of temperate lakes or perennially in enriched tropical lakes	Blue-green algae, especially *Anacystis* (=*Microcystis*), *Aphanizomenon, Anabaena*	Other blue-green algae; euglenophytes if organically enriched or polluted

SOURCE: Wetzel (1975).

The total phosphorous and total nitrogen measures provide a fairly direct indication of lake productivity, while the remaining three measures, light extinction, TOC, and inorganic solids, are measures that correlate with lake productivity as well as other factors [algae, taste, and odor (see Chapter 17)].

Algal Endotoxins

Endotoxins are present in gram-negative bacteria and in some strains of at least four species of blue-green algae. *Anabaena flos-aquae, Microcystis aeruginosa, Schizothrix calcicola,* and *Aphanizomenon flos-aquae* have all been associated with the production of endotoxins. Blooms of these species may or may not include the toxins depending on which strains are present. Poisoning of wild and domestic animals is often the first indication that a bloom is toxic. Birds and animals ingesting water containing the algae die within hours or days. Human deaths have not been reported as due to ingestion of endotoxins; however, treated water containing residual endotoxin levels have been associated with gastroenteritis outbreaks. The largest to date occurred in 1975 in Sewickly, Pennsylvania, where 62 percent of the population became ill.

Algae and Filter Clogging

Algae that pass through preliminary treatment processes and become trapped among the spaces in a filter bed can cause gradual or rapid loss of head. Although effective coagulation and sedimentation can remove up to 90 or 95 percent of the incoming algae, the remainder may be sufficient to significantly shorten filter runs, even to the extent that the amount of water required to backwash the filter is greater than the amount of filtered water produced.

TABLE 3-24. General Ranges of Primary Productivity of Phytoplankton and Related Characteristics of Lakes of Different Trophic Categories

Trophic Type	Mean Primary Productivity[a] (mg C · m⁻²/day⁻¹)	Phytoplankton Density (cm³/m³)	Phytoplankton Biomass (mg C/m³)	Chlorophyll a (mg/m³)	Dominant Phytoplankton[b]	Total P (μg/L)	Total N (μg/L)	Light Extinction Coefficients (η/m)	Total Organic Carbon (mg/L)	Total Inorganic Solids (mg/L)
Ultraoligotrophic	<50	<1	<50	0.01–0.5		<1–5	<1–250	0.03–0.8	—	2–15
Oligotrophic	50–300	—	20–100	0.3–3	Chrysophyceae, Cryptophyceae	—	—	0.05–1.0	<1–3	—
Oligomesotrophic	—	1–3	—	—	Dinophyceae, Bacillariophyceae	5–10	250–600	—	—	10–200
Mesotrophic	250–1000	—	100–300	2–15		—	—	0.1–2.0	<1–5	—
Mesoeutrophic	—	3–5	—	—		10–30	500–1100	—	—	100–500
Eutrophic	>1000	—	>300	10–500	Bacillariophyceae, Cyanophyceae	—	—	0.5–4.0	5–30	—
Hypereutrophic	—	>10	—	—	Chlorophyceae, Euglenophyceae	30 to >5000	500 to >15,000	—	—	400–6000
Dystrophic	<50–500	—	<50–200	0.1–10		<1–10	<1–500	1.0–4.0	3–30	5–200

[a] Referring to approximately net primary productivity, such as measured by the ^{14}C method.
[b] Dominant phytoplankton associated with oligotrophy and eutrophy.

SOURCE: Wetzel (1975).

52

FIGURE 3-6. Filter-clogging algae (linear magnifications in parentheses): *Anabaena flos-aquae* (500); *Anacystis dimidiata* (1000); *Asterionella formosa* (1000); *Chlorella pyrenoidosa* (5000); *Closterium moniliferum* (250); *Cyclotella meneghiniana* (1500); *Cymbella ventricosa* (1500); *Diatoma vulgare* (1500); *Dinobryonsertularia* (1500); *Fragilaria crotonensis* (1000); *Melosira granulata* (1000); *Navicula graciloides* (1500); *Oscillatoria princeps* (top) (250), *O. chalybea* (middle) (250), *O. splendida* (bottom) (500); *Palmella mucosa* (1000); *Rivularia dura* (250); *Spirogyra porticalis* (125); *Synedra acus* (500); *Tabellaria flocculosa* (1500); *Trachelomonas crebea* (1500); *Tribonema bombycinum* (500) (Palmer, 1959).

It should be noted that in slow sand filters, algae and their microorganisms layered over the surface may actually aid in treatment. The algae photosynthesize and release oxygen, which is utilized by associated bacteria, fungi, and protozoans. These latter organisms can then break down incoming organic matter and hence reduce filtered water organic levels (Palmer, 1959).

Nuisance filter-clogging algae include diatoms whose rigid cell walls prevent easy passage through filter media. Common problem diatoms include *Asterionella, Fragillaria, Tabellaria,* and *Synedra.* Various members of the blue-green algae, the green algae, and the golden browns also cause problems. *Palmella,* one of the green algae, forms copious mucilaginous material around its cells and literally gums up the filter bed. Figure 3-6 illustrates some of the more common filter-clogging algae.

PROTOZOANS

The protozoans are a group of unicellular, nonphotosynthetic organisms, probably derived from various groups of unicellular algae. Most are motile, moving via use of flagellae, amoeboid locomotion, or cilia. Figure 3-7 illustrates typical flagellate, amoebic, and ciliated protozoans. Several protozoans are parasites, and five are human parasites potentially transmitted via contaminated water. Table 3-25 lists these five protozoans and describes their associated diseases and modes of transmission.

Protozoans in Water

As shown in Table 3-25, two of the pathogenic protozoans have water and soil as their natural habitats. Their survival time in water is thus not an issue. *Naegleria fowleri* is a free-living amoeba found in soil, water, and decaying vegetation. Surveys have shown this organism to be ubiquitous in southeastern freshwater lakes, although the risk of infection from swimming in such lakes has been estimated at <1 in 2.5 million exposures. *Acanthamoeba* is another free-living amoeba found in both freshwater and sewage effluents.

Two others of the protozoans in Table 3-25 are able to survive long periods of time in water as cysts. *Giardia lamblia* and *Entamoeba* cysts can survive up to 3 months in water. While the last outbreak of *Entamoeba histolytica* in the United States was in 1953, giardiasis appears to be on the in-

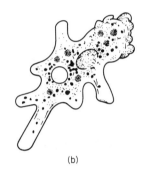

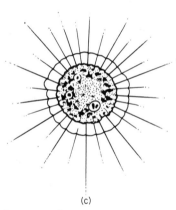

FIGURE 3-7. Representative protozoans. (a) Flagellated protozoan of the trypanosome group. One strain is the agent of African sleeping sickness, transmitted by the tsetse fly. (b) Ameboid protozoan. (c) Ciliate protozoan (Stanier, 1970).

crease. During 1951–1970 only 12 cases of giardiasis per year were reported. From 1971 through 1981, however, 50 outbreaks comprising nearly 20,000 cases have been recorded. While some fraction of this increase is due to increased awareness of the problem and better reporting, giardiasis is in fact more common most likely because of increased pressure on watershed areas without concurrent upgrading of the systems providing treated water from these watersheds. Most giardiasis outbreaks have been associated with small municipal systems or semipublic utilities where treatment may be inade-

TABLE 3-25.   Human Parasitic Protozoans

Protozoan	Host(s)	Disease	Transmission	Occurrence
Acanthamoeba castellani	Fresh water, sewage, man	Amebic meningoencephalitis	Gains entry through abrasions, ulcers, and as secondary invader during other infections	
Balantidium coli	Pigs, humans	Balantidiasis (dysentery)	Contaminated water (pig feces)	Micronesia has been the only known site of an outbreak
Entamoeba histolytica	Humans	Amoebic dysentery	Contaminated water	Last United States outbreak 1953
Giardia lamblia	Animals, humans	Giardiasis (gastroenteritis)	Contaminated water	Mexico, United States, USSR
Nagleria fowleri	Soil water, decaying vegetation	Primary amoebic meningoencephalitis	Nasal inhalation with subsequent penetration of nasopharynx; exposure from swimming in freshwater lakes	

quate or less consistent than in large systems associated with more developed areas. Figure 3-8 shows some scanning electron microscopy and phase micrographs of *Giardia* cysts taken from humans.

The fifth pathogenic protozoan listed in Table 3-25 has been associated with only one waterborne disease outbreak. *Balantidium coli* is a ciliated protozoan and primarily a parasite of pigs. The known waterborne outbreak occurred in Micronesia and was attributed to contamination of a water supply with pig feces after a typhoon.

Protozoans and Water Treatment

Drinking water does not seem to be a significant route of transmission for *Naegleria*, *Balantidium*, and *Acanthamoeba* (Berger, 1983). On the other hand, the pattern of giardiasis outbreaks in the United States suggests that consumption of untreated or inadequately treated drinking water is an important route of infection. Similarly, *Entamoeba histolytica* is widely distributed in the population, with prevalence rates ranging from 3 to 10 percent in the United States and an annual reported case rate of amoebiasis of 2000–3000. Such cases have also been associated with untreated or less than adequately treated supplies.

Both *Giardia* and *Entamoeba* can be controlled through water treatment. Filtration can remove >99 percent of *Entamoeba histolytica* and *Giardia* cysts

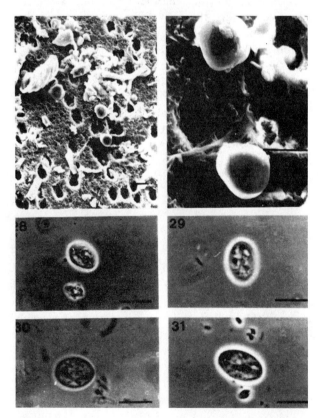

FIGURE 3-8.   *Giardia* cysts from humans. 26, SEM micrograph of several *Giardia* cysts from human material on an 8-μm pore membrane at 750×; 27, same material magnified to 3000×; 28, 29, 30, 31, phase micrographs of cysts of the same sources at 2000× (EPA, 1979).

TABLE 3-26. USPHS Minimum Cysticidal and
Bactericidal Residuals[a] (after 30 mins Contact)

pH Value	Free Chlorine, HOCl + OCl⁻		
	Bactericidal, 0–25°C	Cysticidal, 22–25°C	Cysticidal, 2–5°C
6.0	0.2	2.0	7.5
7.0	0.2	2.5	10.0
8.0	0.2	5.0	20.0
9.0	0.6	20.0	70.0

[a] Concentration required for inactivation of 99.999% of cysts.
(See Palin, A. T., as reported by J. C. Hoff in the EPA book on
giardiasis (1979).

with optimal coagulant doses and stable filter operation. However, poor cyst removal has been noted when coagulant doses are low or less than optimal, when filtered water turbidities increase during filter runs, and during filter ripening periods (Logsdon, 1981).

Similarly, chlorine and other disinfectants can be effective against these cysts; however, typically longer contract time and/or higher doses are required than are necessary to meet coliform drinking water standards. Table 3-26 is illustrative. As shown, at pH 6 a chlorine residual of 0.2 mg/L after 30 minutes of contact is sufficient to ensure bacteria-free water, while a residual of 2.0 mg/L at 22–25°C and a residual of 7.5 mg/L at 2–5°C is required to provide cyst-free water. As the pH increases, the differences between the chlorine required for bactericidal and cysticidal action increases dramatically. (See Chapter 12.)

HELMINTHS

Parasitic worm infections are not a widespread problem in the United States. However, some types do still regularly occur, particularly in southeastern United States. Ascariasis, an infection of the small intestine, occurs worldwide and is common, especially among children in southern United States. In more tropical countries and/or in countries with less developed sewage treatment systems, certain types of helminth infections are endemic. Schistosomiasis, a disease caused by *Schistosoma* infections of the liver or urinary system is endemic in parts of Africa and also occurs in the Arabian peninsula, South America, the Middle East, India, and the Orient.

Transmission of helminths occurs in a variety of ways: via contaminated drinking water or vegetables and through body contact with contaminated irrigation water and/or soils. Table 3-27 summarizes information on some of the common helminths in the United States and worldwide, and Figure 3-9 shows the eggs of two varieties. The primary mode of control is through effective sewage treatment and disposal. If contamination of a drinking water supply does occur, conventional treatment will generally remove the helminth eggs, which are denser than water and larger than filter pores. However, some of the adult worm stages can pass through the pores of filters. For schistosomiasis, control of the intermediate snail host population can be effective.

CHARACTERIZATION OF MICROBIAL QUALITY

A variety of methods exist to characterize or measure microbial water quality. They include methods to measure incidence of waterborne disease, indicate the safety of a potable water supply, ascertain reservoir trophic status, and so on. Techniques and issues associated with common microbial water quality measures are described below.

Measurement of Waterborne Disease Outbreaks

In the United States the Centers for Disease Control (CDC) keeps track of outbreaks of waterborne diseases, as reported by local and state public health agencies. The Health Effects Research Laboratory of the Environmental Protection Agency and its predecessor, the U.S. Public Health Service (USPHS), maintain similar records. Figure 3-10 shows the reported incidence of waterborne disease outbreaks since 1920. By definition, an outbreak consists of two or more cases of a disease. This indicates a common source that can be investigated. In most of the waterborne outbreaks reported, the suspect water can be found to be bacteriologically or chemically contaminated, but in only a few can the etiologic agent be isolated (Craun, Waltrip, and Hammonds, 1983).

It is significant to note an apparent steady increase in the number of outbreaks that have occurred since 1950. Without doubt some portion of this rise is due to increased awareness and consequent reporting on the part of the general public as well as public health officials of the potential role of water in transmitting disease. Associated improve-

TABLE 3-27.   Characteristics of Helminths which Infect Man

Agent	Host(s)	Disease	Transmission	Occurrence
Ascaris lumbricoides (intestinal roundworm)	Man	Ascariasis—moderate infections cause digestive and nutritional problems, abdominal pain, and vomiting; live worms passed in stools or vomited; serious cases involving liver can cause death	Ingestion of infected eggs from soil, salads, and vegetables contaminated with eggs from human feces	Worldwide—especially in moist tropical areas, where prevalence can exceed 50%; in United States disease is most common in the south
Schistosoma mansoni, S. haematobium, S. intercalatum, S. japonicum	Man, domestic animals, and rats serve as primary hosts; snails act as necessary intermediate host	Schistosomiasis—a debilitating infection where worms inhabit veins of host; chronic infection affects liver or urinary system	Water infected with cercariae (larvae) which have developed in snails; penetration through human skin; eggs excreted via urine or feces and miricidium (larvae) develop in water and reinfect snails	Africa, Arabian Peninsula, South America, Middle East, the Orient, parts of India; in United States immigrants from Middle East may carry disease
Various schistosomes	Birds and rodents; man is a nonnormal host	Schistosome dermatitis (swimmer's itch); local skin dermatitis caused by penetration of larvae; larvae die in skin	Same	Widely distributed, but only locally endemic
Necator americanus or *Ancylostoma duodenale* (hookworm)	Man	Ancylostomiasis—hookworm disease; debilitating disease associated with anemia; heavy infestations can result in retardation; light infections produce few effects	Eggs from deposited feces develop into larvae which penetrates the skin; ancyclostoma can be acquired orally	Widely endemic in moist tropical and subtropical areas where disposal of human feces not adequate
Strongyloides stercoralis	Man and possibly dogs	Strongyloidiatis—intestinal infection causing cramps, nausea, weight loss, vomiting, and weakness; rarely results in death	Infective larvae in moist soils resulting from fecal contamination penetrate skin and reach digestive system via venous and respiratory systems	Similar distribution to hookworm
Trichuris trichiura (whipworm)	Man	Trichuriasis—a nematode infection of the large intestine; often without symptoms, but heavy infestations result in abdominal pain, weight loss, and diarrhea	Ingestion of eggs in soil and/or vegetables contaminated with fecal material	Worldwide—especially in warm moist environments

ments in sampling and analytical techniques allow disease outbreak sources (water, food, human contact) to be better defined. Finally, it is likely, although not demonstrated, that increasing public usage of watershed areas coupled with great variability in the degree of water treatment provided has resulted in an actual increase in waterborne disease outbreaks.

Microbial Indicators of Waterborne Disease

Isolating and identifying specific pathogenic microorganisms in water is beyond the scope of most laboratories. Numerous viruses associated with waterborne disease have yet to be isolated from water supplies at all. Because of these technical difficulties and also because the numbers of patho-

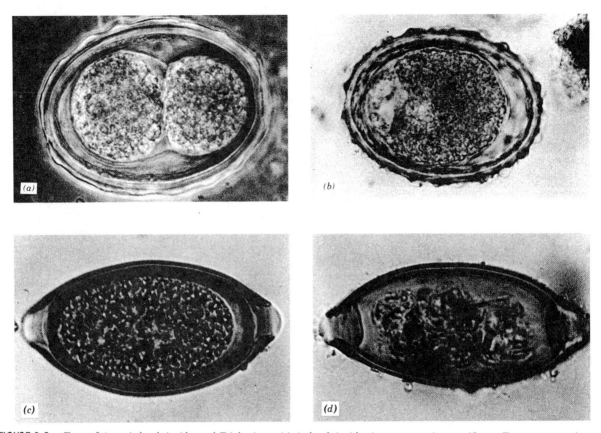

FIGURE 3-9. Eggs of *Ascaris lumbricoides* and *Trichuris* sp. (a) *A. lumbricoides* (or var. *suum*) eggs, 65 μm. From west-southwest anaerobically digested sludge; typical; two-cell stage; isolated by sucrose flotation; phase contrast; two blue filters; unstained. (b) *A. lumbricoides* (or var. *suum*) egg, 65 μm. From west-southwest anaerobically digested sludge; typical; unembryonated; isolated by sucrose flotation; bright field; two blue filters; unstained. (c) *Trichuris* sp. egg, 80 μm. From District Nu Earth; typical; unembryonated; isolated by sucrose flotation; bright field; two blue filters; unstained. (d) *Trichuris* sp. egg, 80 μm. From District Nu Earth; degenerate; unembryonated; isolated by sucrose flotation; bright field; 2 blue filters; unstained (Fox, 1981).

gens relative to other microorganisms in water can be very small, indicator organisms are utilized to measure the potential of a water to transmit disease.

Since most pathogens likely to be transmitted via the water route are shed in human and/or animal feces, the definition of an indicator organism is a microorganism whose presence is evidence of fecal contamination of warm-blooded animals. Indicators may be accompanied by pathogens but typically do not cause disease themselves. The ideal indicator organism should have the following characteristics (NAS, 1977):

1. Be applicable to all waters.
2. Be present in sewage and polluted waters when pathogens are present.
3. Numbers should correlate with the degree of pollution.
4. Be present in greater numbers than pathogens.
5. There should be no aftergrowth or regrowth in water.
6. There should be greater or equal survival time than pathogens.
7. Be absent from unpolluted waters.
8. Be easily and quickly detected by simple laboratory tests.
9. Should have constant biochemical and identifying characteristics.
10. Hamless to man and animals.

No organism or group of organisms currently meets all of these criteria; but the coliform bacteria fulfill

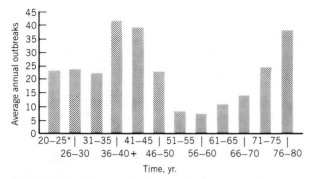

FIGURE 3-10. Reported incidence of waterborne disease outbreaks in the United States since 1920. *, Average of 6 yr; +, average of 4 yr, 1937 missing.

most of them, and this group is the most common indicator in use today. The group is operationally defined as gram-negative, nonsporing rods capable of fermenting lactose with the production of acid and gas within 48 hr at 35°C.

Drawbacks to using the coliforms as indicators of the presence or absence of disease organisms include the fact that under certain conditions coliforms can regrow in water and become part of the natural aquatic flora. Their detection then represents a false positive. False positives can also be obtained from the bacterial genus *Aeromonas*, which can biochemically mimic the coliform group. On the other hand, false-negative results can be obtained when coliforms are present along with high populations of other plate count bacteria. The latter microorganisms can act to suppress coliform activity (Allen and Geldreich, 1975). Finally, a number of pathogens have been shown to survive longer in natural waters and/or through various treatment processes than coliforms. Table 3-28 illustrates the comparative die-off rates for coliforms, enterococci (another potential indicator), and some enteric pathogens.

Because of these problems, other organisms have been suggested as alternative indicators for drinking water as well as other activities. Table 3-29 summarizes some of these bacteria and their potential indicator applications.

Enumeration of Algae in Water Supplies

Enumeration of algae within a water supply can be useful in a variety of contexts. Maintaining regular records of numbers of taste and odor algae can aid a water supply manager in determining when potential problems are likely to occur. The total area, or volume, of algae, especially of diatoms, can aid in establishing a relationship between phytoplankton and length of filter runs. For recreational reservoirs, keeping track of the trophic status via measurement of indicator algae or other parameters (see Table 3-24) can help determine appropriate level of usage for the reservoir. Finally, scanning for sewage organisms can identify problems of contamination.

As summarized by Palmer (1959), an adequate routine procedure for treatment plants using surface water supplies would include regular inspection of the raw water supply, treatment plant, and distribution system for attached growths, floating mats, and blooms. In visible growths, identification and enumeration of dominant organisms should be made. In addition, regular plankton analyses of water samples from these areas should be made. When supplemented with physical and chemical water quality data, a useful record of algal activity and/or problems can be established.

TABLE 3-28. Comparative Die-off Rates of Indicator Bacteria and Enteric Pathogens in Water

Bacteria	Half-Time[a] (hr)	Number of Strains Analyzed
Indicator bacteria	17.0	29
Coliform bacteria (avg.)	22.0	20
Enterococci (avg.)	19.5	
Coliform from raw sewage	19.5	
Streptococci from raw sewage	19.5	
Pathogenic bacteria		
Shigella dysenteriae	22.4	1
Sh. sonnei	24.5	1
Sh. flexneri	26.8	1
Salmonella enteritidis ser. paratyphi A	16.0	1
S. enteritidis ser. typhimurium	16.0	1
S. typhi	6.0	2
S. enteritidis ser. paratyphi B	2.4	1

[a] The time–time was determined graphically from the time required for a 50% reduction in the initial population.

SOURCE: McFeters et al. (1974).

TABLE 3-29.   Relationship among Tests for Various Indicators and Pathogens and Water Sources[a]

Indicator/ Pathogen	Drinking Water	Effluent Discharge	Swimming Pool	Recreational Beach Waters	Shellfish Waters	Irrigation Waters	Comments
Coliform	****						Still one of the best indicators for drinking water safety, but should be supported by heterotroph plate counts
Fecal coliform (*E. coli*)	*[b]	****	**	****	****	****	Indicator of stream and lake pollution. *E. coli* proposed to replace fecal coliform for primary contact water quality standard
Fecal streptococci		*		***	***		Main purpose is to try to establish source of pollution
Pseudomonas aeruginosa		*	***	***			Simple reliable test; organism is man and sewage related
Clostridium perfringens	*[b]			*[c]	**		Historical pollution and tracer
Bifidobacterium				**	**[d]		New indicator—to establish validity and specificity of pollution source.
Coagulase positive staphylococci			***	**			Pathogen which may occur in grossly contaminated waters.
Salmonella	*[b]	*			**[d]	*	Pathogen which may occur in grossly contaminated waters. Levels are significantly lower than indicator organisms.
Vibrios					***[d]		*Vibrio parahaemolyticus* emphasized, NAG vibrios considered
Fecal sterols	*[b]	**		**	**	*	Absolute indicator of fecal material
Candida albicans		*	**	**			Simple reliable test—organisms related to human activity

60

TABLE 3-29. (*Continued*)

Indicator/ Pathogen	Drinking Water	Effluent Discharge	Swimming Pool	Recreational Beach Waters	Shellfish Waters	Irrigation Waters	Comments
Fungi				*c			To collect data on relationship between occurrence of organisms and skin infections
Viruses	**	*	*	*	***d	*	May have greater role in drinking water safety when technology is simplified
Phage (bacteria)	*e	*		*			May be relationship between trophic status and phage.
Heterotrophs	****	*		**			Indicator of productivity of water.

a Indicator rating: ****, regularly = daily; ***, routinely = once weekly; **, occasionally = 10–20 times per year; *, special problems studies.
b Well waters or tracing infection.
c Sediments and/or sands.
d Shellfish, water, and sediments.
e Ensuring water main safety after breakdowns.

REFERENCES

Acton, J. D., Kuera, L. S., Myrvik, Q. N., and Weiser, R. S., *Fundamentals of Medial Virology,* Lea and Febiger, Philadelphia (1974).

AWWA, *Handbook of Taste and Odor Control Experiences in the US and Canada,* AWWARF, Denver CO (1976).

Berger, P. S., and Argaman, Y., "Assessment of Microbiology and Turbidity Standards of Water," USEPA 570/9-83-001 (July 1983).

Bitton, G., *Introduction to Environmental Virology,* Wiley, New York (1980).

Bitton, G., Gifford, G. E., and Pancordo, O. C., Publication No. 40, Florida Water Resources Research Center (1976).

Cabelli, V., "New Standards for Enteric Bacteria." In: R. Mitchell, Ed. *Water Pollution Microbiology,* Vol. 2. Wiley, New York (1978).

California State Department of Health, *A Manual for the Control of Communicable Diseases in California,* California State Department of Health (1975).

Craun, G. F., Waltrip, S. C., and Hammonds, A. F., "Waterborne Outbreaks in the United States 1971–81," USEPA, MERL, 1983.

EPA, *Waterborne Transmission of Giardiasis,* EPA-600/9-79-001 (June 1979).

Fox, C. J., Fitzgerald, P. R., and Lue-Hing, C., *Sewage Organisms: A Color Atlas,* The Metropolitan Sanitary District of Greater Chicago, Chicago, IL (1981).

Geldreich, E., et al., "Bacterial Dynamics in a Water Supply Reservoir: A Case Study," *JAWWA,* **xx,** 31–40 (1980).

Goldman, C. R., and Horne, A. J. *Limnology,* McGraw-Hill, New York (1983).

Guy, M. D., McIver, J. D., and Lewis, M. J., "The Removal of Virus by a Pilot Treatment Plant," *Water Res.,* **11,** 421–428 (1977).

Grabow, W. D. K., "The Virology of Waste Water Treatment," *Water Res.,* **2**(10), 675 (1968).

Hoadley, A. W., and Dutka, B. J., *Bacterial Indicators, Health Hazards Associated with Water,* ASTM Special Technical Publication 635, American Society for Testing and Materials, Philadelphia (1977).

Hutchinson, G. E., *A Treatise on Limnology,* John Wiley & Sons, New York (1957).

Larson, T. E., "Deterioration of Water Quality in Distribution Systems," *JAWWA,* **58,** 1307–1316 (1966).

Leong, L. Y. C., "Removal and Inactivation of Viruses by Treatment Processes for Potable Water and Wastewater: A Review," *Water Sci. Tech.,* **15,** 91–114 (1983).

Logsdon, G. S., Symons, J. M., Haye, R. L., and Arozarena, M. M., "Alternative Filtration Methods for Removal of *Giardia* Cysts and Cyst Models." *JAWWA,* **73,** 111–118 (1981).

Liu, O. C., et al. *Virus and Water Quality; Occurrence and Control,* 13th Water Quality Conference, University of Illinois, Urbana-Champaign (1971).

McFeters, G. A., et al., "Comparative Survival of Indicator Bacteria and Enteric Pathogens in Well Water," *Appl. Microbiol.*, **27**, 823–829 (1974).

Mitchell, R., *Water Pollution Microbiology*, Wiley, New York (1972).

Mojadjer S., and Mehrabian, S., "Studies on the Survival of *S. flexeri* in River and Tap Waters," *Arch. Roum. Exp. Microbiol.*, **34**, 307–312 (1975).

Moore, B. E., Sagik, B. P., and Malina, J. F., Jr., *Water Res.*, **9**, 197–203 (1975).

National Research Council, *Drinking Water and Health,* Vol. 2, National Academy Press, Washington, DC (1980).

National Research Council, *Drinking Water and Health,* Vol. 4, National Academy Press, Washington DC (1982).

Palmer, C. M., "Algae in Water Supplies," U.S. Public Health Service Publication No. 657 (1959).

Rheinheimer, G. *Aquatic Microbiology,* 2nd ed., Wiley, New York (1980).

Robeck, G. G., Clark, N. A., and Dostal, K. A., "Effectiveness of Water Treatment Processes in Virus Removal" *JAWWA,* **54**, 1275–1290 (1962).

Schaub, S. A., and Sagik, B. P., *Appl. Microbiol.*, **30**, 212–222 (1975).

Sobsey, M. D., "Enteric Viruses and Drinking Water Supplies," *JAWWA,* **67**, 414–418 (1975).

Stanier, R. Y., Doudorff, M., and Adelberg, E. A., *The Microbial World,* 3rd ed. Prentice-Hall, Englewood Cliffs, NJ (1970).

Starkey, R. L., "Transformations of Iron by Bacteria in Water," *JAWWA,* **37**, 963–984 (1945).

Sykora, J. L., and Keliti, G., "Cyanobacteria and Endotoxins in Drinking Water Supplies," in W. W. Carmichael (ed.), *The Water Environmental Algal Toxins and Health,* Plenum Press, New York (1981).

Wetzel, R. G., *Limnology,* W. B. Saunders Co., Philadelphia (1975).

—4—

Water Quality Criteria and Standards

Water quality criteria have become an important and sometimes controversial segment of the water supply field. Much of this emphasis stems from findings in the past decade linking low levels of contaminants to higher incidence of diseases such as cancer. Following the passage of the Safe Drinking Water Act in 1974, the principal responsibility for setting water quality standards shifted from state and local agencies to the federal government. As our ability to measure trace quantities of contaminants in water improves and our knowledge of the health effects of these compounds increases, we are faced with increasingly complex water quality regulations.

These regulations are vitally important to environmental engineers for a number of reasons. Standards affect the selection of raw water sources, choice of treatment processes and design criteria, range of alternatives for modifying existing treatment plants to meet current or future standards, and treatment costs. Engineers in the drinking water field should know what standards are currently applicable and what changes can be expected in the future so treatment plants can be designed and operated in compliance with the regulations and so

consumers can be assured of an acceptable quality water.

This chapter describes the mechanisms of the regulatory process, including beneficial use designation, criteria development, standard promulgation, and goal selection. Resulting water quality regulations are then presented by discussing historical background, current regulations, and future expectations. Standards described herein are current as of early 1985, but are subject to change.

THE REGULATORY PROCESS

A brief review of the regulatory process highlights the mechanisms behind it. Water quality criteria, standards, and goals are discussed. Although often used interchangeably in everyday speech, there are actual differences in the terms *criteria, standards,* and *goals*. All fit under the general category of water quality "regulation."

Water quality regulation typically proceeds in a logical stepwise fashion:

1. Beneficial uses are designated.
2. Criteria are developed.

3. Standards are promulgated.

4. Goals are set.

For various reasons, these four steps are not always followed in exactly this progression, but the above process is certainly a common procedure. Figure 4-1 illustrates the relationships of the various regulatory process steps in determining treatment for drinking water.

Beneficial Use Designation

The first step in the regulatory process is designating beneficial uses for individual water sources. Surface waters and groundwaters are typically designated by a state water pollution control agency for beneficial uses such as municipal water supply, industrial water supply, recreation, agricultural irrigation, power and navigation, and protection or enhancement of fish and wildlife. These beneficial uses are based on the quality of the water, present and future pollution sources, availability of suitable alternative sources, historical practice, and availability of treatment processes to remove undesirable constituents for a given end use.

For the purposes of this chapter, the only beneficial use that will be discussed is municipal water supply. Once a supply is designated for municipal use, the water quality criteria that apply are based on protecting human health, providing aesthetically acceptable water, and related considerations.

The designation of a source for municipal use can affect subsequent activities therein. For example, the federal Safe Drinking Water Act contains provisions for regulating underground injection, such as from brine or other fluids associated with oil or gas production, if an underground drinking water source would be endangered.

Subsequent regulations provide more stringent control where an aquifer is designated as the sole source of local drinking water. On the other hand, a state agency may determine that it is not technically or economically feasible to use a given source for drinking water; an example would be a confined aquifer heavily contaminated by landfill leachate or agricultural wastes. In that case, conscious further source degradation may be allowed, and efforts would be focused on protecting an alternative source better suited for drinking water.

Criteria Development

To protect given beneficial uses, water quality criteria have been developed by various groups to define contaminant concentrations which should not be exceeded. Until they are translated into standards through rule making or adjudication, criteria are in the form of recommendations or suggestions only and do not have the force of regulation behind them. Criteria are developed solely on the basis of data and scientific judgment without consideration of technical or economic feasibility.

Criteria are developed for different beneficial uses. For a single constituent, separate criteria could be set for drinking water (based on health effects or appearance), for waters used for fish and shellfish propagation (based on toxic effects), or for industry (based on curtailing interference with specific industrial processes).

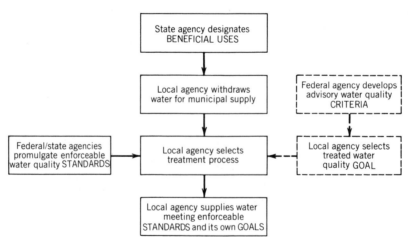

FIGURE 4-1.   Water quality regulatory process interactions.

Standard Promulgation

Water quality standards, in contrast to criteria, have direct regulatory force. Examples include the United States Environmental Protection Agency (EPA) and state standards for drinking water, which are described in more detail in the following section.

Once designation of water bodies for specific beneficial uses has been made and water quality criteria have been developed for those beneficial uses, the regulatory agency is ready to set standards. To do so, it considers water quality criteria plus other factors such as economics, technical feasibility, and political realities, as shown in Figure 4-2.

Standards may take the form of (1) process specification or treatment (such as requiring granular activated carbon or filtration), (2) product quality specification (such as maximum contaminant levels), or (3) a combination of the above (Tate and Trussell, 1977). Quality standards in the past have been based on a number of considerations, including background levels in natural waters, analytical detection limits, technological feasibility, aesthetics, and health effects.

The ideal method for establishing standards involves a scientific determination of health risks or benefits, a technical/engineering decision of costs to meet various water quality levels, and a regulatory/ political decision that weighs benefits and costs.

Goal Selection

Water quality goals are contaminant concentrations which an agency or water supplier chooses to achieve in order to ensure it consistently meets regulated levels. Goals are typically more stringent than standards and include constituents not covered by regulations but of particular importance to the goal-setting entity. For example, if the turbidity standard is 1 NTU, a utility might choose an operating goal of 0.5 NTU to ensure satisfactory performance. The American Water Works Association goals (AWWA, 1968, 1974) are examples of a professional group effort. Alternatively, an individual water supplier may elect to provide water quality that is better than required by the applicable standards or for constituents not regulated by standards. Examples include goals for turbidity or trihalomethane (THM) in treated water lower than required by regulation or goals for unregulated parameters like total organic carbon or standard plate

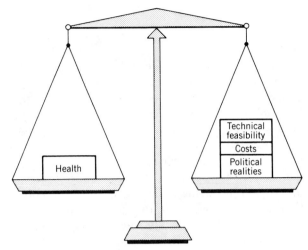

FIGURE 4-2. Factors in setting water quality standards.

counts. Decisions on setting goals involve determination of costs, benefits, and the overall philosophy or posture of a supplier.

WATER QUALITY REGULATIONS: PAST, PRESENT, AND FUTURE

The evolution of water quality regulations has followed a pattern that sheds light on current requirements. Existing regulations, both in the United States and abroad, are detailed for easy reference. Some projected trends and anticipated regulations for the future are presented.

Historical Background

The development of water quality criteria and standards, at least in a quantifiable sense, is a recent phenomenon. A history of water quality standards through 1971 has been described by the AWWA (1971). The reader is referred to that text for more detail.

The first standards in the United States were written in 1914, but there have been numerous developments since then, particularly in the last decade. Key developments have been activities prior to 1900, the actions of the U.S. Public Health Service in establishing limits that were widely followed voluntarily, water quality criteria development, and the entry of the federal government into a standards-setting role for community water supplies.

Events Prior to 1900

Historical records indicate that water quality standards, except for infrequent references to aesthetics, were notably absent from the time of ancient civilization through most of the nineteenth century. Typically, consumers relied on sensory perceptions of taste, odor, and visual clarity to judge the quality of the supply. The deficiency of this system was clearly pointed out during the London cholera epidemic in 1854, when John Snow did epidemiological investigations tracing cholera to sewage contamination in a single well, the Broad Street well. Even though the well was contaminated, some consumers traveled there specifically because they preferred its water, presumably on the basis of taste, appearance, or smell. From this example it is clear that standards need to be quantifiable and related directly to measurable water quality contaminants that could have health effects and not just the appearance or aesthetics of a supply.

After the germ theory of disease developed by Pasteur in the 1840s was recognized, the issue of drinking water contaminated from sewage was explored. Because bacteriological tests were not available until the end of the nineteenth century, the earliest quantitative measurements were chemical tests. Since it was recognized that ammonia and albumoid nitrogen from fresh sewage were gradually oxidized in receiving water to nitrites and nitrates, these forms of nitrogen were measured in drinking water in an attempt to ensure that contamination, if present, was not recent. However, this method was an indirect measure of bacterial contamination and did not serve to curtail outbreaks of waterborne disease, particularly typhoid, in the United States.

The development of a bacterial test for water supplies by Theobald Smith made is possible to directly analyze bacterial water quality. In 1892, the New York State Board of Health first applied this technique to study pollution in the Mohawk and Hudson Rivers.

Actions of the U.S. Public Health Service

The Public Health Service has had an indirect but nevertheless key role in setting water quality standards in the United States. The ability to detect bacteria, coupled with the introduction of chlorination as a disinfectant in 1907, led to the first quantitative water quality standards. In 1914, the U.S. Public Health Service, a part of the Treasury Department, adopted the first standards for drinking water supplied to the public by any common carrier engaged in interstate commerce. Maximum permissible limits were specified for bacterial plate count and *Bacillus coli* (a coliform bacteria).

Following the entry of the Public Health Service into the regulatory field, standards development proceeded rapidly. Revised standards were issued in 1925, 1942, 1946, and 1962 (U.S. Public Health Service, 1962), as shown in Table 4-1. Development of these standards, as indicated by the types of parameters included plus the maximum allowable levels, provides insight into progress in the sanitary engineering field plus response to the pertinent issues of the times.

After the initial emphasis on controlling waterborne bacteria, new parameters were added to limit exposure to other contaminants that cause acute effects, such as arsenic, or that adversely affect the aesthetic quality of the water. In 1925, a number of aesthetic parameters (color, odor, and taste) were added, along with certain minerals (chloride, copper, iron, lead, magnesium, sulfate, and zinc). Except for lead, these minerals are related to taste or aesthetics. In 1942, a number of constituents were added, including selenium, residue (dissolved solids), turbidity, fluoride, manganese, alkyl benzene sulfonate, and phenols. The latter two compounds marked the first time that specific organic constituents were covered by regulations. The 1946 standards were similar to the 1942 set except that a limit was set for another toxic constituent, chromium. Following the dawn of the atomic age, standards in 1962 included radium-226, strontium-90 and gross beta activity. Addition of an indicator of organics (carbon chloroform extract) plus additional toxic constituents (cadmium, cyanide, nitrate) reflected an awareness of the rapid postwar development of the chemical industry plus new data on toxicological effects. The last action of the Public Health Service, before its standards-setting function was transferred to the newly formed EPA in 1970, was to recommend additional parameters such as pesticides, boron, and the uranyl ion.

It is clear from reviewing the history of regulations, at least in the United States, that:

1. The number of regulated contaminants has continued to increase as toxicological evidence and analytical techniques improve.

2. Maximum permissible concentrations are usually lowered over time and seldom raised.

TABLE 4-1. History of U.S. Public Health Service Drinking Water Regulations

Constituent					

Constituent	1914	1925	1942	1946	1962
Bacteriological					
Plate count	X				
Coliform bacteria	X	X	X	X	X
Physical					
Color		X	X	X	X
Odor		X	X	X	X
Residue			X	X	X
Taste		X	X	X	X
Turbidity			X	X	X
Inorganic					
Arsenic			X	X	X
Barium					X
Cadmium					X
Chloride		X	X	X	X
Chromium				X	X
Copper		X	X	X	X
Cyanide					X
Fluoride			X	X	X
Iron		X	X	X	X
Lead	X	X	X	X	X
Manganese			X	X	X
Magnesium		X	X	X	X
Nitrate					X
Selenium			X	X	X
Silver					X
Sulfate		X	X	X	X
Zinc		X	X	X	X
Organic					
Alkyl benzene sulfonate			X	X	X
Carbon chloroform extract					X
Phenols			X	X	X
Radioactivity					
Radium 226					X
Strontium 90					X
Gross beta activity					X

SOURCE: Adapted from AWWA (1971).

Another significant feature of the Public Health Service standards is the development of a two-tiered system, which began in 1925. Water quality contaminants were controlled by either tolerance limits or recommended limits depending on how the effect of the contaminant was viewed. Tolerance limits were set for substances which, if present in excess of specified concentrations, constituted grounds for rejecting the supply; examples include arsenic, chromium, and lead. On the other hand, recommended limits were constituent concentrations that should not be exceeded if other more suitable supplies were or could be made available; examples included chloride, iron, and sulfate. This type of differentiation was the forerunner of present regulations, wherein the tolerance limits correspond to primary regulations intended for public health protection and recommended limits are analogous to secondary standards for public welfare or aesthetics.

During the time the Public Health Service was preparing standards, they strictly applied only to suppliers of water to interstate commerce since the original intent was to protect the health of the traveling public. Thus, standards applied to water used on commercial trains, airplanes, buses, and similar vehicles.

What actually happened is that the Public Health Service standards became recognized informally as water quality criteria and were adopted or adapted by many regulatory agencies at the state or local level as standards.

Thus, prior to the entry of the EPA into the role of regulating community water supplies, many water suppliers were producing water in accordance with the levels shown in the Public Health Service standards. A similar response occurred internationally, with agencies such as the World Health Organization using the Public Health Service Standards as a guideline in developing their own standards.

Water Quality Criteria Development

A number of publications summarize water quality criteria for various beneficial uses, including drinking water. In 1952, the California State Water Pollution Control Board in conjunction with the California Institute of Technology published a report titled *Water Quality Criteria*, which summarized scientific and technical literature on water quality for various beneficial uses. The report was revised in 1963 (McKee and Wolf, 1963) and republished by the California State Water Resources Control Board (McKee and Wolf, 1971).

Federal agencies have also developed water quality criteria documents in response to the Federal Water Pollution Control Act and Safe Drinking Water Act. These serve as references for judgments concerning the suitability of water quality for designated uses, including drinking water. Some treatment is presumed before use. These references include:

- *Water Quality Criteria* (National Technical Advisory Committee to the Secretary of the Interior, 1968), reprinted by the EPA (1972).
- *Water Quality Criteria,* 1972, prepared by the National Academy of Sciences and the National Academy of Engineering (1972) for the EPA.
- *Quality Criteria for Water,* published by the EPA (1976b).

The three documents are often referred to as "the green book," "the blue book," and "the red book," respectively.

The National Academy of Sciences (1977, 1980) developed a systematic approach to establishing quantitative criteria and made a major contribution to the field. In doing so, the academy iterated four principles for safety and risk assessment of chemical contaminants in drinking water:

- Effects in animals, properly qualified, are applicable to man.
- Methods do not now exist to establish a threshold for long-term effects of toxic agents.
- The exposure of experimental animals to toxic agents in high doses is a necessary and valid method of discovering possible carcinogenic hazards in man.
- Material should be assessed in terms of human risk, rather than as "safe" or "unsafe."

The academy divided criteria development into two different methodological approaches, depending on whether the compound in question was a carcinogen or a noncarcinogen. For carcinogens the academy used a probabilistic multistage model to estimate risk from exposure to low doses. This model is equivalent to a linear model at very low dosages. In selecting a risk estimation model, the academy (1980) evaluated a number of quantitative models to describe carcinogenic responses at varying doses of a given model. These models included:

1. *Probabilistic multistage model.* This model assumes that carcinogenesis consists of one or more stages at the cellular level beginning with a *single* cell mutation, at which point cancer is initiated. The model relates doses, d, to the probability of response, $P(d)$, as follows:

$$P(d) = 1 - \exp[-(\lambda_0 + \lambda_1 d + \lambda_2 d^2 + \cdots + \lambda_k d^k)]$$

2. *Linear, no threshold model.* This model assumes that carcinogenic risk, $P(d)$, is directly proportional to dose d as follows:

$$P(d) = \alpha d$$

3. *Tolerance distribution model.* This model assumes that each member of the population at risk has an individual tolerance for the toxic agent below which a dose will produce no response; higher doses will produce a response. Tolerances vary among the population according to some probability distribution F. Toxicity tests have frequently shown an approximately sigmoid relationship with the logarithm of dose, leading to development of the log-normal or log-probit model. The distribution of F is normal against the logarithm of dose. Modified versions of this model have been developed.

4. *Logistic model.* The logistic model assumes a logistic distribution of the logarithms of the individual tolerance and a theoretical description of certain chemical reactions.

5. *"Hitness" model.* Based on radiation-induced carcinogenesis, this model assumes that the site of action has some number ($N \geq 1$) of critical "targets" and that an event occurs if some number ($n \leq N$) of them are "hit" by k or more radiation particles. Single-hit ($N = 1$, $k = 1$), two-hit ($N = 1$, $k = 2$), and two-target ($N = 2$, $k = 1$) models are the more commonly used. The single-hit model is similar to the linear, no-threshold model.

6. *Models of time to tumor occurrence.* Unlike the previous models, this assumes a latency period between exposure and carcinogenesis such that higher doses produce a shorter time to occurrence.

The difficulty of using any of the above models is the inability to determine whether predictions of risk at very low dosages are accurate. It is simply impractical to test the large number of animals needed to observe a response at very low dosage.

Figure 4-3 illustrates the effect of model selection on predicted response at low dosages for three different models (NAS, 1980). Note that in the range of doses that can be measured (5–95 percent response rate), the three models give similar estimates of response. However, on extrapolation to very low doses, predicted responses differ by several orders of magnitude, with the single-hit model giving the most conservative estimate.

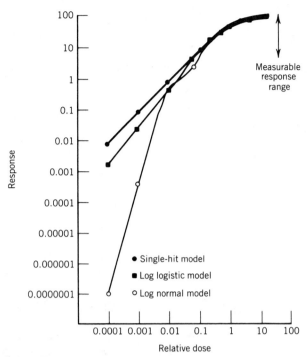

FIGURE 4-3. Effect of model selection on predicted response at low dosage (after NAS, 1980).

The academy's Safe Drinking Water Committee selected the probabilistic multistage model to estimate carcinogenic risk at low doses because (1) it was based on a plausible biological mechanism of carcinogens, a single cell mutation, and (2) other models were empirical.

For the carcinogenic compounds, no "safe" level could be estimated. However, estimates were made such that concentrations of a compound in water could be correlated with an incremental lifetime cancer risk, assuming a person consumed 2 L/day of water for 70 yr. For example, a chloroform concentration of 0.29 μg/L corresponded to an incremental lifetime cancer risk of 10^{-6}. Thus, an individual's risk of cancer would increase by 1 in 1 million by drinking 2 L/day of water with 0.29 μg/L chloroform for 70 yr; alternatively, in a population of 1 million one person would get cancer who otherwise would not have. The academy provided the criteria to allow correlations of contaminant levels and risks but made no judgment on an appropriate risk level. The latter decision properly falls in the sociopolitical realm of standards setting.

For noncarcinogens, data from human or animal exposure to a toxic agent were reviewed and calculations made to determine the no-adverse-effect

doses in humans. Then, depending on the type and reliability of data, a safety factor was applied. This factor ranged from 10 (where good human chronic exposure data was available and supported by chronic oral toxicity data in other species) to 1000 (where limited chronic toxicity data was available). Based on these levels and estimates of the fraction of a substance ingested from water (compared to food, air, or other sources), the academy method allowed calculations of acceptable daily intake and a suggested no-adverse-effect level in drinking water.

The EPA Office of Drinking Water has used criteria developed by the National Academy of Sciences to develop advisory documents, collectively known as "health advisories." Suggested no-adverse-effect response levels (SNARLs) have been established for individual compounds for short-term exposure situations. For long-term exposure to carcinogens, the documents indicate a range of concentrations corresponding to 10^{-5}, 10^{-6}, and 10^{-7} incremental lifetime cancer risks. Examples of these criteria are shown in Table 4-2 and in individual SNARL documents (EPA December 21, 1979; February 6, 1980; March 19, 1980; May 9, 1980). Discretion on applications of these criteria, collectively known as SNARLs, has been left to individual state agencies since the criteria documents carry no regulatory force.

Other criteria for 64 toxic pollutants or pollutant categories in drinking water have been published by the EPA, Office of Water Planning and Standards (November 1980, August 1981), pursuant to the Federal Water Pollution Control Act. Criteria include concentrations that either are protective against toxic effects or correspond to estimated risk levels for carcinogens. For the latter group the criteria indicate levels that may result in an incremental increase over the lifetime of 10^{-5}, 10^{-6}, and 10^{-7}. While the criteria establish levels to protect people from toxic constituents, they only show the risks associated with levels of carcinogens but do not select an appropriate risk level.

Although recent steps have been taken to systematically develop water quality criteria based on health effects, not all estimates are in agreement. Table 4-3 illustrates concentrations for four compounds corresponding to a 10^{-6} incremental lifetime cancer risk, as developed by the National Academy of Sciences (1977, 1980) and the EPA Cancer Assessment Group for the Office of Water Planning and Standards (November 1980, August 1981). Use

TABLE 4-2. Water Quality Criteria Examples—EPA SNARLS

Constituent	Carcinogenic Levels μg/L for 1/1,000,000 Incremental Cancer Risk		Noncarcinogenic Levels SNARL in mg/L for			
	NAS	EPA–CAG	1 Day	7 Days	10 Days	Long-term
Trichloroethylene	4.5	—	2,000	—	200	75
Tetrachloroethylene	3.5	—	2,300	—	175	20
Fuel oil #2 (kerosene)	—	—	—	100	—	—
Benzene	—	1.5	—	350	—	—
Benzo[a]pyrene	PAH mixture carcinogenic: Exact role of benzo[a]pyrene uncertain		—	25	—	—
1,1,1-Trichloroethane	Not considered to be carcinogen		—	—	—	1,000
Dichloromethane (methylene chloride)	Inconclusive evidence to date		13,000	—	1,300	150
Chlordane	0.028	0.023	63	—	63	8
Formaldehyde	Some evidence of carcinogenicity		—	—	30	—
Carbon tetrachloride	4.5	—	200	—	20	—
1,2-Dichloroethane	0.7	0.95	None developed; no satisfactory data			
trans-1,2-Dichloroethylene	No studies completed		2,700	—	270	—
cis-1,2-Dichloroethylene	No studies completed		4,000	—	400	—
1,1-Dichloroethylene (vinylidene chloride)	—	0.033	1,000	—	—	70
Xylenes	No tests available		12,000	—	1,400	620
Methyl ethyl ketone	No studies found		7,500	—	750	—
n-Hexane	No studies found		13,000	—	4,000	—
Ethylene glycol	No evidence to date		19,000	—	—	5,500

of different assumptions and methods of extrapolation leads to differences shown in the table.

Current Regulations

Federal, state, and international drinking water regulations are described.

Federal. With the passage of the Safe Drinking Water Act (U.S. Congress, 1974), the federal government through the EPA was given the authority to set standards for drinking water quality delivered by community (public) water suppliers. Thus, direct federal influence on water quality was authorized, as opposed to the indirect influence exerted by the Public Health Service.

Following the example of the two-tiered standards established by the Public Health Service, the Safe Drinking Water Act requires two types of standards: primary and secondary. Primary standards protect public health, to the extent feasible, using technology, treatment techniques, and other means, which the administrator of the EPA determined as generally available (taking costs into consideration) on the date of enactment of the act. Standards can include performance requirements (maximum contaminant levels, or MCLs) and/or treatment requirements. The act also contains provisions for secondary drinking water regulations for MCLs on contaminants that may adversely affect the odor or appearance of water or the public welfare.

The Safe Drinking Water Act contained a series of steps and timetables for developing the regulations. Procedures were established for interim primary drinking water regulations, revised national

TABLE 4-3. Examples of Risk Estimates by Two Different Methods

Compound	Concentration (μg/L) for 1/1,000,000 Incremental Lifetime Cancer Risk	
	EPA–ODW[a]	EPA–OWPS[b]
Carbon tetrachloride	4.5	0.40
Chloroform	0.29	0.19
Polychlorinated biphenyl	0.16	0.00079
Tetrachloroethylene	4.5	2.7

[a] From NAS, 1977, 1980.

[b] From EPA, November 1980, August 1981.

primary drinking water regulations, national secondary drinking water regulations, and periodic review and update of the regulations. With each step, proposed regulations were to be developed by the EPA, published in the *Federal Register,* discussed at public hearings, commented on by interested parties, and revised as necessary before final promulgation.

In the United States, there are a number of drinking water regulations currently in force, including:

1. The National Interim Primary Drinking Water Regulations (NIPDWR) (EPA, 1975), which are similar in content to the 1962 Public Health Service standards.
2. The trihalomethane regulations (EPA, November 29, 1979, March 11, 1980, and March 5, 1982), an amendment to the NIPDWR.
3. Requirements for monitoring sodium and corrosion (EPA, August 27, 1980), another amendment to the NIPDWR.
4. Secondary drinking water regulations (EPA, July 19, 1979), which are advisory in nature, to be applied by the states as they see fit.

The EPA NIPDWR (EPA, 1975) were published December 24, 1975, and became effective June 24, 1977. As shown in Table 4-4, they contain MCLs for a number of inorganic chemicals, organic chemicals, physical parameters, radioactivity, and bacteriological factors. Since all the primary standards are based on health effects to the consumer, they are mandatory standards. MCLs are set as limits never to be exceeded (with some minor exceptions). All standards, except turbidity, apply at the consumer's tap. Turbidity is to be measured at a

representative entry point to the distribution system; this is typically measured in finished water at a water treatment plant.

Perhaps the most substantial change of the EPA NIPDWR compared to the Public Health Service standards was the designation of turbidity as a health-related rather than an aesthetic parameter. This change was based on new information that the particulates responsible for turbidity do more than make the water cloudy and aesthetically unpleasing; they could also, under some circumstances, protect pathogenic microorganisms from the action of a disinfecting agent (Symons and Hoff, 1975).

The inclusion of turbidity as a health-related standard has led many water suppliers to design and

TABLE 4-4. Environmental Protection Agency National Interim Primary Drinking Water Regulations (1975)

Constituent	Maximum Contaminant Level[a]
Inorganic Chemicals	
Arsenic	0.05
Barium	1
Cadmium	0.010
Chromium (total)	0.05
Fluoride	1.4–2.4[b]
Lead	0.05
Mercury	0.002
Nitrate (as N)	10
Selenium	0.01
Silver	0.05
Organic Chemicals	
Chlorinated hydrocarbons	
Endrin	0.0002
Lindane	0.004
Methoxychlor	0.1
Toxaphene	0.005
Chlorophenoxys	
2,4-D	0.1
2,4,5-TP Silvex	0.01
Physical Parameters	
Turbidity (TU)	1[c]
Radioactivity	
Gross alpha (pCi/L)	15
Radium-226 and 228 (pCi/L)	5
Tritium (pCi/L)	20,000
Strontium-90 (pCi/L)	8
Bacteriological factors	
Coliform bacteria (per 100 mL)	1

[a] In mg/L unless otherwise noted.

[b] Depends on temperature.

[c] Under certain circumstances up to 5 TU may be allowed.

construct filtration facilities for supplies that were previously unfiltered. A notable example is the city of Los Angeles, which had relied on a mountain watershed supply that typically meets all primary MCLs except turbidity.

The next action affecting the NIPDWR was an amendment to include MCLs for total trihalomethanes (THMs). Regulations were proposed in 1978 (EPA, February 9, 1978, and July 6, 1978), promulgated in 1979 (EPA, November 29, 1979), and corrected in 1980 (EPA, March 11, 1980). For community water supplies serving 75,000 persons or more, the effective dates were November 29, 1980, for monitoring and November 29, 1981, for MCLs. Suppliers serving a population of 10,000–74,999 had until 1982 and 1983 to comply with monitoring requirements and MCLs, respectively.

The THM regulations set a maximum contaminant level of 0.10 mg/L (100 μg/L) for total THMs, defined as the sum of the concentrations for trichloromethane (chloroform), bromodichloromethane, dibromochloromethane, and tribromomethane (bromoform). Since THMs can continue to form after chlorine is applied, MCLs are applicable in the distribution system, measured by a running annual average of at least four points in the system at no less than quarterly intervals.

The development of the THM regulation is an example of the interaction among technical, technological, and political factors. With the first findings of chlorinated organics in drinking water in New Orleans and postulated links between organics and higher cancer incidence (Harris, 1974), there was a widespread interest in the United States, particularly in the EPA, to begin to control exposure to these compounds. The Environmental Defense Fund sued the EPA, seeking more comprehensive EPA control over organic chemical contaminants than were contained in the NIPDWR (EPA, 1975). Surveys by the EPA (Symons et al., 1975; EPA, 1976a) found that the most prevalent chlorinated organics in water, both in terms of frequency and concentration, were THMs. Next, an animal feeding study (NCI, 1976) found that chloroform was carcinogenic to mice and rats. Several epidemiological studies (summarized by EPA, November 29, 1979, and NAS, 1980) of THM concentrations in drinking water and cancer incidence showed some positive associations, lending credence to the animal feeding data. The EPA published an Advance Notice of Proposed Rulemaking (EPA, July 14, 1976) and later proposed THM regulations (EPA, February 9,

1978, and July 6, 1978). After an extensive public comment period, the EPA promulgated regulations to control THMs (EPA, November 29, 1979, and March 11, 1980). The final regulations for THM control were challenged by AWWA (1980) in a lawsuit that was settled in 1982. The EPA (March 5, 1982) proposed new rules for THM control that established procedures for selecting treatment methods and finalized the rule (EPA, Feb. 28, 1983) identifying five treatment techniques generally available for THM control and five that may need to be examined if the first set does not solve the problem.

Treatment to meet the THM regulations and the turbidity MCL appears to be the most substantive challenge to water utilities to result from the National Interim Primary Drinking Water regulations to date.

A second amendment to the NIPDWR (EPA, August 27, 1980) made some minor changes by requiring monitoring for sodium and corrosivity characteristics. Although no new MCLs were set, the amendment may be the forerunner of future regulations to control parameters linked to health effects. Increased sodium levels have been linked to hypertension, and there is some evidence relating corrosive water to release of toxic heavy metals (such as lead) from distribution systems and home plumbing materials. Hard water has, in some cases, been linked to lower cardiovascular disease rates. At this point, the regulatory stance has been to gather more information through monitoring rather than to set MCLs, while additional health effects data is being assembled.

The EPA has also promulgated secondary drinking water regulations (EPA, July 19, 1979), as summarized in Table 4-5. The national secondary drinking water regulations pertain to those contaminants such as taste, odor, and color that may adversely affect the aesthetic quality of drinking water. These secondary levels represent reasonable goals for drinking water quality but are not federally enforceable; rather, they are intended as guidelines for the states. The states may establish higher or lower levels as appropriate to their particular circumstances, provided that public health and welfare are adequately protected.

The EPA (March 4, 1982) is considering regulating volatile organic chemicals (VOCs) in drinking water through (1) providing health and treatment guidance to the states, (2) requiring monitoring followed by state action as appropriate, or (3) requiring monitoring and setting maximum contaminant

TABLE 4-5. EPA National Secondary Drinking Water Regulations (1979)

Constituent	Maximum Contaminant Level	Effect on Water Quality
Chloride	250 mg/L	Salty taste
Color	15 color units	Objectionable appearance
Copper	1 mg/L	Undesirable taste
Corrosivity	Noncorrosive	Stains; dissolution of metals; economic losses
Foaming agents	0.5 mg/L	Undesirable appearance
Iron	0.3 mg/L	Bitter, astringent taste; stained laundry
Manganese	0.05 mg/L	Impaired taste; discolored laundry
Odor	3 threshold odor number	Undesirable smell
pH	6.5–8.5	Corrosivity; impaired taste
Sulfate	250 mg/L	Detectable taste; laxative at high concentrations
TDS	500 mg/L	Objectional taste
Zinc	5 mg/L	Undesirable taste, milky appearance at high concentrations

levels. Possible ranges for MCLs for six of the most frequently detected volatile organics are shown in Table 4-6. The EPA (July 12, 1984) has proposed recommended MCLs (RMCLs) of zero for seven substances considered carcinogenic: tetrachloroethylene, trichloroethylene, carbon tetrachloride, 1,2-dichloroethane, vinyl chloride, benzene, and 1,1-dichloroethylene. For noncarcinogens, proposed RMCLs were 0.2 mg/L for 1,1,1-trichloroethane and 0.75 mg/L for p-dichlorobenzene. Proposed MCLs are planned in 1985.

State. Although the EPA sets national regulations, the Safe Drinking Water Act gives states the opportunity to obtain primary enforcement responsibility (primacy). States with primacy must develop their own drinking water standards, which must be at

TABLE 4-6. Potential MCLs for Volatile Organics

Compound	Potential MCL, μg/L
Trichloroethylene	5–500
Tetrachloroethylene	5–500
Carbon tetrachloride	5–500
1,1,1-Trichloroethane	1000
1,2-Dichloroethane	1–100
Vinyl chloride	1–100

SOURCE: EPA (March 4, 1982).

least as stringent as the EPA standards. Over 40 states have applied for and have been granted primacy.

In many instances, the state water quality standards are identical to the EPA NIPDWR and amendments thereto. Individual state health departments should be consulted for applicable regulations.

The California Department of Health Services (1977) regulations serve as a case in point for how federally derived regulations are applied in one state. The California regulations are the same as EPA regulations shown in Tables 4-4 and 4-5 with the following exceptions:

1. The state requires that "surface waters exposed to significant sewage hazards or significant recreational use shall receive, as a minimum, pretreatment, filtration, and disinfection. The filtered water turbidity, measured daily, shall be less than 0.5 turbidity units on a monthly average for an acceptable level of public health protection." Thus, the state MCL for turbidity on certain sources is one-half the EPA MCL. In addition, this regulation is an example of a combined process (pretreatment, filtration, disinfection) and product quality (0.5 TU) standard.

2. The radioactivity MCLs include 50 pCi/L of gross beta particle activity.

3. If secondary standards, which differ slightly from the EPA standards, are not met, the state

may require the supplier to conduct a study, correct the situation, or use new supplies of better quality.

4. Specific treatment design or operating criteria are applied on a case-by-case basis.

5. Utilities serving less than 10,000 persons are required to analyze distribution system samples for THMs.

6. The state is using criteria developed in the EPA SNARL documents to regulate use of well water supplies if concentrations of trichloroethylene (TCE), perchloroethylene (PCE), or other designated organics exceed state-defined "action levels," which are based on the 10^{-6} incremental lifetime cancer risk levels or acceptable daily intake.

International. A number of agencies outside the United States have developed drinking water regulations. These include standards for individual countries or groups of countries.

The World Health Organization (WHO) has been at the forefront of developing standards. European standards (WHO, 1970) and international standards (WHO, 1971) are both available; revisions were published in 1984. The European standards are stricter than the international standards, since the WHO recognized different economic and technological capabilities of the various countries. WHO standards are meant for guidance only and are recommendations, not mandatory requirements. Therefore, strictly speaking they are criteria rather than standards. However, the WHO standards have been adopted in whole or in part by a number of countries as a basis of formulation for national standards.

The WHO European standards contain recommendations for bacteriological examination, virus concentration, larger organisms, radioactive factors, toxic chemicals, extractable organics, polycyclic aromatic hydrocarbons, pesticides, chemical substances that may give rise to trouble in piped supplies of drinking water (typically aesthetic factors), and general physical, chemical, and aesthetic characteristics. International standards cover a similar array of topics.

The European Community, which represents nine countries with a combined population of about 250 million, has developed drinking water standards that are being implemented in national laws (Knoppert, 1980). All parameters, whether they have a direct influence on human health or not, have a mandatory character. The standards include organoleptic factors (such as color, palatability, turbidity, odor), physiochemical factors (such as chloride, calcium, sodium, TDS), undesirable factors (such as nitrates, hydrogen sulfide, TOC, other organochlorine compounds, zinc, fluoride), toxic factors (such as arsenic, chromium, lead, pesticides, polycyclic aromatic hydrocarbons), and microbiological factors (such as total and fecal coliforms, clostridium, and total count).

Drinking water criteria and standards have been developed by a number of countries including Japan, the United Kingdom, the USSR, South Africa, Brazil, and Australia. Their national regulations reflect the strong influence of the U.S. Public Health Service and WHO standards. Many requirements are similar from country to country.

A selected set of international drinking water regulations, representing a range of constituents with health or aesthetic impacts, is shown in Table 4-7. For some of the parameters, such as selenium, the regulations are fairly consistent from country to country. For others, such as nitrate and zinc, there is a wider diversity of allowable levels. Several recent regulations address organics, including (1) U.S. MCL of 0.10 mg/L total trihalomethanes (EPA, 1979), (2) Canadian maximum acceptable concentrations for 14 pesticides and total trihalomethanes (0.35 mg/L) (Minister of National Health and Welfare, 1978), and (3) WHO guidelines for 20 organics, including 0.030 mg/L chloroform (Ozolins, 1983).

Future Expectations

The continued process of water quality regulation is expected to produce additional standards in the future. It is clear from Figure 4-4 that the number of federally regulated constituents has increased historically, a trend that is expected to continue. (*Note:* the decline in number in 1975 is due to the change in federal regulatory responsibility from the Public Health Service to the EPA; the 1975 EPA regulations included only primary standards while secondary standards were added in 1979. The number of primary standards has continued to increase.)

Other possible modifications include (1) standards for inorganics based on speciation or oxidation state, (2) more stringent microbial standards, (3) emphasis on tastes and odors, (4) control of water quality degradation in distribution system and

TABLE 4-7. Selected International Drinking Water Regulations

Parameter	Units	U.S. EPA (1980) Recommended	U.S. EPA (1980) Maximum Contaminant Level	WHO, International (1971) Recommended	WHO, International (1971) Maximum Permissible Level	WHO, European (1970)	European Community Guide Levels	European Community Maximum Admissible	Netherlands (1981)	USSR (1975)	Norway
Aluminum	mg/L	—	—	—	—	—	0.05	0.2	—	0.5	<0.1
Arsenic	mg/L	—	0.05	—	0.05	0.05	—	0.05	0.2	0.05	<0.01
Chloride	mg/L	250	—	200	600	200	25	>200	—	—	—
Copper	mg/L	1	—	0.05	1.5	0.05	0.1	—	3.0[b]	—	<0.05
Fluoride	mg/L	—	1.4–2.4[a]	—	0.6–1.7[a]	0.7–1.7[a]	—	0.7–1.5[a]	1.2	0.7–1.5[a]	<1.5
Iron	mg/L	0.3	—	0.1	1.0	0.1	0.05	0.2	1	0.3	<0.1
Lead	mg/L	—	0.05	—	0.1	0.1	—	0.05	0.3	0.1	<0.05
Magnesium	mg/L	—	—	—	150[c]	125[c]	30	50	—	—	<10
Manganese	mg/L	0.05	—	0.05	0.5	0.05	0.02	0.05	0.05	0.1	<0.1
Mercury	mg/L	—	0.002	—	0.001	—	—	0.001	—	—	<0.0005
Nitrate	mg/L-N	—	10	—	—	23	6	11	23	10	<2.5
Phosphorus	mg/L	—	—	—	—	—	0.15	2.0	—	0.0035	—
Sodium	mg/L	20	—	—	—	—	20	175	20	—	—
Sulfate	mg/L	250	—	200	400	250	25	250	—	—	—
Zinc	mg/L	5	—	5	15	5.0	0.1	5	1.5[b]	—	<0.3
TDS	mg/L	500	—	500	1500	100–500	—	1500	—	—	—
Total hardness	mg/L $CaCO_3$	—	—	100	500	—	—	—	—	—	—
Selenium	mg/L	—	0.01	—	0.01	0.01	—	0.01	0.05	0.001	<0.01
Color	mgPt/L	15	—	5	50	—	1	20	20	—	<5
Turbidity	TU	—	1	5	25	—	0.4	4	0.5	—	<0.5
pH		6.5–8.5	—	7.0–8.5	6.5–9.2	—	6.5–8.5	9.5	—	—	8.0–8.5
Coliform	org/100 mL	—	1	—	1	1	—	<1	0–2	—	<1
Foaming agents (MBAS)	mg/L	0.5	—	0.2	1.0	0.2	—	0.2	—	—	—
Other organochlorine compounds	μg/L	—	—	—	—	—	1	—	—	—	—
Total organic carbon (TOC)	mg/L	—	—	—	—	—	—	[d]	—	—	—
Trihalomethane (THM)	μg/L	—	100	—	—	—	[e]	—	—	—	—

[a] Concentration is described as a function of ambient temperature.
[b] After 16 hr lead or copper pipe.
[c] If there is less than 250 mg/L of sulfate.
[d] The reason for any increase in the usual concentration must be investigated.
[e] Concentration of THM as low as possible.

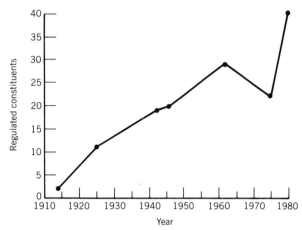

FIGURE 4-4. Increase in number of federally regulated water quality constituents.

home plumbing, and (5) additional inorganic constituents (Tate, 1981). Many of these are expected to occur as EPA produces revised primary drinking water regulations; others are more long-term in nature.

Revised Drinking Water Regulations. In the near-term, substantial changes from current regulations are expected in revised national primary drinking water regulations, developed by the EPA to comply with the requirements of the Safe Drinking Water Act (U.S. Congress, 1974). The act requires revisions following promulgation of the interim standards (EPA, 1975) and review of health effects data by the National Academy of Sciences. Several health effects reviews have been published (NAS 1977, 1980, 1982, 1983) and revised regulations are still pending. After revised regulations are issued, they are to be reviewed and revised as needed at 3-yr intervals.

The EPA announced (EPA, Oct. 5, 1983) that revised primary regulations would be developed in four phases:

- Phase I—volatile synthetic organics (VOCs) (previously described).
- Phase II—synthetic organic chemicals (SOCs), inorganic chemicals, and microbiological contaminants.
- Phase III—radionuclides.
- Phase IV—disinfectant by-products including trihalomethanes.

Revised standards, expected to require several years for completion, may deviate in form from present interim standards, which are applicable to all suppliers above a minimum population. In new regulations, contaminants may fall into one of three tiers: (1) Category I—frequent occurrence justifying national regulation by MCLs and national monitoring (such as turbidity and coliform), (2) Category II—sufficient frequency to justify national MCLs but flexible monitoring determined by states, and (3) Category III—health concern, but insufficient frequency to require regulation, leading to publication of health advisories.

Some of the major issues that could affect water quality and treatment are:

- Microbiology and turbidity—lower turbidity MCL, mandatory filtration (with some possible exceptions) of all surface water supplies, mandatory disinfection of all services, possible additional regulations for *Giardia,* viruses, and Legionella.
- Inorganic chemicals—modification of existing MCIs plus consideration of regulations for aluminum, antimony, molybdenum, asbestos, sulfate, copper, vanadium, sodium, nickel, zinc, thallium, beryllium, and cyanide.
- Corrosion—monitoring for corrosion by-products.
- Synthetic organic chemicals—modification of MCLs for compounds currently regulated plus MCLs or monitoring for a long list of other pesticides, polynuclear aromatic hydrocarbons, phthalates and adipates, and acrylamide.
- Radionuclides—revision of existing regulations plus MCLs for uranium and radon.
- Disinfection by-products—to be determined, but likely to include lower THM MCL, limitations on disinfectant use or by-products, depending on findings of health effects studies.

Long-Term Trends. *Standards for Inorganics Based on Speciation (Oxidation State).* It is well known that for certain inorganics constituents, the health effects are only related to one species or oxidation state of the element. For example, hexavalent chromium [Cr(VI)] is known to have adverse health effects, while trivalent chromium [Cr(III)] is generally believed to be relatively innocuous. Setting the standard for hexavalent chromium instead of total chromium would be a logical consequence of these findings. Similar developments may occur for arsenic and selenium.

More Stringent Microbial Standards. With the current emphasis on use of alternative disinfectants for control of THMs, there is concern that the microbiological quality of the water not be degraded by using disinfectants other than chlorine. One way of controlling this would be to set more stringent standards for microorganisms (such as 0.1 coliform/100 mL rather than 1 coliform/100 mL). Another would be adding requirements for maximum contaminant levels for additional microorganisms, such as specific types of bacteria in addition to the coliforms.

Emphasis on Tastes and Odors. Recently, analytical chemists have been able to identify geosmin, methylisoborneol (MIB), and other odor-causing compounds at very low levels in water supplies. With identification and quantification, the use of systematic control procedures becomes more viable. Therefore, it can be expected that controlling tastes and odors may receive more attention in the future. Even though tastes and odors are typically not directly related to health effects, they are still a principal cause of consumer complaints in many water systems.

Control of Deterioration in Distribution Systems and Home Plumbing. It is likely that in the future standards will place more emphasis on preventing or curtailing deterioration of water quality in distribution systems and home plumbing. Once the water leaves a water treatment plant, it may be subject to pipeline corrosion and the input of metals; leaching of substances from asbestos/cement pipe, plastic pipe, or lining materials; bacterial growth on pipe walls and by-products thereof; contamination from open treated water storage reservoirs; and increasing mutagenic activity from chemical or biological activity. This topic has begun to receive emphasis (NAS, 1982), and it can be expected that there will be more standards related to preventing water quality degradation between the water treatment plant and the consumer's tap in the future.

Additional Inorganic Constituents. Additional inorganic constituents are candidates for regulation in the future based on health effects data that is currently accumulating from toxicological and epidemiological tests. Hardness is an example. Health effects information, although subject to limitations and contradictions, will likely play a significant role in the development of any future standards.

REFERENCES

AWWA, "Quality Control for Potable Water," *JAWWA*, **60**(12), 1317–1322 (1968).

AWWA, *Water Quality and Treatment*, 3rd ed., McGraw-Hill, New York (1971).

AWWA, "Water Quality Goals," *JAWWA*, **66**(4), 221–230 (1974).

AWWA, "AWWA Files TTHMs Appeal in Federal Court," *Willing Water*, **24**(3), 1, 3 (1980).

California State Department of Health, "Domestic Water Quality and Monitoring," Title 22, Division 4, California Administrative Code (October 20, 1977).

EPA, "National Interim Primary Drinking Water Regulations," *Fed. Reg.*, **40**(248), 59566–59587 (December 24, 1975).

EPA, "Advance Notice of Proposed Rule-Making," *Fed. Reg.*, **41**(136), 28991–28999 (July 14, 1976).

EPA, "National Organics Monitoring Survey," Unpublished (1976a).

EPA, *Quality Criteria for Water*, Washington, DC (1976b).

EPA, "National Interim Primary Drinking Water Regulations: Control of Organic Chemical Contaminants in Drinking Water, Proposed Rule," *Fed. Reg.*, **43**(28), 5757–5780 (February 9, 1978).

EPA, "Interim Primary Drinking Water Regulations, Proposed Regulations for Control of Organic Chemical Contaminants in Drinking Water," *Fed. Reg.*, **43**(136), 29135–29150 (July 6, 1978).

EPA, "National Secondary Drinking Water Regulations," *Fed. Reg.*, **44**(153), 42195–42202 (July 19, 1979).

EPA, "National Interim Primary Drinking Water Regulations; Control of Trihalomethanes in Drinking Water; Final Rule," *Fed. Reg.*, **44**(231), 68624–68707 (November 29, 1979).

EPA, Health Effects Branch, Criteria and Standards Division, Office of Drinking Water, "SNARL for Trichloroethylene" (December 21, 1979).

EPA, Office of Drinking Water, "SNARL for Tetrachloroethylene" (February 6, 1980).

EPA, "National Interim Primary Drinking Water Regulations; Control of Trihalomethanes in Water; Correction," *Fed. Reg.*, **45**(49), 15542–15547 (March 11, 1980).

EPA, Health Effects Branch, Criteria and Standards Division, Office of Drinking Water, "SNARL for Fuel Oil #2 or Kerosene" (March 19, 1980).

EPA, Office of Drinking Water, "SNARL for 1,1,1-Trichloroethane" (May 9, 1980).

EPA, "Interim Primary Drinking Water Regulations; Amendments," *Fed. Reg.*, **45**(168), 57332–57357 (August 27, 1980).

EPA, Office of Water Planning and Standards, Criteria and Standards Division, "Water Quality Criteria Documents; Availability," *Fed. Reg.*, **45**(231), 79318–79379 (November 28, 1980).

EPA, "Water Quality Criteria; Corrections," *Fed. Reg.*, **46**(156), 40919 (August 13, 1981).

EPA, "National Revised Primary Drinking Water Regulations, Volatile Synthetic Organic Chemicals in Drinking Water," *Fed. Reg.*, **47**(43), 9350–9358 (March 4, 1982).

EPA, "National Interim Primary Drinking Water Regulations; Trihalomethanes," *Fed. Reg.*, **47**(44), 9796–9799 (March 5, 1982).

EPA, "National Interim Primary Drinking Water Regulations; Trihalomethanes," *Fed. Reg.*, **48**(40), 8406–8414 (Feb. 28, 1983).

EPA, "National Revised Primary Drinking Water Regulation; "Advance Notice of Proposed Rulemaking," *Fed. Reg.*, **48**(194), 45503–45521 (Oct. 5, 1983).

EPA, "National Primary Drinking Water Regulations; Volatile Synthetic Organic Chemicals," *Fed. Reg.*, **49**(114), 24330–24355 (July 12, 1984).

Harris, R., "The Implications of Cancer-Causing Substances in Mississippi River Water," Environmental Defense Fund (November 1974).

Knoppert, P. L., "European Communities Drinking Water Standards," *1980 AWWA Annual Conference Proceedings,* June 15–20, 1980, Atlanta, AWWA, Denver, CO, pp. 1121–1133.

McKee, J., and Wolf, H. W., *Water Quality Criteria,* 2nd ed., Publication No. 3-A California State Water Quality Control Board, Sacramento (1963).

McKee, J., and Wolf, H. W., *Water Quality Criteria,* 2nd ed., Publication No. 3-A California State Water Quality Control Board, Sacramento (1971).

Minister of National Health and Welfare, "Guides for Canadian Drinking Water Quality," Ottawa, 1978.

National Cancer Institute, Carcinogenesis Program, Division of Cancer Cause and Prevention, "Report on Carcinogenesis Bioassay of Chloroform," Bethesda, MD (March 1, 1976).

National Academy of Sciences and National Academy of Engineering, *Water Quality Criteria, 1972,* Prepared for EPA, Washington, DC (1972).

National Academy of Sciences, Safe Drinking Water Committee, *Drinking Water and Health,* Washington, DC (1977).

National Academy of Sciences, Safe Drinking Water Committee, *Drinking Water and Health,* Vol. 2 and 3, National Academy Press, Washington, DC (1980).

National Technical Advisory Committee to the Secretary of the Interior, *Water Quality Criteria,* Washington, DC (1968).

National Academy of Sciences, Safe Drinking Water Committee, *Drinking Water and Health,* Vol. 4, National Academy Press, Washington, DC (1982).

National Academy of Sciences, Safe Drinking Water Committee, *Drinking Water and Health,* Vol. 5, National Academy Press, Washington, DC (1983).

Ozolins, G., World Health Organization, "Guidelines for Drinking Water Quality," Volume I—Recommendations, Unedited final draft, Geneva (Jan. 14, 1983).

Symons, J. M., et al., "National Organics Reconnaissance Survey for Halogenated Organics in Drinking Water," *JAWWA* **67,** 634–647 (November 1975).

Symons, J. M., and Hoff, J. C., "Rationale for Turbidity Maximum Contaminant Level," *AWWA Third Water Quality Technology Conference,* Atlanta (December 8–10, 1975).

Tate, C. H., and R. R. Trussell, "Developing Drinking Water Standards," *JAWWA* **69,** 9, 486–498 (September 1977).

Tate, C. H., "Water Quality for the Future," *AWWA 1981 Annual Conference Proceedings,* St. Louis, MO (1981), pp. 135–144.

U. S. Congress, *Safe Drinking Water Act,* PL 93-523 (December 17, 1974).

U.S. Public Health Service, *Public Health Service Drinking Water Standards: 1962,* USPHS Publication 956, U.S. Government Printing Office, Washington, DC (1962).

World Health Organization, *European Standards for Drinking Water,* 2nd ed., Geneva (1970).

World Health Organization, *International Standards for Drinking Water,* 3rd ed., Geneva (1971).

— 5 —

Principles of Process Engineering

The principal objective of a water treatment facility is to provide a product that satisfies a set of quality standards at a reasonable price to the consumer. Undesirable constituents in the water source must either be removed or rendered innocuous by appropriate physical and chemical changes carried out in well-designed engineered systems. The creation and assessment of alternative systems to meet this objective comprises the scope of *process engineering and design* (Rudd, 1968). The alternative systems created are assessed based on standards of technical feasibility, operability, and cost.

Process engineering generally represents a small proportion of the total cost of a water treatment project, ranging from about 5 to 20 percent depending on the size, complexity, and innovative design aspects of the project. However, engineering decisions made during this stage can have a major impact on total project cost. Errors in process design can lead to selection of a system that is not cost-effective or that may require expensive change orders during or following construction to satisfy water quality standards.

SEPARATION PROCESSES

All water sources contain a broad spectrum of inorganic and organic constituents of diverse origins. Dissolved minerals, for example, enter the water source by natural weathering processes of erosion and dissolution. Phenol, on the other hand, is discharged into water supplies as a waste product of industrial activity. These constituents are converted from a concentrated to a dilute state as they are mixed in water. Water treatment reverses this process, at the cost of expending thermodynamic work or energy.

The minimum amount of work required to "unmix," or remove, dissolved constituents from water is given by the change in Gibbs free energy at constant temperature (T) and pressure (P) for a reversible process (Denbigh, 1961).

$$W_{min} = -\Delta G_{reversible} \qquad (5\text{-}1)$$

The free-energy change due to mixing of i constituents for nonideal solutions is found (Denbigh, 1961)

TABLE 5-1. Alternative Separation Processes for Removal of Constituents from Water

Constituent	Separation Processes
Algae	Straining, coagulation–sedimentation, coagulation–flotation, oxidation–filtration
Bacteria, pathogenic	Coagulation–sedimentation, adsorption, ultrafiltration
Calcium	Precipitation–sedimentation, ion exchange, reverse osmosis
Chloroform	Adsorption, gas stripping, reverse osmosis
Clays	Coagulation–sedimentation, ultrafiltration
Fluoride	Precipitation–sedimentation, adsorption
Humic acids	Coagulation–sedimentation, adsorption, ultrafiltration, reverse osmosis
Iron, ferrous	Oxidation–filtration, reverse osmosis
Mercury	Coagulation–sedimentation, adsorption, ion exchange
Nitrate	Ion exchange, biological reduction, reverse osmosis
Phenol	Oxidation, adsorption
Salts, dissolved	Distillation, freezing, ion exchange, reverse osmosis
Sulfate	Ion exchange, precipitation–sedimentation

to be

$$\Delta G = RT \sum_i n_i \ln a_i \qquad (5\text{-}2)$$

where R is the universal gas constant, T the absolute temperature, n_i the number of moles of constituent i in the system, and a_i the activity of the ith constituent in the mixture. The activity is the product of the mole fraction and the activity coefficient f_i of the ith component, $a_i = x_i f_i$.

Seawater contains approximately 30 g/L of sodium chloride (NaCl). Removal of 1 g-mole of this salt will require a minimum of roughly 11.5 kJ (10.9 Btu) of energy, according to Eq. 5-2. By contrast, gold is present in seawater at a concentration of 4×10^{-6} mg/L. Recovery of 197 gm (1 g-mole) of gold from seawater would require at least 57 kJ (54 Btu) of energy. In general, the more dilute a constituent, the more energy is required to separate it into its pure state.

In most water treatment and reuse applications, however, we need not recover the undesirable constituents; we simply wish to remove them in a form that can be easily and safely disposed. To achieve this goal, we can use a variety of *separation processes* that exploit various physicochemical phenomena to remove undesirable constituents from solution. Some of these constituents and the feasible separation processes for their removal are shown in Table 5-1.

TABLE 5-2. Selected Physicochemical Separation Processes

Process	Separating Agent	Principle of Separation	Examples in Water Treatment/Reuse
Distillation	Heat	Vapor pressure	Desalination
Stripping	Noncondensible gas (e.g., air)	Phase equilibria	Removal of dissolved gases (H_2S, CH_4, NH_3)
Absorption	Nonvolatile liquid (e.g., H_2O)	Phase equilibria	Addition of CO_2, Cl_2, O_3 to water
Adsorption	Solid adsorbent	Difference in Gibbs free energy	Removal of organics, trace metals
Ion exchange	Solid exchange resin	Chemical equilibria	Water softening, removal of nitrate
Reverse osmosis	Semipermeable membrane and pressure gradient	Diffusion	Demineralization
Drying of solids	Heat	Water evaporation and diffusion	Dewatering of sludge
Precipitation	Chemical oxidant, excess precipitant, pH	Nucleation, insolubility of solid	Lime–soda softening, Fe and Mn removal
Granular media filtration	Unconsolidated inert solids	Reduction of Gibbs interfacial free energy, size	Separation of clays, bacteria, algae

Separation processes are one of the keys to successful performance of water treatment plants. Table 5-2 lists several commonly used physicochemical separation processes. These processes employ a variety of separating agents to provide the driving force for separation. Some rely on the phenomenon that matter will pass from one homogeneous phase to another if a concentration difference exists, analogous to the transfer of heat across an interface separating materials at different temperatures. Concentration gradients arise due to differences in solubility, relative vapor pressures (volatility), or interfacial attractions. Heat is the separating agent used to desalinate seawater by distillation or to dewater solids suspensions by drying. Undesirable dissolved gases such as methane (CH_4) or hydrogen sulfide (H_2S) can be removed or stripped from water by transferring the constituents from water to a gas phase (e.g., air). Certain dissolved inorganic molecules are separated by contacting the water with an ion exchange resin which exploits the resin's preferential equilibria for the undesirable ion.

Other separation processes depend primarily on physical characteristics of constituents to achieve separation. Particulates will separate from aqueous suspensions if their density is either greater than water (sedimentation) or less (flotation) and the particulate size exceeds about 1 μm. Metal screens with openings as small as 20 μm can effectively remove larger algae and other particulates whose size exceed the minimum mesh dimension. Physical separation processes used in water treatment are listed in Table 5-3.

A water treatment plant in a municipality or at an industrial facility also requires various physical operations. Fluids must be transported between separation processes, and various chemicals must be thoroughly mixed with process streams. One mixing operation, called *flocculation,* accelerates the rate of particulate removal by gravity sedimentation through aggregation of the fine particulates into larger agglomerates. The process engineering tasks in plant design thus entail designing alternative systems with both separation processes and physical operations.

PROCESS DESIGN

Selection of the apparent best alternative, including assessment of the relative cost of each alternative, requires the preliminary process design of all potential system components. Various tools are available for this task. For several separation processes, quantitative models of system performance, personal experience, or literature data can be used to predict efficiency of separation. If no models exist, pilot studies of the alternative combinations may be necessary to develop the design data or empirical models of performance.

Alternatively, models can be developed based on the physical-chemical phenomena controlling the separation process (Weber, 1972; Keinath and Wanielista, 1975). The engineer can often develop mathematical descriptions of the process based on a materials balance and models of physical and chemical transformation rates. The objective of this chapter is to establish the framework for developing models of individual separation processes. In later chapters, we will apply these techniques to analyze widely used separation processes in water treatment.

TABLE 5-3. Selected Mechanical Separation Processes

Processes	Separating Agent	Principle of Separation	Examples in Water Treatment/Reuse
Ultrafiltration	Membrane and pressure	Molecular size	Organics removal
Sedimentation	Gravity	Size, density	Solids–liquid separation
Flotation	Gravity, rising or attached air bubbles	Size, density	Solids–liquid separation
Thickening	Gravity	Size, density, and structure	Liquid–solids separation, e.g, alum sludge
Centrifuge	Centrifugal force	Size, density	Dewatering of sludges
Cake filtration	Cloth or metal membrane, vacuum or mechanical pressure	Size	Dewatering of sludges
Screening	Metal screen, various-size openings	Size	Microstrainers for algae removal

MATERIALS BALANCES

Quantitative description of any separation process begins with an accounting of all materials that enter, leave, accumulate in, or are transformed in the system. The basis for this accounting procedure, known as a *materials balance,* is the law of conservation of mass, which accounts for changes in any component due to fluid flow or chemical transformations. A constituent that passes through a processing system unchanged in total mass (but not perhaps in concentration) is called a *conservative* element. Constituents that undergo reactions are known as *nonconservative* elements. For example, the chloride ion will pass through a water treatment plant unchanged because this inorganic ion does not adsorb on particulate matter or undergo biological transformations. Thus, it acts as a conservative constituent. On the other hand, ferrous ion, an ion often found in ground and surface waters, undergoes oxidation and hydrolysis, leading to the formation of an insoluble precipitate at appropriate chemical conditions. It thus acts as a nonconservative constituent.

Accounting for the fate of some constituent in a water treatment plant, or in an individual separation process, can be approached from various levels. Ideally we want to predict the time and spatial variations of a constituent based on forces operating at the ionic or molecular levels, but this molecular description is seldom used in engineering design. At the opposite extreme, we can ignore all molecular interactions and the internal details of the system or unit and assume no local gradients of mass or temperature. The loss of mechanistic insight with this approach is balanced by its ease of mathematical analysis. The full spectrum of levels of description is discussed in detail by Himmelblau and Bischoff (1968).

For any level of analytical description, a materials balance for a system with defined boundaries can be expressed generally as

$$
\begin{array}{l}
\text{Accumulation} \\
\text{of material} \\
\text{in system}
\end{array}
=
\begin{array}{l}
\text{net transport} \\
\text{into system}
\end{array}
$$

$$
-
\begin{array}{l}
\text{net transport} \\
\text{out of system}
\end{array}
+
\begin{array}{l}
\text{net transformation} \\
\text{in system}
\end{array}
\quad (5\text{-}3)
$$

Mass may be transported across system boundaries by bulk fluid flow (advection) or by molecular or turbulent diffusion. Transformations occur be-

cause of chemical reactions and mass transfer between phases within the system boundaries.

For the purpose of process design and model development of separation processes, we will use three levels of system description as defined by Himmelblau and Bischoff (1968): (1) macroscopic; (2) maximum gradient; and (3) multiple gradient. Each of these is discussed below.

Macroscopic Description

The macroscopic description of the flow systems ignores internal details of processing equipment and assumes that mass, concentration, or temperature are only functions of time. This is the simplest level for quantitative description of the system and is the most widely used in water treatment process design.

Single-Component System. For a pure fluid flowing in a system in which no transformations occur and with a single entrance and exit, the materials balance expression (Eq. 5-3) takes the differential form

$$
\frac{d}{dt}(m) = \rho_1 \bar{v}_1 S_1 - \rho_2 \bar{v}_2 S_2 \quad (5\text{-}4)
$$

where m is the total mass in the system, ρ the fluid density, and v the fluid velocity averaged over the cross sections at the system entrance and exits with cross-sectional areas of S_1 and S_2, respectively.

At steady state, that is, no changes in mass with time, and constant density Eq. 5-4 becomes

$$
\bar{v}_1 S_1 = \bar{v}_2 S_2 \quad (5\text{-}5)
$$

Thus, the rate of mass entering equals that leaving, a result used widely in analyses of fluid flow in both open and closed conduits.

Multicomponent Systems. In a multicomponent system a materials balance can be written for each constituent of interest. Figure 5-1 depicts a hypothetical water treatment plant in which dissolved and particulate constituents are transformed or removed by various separation processes. Internal details of the separation processes are ignored and treated as "black boxes." In the first black box with system boundaries as shown in Figure 5-1, the mass concentration (C_i) of undesirable dissolved gases are reduced by air stripping, or *aeration.* Compo-

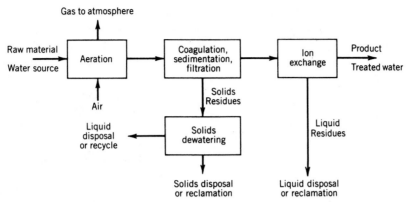

FIGURE 5-1. Macroscopic view of hypothetical water treatment plant.

nent i passes through the separation process by bulk flow. Some fraction is removed by mass transfer to the gas stream. Transformations due to chemical reactions between oxygen and component i are also possible. Using Eq. 5-3, the macroscopic materials balance for component i in the liquid stream at mass concentration C_i, entering at volumetric flowrate Q, becomes

$$\frac{dm_i}{dt} = Q(C_i^{in} - C_i^{out}) + w_i^m + \bar{r}_i V \qquad (5-6)$$

where w_i^m is the net rate of mass transport to the gas phase (negative for removal from system, positive for addition to system). The average rate of mass transformation of component (generation or disappearance) per unit system volume V due to chemical reactions is $\bar{r}_i$. The application of the steady-state form of Eq. 5-6 to a variety of process design problems is presented by Hougen et al. (1954). Example 5-1 illustrates the use of the macroscopic materials balance. Additional balances can be made for

any component for each of the separation processes noted in Figure 5-1.

EXAMPLE 5-1

Problem

A water treatment plant processing a turbid river water consists of the following physical operations and separation processes: coagulation, flocculation, sedimentation, and filtration plus disinfection. During winter, the suspended solids concentration (C_0) in the plant reaches an average peak of 1000 g/m³ with an average flow of 1 m³/sec (22.8 mgd). Tests indicate that alum doses up to 100 g/m³ as alum [$Al_2(SO_4)_3 \cdot 14H_2O$] are required for effective coagulation/sedimentation. Assuming complete reaction of the aluminum ion to aluminum hydroxide [$Al(OH)_3$] and 98 percent removal of total suspended solids by sedimentation and filtration, what maximum rate of solids production must be handled by the solids processing system?

Solution

$Al_2(SO_4)_3 \cdot 14H_2O$

$Q = 1$ m³/sec

Raw water
C_0

Treatment plant

Treated water

q, C_S C_E $C_E = 0.02C_0$

$Al(OH)_3$ + Suspended solids

□

Assume steady state, $dm_i/dt = 0$. Transport by bulk flow only, that is, $w_i^m = 0$, and consider the treatment plant as "black box."

Materials Balance for Suspended Solids, Subscript 1 (Eq. 5-6)

$$-(W_{out} - W_{in}) + r_1 V_{tot} = 0$$
$$\text{where } W = \text{mass/time}$$

But $r_1 = 0$; therefore,

$$W_{out} = W_{in}$$
$$W_{in} = QC_0 \qquad W_{out} = QC_E + qC_S$$

Thus

$$qC_s = QC_0(1 - 0.2)$$
$$= (1 \text{ m}^3/\text{sec})(1000 \text{ g/m}^3)(0.98) = 980 \text{ g/sec}$$

Materials Balance for Aluminum Hydroxide, Al(OH)₃, Subscript 2

$$-(W_{out} - W_{in}) + r_2 V_{tot} = 0$$

Al(III) reacts with H_2O to form $Al(OH)_3$. Thus, each mole of dissolved Al(III) added produces 1 mole of $Al(OH)_3$. Therefore,

$$r_2 V_{tot} = \text{mass of Al(OH)}_3 \text{ produced/unit time}$$
$$= (100 \text{ g/m}^3)(1 \text{ m}^3/\text{sec})$$
$$\left(\frac{\text{mole}}{594 \text{ g alum}}\right)\left(\frac{78 \text{ g Al(OH)}_3}{\text{mole}}\right)$$
$$= 13.13 \text{ g/sec}$$

Since $W_{in} = 0$,

$$W_{out} = 0.98 r_2 V_{tot} \simeq 12.9 \text{ g/sec}$$

Maximum rate of total solids production (on a dry basis)

$$\simeq 992.9 \text{ g/sec}$$
$$\simeq 1430 \text{ kg/day (650 lb/day)}$$

Maximum-Gradient Description

In the macroscopic description of a treatment plant or of an individual separation process, we neglected all spatial variations of concentration. Often, however, it is necessary to describe the change in con-

centration of some constituent in at least one direction. Examples include the change in turbidity or head loss as a function of depth in a granular media filter or the variation in the levels of calcium ion with depth in an ion exchange column. In such cases, it is necessary to include spatial variations of concentration in the materials balance expression. The independent variable is usually chosen to match the direction exhibiting the largest concentration gradient, for example, depth, as in the above examples. The maximum-gradient materials balance, sometimes referred to as a one-dimensional model, finds extensive application in process analysis.

Consider a tubular vessel of radius R as shown in Figure 5-2, through which a fluid flows at an average velocity v_z. Assume that component A in the influent stream decomposes to various products at rate per unit volume of $\bar{r}_A$. We wish to develop quantitative equations describing the space and time variation of A in the vessel. Our natural starting point is a materials balance for component A, using the general expression Eq. 5-3. However, a macroscopic materials balance for component A is impossible because concentration is a function of position in the reactor. Thus, we must evaluate transformations that occur in a hypothetical infinitesimal element of volume $\pi R^2 \Delta Z$ during a small time interval Δt. Although not completely rigorous, this mathematical technique is convenient and conceptually simple. (For additional applications, see Bird, Stewart, and Lightfoot, 1960.)

Between time t and $t + \Delta t$, the mass of component A accumulating in the volume element will be

$$C_A(\pi R^2 \Delta Z)|_{(t+\Delta t)} - C_A(\pi R^2 \Delta Z)|_t$$

where the parallel with subscript refers to the value of the dependent variable at that point in time or space. Net transport of component A by bulk flow in the Z direction during the time interval Δt across

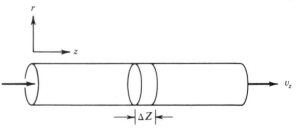

FIGURE 5-2. Notation for tubular vessel.

the volume element bounded at Z and $Z + \Delta Z$ will be

$$C_A v_Z \pi R^2 \, \Delta t|_Z - C_A v_Z \pi R^2 \, \Delta t|_{(Z+\Delta Z)}$$

In addition, the net transformation of A in time W_t will be $r_A \pi R^2 \, \Delta Z \, \Delta t$.

Neglecting all radial gradients and assuming that transfer of mass by diffusional mechanisms is much smaller than advection, the general materials balance becomes

$$\frac{(C_A|_{(t+\Delta t)} - C_A|_t)}{\Delta t}$$
$$+ \frac{(C_A|_{(Z+\Delta Z)} - C_A|_Z v_Z)}{\Delta Z} = r_A \quad (5\text{-}7)$$

Taking the limit of these expressions as Δt and ΔZ approaching zero, we obtain the differential equation

$$\frac{\partial C_A}{\partial t} + \frac{\partial (C_A v_Z)}{\partial Z} = r_A \qquad (5\text{-}8)$$

for the unsteady-state maximum-gradient materials balance in this system. If the system is heterogeneous (e.g, gas–liquid absorption or liquid–solid adsorption), we must also account for mass transfer across phase boundaries. Defining W_i^m as the rate of mass transfer per unit volume (by convention $W_i^m > 0$ when mass is added to a phase), the more general materials balance becomes

$$\frac{\partial C_A}{\partial t} + \frac{\partial (C_A v_Z)}{\partial Z} = r_A + W_A^m \qquad (5\text{-}9)$$

Application of the maximum gradient materials balance is shown in Example 5-2.

EXAMPLE 5-2

Problem

Given the tubular vessel shown in Figure 5-2, develop a steady-state relationship between the length of the reactor and an integral expression of the kinetic decomposition of component A for a homogeneous system.

Solution

We can employ the maximum-gradient model to describe this system. Assuming steady-state operation, $\partial C_A / \partial t = 0$, Eq. 5-9 becomes

$$\frac{\partial (C_A v_Z)}{\partial z} = r_A + W_A^m$$

Because the system is homogeneous, $W_A^m = 0$. Thus, assuming $v_Z \neq f(z)$,

$$\int_0^{C_A} \frac{dC_A}{r_A} = \frac{1}{v_Z} \int_0^L dZ$$

Therefore

$$L = v_Z \int_0^{C_A} \frac{dC_A}{r_A}$$

Therefore, if the kinetics are known, the reactor size can be determined. □

Multiple-Gradient Description

In some industrial processes, notably ion exchange or adsorption separation processes, and in some physical operations, such as fluid mixing, concentration gradients in more than one direction must be considered. Mass transfer by diffusion may also be comparable to mass transfer by bulk flow. The materials balance concept can be applied to develop mathematical models for these systems. Analytical or numerical solution of the resulting differential equations permits the prediction of system performance under varying operating conditions (Bird et al., 1960).

The shell mass balance method used to develop Eq. 5-9 is easily extended to three dimensions. The mass flux (J) of component i by turbulent diffusion in a given direction is defined by analogy with Fick's law of molecular diffusion (Bird et al., 1960),

$$J_i = \frac{-\bar{D} \, \partial C_i}{\partial Z} \qquad (5\text{-}10)$$

where $\bar{D}$ is the average turbulent diffusion coefficient.

Referring to our physical model shown in Figure 5-2, a multiple-gradient materials balance would account for the radial concentration gradient as well as the turbulent diffusion in one or more directions. If we neglect turbulent diffusion in the radial direction, the multiple-gradient materials balance for component A becomes

$$\frac{\partial C_A}{\partial t} + \frac{1}{r}\frac{\partial(rv_rC_A)}{\partial r} + \frac{\partial(v_ZC_A)}{\partial Z}$$
$$= \bar{D}_Z\frac{\partial^2 C_A}{\partial Z^2} + R_A \quad (5\text{-}11)$$

in cylindrical coordinates. Additional examples of the multiple-gradient description are given in Bird et al. (1968), Weber (1972), and other textbooks.

Such differential equations are usually not amenable to analytical solutions, but they can be solved relatively easily using finite-difference techniques. They are routinely used in modeling studies of several separation processes (see, e.g., Keinath and Wanielista, 1975) and of the fate of pollutants in aquatic environments. In the future such models may also find wider application in the design and operation of water treatment separation processes. Thus far, applications have been restricted to research investigations.

INTERFACIAL MASS TRANSFER

Water treatment problems, such as removal of dissolved gases [methane (CH_4) from groundwaters], the dissolution of ozone (O_3) in aqueous solutions, and the precipitation of calcium carbonate in fluid conduits, all transfer material from one phase to another, liquid to gas, gas to liquid, and liquid to solid, respectively. The underlying driving force is the gradient of concentration, with mass transferring in the direction of decreasing concentration.

Interfacial mass transfer plays a crucial role in the separation processes listed in Table 5-2. The engineer's task is to maximize the rate of mass transfer between phases with a minimum expenditure of work. This rate affects the size of equipment needed to separate constituents to a desired concentration.

Phase Equilibria

When a system consists of a single phase, it is known as an *homogeneous* system. A *heterogeneous* system, by contrast, contains two or more

phases in contact. The maximum amount of material that can pass from one phase to another in a heterogeneous system can be estimated using methods of phase equilibria. As with chemical equilibria computations, rates of change cannot be obtained by thermodynamic methods.

When air is mixed with oxygen-free water at constant temperatures and pressure, mass will transfer between the gas and liquid phases until the oxygen concentration in the water reaches its equilibrium value. At this point, as shown by Gibbs (Denbigh, 1961), the chemical potential of oxygen is equal in both phases. In general, for any two phases α and β, mass transfer ceases when

$$\mu_i^\alpha = \mu_i^\beta$$

at constant temperature (T) and pressure (P), analogous to the conditions for thermal ($T^\alpha = T^\beta$) and mechanical ($P^\alpha = P^\beta$) equilibrium. The chemical potential, μ, of component i is defined as the change in Gibbs free energy per mole of i transferred at constant T, P, and number of moles of other components present (Moore, 1962).

Gas–Liquid. Chemical potential is an abstract thermodynamic quantity that can be related to measurable system properties. For a gas–liquid system at equilibrium, assuming ideal fluid behavior in each phase, the partial pressure of component i, p_i, is related to its mole fraction in the liquid phase by the equation

$$p_i = H_ix_i \quad (5\text{-}12)$$

This expression is known as Henry's law (see, e.g., Moore, 1962), with H_i the Henry constant. The mole fraction of i is given as $x_i = n_i/(n_i + n_j)$. Henry's coefficients are tabulated for insoluble gases and volatile liquids in water (Perry, 1973). Gas–liquid systems have many important applications in water treatment, as will be discussed in Chapter 11. Use of Henry's law is illustrated in Example 5-3.

Henry's law is a good approximation of gas–liquid equilibria when the total system pressure is 101 kPa (1 atm) or less, and the mole fraction is less than about 5×10^{-2}. Outside of this range, corrections must be made for nonideal behavior and pressure effects on the Henry coefficient. The recommended procedures are covered in several sources (Reid, 1977; Perry, 1973).

EXAMPLE 5-3

Problem

Chloroform is a'volatile organic liquid (vapor pressure = 175 mm Hg at 20°C) that is an undesirable by-product of water disinfection with chlorine in the presence of certain naturally occurring organic compounds. At 20°C and a total pressure of 1 atm, the Henry constant for chloroform is 172 atm. What would be the equilibrium concentration of $CHCl_3$ in water under these conditions, assuming that the partial pressure of $CHCl_3$ is at saturation?

Solution

Assume that gas and liquid phases are ideal. Then, Henry's law holds,

$$x_{CHCl_3} = \frac{p_{CHCl_3}}{H_{CHCl_3}} \quad \text{where } p_{CHCl_3}$$
$$= \text{saturated vapor pressure at 20°C}$$

or

$$x_{CHCl_3} = \frac{175 \text{ mm Hg}}{760 \text{ mm Hg}} \frac{1 \text{ atm}}{172 \text{ atm}} = 1.34 \times 10^{-3}$$

Now convert mole fraction to mass concentration. In 1 L of water, there are 1000/18 = 55.6 g-mole H_2O. Thus,

$$x = \frac{n}{n + n_{H_2O}}$$

and

$$n_{CHCl_3} = \frac{n_{H_2O}x}{1 - x} = 7.45 \times 10^{-2} \text{ g-mole}$$

$MW_{CHCl_3} = 119$. Therefore,

$$C_{CHCl_3} = 7.45 \times 10^{-2} \times 119 = 8.9 \text{ g/L}$$

This concentration is, in fact, the solubility of $CHCl_3$ in water at 20°C. Because treated water is open to the atmosphere, the partial pressure of $CHCl_3$ in the air will be negligible, and thus the concentration of $CHCl_3$ will tend toward zero. In a closed system, however, such as an enclosed reservoir or tank, a slight vapor pressure might develop. Concentrations of $CHCl_3$ up to 200 μg/L in tap water have been reported. This is equivalent to a partial pressure of only 4×10^{-3} mm Hg (0.53 Pa). $\square$

Liquid–Liquid. Systems of two immiscible or slightly miscible fluids are used to extract a component from one liquid phase to another. If the component is present in small amounts, ideal fluid behavior can be assumed. The general phase equilibria condition then becomes

$$\frac{x_i^{\alpha}}{x_i^{\beta}} = N(T) \qquad (5\text{-}13)$$

where $N(T)$ is the distribution coefficient for component i between the two phases α and β and is a function of temperature only (Denbigh, 1961). Equation 5-13 is known as the Nerst distribution law and can be used to predict the equilibrium concentrations of a component between two partially miscible liquids. Values of N are available for many water–liquid systems (Perry, 1973).

Liquid–liquid extraction, although extensively used in the chemical process industries, has few applications in water treatment. One potential application, however, is a proposed method that uses a two-stage extraction procedure for aluminum recovery from alum sludges generated in the coagulation/sedimentation separation process. The aluminum is first transferred to an organic liquid phase and then reextracted into the aqueous phase. Several analytic methods also rely on liquid–liquid extraction, for example the analysis of chloroform in water. Chloroform is first extracted from the water with a suitable organic liquid (e.g., pentane) and then measured by gas chromatography.

Solution–Solid Interfacial Equilibria. Consider a very dilute organic solute (e.g., phenol in water). If an insoluble solid phase is added to this system, phenol may be attracted to the solid interface. Mass transfer between the liquid and solid phase will occur until thermodynamic equilibrium is satisfied. In this case, however, the adsorbed phenol is concentrated on the surface rather than in the bulk of the solid phase. Thus, we wish to know the liquid bulk concentration and the solid surface concentration at equilibrium.

The surface concentration or density, Γ, is defined as the quantity of adsorbed component per unit surface area, given as

$$\Gamma_i \equiv \frac{n_i^s}{A}$$

where $n_i{}^s$ refers to the number of moles of i attached to the total surface area A. For a single solute at low concentration (ideal solution) in water at constant T and P, Gibbs showed (see, e.g., Adamson, 1967) that the equilibrium surface density is proportional to the change in surface tension, δ, due to the change in solution concentration of the solute. Expressed mathematically,

$$\Gamma = -\frac{1}{RT}\frac{d\delta}{d\ln C} \qquad (5\text{-}14)$$

where C is the equilibrium molar concentration of the solute. Thus, when a solute decreases the surface tension of a solid phase, it will tend to accumulate on that surface. The above expression, known as the Gibbs isotherm because it applies for a fixed temperature, is valid for determining the equilibrium distribution of solutes between any two phases. It can describe the accumulation of a detergent at the water–air interface or the adsorption of phenol on activated carbon or other adsorbent. To apply the Gibbs isotherm, experimental data are required on surface tension changes with changes in bulk solute concentrations. Application of the Gibbs isotherm will be discussed in Chapter 9.

Mass Transfer Between Phases

The separation processes used in water treatment and reuse that depend on efficient mass transfer between phases include gas absorption and its converse gas-desorption or stripping, drying of solids, precipitation, coagulation, adsorption, and ion exchange. Mass transfer also affects the performance of membrane processes such as reverse osmosis, electrodialysis, and hyperfiltration.

The rate at which equilibrium conditions are reached in each of these diffusion processes depends on the magnitude of interfacial mass transfer due to both molecular and turbulent diffusion. Data are available on molecular diffusion coefficients of various solutes in water (see, e.g., Perry, 1973), some of which are listed in Table 5-4. Most molecular diffusion coefficients exhibit values of about 1×10^{-9} m²/sec. Turbulent diffusion coefficients must, however, be experimentally determined or estimated from empirical correlations. Measured values of turbulent diffusion coefficients in water range from 10^{-5} to 10^2 m²/sec. Thus, the rate of turbulent mass transfer generally far exceeds that due to molecular motion.

TABLE 5.4. Molecular Diffusion Coefficients of Common Solutes in Water

Constituent	Molecular Diffusion Coefficient (25°C) (m²/sec $\times 10^9$)
Dissolved Gases	
O_2	2.5
CO_2	2.0
SO_2	1.7
Cl_2	1.4
NH_3	2.0
Weak electrolytes	
Formic acid	1.4
Ethanol	1.3
Glucose	0.7
Lactose	0.5
Urea	1.4
Humic acid molecule (5 nm in diameter)	0.1
Strong electrolytes (at infinite dilution)	
NaCl	0.8
KCl	1.0
$CaCl_2$	1.33
$MgCl_2$	1.25
$NaHCO_3$	1.63
Na_2SO_4	1.2

Mass Transfer Coefficients.   Mass transfer of a component between two phases consists of three sequential steps: (1) transport of the component from the bulk of one phase to the interface; (2) an exchange at or transport through the phase boundary; and (3) transfer from the interface to the bulk of the second phase. This sequence is shown schematically in Figure 5-3 for a gas–liquid system in which component A in the gas phase at a partial pressure of p_A is being absorbed into the liquid phase, containing a bulk concentration of A, C_A, in any appropriate units.

At the interface, it is assumed that the liquid concentration C_{A_i} is in equilibrium with the gas phase partial pressure p_{A_i}. Because the gas is being absorbed, p_{A_i} will be less than bulk partial pressure p_A. Similarly $C_{A_i} > C_A$.

The rate of mass transfer of A per unit interfacial area N_A will depend on numerous factors including temperature, the molecular diffusivity of A, solubility of A in the liquid, and hydrodynamic conditions. By analogy with Fick's law of molecular diffusion, the turbulent rate of mass transfer from gas to liquid

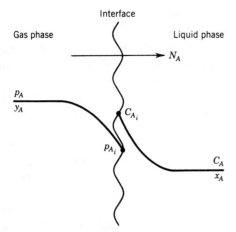

FIGURE 5-3. Mass transfer from gas to liquid phase.

is defined as proportional to a linear concentration gradient,

$$N_A = k_G(p_A - p_{A_i}) \qquad (5\text{-}15)$$

where k_G is a rate constant called the gas phase mass transfer coefficient (units of moles area^{-1} time^{-1} pressure^{-1}). The mass transfer into the liquid phase then becomes

$$N_A = k_L(C_{A_i} - C_A) \qquad (5\text{-}16)$$

where k_L refers to the liquid phase mass transfer coefficient (units of length time^{-1}). Similar expressions can be written for any set of units (see Treybal, 1980).

The coefficients k_G and k_L depend on conditions at the interface and are termed *local mass transfer coefficients*. The interfacial concentrations of A, p_{A_i}, and C_{A_i} are generally not easily measured, however. If we define overall coefficients K_G and K_L (the subscripts referring again to gas and liquid phases, respectively), then by analogy, we define the following:

$$N_A = K_G(p_A - p_{AE}) \qquad (5\text{-}17)$$

and

$$N_A = K_L(C_{AE} - C_A) \qquad (5\text{-}18)$$

In these expressions, p_{AE} is the partial pressure of A in equilibrium with c_A at the system temperature and pressure. Similarly c_{AE} is the concentration of A in equilibrium with the bulk partial pressure of A.

The overall mass transfer coefficients depend on local mass transfer coefficients for both gas and liquid. By definition, the rate of mass transfer per interfacial area from gas to liquid is equivalent for both local and overall mass transfer coefficients. Thus, from Eqs. 5-15 and 5-17, we obtain

$$K_G(p_A - p_{AE}) = k_G(p_A - p_{A_i}) \qquad (5\text{-}19)$$

Assuming Henry's law (Eq. 5-12) holds, then $p_A = H_A x_A$. In dilute solutions $p_A = Hc_A$ because $x_A = c_A/(c_A + c_B) = c_A/c_B$. Thus $\mathbf{H} = H_A/c_B$ where $c_B = 55.6\,\text{kg}$ mole/m^3 for water. Substituting Eq. 5-16, we obtain

$$\frac{1}{K_G} = \frac{1}{k_G} + \frac{\mathbf{H}}{k_L} \qquad (5\text{-}20)$$

Similarly, for liquid phase mass transfer,

$$\frac{1}{K_L} = \frac{1}{k_L} + \frac{1}{\mathbf{H}k_G} \qquad (5\text{-}21)$$

Note that the units of the mass transfer coefficients are defined by Eqs. 5-15 and 5-16. Any consistent set of units may be used, including mole fractions, mass, or moles.

As can be seen from the form of Eqs. 5-20 and 5-21, the overall mass transfer coefficient may be controlled by either liquid or gas mass transfer resistances. For sparingly soluble constituents in water with high values of the Henry constant, the rate of mass transfer will be controlled by resistances at the liquid side of the gas–water interface.

For a more water-soluble gas, such as ammonia, with a Henry constant at 20°C of about 71 kPa (0.7 atm), resistance at the gas side of the interface may control the rate of gas transfer. For the sparingly soluble gas, mass transfer rates could be improved by increasing turbulence on the liquid side of the interface, for example, by intensifying mixing. Between the two solubility extremes, the rate of interfacial mass transfer may depend on resistances at both sides of the interface, as in the SO$_2$–air–water system (Dankwerts, 1970). See Chapter 11 for a more detailed evaluation of this problem.

Estimating Mass Transfer Coefficients. The overall liquid and gas phase mass transfer coefficients defined by Eqs. 5-20 and 5-21 cannot be estimated from fluid properties and must be determined experimentally. One procedure is to measure the transfer coefficients in small-scale models of the separation process; full-scale design can be based on these data. When such data are unavailable, empirical correlations that relate mass transfer coefficients to appropriate dimensionless groups can be used. Many such compilations are available (Treybal, 1968; Perry, 1973) for a variety of mass transfer equipment.

For most diffusional separation equipment, the interfacial area is neither known nor easily measured. Thus, the rate of mass transfer is usually reported volumetrically rather than by interfacial area. The overall mass transfer coefficients for gas–liquid systems are then defined by

$$M_A = K_G a(p_A - p_{AE}) = K_L a(c_{AE} - c_A) \quad (5\text{-}22)$$

where M_A is the mass transfer rate per unit volume (mol time^{-1} length^{-3}) and a is the interfacial area per unit system volume (specific interfacial area) (units of length^{-1}). Similarly the relation between overall and local transfer coefficients becomes

$$\frac{1}{K_G a} = \frac{1}{k_G a} + \frac{\mathbf{H}}{k_L a}$$

and

$$\frac{1}{K_L a} = \frac{1}{k_L a} + \frac{1}{\mathbf{H} k_g a} \quad (5\text{-}23)$$

Mass transfer correlations generally provide empirical data on the volumetric mass transfer coefficients (e.g., $K_L a$).

Process Design of Mass Transfer Processes

Interfacial mass transfer separation processes are conducted with two types of equipment: multiple-stage contacting systems and continuous contacting systems (King, 1971). Multiple-stage systems consist of a succession of discrete regions or stages where intense mixing of the two phases produces rapid rates of mass transfer. Examples common to water treatment include mixing chambers in series for purposes of particulate aggregation or for disinfection.

Continuous contacting systems provide uninterrupted contact between phases. Examples of these systems are columns or towers containing a variety of packing materials or a single-stage agitated vessel. Mass transfer from the liquid to the gas phase, such as the removal of ammonia from water, can be conducted in structures similar to cooling towers. Dissolved constituents are most efficiently transported from the liquid phase to the solid phase in packed columns, using, for example, ion exchange resin or other granular adsorbents.

Process design of packed columns or towers is based on the concept of *transfer units,* whose height and number determine the total height of the contacting unit required to achieve a specified separation efficiency. The height of a transfer unit (HTU) reflects the rate of mass transfer for the particular packing materials. The number of transfer units (NTU) is a measure of the mass transfer driving force, determined by the difference between actual and equilibrium phase concentrations.

Consider a gas–liquid contacting system as shown in Figure 5-4. Flow is countercurrent, and contact between the gas stream G and the liquid stream L can be achieved by a variety of mixing systems, each with a different rate of mass transfer. Removal of an undesirable dissolved gas in the liquid phase will require system height Z and a gas flow selected to reduce the mole fraction of the dissolved gas from x_1 to x_2.

The required gas flowrate is determined from a steady-state macroscopic materials balance for the dissolved gas. Thus, we neglect all internal details of the contacting system. The overall materials balance becomes

$$L x_1 + G y_1 = L x_2 + G y_2 \quad (5\text{-}24)$$

where x and y denote the mole fraction of the dissolved gas in the liquid and gas streams, respec-

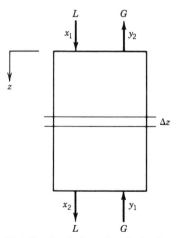

FIGURE 5-4. Notation for design of packed-column, gas–liquid transfer.

tively, and L and G are the respective liquid and gas loading rates (mole area^{-1} time^{-1}). Equation 5-24 assumes that the solutions are dilute and thus neglects any mass transfer within L and G. Solving for G and assuming that the desorbing or stripping gas contains no undesirable gas ($y_1 = 0$), Eq. 5-24 becomes

$$\frac{G}{L} = \frac{x_1 - x_2}{y_2} \tag{5-25}$$

For a specified degree of removal, the required gas flow reaches a minimum when y_2 is at a maximum, when y_2 is in equilibrium with the influent gas concentration x_1. Assuming that Henry's law is valid (dilute solutions) and that the gas phase is ideal, Eq. 5-25 becomes

$$\left(\frac{G}{L}\right)_{min} = \frac{(x_1 - x_2)P_T}{(x_1)H} \tag{5-26}$$

The use of Eq. 5-26 is illustrated in Example 5-4.

EXAMPLE 5-4

Problem

Desorption or stripping of ammonia gas is a potential separation process for the reduction of nitrogen levels in wastewaters. A typical domestic wastewater contains 30 mg/L NH_3–N. What is the minimum gas–liquid ratio required to achieve 90% removal of the ammonia at 5 and 20°C, assuming that the dissolved ammonia is undissociated and the system pressure is 1 atm? Given: H at 5°C = 1.8 atm; H at 20°C = 4.0 atm.

Solution

x_1 = influent mole fraction of NH_3

C_{NH_3} = moles NH_3–N/L

$$x_1 = \frac{C_{NH_3}}{C_{NH_3} + C_j}$$

$C_j = C_{H_2O} = 1000/18 = 55.6$ mole/L

$C_{NH_3} = 30$ mg/L NH_3–N = 2.1×10^{-3} mole/L

Therefore

$$x_1 = 3.77 \times 10^{-5}$$

From Eq. 5-26 the minimum gas–liquid ratio is

$$\left(\frac{G}{L}\right)_{min} = \frac{x_1 - x_2}{Hx_1}P_T = \left[1 - \frac{x_2}{x_1}\right]\frac{P_T}{H}$$

[Notice that $(G/L)_{min}$ is independent of x_1.] Thus, at 5°C

$$\left(\frac{G}{L}\right)_{min} = 0.9 \times \frac{1}{1.8} = 0.50$$

$$= 0.6 \text{ m}^3, \text{ air (STP)/} \\ \text{kg, H}_2\text{O (10 ft}^3\text{/lb)}$$

and at 20°C

$$= 0.9 \times \tfrac{1}{4} = 0.23$$

$$\simeq 0.3 \text{ m}^3, \text{ air (STP)/} \\ \text{kg, H}_2\text{O (4.6 ft}^3\text{/lb)}$$

These minimum gas–liquid ratios can be compared with typical reported air–wastewater ratios of approximately 3 m^3, air (STP)/kg, wastewater (50 ft^3/lb) to achieve 90% removal at temperatures of about 20°C.   ☐

The required gas–liquid contact time and the corresponding height of the contacting equipment cannot be determined using the macroscopic materials balance. An additional level of detail is required, available from the maximum gradient model (see Eq. 5-9) which incorporates the rate of mass transfer.

Consider an elemental depth ΔZ of cross-sectional area A in the contacting system. A liquid phase materials balance for the dissolved gas can be written as

In − out − loss to gas phase = accumulation

Defining the volume rate of mass transfer from liquid to gas as M_L (mole volume^{-1} time^{-1}) and using the shell balance method, the materials balance becomes

$$L(x_z - x_{z+\Delta z})A\,\Delta t - M_L A\,\Delta Z\,\Delta t = 0 \tag{5-27}$$

for steady-state conditions. Dividing by $A\,\Delta Z\,\Delta t$, and taking the limit as ΔZ, the maximum gradient model leads to

$$L\frac{dx}{dZ} + M_L = 0 \qquad (5\text{-}28)$$

A similar materials balance for the gas phase gives

$$G\frac{dy}{dZ} + M_G = 0 \qquad (5\text{-}29)$$

From Eq. 5-22, the volume rate of mass transfer from the liquid to the gas is given as

$$M_L = K_L a(c_E - c) \qquad (5\text{-}30)$$

or on a mole fraction basis,

$$M_L = \frac{K_L a}{C_0}(x_E - x) \qquad (5\text{-}31)$$

where C_0 is the molar volume of water.

Combining Eqs. 5-28 and 5-31 and integrating, the total height of the contacting system becomes

$$Z = \frac{L}{K_L a C_0}\int_{x_1}^{x_2}\frac{dx}{x_E - x} \qquad (5\text{-}32)$$

assuming that the overall liquid phase mass transfer coefficient $K_L a$ is independent of depth.

The first term in this equation is defined as the height of a transfer unit (HTU), given by

$$(\text{HTU})_{\text{OL}} = \frac{L}{K_L a C_0} \qquad (5\text{-}33)$$

the subscript OL indicating that the HTU is based on the overall mass transfer coefficient $K_L a$ for the liquid phase. The second term is the number of transfer units given by

$$(\text{NTU})_{\text{OL}} = \int_{x_1}^{x_2}\frac{dx}{x - x_E} \qquad (5\text{-}34)$$

The magnitude of the HTU is seen to depend inversely on $K_L a$. Thus, increasing the rate of mass transfer by increasing the interfacial area a or by decreasing the liquid or gas phase mass transfer resistances reduces the required column height. Use of the transfer unit concept to design mass transfer processes will be illustrated in subsequent chapters (see Chapters 9–11).

REACTION KINETICS

In the previous section, we have covered quantitative descriptions of the rate of mass transfer between phases and its application to process design of separation processes. In addition to knowledge of the rates of mass transfer, process design of many separation processes requires information on the rates of chemical conversions, such as the rate of precipitation of calcium carbonate or the rate of viral kill with ozone (O_3). While chemical thermodynamics provides estimates of the maximum possible extent of a given chemical transformation, an alternative approach must be used to determine the rate of these transformations. Thus arises the need for *reaction kinetics*.

Reaction kinetics is a branch of physical chemistry concerned with the rates at which molecules collide and are transferred into new compounds. On the basis of rate data, models can be developed to describe quantitatively the rates of the chemical reactions. The application of rate data for the design of structures or reactors in which chemical transformations occur is known as *reactor engineering*. The following sections discuss various models of chemical reactions common to water treatment problems and their application to reactor selection and design.

Rates of Chemical Reactions

Consider the reaction between hypochlorous acid (HOCl) and ammonia (NH_3) in water, shown below in stoichiometric form:

$$NH_3 + HOCl \rightarrow NH_2Cl + H_2O \qquad (5\text{-}35)$$

This expression states that a mole of NH_3 will react with a mole of HOCl to form a mole of monochloramine (NH_2Cl) and a mole of H_2O. NH_3 and HOCl are called the *reactants* while NH_2Cl and H_2O are the reaction *products*. The rate of transformation of any ith reactant or product is defined as the quantity of material changing per unit time per unit volume, given as

$$r_i = \text{rate of reaction of } i = \frac{1}{V}\frac{dN_i}{dt} \qquad (5\text{-}36)$$

where N_i is the number of moles of i and V the system volume. In water treatment applications, most reactions occur in solution at constant density. Thus, $N_i = VC_i$, and the rate of reaction is given by

$$r_i = \frac{dC_i}{dt} \qquad (5\text{-}37)$$

The fundamental problem of reaction kinetics is describing r_i mathematically with respect to the concentration of reactants, products, and parameters of water quality (e.g., temperature, pH, and ionic strength).

Types of Reactions

Chemical reactions occurring in a single phase are known as *homogeneous* reactions. A typical example is the reaction between NH_3 and $HOCl$. In contrast, *heterogeneous* reactions occur when two or more phases participate in the reaction, for example the precipitation of $CaCO_3$ or aluminum hydroxide. The rate of transformation in heterogeneous reactions may also depend on mass transfer of reactants and products between phases.

The rates of some chemical transformations, called *simple* reactions, can be described by an expression for r_i. Complex reactions consist of numerous reaction paths and require multiple reaction rate expressions to describe the overall reaction rate. Such reactions are common to water treatment technology.

Finally, reactions that proceed in one direction only are designated *irreversible*. Reactions occurring in both forward and reverse directions are known as *reversible*.

Reaction Models, Mechanisms, and Order

Consider a simple, irreversible reaction between two reactants A and B producing a product C occurring at constant density in solution. This reaction can first be described with a stoichiometric expression.

$$A + B \rightarrow C \qquad (5\text{-}38)$$

It can be hypothesized that the rate of reaction of A, r_A, is proportional to the product of the reactant molar concentrations C for each reactant A and B. The rate expression would then be

$$-r_A = -\frac{dC_A}{dt} = kC_A C_B \qquad (5\text{-}39)$$

the minus sign indicating that A is disappearing due to the reaction. The proportionality constant between r_A and the reactant molar concentrations is designated as the *reaction rate constant* and by definition is independent of reactant concentrations.

Similarly, the rates of reaction of B and C are given by

$$-r_B = -\frac{dC_B}{dt} = kC_A C_B \qquad (5\text{-}40)$$

$$r_c = \frac{dC_c}{dt} = kC_A C_B \qquad (5\text{-}41)$$

From these expressions, the reaction rates of reactants and products are related by

$$r_C = -r_A = -r_B$$

Often, however, the rate of reaction is not linearly proportional to reactant concentrations. The stoichiometric equation for the general reaction scheme is

$$aA + bB \rightarrow cC + dD \qquad (5\text{-}42)$$

where a, b, c, and d are stoichiometric coefficients. An empirical reaction rate expression may take the form

$$r_A = -kC_A^\alpha C_B^\beta \qquad (5\text{-}43)$$

The coefficients α, β in this rate expression specify the *order* of the reaction with respect to a given constituent and are not necessarily equal to the stoichiometric coefficients a and b. The latter coefficients, however, determine the relation between the individual reaction rates of the reactants and products and is given by

$$-\frac{1}{a}\frac{dC_A}{dt} = -\frac{1}{b}\frac{dC_B}{dt} = \frac{1}{c}\frac{dC_c}{dt} = \frac{1}{d}\frac{dD_d}{dt} \qquad (5\text{-}44)$$

For example, in the reaction

$$2A \rightarrow B$$

each mole of B formed requires 2 moles of A, and thus the rate of disappearance of A is twice as fast as the rate of formation of B, or

$$-r_A = 2r_B \qquad (5\text{-}45)$$

The order of a chemical reaction is a useful empirical parameter characterizing reaction types. The *overall* reaction order is defined as the sum of the power coefficients of the reactant concentrations. In the formation of monochloramine (see Eq. 5-35) the rate of formation has been shown (Weil and Morris, 1949) to be

$$r_{NH_2Cl} = k_{obs}C_{NH_3}C_{HOCl} \qquad (5\text{-}46)$$

This reaction is first order with respect to the reactants and second order overall. We shall see shortly the importance of the reaction order in reactor design.

In the case of monochloramine formation, the overall reaction order is simply the sum of the stoichiometric coefficients, and the reaction rate expression can be written directly from the stoichiometric equation. For all chemical reactions, however, the rate expression and the reaction orders must be determined experimentally. Agreement with stoichiometric equations is usually fortuitous. An empirical model of an overall reaction rate, however, is sufficient for the purposes of reactor design. The actual reaction mechanism can be overlooked, although such information is often useful and can permit extrapolations of the empirical rate law. (See Frost and Pearson, 1961).

Rate Models of Simple Homogeneous Reactions

Examples of simple homogeneous reactions in water treatment systems include oxidation of Fe(II) to Fe(III) and the formation of monochloramine. Rate expressions for these reactions are easily developed and manipulated mathematically. Thus, they can be used to describe the overall kinetics of more complex reactions, which consist of a sequence of simple or elementary reactions.

Irreversible Reactions. Consider the following first-order irreversible homogeneous reaction occurring in solution

$$A \xrightarrow{k} \text{products}$$

The rate expression for the disappearance of A becomes

$$r_A = kf(C_A) \qquad (5\text{-}47)$$

where k is specified to be independent of C_A but a function of temperature, pH, and other water quality parameters. For a first-order reaction Eq. 5-47 becomes

$$-r_A = -\frac{dC_A}{dt} = kC_A \qquad (5\text{-}48)$$

Initially the concentration of A is C_{A_0}, which decreases to C_A as the reaction proceeds. The expression can be integrated by separating variables to give

$$-\ln \frac{C_A}{C_{A_0}} = kt \qquad (5\text{-}49)$$

If a plot of C_A/C_{A_0} versus time on semi log paper is linear, then the observed kinetics are first order, with the rate constant k obtained from the slope of the line. Figure 5-5 shows first-order kinetics for the ozone decomposition in water at various pH levels (Stumm, 1956).

Second-order irreversible homogeneous reactions are more commonly encountered in aqueous

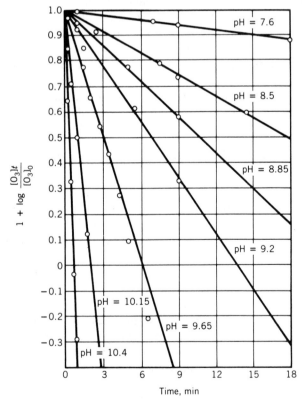

FIGURE 5-5. Ozone decomposition in water ($I = 5 \times 10^{-2}$) illustrating first-order kinetics (Stumm, 1956).

solutions, for example the formation of chloramine. The stoichiometric equation is

$$A + B \xrightarrow{k} \text{products}$$

with the corresponding rate expressions given as

$$-r_A = -r_B = -\frac{dC_A}{dt} = kC_A C_B \qquad (5\text{-}50)$$

When the initial concentrations of A and B are equal, this rate expression becomes

$$-\frac{dC_A}{dt} = kC_A^2 \qquad (5\text{-}51)$$

which on integration yields

$$\frac{1}{C_A} - \frac{1}{C_{A_0}} = kt \qquad (5\text{-}52)$$

A plot of $1/C_A$ versus time of reaction tests the second-order reaction model with the rate constant computed from the slope.

For the more general case, with $C_{A_0} \neq C_{B_0}$, the integrated form of the rate expression can be shown to be

$$\ln \frac{C_B C_{A_0}}{C_A C_{B_0}} = (C_{B_0} - C_{A_0})\, kt \qquad (5\text{-}53)$$

Often, irreversible complex homogeneous reactions with unknown mechanisms can be modeled with an n-order rate expression

$$-r_A = -kC_A^n \qquad (5\text{-}54)$$

where n is an empirically determined reaction order. Integration yields the expression

$$C_A^{(1-n)} - C_{A_0}^{(1-n)} = (n-1)kt \qquad n \neq 1 \quad (5\text{-}55)$$

which can be tested by trial-and-error techniques for various values of n. Rate expressions for different reaction orders are tabulated in several sources (e.g., Smith, 1970; Perry, 1973; Levenspiel, 1972).

Reversible Reactions. Many reactions occurring in water treatment are irreversible because an insoluble product is formed or because the rate constant is large. In some cases, however, the rate of reactant

formation due to reactions between products cannot be neglected. The reaction rate expression must then account for the reverse reactions.

Consider a reversible first-order homogeneous reaction

$$A \underset{k_r}{\overset{k_f}{\rightleftharpoons}} B \qquad (5\text{-}56)$$

where k_f is the rate constant of the forward reaction and k_r the rate constant of the reverse reaction. Under reversible conditions, the rate of reaction for component A depends on both the forward and reverse reactions and is given by

$$-\frac{dC_A}{dt} = k_f C_A - k_r C_B \qquad (5\text{-}57)$$

The concentration of A will thus decrease with time (assuming $k_f > k_r$) until it reaches a constant value corresponding to the equilibrium concentration. At this point, $dC_A/dt = 0$ and

$$\frac{C_B}{C_A} = \frac{k_f}{k_r} = K_{eq.} \qquad (5\text{-}58)$$

which is identical to the expression for the equilibrium constant. Examining the kinetics of the reaction, then, the equilibrium constant is simply the ratio of the rate constants of the forward to the reverse reactions of a reversible reaction sequence. Rate expressions for various types of reversible reactions are tabulated in several sources (Smith, 1970; Frost and Pearson, 1961).

Comparison of Reaction Rates: Irreversible Reactions. The integrated rate expressions discussed above permit prediction of the time required for a desired level of change during a chemical reaction. A widely used parameter to compare reaction rates of irreversible reactions is the half-life, the time within which half of the initial concentration of a reactant has disappeared, that is $C_A/C_{A_0} = 0.5$.

For first-order irreversible reactions, the half-life is obtained by integrating Eq. 5-49:

$$t_{1/2} = \frac{0.693}{k} \qquad (5\text{-}59)$$

which is independent of the initial reactant concentration and inversely proportional to the reaction rate constant. From Eq. 5-52 the half-life for a second-order irreversible reaction is seen to be inversely proportional to both the rate constant and the initial reactant concentration, given as

$$t_{1/2} = \frac{1}{kC_{A_0}} \qquad (5\text{-}60)$$

Values of $t_{1/2}$ as a function of k and C_{A_0} are shown in Figure 5-6 for first- and second-order reactions. First-order rate constants greater than 1 sec^{-1} are characteristic of "fast" reactions. For example, the hydrolysis of chlorine in water has a first-order rate constant of 11 sec^{-1} at 20°C (Eigen, 1962). The reaction is essentially complete (99 percent conversion) in less than 0.5 sec at pH greater than 5.

Rate constants for typical first- and second-order reactions seen in water treatment applications are listed in Table 5-5. Additional compendiums of rate constants for reactions of interest in water treatment can be found in the literature (e.g., see Stumm and Morgan, 1981; Pankow and Morgan, 1982; Hoigne and Bader, 1983). Many of the rate constants are high, which is typical of ionic reactions in solution (Frost and Pearson, 1961). An interesting exception is the reaction of ozone with ammonia. Assuming equimolar concentrations of NH_3 (1.5 mg-N/L) and O_3, the maximum half-life of NH_3 at

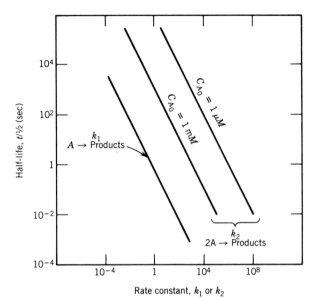

FIGURE 5-6. Comparison of reaction half-life for first- and second-order reactions.

25°C can be shown to be approximately 8 min. The effect of initial reactant concentration on the half-life of second-order reactions is illustrated in Figure 5-6 as a function of the second-order rate constant. If the rate constant is greater than 10^4 M^{-1}/sec, with 1 mM initial concentration, the reactions would be

TABLE 5-5. Selected Rate Constants for Reactions Common to Water Treatment

Reaction	Conditions of Measurement	Rate Constant
Second Order		
$NH_3 + HOCl \rightarrow NH_2Cl + H_2O$[a]	25°C, pH 8.3	5.1×10^6 M^{-1}/sec
Phenol + HOCl → 2-Chlorophenol[b]	25°C, pH 7	2.2×10^3 M^{-1}/sec
$NH_3 + OH° \rightarrow$ products[c]	22–25°C	8.7×10^7 M^{-1}/sec
$NH_3 + O_3 \rightarrow$ products[c]	22–25°C	20 M^{-1}/sec
Phenol + OH° → products[d]	22–25°C	10^{10} M^{-1}/sec
First Order		
$Cl_2(aq) + H_2O \rightarrow HOCl + H^+ + Cl^-$ [e]	20°C	11 sec^{-1}
$SO_2(aq) + H_2O \rightarrow HSO_3 + H^+$ [f]	20°C	3.4×10^6 sec^{-1}
$CO_2(aq) + H_2O \rightarrow HCO_3 + H^+$ [g]	20°C	0.02 sec^{-1}
Higher Order		
Oxidation of Fe(II)[h]	20°C, pH 5	1.3×10^{12} M^{-2}/sec atm

[a] From Weil and Morris (1949).
[b] From Soper and Smith (1926).
[c] From Hoigne and Bader (1978).
[d] From Hoigne and Bader (1976).
[e] From Eigen and Kustin (1962).
[f] From Eigen, Kustin, Maass (1961).
[g] From Kern (1960).
[h] From Stumm and Lee (1961).

considered "fast." As seen in Table 5-5, many ionic reactions will fall in this category.

Rate Models: Complex Reactions

Many reactions in water treatment consist of multiple-reaction paths, with competing sequential or parallel reactions. Complex reactions, however, consist of several elementary reaction steps and are amenable to analysis.

Series Reactions. Consider the first-order irreversible decomposition of reactant A to form a species R which in turn transforms into the final product S, shown by

$$A \xrightarrow{k_1} R \xrightarrow{k_2} S$$

with k_1 and k_2 the respective rate constants of the sequential reactions. The rate expressions for each constituent are

$$-\frac{dC_A}{dt} = k_1 C_A$$

$$\frac{dC_R}{dt} = k_1 C_A - k_2 C_s \qquad (5\text{-}61)$$

$$\frac{dC_S}{dt} = k_2 C_R$$

which can be integrated to yield models for each component as a function of time. For example, the concentration of S is given by Levenspiel (1972):

$$C_S = C_{A_0}\left[1 + \frac{k_2}{k_1 - k_2} \exp(-k_1 t)\right.$$

$$\left. + \frac{k_1}{k_2 - k_1} \exp(-k_2 t)\right] \qquad (5\text{-}62)$$

The relative rate of formation of S, and analogously R, depends on the rate constants k_1 and k_2. When $k_2 \gg k_1$, Eq. 5-62 reduces to

$$C_S \cong C_{A_0}[1 - \exp(-k_1 t)]$$

Similarly, for reverse conditions, $k_1 \gg k_2$,

$$C_S \cong C_{A_0}[1 - \exp(-k_2 t)]$$

Figure 5-7 illustrates the effect of the relative reaction rate constants k_1/k_2 on the formation of the final

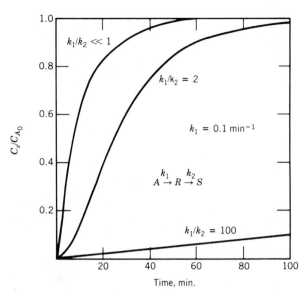

FIGURE 5-7. Rates of formation of reaction products in series reactions as function of relative rate constants.

product for the series reaction. The rate of formation of the final product for irreversible reactions in series is controlled by the *slowest* reaction in the sequence. This reaction becomes the rate-limiting or rate-determining step in the overall reaction sequence.

Parallel Reactions. Strong oxidants such as chlorine and ozone are routinely used in water treatment for the reduction of tastes and odors, color, and bacterial levels. Because of their reactivity, these oxidants simultaneously combine with several substrates present in the water in addition to the one of interest. This leads to increased chemical consumption. Chemical costs are reduced if the undesirable parallel reactions of the oxidant can be suppressed.

Consider the following sequence of irreversible second-order parallel reactions in which reactant B combines with reactants to form a desirable by-product R and an undesirable by-product S:

$$A + B \xrightarrow{k_1} R$$
$$C + B \xrightarrow{k_2} S$$

We would like to control the environmental factors such that the ratio of moles of R formed per mole of B consumed, defined as the yield of R, is a maximum. In other words, we would like to increase the *selectivity* of B for reactant A.

The rates of change for R, S, and B are given by

$$\frac{dC_R}{dt} = k_1 C_A C_B$$

$$\frac{dC_S}{dt} = k_2 C_c C_B$$

$$-\frac{dC_B}{dt} = k_1 C_A C_B + k_2 C_c C_B$$

and the yield of product R becomes

$$\frac{dC_R}{-dC_B} = \frac{k_1 C_A}{k_1 C_A + k_2 C_c} = \frac{1}{1 + (k_2/k_1)(C_c/C_A)} \quad (5\text{-}63)$$

This ratio reaches a maximum when $k_1 \gg k_2$ assuming similar initial concentrations of the reactants A and C, or if $C_C \ll C_A$ with similar rate constants. The rate of formation of R compared to S is shown schematically in Figure 5-8 for different relative values of k_1/k_2.

Control of the yield and selectivity of parallel reactions pose challenging engineering problems. Options include altering the rate constants by varying suitable water quality parameters (e.g., pH), removing competing reactants, or selecting the appropriate reactor system.

Series–Parallel Reactions.   Reactions consisting of both series and parallel reaction paths are classified as series–parallel reactions. An important example in water treatment is the breakpoint chlorination reaction in which ammonia is progressively oxidized with chlorine to nitrogen gas and other products. Analysis of series–parallel reactions is beyond the scope of this text. Readers are referred to other sources for discussions of these reactions. (See, e.g., Levenspiel, 1972.)

Heterogeneous Reactions

In the rate expressions for simple and complex reactions discussed thus far, we have assumed that reactions are homogeneous. If one or more additional phases are present, however, mass transfer phenomena can no longer be neglected.

Rate expressions for heterogeneous reactions must account for both the reaction kinetics and the rate at which reactants or products are transported between phases. In a gas–liquid system, for example, O_3 bubbled into water, the ozone molecules must first be transferred from the gas phase to the solution before reactions between O_3 and dissolved reactants can occur. Because these processes occur in series, either one can become rate limiting. As we have seen, the rate of mass transfer depends on the interfacial area of the gas phase, and thus, the manner in which the gas and the liquid are contacted. These three factors, the kinetics, the rate of mass transfer, and the type of phase contacting device, are interdependent and may control the overall rate of reaction.

Some typical examples of heterogeneous reaction systems common to water treatment are listed in Table 5-6. As we shall see for the gas–liquid examples, the gas–liquid contacting system will be selected to minimize the rate-limiting step, either the reaction rate or the rate of mass transfer in the overall reaction sequence. The precipitation of insoluble solids is a typical liquid–solid system in which a new phase, the precipitant, is formed as the reactants pass from the solution to the surface of the new phase by diffusion.

Mass Transfer and Reaction Kinetics.   Consider a gas A dispersed in water reacting with water molecules to form products that combine with dissolved

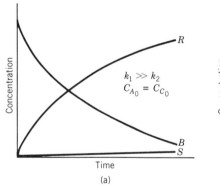
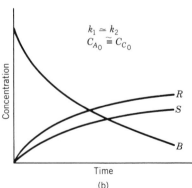

FIGURE 5-8.   Examples of product selectivity for second-order parallel reactions.

TABLE 5-6. Examples of Heterogeneous Reaction Systems Common to Water Treatment

Type	Examples
Gas–liquid	Oxidation–reduction reactions with $O_3(g)$, $Cl_2(g)$, and $ClO_2(g)$
	$CO_2(g)$ absorption for pH control and dechlorination with $SO_2(g)$
	Ammonia removal by air stripping
Liquid–solid	Disinfection with HOCl or $O_3(aq)$
	Precipitation of solids [e.g., $Al(OH)_3$, $CaCO_3$, $Ca_3(PO_4)_2$, and others]
	Corrosion of metallic surfaces
	Adsorption (on activated carbon, alumina)

materials in the solution. According to the two-film model of this process (Lewis and Whitman, 1924), the overall rate of reaction of gas A depends on the following sequence of events:

1. *Diffusion in the gas phase.* Molecules of A are transported by molecular diffusion through a hypothetical gas film or zone of mass transfer resistance to the gas–liquid interface.

2. *Interfacial equilibrium.* Gas A at the interface is assumed to be at an equilibrium concentration with water.

3. *Reaction in film.* Gas A reacts with H_2O at the interface or at some location in a hypothetical liquid film on the liquid side of the interface. The reaction products react with dissolved materials in the film.

4. *Diffusion into liquid phase.* The reaction products are transported by molecular diffusion through the liquid film and then mixed completely with the bulk phase by turbulent diffusion.

The overall rate expression for the disappearance of A depends on the type of gas, the rate of reaction with water, and the type of contacting system used. If the reaction is extremely fast, as in the case of SO_2 or Cl_2, only gas or liquid phase mass transfer will control the overall rate. Under these conditions, the reaction is said to be diffusion controlled. At the other extreme, very slow kinetics lead to kinetically controlled reactions. Examples include disinfection and often precipitation of solids.

The behavior of gases dispersed in water according to the two-film model is shown in Figure 5-9 for the cases of a very fast reaction between (Cl_2) and water and a relatively slow reaction (CO_2). In the latter case, free carbon dioxide exists in the liquid phase, whereas in the former no free chlorine is found, having reacted instantaneously at the interface to form HOCl if the pH > 5.

Reaction Rate Models. Models of overall rate expressions for heterogeneous reactions have been developed for several different cases representing various rates of mass transfer and chemical reactions (Levenspiel, 1972; Dankwerts, 1970; Saunier, 1976). Each case requires an independent analysis, and because of analytical complexities, few systems have been quantitatively modeled. However, defining the kinetic regime in which a heterogeneous reaction falls determines the type of contacting device that will maximize the overall reaction rate.

In the case of gas–liquid reactions, very fast reactions will occur at the interphase boundary. To maximize the overall reaction rate, a contactor system should be chosen that provides large interfacial areas (Levenspiel, 1972). For a gas bubbled into

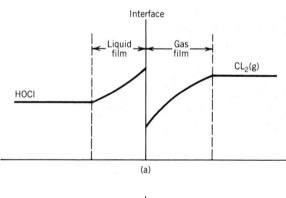

(a)

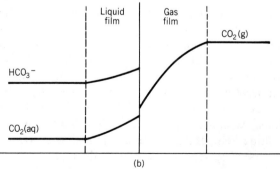

(b)

FIGURE 5-9. Gas–liquid mass transfer with chemical reaction. (a) Chlorine gas in water, "fast" reaction. (b) Carbon dioxide in water, "slow" reaction.

water, the smallest possible bubbles would provide the largest interfacial area for a given gas flowrate.

Slow reactions, on the other hand, cannot be influenced by varying the rate of mass transfer, and thus large interfacial areas per volume of reactor will not increase the rate of reaction. Such reactions therefore need large volumes of liquid relative to the gas volume. Bubble contactors can produce a liquid volume fraction that approaches 98 percent (Kramer and Westererp, 1963).

Additional examples of fast and slow heterogeneous reactions will be discussed in Chapter 11.

Dependence of Rate Constants on Intensive Variables

A distinctive feature of water treatment is the varying character of each water source. A wide range of values of temperature, pH, and ionic composition are encountered in water treatment practice. In addition, environmental conditions often produce wide seasonable fluctuations in these intensive variables. Quantitative estimates of the dependence of empirical rate constants on temperature, pH, ionic composition, and other factors are essential for proper control of the reaction of interest.

Temperature. Chemical reactions occur when reactant molecules collide. Because the frequency of these collisions will increase as the temperature increases, a temperature rise increases the rate constants. It has been found (see, e.g., Moore, 1962) that the temperature dependence of reaction rate constants can be well represented by an equation of the form

$$k = k_0 \exp \left(\frac{-E}{RT} \right) \qquad (5\text{-}64)$$

where k_0 is the frequency factor representing the number of molecular collisions per time and E the activation energy. According to this expression, known as Arrhenius' law, a semilog plot of the rate constant as a function of the inverse of the absolute temperature should be linear. Examples of this behavior are shown in Figure 5-10. For microbial reactions, Arrhenius-type behavior is exhibited up to a limiting temperature where enzyme deactivation occurs. The reaction rate then decreases rapidly with further temperature increases.

Based on the Arrhenius model, it can be seen that the temperature sensitivity of the rate con-

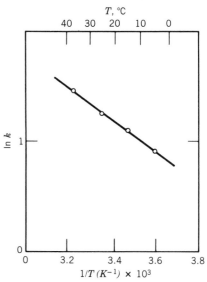

FIGURE 5-10. Temperature dependence of second-order rate constant, formation of monochloramine (Weil and Morris, 1949).

stants depends on the magnitude of the activation energy. Values of E for reactions in solution range from 4 to 125 kJ/mole (1–30 kcal/mole) (Moelwyn-Hughes, 1947). For example, the formation of chloramine (see Figure 5-10) has an activation energy of 10.4 kJ/mole (2.5 kcal/mole) (Weil and Morris, 1949), whereas the hydrolysis of aqueous CO_2 has an E of about 55 kJ/mole (13 kcal/mole) (Kern, 1960). Reactions with high E values show a much greater sensitivity to temperature increases compared to low E reactions, as shown in Figure 5-11. A common formula states that rate constants generally double for every 10°C increase in temperature.

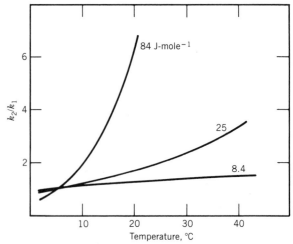

FIGURE 5-11. Sensitivity of temperature dependence of rate constants to activation energy.

As can be seen from Figure 5-11, this rule only holds for an activation energy of approximately 50 kJ/mole (12 kcal/mole), which is in the median range of most reactions in solution (Frost and Pearson, 1961).

Ionic Strength. We have previously used molar concentration terms in reaction rate expressions. For reactions in nonideal solution, activities should replace these concentrations (Moore, 1962). Thus, for an irreversible second-order reaction the rate of disappearance of either reactant would be

$$-r_A = -r_B = k(a_A)(a_B)$$

where $a_i = f_i C_i$ and f_i is the component activity coefficient. The observed rate "constant" then becomes a function of the activity coefficients, given by

$$k_{obs} = k f_A f_B$$

for the example above.

For waters with ionic strength of less than 0.01 M, the observed rate constant will vary with both ionic charge and ionic strength, based on the Debye–Hückel law (Frost and Pearson, 1961) and given by

$$\ln \left(\frac{k}{k_{I=0}} \right) = 2 Z_A Z_B \alpha \sqrt{I} \qquad (5\text{-}65)$$

where Z_i is the charge on the ionic reactant and α is a constant from the extended Debye–Hückel theory given as $A = 1.82 \times 10^6 (\varepsilon T)^{2/3}$ where ε is the dielectric constant for water and T is the absolute temperature (°K). ($\alpha = 1.17$ at 25°C). This relation shows that increases in ionic strength or salt content can cause either a decrease or increase in the rate constant relative to its value at zero ionic strength. For reactions between neutral molecules, no effect of ionic strength on the rate constant is predicted. Weil and Morris (1949) demonstrated that the rate constant for the formation of chloramine was independent of ionic strength, supporting the reaction mechanism hypothesis that the neutral molecules NH_3 and $HOCl$ were the principal reactants.

Generally, potable water sources exhibit ionic strengths less than 10^{-2} M [approximately 400 mg/L TDS, or 60 mS/m (600 μmho/cm)], and the effect of ionic strength on rate constants is small. When treating saline waters or seawater ($I = 0.65$ M), however, the effect of ionic strength cannot be neglected. Rate constants determined in low-ionic-strength waters for ionic reactions should not be extrapolated to high-ionic-strength systems.

pH. The hydrogen ion concentration (pH) is a process variable that plays a major role in the control of reaction selectivity and product distribution. The pH influences reaction rates through direct reaction pathways (e.g., precipitation of aluminum or magnesium hydroxide), control of reactant species, or as a reaction catalyst. It is only in the latter case, however, that the actual rate constant is affected. Often, control of the pH will permit acceleration of desired reaction pathways. A classic example is the oxidation of manganese (Morgan and Stumm, 1964).

The influence of pH on reaction rates can be illustrated by the chloramine formation reaction discussed previously. The rate expression for this second-order, nearly irreversible reaction is given as

$$\frac{d NH_3Cl}{dt} = k_{obs}(NH_3)(HOCl) \qquad (5\text{-}66)$$

The observed role constant depends strongly on pH, as shown in Figure 5-12, with a maximum rate occurring at pH 8.3, corresponding to the pH at which the product of the concentrations of the undissociated reactants is also a maximum.

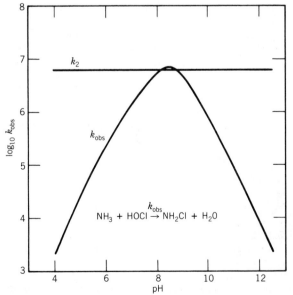

FIGURE 5-12. pH dependence of observed rate constant, formation of monochloramine.

The reaction rate expression for the formation of monochloramine can also be written in terms of the total concentrations of ammonia and chlorine species, as developed from the respective ionic equilibria,

$$HOCl \rightleftharpoons H^+ + OCl^- \qquad (K_1)$$

$$NH_3 + H_2O \rightleftharpoons NH_4^+ + OH^- \qquad (K_2)$$

and given by

$$\frac{d\,NH_2Cl}{dt} = K_{obs}\left[\frac{y}{1 + K_2/OH^-}\right]\left[\frac{x}{1 + K_1/H^+}\right] \quad (5\text{-}67)$$

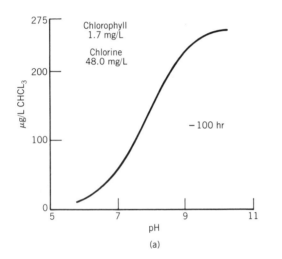

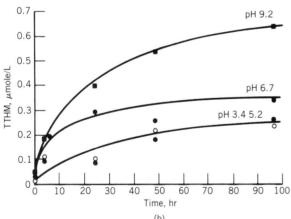

FIGURE 5-13. Acid–base catalysis; effect of pH on rate formation of chloroform by chlorination of organic compounds. (a) Chlorination of chlorophyll at varied pH. Initial chlorophyll concentration (nominal), 1.7 mg/L; initial chlorine, 5.7×10^{-4} M (40 mg/L). pH values, 5.8, 6.6, 7.0, 9.2, 10.0. Chlorine demand after 100 hr 36 mg/L in all solutions. (b) Effect of pH on THM production from 1 mg/L humic acid, 25°C, chlorine dose 10 mg/L.

where $x = HOCl + OCl^-$ and $y = NH_3 + NH_4^+$. The rate constant then becomes

$$k_2 = \frac{k_{obs}}{(1 + K_2/OH^-)(1 + K_1/H^+)}$$

and is found to be independent of pH (see Figure 5-12) and a function of temperature and ionic strength only. As we shall see, control of the system pH is an important factor in proper design and operation of water treatment plants.

In many cases, however, the pH controls the magnitude of the rate constant by acting as a reaction catalyst (Frost and Pearson, 1961). The hydrogen or hydroxyl ions can form reactive intermediates with various reactants, which accelerate the overall reaction rate. A significant example in water treatment is the reaction pathway for producing trihalomethanes during chlorination of drinking water. The reaction rate increases rapidly with rising pH, as shown in Figure 5-13 (Morris and Baum, 1978). The rates of all oxidation–reduction reactions occurring in water also depend on the catalytic effect of the hydroxyl or hydrogen concentrations. Examples of these reactions will be discussed in subsequent chapters.

REACTOR DESIGN

The final stage in the process design of a chemical conversion system is the selection and sizing of the chemical reactor. Chemical reactions can be carried out in a variety of structures including large rectangular tanks, pipes, long channels, columns, or towers. Stoichiometric and kinetic descriptions of chemical reactions combined with knowledge of practical flow patterns provide the basis for reactor selection and design. Other factors to consider include the quantity of material being processed and the structural requirements of the reactor selected (Levenspiel, 1972; Perry, 1973; Kramer and Westerterp, 1963; Smith, 1970; Froment and Bischoff; 1979).

Application of reaction engineering in the chemical processing industries has been widespread (Shreve and Brink, 1977) and has significantly increased productivity. In the treatment of water for domestic, agricultural, or industrial use, however, the impact of reaction engineering has been less dramatic. The principal constraint has been the lack of fundamental data on reaction mechanisms and reaction rate constants in solution. As these data are generated, use of reaction engineering in water treatment design should increase, especially because reactions in water are generally isothermal and occur without significant volume changes. These factors simplify the reactor analysis considerably.

Reactor types

Batch and *flow* reactors are the principal types of chemical reactors. Batch reactors are characterized by noncontinuous and thus non-steady-state operation. Reactants are mixed together, and the reaction is allowed to go to completion. Batch reactors are widely used in the production of small-volume, specialty chemicals in the chemical processing industries. Their use in water treatment applications however is restricted to laboratory-scale investigations. Flow reactors operate on a continuous basis and are used exclusively in water treatment applications because of the large volumes of water processed.

With few exceptions, most reactions of importance in water treatment are heterogeneous; that is, they consist of reactions occurring in more than a single phase, so-called multiphase reactions. Typically, a gas is mixed with water to achieve the transformation of some undesired constituent, for example, oxidation of manganese. In other cases, a solid precipitate is formed, which removes the contaminant.

Various types of reactors can be used to carry out multiphase reactions. Examples, characteristic of water treatment problems include stirred tanks, venturi mixers, several tanks in series, packed columns, spray towers, and recycle reactors.

Stirred tanks provide intense or slow mixing of reactants, depending on the reactions desired. In the mixing of coagulants, intense mixing is desirable to disperse the reactants quickly. Flocculation, on the other hand, requires moderate agitation to increase the rate of formation of solids that can be subsequently removed. Venturi reactors consist of venturi nozzles through which reactants are sprayed into the water, thereby provided intense mixing with low pressure drops. They are often used for injection of chlorine and other higher-soluble gases into water.

Packed columns consist of a cylindrical column containing appropriate packing materials that provide high interfacial areas between water and the gas, usually air. They are used for stripping of undesirable gases or volatile organic compounds from water. Two-phase flow in packed towers is countercurrent. Other reactor types used to contact water and gases include bubble tanks, where a gas is bubbled into the water in tanks, and spray towers, where water is sprayed into the air, used primarily for removal of volatile materials. Recycle reactors operate with a portion of the flow returned to the reactor inlet. Such reactors are used principally for precipitation reactions in which a portion of the precipitated solids is recycled to accelerate the rate of contaminant removal.

Examples of multiphase reactor types used in water treatment processes are summarized in Table 5-7.

Fluid behavior in real reaction vessels is complex and difficult to describe mathematically. This problem is resolved through the use of two ideal models that represent the extreme fluid conditions in a reactor, complete mixing or no mixing of reactants. The contents of a continuous-stirred tank reactor (CSTR), sometimes referred to as a backmix reactor or a mixed-flow reactor, are defined as thoroughly mixed. The concentration of reactants and products is assumed to be uniform at all points in the reactor. Thus, the concentration of any compo-

TABLE 5-7. Examples of Reactors in Water Treatment Multiphase Reactions

Process	Reactor Type	Example
Oxidation	Stirred tanks	Oxidation of iron
	Tanks in series	Oxidation of manganese
	Bubble tanks	Dechlorination by SO_2
	Venturi	Ozone reactions
Disinfection	Stirred tanks	Ozone
	Tanks in series	Chlorine
	Venturi	Chloramination
Coagulation–flocculation	Stirred tanks, tanks in series	Removal of particulates with alum
	Recycle reactor	Removal of humic compounds with ferric chloride
Lime softening	Stirred tanks, recycle reactor	Removal of calcium by lime
Air stripping	Packed column	CO_2 removal
	Bubble tanks	NH_3 removal

nent leaving the reactor is identical with the concentration at any point inside the reactor. Furthermore, reactants in the influent of the reactor are assumed to be instantaneously mixed, and their concentration decreases immediately to the bulk concentration in the reactor. The opposite extreme is simulated by a *plug-flow (PF) reactor*. Fluid passes through the reactor as if it were plug, with no longitudinal or latitudinal mixing. Each fluid element thus has the same residence time in the reactor. Each fluid element in the reactor acts as a small batch reactor, with the batch reaction time equivalent to residence time in the PF reactor.

As would be expected, no real reactors meet exactly the criteria of either the CSTR or PF models. However, the models often closely simulate fluid behavior in real systems and have the important advantage of mathematical simplicity. The effect of flow patterns that deviate from these models on reactor design will be discussed subsequently.

Conversion in Ideal Reactors: Simple Reactions

Consider a simple, irreversible first-order reaction *A* yielding products with rate constant *k*. Reactor volume often has a dominant influence on capital cost. As a result, we wish to select the reactor type and flow pattern that will produce maximum removal of an undesirable contaminant with a minimum of reactor volume for a given flow. We apply the principles of process design discussed earlier, materials balances and reaction kinetics.

Because the flow patterns in the ideal reactor models are fixed, we can neglect local concentration variations and use the macroscopic materials balance approach (see Eq. 5-3).

Batch Reactor. No material enters or leaves the batch reactor during the course of the reaction and all the components of reaction are assumed to be instantaneously and uniformly distributed throughout the reactor at the beginning of the reaction. Thus, the materials balance for component *i* becomes

$$\frac{d(C_i V)}{dt} = r_i V \qquad (5\text{-}68)$$

where *V* is the batch reactor volume, C_i the molar concentration, and r_i the reaction rate of *i*. As the volume stays constant for most reactions in solu-

tion, this expression simplifies to

$$\frac{dC_i}{dt} = r_i$$

For the first-order disappearance of reactant *A*, this equation becomes

$$\frac{dC_A}{dt} = kC_A$$

which integrates to

$$\frac{C_A}{C_{A_0}} = \exp(-kt)$$

This expression is identical to Eq. 5-49, and thus conversion in a batch reactor can be computed directly from the integrated forms of the reaction rate expressions. In the case of a first-order reaction a semilog plot can be used.

Continuous-Stirred Tank Reactor. In a CSTR the fluid element is by definition the total reactor volume, that is, uniform concentration throughout the reactor. Water enters and leaves the reactor at a given volumetric flowrate of $Q[L^3 t^{-1}]$, and the quantity of any component in the reactor is given by $C_i V$. Notation for the CSTR is shown in Figure 5-14. The concentration of component *i* is C_{i_0} entering the re-

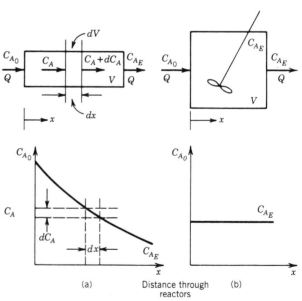

FIGURE 5-14. Notation for ideal reactors: (a) Plug flow; (b) mixed flow.

actor and C_i leaving the reactor. In symbols, the materials balance becomes

$$\frac{d(C_i V)}{dt} = QC_{i_0} - QC_i + r_i V \qquad (5\text{-}69)$$

Usually continuous reactors are operated near steady state, that is, no change in C_i with time, or $dC_i/dt = 0$. Making this assumption and rearranging, Eq. 5-69 becomes

$$\theta = \frac{V}{Q} = \frac{C_{i_0} - C_i}{-r_i} \qquad (5\text{-}70)$$

where θ is defined as the mean hydraulic residence time in the reactor of volume V. The volume of a CSTR for any desired conversion of component i can be computed directly from Eq. 5-70 if the reaction kinetics and rate constants are known.

Plug-Flow Reactor. The fluid elements in a PF reactor flow in an orderly fashion without intermixing. As is shown in Figure 5-14, the reactant influent concentration will decrease as the fluid element progresses through the tubular reactor. For each element the concentration of a reactant entering is C_i, and exiting concentration is $C_i + dC_i$, where dC_i is the differential change in concentration in the fluid element.

For steady-state operation, the materials balance around the volume element dV becomes

$$QC_i - Q(C_i + dC_i) + r_i\, dV = 0 \qquad (5\text{-}71)$$

Rearranging, separating variables, and integrating with the boundary conditions, this expression becomes

$$\theta = \frac{V}{Q} = \int_{C_{i_0}}^{C_i} \frac{dC_i}{r_i} \qquad (5\text{-}72)$$

which can be integrated for any known kinetics. Equation 5-72 is the general design equation for a PF reactor. The reactor size needed for a particular conversion can be predicted if the kinetics are known. A tank or reactor of this size provides a hydraulic residence time equal to the time required to accomplish the conversion in a batch reactor.

Comparison of CST and PF Reactors: Simple Reactions. For a first-order homogeneous irreversible reaction in solution in which contaminant A is con-

verted to an innocuous form, the mean hydraulic residence times for CST and PF reactors are obtained from the respective design equations, as follows:

$$r_A = -kC_A$$

and

$$\theta_{\text{CSTR}} = \frac{1}{k}\left[\frac{C_{A_0} - C_A}{C_A}\right] \qquad (5\text{-}73)$$

$$\theta_{\text{PF}} = \frac{1}{k}\ln\frac{C_{A_0}}{C_A} \qquad (5\text{-}74)$$

If the desired conversion of reactant A is 95 percent ($C_A/C_{A_0} = 0.05$), it is easily seen that the CSTR requires a residence time approximately six times larger than a PF reactor. For larger conversions, the ratio of the residence time of the CSTR to PF reactor increases, as shown in Figure 5-15.

For a second-order irreversible reaction with kinetics given by $r_A = -kC_A^2$, the ratio of the CSTR and PF reactor residence times becomes

$$\theta_{\text{CSTR}}/\theta_{\text{PF}} = \frac{C_{A_0}/C_A - 1}{1 - C_A/C_{A_0}} \qquad (5\text{-}75)$$

Compared to first-order kinetics, the residence time ratio increases to 20 at a 95 percent conversion level, as compared in Figure 5-15.

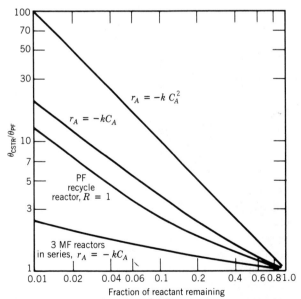

FIGURE 5-15. Comparison of reactor volumes for different reactor types, reaction kinetics, and extent of conversion.

In general, it can be demonstrated that for all reaction orders greater than zero, the volume of a CSTR will always be larger than that of the PF reactor for equivalent conversion (Levenspiel, 1972). Increasing the desired conversion enlarges the difference in reactor size. From an economic viewpoint, then, a PF reactor is superior to a CSTR whenever the reaction order is greater than zero.

Tanks-in-Series Reactors. In water treatment applications PF or tubular reactors are usually impractical from a structural standpoint. Also, most reactions require intense mixing of reactants with the water as the initial step in the reactor. A common alternative to the PF reactor is a series of CSTRs constructed in an appropriate manner, which provides the advantages of the PF reactor and is easily constructed.

Consider three CSTRs of equal volume in series. For first-order kinetics, a materials balance around the first CSTR gives, for steady-state operation,

$$\theta_i = \frac{C_0 - C_1}{kC_1}$$

or

$$\frac{C_0}{C_1} = 1 + k\theta_1$$

Similarly

$$\frac{C_1}{C_2} = 1 + k\theta_2 \qquad \frac{C_2}{C_3} = 1 + k\theta_3$$

Thus, overall conversion for the reactor system is

$$\frac{C_0}{C_3} = \prod_{i=1}^{3} (1 + k\theta_i) \qquad (5\text{-}76)$$

For reactors of equal volume

$$\theta = \frac{1}{k}\left(\frac{C_0}{C_3}\right)^{1/3} - 1 \qquad (5\text{-}77)$$

Comparing the residence time for three CSTRs in series to that of a PF reactor, we find that at a 95 percent conversion level and first-order kinetics, the residence time ratio is approximately 1.7 compared to 6 for a single CSTR. Thus, significant reductions in required reactor volume can be achieved by using a tanks-in-series design whenever the reaction order is greater than zero, as illustrated in Figure 5-15.

For N CSTRs of equal volume in series, the design equation can be shown to be

$$\theta_N = \frac{1}{k}\left[\left(\frac{C_0}{C_N}\right)^{1/N} - 1\right] \qquad (5\text{-}78)$$

with θ_N the mean hydraulic residence time in a single reactor and the total residence time in the system given by $\theta = N\theta_N$. As the number of tanks increases, the tanks-in-series reactor approaches the performance of a PF reactor. Thus,

$$\lim_{N\to\infty} \theta_N = \theta_{PF} = \frac{1}{k} \ln \frac{C_0}{C}$$

Recycle Reactor. The reuse of treated wastewater for industrial cooling purposes, which is increasing in water-short areas worldwide, requires removal of phosphorus to minimize scaling and biological growth on heat exchanger surfaces. One phosphorus removal process relies on the reaction with calcium to form insoluble calcium phosphate solids. Very slow precipitation will occur, however, if phosphate solids are not initially present in the system, and thus seeding is required for economically acceptable reaction rates.

If the reaction is carried out in a CSTR or PF reactor with an initial seeding of phosphate solids, high level of seed is required. To accomplish this, a portion of the solids are separated from the reactor effluent and returned to the influent. As a result phosphorus can be efficiently removed. This type of reactor is known as a *recycle reactor* and is used to improve the yields of heterogeneous reactions. In addition to phosphorus precipitation with calcium, water treatment applications include lime softening, lime coagulation, and organic carbon reduction in microbial systems (activated sludge).

Notation for determining chemical conversion in recycle reactors is shown in Figure 5-16. A portion of the throughput Q is separated from the reactor effluent and recirculated to the influent at a rate q. We pose the question, how does the recycle ratio $R = q/Q$ affect the size of a CSTR or PF reactor needed to achieve a desired conversion for a heterogeneous reaction exhibiting first-order kinetics?

Continuous-Stirred Tank Reactor. A steady-state materials balance around the CSTR of volume V (see Figure 5-15) gives

$$QC_{A0} + qC_A - C_A(Q + q) - kC_AV = 0$$

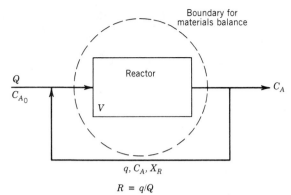

Boundary for
materials balance

Reactor

Q
C_{A_0}

C_A

V

q, C_A, X_R

$R = q/Q$

FIGURE 5-16. Notation for materials balance for a recycle reactor.

This simplifies to

$$\theta_{CSTR} = \frac{1}{k}\left(\frac{C_{A_0}}{C_A} - 1\right) \qquad (5\text{-}79)$$

which is identical with Eq. 5-70 for a CSTR. However, the rate constant k for this heterogeneous reaction depends on the concentration of solids in the reactor. This concentration can be controlled by recycling the separated solids that are present at a concentration of X_R. Therefore, the recycle ratio R can be controlled to produce the desired level of solids in the reactor and thus a maximum rate constant.

Plug-Flow Reactor. The steady-state materials balance for the PF reactor with recycle (PFR) around the fluid element dV becomes

$$(Q + q)\, dC_A = r_A\, dV$$

Separating variables and integrating between the influent concentration C_{A_I} and C_A, one obtains, for first-order kinetics,

$$\theta_{PFR} = \left[\frac{1 + R}{k}\right] \ln \left[\frac{C_{A_I}}{C_A}\right] \qquad (5\text{-}80)$$

A materials balance shows that the influent concentration is given by

$$C_{A_I} = \left[\frac{C_{A_0}Q + C_{A}q}{Q + q}\right]$$

Thus, the residence time for a PF reactor with recycle becomes

$$\theta_{PFR} = \frac{1 + R}{k} \ln \left(\frac{C_{A_0}/C_A + R}{1 + R}\right) \qquad (5\text{-}81)$$

With no recycle ($R = 0$), the expected design equation for a PF reactor is obtained. At the other extreme, with $R \to \infty$, the reactor approaches the CSTR model. The recycle reactor is thus a model of fluid behavior lying between the two extremes of complete mixing and no mixing. In Figure 5-15, the mean residence time, θ_{PFR}, for a PF reactor with 100 percent recycle ($R = 1$) is compared to θ_{PF}. At the 95 percent conversion level, 100 percent recycle increases the required reactor volume by a factor of 4. As R decreases, a smaller reactor volume is required. Thus, a PF recycle reactor should be designed with an efficient solids separator to ensure minimum recycle ratios that are consistent with the levels of solids required in the reactor. Complete process design also requires a solids balance (see, e.g., Ferguson, 1973).

Conversion in Ideal Reactors: Complex Reactions

Simple, elementary reactions are the exception rather than the rule in water treatment applications. Usually chemicals added to water undergo series, parallel, or series–parallel reactions that often compete with a desired reaction path or paths and thereby decrease reaction selectivity. Because rate expressions and rate constants are rarely available for all elementary paths in a complex reaction sequence, quantitative reactor design is difficult. Nonetheless, for complex reactions we can at best obtain a basis for selecting the type of reactor that maximizes formation of the desired reaction products.

Parallel Reactions. Consider the hypothetical first-order irreversible decomposition of reactant A by the following parallel reactions:

in which R and S are desirable and undesirable reaction pathways, respectively. We can define an instantaneous yield ϕ for the desired reaction as the quantity of AR formed for each unit of reactant consumed. In symbols,

$$\phi_R = \frac{dC_R}{-dC_A} = \frac{k_1 C_A}{+(k_1 C_A + k_2 C_A)} = \frac{1}{1 + k_2/k_1} \qquad (5\text{-}82)$$

Thus, the instantaneous yield for the parallel reactions is dependent only on the rate constants of the individual reactions, regardless of the reactor type. The required reactor volume depends on the reaction rate of the desired reaction pathway. In general, for several reactions in parallel, the reactor volume is determined by the slowest reaction pathway of the desired pathways (Kramer and Westerterp, 1963).

Reaction Selectivity. We wish to choose a reactor type that maximizes the selective reaction between reactants leading to a desired level of removal or transformation of the undesirable constituent. In other words, we wish to minimize the concentration of the contaminant in water. As we have seen for first-order reactions, reductions of the contaminant are dependent only on the relative reaction rate constants. Contaminant reduction can only be controlled by altering the rate constants, for example, changing pH until $k_1 > k_2$. Catalysts may also increase the rate of destruction of the undesirable constituent. Temperature, a widely used control parameter in the chemical processing industries, is generally not a controllable process parameter in potable water treatment. In certain hot-water systems, however, temperature control may permit optimization of reactor performance because the temperature sensitivity of each rate constant varies depending on activation energies.

If the parallel reactions are second order overall, however, the relative rates of parallel pathways will also depend on the concentration of reactants. An important example in water treatment is disinfection with chlorine. The reactant chlorine will react with several constituents in water, which leads to consumption of the reactant, formation of undesirable by-products such as chlorinated organic compounds, and a reduced rate of bacterial deactivation.

Consider the simple case of reactant B (e.g., Cl_2), which follows two reaction pathways, one with constituents A (bacteria) and one with D (natural organic matter),

$$A + B \xrightarrow{k_1} \text{products } (R)$$

$$D + B \xrightarrow{k_2} \text{products } (T)$$

The second reaction pathway is undesirable, and the reactor should be designed to minimize the consumption of B via this pathway.

The instantaneous consumption of B relative to the formation of desired products is given by

$$\phi_B = \text{consumption of } B$$
$$= \frac{dC_b}{dC_A} = \frac{k_1 C_A C_B + k_2 C_D C_B}{k_1 C_A C_B}$$
$$= 1 + \frac{k_2}{k_1} \frac{C_D}{C_A} \qquad (5\text{-}83)$$

The influence of the relative reaction rates and initial concentrations of reactants on the consumption of reactant B is illustrated in Figure 5-17. Note that increasing k_1 relative to k_2 will increase reactant selectivity.

How does the reactor type affect the consumption of reactant for the second-order reactions? The formation of R relative to T, which we wish to maximize, is given by

$$\frac{dC_R}{dC_T} = \frac{k_1 C_A}{k_2 C_D}$$

The reactor type will affect only the reactant concentrations and not the rate constants. Thus, neither type of ideal reactor can control the relative concentrations of C_A and C_D. However, if the reaction with D were third order overall, the instantaneous yield would become

$$\frac{dC_R}{dC_T} = \frac{k_1 C_A}{k_2 C_D^2}$$

Under these conditions, the yield could be im-

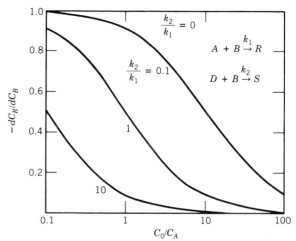

FIGURE 5-17. Instantaneous product yield in series–parallel reactions for varying rate constants and concentrations of reactants.

proved by keeping the concentration of D as low as possible by using a CSTR.

In general, a PF reactor will favor the reaction of highest order, while a CST reactor favors the reaction of lower order (Levenspiel, 1972). Thus, the reactor type provides some degree of control over reaction selectivity. Once the reactor has been constructed, however, only control of relative rate constants can effectively maintain the reaction selectivity in parallel reaction schemes.

Series Reactions. In the series reaction

$$A \xrightarrow{k_1} R \xrightarrow{k_2} S$$

with R a desirable product and S undesirable, it was shown earlier that product distribution in a batch reactor depends on the relative rate constants k_1 and k_2 (see Eq. 5-62). Thus, as with parallel reactions, control of reaction selectivity for series reactions depends on control of the relative rate constants regardless of the type of reactor used. Again, reactions of higher order will be favored by PF reactors. A PF reactor will give higher yields of R compared to a CSTR reactor because the reaction order is greater than zero. Example 5-5 illustrates these concepts for a second-order series–parallel reaction (see also Levenspiel, 1972).

EXAMPLE 5-5
Reaction Kinetics

Chlorine, which is commonly used for water disinfection, is a strong oxidant and will react with numerous substances in water. In some cases, reaction products can cause undesirable taste and odor problems in the treated water, for example, the formation of chlorinated phenols.

Estimate the relative rate of formation of 2-chlorophenol compared to the rate of formation of monochloramine (NH_2Cl) in a surface water containing 0.5 mg/L NH_3–N and 50 μg/L phenol at pH 6 and 8 and TDS of 320 mg/L using the following data and neglecting other possible chlorine reactions.

$$T = 25°C$$

$$pK_1 \text{ (HOCl)} = 7.537$$

$$pK_2 \text{ (NH}_4^+\text{)} = 9.246$$

$$pK_3 \text{ (phenol)} = 9.87$$

Reactions

$$NH_3 + HOCl \xrightarrow{k_1} NH_2Cl + H_2O$$

$$C_6H_5OH + HOCl \xrightarrow{k_2} C_6H_5OCl + H_2O$$

$$k_1 = 6.2 \times 10^6 \, M^{-1} \, S^{-1} \quad \text{(Weil and Morris, 1949)}$$

$$k_2 = 5 \times 10^3 \, M^{-1} \, S^{-1} \quad \text{(Soper and Smith, 1926)}$$

Both reactions are second order and depend on pH, which controls the concentration of the chemical species participating in the reactions.

Rates of Formation

$$\frac{d(NH_2Cl)}{dt} = k_1(NH_3)(HOCl)$$

$$\frac{d(PhOCl)}{dt} = k_2(PhOH)(HOCl)$$

Relative Rates of Formation

$$\phi = \frac{d(PhOCl)}{d(NH_2Cl)} = \frac{k_2(PhOH)}{k_1(NH_3)}$$

The concentrations of phenol and ammonia can be estimated from the dissociation constants and the concentrations of the constituents measured analytically.

Let

$$x = NH_4^+ + NH_3 = 0.5 \text{ mg/L NH}_3\text{–N}$$
$$(3.57 \times 10^{-2} \text{ m}M)$$

$$y = PhOH + PhO^- = 50 \text{ }\mu\text{g/L}$$
$$(5.32 \times 10^{-4} \text{ m}M)$$

But

$$K_2 = \frac{(NH_3)(H^+)}{NH_4^+} = 5.67 \times 10^{-10}$$

$$K_3 = \frac{(PhO^-)(H^+)}{PhOH} = 1.35 \times 10^{-10}$$

assuming the activity coefficients are unity, a reasonable assumption at low TDS. Therefore,

$$NH_4^+ = \frac{(NH_3)(H^+)}{K_2}$$

$$\text{and } NH_3 = \frac{x}{1 + (H^+)/K_2}$$

Similarly,

$$PhOH = \frac{y}{1 + K_3/(H^+)}$$

Results

	pH 6	pH 8
NH_3 (mM)	2.02×10^{-5}	1.92×10^{-3}
PhOH (mM)	5.32×10^{-4}	5.26×10^{-4}
ϕ	2.12×10^{-2}	2.21×10^{-4}

Therefore, at pH 6, relative rate of formation of the odor causing 2-chlorophenol is 40 times slower than the rate of formation of monochloramine. Increasing the pH increases the relative rate of formation of monochloramine. Thus, one possible technique for control of chlorinated phenols is increasing the pH before chlorination.

Rate constants in this example differ by 3 orders of magnitude. In those cases where rate constants of competing parallel reactions are similar, control of the reaction pathways can be more difficult. □

Conversion in Nonideal Reactors

Thus far, we have discussed conversion of chemical reactants in two model reactor types, the CSTR, or mixed-flow reactor, and the PF reactor. The ideal reactor models represent the extremes of flow behavior in process equipment. Often flow behavior deviates substantially from the assumptions of ideal flow, which can cause decreased removal efficiencies or the formation of undesirable by-products.

There are three principal types of nonideal fluid behavior in processing equipment: short-circuiting, dead space, and dispersion. *Short-circuiting* is characterized by a segment of the fluid stream having a residence time considerably shorter than the mean hydraulic residence time. This is a common problem in CSTRs, rectangular basins, and packed columns or towers caused usually by poor fluid mechanical design of inlet structures or internal packing. In contrast, when a segment of fluid has a significantly longer residence time than the mean, this indicates the presence of a hydraulic *dead space*. This often occurs in long rectangular basins used for sedimentation. This fluid segment shows little mixing with the bulk of the fluid stream. *Dispersion* occurs mainly in PF-type reactors (channels, pipes, or columns) and is characterized by longitudinal

mixing due to turbulent diffusion, which distorts the flat velocity profile assumed for a PF reactor.

Quantitative models of nonideal fluid behavior and its impact on the efficiency of contaminant removal can be applied to reactor design in water treatment applications. Causes of poor reactor performance can be evaluated using nonideal flow models. Safety factors for reactor sizing can be estimated analytically. In addition, a qualitative understanding of fluid mixing will aid the designer in selection of appropriate equipment for addition of chemicals in water treatment applications.

Mixing of Fluids. In the development of reactor design equations for the CSTR, we have assumed that all molecules move about freely in the fluid with no interference from surrounding molecules, analogous to an ideal gas. Mixing of fluids is assumed to be instantaneous. Molecules just added to the reactor are not distinguished from those already present.

For real fluids, however, the time required to reach uniform concentration in a CSTR depends on the intensity of turbulence and may not be instantaneous. Chemical solutions added to water are first dispersed into clumps of fluid and then broken into smaller aggregates by turbulent motion (Uhl and Gray, 1966). Molecular diffusion completes the mixing process. If chemical reactions are rapid or heterogeneous, as is usually the case in water treatment applications, the time required for complete mixing may be comparable to rates of reaction. For some reactions, this may lead to formation of undesirable reaction by-products, resulting in increased chemical costs.

When two fluids are mixed together, the molecular behavior of the dispersed fluid will fall between two extremes (Levenspiel, 1972). If molecules are completely free to move about, the dispersed fluid behaves as a *microfluid* and exhibits no fluid segregation. At the opposite extreme, the dispersed fluid remains in clumps containing a large number of molecules. Clump size may range from roughly 100 to 500 μm. Such a fluid is known as a *macrofluid*. All heterogeneous systems, such as a gas dispersed in water or a suspension of precipitated solids or bacterial cells, are classified as macrofluids. As the macrofluid is transformed to a microfluid by physical mixing processes (turbulence, molecular diffusion), the *degree of segregation* and the *scale of segregation* (i.e., the average size of the segregated clumps) decreases.

In a PF reactor, each fluid element is considered to be a small batch reactor passing through the reaction vessel. Consequently, the type of molecular fluid behavior will not affect reactor performance. Fluid segregation in a CSTR, however, will cause a deviation from ideal conversion because the reactant concentration does not immediately decrease to a uniform value in the reactor. The macrofluid dispersed in the CSTR will behave as a PF reactor until the degree of segregation decreases to zero. The extent of reaction, and thus the efficiency of the reactor, will depend on the distribution of residence times of the macrofluid elements (e.g., gas bubbles, clumps of cells, precipitated solids) in the CSTR.

As would be anticipated, the extent of conversion of a reactant acting as a macrofluid in a CSTR depends on the order and type of reaction. Levenspiel (1972) has computed the effect of molecular fluid behavior on the reactor volume of a CSTR compared to a PF reactor for several reaction orders. For 95 percent conversion of reactant this ratio of reactor volumes increases as reaction order increases, assuming microfluid behavior. For a macrofluid in a CSTR, the reverse is true. For a linear kinetics, with $n = 1$, molecular fluid behavior does not affect the volume ratios. When reaction order is greater than 1, however, segregation or macrofluid behavior decreases the required volume of the CSTR relative to a PF reactor. For reaction orders less than 1, the reverse occurs and segregation (macrofluid behavior) is detrimental to performance.

Thus, the intensity of fluid mixing on a molecular level can influence reactor performance in several ways depending on the kinetics. Quantitative models relating fluid mixing to reactor performance have been developed but have had only limited success and are beyond the scope of this text. (See Levenspiel, 1972; Smith, 1970.) Application of these concepts to design of reactors for oxidation–reduction reactions in water treatment applications is limited. Important examples where additional research work is needed include (1) mixing of coagulants for removal of particulates and dissolved organic carbon, (2) oxidation of organic compounds in water, (3) oxidation (disinfection) of bacterial cells, and (4) precipitation reactions for removal of undesirable inorganic contaminants (e.g., barium, calcium, manganese, and arsenic).

Residence Time Distributions. In addition to the effects of initial mixing of reacting fluids and the relative micro- or macrofluid behavior on reactor design, the characteristics of macroscopic fluid behavior in processing equipment must also be evaluated. Both a microfluid or a macrofluid may show deviations from the ideal reactor flow models. In contrast to the qualitative descriptions of micromixing, macroscopic behavior of fluids can be conveniently and readily described by knowledge of the residence times of fluid elements in a reactor.

The *residence time* is defined as the length of time required for a fluid element to pass through a reactor. We wish to estimate the effect of the distribution of residence times on the conversion of a constituent in a chemical reactor. We have solved this problem for the two ideal flow models for various reaction kinetic schemes but must now evaluate the effects of deviations from the ideal models due to short-circuiting, dead space, longitudinal dispersion, or other causes of nonideal flow behavior.

Consider a nonreacting fluid flowing through a reactor of any shape. Fluid elements will reach the effluent by various pathways, resulting in a distribution of residence times for the various fluid elements passing through the reactor. We seek a distribution function E by which the fraction of fluid elements in the effluent with residence time between t and $t + dt$ is given by $E \, dt$. It is convenient to normalize the function E such that

$$\int_0^\infty E \, dt = 1 \qquad (5\text{-}84)$$

where the function E can be thought of as the exit age distribution function of fluid elements.

A plot of E versus time will represent the residence time distribution (RTD) of elements in the reactor. The average residence time of fluid elements in the reactor $\bar{\theta}$ is simply the first moment of the E curve, given mathematically as

$$\bar{\theta} = \int_0^\infty tE \, dt \qquad (5\text{-}85)$$

It can be shown that $\bar{\theta}$ is equal to the mean hydraulic residence time (Trussell et al., 1979).

For an existing reactor, such as a tank, packed column, or basin, the exit age function E or RTD can be easily determined using stimulus–response (tracer) methods. A dye or nonreactive substance is injected into the reactor influent and its concentration measured as it leaves the reactor. Two principal types of stimulus methods are most frequently used for this purpose, the pulse-function and the step-function input, although any type of stimulus can be used.

A pulse input is an instantaneous addition of tracer material into the system influent. The E function can be determined by measuring the effluent concentration of tracer then normalizing. This is known as the $\bar{C}$ function and is given by

$$\bar{C}(t) = \frac{C(t)}{\int_0^\infty C \, dt} = \frac{C(t)}{M} \qquad (5\text{-}86)$$

such that $\int_0^\infty \bar{C} \, dt = 1$, where M = mass of tracer added. The $\bar{C}$ function is illustrated in Figure 5-18a.

For a step-function input the concentration of some suitable tracer is suddenly increased to C_0 at time t. The effluent concentration of the tracer is measured and the normalized concentration C/C_0 plotted versus time, as shown in Figure 5-18b, giving the so-called F function. The F function is a cumulative frequency distribution and is related mathematically to the E function by

$$F = \frac{C}{C_0} = \int_0^t E \, dt \quad \text{or} \quad E = \frac{dF}{dt} \qquad (5\text{-}87)$$

Thus, the F function gives the fraction of fluid ele-

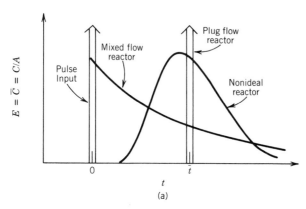

(a)

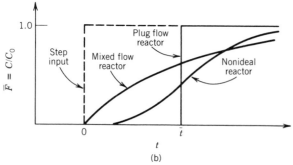

(b)

FIGURE 5-18. Response curves for mixed-flow, plug-flow, non-ideal reactor. (a) $\bar{C}$ curve, response to pulse input. (b) $\bar{F}$ curve, response to step-function input.

ments in the effluent with residence times less than t. The $\bar{C}$ and F functions are related to the E function mathematically. Either type of function can thus be used to define the exit age distribution (the RTD).

The RTDs for the ideal CSTRs can be developed by materials balance methods. Consider a pulse or delta function stimulus added to a CSTR. The concentration of tracer in the reactor, and therefore in the reactor effluent, at time $t = 0$ is the total tracer amount added divided by the reactor volume V. Because no additional tracer is added, the concentration C will decrease with time. From a non-steady-state macroscopic mass balance for a volumetric flowrate Q,

$$Q(C_0 - C) = V \frac{dC}{dt}$$

Integrating, this becomes

$$C = C_0 \exp\left(-\frac{t}{\bar{\theta}}\right) \qquad (5\text{-}88)$$

where $\bar{\theta} = V/Q$ and C_0 is the initial reactor concentration. The $\bar{C}$ function or RTD is found by normalizing the effluent tracer concentration such that

$$\int_0^\infty \bar{C} \, dt = 1$$

The area under the C curve is given as

$$A = \int_0^\infty C \, dt = C_0\bar{\theta}$$

and therefore

$$\bar{C} = \frac{C}{A} = \frac{1}{\bar{\theta}} \exp\left(-\frac{t}{\bar{\theta}}\right) \qquad (5\text{-}89)$$

The response of a CSTR to a step-function input is given by the F function, which is obtained from Eqs. 5-87 and 5-89.

$$F = \int_0^t \bar{C} \, dt = 1 - \exp\left(-\frac{t}{\bar{\theta}}\right) \qquad (5\text{-}90)$$

For a PF reactor the stimulus and response (RTD) functions will be identical.

Estimating Removal from Tracer Data. The prediction of nonideal flow patterns in chemical reactors has two principal applications to reactor design. First, when the kinetics of a reaction can be described analytically, we can determine the size of a given reactor if the flow pattern deviates from the ideal model of either mixed (CSTR) or plug flow. Various degrees of nonideal flow behavior can be stimulated to determine the effects on reactor size.

We can also determine if a given reactor is large enough to achieve a specified degree of removal. Here, the residence time distribution (RTD) can be determined directly from tracer studies. We then wish to predict the reactant conversion from the tracer data. If the RTD matches the ideal flow models, the reactor design methods presented can be used. For the general case, the RTD specifies the fraction of fluid elements that have remained in the reactor for time t. Assuming that no intermixing of fluid elements occurs, each element remaining completely segregated as it travels through the reactor, the conversion in each element with time is given by the integrated batch kinetics expression.

Thus, the average concentration $\bar{C}_A$ of a reactant A in the effluent is the sum of conversion in all fluid elements, given by

$$\bar{C}_A = \Sigma \begin{pmatrix} \text{concentration of} \\ \text{reactant remaining} \\ \text{in element of age} \\ \text{between } t \text{ and } t + dt \end{pmatrix} \begin{pmatrix} \text{fraction of ele-} \\ \text{ments in effluent} \\ \text{with age between} \\ t \text{ and } t + dt \end{pmatrix}$$

or in symbols

$$\bar{C}_A = \int_0^\infty (C_A)_{\text{BATCH}} E \, dt \qquad (5\text{-}91)$$

Given an integrated rate expression and the exit age distribution function (E function) determined by either a pulse or step input stimulus response, the average concentration of any constituent in the reactor effluent can be estimated.

The above equation has certain limits. An RTD is not unique to a single flow pattern in processing equipment. For example, the RTD for a PF reactor followed by a CSTR and the RTD for the opposite flow pattern are identical. Reactant conversion, however, is only the same if the reaction kinetics are first order. For nonlinear processes (reaction order $\neq 1$) the equation gives the upper bound of reactant conversion for $n > 1$ and the lower bound for $n < 1$ (Levenspiel, 1972).

Conversion from RTD Models. When tracer data is unavailable, estimating the effect of nonideal flow

patterns on reactant conversion can be based on models of the expected RTD. Numerous single- and multiparameter models have been proposed for this purpose, but complexity has restricted their practical applications. However, two one-parameter models have been widely used in water and wastewater treatment applications, the dispersion model and the tanks-in-series model (Levenspiel, 1970; Kramer and Westerterp, 1963; Froment and Bischoff, 1979).

Dispersion Model. The dispersion model draws on the analogy of diffusional mixing in tubular or PF reactors, which causes deviations from the assumed flat velocity profile. Consider the flow of a fluid in a tubular reactor without chemical reaction. Flow occurs into and out of a fluid element by convection and axial or longitudinal diffusion. The one-dimensional, maximum-gradient mass balance around a fluid element of volume dV for non-steady-state becomes

$$D_L \frac{\partial^2 C}{\partial x^2} - u \frac{\partial C}{\partial x} = \frac{\partial C}{\partial t} \qquad (5\text{-}92)$$

where D_L is the longitude dispersion coefficient, x is the distance into the reactor in the direction of flow, and u is the average fluid velocity in the x direction. Converting this to dimensionless form by substituting $z = x/L$ and $\theta = t/\bar{\theta} = tu/L$ with L the length of the reactor, the partial differential equation representing the dispersion model becomes

$$\left(\frac{D_L}{u_L}\right) \frac{\partial^2 C}{\partial X^2} - \frac{\partial C}{\partial Z} = \frac{\partial C}{\partial \theta} \qquad (5\text{-}93)$$

The dimensionless group D_L/uL is termed the dispersion number and is the inverse of the Peclet number (Pe). When no axial dispersion occurs, $Pe^{-1} = 0$, corresponding to a PF reactor. For very large dispersion, $Pe^{-1} \to \infty$, representing a CSTR.

For an open vessel with undisturbed flow at the vessel boundaries, the exit age distribution function or RTD is found to be (Levenspiel and Smith, 1957)

$$E = \frac{1}{(\sqrt{\pi \bar{\theta} Pe^{-1}})} \exp\left[\frac{-(1 - \bar{\theta})^2}{4\bar{\theta} Pe^{-1}}\right] \qquad (5\text{-}94)$$

where $\tau = t/\bar{\theta}$. Thus, the dispersion model provides a RTD that depends on a single parameter, the dimensionless dispersion Pe^{-1}, which can be estimated for various types of reactors. Reactant conversion can then be estimated numerically or graphically using Eq. 5-91 and known kinetics. (For an example of this approach see Trussell and Chao, 1977.)

Reactant conversion in a PF reactor with dispersion has been computed analytically for first- and second-order kinetics. For small deviations from plug flow ($Pe^{-1} \leq 0.1$), the conversion for first-order decomposition of A with rate constant k can be shown (Wehner and Wilhelm, 1956) to be

$$\frac{C_A}{C_{A_0}} = \exp(-k\bar{\theta} + (k\bar{\theta})^2) \qquad (5\text{-}95)$$

which can be compared to conversion in an ideal PF reactor:

$$\frac{C_A}{C_{A_0}} = \exp(-k\bar{\theta})$$

Thus, the size ratio for identical conversion is

$$\frac{V}{V_{PF}} = 1 + k\,Pe^{-1} \qquad (5\text{-}96)$$

where V is the volume of the reactor with dispersion and V_{PF} the volume of an ideal plug PF reactor. Table 5-8 summarizes this ratio for "slow" and "fast" first-order reactions.

For slow reactions, with $k < 10^{-3}$ sec^{-1}, a PF assumption is valid for dispersion numbers less than approximately 0.01. This has been demonstrated by Trussell and Chao (1977) for the kinetics of disinfection. Fast reactions, however, with $k > 1$ sec^{-1} (half-life < 1 sec), are sensitive to dispersion, and small deviations from plug flow can dramatically decrease removal efficiencies in a PF reactor. In cases of fast reactions, the tubular reactor should be designed structurally to minimize any axial mixing.

Tanks-in-Series Model. This model assumes that the RTD from any reactor can be simulated by a series of equal-volume CSTRs. The exit age distribution function for a single CSTR was shown to be

$$E(t) = \frac{1}{\theta} \exp -\frac{t}{\theta}$$

For N tanks in series the RTD can be shown to be

$$E(t) = \frac{N(Nt/\bar{\theta})^{N-1}}{(N-1)} \exp\left(-\frac{Nt}{\bar{\theta}}\right) \qquad (5\text{-}97)$$

where $\bar{\theta} = N\bar{\theta}_i$ = mean residence time in the N tank system. Thus, the RTD is characterized by the single parameter N, which can be assumed or obtained from tracer data by statistical fitting techniques. For $N \to \infty$, this model approaches plug flow. As N decreases, deviation from ideal plug flow becomes more pronounced. Conversion using the tanks-in-series models can be estimated from the RTD model and Eq. 5-91 with the restrictions as noted (Kramer and Westerterp, 1963).

Other Models. When the dispersion or tanks-in-series models do not satisfactorily model the deviations from ideal fluid behavior, multiparameter models are used. (For a review of this approach, see Levenspiel, 1972; Chapter 15.) These models are rather cumbersome, require tracer data to provide estimates of the model parameters, and are not widely used in water treatment process design.

REFERENCES

Adamson, A. W., *Physical Chemistry of Surfaces*, 2nd ed., Wiley-Interscience, New York (1967).

Bird, R. B., Stewart, W. E., and Lightfoot, E. N., *Transport Phenomena*, Wiley, New York (1960).

Dankwerts, P. V., *Gas-Liquid Reactions*, McGraw-Hill, New York (1970).

Denbigh, K., *Principles of Chemical Equilibrium*, Cambridge University Press, London (1961).

Eigen, M., and Kustin, K., *J. Amer. Chem. Soc.*, **84**, 1355–1361 (1962).

Eigen, M., Kustin, M., and Maas, G., *Z. Physik. Chem.*, **30**, 130 (1961).

Ferguson, J. F., Jenkins, D., and Eastman, J., *J. Water Poll. Control Fed.*, **45**(4), 620–631 (1973).

Froment, G. F., and Bischoff, K. B., *Chemical Reactor Analysis and Design*, Wiley, New York (1979).

TABLE 5-8. Effect of Reaction Rate and Dispersion Number (Deviation from Ideal Plug Flow) on Relative Reactor Volume, First-Order Kinetics

	V/V_{PF}	
Dispersion Number	"Fast" Reaction ($k = 1.8 \times 10^3$ sec^{-1})	"Slow" Reaction ($k = 1.8 \times 10^{-3}$ sec^{-1})
0.001	2.8	1.0
0.01	19	1.02
0.1	181	1.18

Frost, A. A., and Pearson, R. G., *Kinetics and Mechanism,* 2nd ed., Wiley, New York (1961).

Himmelblau, D. M., and Bischoff, K. B., *Process Analysis and Simulation,* Wiley, New York (1968).

Hoigne, J., and Bader, H., *Water Res.,* **10,** 377 (1976).

Hoigne, J., and Bader, H., *Environ Sci. Tech.,* **12,** 79 (1978).

Hoigne, J., and Bader, H., *Science,* **190,** 782 (1975).

Hoigne, J., and Bader, H., *Water Res.,* **17,** 173–185 (1983).

Hougen, O. A., Watson, K. M., and Ragatz, R. A., *Chemical Process Principles,* 2nd ed., Part I, Wiley, New York (1954).

Keinath, T. M., and Wanielista, M., eds., *Mathematical Modeling for Water Pollution Control Processes,* Ann Arbor Science, Ann Arbor, MI (1975).

Kern, D. M., *J. Chem. Educ.,* **37**(1), 14 (1960).

King, C. J., *Separation Processes,* McGraw-Hill, New York (1971).

Kramer, H., and Westerterp, K. R., *Elements of Chemical Reactor Design and Operation,* Academic Press, New York (1963).

Lee, G. F., and Morris, J. C., "Kinetics of Chlorination of Phenol: Chlorophenolic Tastes and Odors," *Int. J. Air Water Poll.,* **6,** 419–431 (1962).

Levenspiel, O., *Chemical Reaction Engineering,* 2nd ed., Wiley, New York (1972).

Levenspiel, O., and Smith, W. K., *Chem. Eng. Sci.,* **6,** 227–233 (1957).

Lewis, W. K., and Whitman, W. G., *Ind. Eng. Chem.,* **16,** 1215–1220 (1924).

Moelwyn-Hughes, E. A., *Kinetics of Reactions in Solution,* Clarendon Press, Oxford (1947).

Moore, W. J., *Physical Chemistry,* 3rd ed., Prentice-Hall, Englewood Cliffs, NJ (1962).

Morgan, J. J., and Stumm, W., "Manganese Oxidation," *Proceedings of the Second Conference on Water Pollution Research,* Pergamon, Tokyo (1964).

Morris, J. C., and Baum, B., in E. G. Jolley et al. (eds.), *Water Chlorination, Environmental Impact and Health Effects,* Vol. 2, Ann Arbor Science, Ann Arbor, MI (1978), pp. 24–49.

Perry, R. H., *Chemical Engineers Handbook,* 5th ed., McGraw-Hill, New York (1973).

Reid, R. C., Prausnitz, J. M., and Sherwood, T. K., *Properties of Gases and Liquids,* 3rd ed., McGraw-Hill, New York (1977).

Rudd, D. F., and Watson, C. C., *Strategy of Process Engineering,* Wiley, New York (1968).

Saunier, B., "Reaction Kinetics of Chlorine with Nitrogen Species," Ph.D. Thesis, University of California, Berkeley (1976).

Shreve, R. N., and Brink, J. A., *Chemical Process Industries,* 4th ed., McGraw-Hill, New York (1977).

Singer, P. C., and Stumm, W., *Science,* **167–168,** 3291 (1970).

Smith, J. M., *Chemical Engineering Kinetics,* 2nd ed., McGraw-Hill, New York (1970).

Soper, F. G., and Smith, G. F., *J. Chem. Soc. Lond.,* **46,** 1582–1591 (1926).

Stumm, W., "Chemical Aspects of Water Ozonation," *Schw. Z. Hydrol.,* **18,** 201 (1956).

Stumm, W., and Lee, G. F., *Ind. Eng. Chem.,* **53,** 143 (1961).

Stumm W., and Morgan, J. J., *Aquatic Chemistry,* 2nd ed., Wiley, New York (1981).

Treybal, R. E., *Mass Transfer Operations,* 2nd ed., McGraw-Hill, New York (1968).

Trussell, R. R., and Chao, J. L., "Rational Design of Chlorine Contact Facilities," *J. Water Poll. Control Fed.,* **49**(4), 659–667 (1977).

Trussell, R. R., Selleck, R. E., and Chao, J. L., "Hydraulic Analysis of Model Treatment Units," *J. Env. Eng. Div.,* *ASCE,* **105**(EE4), 796–798 (1979).

Uhl, V. W., and Gray, J. B., eds., *Mixing, Theory and Practice,* Academic Press, New York (1966).

Weber, W. J., *Physiochemical Processes,* Wiley-Interscience, New York (1972).

Wehner, J. F., and Wilhelm, R. H., *Chem. Eng. Sci.,* **6,** 89–93 (1956).

Weil, I., and Morris, J. C., *J. Amer. Chem. Soc.,* **71,** 1664–1671 (1949).

— 6 —

Precipitation, Coagulation, Flocculation

INTRODUCTION

Of the many unit processes and operations used in water treatment, coagulation and flocculation require a unique combination of chemical and physical phenomena for producing a water acceptable for human consumption. These are essential pretreatment processes for the removal of finely divided particulate matter which, due to its small size (usually less than 10 μm), will not settle out of suspension by gravity in an economical time frame. Aggregation of fine particulate matter into larger particulates by the use of coagulation and flocculation facilities permits cost-effective removal in subsequent solids separation processes.

The characteristics of particulate contaminants found in water supply sources were described in Chapter 2. Particulates of inorganic origin, such as clay, silt, and mineral oxides, generally enter a surface water by natural erosion processes and can decrease the clarity of the water to unacceptable levels. Organic particulates, such as colloidal humic and fulvic acids, are a product of decay and leaching of organic debris and litter which have fallen in the water source. These particulates impart a color

to the water which in some extreme cases can be nearly opaque. Removal of these particulate contaminants is required both for aesthetic and health concerns.

As discussed in Chapter 3, a finished drinking water of acceptable quality for potable use must meet various standards defined in drinking regulations developed by designated regulatory agencies. In the United States, the national interim primary and secondary standards define a turbidity standard of 1 TU and a color standard of 3 color units. Meeting these standards often requires well-designed and -operated coagulation and flocculation facilities. Removal of particulate contaminants may also assist in the removal of toxic contaminants, such as heavy metals, pesticides, and viruses, known to be associated with inorganic and organic particulate matter in water. Removal of colloidal organic particulates that produce color has gained increased attention because of the discovery that these particulates, although posing no known health hazards to consumers, will react with chemical oxidants used in water treatment, such as chlorine and ozone, to form organic by-products with known or potential health hazards for consumers. In the case of chlorine, humic and fulvic acids react with free chlorine to form chloroform, a suspected animal carcinogen, and other chlorinated compounds of unknown toxicity. Maximum economical removal of these organic precursors prior to the addition of chlorine has thus become a major goal of water treatment plants.

Fine particulate material is removed from water by addition of inorganic or organic chemicals that accelerate the aggregation of the particulates into larger aggregates. The chemicals used in this process include metal ions such as aluminum or iron, which hydrolyze rapidly to form insoluble precipitates, and natural or synthetic organic polyelectrolytes, which rapidly adsorb on the surface of the particulates, thereby accelerating the rate at which the particulates aggregate. These aggregates are then removed from the water by physical means such as gravity sedimentation, flotation, or filtration through granular media.

PROCESS OVERVIEW

Conceptually, the aggregation of particulate material is a two-step sequential process. In the initial step, the interparticulate forces responsible for the stability of the particulates are reduced or eliminated by addition of suitable chemicals. Subsequentially, particulate collisions occur due to transport by molecular motion or mechanical mixing. If these collisions are successful, aggregation occurs.

Table 6-1 provides an overview of the coagulation–flocculation process with respect to (1) the principle phenomenon occurring, (2) the component actions associated with each phenomenon, (3) the standard terminology for those actions, and (4) the facilities required to complete the process (modified after Fiessinger, 1978).

The chemicals used to destabilize particulates are known as *coagulants*. Chemical handling and feeding equipment must be designed for preparation of the chemical coagulant prior to addition. The coagulant is then injected into the process stream through a mixing device that should provide rapid and thorough dispersion of the coagulant in the water. This *rapid, flash,* or *initial mixing* stage, which occurs over a short time frame (usually less than 1 min), serves to optimize the effectiveness of the coagulant for particulate destabilization. In this text, this phenomenon is called *coagulation*.

Following destabilization, less intense mixing of the particulates must be provided to increase the rate of particulate encounters or collisions without

TABLE 6-1. Overview of Coagulation–Flocculation Process Aggregation of Particulates

Phenomenon	Action	Terminology	Facility Involved
Formation of active coagulant species	Preparation of coagulant (dilution, dissolution)	Chemical handling (coagulant preparation)	Chemical handling and feeding equipment
	Treatment dispersion of coagulant chemical reactions with ligands (OH^-, SO_4^-) e.g., hydrolysis, polymerization, complex formation	Rapid mixing, Flash mixing, Initial mixing	Mixing device for rapid and thorough dispersion of chemical; high turbulence, high-shear environment
Particulate destabilization	Compression of double layer by indifferent electrolytes	Aggregation 1. Coagulation	
	Charge neutralization by specifically adsorbed charged species		
	Surface precipitation and formation of interparticle "bridges"		
	Coagulant precipitation and entrapment (sweep floc) of particulates		
Particulate transport	Random collisions due to thermal motion of water molecules (Brownian motion)	2. Flocculation a. Perikinetic flocculation	
	Ordered collisions due to differential relative particulate velocities achieved by mixing and differential settling	b. Orthokinetic flocculation	Flocculation basins and mixing devices for low-shear turbulence

breaking up or disrupting the aggregates being formed. This phenomenon is called *flocculation*. For colloidal particulates (less than 1 μm) Brownian motion provides some degree of particulate transport. This is known as *perikinetic flocculation* (Kruyt, 1952). For larger particulates, Brownian motion is very slow, and transport requires mixing by mechanical means. Mechanical mixing devices such as paddles or turbine mixers are used to provide ordered collisions due to differential particulate velocities in the mixing basin. This subprocess is termed *orthokinetic* (Kruyt, 1952) flocculation. Proper design of both the mixing device and the mixing basin (flocculation reactor) is necessary to achieve optimum aggregation of the finely divided particulate contaminants prior to removal in the solids separation process.

This chapter will review the chemical and physical bases for the phenomena occurring in the coagulation and flocculation process and the process design of the main facilities used to complete particulate aggregation. Design details of coagulation and flocculation equipment are presented in Chapter 23.

STABILITY OF PARTICULATES

The principal characteristic of fine particulate matter suspended in water is its relative stability, causing it to remain in suspension for long periods of time. The particulate suspensions are thermodynamically unstable, and given sufficient time colloids and fine particles will settle. However, this process is not economically feasible. Coagulation–flocculation facilities must eliminate particulate stability and thereby increase the rate of subsequent particulate removal by solids–liquid separation processes. A review of the causes of particulate stability will provide an understanding of the techniques that can be used to destabilize particulates.

Particulate Characteristics

The particulates in water are broadly categorized into two major size groups: colloidal material, with an upper limit of approximately 1 μm and a lower limit of approximately 5nm, and suspended solids, consisting of particulates larger than approximately 0.5 μm. Particles smaller than 5 nm are considered to be in true solution. This size spectrum covers roughly 6–7 orders of magnitude and encompasses a heterogenous mixture of particulates with a wide range of physical, chemical, and biological properties.

In addition to classification by size, colloidal and coarse particulate suspensions are also characterized according to the nature of their water–solid interface. *Hydrophobic* particulates have a well-defined interface between the water and solid phases and have a low affinity for water molecules. They are thermodynamically unstable and will aggregate irreversibly over time. *Hydrophilic* particulates are characterized by the lack of a clear phase boundary and are generally solutions of macromolecular organic compounds, such as proteins or humic acids. The particulate suspensions can be reconstituted after aggregation and are thus reversible.

Hydrophobic particulates in water are primarily of an inorganic origin and include some clay particulates and nonhydrated metal oxides. Many inorganic particulates in natural waters show some hydrophilic properties because water molecules will bind to the surface (Stumm and Morgan, 1981). Examples include hydrated metal oxides (iron or aluminum oxides) and silica (SiO_2) and asbestos fibers.

In contrast, hydrophilic particulates are primarily of organic origin and include a wide diversity of biocolloids (humic and fulvic acids, viruses) and suspended living or dead microorganisms (bacteria, algae). Because biocolloids may adsorb on the surfaces of inorganic particulates, particulates generally exhibit heterogeneous surface properties in natural water systems.

Mechanisms of Stability

The principal mechanism controlling the stability of hydrophobic and hydrophilic particulates is electrostatic repulsion. In the case of hydrophobic surfaces, an excess of anions or cations may accumulate at the interface, producing an electrical potential that can repulse particulates of similar surface potential. For hydrophilic surfaces, typically electrical charges arise from dissociation of inorganic groups, for example, a carboxyl or other organic acid group located on the particulate surface or interface.

In addition to this electrostatic repulsion, particulates may also be quite stable due to the presence of adsorbed water molecules that provide a liquid barrier to successful particulate collisions.

The electrical charges existing at particulate surfaces arise in three principal ways (Stumm and Morgan, 1981):

1. *Crystal imperfections.* Under geologic conditions, silicon atoms in crystalline materials can be replaced by atoms with lower valence, such as an aluminum ion, giving an excess negative charge to the crystal material. This process, known as isomorphous substitution, produces negative charges on the surface of clay particles (Olphen, 1963).

2. *Preferential adsorption of specific ions.* When some particulates are dispersed in water, soluble polyelectrolytes of natural origin may adsorb on their surfaces. Typically, a negatively charged polymer, such as a fulvic acid molecule, may adsorb on the surface of a positively charged particulate (e.g., $CaCO_3$).

3. *Specific chemical reactions of ionogenic groups on particulate surfaces.* Many particulate surfaces contain ionogenic groups, such as hydroxyl or carboxyl functional groups, which dissociate in water, producing a surface electrical charge that depends on the solution pH. Typical examples include hydrolyzable metal oxides (e.g., iron oxides) and bacteria (carboxyl groups on bacterial surface).

As a rule, however, most particulates have complex surface chemistry, and electrical surface charges may arise from several sources.

Origin of the Double Layer

Further insights into the mechanism of stability can be obtained from a closer evaluation of particulate surface chemistry. When particulates are dispersed in water, ions of opposite charge to the surface charges accumulate close to the particulate surface to satisfy electroneutrality. This accumulation of ions is opposed by the tendency of ions to diffuse in the direction of decreasing concentration (Fick's law). These two opposing forces, electrostatic attraction and diffusion, produce a diffuse cloud of ions surrounding the particulate which can extend up to 300 nm into the solution. This is known as the *electrical double layer* (Kruyt, 1952).

The structure of the double layer is shown schematically in Figure 6-1. Because of the excess of cations in this case near the surface, an electrical potential difference arises, which decreases exponentially with distance from the surface from a maximum value at the particle surface. A schematic of

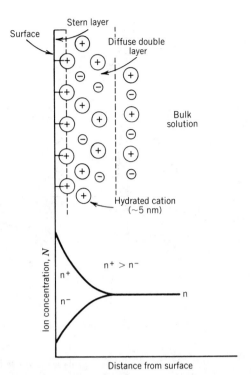

FIGURE 6-1. Schematic of electrical double layer at solid–liquid interface.

the shape of the electrical potential curves is shown in Figure 6-2.

When a particle moves in an electric field, some portion of the water near the surface of the particle moves with it. This gives rise to the shear plane, shown in Figure 6-2. The electrical potential between the shear plane and the bulk solution can be determined by electrophoresis measurements (measurement of rate of movement of particulates in an electric field) and is called the *zeta potential.*

The double layer consists of two major regions, an inner layer where adsorbed molecules are found and an outer layer of the oppositely charged, so-called counterions. The inner, or Stern layer, is approximately 5 nm deep, corresponding in size to the hydrated cation or nonhydrated anion. The extent of the outer or diffuse layer depends on the ionic strength.

Both the diffuse double layer and the inner layer have several important impacts on the behavior of particulate surfaces in water. For example, a difference may exist between the bulk and surface pH value at the solid–liquid interface. Also, numerous ions can accumulate in the inner layer, leading to adsorption, ion exchange, or precipitation reactions not normally predicted on the basis of equilibrium models (Stumm and Morgan, 1981).

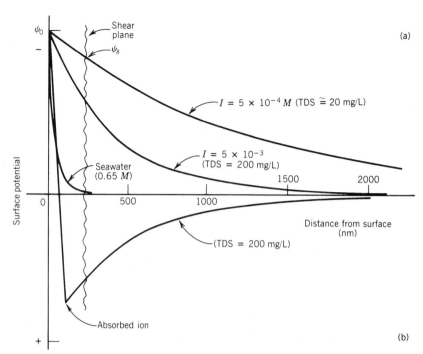

FIGURE 6-2. Schematic of double layer. (a) Effect of increasing concentration of indifferent electrolyte. (b) Reversal of potential by specific adsorption.

As shown in Figure 6-2, increasing the ionic strength of the solution as represented by an increase in the total dissolved solids (TDS) will compress the diffuse layer. Three curves are shown for TDS values ranging from 20 to 35,000 mg/L, the TDS level of seawater. Particulates in seawater exhibit a low zeta potential due to suppression of the double layer. In addition, some ionic species will adsorb on the surface of the particulate, thus reversing the surface potential. A cationic (positive-charge) polyelectrolyte adsorbed on a negative surface would produce a charge reversal as seen in Figure 6-2.

Mechanisms of Destabilization

The effective removal of the colloidal and suspended particulates from water depends on a reduction in particulate stability. Destabilization mechanisms that can be exploited to achieve particulate destabilization include compression of the electrical double layer, electrostatic attraction, interparticle bridging, and enmeshment, or "sweep floc."

Double-Layer Compression. As was shown schematically in Figure 6-2, increasing the ionic strength

compresses the double layer, causing a decrease in its thickness. Empirically, when the zeta potential is reduced below approximately ±20 mV, rapid coagulation is likely to occur (Kruyt, 1952). This varies from system to system, however. The amount of dissolved ions that produce this rapid coagulation, defined as the *critical coagulation concentration (CCC)*, will depend on the type of particulate as well as the type of dissolved ion.

For hydrophobic particulates, models of the electrical double layer predict that the CCC is inversely proportional to the sixth power of the charge on the ion (Schulz–Hardy rule). Thus, the CCC values for mono-, di-, and trivalent ions are, respectively, in the ratio

$$1 : \tfrac{1}{2^6} : \tfrac{1}{3^6} \quad \text{or} \quad 100 : 1.6 : 0.14$$

(Kruyt, 1952), with concentrations in milliequivalents/liter. Thus, if 3000 mg/L of NaCl will produce rapid coagulation of an hydrophobic particulate, 44 mg/L of $CaCl_2$ will achieve similar results.

Reduction of the zeta potential permits close approach of hydrophobic particulates such that short-range attractive forces (primarily London–van der Waals forces of molecular origin; see Kruyt, 1952) can act to produce successful collisions.

Electrostatic Attraction. In addition to double-layer compression, particulates can be destabilized by electrostatic attraction, which occurs when surfaces are oppositely charged. This can be promoted by the adsorption of specific ions on the surface of the particulates, as shown schematically in Figure 6-2. Many particulates found in natural waters have surface charges dependent on the solution pH and can exhibit both positive and negative surface charges. The pH corresponding to a surface charge of zero is defined as the *zero point of charge (ZPC)*. Above the ZPC the surface charge will be negative (anionic), and below the ZPC the charge will be positive. The ZPCs of a number of inorganic and organic materials found in natural waters are summarized in Table 6-2 (data from Parks, 1967, and Stumm and Morgan, 1981). As can be seen, clays, silica, and most organic particulates are negatively charged at a neutral pH (pH 7). These represent the predominant types of particulates found in natural waters.

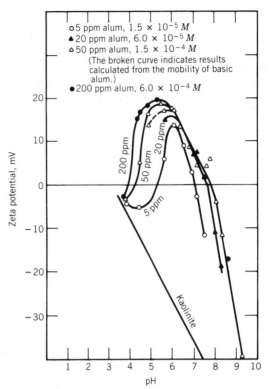

FIGURE 6-3. Effect of pH and alum dose on zeta potential of kaolin (after Packham, 1965).

A reduction of the surface charge by alteration of the pH or addition of specifically adsorbed ions can also lead to charge neutralization and destabilization of the particulates. In Figure 6-3, the effect of dissolved aluminum species on the electrophoretic mobility of kaolin solutions is depicted as a function of pH (Packham, 1962).

TABLE 6-2. Surface Characteristics of Particulates Commonly Found in Natural Waters and Wastewaters

	Zero Point of Charge, pH $_{ZPC}$
INORGANIC (hydrophobic)	
"Al(OH)$_3$" (amorph)	7.5–8.5
Al$_2$O$_3$	9.1
CuO$_3$	9.5
"Fe(OH)$_3$" (amorph)	8.5
MgO	12.4
MnO$_2$	2–4.5
SiO$_2$	2–3.5
Clays	
Kaolinite	3.3–4.6
Montmorillonite	2.5
Asbestos	
Chrysotile	10–12
Crocidolite	5–6
CaCO$_3$	8–9
Ca$_5$(PO$_4$)$_3$OH	6–7
FePO$_4$	3
AlPO$_4$	4
ORGANIC (hydrophilic)	
Algae	3–5
Bacteria	2–4
Humic acid	3
Oil droplets	2–5

SOURCE: Parks (1967) and Stumm and Morgan (1981).

Interparticle Bridging. Although the electrostatic model describes the behavior of many particulate suspensions, it is unable to predict the destabilization of biocolloids and other particulate systems with the same surface charge. Evidence indicates (Ruehrwein and Ward, 1952; LaMer and Healy, 1963) that long-chain polymers carrying negative charges can form bridges between particulates, thus destabilizing the suspension. This mechanism has also been shown to be the major mechanism controlling the aggregation of bacterial and algae suspensions (Tenney and Stumm, 1965).

Enmeshment (Sweep Floc). The destabilization of particulates with sodium or calcium ions is not a viable strategy in water treatment. Some soluble cations, however, such as aluminum, iron, or magnesium, hydrolyze and form an insoluble precipitate, thereby minimizing the concentration of ions added to the water. This type of destabilization has been described as an *enmeshment mechanism* or *sweep floc* (Packham, 1965; Stumm and O'Melia, 1968) in which finely divided particulates are entrapped in the amorphous precipitate formed. The molecular events leading to sweep floc have not been clearly defined. It is likely that nucleation of the precipitate may occur on the surface of particulates, leading to the growth of an amorphous precipitate and the entrapment of particles in this amorphous structure. This mechanism predominates in water treatment applications where pH values are generally maintained between pH 6 and 8 and aluminum or iron salts are used at concentrations exceeding saturation with respect to the amorphous metal hydroxide solid that is formed. (For a thorough review of destabilization mechanisms see Stumm and O'Melia, 1968).

CHEMISTRY OF COAGULATION

The two major operating costs of coagulation–flocculation are amortization of capital costs and costs for coagulation chemicals. In the United States, annual costs for water treatment coagulation chemicals may be on the order of 120×10^6. Consequently, treatment plant managers are continually seeking ways to reduce these costs.

The two primary functions of coagulant chemicals are particle destabilization and strengthening of flocs to reduce floc breakup. Chemicals serving one or both of these purposes must also satisfy several practical constraints, including low cost, ease of handling, availability, and chemical stability during storage. In addition, the coagulant must form highly insoluble compounds or be strongly adsorbed on particulate surfaces, thus minimizing the concentration of soluble residuals that might pass through the treatment plant.

Selection of the type and dose of coagulant depends on the characteristics of the coagulant, the particulates, and the water quality. As we shall see, the interdependence between these three elements is understood qualitatively; however, prediction of the optimum coagulant combination from characteristics of the particulates and the water quality is not yet possible. As a consequence, each coagulation problem must be solved empirically.

The standard procedure for bench-scale testing of coagulant doses and types is the use of a "jar test" apparatus. This test, originally developed by Langelier (1921) and refined over the years (see Black et al., 1957; TeKippe and Ham, 1970) permits rapid evaluation of a range of coagulant types and doses in a simple, commercially available apparatus (Phipps-Bird†).

The apparatus consists of six independently controlled paddles arranged in a row that can mix six beakers containing the particulate suspension and the coagulants of interest. The test attempts to simulate expected or desired conditions in the coagulation-flocculation facilities. Refer to the ASTM procedure (ASTM, 1976) for a review of the experimental protocols. Generally the test consists of a rapid mix phase (high mixing intensity) with simple batch addition of the coagulant or coagulants, followed by a slow mix period to simulate flocculation. Flocs are allowed to separate from the water and samples are taken from the supernatant. Turbidity or suspended solids removal can then be plotted as a function of coagulant dose.

Figure 6-4 (Trussell, 1978) shows a typical example summarizing jar test data for selection of the optimum alum dose and system pH for turbidity removal. From this figure the optimum dose pH would be approximately a pH of 7 and an alum dose of 8 mg/L.

In addition to performance, coagulant selection will depend on cost and the quantity and dewatering characteristics of the solids produced. Often, combinations of inorganic coagulants and polyelectrolytes provide the lowest-cost solutions to coagulation problems (James and O'Melia, 1982). Because of the many available coagulant–polymer combinations, a preliminary cost analysis is suggested to select viable combinations for jar testing.

Finally, full-scale testing is usually required to refine the optimum coagulant combinations and doses because of the limits of the jar test in simulating the hydraulic conditions in full-scale facilities.

Inorganic Coagulants

The two principal inorganic coagulants used in water treatment are salts of aluminum and ferric ions. These hydrolyzable cations are readily available as sulfate or chloride salts in both liquid and solid (dry)

† Phipps-Bird jar tester, Phipps-Bird Co., Richmond, Virginia.

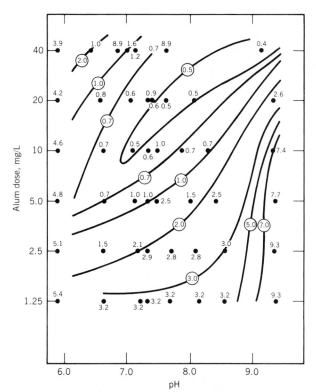

FIGURE 6-4. Turbidity topogram (Trussell, 1978).

particulate surfaces or other ligands in complex and poorly understood ways. Thus, the optimum dose of a coagulant depends strongly on the particular water chemistry and the types of particulates.

Kinetics. The overall stoichiometric reaction greatly oversimplifies the successive reaction steps that occur when these cations are added to water. Both aluminum and ferric ions undergo rapid hydrolysis reactions. For example, the first hydrolysis reaction of the aluminum ion is

$$Al(H_2O)_6^{3+} \leftrightharpoons [Al(H_2O)_5OH]^{2+} + H^+ \quad (6\text{-}1)$$

where the aquo ion, $Al(H_2O)_6^{3+}$, representing the actual species in water, is shown. At a pH of 4, half of the Al^{3+} hydrolyzes to $Al(OH)^{2+}$ within 10^{-5} sec (Baes and Mesmer, 1976). Subsequent hydrolysis reactions lead to the formation of monomeric and polynuclear aluminum species. Many such polynuclear species have been reported, containing up to 13 aluminum ions (i.e., $Al_{13}(OH)_{34}^{5+}$) with a multivalent charge. (See Baes and Mesmer, 1976, for a recent summary; also, Brossett et al., 1954; Matijevic, 1961; Sullivan and Singley, 1968; Singley and Sullivan, 1969.) The hydrolysis products, not the aluminum ion, cause particulate destabilization through charge neutralization (Mattson, 1928; Stumm and O'Melia, 1968).

As seen in Eq. 6-1, the rate of hydrolysis will be strongly dependent on both pH and concentration of aluminum. The hydrolysis products can also react with particulate surfaces and other liquids, thus exerting a coagulant demand. For example, phosphates found in synthetic detergents can significantly increase coagulant demand of a surface supply subject to waste discharges (Morgan and Engelbrecht, 1960).

Equilibrium. In most water treatment applications for removal of both turbidity and color (TOC), the pH during coagulation ranges between 6 and 8, the lower limit imposed by accelerated corrosion rates that occur at values below pH 6. In this pH range the soluble hydrolysis products that may be primarily responsible for particulate destabilization have a very short existence, on the order of microseconds. Using estimated equilibrium constants for the major hydrolysis reactions, the expected distribution of hydrolysis products after approximately 1 hr of reaction (upper limit of coagulation–flocculation de-

form. In the United States the predominant water treatment coagulant is aluminum sulfate, or "alum," sold in a hydrated form as $Al_2(SO_4)_3 \cdot xH_2O$, where x is usually 14. Lime (CaO) is also used for coagulation in those instances where high pH values (>pH 10) are desired. At these pH values magnesium ion will hydrolyze, forming an insoluble precipitate that aids in particulate destabilization.

The properties of commonly used water treatment coagulants are described in detail in Chapter 21. Details on design of chemical handling facilities are also covered in that chapter.

Chemistry. The aqueous chemistry of aluminum and ferric ions is complex. Characteristically, as with all cations in water, these ions react with various ligands (e.g., OH^-, SO_4^{2-}, PO_4^{3-}), forming both soluble and insoluble products that will influence the quantity or dose of the coagulant required to achieve a desired level of particulate destabilization. Both aluminum and ferric ions can undergo a series of reactions with the hydroxyl ion (OH^-), forming both monomeric and polynuclear (more than one cation) species, which in turn react with

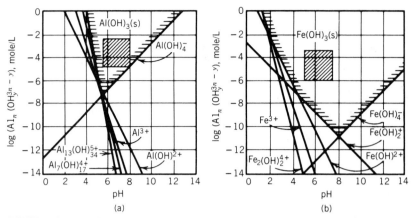

FIGURE 6-5. Equilibrium composition of solutions in contact with freshly precipitated, $Al(OH)_3$ and $Fe(OH)_3$. Calculated using representative values for the equilibrium constants for solubility and hydrolysis equilibria. Shaded areas are approximate operating regions in water treatment practice; coagulation in these systems occurs under conditions of oversaturation with respect to the metal hydroxide (Stumm and O'Melia, 1968).

tention times) can be defined. Figure 6-5 (from Stumm and O'Melia, 1968) illustrates the equilibrium composition of aluminum and iron species in contact with the freshly precipitated hydroxide using estimated values of the appropriate equilibrium reactions as noted.

These diagrams show clearly that in the pH range normally encountered in water treatment and using typical alum or ferric sulfate doses (10–100 mg/L), the predominant metal species will be the insoluble hydroxide. The equilibrium diagrams also illustrate the importance of pH in controlling the level of soluble metal species that will pass through the treatment process. As shown, ferric species are more insoluble than aluminum species and are also insoluble over a wider pH range. Thus, ferric ion is often the coagulant of choice to aid destabilization in the lime-softening process, which is carried out at higher pH values (pH > 9).

Stoichiometry. For engineering design purposes, the stoichiometric hydrolysis reactions of aluminum and ferric ions provide a useful estimate of coagulant quantities. The overall reactions with water, shown in Eq. 6-2, indicate that each mole of trivalent ion should produce 1 mole of the metal hydroxide and 3 moles of hydrogen ions if the reaction goes to completion.

$$Al^{3+} + 3H_2O \leftrightharpoons Al(OH)_{3(s)} + 3H^+$$
$$Fe^{3+} + 3H_2O \leftrightharpoons Fe(OH)_{3(s)} + 3H^+ \tag{6-2}$$

Thus, 1 mg of aluminum sulfate, or alum $[Al_2(SO_4)_3 \cdot 14H_2O]$, may produce approximately 0.26 mg of insoluble $Al(OH)_3$ and may consume approximately 0.5 mg of alkalinity (expressed as $CaCO_3$). Similarly, 1 mg of ferric sulfate $[Fe_2(SO_4)_3]$ may produce approximately 0.5 mg of $Fe(OH)_3$ as precipitate, consuming approximately 0.75 mg of alkalinity. The actual amounts of precipitate and acidity (H^+) formed will depend on system chemistry, particularly the pH, and concentrations of reactive liquids.

Interactions with Particulates. When ferric or aluminum ions are added to water, parallel and sequential reactions occur, the rates of which depend on several solution parameters, including pH, ionic species in the water, temperature, type and concentration of particulates, concentration of the coagulant, and the mixing conditions at the point of coagulant addition. All of these factors will influence the quantity of inorganic coagulant needed to achieve destabilization of the particulates.

Of these solution parameters affecting coagulant–particulate interactions, the solution pH plays a dominant role. For aluminum, at pH values less than pH 6, it appears that positively charged Al species remain in solution long enough to interact with particulates and cause destabilization by charge neutralization. Turbidity-causing particulates are destabilized by adsorption while organic particulates, such as humic and fulvic acids, appear to form insoluble precipitates with the charged Al species. A similar picture of this phenomenon applies to ferric ion, but at a pH below about pH 4. A characteristic of this pH region is that a stoichiometric relation between coagulant dose and particulate concentration occurs (Stumm and O'Melia, 1968).

Above pH 6 for aluminum and pH 4 for ferric ion, the formation of an amorphous precipitate occurs rapidly, causing entrapment of the particulates. This "sweep floc" mechanism usually requires a greater quantity of coagulant than charge neutralization, resulting in the formation of larger quantities of solids (sludge). For turbidity-causing particulates, no stoichiometric relation is observed. An increase in clay particulates, for example, may actually accelerate the aggregation of the coagulant–clay floc particles.

Removal of color-producing particulates by sweep floc appears to occur by adsorption of the organic molecules on the amorphous precipitate. Under these conditions, some stoichiometry is observed (Dempsey et al., 1984). Characteristically, coagulant requirements for sweep floc are not influenced by the type of inorganic particulates present (see Figure 6-6).

Current research on the interactions between inorganic coagulant species and particulate surfaces suggests that other forms of inorganic coagulants may provide improved performance. One such coagulant, so-called polyaluminum chloride (PAC), has been shown to have some advantages compared to alum for the removal of naturally occurring organic compounds when charge neutralization is the predominant destabilization mechanism (Dempsey et al., 1984). PAC is used in Japan, Germany, and France and may have potential in the United States in certain applications.

Organic Coagulants

Organic polymers have gained widespread use as water treatment coagulants in the United States since their introduction in the early 1950s. Polymers are long-chain molecules consisting of repeating chemical units with a structure designed to provide distinctive physicochemical properties to the polymer. The chemical units usually have an ionic nature that imparts an electrical charge to the polymer chain. Hence, organic polymers are often termed *polyelectrolytes*.

Polymers are used as coagulants in numerous technological fields, notably the mining industry, papermaking, and water and wastewater treatment. In water treatment applications, organic polymers are generally designed to be water soluble, to adsorb completely on, or react rapidly with, particulates, and to contain a chemical structure suitable for the intended use. This includes use as primary coagulants, coagulant or filter aids, and sludge conditioning. When used as primary coagulants, polymers, in contrast to aluminum or ferric ions, do not produce voluminous floc volumes. This is especially advantageous in applications where low floc volumes are desirable, for example, when granular media filters are used for particulate removal.

Although polymers have significant potential in water treatment, their use has been restricted due to high cost and uncertainties regarding chemical impurities associated with polymer synthesis. Such

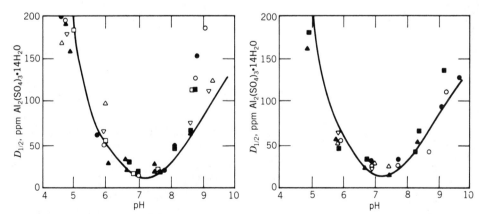

FIGURE 6-6. Coagulation of clays and organics, which supports the hypothesis that sweep floc is not influenced by the type of particles present. (Packman, 1962). ×, Mole; ▽, Great Ouse; ●, Chelmer; ⊙, Severn; □, Nar; ——, Kaolinite (Speswhite); △, Kennet; ◇, Itchen.

uncertainties have restricted the maximum concentration levels recommended or allowed by regulatory agencies in the United States and elsewhere. In France and Japan, synthetic organic polymers are currently not permitted in water treatment. In the United States, however, over 600 polymers have been approved for use in potable water treatment by the U.S. EPA (1979).

Because of the complex interactions between polymers and particulates and the uncertain influence of water quality on these interactions, polymer selection is empirical. A brief review of polymer properties and the mechanisms of polymer action will provide a basis for rational selection of polymers to be used in screening tests.

Types of Polymers. Polymers are broadly classified into natural and synthetic origin, the latter type predominating in water treatment applications. Some natural organic polymers, such as sodium alginate, a compound extracted from brown seaweed, and chitosan, obtained from chitin shells, show promise as coagulants, but costs are currently too high relative to synthetic polymers.

Synthetic organic polymers are made either by homopolymerization of the monomer or by copolymerization of two monomers. Polymer synthesis can be manipulated to produce polymers of varying size (molecular weight), charge groups, number of charge groups per polymer chain (charge density), and structure (linear or branched). A typical example is the production of polyacrylamide in which the monomer, acrylamide, homopolymerizes under appropriate conditions to form the polymer, as shown below, with n monomer units.

$$
\begin{array}{c}
CH_2{=}CH \\
| \\
C{=}O \\
| \\
NH_2
\end{array}
\xrightarrow{\text{initiation}}
\left[
\begin{array}{c}
CH_2{-}CH \\
| \\
C{=}O \\
| \\
NH_2
\end{array}
\right]_n
$$

Acrylamide　　　　　　　Polyacrylamide

This polymer carries no ionic charge and is referred to as nonionic polymer. Subsequent hydrolysis of polyacrylamide under basic pH conditions produces a polymer with anionic charges, as shown below.

$$
\left[
\begin{array}{c}
CH_2{-}CH \\
| \\
C{=}O \\
| \\
NH_2
\end{array}
\right]_x
\left[
\begin{array}{c}
CH_2{-}CH \\
| \\
C{=}O \\
| \\
O^-Na^+
\end{array}
\right]_y
$$

Thus, the number of anionic groups, in this case a carboxyl group, can be controlled providing anionic polymers of different molecular weights and charge density.

The third type of polymer has a cationic, or positive-charge, group incorporated in the polymer chain, usually by a copolymerization process. Table 6-3 describes the characteristics and uses of several examples of nonionic, anionic, and cationic polymers widely used in water treatment applications.

Mechanisms of Action. Organic polymers have two principal objectives in water treatment, destabilization of particulates and formation of larger and more shear-resistant flocs. Destabilization occurs primarily through charge neutralization. Thus, negatively charged particulates, such as clay suspensions, can be destabilized with a cationic polymer. In addition, however, destabilization of particulates with negative charges can be achieved with anionic polymers under appropriate conditions (Sommerauer et al., 1968). This has been described as a polymer-bridging mechanism, and its existence has been confirmed by a variety of techniques (LaMer and Healy, 1963; Ruehrwein and Ward, 1952; Michaels and Morelos, 1955; Evans and Napper, 1973; Ries and Myers, 1968). Increase in floc size and strength is a result of the bridging mechanism.

Successful use of polymers in water treatment requires adequate dispersion of the polymer to promote more uniform polymer adsorption and selection of the appropriate type and dose of polymer to achieve the objective. Depending on the predominant mechanism, this selection will be influenced in different ways by the physicochemical properties of the particulates, and the solution.

Charge Neutralization

Because most particulates in natural waters are negatively charged (clays, humic acids, bacteria) in the neutral pH range (pH 6–8), cationic polymers find the widest use to achieve particulate destabilization through charge neutralization. Cationic polymers can be used in conjunction or in lieu of inorganic coagulants for this purpose. With hydrophobic particulates having a defined phase boundary, cationic polymers appear to adsorb on particulate surfaces in a patchlike pattern (Kasper, 1971; Gregory, 1973), reducing electrostatic repulsion such that particulates may attach to each other following induced collisions. Generally, the optimum dose occurs when the particulate surface is only partially

TABLE 6-3. Synthetic Organic Coagulants Used in Water Treatment

Type	Examples	Molecular Weight Range	Uses	Observations
Nonionic	Polyacrylamide $-[CH_2-CH]_n-$ with $C=O$, NH_2	10^5-10^7	Coagulant aid, filter aid	Used to increase floc strength, available as powder or emulsion, used mostly as filter aid
Anionic	Partially hydrolyzed polyacrylamide $-[CH_2-CH]_x-[CH_2-CH]_y-$ with $C=O$, NH_2 and $C=O$, ONa	10^4-10^7	Coagulant aid, filter aid, sludge conditioning	Produced by controlled hydrolysis of polyacrylamide; range of MW, charge density available, charge depends on pH
Cationic	Poly(DADMAC) or poly(DMDAAC) polymers $-[CH_2-CH-CH-CH_2]_n-$ with CH_2 CH_2, N^+ Cl^-, CH_3 CH_3	10^4-10^6	Primary coagulant, turbidity/color removal, sludge conditioning	Most widely used primary coagulant; may be used in conjunction with inorganic coagulant; chlorine resistant; charge density not pH sensitive; available in liquid form
Cationic	Quarternized polyamines $-[CH_2-CH-CH_2-N^+]_n-$ with OH, CH_3, CH_3	10^4-10^5	Primary coagulant, color/turbidity removal	Mostly widely used primary coagulant for color removal; properties similar to poly(DADMAC)
Cationic	Polyamines $-[CH_2-CH_2-NH]_n-$	10^4-10^7	Primary coagulant, also coagulant aid (high MW)	Includes several types of polymers; less widely used as primary coagulant, reacts with chlorine; charge density depends on pH

covered (less than 50 percent) with polymer. Overdosing will cause restabilization. The optimum dose appears to increase in proportion to the surface area concentration of the particulates.

For hydrophilic particulates, such as humic acids, a precipitation mechanism is postulated (Glaser and Edzwald, 1979) as evidenced by the stoichiometry observed between optimum dose and concentration of humic acid.

For this mechanism, charge density is a more important polymer characteristic than molecular weight in controlling the magnitude of the optimum polymer dose. This depends, however, on the concentration of particulates (Gregory, 1974). At "high" concentrations ($>10^{14}$ particles/L, or about

100 mg/L, clay particles with $d = 0.1$ μm), increasing molecular weight may reduce polymer dose requirements due to the bridging mechanism. This would require evaluation on a case-by-case basis.

Solution parameters will also impact polymer dose. If the polymer charge density depends on pH [e.g., with nonquarternized polyamines (see Table 6-3)], then the optimum polymer dose will vary with pH, generally decreasing as the pH decreases. The charge density of quarternized polymers such as poly(DADMAC) are only slightly affected by pH. Changes in ionic composition do not appear to affect polymer dose strongly over typical ranges encountered in water treatment (TDS between 50 and 500 mg/L).

Polymer Bridging

Polymer bridging is complex and has not been adequately described analytically. Schematically, polymer chains adsorb on particulate surfaces at one or more sites along the polymer chain. The remainder of the polymer may remain extended into the solution and adsorb on available surface sites of other particulates, thus creating a "bridge" between the surfaces. If the extended polymer cannot find vacant sites on the surface of other particulates, no bridging will occur. Thus, there is an optimum degree of coverage or extent of polymer adsorption at which the rate of aggregation will be a maximum. In mineral slurries of high solids concentration (>1000 mg/L), an optimum coverage of 50 percent has been observed (LaMer and Healy, 1963).

Because polymer bridging is an adsorption phenomenon, the optimum dose will generally be proportional to the concentration of particulates present. Anionic, nonionic, and cationic polymers may function as bridging polymers; however, anionic and nonionic polymers are more widely used due to inherently higher molecular weights.

Increases in molecular weight are advantageous because of the increase in polymer size and thus the potential extent of bridging. For anionic polymers an optimum charge density may exist because low charge densities exhibit slow rates of adsorption, while higher charge densities may adsorb in a planar configuration, thus defeating the objective of bridge formation.

Solution properties (pH, ionic content) affect the polymer configuration in solution and at the interface. High ionic strength (high conductivity) tends to cause polymers to coil, thus decreasing their radius of gyration or length of extension. As noted earlier, the presence of specific cations may promote the rate of polymer adsorption (Sommerauer et al., 1968).

Because of these complex interactions, polymer selection requires empirical testing. In general, though, anionic polymers have been shown to be effective coagulant aids, while nonionic polymers have been effective as filter aids. Doses for coagulant aids range from 0.1 to 5 mg/L, while filter aids are used at dose levels ranging from 0.01 to 0.1 mg/L. Polymer selection for sludge conditioning is dependent on sludge properties, polymer properties, and the mixing environment (O'Brien and Novak, 1977). Polymer bridging is the dominant mechanism in sludge conditioning, and thus polymer molecular weight is the dominant property of interest. For each system, an optimum polymer dose, mixing conditions, and pH must be determined empirically.

KINETICS OF PARTICULATE AGGREGATION: GENERAL MODEL

A schematic of the subprocesses controlling the rate of particulate aggregation during coagulation and flocculation is shown in Figure 6-7. Initially, the distribution of particulates consists of primary particles not yet destabilized. Upon the addition of an inorganic or organic coagulant, the particulates are rapidly destabilized. This leads to the formation of unstable microflocs ranging from 1 to about 100 μm (Argaman and Kaufman, 1970). Further aggregation and growth of the flocs occur due to particulate collisions caused by transport. During the transport process the flocs are subjected to unequal shearing forces. This leads to erosion and disruption of some of the floc aggregates, or *floc breakup*. Finally, after some periods of mixing, a steady-state floc size distribution is reached, and the growth and disruption of floc particles is roughly equal (Parker et al., 1972). The rate at which the steady-state size distribution is achieved, as well as the form of the size distribution, will depend on the hydrodynamics of the system and the chemistry of the coagulant–particulate interactions. Any of the subprocesses shown in Figure 6-7 may control the rate of particulate aggregation. A quantitative description of this aggregation process will provide a basis for the process design of coagulation–flocculation facilities, as well as identifying the key process parameters.

The fundamental problem in mathematical modeling of the coagulation–flocculation process is predicting the change of the particle size distribution function with time for a given set of chemical and hydrodynamic conditions. The general kinetic model must account for changes in the number of particles found in all size classes.

Particles of size d_i collide with size d_j particles forming particles of size d_k when collisions are successful. At the same time, aggregates of size d_k may break up into smaller aggregates due to hydrodynamic shearing forces. The number of collisions, N_{ij}, occurring per unit time per unit volume between i- and j-sized particles is a product of the particle concentrations and a collision frequency function, β, expressed as

$$N_{ij} = \beta(d_i, d_j)n_i n_j \qquad (6\text{-}3)$$

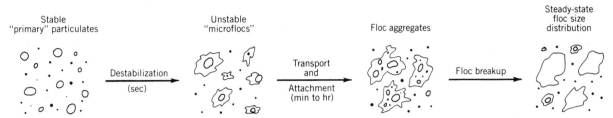

FIGURE 6-7. Subprocesses controlling rate of particulate aggregation.

The *collision frequency* function β depends on the size of the particles, the flocculation transport mechanism, and the efficiency of particulate collisions controlled primarily by the chemistry of coagulation. The formation rate of aggregates in a given size class d_k is the sum of all collisions between i and j particles minus the disappearance of aggregates in this size class due to collisions between k particles and all others. The general model for aggregation, assuming no particle breakup, is given as (Swift and Friedlander, 1964)

$$\frac{dn_k}{dt} = \tfrac{1}{2} \sum \beta(d_i, d_j)n_i n_j - n_k \sum \beta(d_i, d_j)n_j \quad (6\text{-}4)$$

Solution of this equation using appropriate values of β will predict the change in the size distribution of the suspension as aggregation occurs (Lawler et al., 1980). For process design purposes, however, more simplified models are required. The three principal modes of particulate transport are (1) Brownian motion, (2) differential movement due to fluid shear, and (3) differential movement from particle sedimentation.

Brownian (Perikinetic) Flocculation

Brownian motion affects the movement of colloidal particles but has only a minor influence on transport of particles larger than about 1 μm (Smoluchowski, 1917). For a suspension of single-sized spherical particles the collision frequency function for Brownian transport is (Smoluchowski, 1917)

$$\beta = \frac{8}{3} \alpha \frac{kT}{\mu} \quad (6\text{-}5)$$

where k is Boltzmann's constant, T the absolute temperature (K), μ the dynamic viscosity, and α the collision efficiency factor, defined as the ratio of successful to unsuccessful particle collisions with a range of values of $0 \le \alpha \le 1$. Combining Eqs. 6-4

and 6-5 and remembering that the total particle concentration is $N_T = \Sigma \, n_k$, the instantaneous rate of change of N_T due to Brownian or perikinetic flocculation becomes

$$\frac{dN_T}{dt} = -\frac{4}{3} \alpha \frac{kT}{\mu} N_T^2 \quad (6\text{-}6)$$

This reaction is second order with a rate constant $(4/3)\alpha(kT/\mu)$ of 5.4×10^{-15} L/sec at 20°C, assuming $\alpha = 1$. Therefore, compared to most chemical reactions in solution, aggregation of particles by Brownian flocculation is a relatively slow process. For a colloidal suspension of clay particles with a concentration of primary particles of 10^8/L, the half-life following Brownian flocculation at 20°C in a batch or tubular reactor would be about 20 days.

Shear (Orthokinetic) Flocculation

In order to accelerate particle aggregation for colloidal particles and achieve acceptable flocculation rates for coarse-sized particles (>1 μm), mechanical mixing must be employed. Consider particles i and j suspended in water and subjected to a velocity gradient, du/dz, as shown in Figure 6-8, with particles moving in fluid streamlines in the x direction. When the distance between the centers of the particles $R_{ij} \le (d_i + d_j)/2$, a collision will occur.

Laminar Shear. When fluid flow is laminar and steady, the velocity gradients are well defined. Smoluchowski (1917) (see also Swift and Friedlander, 1964) showed that the collision frequency function for spherical particles moving under laminar flow conditions is given by

$$\beta_{\text{sh}} = \frac{8}{\pi} \frac{dv}{dZ} v_i \quad (6\text{-}7)$$

where v_i is the volume of i particles of size d_i. Substitution in Eq. 6-4 and neglecting breakup, the ag-

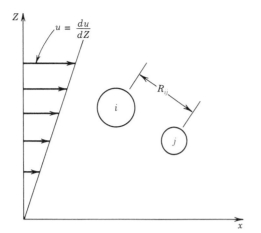

FIGURE 6-8. Orthokinetic flocculation. Particles i and j are subjected to a velocity gradient du/dz. A collision will occur when the distance between the centers becomes $R_{ij} \leq (d_i + d_j)/2$.

gregation rate of particles due to laminar shear is

$$\frac{dN_T}{dt} = \frac{4}{\pi}\frac{du}{dZ}\alpha\phi N_T \qquad (6\text{-}8)$$

where ϕ is the volume fraction of the dispersed phase, that is, the volume of the particulates per volume of solution, and is given by $\Sigma(n_i\pi d_i^3/6)$ for spherical particles, and α is the collision efficiency factor.

Thus, orthokinetic flocculation is a first-order rate expression with respect to N_T, with a rate constant directly proportional to the velocity gradient (see Figure 6-9) and the floc volume fraction. When the particulates are heterogeneous, with a wide size distribution, the rate of aggregation is increased (Swift and Friedlander, 1964) to some extent. Thus, the kinetic model for monodispersed suspensions is conservative.

Turbulent Shear. Fluid flow in mechanically mixed flocculation systems is rarely laminar. Under turbulent flow conditions, the velocity gradient is not well defined and can vary locally in the flocculation reactor. These velocity gradients relate to the dissipation of mechanical energy into heat as the water is mixed. Using fluid mechanical arguments, it has been shown (Camp and Stein, 1943) that the power P dissipated per unit volume V of fluid is given by

$$\frac{P}{V} = \mu G^2 \qquad (6\text{-}9)$$

where G is defined as the root-mean-square (rms) velocity gradient. In turbulent flow the rate of particle aggregation becomes proportional to G (Harris et al., 1966), or

$$\frac{dN_T}{dt} = -KG\phi N_T \qquad (6\text{-}10)$$

In this equation K is an empirical aggregation constant that depends on system chemistry, the heterogeneity of the suspensions, and variations in the scale and intensity of turbulence, which are not incorporated in the rms velocity gradient. Because of the different flow patterns and distributions of velocity gradients promoted by various mixing devices, K must be determined experimentally.

When flow conditions are turbulent, floc breakup cannot be neglected. Small particles are sheared from larger aggregates when the local shear stress exceeds the internal binding forces of the aggregate. The principal mechanisms of aggregate or floc breakup are surface erosion (Argaman and Kaufman, 1970) and floc splitting (Thomas, 1964).

Based on the surface erosion model, it has been shown (Argaman and Kaufman, 1970; Parker et al., 1972) that the formation rate of particle fragments, called primary particles, due to breakup is depen-

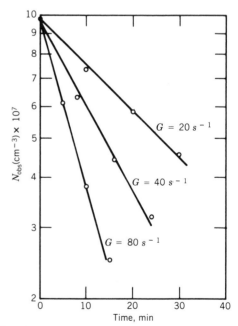

FIGURE 6-9. First-order kinetics; flocculation under orthokinetic conditions (Swift and Friedlander, 1964).

dent on the rms velocity gradient given by

$$\frac{dn_1}{dt} = K_B G^\delta \qquad (6\text{-}11)$$

In this expression n_1 is the number concentration of the primary particles, K_B a floc breakup constant dependent on the internal binding forces or floc strength of the aggregate, and δ a constant that varies between 2 and 4 depending on the hydraulic regime of turbulence (Parker et al., 1972). Thus, the net rate of disappearance of primary particles becomes

$$\frac{dn_1}{dt} = -K_A G n_1 + K_B G^\delta \qquad (6\text{-}12)$$

under turbulent mixing conditions. The aggregation constant K_A and the breakup constant K_B can be determined empirically in laboratory or pilot-scale tests (Argaman, 1971; Parker et al., 1972; Bratby, 1977; Odegaard, 1979; Kavanaugh et al., 1978). Range of reported values are shown in Table 6-4. Note that K_A incorporates K and ϕ.

Differential Sedimentation. The velocity of particles of similar densities settling in a water column is proportional to the size squared. Thus, differential particle motion occurs in heterogeneous suspensions during sedimentation, providing an additional transport mechanism for promoting flocculation. The collision frequency function for this transport mechanism is given by (Friedlander, 1977)

$$\beta(d_i, d_j) = \frac{\pi \, \Delta\rho \, g\alpha}{72\mu} [(d_i + d_j)^3][d_i - d_j] \quad (6\text{-}13)$$

where $\Delta\rho$ is the difference in density between the particle and the fluid and g is the acceleration due to gravity. For suspensions containing a wide range of particle sizes, differential sedimentation can be a significant transport mechanism (O'Melia, 1979). This is illustrated in Example 6-1.

EXAMPLE 6-1

Problem

Consider a particulate suspension of spherical particles with only two sizes, 100 and 1 μm. Compare the collision frequency factor for transport by Brownian motion, fluid shear, and differential sedimentation in a low fluid-shear environment, using the following data:

$$G = 1 \text{ sec}^{-1}$$

$$\Delta\rho = 10 \text{ kg/m}^{-3}$$

$$\mu = 0.001 \text{ kg/m-sec} \quad (T = 20°C)$$

$$\alpha = 1.0$$

Solution

From Eq. 6-5 for $d = 1$ μm, we have

$$\beta_{\text{Br}} = 1.08 \times 10^{-17} \text{ m}^3/\text{sec}$$

From Eq. 6-7

$$\beta_{\text{sh}} = \frac{8Gv_i}{\pi}$$

Since $v_{100\,\mu m} \gg v_{1\,\mu m}$, the 100-$\mu$m particle dominates β due to shear.

TABLE 6-4.   Reported Kinetic Parameters: Flocculation Kinetics

| | Kinetic Parameters | | |
System	Aggregation, K_A	Breakup, K_B (sec)	Reference
Kaolin-alum	4.5×10^{-5}	1×10^{-7}	Argaman (1970)
Kaolin-alum	2.5×10^{-4}	4.5×10^{-7}	Bratby (1977)
Natural particulates-alum	1.8×10^{-5}	0.8×10^{-7}	Argaman (1971)
Alum-phosphate precipitate	2.8×10^{-4}	3.4×10^{-7}	
Alum-phosphate plus polymer	2.7×10^{-4}	1×10^{-7}	Odegaard (1979)
Lime-phosphate, pH 11	5.6×10^{-4}	2.4×10^{-7}	

Thus,

$$\beta_{sh} = 1.33 \times 10^{-12} \text{ m}^3/\text{sec}$$

For the collision frequency of differential sedimentation, we obtain from Eq. 6-13

$$\beta_{sed} = 4.36 \times 10^{-13} \text{ m}^3/\text{sec}$$

which is roughly 30 percent of the shear collision frequency factor. Thus, in low-shear systems such as sedimentation basins, differential sedimentation can become an important transport mechanism. □

APPLICATIONS TO PROCESS DESIGN

Simplified kinetic models of particle aggregation, Eqs. 6-10 and 6-12 provide a basis for process design of coagulation–flocculation systems. Where possible, pilot studies are best suited to determine model parameters. Otherwise, literature data can be used to prepare a preliminary process design, including selection of the flocculation configuration (number of tanks), type and intensity of mixing, and flocculation residence times to achieve the desired removal efficiency.

Reactor Configuration

Because shear flocculation is a first-order reaction with respect to total particle number, the flocculation reactor should be designed to exhibit a resi-

dence time distribution approximating plug flow (see Chapter 5). Since flocculation reactors must be mixed, a combination of continuous-stirred tank reactors (CSTR) in series is the most appropriate flocculation configuration.

For m equal-volume CSTR in series, and neglecting floc breakup, the flocculator performance equation for removal of primary particles, n_1, becomes

$$\frac{n_1^m}{n_1^\circ} = \left(\frac{1}{1 + K_A G(\bar{\theta}/m)} \right)^m \qquad (6\text{-}14)$$

where $\bar{\theta}/m = V_T/mQ$, or the hydraulic residence time in a single reactor of volume V_T/m, where V_T is the total volume of the flocculator. Figure 6-10 illustrates the effect on flocculation performance of increasing the number of reactors in series at various G values as determined by changes in the turbidity (concentration of primary particles). As m increases, the optimum G value decreases.

If floc breakup is included (Eq. 6-12), the flocculation performance equation becomes

$$\frac{n_1^\circ}{n_1^m} = \frac{(1 + K_A G \tau_i)^m}{1 + K_B G^2 \tau_i \Sigma_{i=1}^{m-1}(1 + K_A G \tau_i)^i} \qquad (6\text{-}15)$$

as developed by Argaman (1971), where $\tau_i = \bar{\theta}/m$.

The importance of this equation for process design of flocculation basins is illustrated in Figure 6-11. Shown are theoretical curves using experimental values of K_A and K_B, relating the total residence time $\bar{\theta}$ and the rms velocity gradient G to the system performance as characterized by n_1°/n_1^m for $m = 1$ and $m = 4$ CSTR in series.

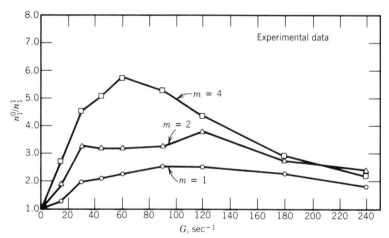

FIGURE 6-10. Performance of multicompartment systems, $\bar{\theta}/m = 8$ min (Argaman, 1971).

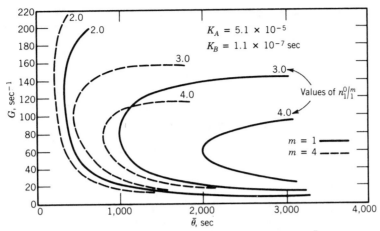

FIGURE 6-11. Performance of flocculator as related to G, $\bar{\theta}$, and m.

As the performance curves illustrate, for any desired performance there is a minimum hydraulic residence time required to achieve the removal goal, regardless of G. The curves also show that for each configuration and performance goal, an optimum value of G can be determined. Finally, and of particular importance, several CSTRs in series decrease the required residence time to achieve a given performance goal, as would be predicted by the reactor design principles presented in Chapter 5.

REFERENCES

Argaman, Y. A., "Pilot-Plant Studies of Flocculation," *JAWWA*, **63**(12), 775–777 (1971).

Argaman, Y. A., and Kaufman, W. J., "Turbulence and Flocculation," *J. Div. Sanit. Eng., Proc. Am. Soc. Civil Eng.*, **96**(SA2), 223–241 (1970).

ASTM D2035-74, "Standard Recommended Practice for Coagulation–Flocculation Jar Test of Water," *ASTM 1976 Annual Book of Standards*, Part 31, WATSR, Philadelphia (1976), p. 865.

ASTM D2035-80, "Standard Practice for Coagulation–Flocculation Jar Test of Water," *ASTM 1984 Annual Book of Standards*, Section 11.02, Philadelphia (1984), pp. 799–803.

Audsley, A., *Flocculation of Suspensions of Solids with Organic Polymers: A Literature Survey*, Warren Spring Laboratory, Ministry of Technology, Mineral Processing Information Note No. 5, Herts, England (1965).

Baes, C. F., and Mesmer, R. E., *The Hydrolysis of Cations*, Wiley-Interscience, New York (1976).

Birkner, F. B., and Edzwald, J. K., "Nonionic Polymer Flocculation of Dilute Clay Suspension," *JAWWA*, **61**(12), 645–651 (1969).

Birkner, F. B., and Morgan, J. J., "Polymer Flocculation Kinetics of Dilute Colloidal Suspensions," *JAWWA*, **60**(2), 175–191 (1968).

Black, A. P., Birkner, F. B., and Morgan, J. J., "Destabilization of Dilute Clay Suspensions with Labeled Polymers," *JAWWA*, **57**(12), 1547–1560 (1965).

Black, A. P., Arthur M. Buswell, Fred A. Eidsniss, "Review of the Jar Test," *JAWWA*, **49**, 1414–1424 (1957).

Bratby, J., Miller, M. W., and Marais, G. V. R., "Design of Flocculation Systems from Batch Test Data," *Water S. Afr.*, **3**(4), 173–182 (1977).

Brossett, C., Biedermann, G., and Gillen, L. G., "Studies on the Hydrolysis of Metal Ions," *Acta Chemica Scandinavica*, **8**, 1917–1926 (1954).

Camp, T. R., and Stein, P. C., "Velocity Gradients and Internal Work in Fluid Motion," *J. Boston Soc. Civil Eng.*, **30**(4), 219–237 (1943).

Dempsey, Brian A., Ganho, Rui M., and O'Melia, Charles R., "The Coagulation of Humic Substances by Means of Aluminum Salts," *JAWWA*, **76**(4), 141–150 (1984).

EPA, "Report on Coagulant Aids for Water Treatment," Office of Water and Waste Management, Cincinnati, OH (April 1979).

Evans, R., and Napper, D. H., "Flocculation of Lattices by Low Molecular Weight Polymers," *Nature*, **246**, 34 (1973).

Fiessinger, F., "Coagulation and Flocculation," Part I, "Coagulation," Paper presented at Congress of International Water Supply Assoc., Kyoto, Japan, Special Subject No. 3 (1978).

Friedlander, S. K., *Smoke, Dust and Haze*, Wiley-Interscience, New York (1977).

Glaser, H. T., and Edzwald, J. K., "Coagulation and Direct Filtration of Humic Substances with Polyethylenimine," *Env. Sci. Technol.*, **13**(3), 299–305 (1979).

Gregory, J., and Sheiham, I., "Kinetic Aspects of Flocculation by Cationic Polymers," *Br. Poly. J.*, **6**, 47–59 (1974).

Gregory, J., "Effects of Polymers on Colloid Stability," in K. J. Ives (ed.), *The Scientific Basis of Flocculation*, Sijthoff & Noordhoff, Rockville, MD (1978).

Harris, H. S., Kaufman, W. F., and Krone, R. B., "Orthokinetic Flocculation in Water Purification," *J. Div. Sanit. Eng., Proc. Amer. Soc. Civil Eng.*, **92**(SA6), 95–111 (1966).

James, C. R., and O'Melia, C. R., *JAWWA*, **79**(3), 6–26 (1982).

Kasper, D. R., "Theoretical and Experimental Investigations of the Flocculation of Charged Particles in Aqueous Solutions by Polyelectrolytes of Opposite Charge," Ph.D. thesis, California Institute of Technology (1971).

Kruyt, H. R., *Colloid Science*, Vol 1. Elsevier, New York (1952).

Langelier, W. F., "Coagulation of Water with Alum by Prolonged Agitation," *Eng. News-Record*, **86**, 924 (1921).

LaMer, V. K., and Healy, T. W., "Adsorption-Flocculation Reactions of Macromolecules at the Solid-Liquid Interface," *Rev. Pure Appl. Chem.*, **13**, 112–133 (1963).

Lawler, D. F., O'Melia, C. R., and Tobiason, J. E., "Integral Water Treatment Plant Design; From Particle Size to Plant Performance," in M. C. Kavanaugh and J. E. Leckie, (eds.), *Particulates in Water*, Advances in Chemistry Series, No. 189, American Chemical Society, Washington, DC (1980).

Matijevic, E., K. M. Mathal, R. H. Ottewill, M. Kerber, "Detection of Metal Ion Hydrolysis by Coagulation. III. Aluminum," *J. Phys. Chem.*, **65**(5), 826–830 (1961).

Mattson, Sante. "Catophoresis and the Electrical Neutralization of Colloidal Material," *J. Phys. Chem.*, **32**, 1532–1552 (1928).

Michaels, A. S., and Morelos, O., "Polyelectrolyte Adsorption by Kaolinite," *IEC*, **47**(9), 1801–1809 (1955).

Morgan, J. J., and Engelbrecht, R. S., "Effects of Phosphates on Coagulation and Sedimentation of Turbid Waters," *JAWWA*, **52**(9), 1303–1314 (1960).

O'Brien, J. H., and Novak, J. T., "Effects of pH and Mixing on Polymer Conditioning of Chemical Sludges," *JAWWA*, **69**(11), 600–605 (1977).

Odegaard, H., "Orthokinetic Flocculation of Phosphate Precipitates in a Multicomponent Reactor with Non-ideal Flow," *Prog. Water Tech.*, **11**, Suppl. 1, 61–88 (1979).

O'Melia, C. R., "Coagulation in Wastewater Treatment," in K. J. Ives (ed.), *Scientific Basis of Flocculation*, Noordhoff International Publishing, Leyden, The Netherlands (1978).

Packham, R. F. "The Coagulation Process," *J. Appl. Chem.*, **12**, 556–568 (1962).

Packham, R. F., "Some Studies of the Coagulation of Dispersed Clays with Hydrolyzing Salts," *J. Coll. Science*, **20**, 81–92 (1965).

Parker, D. S., Kaufmann, W. J., and Jenkins, D., "Floc Breakup in Turbulent Flocculation Processes," *J. Sanit. Eng. Div.*, **98**(SAI), 79–99 (1972).

Parks, G. A., "Aqueous Surface Chemistry of Oxides and Complex Oxide Minerals; Isoelectric Point and Zero Point of Charge," in *Equilibrium Concepts in Natural Water Systems*, Advances in Chemistry Series, No. 67, American Chemical Society, Washington, DC (1967).

Ries, H. E., and Myers, B. L., "Flocculation Mechanism: Charge Neutralization and Bridging," *Science*, **160**, 1449–1450 (1968).

Ruehrwein, R. A., and Ward, D. W., "Mechanism of Clay Aggregation by Polyelectrolytes," *Soil Sci.*, **73**(6), 485–492 (1952).

Singley, J. E., and Sullivan, J. H., "Reactions of Metal Ions in Dilute Solution: Recalculation of Hydrolysis of Ion (III) Data," *JAWWA*, **61**(4), 190–192 (1969).

Smoluchowski, M., *Zeitschrift Physik Chem.*, **92**, 129–168 (1917).

Sommerauer, A., Sussman, D. L., and Stumm, W., "The Role of Complex Formation in the Flocculation of Negatively Charged Soils with Anionic Polyelectrolytes," *Kolloid-Zeit. & Zeit. Polymere*, **225**(2), 147–154 (1968).

Stumm, W., and Morgan, J. J., "Chemical Aspects of Coagulation," *JAWWA*, **54**(8), 971–992 (1962).

Stumm, W., and Morgan, J. J., *Aquatic Chemistry*, 2nd ed., Wiley-Interscience, New York (1981).

Stumm, W., and O'Melia, C. R., "Stoichiometry of Coagulation," *JAWWA*, **60**(5), 514–539 (1968).

Sullivan, J. H., and Singley, J. E., "Reactions of Metal Ions in Dilute Aqueous Solution: Hydrolysis of Aluminum," *JAWWA*, **60**(11), 1280–1287 (1968).

Swift, D. L., and Friedlander, S. K., "The Coagulation of Hydrosols by Brownian Motion and Laminar Shear Flow," *J. Colloid Sci.*, **19**, 621–647 (1964).

TeKippe, R. J., and Ham, R. K., "Coagulation Testing: A Comparison of Techniques," *JAWWA*, **62**(9), 594–628 (1970).

Tenney, M. W., and Stumm, W., "Chemical Flocculation of Microorganisms in Biological Waste Treatment," *JWPCF*, **37**(10), 1370–1388 (1965).

Thomas, D. G., "Turbulent Disruption of Flocs in Small Particle Size Suspensions," *AIChE J.*, **10**, 517–523 (1964).

Trussell, R. R., "Predesign Studies," in R. Sanks (ed.), *Water Treatment Plant Design*, Ann Arbor Science, Ann Arbor, MI (1978).

van Olphen, H., *An introduction to Clay-Colloid Chemistry*, Wiley-Interscience, New York (1963).

— 7 —

Gravity Separation

Gravity separation of suspended material from aqueous solution is the oldest and most widely used process in water treatment and wastewater reclamation. Suspensions in which particulate matter is heavier than water tend to settle to the bottom as a result of gravity forces in the process of sedimentation. Particles lighter than the density of water conversely tend to float in a process designated flotation. Since nearly all particulate matter in natural waters and reclaimed water is as heavy or heavier than water itself, the most widespread process is sedimentation. However, the density of particles can be made lighter than water by the attachment of bubbles to the particulates. Process units used to achieve this latter phenomenon are known as dissolved air flotation or foam fractionation flotation devices. Flotation processes are relatively new in the water treatment field; however, they are under extensive research and have been found effective in removing algae and other organic materials.

This chapter discusses theoretical concepts, practical considerations, and full-scale design features of sedimentation, flotation processes, and process selection.

SEDIMENTATION

The removal by sedimentation of suspended matter from water at low cost and low energy consumption is conceptually simple but often involves complications that render proper basin design a challenge for many engineers.

Prediction of performance of a typical sedimentation tank design for a given quality of raw water or wastewater effluent can, to some extent, be understood with the help of sedimentation theories. When supplemented by understanding of practical aspects of clarifier design, the units can be designed to perform reliably and consistently.

Theories of Sedimentation

In raw water and reclaimed water treatment a wide variety of suspensions ranging from a very low concentration of nearly discrete particles to a high concentration of flocculent solids can be treated by the sedimentation process. To define the different settling characteristics of aqueous suspensions more clearly, particulates have been categorized into three general classes (Katz et al., 1962):

135

Class 1. Discrete particles that will not readily flocculate predominate in relatively low concentrations. An example of this type of suspension is encountered in wastewater grit chamber design and in clarification of certain industrial wastes (e.g., sand and gravel washings).

Class 2. Relatively low solids concentrations of flocculent material. An example of this type of material is found in water subjected to flocculation by chemical addition.

Class 3. Relatively high concentrations of material. This material may be flocculent, but not necessarily so. The term *hindered settling* is generally used to describe separation of this type of solids. Examples of this type of separation are found in sludge thickening.

These are illustrated in Figure 7-1.

Discrete Particles. The beginning of modern theory in sedimentation dates back to the early work of Stokes, Hazen, and Newton. These classical theories coupled with more recent literature in this field have been covered clearly and concisely in a 1956 publication (McGauhey, 1956). The discussion herein presents a brief summary of sedimentation theory.

For aqueous suspensions, particles with densities greater than water will be accelerated down- ward under the force of gravity until the resistance of the liquid equals the effective weight of the particle. The gravity forces are eventually balanced by particle drag forces, at which time the particle will reach an equilibrium settling velocity that is approximately constant and depends on the size, shape, and density of the particle and the density and viscosity of the water.

Classical settling theories have been based on the assumption that particles are spherical in nature. The results are transferable to some nonspherical particulate solutions by the application of coefficients relating other shapes to that of a settling sphere. The general equation used to relate these variables for spheres is

$$V = \sqrt{\frac{4}{3}\frac{g}{C_D}\frac{(\rho_1 - \rho)D}{\rho}} \qquad (7\text{-}1)$$

where C_D = drag coefficient
g = gravity constant
ρ_1 = mass density of particles
ρ = density of water
D = diameter of particle (spherical shape)
V = velocity of settling particle

The drag coefficient varies as a function of density, relative velocity, particle diameter, and fluid

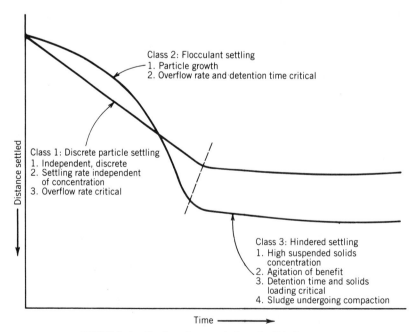

FIGURE 7-1. Setting characteristics of solids in water.

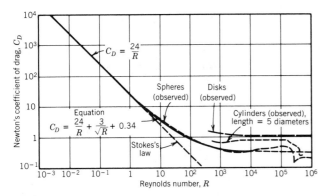

FIGURE 7-2. Newton's coefficient of drag for varying magnitudes of Reynolds number. (Observed curves after Camp, 1946.)

viscosity, which are expressed by the dimensionless Reynolds number, R:

$$R = \frac{\rho V D}{\mu} \tag{7-2}$$

where ρ = density of fluid
μ = viscosity of fluid
D = diameter of particle
V = relative velocity

and related by the Moody diagram (shown in Figure 7-2).

For Reynolds numbers less than 2, C_D is related to Re by the linear expression

$$C_D = \frac{24}{R} \tag{7-3}$$

which was formulated by Stokes. At higher Reynolds numbers, the relationship is nonlinear but approximated by the equation

$$C_D = \frac{24}{R} + \frac{3}{\sqrt{R}} + 0.34 \tag{7-4}$$

By this equation, as R becomes large, C_D approaches the constant value of 0.34. Data in Figure 7-2 show that in the range of $R = 10^3–10^{5.2}$, C_D is approximately 0.4 and then drops to 0.2 for higher values. Equation 7-4 is, therefore, to be considered an overall useful relationship for a broad range of R values but not exact for all values of R.

At low velocities where $C_D = 24/R$ is sufficiently accurate, Eq. 7-3 can be substituted into Eq. 7-1 to give

$$V = \frac{g(\rho_1 - \rho)D^2}{18\mu} \tag{7-5}$$

This expression is known as Stokes's law. At higher Reynolds numbers where $C_D = 0.4$, the velocity relationship becomes

$$V = \sqrt{\frac{10}{3} g \frac{\rho_1 - \rho}{\rho} D} \tag{7-6}$$

Camp (1936) developed a rational theory for relating the removal of discrete particulate suspended matter in an ideal sedimentation basin. As illustrated in Figure 7-3, Camp divided a settling tank into four zones and studied removals in the ideal settling zone. The other parts of the tank, namely inlet zone, sludge zone, and outlet zone, are considered special tank areas that permit ideal settling in the settling zone but do not in themselves achieve particulate removal.

The following assumptions were made by Camp to develop tank removal efficiency equations:

- Horizontal flow in the settling zone.
- Uniform horizontal velocity in the settling zone.
- Uniform concentration of all-size particles across a vertical plane at the inlet end of the settling zone.
- Particles were removed once they reached the bottom of the settling zone.

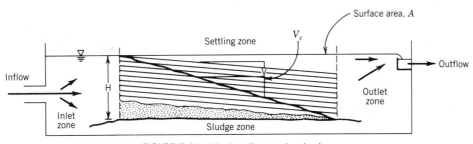

FIGURE 7-3. Ideal sedimentation basin.

• Particles settle discretely without interference from other particles at any depth.

On the basis of these assumptions, all particles in the settling zone travel in a straight-line path, as shown in Figure 7-3. In the calculations by Camp, particle removal efficiency then reduces to a function of flowrate divided by surface area of the clarifier. This ratio, known as the surface overflow rate, was determined to be a useful design criterion. This same conclusion was reached in 1904 by Hazen (1904) but was not generally accepted for design until publication of the work by Camp.

Camp understood that the ideal settling concept was a severe simplification of actual water treatment plant clarifier operation. Particle interaction and currents in the settling area are considered to be the most significant departures from ideality. Currents have the effect of altering the particle-settling paths and also may scour particles already settled. Some degree of particle interaction, including flocculation and perhaps even breakup, occur in real settling basins.

Since surface loading rates do not specify surface shapes, Camp was concerned about developing guidelines for this factor as well. Assuming a rectangular shape, the problem relates to identifying the length–width ratio. This in turn relates to the travel velocity of the water, for excess velocity values would result in scour. The critical scour velocity for discrete-particle suspensions was formulated by Camp as

$$V_c = \sqrt{\frac{8\beta}{f} g(s - 1)D} \qquad (7\text{-}7)$$

where V_c = critical velocity
$\beta \cong 0.04$, a constant
$f \cong 0.03$, a friction factor
s = specific gravity of particles
D = diameter of the particle

In time, the concern for currents, scour velocities, and other nonideal conditions led to design provisions for length/width ratios of 5 or more but with horizontal velocities of less than 30 cm/min (1.0 ft/min) and adequate depths.

To overcome the practical problems of determining particle sizes, settling velocities, and such data for suspensions in water treatment, researchers have gone to experimental techniques using settling columns to simulate prototype clarifiers. In brief, a column is filled with a suspension of particles and samples are removed from selected depths at certain time intervals and analyzed for total suspended solids (TSS) removal.

Referring to Figure 7-3, particles with velocities less than V_c will be removed according to the ratio

$$X_r = \frac{V}{V_c} \qquad (7\text{-}8)$$

where X_r = fraction removed
V = velocity of specified particle size
V_c = critical velocity defined as ideal clarifier overflow rate (i.e., tank inflow/ surface area).

Thus, only particles with $V > V_c$ will be completely removed. Smaller particles will be fractionally removed. The following graphic procedure is often used to determine an approximation of the overall removals:

1. Prepare a discrete settling curve as shown in Figure 7-4.

2. Integrate the area to the left of the curve representing removal fractions by

$$X_r = (1 - X_c) + \int_0^{X_c} \frac{V}{V_c} dx \qquad (7\text{-}9)$$

where $1 - X_c$ = fraction of particles with velocity greater than V_c

$\int_0^{X_c} \dfrac{V}{V_c} dx$ = fraction of particles removed with velocities less than V_c

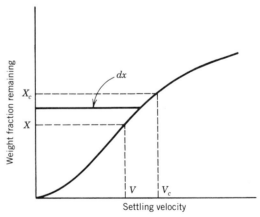

FIGURE 7-4. Typical discrete-particle settling curve.

Flocculant Particles. Flocculant particles are those with a tendency to coalesce during the sedimentation process. There are two principal causes of flocculation during sedimentation:

1. Differences in the settling velocities of particles whereby faster-settling particles overtake those that settle more slowly and coalesce with them.
2. Velocity gradients within the liquid that cause particles in a region of a higher velocity to overtake those in adjacent stream paths moving at slower velocities.

The concept of flocculation within a settling basin is considered beneficial for two principal reasons. First, the combination of smaller particles to form larger ones results in a faster-settling sludge particle because of its increase in diameter. Second, flocculation tends to have a sweeping effect in which large particles settling at a velocity faster than slow particles tends to sweep some of them from suspension. Thereby many tiny particles that would otherwise be lost and appear as turbidity in the effluent are removed. Thus, the net effect of flocculation during settling permits design of smaller clarifiers and improved effluent quality.

Camp and Stein (1943) attempted to quantify the number of collisions leading to flocculation by the following expression:

$$N_{ij} = N_i N_j \frac{\pi}{4} (D_i + D_j)^2 (V_i - V_j) \qquad (7\text{-}10)$$

where N = frequency of particle collision
i, j = particle designation indices
D = diameter of particle
V = particle settling velocity

Application of such equations to sedimentation or flotation tank influent has not been found valuable in practical process unit design or performance prediction. A number of investigators have tried to improve on the design by conducting experiments with suspensions in settling columns. For example, O'Connor and Eckenfelder (1958) proposed a technique involving use of a quiescent settling column equal to the depth of the proposed tank. The test results were then employed to obtain the required overflow rate and detention time for a clarifier.

This technique involves use of a settling column of sufficiently large diameter to prevent excessive wall effects. Approximately 15 cm is generally con-

sidered adequate. The height of the column should be equal to or greater than that anticipated for the prototype settling tank design. Sampling ports should be constructed at approximately 50-cm intervals.

To begin the experiments, a suspension is poured into the top and gently mixed with a perforated plunger to obtain a uniform dispersion of particles. At predetermined time intervals, samples are removed from the ports and analyzed for suspended solids concentrations. Care must be taken to ensure that removal of the samples has a minimal effect on the settling within the column. It is also necessary to ensure uniform temperature control and lack of any vibration that may upset the settling process. Experimental data of suspended solids removal are then plotted on a time–depth graph as illustrated in Figure 7-5.

The overall percent removal, X_r, can be calculated in a manner similar to that used above for discrete particles. Specifically, an approximation can be determined by use of Eq. 7-11.

$$X_r \cong \frac{\Delta Z_1}{Z_6}\left(\frac{R_1 + R_2}{2}\right) + \frac{\Delta Z_2}{Z_6}\left(\frac{R_2 + R_3}{2}\right)$$
$$+ \frac{\Delta Z_3}{Z_6}\left(\frac{R_3 + R_4}{2}\right) + \frac{\Delta Z_4}{Z_6}\left(\frac{R_4 + R_5}{2}\right) \qquad (7\text{-}11)$$

where X_r = percent removal of TSS
Z = depth
R = removal percentages

The terms of this equation are illustrated in Figure 7-5. The accuracy of estimation can be improved by decreasing the interval between isoconcentration lines and adding more terms to Eq. 7-11.

Such use of ideal settling columns is helpful in predicting removals from aqueous suspensions provided that the limitations of such testing are understood. Short-circuiting, inlet and outlet turbulence, density, and temperature-induced currents are typical of nonideal conditions in prototype clarifiers resulting in reduced removal efficiency. To compensate for the nonideal conditions, various authors (Schroeder, 1976; Metcalf and Eddy, 1979) recommend applying a safety factor equal to 0.65–0.85 for the overflow rate and 1.75 to 2.0 for the detention time. If the data are available, the hydraulic efficiency, defined as the detention time for the center of mass of a dye tracer study curve divided by the theoretical detention time, has been used as an approximation for the factor of safety.

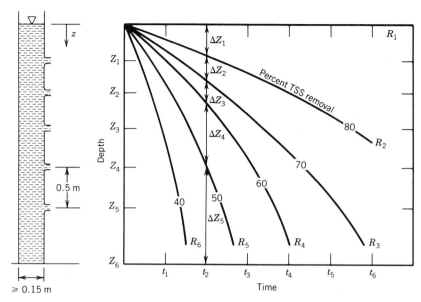

FIGURE 7-5. Typical pilot settling test results for flocculent suspensions.

Wallace (1965) took the settling column technique a step further by combining column settling test results with clarifier hydraulic dye test curves to predict clarifier performance. His procedure involved a multigraph procedure which related settling column test data to dye curve data on a step-by-step basis to determine an integrated overall estimating settling tank process performance efficiency.

The fact that such experimental work is felt necessary by many designers is testimony to the difficulty in using mathematical models to predict particle interaction for quantifying the effect of flocculation in clarifier design. Study of flocculent particle removals has established several fundamentals that are different from those of discrete particles:

1. Importance of depth.
2. Reduced value of complete quiescence in settling.
3. Importance of velocity gradient patterns in the settling regime.

Since flocculent particles tend to grow in size during their downward movement in a settling basin, clarifier depth and detention time are important design considerations. Deeper tanks provide more opportunity for larger particles to "sweep" smaller particles from suspension. All other factors being equal, therefore, deeper tanks are expected to perform better in settling of flocculent particles.

The concept of ideal settling basins with a zone of complete quiescence is not necessarily valid for flocculating particles. Some degree of induced currents, which make a clarifier nonideal, may in fact improve sedimentation because of its enhanced degree of flocculation.

These observations point out that coagulation, flocculation, and sedimentation need to be considered as an integral system. Gradually decreasing velocity gradient paths ending with ideal settling has been considered optimal for water treatment plant design by many engineers for years. Data by TeKippe and Ham (1971) have shown that changes in the velocity gradient path can affect the clarifier design criteria. For example, in a water treatment plant where rapid sand filtration is provided, the bulk of suspended solids can be removed during a very short period of sedimentation provided that the proper coagulation and flocculation velocity gradient path was followed first.

Zone Settling. Zone settling occurs in basins when the interaction of particles tends to decrease the settling velocities of individual particles. When zone settling occurs, particle aggregates tend to form a blanket with a distinct interface between the settling sludge and the basin supernatant. Settling tank influents are rarely of sufficient particle concentration to encounter Class 3 type or zone settling. This concept is much more likely to occur only in the lower regions of a clarifier where the concentration of suspended solids is highest. In water treatment, therefore, zone settling is of primary importance as related to sludge thickening rather than settling tank performance. For completeness, however, the concept is discussed in this chapter and referred to later in the chapter on sludge dewatering.

The concept of zone settling can be illustrated by observations of settling in a quiescent settling column and a graphical plot of the interface height versus time. Such information is presented in Figure 7-6 for a typical zone settling column test. Initially, the uniform concentration of suspended material (B) has its interface at the top of the settling column. As time passes, an interface forms and begins to move downward, forming a supernatant (A) at the top. Shortly thereafter, a layer of dense sludge forms at the bottom (D) and an intermediate transition zone of sludge, changing from hindered settling to compaction (C), forms. In time the settling sludge makes a transition from hindered settling to complete compaction with supernatant at the top.

If ideal settling can be assumed, an estimate of clarifier capacity can be made by calculating the slope of the hindered settling portion of the curve shown in Figure 7-6. The limiting overflow rate of the clarifier must be less than the settling velocity of the suspension for the solids to remain in the clarifier. This limitation can be expressed mathematically by

$$A \geq \frac{Q}{V_s} \tag{7-12}$$

where A = tank surface area
Q = volumetric flowrate
V_s = hindered settling velocity

The solids flux, which consists of the mass of solids movement in a basin divided by the area of the basin per unit time, is defined by

$$G_L = CV_s \tag{7-13}$$

where G_L = limiting solids flux, mass per unit area per unit time
C = solids concentration

For a basin without sludge withdrawal, a limiting solids flux would be equal to the slope of the hindered settling curve times the initial solids concentration.

When sludge is withdrawn at a significant rate from the bottom of the clarifier in a flowthrough system, the limiting solids flux can be increased by a factor equal to the solids removal rate or underflow rate. Further description of this theory is presented by Dick and Young (1972).

Under equilibrium conditions, a limiting solids flux of a clarifier can be expressed as

$$G_L = C_L V_L + \frac{Q_u C_L}{A_T} = \frac{Q_u C_u}{A_T} = \frac{C_0 Q_0}{A_T} \tag{7-14}$$

where G_L = limiting solids flux, mass per unit area per unit time
C_L = limiting solids concentration
V_L = zone settling velocity of sludge at concentration C_L
Q_u = volumetric underflow rate (return rate plus wastage rate if waste sludge is

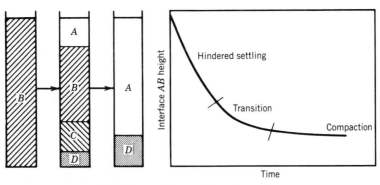

FIGURE 7-6. Zone-settling observations.

taken from underflow)
A_T = thickening area of tank
C_u = solids concentration of underflow
C_0 = mixed liquor concentration
Q_0 = volumetric inflow rate (wastewater feed plus return sludge less wastage rate)

The first term in Eq. 7-14 ($C_L V_L$) is the solids flux due to gravity settling. The second term ($Q_u V_L / A_T$) is the flux due to withdrawal of solids in the underflow. The wastage rate in effluent solids are normally neglible in comparison to other flowrates and can be ignored.

The limiting total flux, G_L, may be determined from graphical or mathematical differentiation of the relationship for total flux at any solids concentration:

$$G_{L_i} = C_i V_i + \frac{Q_u C_i}{A_T} \qquad (7\text{-}15)$$

Mathematical differentiation requires a defined relationship between the zone settling velocity and the initial solids concentration. This may be derived from the data collected in pilot-plant thickening studies analyzed according to the empirical relationship

$$V_i = m C_i^{-n} \qquad (7\text{-}16)$$

where V_i = zone settling velocity of solids at initial concentration C_i
C_i = initial solids concentration, g/L

The coefficients m and n are, respectively, the intercept and slope of the linear portion of a log plot of zone settling velocity against initial solids concentration. Figure 7-7 illustrates such a plot. Other models of the velocity–concentration relationship may be necessary to represent certain types of sludges if the relationship expressed above is nonlinear.

A graphical representation of Eq. 7-14 is shown in Figure 7-8. The batch flux curve is constructed from the settling data such as in Figure 7-7 as a product of concentration and zone settling velocity plotted at the initial concentration. The underflow solids flux is represented by a line tangent to the batch flux curve with a slope equal to the underflow rate per unit area (Q_u / A_T). A line corresponding to the overflow rate, drawn with a slope equal to the

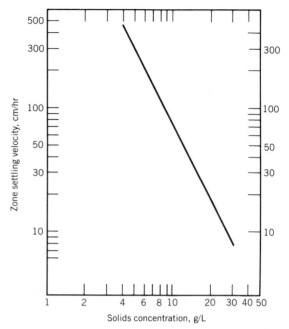

FIGURE 7-7. Relationship between initial settling velocity and zone-settling velocity.

wastewater feed rate divided by the thickening area (Q / A_T), is also shown.

The intersection of the underflow line with the vertical axis is the limiting solids flux. The corresponding underflow concentration (C_u) is given by the intersection of the underflow line on the horizontal axis. The limiting concentration (C_L) is given by the point of tangency of the underflow line to the batch flux curve, and the initial solids concentration (C_0) is given by the intersection of the underflow and overflow lines. This intersection is defined as the "state point" by Keinath, et al. (1976). It is useful to quantify blanket level changes resulting from changes in feed or changes in sludge withdrawal rates. The maximum feed solids concentration permissible is given by the intersection of the overflow line with the batch flux curve.

The limiting flux defines the maximum rate at which solids may be added to the tank and still maintain equilibrium at a given underflow rate. For a given clarifier, the limiting solids flux is not an absolute maximum loading. The limiting flux can be increased by increasing the underflow or sludge withdrawal rate. Thus, the influent solids concentration may be increased at the same time but at the cost of a decreasing underflow concentration, as shown by the dashed line in Figure 7-8. Higher solids loading without corresponding increases in

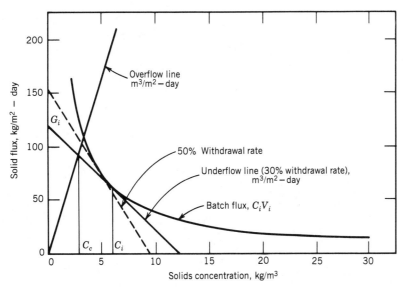

FIGURE 7-8.   Graphical solids flux analysis.

the underflow rate would overload the thickening capacity of the basin. In effect, solids are added to the tank faster than they are withdrawn and solids will be stored in the tank at the limiting concentration (C_L). A solids loading less than G_L is called an underloading condition and requires a reduced underflow concentration for equilibrium if Q_u remains constant.

The solids flux analysis may be performed mathematically with the settling velocity–concentration relationship given by Eq. 7-16. Substitution of this equation into Eq. 7-15, differentiating, and solving for C_L with $dG_i/dC_i = 0$ gives the following:

$$C_L = \left[\frac{m(n-1)A_T}{Q_u}\right]^{1/n} \qquad (7\text{-}17)$$

$$G_L = [m(n-1)]^{1/n}\left[\frac{n}{n-1}\right]\left[\frac{Q_u}{A_T}\right]^{(n-1)/n} \qquad (7\text{-}18)$$

The corresponding influent suspended solids concentration and underflow solids concentrations are given by Eq. 7-19 and 7-20:

$$C_u = \frac{G_L A_T}{Q_u} \qquad (7\text{-}19)$$

$$C_0 = \frac{r}{r+1}\,C_u \qquad (7\text{-}20)$$

where r = ratio of underflow volumetric flowrate to clarifier or thickener inflow rate

These latter two equations are valid only when the absolute value of n is greater than 1.0 and in the range of solids concentrations where the log–velocity concentration relationship is linear. Nonlinearity of these data would necessitate development and use of a different equation to substitute for Eq. 7-16. Additional discussion and practical application of zone settling theories are presented in the section of solids processing.

TYPES OF SEDIMENTATION BASINS

Sedimentation basins are commonly thought of as large rectangular or circular tanks of about 3 m in depth with inlet baffles and effluent weirs. In recent years, the field has been expanded greatly by the innovations of tube settlers, plate separators, upflow clarifiers, and so on, which aim to achieve an equivalent degree of clarification for less money and/or on less land. This section summarizes the classical basin designs and the innovative units. For more detailed design information refer to Chapter 21.

Conventional Basin Designs

Conventional clarifiers used in water treatment and reclamation are of rectangular, circular, or square configuration (see Figure 7-9).

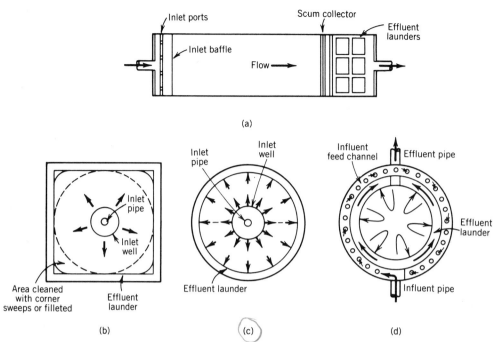

FIGURE 7-9. Types of sedimentation basins. (a) Rectangular. (b) Square, center feed. (c) Circular, center feed. (d) Circular, peripheral feed.

Rectangular Basins. The influent flow to a rectangular basin is normally distributed across the end of the tank by baffle or reaction jet structures, which provide energy dissipation and flow distribution. The concept of distributing flow by geometric similarity, such as duplicated multiple-inlet pipes or ports, has often led to adequate distribution at some flowrates, such as at average conditions, but poor distribution at others (e.g., peak flows). This is principally due to the fact that head loss varies as the square of velocity and flowrate.

Rectangular tank outlet structures are generally composed of finger launders running parallel to the length of the tank. Some design engineers also provide cross baffles in the vicinity of the effluent launders to prevent the return of reverse surface currents toward the tank inlet. This is discussed further under the heading of density currents in this chapter.

Circular Basins. Circular settling tanks have been chosen in many cases because they provide an opportunity to use relatively trouble-free circular sludge removal mechanisms and, for small plants, can be constructed at a lower capital cost per unit surface area. Diameters are calculated on the basis

of overflow rates using approximately the same criteria as common for rectangular basins.

Circular basins can be fed from a central inlet or from multiple peripheral ports or a peripheral skirt launder. Most circular basins provide a single circular baffle around the influent vertical rise pipe for center feed tanks and around the perimeter of the tank for peripheral feed units. These baffles provide space for energy dissipation and direct the flow downward into the depths of the settling tank. The net effect on tank hydraulics is the formation of a doughnut-shaped roll within the tank. Return surface currents flow radially inward for center-feed tanks and outward for peripheral-feed tanks.

Effluent structures for circular tanks' normally consist of a single, V-notch weir constructed at the outside perimeter of the tank. Multiple-weir launders and serpentine weirs are sometimes used to increase weir length, especially on large circular tanks. Unless the influent water has problems with debris and flotable material, baffles near the outlet and surface-skimming devices are not provided.

Square Basins. The development of square clarifiers was done in an effort to combine the advantages of common-wall construction of rectangular basins with the simplicity of circular sludge collectors. This attempt generally has not been successful. Since the effluent launders are constructed along the perimeter of the basins, the corners have more weir length per degree of radial arc. Thus, flows are not distributed equally, resulting in a larger deposition of sludge in the corners of the basin. Corner sweeps, added to the circular sludge collector mechanisms to remove sludge settling in the corners, have been a source of mechanical difficulty. Because of this problem, there are relatively few square basins constructed for water treatment.

Sedimentation Basin Innovations

High-Rate Clarifiers. Sedimentation of aqueous suspensions can be accelerated by increasing particle size or decreasing the distance a particle must fall prior to removal. The first is achieved by coagulation and flocculation prior to sedimentation. The second can be achieved by making the basin shallower, but this is limited by practical aspects of sludge storage, equipment movement, wind effects on the surface, and so on. To pursue this concept, therefore, it is necessary to separate clarification in shallow compartments from the process of sludge withdrawal and surface current effects. The first approach was to provide parallel plates that permit solids to reach the bottom after only short distances of settling. If these plates are oriented in a horizontal direction, they would eventually fill with solids, which would increase the head loss and eventually increase velocities to a point that the suspended materials would be scoured back into suspension. Inclining such parallel plates to a degree that the sludge can flow in a direction opposite that of the suspended liquid was found feasible and led to the development of parallel-plate and tube settlers.

Early development studies were conducted using flat parallel plates, shallow trays, and circular pipes or tubes. The plate spacing or tube sizes and feed rates were set so as to maintain laminar flow at all times. Tests indicated that for alum-coagulated sludge, solids would remain deposited in the tubes until the angle of inclination was increased to 60° or more from horizontal. If, however, flow was arranged such that the movement of the liquid and sludge in the tube were in the same direction, the angle of inclination could be reduced to approximately 30°. Figure 7-10 shows an essentially horizontal orientation and a steeply inclined design.

The nearly horizontal tube settlers theoretically require less tube volume because the depth of particle fall, which varies inversely with the cosine of the inclination angle, is smallest. An angle of inclination of 60° would effectively double the maximum fall distance for particles entering the tube. Thus, nearly horizontal tubes offer the advantage of lower construction costs, but special cleaning procedures are required to remove suspended material that settle in the tubes. Because of the complexity of the cleaning operation, the nearly horizontal tube settlers are generally restricted to use in relatively small plants, usually less than 0.05 m³/sec (1 mgd) capacity.

The most popular, commercially available tube settler is the steeply inclined tube settler. The angle of inclination is steep enough to enable the sludge to flow in a countercurrent direction from the suspension flow passing upward through the tube. Thus solids drop to the bottom of the clarifier and are removed by conventional sludge removal mechanisms.

The Lamella separator offers the advantage of concurrent flow of liquid and sludge. Flow entering a Lamella separator flows downward between the

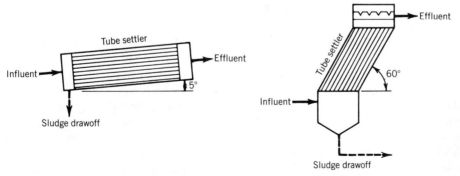

FIGURE 7-10. Microfloc tube settler system alternatives.

plates depositing the sludge as it travels. Laminar flow is established by the parallel plates. At the bottom of each plate, tubes are located near the top of the opening to catch the effluent and transport it upward between the plates to the module separator effluent pipeline. Sludge falls off the bottom of the plate into a hopper for removal. The net effect is that a smaller angle of inclination (normally 30° or more) can be used and still obtain adequate sludge removal from the tube. The Lamella separators require relatively little surface area and have been used extensively in Europe and in industrial applications in the United States. To date, however, they have received little use in municipal treatment of water or wastewater in the United States.

Other design developments include the Chevron tube settler and the Pielkenroad separator. The Chevron design has the cross-sectional area of each rhomboidal tube formed in a V shape. The Pielkenroad separator offers some of the advantages of both the Chevron tube settler and the parallel-plate Lamella separator. The Pielkenroad separator has corrugated plates that provide for some sludge thickening and appear to be less costly to construct than tube settlers.

Gravel-Bed Clarifiers.   Another attempt at providing laminar flow during clarification is the development of gravel-bed clarifiers. Similar in objective to the tube settlers, a bed of rock serves to establish a zone of laminar flow. Suspended material deposits on the rock during these conditions. To remove the sludge, rock must be agitated to loosen the particles, which in turn fall to the bottom of the tank. Conventional sludge removal mechanisms are then employed.

Upflow Clarifiers.   Upflow clarifiers, commonly called solids contact units, have been used for years to achieve good suspended solids removal in less space than conventional clarifiers. The solids contact unit employs flocculation with sedimentation in a single tank. The flocculation portion of the tank is designed to provide the mixing necessary for good floc formation while the sedimentation portion acts as a true upflow-type clarifier with surface overflow rate controlling particle removal.

Upflow clarifiers may be classified into three basic groups: (1) simple upflow clarifiers; (2) reactor clarifiers, with or without sludge recirculation; and (3) sludge blanket clarifiers (see Chapter 21).

Simple upflow clarifiers are essentially classical

separate sedimentation tanks in which inflow enters the tank near the bottom and flows upward toward the weirs located at the surface. The hydraulic regime is established by the orientation of the inlet baffles.

A reactor clarifier is a solids contact unit in which the coagulation and flocculation step occurs in the same basin as sedimentation but flow does not pass upward through the sludge blanket. A typical reactor clarifier is a circular basin designed with both a center feed and flocculation zone employing mechanical mixing in a central conically shaped compartment.

Sludge blanket clarifiers have a distinct suspended sludge blanket layer maintained as a suspended filter through which the flow passes. During this upward passage, small particles are removed by adsorption and filtration onto other particles of the suspended blanket.

MECHANISMS AFFECTING SEDIMENTATION

Accurate prediction of settling tank performance by mathematical and experimental methods is a challenge to the best of design engineers. Model testing using tracers and settling columns is limited by scale-up, which cannot be adequately expressed by principles of similitude, primarily because solids particles are not easily scaled down. In addition, many of the simplifying assumptions of modeling do not hold true in prototype units. Such factors as temperature gradients, wind effects, inlet energy dissipation, outlet currents, and equipment movement affect tank performance but are not easily modeled. This section briefly discusses these influences. Operational or design adjustments to alleviate the resulting performance limitations are developed in Chapters 21 and 22.

Density Currents

Temperature Differentials.   The addition of warm influent water to a sedimentation basin containing cooler water can lead to a short-circuiting phenomena in which the warm water rises to the surface and reaches the effluent launders in a fraction of the nominal detention time. Conversely, cold water added to a basin containing warm water tends to force the incoming water to dive to the bottom of the basin, flow along the bottom, and rise at the basin outlet, as shown in Figure 7-11. This phenom-

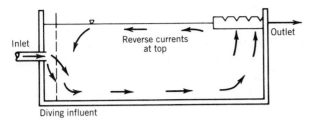

FIGURE 7-11. Influence of adding cooler, more dense water to a settling tank containing warmer water.

enon has been summarized by Hudson (1972). Density differences relating to temperature may be tabulated as follows:

Temperature (°C)	Density (g/mL)
4	1.00000
18	0.99862
20	0.99823

A density change caused by a 2°C variation takes place in the fourth significant figure, which may appear to be insignificant. When considering a clarifier containing 1 million gallons of water, however, the weight differential corresponding to the 2°C difference is approximately 1.6 tons. For example, cooler water entering a 3-m-deep basin at middepth would drop about 1.5 m (5 ft). This phenomenon would induce 1.6×10^4 ft-lb (2.2×10^4 N-m) more potential energy to be dissipated in the basin than an equivalent volume of water entering at 20°C.

This concept can be very critical. In the conduct of dye-dispersion testing in pilot-plant investigations, TeKippe and Cleasby (1968) found minor temperature differences to greatly reduce the reproducibility of experiments and overshadow minor differences in inlet and outlet design variables.

Performance in settling basins constructed with metallic walls exposed to sunlight is also known to be somewhat unpredictable. The heat transmitted through the wall on the sunny side of the basin tends to warm the water, making it less dense than water on the shaded side. The warm water in turn rises, forming a density current which, if sufficiently severe, can turn over the contents of a clarifier.

Turbidity Effects. Density current problems similar to those discussed above may also be caused by changes in influent turbidity resulting from flash

floods or strong winds on lake water surfaces. A rapid increase in turbidity increases the density of the influent and causes it to plunge as it enters the clarifier. In a center-feed circular tank a more dense influent would tend to plummet and flow radially outward from the center, rising toward the effluent launders near the perimeter. Water not leaving the launders would return radially inward along the surface to establish a doughnut-shaped roll. These forces tend to leave the center of the roll in a relatively stagnant position, thereby reducing the effective volume and detention time of sedimentation.

To quantify the effect of varying influent turbidity, Hudson presented the data shown in Figure 7-12. Using these data and a specific gravity of suspended matter of 2.5, the densities of various solids concentrations were estimated to be the following:

Solids Concentration (mg/L)	Density of Mixture (g/mL)
100	1.00006
1,000	1.0006
10,000	1.006

For a 3-m-deep basin, an abrupt change in raw water suspended solids concentration from 10 to 1,000 JTU would impart an increased energy of

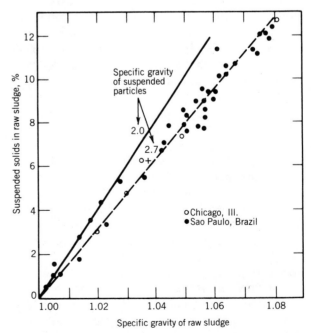

FIGURE 7-12. Estimation of density of suspended particles in Floc (Hudson, 1972).

1.1×10^{-2} N-m/L (3×10^4 ft-lb/mil gal). This effect is comparable to that described earlier for a 2°C change in incoming water temperature.

The solution to problems of varying influent turbidity are similar to those for incoming temperature differences. Specifically the source of water should be carefully selected and the method of removing water from the source should be as immune as possible from quality variations.

Salinity Effects. Density currents leading to short-circuiting can also be caused by changes in influent salinity. A 500-mg/L change in salinity (NaCl) would increase the upsetting energy of water by approximately 3.6×10^{-3} N-m/L (1×10^4 ft-lb/mil gal). This order of change is similar to that described for temperature changes and for a 100-fold increase in turbidity.

Wind Effects

Wind can have a pronounced effect on the performance of large-diameter open-gravity settling basins. High wind velocity tends to push the surface of the water to the downwind side of a basin and causes more water to overflow the weirs on that side. The net effect is a surface current moving in the direction of the wind. To compensate, underflow current in the opposite direction would be created along the bottom of the tank. The resulting circulating current could lead to short-circuiting of the influent to the effluent weir and scouring of settled particles from the bottom. For circular or square basins with diameters greater than 30–35 m this phenomenon can be quite noticeable by degradation of effluent quality.

Inlet Energy Dissipation

Clarifier performance is strongly influenced by inlet energy dissipation. Influent waters are normally carried in a pipe at sufficient velocities to keep solids in suspension. Suddenly the fluid must be slowed down and distributed over a broad area to begin the sedimentation process. In rectangular basins flow is often distributed with a channel across the end of the basin. Baffles are then used to distribute the flow across the tank and in a horizontal and vertical direction simultaneously. If this energy dissipation is not carefully engineered, density and eddy currents and excessive velocity vectors will be created.

Outlet Currents

Outlet currents of a clarifier are often related to design details of effluent weirs and launders. Initially, these weirs were simply flat plates across the end of a rectangular basin. The width of the basin established the length of the weir. When tanks were designed in a long narrow configuration, the weir length was relatively short and was believed to contribute to formation of outlet currents which, if severe, could sweep settleable particles into the tank effluent.

The problem of currents was compounded in early designs because the flat weir plates were sometimes out of level. Concern for this led to the development of V-notch weirs, which provided better lateral distribution of outlet flow when leveling was imperfect.

For upflow clarifiers such as solids contact basins used in water treatment, launders carefully spaced across the surface are considered of vital importance to good process performance. The launders often constructed in a radial pattern serve an important role in directing the vertical flow through the solids contact zone. As solids contact tanks become large, strategic location of the radial weirs becomes more critical.

In general, for most water treatment clarifiers, tank performance is primarily a function of density currents and inlet energy dissipation rather than outlet currents. Careful design of effluent weirs cannot be taken for granted but will not overcome density currents created by other design deficiencies.

Equipment Movement

Another potential effect of clarifier performance is the movement of equipment within the basin. Sludge collection mechanisms, normally consisting of chain-and-flight scrapers, bridge-mounted scrapers, or hydraulic suction units, must move through the contents of the tank to completely remove settled sludge. If such movement is excessive, it can introduce currents that upset the sedimentation process. Most equipment moves on the order of 25–50 cm/min and, as such, has a minimal effect. Equipment movement in the vicinity of the effluent launders, however, is important in that it can stir up solids that can be caught in the effluent currents.

The trend in recent years has been construction of traveling bridge, hydraulic removal systems. These have the advantage of removing sludge near the point of deposition, thus reducing scouring potential.

FLOTATION SYSTEMS

Flotation, in general, is a very effective means of separating solids and liquids. It has been employed outside of the water treatment field since the early 1900s, principally in the area of mineral processing where it is used to concentrate metals. It has also been used in waste treatment for sludge thickening for about 25 years. Recently flotation has been applied in the water and wastewater treatment fields to concentrate and remove algae to clarify water (Hyde et al., 1975).

Theories of Flotation

Flotation is a solids–liquid separation process that transfers solids to the liquid surface through attachment of bubbles to solid particles. The phenomenon of flotation consists of three processes: bubble generation, attachment of solids to the bubbles, and solids separation. In theory, many gases are suitable for flotation systems, but in practice air is used almost exclusively since it is readily available and inexpensive.

Bubble Attachment Phenomena.   Bubbles can be generated in a number of ways, some of which are described in greater detail below. Once formed, the bubble rises to the surface at a rate dependent on the fluid properties, such as the viscosity and density of the water, and the properties of the bubble volume, shape, surface area, and so on. Solids that come in contact with the rising bubble can attach themselves to the bubble and thereby be removed from the liquid. The attachment process is dependent on the hydrodynamic properties of the rising bubble as discussed by Richmond (1977) and the surface forces acting at the bubble–liquid interface as discussed by Perry and Chilton (1973) and Richmond (1977).

Process Variables and Their Interaction.   Process variables can be grouped into two categories; operation and design. Operational variables can be divided into physical and chemical variables. Design process variables include hydraulic loading, solids loading, and detention time (see Table 7-1). These variables are not mutually exclusive and the design-

TABLE 7-1.   Process Variables

| Design | Operational | |
	Physical	Chemical
Hydraulic loading	Recycled flow	Coagulant dose
Solids loading	Operating pressure air quantity	pH
	Solids removal frequency	Influent solids concentration

operational differentiation is somewhat arbitrary. A change in one can affect several others. For example, the designer should provide the capability to run at various air flowrates and recycle flows, but it is an operational problem to determine the optimum conditions once the system is on-line.

For a given system the coagulant dose, pH, and influent solids concentration will affect performance. Zabel (1978), Bare et al. (1975), Zabel and Hyde (1976), and Steiner et al. (1978) have demonstrated that for a given suspension, coagulant dosage affects removal efficiencies. This is true of both inorganic coagulants and organic polymers. For most coagulants removal efficiency increases with increasing coagulant doses up to a certain value, beyond which performance is unaffected by dosage or may even decrease slightly. Zabel (1978) has also indicated that standard sedimentation jar tests will predict the proper optimum coagulant dose to be used for flotation. When coagulant doses in excess of the optimum as determined by sedimentation jar tests are applied to flotation systems, a deterioration in effluent quality was noted. This can be attributed to the production of a weaker floc at higher doses and its breakdown when air is added to the system. This jar test method should only be used for preliminary screening of chemicals. Prudent design should incorporate the data from actual flotation tests.

The effects of pH on the flotation process have not been studied per se. However, data presented by Bare et al. (1975) indicate that for ferric sulfate and, by inference, aluminum sulfate, performance will be enhanced as pH is increased for low coagulant doses, will deteriorate slightly at moderate doses, and will increase again at high coagulant doses. The effect of pH while utilizing organic polymers has not been investigated. In light of these facts, bench-scale or pilot-scale systems should be operated at the anticipated pH of the flow stream while investigating other variables.

Influent solids concentration also has an adverse effect on effluent quality, hindering performance at higher concentration. Zabel and Hyde (1976) observed a qualitative increase in effluent turbidity matching periods of increased influent turbidity in a pilot plant using dissolved air flotation to remove suspended solids from a water supply. More quantitative relationships were obtained by Steiner et al. (1978), who showed nearly proportional increases in effluent quality as influent quality changed. However, the researchers dealt with an emulsified oil waste.

Recycle flow, operating pressure, air quantity, and solids removal frequency are adjustable variables under the direct control of the operator. The first three are interrelated to a certain degree. As the amount of recycle increases, a greater mass of air is available for flotation. For example, twice as much air can be dissolved at 10 percent recycle than at 5 percent. The effect of this increased air is twofold. First, more bubbles are available thereby enhancing performance. Second, more turbulence is created, which tends to shear the floc and cause poorer performance. Bare et al. (1975) have reported an optimum of 25 percent recycle for algae removal, while Zabel (1978) reported optimums of 8–10 percent for raw water treatment.

A similar effect is demonstrated by operating pressure. The work of Childs et al. (1976), Bare et al. (1975), and Zabel and Hyde (1976) indicate better performance as pressure is increased. Childs, however, increased recycle while increasing pressure and by doing so did not isolate the pressure effect. Bare and Zabel show generally increasing performance as pressure is raised, but Bare only studied systems up to 45 psig and Zabel up to 70 psig.

Air quantity is a variable that incorporates the ideas expressed in the previous two paragraphs. If either the percent recycle or pressure is increased (or if both are), the quantity of air available for flotation will increase. Zabel showed that the optimum air quantity for raw water turbidity removed was about 6 g/m^3 of water. Beyond that value, no change in performance was noted. These results were shown to be independent of pressure or recycle as would be expected but were based on a constant influent raw water quality.

Some investigators have promoted the value of air–solids ratio (A/S) on the performance of flota-

tion units. It is defined as the ratio of air in solution to the concentration of suspended solids in the water to be treated. Based on the previous discussions of air quantity and influent solids concentration, increasing the A/S ratio would be expected to have a positive effect on performance below the optimum air quantity and an adverse effect above that value. This quantity is of no importance in relatively dilute waste streams because the amount of air present is usually far in excess of the required amount. However, as reported by Walzer (1978), it can improve performance of units treating concentration water, waste streams, or sludges within limited A/S ranges and at air flowrates below the optimum.

Solids removal frequency, or the period of time between removal of float from the surface, affects performance of a flotation unit. As more time elapses between float removal, performance decreases in terms of treated water quality. Zabel and Hyde (1976) found that 40 hr was the maximum allowable period for waters with low initial turbidities (5–6 FTU). Poorer raw water quality would require more frequent skimming. Continuous skimming has little or no effect on flotation units used for water treatment.

The two variables under the control of the designers, hydraulic and solids loading, also affect the operation of a flotation system. Hydraulic loading has a similar effect in flotation to that in sedimentation. Walzer (1978) shows that the design concept of sedimentation tanks in terms of terminal settling velocities is analogous to particle rise rate in flotation. This is based on ideal, noncoalescing particles. Any deviation toward nonideality necessitates an empirical approach specific to the water or sludge to be treated by flotation. In general, performance can be expected to decrease dramatically at high hydraulic loadings.

Types of Flotation Systems

There are several types of flotation systems. Each differs principally in the manner in which bubbles are generated. Major types include dissolved air flotation, dispersed air flotation, and electrolytic fractionation. Each type has found applications in specific industries, water treatment processes, or waste treatment processes. Dissolved air flotation is by far the most common system. Refer to Chapter 21 for more detailed design discussion.

In the dissolved air flotation process, air is added to a pressurized process stream and is released from solution when the pressure is removed. The small bubbles formed adhere to or enmesh in the solids present, thereby floating them to the surface for removal.

In contrast to dissolved air systems, dispersed air flotation systems do not dissolve the air in the water. Rather, air and water are mixed in such a way that relatively small and widely dispersed air bubbles are created. The major types of dispersed air flotation systems are differentiated by the means in which the air–water mixture is created.

Electrolytic flotation, or electroflotation, which involves generating the bubbles by electrolysis, claims to provide a profuse number of very small bubbles with minimum turbulence and to aid the flocculation of existing solids due to the electric field gradient between the electrodes (Chambers et al., 1976).

REFERENCES

Bare, W. F. Rance, Jones, N. B., and Middlebrooks, E. J., "Algae Removal Using Dissolved Air Flotation," *J. Water Pollut. Control Fed.* **47**(1), 153–169 (January 1975).

Camp, T. R., "A Study of the Rational Design of Settling Tanks," *Sewage Works J.,* **8,** 742–758 (1936).

Camp, T. R., "Sedimentation and The Design of Settling Tanks," *Trans. Amer. Soc. Civil Eng.,* **111,** 895–936 (1946).

Camp, T. R., and Stein, P. C., "Velocity Gradients and Hydraulic Work in Fluid Motion," *J. Boston Soc. Civil Eng.,* **30,** 219 (1943).

Chambers, D. B., and Cottrell, W. R. T., "Flotation: Two Fresh Ways to Treat Effluents," *Chem. Eng.,* 95–98 (August 2, 1976).

Childs, A. R., Burfield, I., and Rees, A. J., "Operational Experience with the 2300 m³/d Pilot Plant of the Essex Water Company," Paper 9 (1976).

Dick, R. I., and Young, K. W., "Analysis of Thickening Performance of Final Settling Tanks," *Proceedings of the 27th Industrial Waste Conference,* Purdue University, Lafayette, IN (1972).

Hazen, A., "On Sedimentation," *Trans. Amer. Soc. Civil Eng.,* **53,** 45–71 (1904).

Hudson, H. E., Jr., "Density Considerations in Sedimentation," *JAWWA,* **64,** 382 (1972).

Hyde, R. A., Miller, D. B., and Packham, R. F., "Water Clarification by Flotation," Water Research Association, U.K., Paper No. 14-5 (1975).

Katz, W. J., Geinopolos, A., and Mancini, J. L., "Concepts of Sedimentation Applied to Design," *Water and Sewage Works,* **109,** 118–129 (1962).

Keinath, T. M., *et al.,* "A Unified Approach to the Design and Operation of the Activated Sludge System," *Proceedings of the 31st Purdue Industrial Waste Conference,* Purdue University, West Lafayette, IN, p. 914 (1976).

McGauhey, P. H., "Theory of Sedimentation," *JAWWA,* **48,** 437–454 (1956).

Metcalf & Eddy, Inc., *Wastewater Engineering, Treatment, Disposal, and Reuse,* 2nd ed., McGraw-Hill, New York (1979).

O'Connor, D. J., and Eckenfelder, W. W., Jr., "Evaluation of Laboratory Settling Data for Process Design," in W. W. Eckenfelder and Brother Joseph McCabe, F.S.C. (eds.), *Biologic Treatment of Sewage and Industrial Wastes,* Vol. 2, Reinhold Publication Corporation, New York (1958).

Perry, Robert H., and Chilton, Cecil H., *Chemical Engineers Handbook,* 5th ed., McGraw-Hill, New York (1973).

Richmond, Peter, "Some Fundamental Concepts in Flotation," Paper Delivered at SCI Meeting, Principals and Practice of Flotation, January 11, 1977.

Schroeder, E. D., "Water and Wastewater Treatment," *Environmental Engineering and Water Resources Series,* McGraw-Hill, New York (1976).

Steiner, J. L., Bennett, G. F., Mohler, E. F., and Clere, L. T., "Air Flotation Treatment of Refinery Waste Water," *Chem. Eng. Progress,* 39–45 (December 1978).

TeKippe, R. J., and Cleasby, J. L., "Model Studies of Peripheral Feed Settling Tank," *J. Amer. Soc. Civil Eng.,* **94,** No. SA1, 85–102 (1968).

TeKippe, R. J., and Ham, R. K., "Velocity-Gradient as in Coagulation," *JAWWA,* **63,** 439 (1971).

Wallace, A. T., "Hydraulic and Removal Efficiencies in Sedimentation Basins," Ph.D. Thesis, University of Wisconsin (1965).

Walzer, James G., "Design Criteria for Dissolved Air Flotation," *Pollut. Eng.,* **10,** 46–48 (February 1978).

Zabel, R., *Flotation,* Water Research Center, U.K., pp. F1–F10 (1978).

Zabel, T. F., and Hyde, R. A., "Factors Influencing Dissolved Air Flotation as Applied to Water Clarification," Paper 8, Water Research Center, U.K. (1976).

—8—

Filtration

INTRODUCTION

Filtration is a unit process widely used in water and wastewater treatment for the removal of particulate materials commonly found in water. In this process, water passes through a filter medium, and particulate materials either accumulate on the surface of the medium or are collected through its depth. Filters have been found effective for removing particulates of all size ranges including algae, colloidal humic compounds, viruses, asbestos fibers, and colloidal clay particulates, provided that proper design parameters are used.

A wide range of media is utilized in filtration systems as summarized in Table 8-1. Typical examples include screens with openings of $1-100$ μm and granular materials, usually sand, anthracite coal, or magnetite, with sizes ranging from 0.1 to 10 mm. Diatomaceous earth, a deposit formed from siliceous fossil remains of diatoms, is also employed as a filtering medium in certain filtration applications.

Particulate matter in water causes a decrease in water clarity measured generally by light-scattering techniques and commonly defined in units of turbidity. The National Interim Primary Drinking Water Standards specify a maximum contaminant level of one turbidity unit. This turbidity level, selected to ensure adequate disinfection (Symons and Hoff, 1975), can rarely be achieved by coagulation–sedimentation alone. Consequently, filtration assumes the role of the final treatment barrier for removal of undesirable particulates in water treatment. In addition, many contaminants, such as viruses, heavy metals, or some pesticides, may be associated with particulates, and thus efficient removal of particulates can improve overall water quality (see e.g., Kavanaugh et al., 1978; Schaub and Sagib, 1975).

Whether the filter medium is a screen or granular material, the operating characteristics of the process are similar. As particulate matter accumulates in or on the filter medium, the pressure drop increases until the available hydraulic head is exhausted. The medium is then cleaned hydraulically during a short regeneration cycle and placed back in service. This discontinuous operating mode of filtration is analogous to other common unit processes, such as granular activated carbon and ion exchange, which will be discussed in subsequent chapters.

HISTORICAL OVERVIEW

Filtration was recognized quite early in recorded technological history as a unique process for improving the clarity of water. As summarized by

TABLE 8-1. Types of Filter Media

Types	Examples	Size Range
Screens	Polyethylene, stainless steel, cloth	1–100 μm effective size opening
Diatomaceous earth	Siliceous fossil remains	Mean size 7–50 μm
Granular	Sand, anthracite coal, magnetite, garnet sand, coconut shells	0.1–10 mm

Baker (1949), the earliest recorded reference to the use of filters for water treatment occurred about 3000 years ago in India. Various types of filters were utilized in small-scale water-purifying units including wick syphons and various kinds of cloth strainers. In China, as early as the tenth century B.C., granular materials were apparently placed in the bottom of wells to improve the clarity of the well water.

Modern engineering applications of filters for the purification of water supplies dates from the eighteenth century. The first patent for a filter was granted in Paris, France, in 1746. The earliest use of filters for a domestic water supply occurred in Scotland in 1804, followed thereafter by the installation of sand filters in England in 1829. The first attempt at filtering a municipal supply in the United States occurred at Richmond, Virginia, in 1832 under the direction of Albert Stein (Baker, 1949). The sand filters were operated in an upflow manner without any form of pretreatment. This system failed, however, as little particulate removal was achieved without coagulation.

Considerable controversy surrounded the use of filters for the removal of bacteria in the nineteenth century. It was not until the germ theory of disease had been discovered, however, that the utility of filtration for the prevention of disease was demonstrated. In 1892, a cholera epidemic struck the city of Hamburg, Germany. The neighboring city of Altona, which treated its water by slow sand filtration, escaped the epidemic (Baker, 1949). This incident conclusively demonstrated the efficacy of slow sand filters for removal of harmful pathogens. Since then, the value of granular media filtration has been well recognized. The majority of treatment plants treating surface waters have installed filters for the purposes of meeting drinking water standards and providing water of reasonable aesthetic and microbiological quality. The advent of disinfection in the early twentieth century, however, was the final milestone that assured the production of a safe drinking water.

FILTRATION APPLICATIONS IN WATER AND WASTEWATER TREATMENT

Filters find many uses in the treatment of water and wastewaters. Treatment of surface sources for potable supplies will normally require filtration to achieve the turbidity standard. Both granular media and diatomaceous earth filters have been employed in this role. In some cases, notably in Western Europe, microscreens have been used for algae removal prior to the use of granular media filters. Depending on the levels of particulate matter in the untreated water, granular filters may be installed without intervening sedimentation. Some common process locations for filters in water treatment are shown in Figure 8-1.

Increasingly stringent discharge requirements imposed on wastewater discharges have led to expanded installation of filtering devices at municipal wastewater treatment plants. Both microscreens and granular filters have been used to increase removal of suspended solids in secondary effluents from waste treatment facilities (Tchobanoglous, 1970; Baumann and Huang, 1974). For phosphorus removal, filters may be obligatory to meet discharge standards of 1 mg P/L for total phosphorus (Kavanaugh et al., 1977; Boller and Kavanaugh, 1975).

Finally, recycling of wastewaters in various industrial sectors, particularly the oil, petrochemical, and steel industries, has increased the key role of filters in industrial water treatment. Granular filters have been shown capable of removing emulsified oil droplets, iron-scale particulates, and other types of suspended solids found in industrial water streams. Increased demand for granular filters and microscreens can be anticipated as water recycling expands in both industrial and domestic markets.

CHARACTERISTICS OF PARTICULATES REMOVED BY FILTERS

Filters are capable of removing a wide range of particulate materials of both natural and human origin. Characteristics of some of these materials are sum-

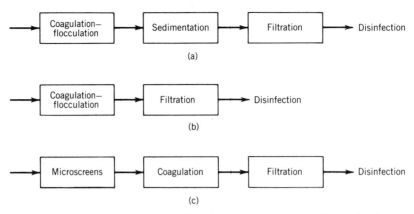

FIGURE 8-1. Common applications of filters in water treatment. (a) Conventional water treatment. (b) Direct filtration. (c) Unusual combinations.

marized in Table 8-2. Typically the size of particulate material ranges from approximately 0.1 to 1000 μm with highly variable shapes.

The density of particulate material ranges from values near unity for flocculant particulates formed by inorganic coagulants, such as iron and aluminum, up to higher values characteristic of particulate oxides and carbonates formed by precipitation processes. Most of the natural particulates found in water supplies, including bacteria, algae, or clay, will be negatively charged in natural waters (pH > 5). Flocculant material formed by iron or aluminum may be negatively or positively charged depending on the pH and the ionic composition of the water (see Chapter 6).

The particulate matter captured within or on a filter can be reentrained in the water due to shearing of the deposited material. The qualitative characteristic of the particulates, which indicates the resistance of flocs to shearing forces, is the floc strength. The floc strength of ferric and aluminum particulates is generally low, which limits the ultimate capacity of granular filters for removal of these solids. This is the principal reason that sedimentation is often required prior to filtration. The addition of polyelectrolytes to destabilize particles and increase floc strength may improve particulate capture and decrease reentrainment (Habibian, 1971).

A final particulate characteristic of importance is the degree to which the solids being removed in the filters can be compacted. In a granular media filter, as the void space is reduced by captured solids, head loss will increase rapidly. As shown in Table 8-2, flocculant solids exhibit a low percent solids under gravity sedimentation conditions while discrete particulates, such as calcium carbonate or ferric oxide particles, can achieve a gravity compaction level of up to 4–6 percent. In general, a larger mass of discrete solids can be retained in granular filters compared to flocculent materials, due to this fact.

The heterogeneous particulate suspensions com-

TABLE 8-2. Characteristics of Particulates Removed by Filtration

Type	Size Range (μm)	Shape	Density (g/cm^3)	Zero Point of Charge (pH)	Floc Strength	Compaction (Percent Solids, Gravity Sedimentation)
Bacterial flocs	0.5–1000	Variable	~1.02	2–3	Medium	0.5–5.0
Algae	1–200	Variable	~1.05	2–3	Medium	1
"Fe(OH)$_3$ or "Al(OH)$_3$"	0.1–1000	Variable	~1.01	5–9	Low	0.5–2
Fe$_2$O$_3$	0.1–50	Plates	5.2	9–10	Low	4–6
CaCO$_3$	0.1–50	Rounded	2.4	8–9	Low	4–6
Oil droplets	10–100	Spherical	0.8	3–5	Medium	?

monly found in water can often be characterized by a size distribution function known as the power law. The power law shown in Eq. 8-1 states that the number of particulates ΔN per size category is an inverse power function of the size, l, of the particulate material.

$$\frac{d\,\Delta N}{dl} = Al^{-\beta} \qquad (8\text{-}1)$$

Several examples of power law size characteristics for typical suspensions are shown in Figure 8-2.

The slope of the power law function is a useful parameter to characterize the type of suspension being treated. Depending on the value of the power law coefficient β, the major portion of the surface area or volume fraction of a suspension will be found in certain size ranges. As summarized in Table 8-3 for power law coefficients less than 3, the majority of the surface area and volume of the suspension is typically found in particulate fractions greater than 2 μm. For power law coefficients greater than 3, a substantial fraction of the surface area of the suspension will be found in the smaller particle sizes, below 2 μm. The characterization of

TABLE 8-3. Influence of Power Law Coefficient on Distribution of Surface Area and Volume of Particulates by Size

Power Law Coefficient, β	Percent of Surface Area in Fraction >2 μm	Percent of Volume in Fraction >2 μm
1	99.95	99.995
2	98.3	99.95
3	73.3	98.3
4	25	73.3

the size distribution of a particulate suspension is useful in determining whether alteration of the size distribution is needed to achieve optimum filter design.

MECHANISMS OF PARTICULATE REMOVAL

For filters in which thin media are used, such as screens or membranes, the principal mechanism of particulate removal is straining, where some characteristic size of particulates is larger than the openings in the filter medium. In the case of granular deep-bed filters, particulates can penetrate into the depth of the filter medium and the mechanisms of removal are more complex. Generally, the particulate matter must be transported from the fluid streamlines to the surface of the media or collector. Particles will deviate from the fluid streamlines due to gravitational forces, diffusion gradients, and inertial effects of momentum (Mintz, 1966; Agrawal, 1966; O'Melia and Stumm, 1967; Ison and Ives, 1969). The relative magnitude of these mechanisms will depend on the water quality and physical characteristics of the particulates and the filter media. When the particulate material closely approaches the surface of the filter media, attachment may occur provided that surface chemical interactions are favorable. At this point, the filtration process becomes analogous to coagulation and the causes of a "successful" collision or capture of particulates are the same as discussed in Chapter 6.

PROCESS SELECTION

A design engineer initiates the selection of the appropriate solids–liquid separation process by conducting a preliminary screening of processes that may be suitable for the design problem. A number

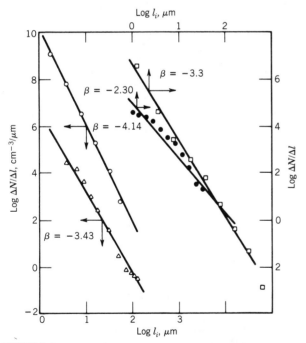

FIGURE 8-2. Examples of particulate size frequency distributions: (△) Effluent, sedimentation basin, pilot-activated sludge plant; (●) Lake Zurich, 40 m; (□) Deer Creek Reservoir, Utah, 20 m; (⊙) digested primary and secondary sludge (Kavanaugh, 1980).

of filtering devices are available as has been discussed. In addition, filtration must be compared to other solid–liquid separation processes available to the designer, including sedimentation, flotation, and possibly membrane processes. This selection process is usually based on the experience of the designer.

Some attempts have been made, however, to develop a methodology for process selection based on the physical characteristics of the particulates. One such approach is shown in Figure 8-3 (Kavanaugh et al., 1980). Expected regions are defined in which various processes are likely to be appropriate for removal of particulate materials depending on the initial number and mass concentration of particulate material and the average size characterizing the distribution. For particulate suspensions with an average size greater than 100 μm and suspended solids concentrations greater than 50 mg/L, gravity sedimentation may be the most cost-effective solid–liquid separation process. When the suspended solids concentration is less than 50 mg/L and the average size of the suspension is on the order of 30 μm, a screening process may be most effective. Over a fairly wide size range and for suspension concentra-

tions generally less than 50 mg/L, (dilute suspensions) filtration using granular media without prior sedimentation may be a feasible process. Above this mass concentration range, it is likely that the solids capacity of filters may be exceeded and some type of coagulation–sedimentation process would be desirable as a pretreatment for the filters.

The designer is then faced with a number of issues related to mechanical and process design of the filtering system. Various techniques are available to guide the designer in the ultimate selection of the design criteria, including analytical models of the process, previous experience, pilot studies, and published literature. For most new applications not previously tested, pilot studies are highly desirable because of the uncertainties in modeling techniques. Figure 8-3 is a useful guide for the selection of suitable processes to be studied in pilot investigations.

SCOPE OF CHAPTER

This chapter will discuss the effect of various filtration design variables on the efficiency of particulate

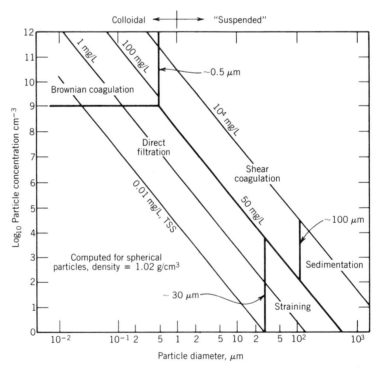

FIGURE 8-3. Solids–liquid separation process selection diagram (Kavanaugh et al., 1980).

removal. The hydraulics of flow through porous media will be presented. Process design considerations for deep-bed granular filters and diatomite filters will then be presented. Mechanical design issues will be covered in Chapter 21.

QUANTITATIVE PREDICTIONS OF PARTICULATE REMOVAL

Particulate removal in filter media can occur by straining or by attachment to the media itself. In addition, material already deposited can be reentrained or detached due to shearing forces that increase as the filter clogs. The relative importance of different mechanisms will depend on physicochemical variables, as will be discussed in the next section.

Collection Efficiency of Filter Media

Straining. When particles are larger than pore openings in granular media or sieve openings in screens, particulate material can be removed by straining. In addition, it is possible for particles smaller than pore sizes to be removed by straining due to the collection of particulates on the surface of the media and closure of existing pore openings. At relatively high concentrations of solids, it is possible for several smaller particles to arrive at an opening simultaneously, thus blocking or clogging the opening. Various investigators have suggested that straining becomes an important removal mechanism when the ratio of the particle size to the media size in porous media is greater than 0.2 (Herzig et al., 1970; Boller, 1980). It has also been demonstrated that the ratio of particle to media size at which straining becomes important will depend to some extent on the number flux of particles approaching the media (*flux* defined as the superficial velocity times the particle number concentration) (Kavanaugh, 1974).

Other investigators (Maroudas and Eisenklam, 1965a,b; Tien et al., 1979) have reported data that indicates that for particle sizes greater than approximately 100 μm, straining in porous media will become a dominant removal mechanism. At the present time, however, no unambiguous criteria are available to determine when straining can occur. In the case of granular media filtration straining is undesirable because head loss will increase rapidly due to the formation of a surface mat (Ives, 1982).

Consequently, in the design of granular filters the size of the filter media should be selected to minimize this straining phenomena (Boller, 1980).

Nonstraining Mechanisms. Estimates of the rate of particulate capture in granular filter media due to nonstraining mechanisms can be made from a knowledge of particulate mechanics in porous media under the influence of hydrodynamic and physicochemical forces. Solution of the equations governing particulate motion in porous media require selection of a geometric model of the porous media and the quantitative description of all forces acting on the particulates as they pass through the granular media. Recent reviews (Ives, 1982; Tien et al., 1979) evaluate this approach for several alternative geometric models of granular media. Geometric models which have been used include (1) the isolated single-sphere model (Yao et al., 1971; Rajagopolan and Tien, 1976; O'Melia and Ali, 1978), (2) the sphere-in-liquid shell model (Spielman and Fitzpatrick, 1973; Payatakes et al., 1974; Rajagopolan and Tien, 1976), (3) the model of parallel capillaries (Payatakes et al., 1974), and (4) the constricted tube model (Payatakes, 1973).

Although quantitative estimates of capture efficiency vary depending on the model, qualitative agreement between results of the investigations cited is apparent (see Tien et al., 1979). For illustrative purposes, only the isolated single-sphere model will be discussed here because of its simplicity and because it gives qualitative agreement with more complex models.

Isolated Single-Sphere Model. A schematic of an isolated spherical collector is shown in Figure 8-4. Particulates are transported past the spherical collector and must deviate from the fluid streamlines to be removed from the suspension. The efficiency of particulate collection is defined as the number of successful collisions for all particulates in the cross-sectional area of the collector divided by the total possible number of collisions between the particulates and the collector, as shown by Eq. 8-2.

$$\eta \equiv \frac{\text{successful number of collisions}}{\text{total number of possible collisions in cross-sectional area perpendicular to isolated collector}} \quad (\text{area} = \pi d_m^2/4) \quad (8\text{-}2)$$

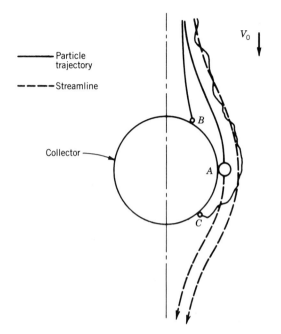

FIGURE 8-4. Modes of action of the basic transport mechanisms. A, interception; B, sedimentation; C, diffusion.

where d_m = media size or diameter.

The collection efficiency throughout the depth of the granular media is thus the summation of the efficiency of all individual collectors in the filter bed. The change in particulate concentration N with depth then becomes

$$-\left(\frac{\partial N}{\partial x}\right)_t = \frac{\psi(1 - \varepsilon_0)}{d_m} \eta N \qquad (8\text{-}3)$$

where x is the media depth, ψ is a shape factor defined as the ratio of area and volume shape factors for granular media ($\psi = 6$ for spherical media), d_m is the collector diameter, and ε_0 is the initial pore volume or porosity of the granular media. Assuming that removal is constant with time, that is, steady-state removal, integration of Eq. 8-3 gives, for the fraction of particulates removed, N/N_0, the following expression

$$\frac{N}{N_0} = \exp\left[\frac{-\psi(1 - \varepsilon_0)}{d_m} L\eta\right] \qquad (8\text{-}4)$$

where L = total depth of the media.

This simple model provides a framework for understanding the effects of various design variables

on the efficiency of filtration. As seen by Eq. 8-4, removal efficiency can be improved by altering the physicochemical nature of the system to increase the individual collection efficiency, η. A decrease in porosity would also produce an increase in particulate removal efficiency. In addition, the model predicts the often observed fact that increasing the filter depth or decreasing the filter media size will improve particle capture.

Figure 8-5 schematically shows the effect of media depth, media size, and individual collector efficiency on the efficiency of particulate capture in granular media. In order to obtain the desired filter performance, the physicochemical conditions in the filtration system can be altered to increase the collector efficiency. The effect of filter design variables on η is discussed in subsequent sections.

Transport Mechanisms

As shown in Figure 8-4, suspended particulates must deviate from fluid streamlines in order to reach the surface of the filter media. Several transport mechanisms have been postulated for achieving this step.

Impaction. Particulate transport by impaction occurs when the inertia of particulates approaching a collector is greater than the hydrodynamic force tending to sweep the particles past the collector. In general, this mechanism is shown to be insignificant in water filtration (Agrawal, 1966; Ison and Ives, 1969) although of major significance in air filtration.

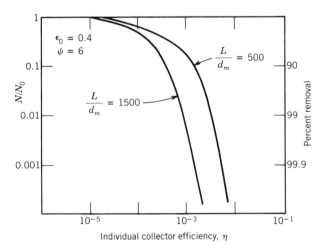

FIGURE 8-5. Effect of media depth (L), media size (d_m), and individual collector efficiency (η) on particulate capture in granular media.

Hydrodynamic Forces. Because of nonuniform shear distribution within the pore spaces and local contraction and expansion effects, some particles may be transported by hydrodynamic forces out of the fluid streamlines to the collector surface. Under low Reynolds number conditions (laminar flow), this is not an important mechanism. As the velocity and turbulence increase, however, this mechanism may contribute to particulate capture.

Interception. Particulates remaining in fluid streamlines that pass within a distance from the collector surface of half the particulate diameter from the collector surface are said to be intercepted by the medium grain. This phenomenon occurs in both air and water filtration. The quantitative estimate of this mechanism is given by (Yao et al., 1971)

$$\eta_I = \frac{3}{2} \left(\frac{d_p}{d_m} \right)^2 \qquad (8\text{-}5)$$

where d_p is the size of the particulate material. This equation is valid for laminar flow and spherical particles. Thus, as the ratio of the particulate size to the media size increases, the collection efficiency due to interception will also increase.

For typical particulate sizes found in surface waters (d_p on the order of 10 μm), and in media of 0.6 mm (typical size used), η_I is on the order of 10^{-3}. Thus, of a thousand possible collisions, only one occurs due to interception.

Sedimentation. Particles with a density significantly greater than water will tend to deviate from the fluid streamlines due to sedimentation. The influence of density, particle size, and superficial velocity of collector efficiency due to sedimentation has been shown to be (Agrawal, 1966; Yao, 1968).

$$\eta_s = \frac{\Delta\rho \, g d_p^2}{18\mu V_0} \qquad (8\text{-}6)$$

where g is the gravitational constant, $\Delta\rho$ is the density difference between the particulate and water, and V_0 is the superficial velocity.

For typical values of these parameters in a filtration system for the removal of flocculent solids ($d_p = 10$ μm, $\Delta\rho = 0.05$ kg/m^3, $T = 20$°C, $V_0 = 0.2$ cm/sec), η_s is also on the order of 10^{-3}. Discrete particulates of higher density (2–5 kg/m^3) (e.g., iron oxide particles) would be captured at much higher

efficiency levels ($\eta_s = 10^{-2}$–10^{-1}) due to this mechanism.

Diffusion. In addition to interception and gravity, particles influenced by Brownian motion will deviate from the fluid streamlines due to diffusion. Diffusion influences only those particles in the colloidal range (size less than approximately 1 μm). The efficiency of a spherical individual collector due to the diffusion mechanism has been shown by Levich (1962) to be

$$\eta_D = 0.9 \left(\frac{kT}{\mu d_p d_m V_0} \right)^{2/3} \qquad (8\text{-}7)$$

where k is the Boltzmann constant (1.38×10^{-23} J °K) and T the absolute temperature (°K). For colloidal particles ($d_p = 0.1$ μm) in water at 20°C and other conditions typical for water filtration ($d_m = 0.06$ cm, $V_0 = 0.2$ cm/sec), η_D is approximately 10^{-3}.

The relative importance of these various mechanisms for transporting the particle to the surface will thus depend on the physical properties of the filtration system.

Other Transport Mechanisms. It has been postulated that surface forces due to London–van der Waals attraction and electrostatic attraction may be sufficient to cause particles to deviate from streamlines. It is unlikely that this occurs under normal filtration conditions, however (Spielman and Goren, 1970).

Attachment Mechanisms

As particles approach the surface of the media, short-range surface forces will begin to influence particle dynamics. If particles have been destabilized, the collision between the particulates and the media or collector surface is likely to be successful. It has been shown both theoretically and experimentally (Stein, 1940; Yao, 1968; Kavanaugh, 1974) that particle capture will only occur when the surface charge of the particulate and media are of opposite sign. In the usual case where the media surface is covered by deposited particles, collection is only effective if the particulate material has been adequately destabilized (Yao et al., 1971). Again, under these conditions this process is exactly analogous to coagulation (Stumm and O'Melia, 1967).

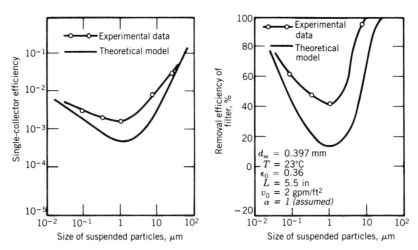

FIGURE 8-6.   Comparison of theoretical model and experimental data.

Detachment Mechanisms

As particulates collect within the pores of the filter medium, hydrodynamic shearing forces tend to increase due to increasing velocity in the pore space. If the increased shearing force exceeds the surface chemical forces holding the particles to the surface, deposited solids can be detached and reentrained in the water. This phenomenon has been observed by a number of authors (Stein, 1940; Mintz, 1966; Kavanaugh, 1974) and has been shown to occur, for example, when hydraulic surges occur in full-scale filters (Hudson, 1959). There is no detailed model of detachment, although some models have incorporated a detachment term into a kinetic description of particle capture, as will be discussed.

Effect of Process Variables on Collection Efficiency

Assuming that the individual transport mechanisms are additive, the individual spherical collector efficiency η can thus be related to the physicochemical properties of the system by summation of the transport efficiencies for the individual mechanisms (Yao et al., 1971). This procedure also assumes that all collisions lead to attachment and that particle destabilization is complete. A comparison of theoretical calculations and experimental results of Yao et al. (1971) are shown in Figure 8-6. More recent models have attempted to account for surface chemical interactions under conditions in which electrostatic repulsion is negligible (Spielman and Fitzpatrick, 1973). The several proposed models for predicting

the individual collector efficiency are shown in Table 8-4.

Effect of Particle Size.   The model of Yao et al. (1971), Tien and co-workers (1979), or Spielman and Fitzpatrick (1973) can be used to predict the effect of particle size on individual collector efficiency. Figure 8-7 is a schematic diagram of the influence of particle size on collector efficiency using typical values as obtained from models and vari-

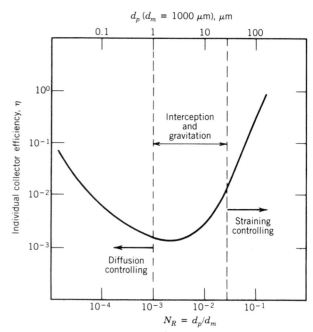

FIGURE 8-7.   Effect of particle size on collector efficiency, typical values from models and experimental results.

TABLE 8-4. Models of Individual Collector Efficiency (No Straining)

Author	Model, Porous Media	Form of Model,* η_{total}
Yao et al. (1971)	Single spherical	$0.9\left(\dfrac{kT}{\mu d_p d_m V_0}\right)^{2/3} + \dfrac{3}{2}\left(\dfrac{d_p}{d_m}\right)^2 + \dfrac{(\rho_p - \rho)g d_p^2}{18\,\mu V_0}$
Rajagopalan and Tien (1976)	Sphere-in-cell $(N_R < 0.18)$	$0.72 A_s N_{L_0}{}^{1/8} N_R{}^{15/8} + 0.0024 A_s N_G{}^{1.2} N_R{}^{-0.4} + 4 A_s{}^{1/3} N_{Pe}$
Fitzpatrick and Spielman (1973)	Sphere-in-cell	$f(N_{Ads} \cdot N_{GR})$ (no closed form)

* See references for definition of dimensionless groups.

ous experimental results. It is assumed that particles are destabilized. As shown in Figure 8-7, for particle sizes above approximately 40 μm, straining may be the controlling removal mechanism. This assumes a media size of approximately 1 mm. For particles below 1 μm, diffusion will control particle transport. Between these two limits, interception and gravity sedimentation should dominate particle capture. As shown in Figure 8-7, removal efficiency exhibits a minimum for particle sizes between 1 and 10 μm. This has been demonstrated experimentally by numerous investigators (Yao et al., 1971; Fitzpatrick and Spielman, 1973; Ghosh et al., 1975).

Other Variables

Two other key design variables are media size and filtration rate or superficial velocity. In general, it has been demonstrated that an increase in superficial velocity will lead to a decrease in removal efficiency. Conversely, a decrease in media size will lead to an increase in removal efficiency. These results have been obtained with ideal suspensions. In practical conditions, however, heterogeneous suspensions with a range of particle sizes are being filtered. Interactive effects are to be anticipated. Comparison of the predictions using the Yao–O'Melia model and those observed in filtration experiments using heterogeneous suspensions is shown in Figure 8-8 (Kavanaugh et al., 1978). As shown, the collector efficiency depends less on particle size than predicted by the models. Individual collector efficiency tends to be relatively constant over the range of particle sizes observed between approximately 0.5 and 50 μm and exhibits values for η between 10^{-3} and 10^{-2}.

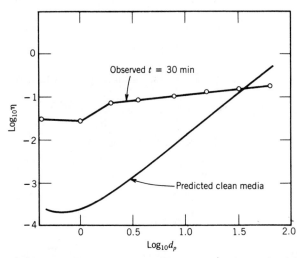

FIGURE 8-8. Comparison of predicted versus observed using Yao–O'Melia model. (Kavanaugh et al., 1978).

Surface Properties

While it is possible to manipulate the filter velocity and the media size, it appears that the predominant mechanism affecting collector efficiency in the absence of straining is the attachment mechanism. Under most conditions, transport is not rate limiting. Adequate destabilization is essential for satisfactory filtration.

HYDRAULICS OF FLOW-THROUGH POROUS MEDIA

Clean Media

When water or any fluid passes through porous materials, either granular or consolidated, energy losses occur due to both form and drag friction at the surface of the media material. In addition, losses occur due to continuous contraction and expansion experienced by the fluid as it passes through pore openings in the media. Flow patterns through porous media are quite complex, and thus the prediction of head loss requires different strategies than used for pipes. Head loss will depend on a wide range of system variables, including the fractional void volume or porosity, the particle shape, roughness, size and size distribution of the granular media, manner of packing, and type of fluid flow, that is, whether it is laminar, transitional, or turbulent. A variety of models have been developed for the prediction of head loss through clean porous media as a function of the relevant physical parameters. Although all models are subject to some shortcomings and are, in general, based on empirical observations, they provide a reasonably accurate estimate of the expected head loss through a bed of clean porous media at the initial stages of filtration.

Laminar Flow

Laminar or viscous flow is characterized by viscous forces dominating inertial forces. For the Reynold's number in porous media defined as

$$N_{Re} = \frac{d_m V_0 \rho_L}{\mu(1 - \varepsilon_0)}$$

flow is observed to be laminar when the Reynold's number is less than 10.

In a classic study by Darcy (1856) he observed that the hydraulic gradient, $\Delta p/L$, under laminar conditions in porous media was given by

$$\frac{\Delta p}{L} = \frac{V_0 \mu}{\kappa} \qquad (8\text{-}8)$$

where Δp is the pressure drop and L the depth of the porous media, with κ is the hydraulic permeability, which must be determined by experiment. This empirical relationship indicates that the pressure drop through porous media is directly proportional to the superficial velocity, the depth of the medium, and the viscosity. The hydraulic permeability can be determined empirically.

Subsequently, Carman (1937) developed a theoretical expression for the hydraulic permeability by postulating a physical model of the porous media, consisting of a series of parallel capillaries with an equivalent hydraulic radius. The fluid path length is greater than the depth of the bed due to the tortuous motion of the fluid passing through the porous media. The Carman–Kozeny equation is given as

$$\frac{\Delta p}{L} = \frac{k\mu S^2 V_0}{\varepsilon_0^3}$$

for laminar flow conditions. The term S is the specific surface of the medium and k is the Kozeny–Carman constant, which has been experimentally shown to be approximately 5 (Fair and Hatch, 1933, see also Happel and Brenner, 1965). For single-size spherical particles S is given by

$$S = \frac{6(1 - \varepsilon_0)}{d_m}$$

Thus, the Kozeny–Carman equation predicts that the hydraulic permeability is

$$\kappa = \frac{\varepsilon_0^3 d_m^2}{(1 - \varepsilon_0)^2 \times 180} \qquad (8\text{-}9)$$

for spherical particles. Combining Eqs. 8-9 and 8-8 and converting to head loss, the Kozeny–Carman equation becomes

$$\frac{\Delta H}{L} = \frac{180(1 - \varepsilon_0)^2 \mu V_0}{\varepsilon_0^3 d_m^2 \rho_L g} \qquad (8\text{-}10)$$

with ΔH as the head loss in units of length and g the gravitational constant. For nonspherical particles the specific surface is given by the following equation, which relates the surface to the volume of media for a given media size:

$$S = \frac{\psi(1 - \varepsilon_0)}{d_m}$$

The shape factor is 6 for spherical particles.

The shape factor will vary depending on the type of media. A range of these values for different commonly used filter media is shown in Table 8-5.

Transition Flow

Traditionally, filtration rates used for water and wastewater filtration have been low, on the order of 5 m/hr (2 gpm/ft^2). Thus, Reynolds numbers are generally less than 10 and laminar flow prevails.

TABLE 8-5. Typical Sphericity, Shape, and Porosity
Factors of Granular Materials

Description	Sphericity, ψ	Shape Factor, S	Typical Porosity, ε_0
Spherical	1.00	6.0	0.38
Rounded	0.98	6.1	0.38
Worn	0.94	6.4	0.39
Sharp	0.81	7.4	0.40
Angular	0.78	7.7	0.43
Crushed	0.70	8.5	0.48

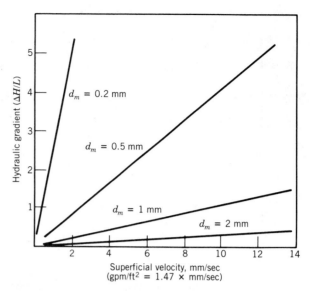

FIGURE 8-9. Effect of velocity on head loss, $T = 20°C$, $\varepsilon_0 = 0.4$, spherical media, single size.

Recent practice, however, has tended toward the use of larger-sized media, deeper filtration beds, and higher filtration rates with appropriate pretreatment. Reynolds numbers as high as 50 are not uncommon. Several approaches have been suggested for prediction of head loss through porous media for transition or turbulent flow. For Reynolds numbers greater than 1000, the Burke–Plumber equation (see Bird et al., 1960) is shown by

$$\frac{\Delta p}{L} = \frac{1.75}{d_m} \rho_L V_0^2 \frac{1 - \varepsilon_0}{\varepsilon_0^3} \qquad (8\text{-}11)$$

Combining Eqs. 8-10 and 8-11, a general expression is obtained, valid over the complete range of Reynolds numbers expected in granular media filtration, assuming spherical media.

$$\frac{\Delta H}{L} = \frac{180 \mu V_0}{\rho_L g d_m^2} \frac{(1 - \varepsilon_0)^2}{\varepsilon_0^3} + \frac{1.75 V_0^2}{d_m g} \frac{1 - \varepsilon_0}{\varepsilon_0^3} \qquad (8\text{-}12)$$

The effect of velocity on head loss per unit depth of media for various media sizes at 20°C is shown in Figure 8-9.

From Eq. 8-12 it can be seen that the head loss in clean media increases linearly as the media depth increases. Head loss will increase linearly with superficial velocity until inertial effects become important at which time the head loss increases as the square of the velocity. Under laminar flow conditions, the head loss is inversely proportional to the square of the media size. As shown previously, smaller media sizes increase particle capture efficiency, but head loss increases rapidly as seen in Figure 8-9. An optimum size can be determined by pilot studies.

Equation 8-12 also defines the dependence of head loss on water temperature. An increase in the

temperature will cause a decrease in fluid viscosity, thereby decreasing the head loss. Finally, the head loss per unit depth is highly sensitive to the void fractional volume or porosity. Under laminar flow conditions, the influence of the porosity on the relative head loss can be illustrated as shown in Figure 8-10. The hydraulic gradient at any given porosity is compared to the hydraulic gradient for $\varepsilon_0 = 0.40$ for the two models of the permeability coefficient, the Carman–Kozeny and the Happel models. The Happel model is based on a sphere-in-fluid shell geometric model of porous media compared to the Carman–Kozeny model, which is based on a parallel capillary model, as discussed. As shown, both models given similar results. Note that a 50 percent decrease in the porosity from 0.4 to 0.2 causes approximately a 12-fold increase in the hydraulic gradient.

Nonuniform Beds

In practical applications the media used in filtration are not uniform or spherical and consist of a range of particle sizes. In addition, recent trends have been toward the use of multimedia filters in order to increase the capture of particulates throughout the filter depth, thereby increasing the length of filtration operation.

Prediction of head loss through clean polydispersed media can be obtained by segmenting the depth of media into a series of layers and summing

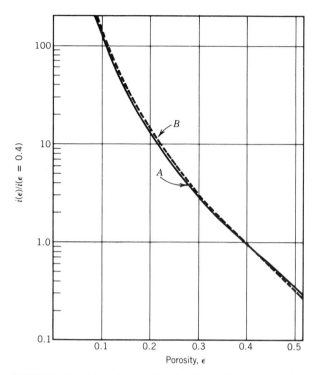

FIGURE 8-10. Dependence of hydraulic gradient on porosity; laminar flow through porous media. Models of: *A*, Carman–Kozeny; *B*, Happel.

where ΔH_0 is the head loss through the clean media at time $t = 0$, σ is the mass specific deposit (mass/volume) or quantity of solids per volume of filter bed, and γ_i are empirical constants in units of volume/mass.

When specific deposits are low and the pore volume occupied by the solids is less than 10 percent, higher terms in Eq. 8-13 can be neglected, giving the expression

$$\frac{\Delta H}{\Delta H_0} = 1 + \gamma_1\sigma \qquad (8\text{-}14)$$

For larger values of the specific deposit, a second-order term may be required to predict head loss (Boller, 1980).

Straining and Cake Formations.   It has been demonstrated that under low-flow conditions ($V_0 < 0.1$ cm/sec) and relatively high solids flux, the formation of a mat of solids on the top of the porous media can occur, even if the size of the particles is considerably smaller than the pore size opening (O'Melia and Ali, 1978). Under these conditions the head loss would be expected to increase dramatically as the fluid would then have to pass through a layer of porous media containing particles considerably smaller than the original media. It has been shown by several authors (see Ives, 1982) that this head loss exhibits an exponential increase pattern, given by

$$H_s = k_1 \exp(k_2 t) \qquad (8\text{-}15)$$

where H_s is the head loss due to straining and k_1 and k_2 are constants.

Effect of Deposit Morphology on Head Loss Buildup.   The impact of the mass of solids collected in the filter on the head loss is a function of the type of solid and the density and the degree to which the solids are compacted in the pore space. In Eq. 8-14 the constant γ can be interpreted as a compaction coefficient with units of volume/mass.

A compaction coefficient or the ratio of the volume-specific deposit to the mass-specific deposit is shown in Figure 8-11 as a function of the porosity of the deposit. Volume and mass balances around an element of filter media demonstrate that this ratio is given by

$$\frac{\sigma^*}{\sigma} = \frac{1/\rho_L - 1/\rho_s}{(\rho_s/\rho_L - 1)(1 - \varepsilon_D)} \qquad (8\text{-}16)$$

the head loss through each layer segment. Ultimately, an equivalent diameter can be used in the head loss correlations whereby this equivalent size of media gives the same head loss as would be obtained by empirical observation. This is the basis for the effective size of the media, equivalent to the 10% fraction by weight, the so-called d_{10} media size. These approaches have been illustrated elsewhere (Fair, Geyer, and Okun, 1968).

Clogging Media

As particulate material is captured in granular media filters, interstitial pore space will decrease. In addition, the media size will change due to accumulation of particulate matter on the media surfaces. This may lead to either an increase or decrease in the specific surface of the media (Herzig et al., 1970). In general, it has been shown that the increase in head loss due to accumulation of solids in the pore space can be approximated by an infinite series as (Herzig et al., 1970; Ives, 1982)

$$\frac{\Delta H}{\Delta H_0} = 1 + \gamma_1\sigma + \gamma_2\sigma^2 + \cdots \qquad (8\text{-}13)$$

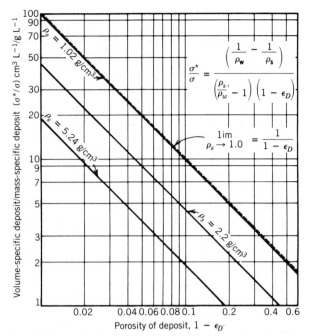

$$\frac{\sigma^*}{\sigma} = \frac{\left(\dfrac{1}{\rho_w} - \dfrac{1}{\rho_s}\right)}{\left(\dfrac{\rho_s}{\rho_w} - 1\right)\left(1 - \epsilon_D\right)}$$

$$\lim_{\rho_s \to 1.0} = \frac{1}{1 - \epsilon_D}$$

FIGURE 8-11. Dependence of volume-specific deposit on porosity and density of deposited solids.

where σ^* is the volume-specific deposit (volume solids–volume filter), ρ_L and ρ_s the density of water and the solid particulates, respectively, and ε_D the porosity of deposit.

For discrete particles, such as clays or ferric oxide scale particles, particulate matter may exhibit a porosity of deposit on the order of 60–80 percent. For flocculent materials, however, with ρ_s close to unity, the porosity of deposit will be high, on the order of 90–95 percent. Thus, the volume of deposit

for an equivalent mass compared to discrete particles may be high.

This has been demonstrated for a number of systems as shown in Table 8-6, which summarizes the compaction coefficients for several particulates determined under a variety of operating conditions. This compaction coefficient appears to be independent of approach velocity (Heertjes and Lerk, 1967; Kavanaugh, 1974) but dependent on the granular media size, the density of the solids, and the porosity of deposit. It is difficult to predict the compaction coefficient theoretically, and measurements of deposit density are also not easily achieved. Several investigators have indicated that the compaction coefficient may vary as an inverse power of the granular media size (Heertjes and Lerk, 1967; Mintz, 1966; Kavanaugh, 1974). For systems in which the pore space does not decrease more than 10–20 percent, it is likely that the head loss is a linear function of the mass-specific deposit in the absence of straining. This equation can then be used to predict the time at which a limiting head loss system will be reached in a constant rate filtration system, as discussed in subsequent sections.

GRANULAR MEDIA FILTERS

General Description

Granular media filtration is the most common type of filter process used for particulate removal in water and wastewater treatment following various pretreatment processes. Diatomaceous earth filters are much less commonly used. Granular filters consist

TABLE 8-6. Compaction Coefficients for Different Filtration Systems

Type Solids	Approximate Density (kg/m³)	Media Size (mm)	Superficial Velocity (m/hr)	Compaction Coefficient (L/g)
Discrete, TiO₂[a]	3.9	0.27	5	0.26
		0.54	5	0.11
		0.94	5	0.05
Discrete, Latex[a]	1.02	0.94	10	0.36
		1.52	10	0.10
Discrete, Kaolin[b]	2.2	1.8	11	0.04–0.28
Flocculant[c]	1.01	2.2	10	0.5
Fe(OH)₃–FePO₄		1.9	10	0.7

[a] From Kavanaugh (1974).
[b] From Tanaka (1982).
[c] From Kavanaugh et al. (1977).

of a bed of porous material contained in a structure that permits regeneration of the solids retention capacity of the filter when either the available head loss has been reached or the effluent quality standard is exceeded. Normally, water flow is by gravity through the granular media in a downflow mode. Other flow arrangements have been used, however, including a reverse flow or upflow filter, a byflow filter in which portions of the flow pass upflow and a portion downflow with the effluent removed at the midpoint of the media. Usually, granular media filtration is a discontinuous process consisting of a filtration and regeneration or backwashing cycle. Some continuous filters, however, are available from manufacturers.

Process design of granular filters requires the selection of several design variables, including the appropriate pretreatment prior to granular media filtration, the type and size of filter media, the depth of the media, the superficial velocity, and the backwash rate. Pretreatment is essential because of the need to achieve particulate destabilization, as has been discussed previously. In addition, pretreatment may increase the strength of the flocculent material to withstand hydraulic stress within the filter, thus permitting higher solids retention prior to breakthrough. There is also some evidence that the reduction of floc size and floc volume can lead to longer filter run times due to reduction of floc volume and a smaller compaction coefficient (Camp, 1968).

The design engineer is faced with numerous choices with respect to media selection, including the type, size, and size range of the media. Theoretically, it is most desirable to decrease the media size in the direction of flow through the filter. Optimally, this is achieved by upflow filtration. Upflow filters have inherent disadvantages, however, including potential bursts of solids due to sudden or local fluidization and the potential contamination of the treated water by backwashing. In order to avoid this potential health problem, designers have developed multimedia filters consisting of two or more layers of filter media of varying density permitting the use of larger-size media on top of a finer grade of material. If a multilayer filter is selected, the designer must determine a media size that minimizes the intermixing of the filter layers and also provides adequate depth to satisfy the run length criteria.

Traditionally, the design flowrate or superficial velocity for granular filters used in water treatment has been regulated by state public health agencies.

The so-called rapid sand filtration rate of (5 m/hr) 2 gpm/ft^2 was based on historical studies conducted at Louisville (Fuller, 1898). Modern filter design, however, has tended toward ever-increasing flowrates because of significant savings achieved in capital costs. Flowrates as high as (30 m/hr) 12 gpm/ft^2 have been used. Such high rates usually require appropriate pretreatment in order to maintain desirable effluent quality.

The sizing of the backwash facilities, including the size of the backwash lines and the flowrates, and whether or not a combination of air and/or water will be used are also critical process design problems. As much as 50 percent of the capital cost may be invested in backwash facilities, including the facilities required for treating or handling backwash solids (Kawamura, 1975).

Mechanical Design

In addition to selection of the process design variables, numerous mechanical design issues must be resolved. This includes the type of flowrate control system, the minimum number of filter units, and consequently the size of each individual unit and the design of numerous appurtenances including washwater troughs, inlet weirs, filter bottoms, and process monitoring equipment. The successful design of a filter unit is heavily dependent on the appropriate selection of these filter elements. These issues will be discussed in Chapter 21.

MATHEMATICAL DESCRIPTION OF GRANULAR MEDIA FILTRATION

The objective of a mathematical model of granular filtration is the prediction of the time profiles of head loss and concentration of particulates for a given set of process design variables. Because of the complex nature of granular filters, including the heterogeneous characteristics of both particulates and the filter media and the ill-defined flow patterns through clogging media, no universal model has yet been developed. Models provide, however, a convenient semiquantitative description of filter performance and are essential tools in the design and interpretation of necessary pilot-plant investigations (Mintz, 1966) which are desirable prior to most filter designs. The following discussion highlights key modeling issues and presents some examples of the use of models for filter process design. Recent detailed reviews of filtration modeling are recommended for further study (Tien et al., 1979; Ives, 1982).

The effluent and head loss history of granular filters is governed by three equations: (1) a macroscopic mass balance; (2) the kinetics of particulate removal; and (3) a general expression for increase of head loss across the media with time and accumulation of solids. The general macroscopic mass balance or continuity equation is given by

$$V_0 \left(\frac{\partial C}{\partial x} \right)_t + \left(\frac{\partial (\sigma + \varepsilon C)}{\partial t} \right)_x = 0 \qquad (8\text{-}17)$$

where x and t are depth and time, respectively; C is the concentration of particulates in the water; and σ is the specific deposit or mass of particulates per volume of media. This equation assumes plug flow of water in an axial direction, no longitudinal dispersion, and constant cross-sectional area. The change of suspension concentration with depth can be expressed by Eq. 8-18 for the isolated collector model; namely

$$-\left(\frac{\partial C}{\partial x} \right)_t = \eta(\sigma) \frac{(1 - \varepsilon_0)C}{d_m} \qquad (8\text{-}18)$$

where $\eta(\sigma)$ indicates that the individual collector efficiency is a function of the specific deposit, σ.

Equation 8-18 is often expressed as

$$-\left(\frac{\partial C}{\partial x} \right)_t = \lambda(\sigma)C \qquad (8\text{-}19)$$

where $\lambda(\sigma)$ is known as the filter coefficient and is also a function of specific deposit as well as the physical properties of the filter media. Numerous empirical models of $\eta(\sigma)$ or $\lambda(\sigma)$ have been proposed (for summaries see Mintz, 1966; Herzig et al., 1970). All models require pilot studies to calibrate the kinetic expressions. They are cumbersome for design purposes and rarely used. They provide, however, an important analytical framework for evaluation of pilot-plant results.

Head loss through the granular media as a function of the specific deposit is given by Eq. 8-14. A combination of Eqs. 8-14, 8-17, and 8-19 can be used to describe the performance of granular filters in the absence of straining.

Process Optimization

The traditional optimization criterion for granular filters operating in a discontinuous mode is that optimum water production will occur when the time to reach a limiting head loss is reached at the same moment that the effluent concentration exceeds the specified standard. A schematic of this is shown in Figure 8-12. However, if the run length is short, filtration efficiency or net water production may be low. The additional constraint on the optimization criterion is that a filter must be operated to give an acceptable production efficiency, defined as the ratio of the effective filtration rate to the design filtration rate. This criterion can be evaluated using the concepts of unit filter run volume (UFRV) and unit backwash volume (UBWV) (Trussell, 1978; Trussell, et al., 1980). The UFRV is the volume of water that passes through the filter during a run and the UBWV is the volume required to backwash the filter. Both are expressed in units of volume per unit area per run, or:

$$\text{UFRV} = V_f/A$$
$$\text{UBWV} = V_b/A$$

where V_f is the volume filtered during one run of a filter of area A and V_b is the volume required to backwash the filter. The effective filtration rate (R_e) depends on these unit volumes, on the duration of the filter run (t_f), and the duration of the filter backwash (t_b), in the following way;

$$R_e = (\text{UFRV} - \text{UBWV})/(t_f + t_b) \qquad (8\text{-}20)$$

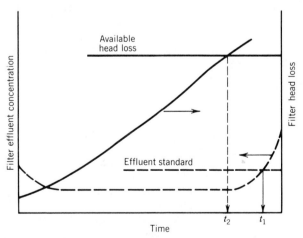

FIGURE 8-12. Schematic description of optimum filtration times. Optimum occurs when $t_1 = t_2$ if t_{opt} provides satisfactory filtration efficiency.

Neglecting the time required to backwash, the ratio of the effective filtration rate to the maximum design filtration rate (R_d) is given by:

$$R_e/R_d = (UFRV - UBWV)/UFRV \quad (8\text{-}21)$$

This relationship is plotted in Figure 8-13 for various UBWV's.

Typically, washwater quantities are approximately 8 m³/m² (200 gal/ft²). Thus, in order to achieve a production efficiency greater than 95 percent, a unit filter run volume of at least 200 m³/m² run (5000 gal/ft² run) is required. Consequently, the filter should be designed to provide this unit filter run volume while meeting the other optimization constraint, namely, the equivalence between the time to reach breakthrough and the time to reach the head loss limit. Several optimization models have been developed based on the concepts discussed above (Ives, 1982; Kavanaugh, 1974; Adin and Rebhem, 1975; Mintz, 1966; Letterman, 1976; Tien et al., 1979).

A schematic of filtration optimization concepts is shown in Figure 8-14, which relates the optimum filter time to the media size and the filtration rate (Kavanaugh, 1977). As can be seen, increasing the filtration velocity will tend to decrease both the time to reach head loss and the time to reach breakthrough. Increasing the media size will tend to decrease the time to reach breakthrough t_1 but increase the time to reach the limiting head loss, t_2. These two planes intersect at a point that shows a global optimum for dependence on media size and filtration rate.

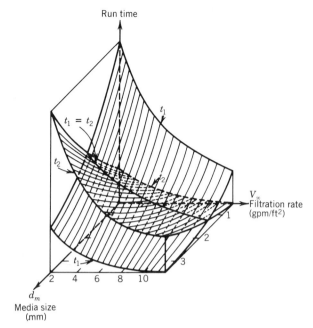

FIGURE 8-14. Filtration optimization for constant bed depth and influent solids concentration.

The effect of significant design parameters on t_1 and t_2 is summarized in Table 8-7. These design variables will influence filter performance and thus t_1 and t_2 based on the rate of particulate capture and on the rate of head loss increase. Media size, media depth, and flowrate are subject to designer selection. Influent solids concentration will depend on the location of the filter in the process scheme. Floc

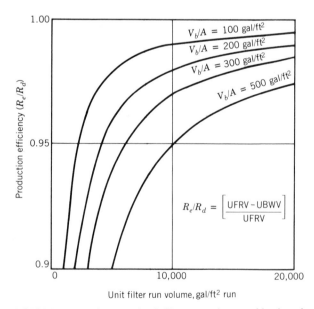

FIGURE 8-13. Influence of unit filter run volume and backwash volume on filter production efficiency.

TABLE 8-7. Dependence of Length of Run on Design Parameters

Parameter	Effect of Parameter Increase On	
	Time to Breakthrough (t_1)	Time to Limiting Head Loss (t_2)
d_m	Decrease	Increase
L	Increase	Decrease
Q/A	Decrease	Decrease
C_0	Decrease	Decrease
Floc strength	Increase	Decrease
Deposit density	Decrease	Decrease
Porosity	Decrease	Increase

strength and deposit density are difficult to control. Use of polymers can be employed to improve floc strength. The density of the deposit and thus the compaction coefficient can be chiefly controlled by the use of polymers for destabilization in place of aluminum or ferric hydroxide flocs. The limits of media size should be chosen to minimize interlayer mixing, which tends to decrease porosity and thereby lead to a rapid increase in head loss (decrease in t_2) (Cleasby and Woods, 1975).

Pilot Studies for Process Design

Based on the current status of modeling of the granular media filtration process, it is not yet possible to select an optimum process design without pilot studies for new applications of granular filters. Pilot studies provide the basis for evaluating the feasibility of the granular media filters for solids–liquid separation (Mintz, 1966). They also provide critical information on such questions as pretreatment requirements, media size, flow rates, and backwashing rates. Once any given model has been calibrated with appropriate pilot-plant studies, sensitivity analyses can be conducted (see Chapter 19).

DIATOMITE (OR PRECOAT) FILTERS

General Description

Diatomaceous earth (DE) or diatomite filters are typical of a group of filters more broadly described as *precoat filtration*. DE (or Fuller's earth) consists of microscopic remnants of the discarded frustules of diatoms. It is the medium most commonly employed in precoat filtration; thus, the term *diatomite filtration* is frequently used to refer to precoat filtration. DE occurs in natural deposits, the largest of which are found near Lompoc, California. These deposits are the primary source of DE in the United States. Other media have also been employed in precoat filters; these include aluminum silicate, carbon, and perlite. None of these have seen widespread use.

Precoat filters have been employed for many years in industrial applications such as preparation of foodstuffs and filtration of beverages, but their most common usage is in filtration of swimming pools. Their use in municipal water treatment in the United States has developed largely since World War II, when they were employed by the military (Black and Spaulding, 1944) in portable filtration systems for removal of the cysts that cause amoebic dysentery.

Diatomite filters have been shown to be very effective for removal of particles from fluids. Specific materials that have been removed successfully include asbestiform fibers, algae, coliforms, and clay suspensions (Hunter et al., 1966; Syrotynski and Stone, 1975; Burns et al., 1970). DE filters have also been employed for iron and manganese removal at a number of locations (Coogan, 1962).

Diatomite filters are typically less expensive to construct than conventional filters and require less space. Thus, when space and capital are limited, diatomite filters may offer distinct advantages. However, design and operation of diatomite filters are more labor intensive than conventional filters. In addition, precoat filters are less capable of dealing with wide variations in influent water quality conditions than conventional deep-bed, granular filters.

Process Description. Particle removal during diatomite filtration, in contrast to most other filtration processes, occurs largely at the surface of the media through formation of a filter cake. Straining plays the key role in precoat filter operation. However, chemical pretreatment of either the media or the raw water particles may be practiced in order to enhance removal by improving particulate capture.

Figure 8-15 shows a schematic of a typical diatomite filtration system. Filter operations are generally a three-step process: precoat, filtration, and backwash. In the first step a precoat or thin layer of filter media approximately 2–5 mm thick is applied to the septum. The septum is a porous plate-screening structure designed to strain out the smallest-diameter particle of precoat or body feed employed. This precoat layer of DE filter aid serves as the initial layer of media for filtration and must be evenly applied to the surface of the septum to achieve uniform, high-quality effluent. Pressure diatomite filters are commonly designed for raw water to enter the bottom of the vessel (as shown in Figure 8-15) and to exit the top. Vertical surface septa, either cylindrical or flat, are frequently employed so high flowrates may be necessary to ensure even coating of the septum. The precoat slurry applied to the septum is composed of 0.3–0.6 percent DE applied at a surface loading of 0.05–0.10 g/cm^2 (0.1–0.2 psf).

After precoat is completed, the valve in the feed

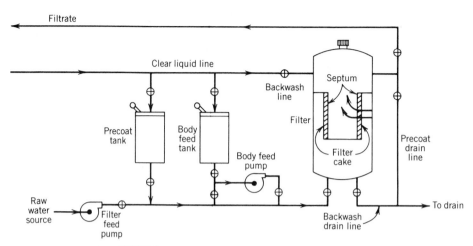

FIGURE 8-15. Typical diatomite filtration system.

line from the precoat slurry tank is shut and filtration begins. Raw water is applied to the septum and filtration occurs across the thin precoat layer. The size of the precoat media largely determines the quality of effluent from the filter, particularly if no chemical destabilization of the influent particles is practiced and no polyelectrolytic coating is applied to the media. Because most naturally occurring particles in water are negatively charged, as is the DE medium, straining is the dominant removal mechanism. By selecting DE media of smaller size, improved effluent quality may be obtained through the concomitant reduction in pore size. If no body feed is applied during filtration, pore space in the filter is rapidly clogged, porosity decreases, interstitial velocities increase (for a constant rate filter), and head loss increases exponentially.

In order to prevent this, *body feed,* consisting of DE, which may be the same as the filter aid applied during precoat, is fed during the filtration cycle. As the filter cycle progresses, the DE filter aid and raw water particles are deposited at the media surface. In this manner, a new filtering surface continuously forms. This process continues until head loss across the filter cake reaches a preset maximum. Body feed during filtration is typically applied at levels of 1–3 mg/L or higher depending on raw water quality, particle size distributions of both the raw water particles and the filter aid, filtration rate, and desired filtration efficiency.

When head loss reaches the specified maximum, the filter inlet line is closed and the backwash line is opened. The filter cake, which has built up during filtration, sloughs off and is discharged to drain. In

some instances it is necessary to employ a surface wash or agitation system to break up an encrusted filter cake, particularly in high-pressure filters. After backwash is complete, the entire cycle begins again. Backwash usually takes about 30 min to complete; filter run lengths of 24–150 hr are common.

A number of variations of the design shown in Figure 8-15 are used. Filters may be designed as a pressurized vessel with a pumped inlet line or as an open filter bed operated through application of a vacuum to the underside of the septum. Pressurized diatomite filters have no upper limit on the amount of pressure that may be lost across the filter, while vacuum-type installations have an approximate practical upper limit of 6 m (20 ft). Advantages of the open, vacuum filters are that noncorrodible materials (such a plastics) can be employed and the septum may be inspected for backwash completion and uniformity of precoat.

Process Models and Mathematics

Much of the developmental work in explaining the performance and process mathematics of diatomite filters has been done at Iowa State University. This work has been summarized in numerous technical publications (e.g., see Baumann et al., 1962; Dillingham et al., 1967; Baumann and Oulman, 1964; Dillingham et al., 1966). A more recent publication that summarizes and updates much of the information compiled and developed in the last 20 years is presented by Baumann (1978).

Much of the work done to date centers on predicting head loss changes during filtration in

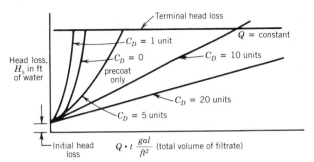

FIGURE 8-16. Effects of body feed rate on head loss development (Baumann, 1978).

diatomite filters. Prior to actual design it is necessary to perform a number of pilot-plant filtration tests in order to determine such factors as the hydraulic characteristics of filter cakes produced during operation. In general, precoat operations are optimized and are most feasible economically when sufficient body feed is applied to allow for linear head loss development with time. This is illustrated in Figure 8-16.

With no body feed, head loss across the cake increases exponentially, as described earlier. If insufficient body feed ($C_D = 1$ unit in Figure 8-16) is applied, head loss development may actually be increased as the existing pore space is rapidly clogged. At some point, however, sufficient body feed is applied to allow for formation of a filter cake with uniform density and porosity such that head loss develops in a linear fashion, proportional to the

thickness of the cake. This point is reached at a body feed rate between 1 and 10 units, as shown in Figure 8-16.

A series of equations has been developed to predict head loss development for various types of septa (e.g., cylindrical or flat) once linear head loss development has been achieved. The mathematics will not be presented here as it is thoroughly documented in the literature (e.g., Baumann, 1978).

As noted previously, mechanical straining is frequently the sole mechanism for removal of suspended particulate matter in a diatomite filter. If monodisperse filter aid is employed (i.e., single, uniform diameter for all media grains) and sphericity is assumed for both the media and suspended particles in the raw water, those suspended particles with a diameter of 0.155 times the media diameter or larger would be removed at the media surface due to straining. Smaller suspended particles would pass through the filter. If media with a heterodisperse size distribution is employed or if conditions other than clean bed are examined, removal due to straining is improved somewhat.

Figure 8-17 demonstrates semiquantitatively the relationships among the various removal mechanisms as a function of suspended particle diameter and defined by single collector efficiency as developed earlier in this chapter. If no particle destabilization is practiced and clean bed conditions are assumed, then Figure 8-17 indicates that particles larger than about 3 μm in diameter are effectively removed from the filtrate by straining with a 15-

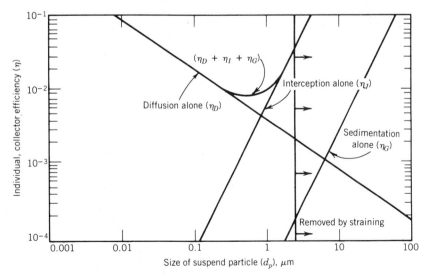

FIGURE 8-17. Removal mechanisms for sample diatomite filter. $T = 25°C$, $d = 15$ μm, $V_0 = 1.5$ gpm/ft², $\rho_p = 1.05$ g/cm³.

μm diameter media. With either ripening of the cake or destabilization of the suspended particles or filter aid, improved removal of particles smaller than 2 μm may be achieved. It is interesting to note for the conditions examined in Figure 8-17 that diffusion and interception are the predominant mechanisms for removal of particles too small to be strained out of suspension; sedimentation in the filter cake is of minor importance. In addition, since straining is the major means of particle removal for large diameters, flocculation is of little or no value with regard to effluent quality. However, body feed rate may be reduced if flocculation is practiced ahead of the diatomite filter.

REFERENCES

Adin, A., and Rebhun, M., "Model for Prediction and Concentration and Headloss Profile in Granular Bed Filtration," *JAWWA*, **66**(2), 109–112 (1979).

Agrawal, G. D., "Electrokinetic Phenomena in Water Filtration," Ph.D. Thesis, University of California, Berkeley (1966).

Baker, M. N., *The Quest for Pure Water,* American Water Works Association, New York (1949).

Baumann, E. R., "Precoat Filtration," in R. L. Sanks (ed.), *Water Treatment Plant Design for the Practicing Engineer,* Ann Arbor Science, Ann Arbor, MI (1978).

Baumann, E. R., and Huang, J. Y., "Granular Filters for Tertiary Wastewater Treatment," *JWPCF,* **46**(8), 1958 (1974).

Baumann, E. R., and Oulman, C. S., "Modified Form of Diatomite Filtration Equation," *JAWWA,* **56,** 330–332 (March 1964).

Baumann, E. R., Cleasby, J. L., and LaFrenz, R. L., "A Theory of Diatomite Filtration," *JAWWA,* **54,** 1109–1119 (September 1962).

Bird, R. B., Stewart, R. E., and Lightfoot, E. N., *Transport Phenomena,* p. 140, John Wiley & Sons, New York (1960).

Black, H. H., and Spaulding, C. A., Diatomite Water Filtration Developed for Field Troops, *JAWWA,* **36,** 1208–1221 (November 1944).

Boller, M., "Flocculation Filtration for Waste Water Treatment," Ph.D. Dissertation, Swiss Technical University, Zurich, Switzerland (1980) (in German).

Boller, M. A., and Kavanaugh, M. C., "Contact Filtration for Additional Removal of Phosphorus in Waste Water Treatment," Presented at International Association for Water Pollution Research (IAWPR) Workshop, Vienna (September 1975).

Burns, D. E., Baumann, E. R., and Oulman, C. S., "Particulate Removal on Coated Filter Media," *JAWWA,* **62,** 121–126 (February 1970).

Camp, T. R., "Floc Volume Concentration," *JAWWA,* **60**(6), 656 (1968).

Carman, P. C., "Fluid Flow Through Granular Beds," *Trans. Inst. Chem. Eng.,* **15,** 150 (1937).

Cleasby, J. L., and Woods, G. F., "Intermixing of Dual Media and Multi-Media Granular Filters," *JAWWA* **67**(4), 197 (1975).

Coogan, G. J., "Diatomite Filtration for Removal of Iron and Manganese," *JAWWA,* **54,** 1507–1517 (December 1962).

Darcy, H. P. G., *Les Fontanier Publiques de la Ville de Dijon.* Paris, Victor Dalmont (1856).

Dillingham, J. H., Cleasby, J. L., and Baumann, E. R., "Diatomite Filtration Equations for Various Septa," *J. Sanit. Eng. Div., ASCE,* **93** (SA1), 41–55 (February 1967).

Dillingham, J. H., Cleasby, J. L., and Baumann, E. R., "Optimum Design and Operation of Diatomite Filtration Plants," *JAWWA,* **58,** 657–672 (June 1966).

Fair, G. M., and Hatch, C. P., "Fundamental Factors Governing the Streamline Flow of Water through Sand," *JAWWA,* **25,** 1551–1565 (1933).

Fair, G. M., Geyer, J. C., and Okun, D. A., *Water and Wastewater Engineering,* Vol. 2, Wiley, New York (1968).

Fitzpatrick, J. A., and Spielman, I. A., "Filtration of Aqueous Latex Suspensions Through Beds of Glass Spheres," *J. Coll. Int. Sci.* **43**(2), 350–369 (1973).

Fuller, G. W., "The Purification of the Ohio River Water at Louisville Kentucky," D. Van Nostrand Co., New York (1898).

Ghosh, M. M., Jordan, T., and Porter, R., "Physicochemical Approach to Water and Wastewater Filtration," *J. Inv. Eng. Div., ASCE,* **101**(EE1), 71 (1975).

Habibian, M., "The Role of Polyelectrolytes in Water Filtration," Ph.D. Thesis, University of North Carolina, Chapel Hill, NC (1971).

Happel, J., and Brenner, H., *Low Reynolds Number Hydrodynamics,* Prentice-Hall, Englewood Cliffs, N.J. (1965).

Heertjes, P. M., and Lerk, C. E., "The Functioning of Deep-Bed Filters, Parts I and II," *Trans. Inst. Chem. Eng.,* **45,** T129–T145 (1967).

Herzig, J. R., et al., "Flow of Suspensions Through Porous Media, Application to Deep Filtration," *I & EC,* **62,** 5, 8 (1970).

Hudson, H. E., "Declining-Rate Filtration" *JAWWA,* **51,** 11 (1959).

Hunter, J. V., Bell, G. R., and Henderson, C. N., "Coliform Organism Removals by Diatomite Filtration," *JAWWA,* **58,** 1160–1169 (September 1966).

Ison, C. R., and Ives, K. J., "Removal Mechanisms in Deep Bed Filtration," *Chem. Eng. Sci.,* **24,** 717–729 (1969).

Ives, K. J., "Deep Filters," *Filtration and Separation,* **4**(3/4), 125 (1967).

Ives, K. M., "Mathematical Models and Design Methods in Solid-Liquid Separation; Deep Bed Filters," Paper presented at NATO Advanced Study Institute (January 1982).

Iwasaki, T. "Some Notes on Sand Filtration," *JAWWA,* **29**(10), 1591–1602 (1937).

Kavanaugh, M., Sigster, K., Weber, A., and Boller, M., "Contact Filtration for Phosphorus Removal," *JWPCF,* **49**(10), 2157 (1977).

Kavanaugh, M., Toregas, G., Chung, M., and Pearson, E. A., "Particulates and Trace Pollutant Removal by Depth Filtration," *Prog. Water Tech.,* **10**(5/6), 197 (1978).

Kavanaugh, M. C., "Mechanisms and Kinetic Parameters in Granular Media Filtration," Ph.D. Thesis, Berkeley (1974).

Kavanaugh, M. C., "Use of Particle Size Distribution Measurements for Selection and Control of Solid/Liquid Separation Processes," in *Particulates in Water,* American Chemical Society Advances in Chemistry, No. 189, (1980), p. 305.

Kawamura, S., "Design and Operation of High-Rate Filters," *JAWWA,* **67**(12), Parts 1–3 (1975).

Letterman, R. D., "Optimizing Deep Bed Water Filters Using a Deposit Distribution Concept," *Filtration and Separation,* **13**(4), 343–350 (July/August 1976).

Levich, V. G., *Physico-Chemical Hydrodynamics,* Prentice-Hall, Englewood Cliffs, NJ (1962).

Maroudas, A., and Eisenklam, P., "Clarification of Suspensions: A Study of Particle Deposition in Granular Media; Part I: Some Observations on Particle Deposition," *Chem. Eng. Sci.,* **20**, 867–873 (1965a).

Maroudas, A., and Eisenklam, P., "Clarification of Suspensions: A Study of Particle Deposition in Granular Media; Part II: A Theory of Clarification," *Chem. Eng. Sci.,* **20,** 875–888 (1965b).

Mintz, D. M. "Modern Theory of Filtration," *International Water Supply Association, 7th Congress,* Barcelona (1966).

O'Melia, C. R., "The Role of Polyelectrolytes in Filtration Processes," Report for EPA, EPA-670/2-74-032 (April 1974).

O'Melia, C. R., and Ali, W., "The Role of Retained Particles in Deep Bed Filtration," *Proj. Water Technol.,* **10**(5,6), 123 (1978).

O'Melia, C. R., and Ali, W., "The Role of Retained Particles in Deep Bed Filtration," *IAWPR Conference,* Stockholm, Sweden (1978).

O'Melia, C. R., and Stumm, W., "Theory of Water Filtration," *JAWWA,* **59**(11), 1393 (1967).

Payatakes, A., "A New Model for Granular Porous Media Application to Filtration Through Packed Beds," Ph.D. Thesis, Syracuse University, Syracuse, New York (1973).

Payatakes, A. C., Tien, C., and Turian, R. M., "Trajectory Calculation of Particle Deposition in Deep Bed Filtration," *J. AIChE,* **20**(5), 889 (1974).

Rajagopolan, R., and Tien, C., "Trajectory Analysis of Deep Bed Filtration With the Sphere-in-Cell Porous Media Model," *J. AIChE,* **22**(3), 523 (1976).

Sakthivadivel, R., "Clogging of a Granular Porous Medium by Sediment," Rep. HEL 15-7, Hydraulic Eng. Lab., University of California, Berkeley (1969).

Schaub, S. A., and Sagib, B. P., "Association of Enteroviruses With Natural and Artificially Introduced Colloidal Solids in Water and Infectivity of Solids: Associated Viruses," Applied Microbiology, 212 August, 1975.

Spielman, L. A., and Cukor, P. M., "Deposition of Non-Brownian Particles Under Colloidal Forces," *J. Coll. Int. Sci.* **43**(1), 51–65 (April 1973).

Spielman, L. A., and Fitzpatrick, J. A., "Theory for Particle Collection under London and Gravity Forces," *J. Coll. Int. Sci.,* **42,** 607 (1973).

Spielman, L. A., and Friedlander, S. K., "Role of the Electrical Double Layer in Particle Deposition by Corrective Diffusion," *J. Coll. Int. Sci.* **46**(1), 22 (1979).

Spielman, L. A., and Cukor, P. M., "Deposition of Non-Brownian Particles under Colloidal Forces," *J. Coll. Int. Sci.,* **43,** 51–62 (1973).

Spielman, L. A., and Goren, S. L., "Capture of Small Particles by London Forces from Low-Speed Liquid Flows," *Env. Sci. Tech.,* **4**(2), 135 (1970).

Stein, P. C., "A Study of the Theory of Rapid Filtration of Water Through Sand," D.Sc. Thesis, Massachusetts Institute of Technology (1940).

Stumm, Werner and O'Melia, Charles R., "Stoichiometry of Coagulation," **JAWWA, 60,** (5), 514 (1968).

Symons, J. M., and Hoff, J. C., "Rationale for Turbidity Maximum Contaminant Level," Paper presented at 3rd Water Quality Technology Conference, AWWA, Atlanta, Georgia (Dec. 8–10, 1975).

Syrotynski ,S., and Stone, D., "Microscreening and Diatomite Filtration," *JAWWA,* **67,** 545–548 (October 1975).

Tanaka, T., "Kinetics of Deep-Bed Filtration," Ph.D. Thesis, University of Southern California, Los Angeles (1982).

Tchobanoglous, G., "Filtration of Treated Sewage Effluent," *J. Sanit. Eng. Div., ASCE,* **96**(SA2), 243 (1970).

Tien, C., Turian, R. M., Pendese, H., "Simulation of the Dynamic Behavior of Deep Bed Filters," *Jr. AIChE,* **25**(3), 385 (1979).

Trussell, R. R., "Predesign Studies," Chapter 3 in *Water Treatment Plant Design,* Sanks, R. L., Editor, Ann Arbor Science, Ann Arbor (1978).

Trussell, R., Trussell, A., Lang, J., and Tate, C., "Recent Developments in Filtration System Design," **AWWA, 73,** 705–710 (December 1980).

Yao, K. M., "Influence of Suspended Particle Size on the Transport Aspect of Water Filtration," Ph.D. Thesis, University of North Carolina, Chapel Hill (1968).

Yao, K. M., Habibian, M. T., and O'Melia, C., "Water and Waste Water Filtration: Concepts and Applications," *Environ. Sci. Tech.,* **5,** 1105–1112 (1971).

—9—

Adsorption

Purification of water supplies by adsorption of undesirable contaminants onto solid adsorbents has a short history compared to other processes. Adsorption was first observed in solution by Lowitz in 1785 and was soon applied as a process for removal of color from sugar during refining (Hassler, 1974). In the latter half of the nineteenth century, unactivated charcoal filters were used in American water treatment plants (Croes, 1883). Large volumes of granular activated carbon (GAC) were manufactured during World War I for use in gas masks. Powdered activated carbon (PAC) was used by Chicago meat packers in the 1920s to control taste and odor in water supplies contaminated by chlorophenols (Baylis, 1929). The first GAC units for treatment of water supplies were constructed in Hamm, Germany, in 1929 and Bay City, Michigan, in 1930. PAC was first used in municipal water treatment in New Milford, New Jersey, in 1930; its use became widespread in the next few decades, primarily for taste and odor control.

During the mid-1970s, interest in adsorption as a process for removal of organics from drinking water heightened as the public became increasingly concerned about water sources contaminated by industrial wastes, agricultural chemicals, and sewage discharges. Another major concern was the observed formation of thrihalomethanes (THMs) and other known or suspected carcinogens during chlorination of water containing organic precursors.

At present, the applications of adsorption in water treatment in the United States are predominantly traditional taste and odor control. However, adsorption is increasingly being considered for removal of synthetic organic chemicals, color-forming organics, and disinfection by-products and their naturally occurring precursors. Some inorganic compounds that represent a health hazard, such as heavy metals, are also removed by adsorption. When water is superchlorinated, adsorption is sometimes used to dechlorinate the final product. By 1977 about a quarter of utilities in the United States used PAC; in 1975 sixty-five plants in the United States used GAC (AWWA Committee, 1977; Robeck, 1975). European water treatment plants have had longer experience using GAC to remove synthetic organic contaminants in water from polluted rivers. Activated carbon remains the principal adsorbent in full-scale water treatment. In the future, it is expected that the application of adsorption to control contamination of drinking water by toxic or carcinogenic compounds at low concentrations will increase.

This chapter discusses the mechanisms and thermodynamics of adsorption, describes the properties of activated carbon and adsorbates, and develops an adsorption process analysis.

Definition of Terms

Adsorption is the physical and/or chemical process in which a substance is accumulated at an interface between phases. For the purposes of water treatment, adsorption from solution occurs when impurities in the water accumulate at a solid–liquid interface. The *adsorbate* is the substance being removed from the liquid phase to the interface. The *adsorbent* is the solid phase onto which the accumulation occurs. Adsorption from *dilute aqueous solution* is said to occur when the concentration of the adsorbate in water is small enough to allow assumptions of ideality. In this limiting case, Henry's law holds: the partial pressure of the adsorbate is proportional to its mole fraction (see Chapter 5), and the ratio of the activity of the adsorbate to its concentration (the activity coefficient) is equal to unity.

MECHANISMS OF ADSORPTION

Adsorption of substances onto adsorbents takes place because there are forces that attract the adsorbate to the solid surface from solution. Alternatively, one can view this thermodynamically as a case where adsorbate has a lower free energy at the surface than in solution. During equilibration, the adsorbate is driven onto the surface to the lower energy state, which it "prefers" in keeping with the second law of thermodynamics. The specific forces or mechanisms by which adsorbate is attracted to the solid solution interface can be physical or chemical.

Physical Adsorption

Electrostatic force is the basic physical principle that describes interactions between molecules of adsorbent and adsorbate. Electrostatic attraction and repulsion based on Coulomb's law are described in Chapter 6. Other physical interactions among molecules, based on the electrostatic force, include *dipole–dipole interactions, dispersion interactions,* and *hydrogen bonding.*

A molecule is said to have a *dipole moment* when there is a net separation of positive and negative charges within it. Molecules such as H_2O and NH_3 have permanent dipoles because of the configuration of atoms and electrons within them (see Chapter 2). They are *polar* compounds. When two dipoles are near each other, they tend to orient their charges to lower their combined free energy: negative charges of one tend to approach positive charges of the other. When electrostatic forces among the charges of the two molecules are summed, the net *dipole–dipole interaction* is an attraction between the two. Polar molecules tend to attract each other. *Hydrogen bonding* is a special case of dipole–dipole interaction in which the hydrogen atom in a molecule has a partial positive charge and attracts an atom on another molecule that has a partial negative charge.

When two neutral molecules that lack permanent dipoles approach each other, a weak polarization is induced in each because of quantum mechanical interactions between their distributions of charge. The net effect is a weak attraction between the molecules, known as the *dispersion interaction* or the *London–van der Waals force* (London, 1930).

In water treatment there is often interest in the adsorption of an organic adsorbate from a polar solvent (water) onto a nonpolar adsorbent (activated carbon). In general, attraction between adsorbate and polar solvent is weaker for adsorbates of a less polar nature; a nonpolar adsorbate is less stabilized by dipole–dipole or hydrogen bonding to water. Nonpolar compounds therefore adsorb more strongly to nonpolar adsorbents. This is known as hydrophobic bonding (Nemethy and Scheraga, 1962); hydrophobic ("disliking water") compounds will adsorb on carbon more strongly. For example, adsorption of fatty acids on carbon from aqueous solution is stronger as the length of the molecule increases because the longer hydrocarbon chain is more nonpolar. This observation is illustrated in Figure 9-1; when generalized to any homologous series of organic molecules, it is known as Traube's rule (Traube, 1891).

Chemisorption

Chemical adsorption, or *chemisorption,* is also based on electrostatic forces. The difference between physical adsorption and chemisorption is not distinct; the former is less specific for which compounds sorb to which surface sites, has weaker forces and energies of bonding, operates over longer distances, and is more reversible. In chemisorption, the attraction between adsorbent and ad-

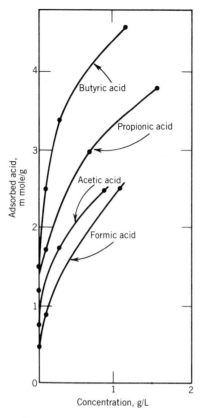

FIGURE 9-1. Illustration of Traube's rule. Mass of carboxylic acids adsorbed on carbon vs. aqueous concentration; degree of adsorption increases with increasing molecular length (Holmes and McKelvey, 1928).

sorbate approaches that of a covalent or electrostatic chemical bond between atoms, with shorter bond length and higher bond energy. Adsorbates bound by chemisorption to a surface generally cannot accumulate at more than one molecular layer, or *monolayer,* because of the specificity of the bond between adsorbate and surface. The bond may also be specific to particular sites or functional groups on the surface of the adsorbent. These properties have modeling implications.

One class of chemical bonding of adsorbate to specific surface sites is acid–base reactions at a functional group. An example is the reaction of hydrated metal ions from solution with hydroxide sites on metal oxides:

$$ROH(aq) + SOH \rightleftharpoons SOR + H_2O(aq)$$

where R is metal ion adsorbate and S is metal oxide adsorbent (Parks, 1975). Parks (1967) and James

and Healy (1972) have reviewed the theory and experimental evidence for this type of chemical bonding in the context of removing heavy metals by adsorption onto silicon and aluminum oxide–based clays and sands for water treatment.

The bond energies of the various mechanisms for adsorption may be approximately ranked from strongest to weakest (Stumm and Morgan, 1981): covalent or electrostatic chemical bonding (> 10 kcal/mole), dispersion interactions and hydrogen bonding (2–10 kcal/mole) and dipole–dipole interactions (<2 kcal/mole).

Adsorption of Electrolytes

For adsorption of ionic species to surfaces, the most important mechanism is electrostatic attraction, which is highly dependent on pH and ionic strength. This is described in Chapter 6 for forces controlling coagulation and in Chapter 10 for the process of ion exchange. Adsorption of electrolytes can be used to control heavy metals, fluoride, and a few other minerals. The use of synthetic resins has been suggested for specific removal of electrolytes.

The adsorption of acids and bases on nonpolar adsorbents such as activated carbon can depend strongly on pH. While both neutral and ionized forms of these compounds may adsorb to carbon from water, the ionized forms tend to be stabilized by interaction with polar water. This means that adsorption of neutral forms is generally much stronger (Getzen and Ward, 1969), and the pH of maximum removal depends on the particular dissociation constant of the acid or base.

THERMODYNAMICS OF ADSORPTION

The adsorption of chemical compounds from solution onto a surface may be viewed as an energetic process driven by thermodynamics. At constant temperature and pressure, a system not in equilibrium will spontaneously change to decrease its Gibbs free energy. At thermodynamic equilibrium, the system of adsorbent, adsorbate, and solvent reaches a minimum free energy level. This section describes models for the energetics of adsorption that allow the engineer to specify the equilibrium state of adsorbent and adsorbate. Later sections discuss the kinetics of adsorption, by which one can estimate the rate of removal of undesired compounds and thereby specify design criteria.

Adsorption Isotherms for Single Solutes

An *adsorption isotherm* specifies the equilibrium surface concentration of adsorbate on adsorbent as a function of bulk concentration of adsorbate in solution. It is called an isotherm because it describes the equilibrium state of adsorbate, adsorbent, and solute at a given temperature.

The *Gibbs equation* (5-14) is an example of a simple equilibrium model. A single adsorbate in dilute solution at concentration C is allowed to equilibrate with a surface on which it has surface density Γ (units of moles/area). The surface tension (γ) is equal to the surface free energy per unit area, which is a function of C. Equilibrium is achieved when total free energy of the system attains a minimum. This equilibrium condition may be expressed (see, e.g., Adamson, 1967) as

$$\Gamma = -\frac{1}{RT}\frac{d\gamma}{d\ln C}$$

For simple systems in which the dependence of surface tension on the bulk concentration is adequately known, this Gibbs adsorption isotherm may be used to describe surface accumulation. This information is not commonly available for adsorbates of interest in water treatment. However, Schay (1975) has shown that the Gibbs equation gives rise to the Langmuir isotherm for a dependence of γ on C observed empirically for a variety of adsorbates.

The *Langmuir adsorption isotherm* describes equilibrium between surface and solution as a reversible chemical equilibrium between species (Langmuir, 1918). The adsorbent surface is considered to be made up of fixed individual sites where molecules of adsorbate may be chemically bound. Denote an unoccupied surface site as $-S$ and the adsorbate in dilute solution as species A, with concentration C, and consider the reaction between the two to form occupied sites ($-SA$):

$$-S + A \rightleftharpoons -SA$$

Assume that this reaction has a fixed free energy of adsorption equal to ΔG_a° that is not dependent on the extent of adsorption and not affected by interaction among sites. Each site is assumed to be capable of binding at most one molecule of adsorbate; the Langmuir model allows accumulation only up to a monolayer.

If there are a total of Γ_∞ moles of surface sites per

square centimeter of adsorbent, and Γ mole/cm^2 of surface sites occupied by adsorbate ($-SA$), then there are $\Gamma_\infty - \Gamma$ mole/cm^2 of unoccupied surface sites ($-S$). Assume that occupied or unoccupied sites may be treated as chemical species with chemical activities equal to their surface concentrations. Then the law of mass action describes an equilibrium constant for the adsorption reaction:

$$K_a = \frac{[-SA]}{[-S][A]} = \frac{\Gamma}{[\Gamma_\infty - \Gamma]C}$$

where $K_a = e^{-G_a^\circ/RT}$. This may be rearranged to obtain

$$\frac{\Gamma}{\Gamma_\infty} = \frac{C}{1/K_a + C} = \frac{K_a C}{1 + K_a C}$$

Strictly speaking, the law of mass action cannot be simply invoked with surface sites representing chemical species, but kinetic arguments to balance the rates of molecules adsorbing to and desorbing from the surface lead to the same result (Adamson, 1967).

In practice, the actual number of sites per surface area and the energy of adsorption are usually unknown. Instead, the variable of interest is q, the number of moles of adsorbate per mass of adsorbent at equilibrium. The Langmuir equation transforms to

$$\frac{q}{Q} = \frac{bC}{1 + bC}$$

where Q is the maximum number of moles adsorbed per mass adsorbent when the surface sites are saturated with adsorbate (i.e., a full monolayer), and b is an empirical constant with units of inverse concentration. An example of a Langmuir isotherm is shown in Figure 9-2 for the adsorption of the three

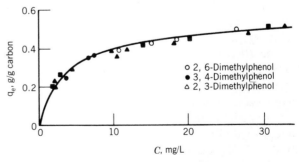

FIGURE 9-2. Langmuir isotherm of dimethylphenols on activated carbon (Singer and Yen, 1980).

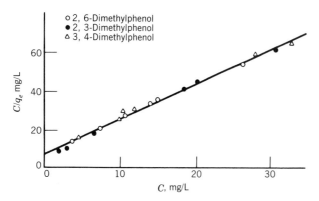

FIGURE 9-3. Isotherms transformed to linearized form of C/q vs. C (Singer and Yen, 1980).

isomers of dimethylphenol. Figure 9-3 shows how experimental data may be plotted to estimate Q and b. With the Langmuir equation rearranged to

$$\frac{C}{q} = \frac{1}{bQ} + \frac{C}{Q}$$

a plot of C/q against C approximates a line with slope $1/Q$ and intercept $1/bQ$. An alternative linearized form is

$$\frac{1}{q} = \frac{1}{Q} + \frac{1}{bQ}\frac{1}{C}$$

so that plot of $1/q$ against $1/C$ has slope $1/bQ$ and intercept $1/Q$.

The Langmuir adsorption isotherm has found wide applicability to adsorption of compounds in water treatment. Its advantages include its simplicity, its foundation in a model with some physical basis, and its ability to fit a broad range of experimental data. Its limitations include (1) the assumption that the energy of adsorption is independent of degree of coverage, (2) reversibility of bonding, and (3) allowance for at most only one monolayer. Mass adsorbed, q, is assumed to approach a saturating value, Q, when C becomes very large.

The *BET adsorption isotherm* (Brunauer et al. 1938) extends the Langmuir model from a monolayer to several molecular layers. Above the monolayer, each additional layer of adsorbate molecules is assumed to equilibrate with the layer below it; layers of different thicknesses are allowed to coexist. The equilibrium for the adsorption onto the new layers is defined by the law of mass action as for the Langmuir model for the first layer. How-

ever, the equilibrium constant is now set equal to $e^{-\Delta G_v^\circ/RT}$, where ΔG_v° is the free energy of precipitation of the adsorbate. The process of sorbing a new layer of adsorbate onto old layers is assumed to be identical to the process of condensing adsorbate from solution to solid or liquid. The resulting isotherm has the form

$$\frac{q}{Q} = \frac{BC}{(C_s - C)[1 + (B - 1)(C/C_s)]}$$

where B is a dimensionless constant related to the difference in free energy between adsorbate on the first and successive layers and C_s is the saturation concentration of the adsorbate in solution. As C approaches C_s, q becomes infinite as adsorbate precipitates onto the surface. The BET isotherm has the general form shown in Figure 9-4, with the surface concentration reaching a plateau as the monolayer is filled, then increasing again with increasing C. Examples of this effect are shown in Figure 9-5 for adsorption of phenol onto various carbon blacks.

The Langmuir and BET models incorporate an assumption that the energy of adsorption is the same for all surface sites and not dependent on degree of coverage. In reality, the energy of adsorption may vary because real surfaces are heterogeneous. The *Freundlich adsorption isotherm* (Freundlich, 1926) attempts to account for this. Assuming that the frequency of sites associated with a free energy of adsorption decreases exponentially

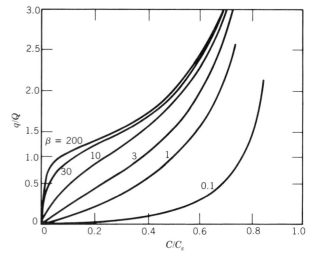

FIGURE 9-4. The BET isotherm for different values of B (Adamson, 1967). q/Q, (mass absorbed/mass carbon)/(monolayer capacity/mass carbon).

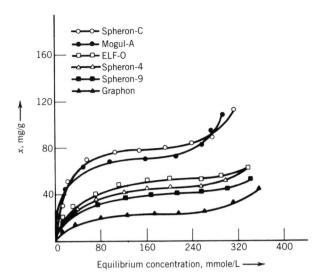

FIGURE 9-5. Example of data for the BET isotherm: adsorption of phenol on different carbon blacks (Puri, 1981).

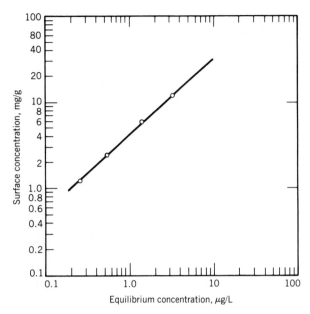

FIGURE 9-6. Freundlich isotherm: adsorption of geosmin on activated carbon (Herzing et al., 1977).

with increasing free energy, one can demonstrate that the isotherm will have the form

$$q = aC^{1/n}$$

where a is a constant. The log–log plot of q against C for this equation is linear. Here the surface concentration of adsorbate does *not* approach a saturating value as C increases, since there are always surface sites with higher free energies of adsorption to fill. The Freundlich isotherm is very widely used to fit observed data empirically even when there is no basis for its underlying assumptions. Figure 9-6 is an example of the use of log–log paper to present linearized data for the adsorption of the odor-causing compound geosmin on activated carbon.

Another approach, which takes into account the heterogeneity of adsorption sites, is the *Polanyi adsorption potential theory* (Polanyi, 1916, 1920; Manes and Hofer, 1969). This model views the region near the surface as a potential energy field caused by London–van der Waals forces. The net potential energy at a point above the surface, ε, is equal to the free energy released in bringing a mole of adsorbate in from solution to the point. The condition for condensation of adsorbate at any point is

$$\varepsilon \geq RT \ln \frac{C_s}{C}$$

where C_s is the saturation concentration of adsorbate; the term on the right is the free energy re-

quired to condense a mole of adsorbate from solution to solid or liquid phase.

The volume adsorbed is equal to the volume below the equipotential surface with this value of free energy. This means that the volume adsorbed is a function of $RT \ln(C_s/C)$ for a given adsorbate and adsorbent. Empirical observations further suggest that for a given adsorbent the energy potential for a variety of adsorbates is proportional to the molar volume of the adsorbate, V. When volume adsorbed is plotted against $(T/V)\log(C_s/C)$, a "correlation curve" emerges that may be used to characterize a single adsorbent for a variety of adsorbates. An example of this is shown in Figure 9-7. Isotherms can be calculated from such a curve.

Multicomponent Adsorption

In water treatment the ideal case of one adsorbate being removed onto an adsorbent is seldom encountered; the objective of adsorption in most real systems is to remove several adsorbates. This complicates both the theoretical picture of equilibrium among adsorbates and adsorbent and the ability of the engineer to apply the theory to practice.

The Langmuir model may be generalized from single- to multicomponent adsorption (Butler and Ockrent, 1930). The assumptions for specific sites, reversible adsorption, and homogeneous free ener-

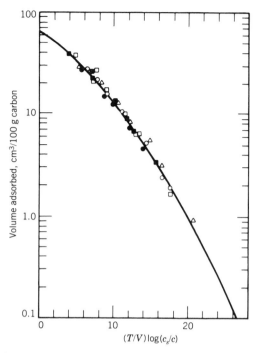

FIGURE 9-7. Correlation curve for adsorption of hydrocarbons on activated carbon: acetamide (○), acetone oxime (□), acrylamide (◯), triazole (●), urethane (△). (Manes, 1980).

gies of adsorption remain the same as for the case of a single adsorbate but are now applied to several adsorbates so that the mass of adsorbate i is given by

$$\frac{q_i}{Q} = \frac{b_i C_i}{1 + \Sigma_{i=1}^{n} b_i C_i}$$

Using this equation, one can in theory estimate the equilibrium capacity of an adsorbent for a complex mixture of compounds from the constants determined from a single solute. It does not allow for changes in free energy due to increased surface coverage or chemical interaction among different adsorbates.

Similarly, the Polanyi potential theory can be extended to multicomponent systems (Rosene and Manes, 1976) by restating the condition for condensation of each adsorbate:

$$\varepsilon_i \geq RT \ln \frac{C_s}{C}$$

in which C_s now represents the saturated concentration of adsorbate in the presence of all other adsorbates at equilibrium. This model has an advantage

over the Langmuir model in that interactions between different adsorbates on the surface are accounted for in the term C_s. One can in theory calculate the order and degree to which adsorbates with different affinities for the adsorbent will be competitively adsorbed.

Because one generally lacks full information on saturation concentrations for the variety of compounds competing for sorption, one may simplify the analysis of adsorption data using the net adsorption energy concept (McGuire and Suffett, 1980). *Net adsorption energy* is defined as the energy associated with the sorption of an adsorbate that displaces solvent from a surface phase:

$$E_T^A = E_{js}^A - E_{is}^A - E_{ji}'$$

where E_{js}^A = energy of affinity of solute with surface
E_{is}^A = energy of affinity of solvent with surface
E_{ji}' = energy of affinity of solute with solvent

The amount of solute adsorbed at equilibrium with a fixed bulk concentration correlates strongly with net adsorption energy for a number of organic compounds (see Figure 9-8). This allows one to predict isotherms for many compounds on the basis of their net adsorption energy and to rank them qualitatively for competitive adsorption.

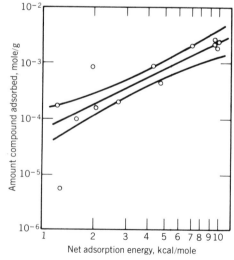

FIGURE 9-8. Amount of organic compound adsorbed vs. net adsorption energy (McGuire and Suffet, 1980).

KINETICS OF ADSORPTION

Although thermodynamic laws specify an equilibrium state between adsorbent and adsorbates, the removal of compounds in water treatment is often determined by the rate of adsorption during contact with adsorbent. The process of adsorption can be categorized as a set of sequential steps with individual rate laws, as described in Chapter 5.

The first step is transport of solute from bulk solution phase to the boundary layer or surface film surrounding the adsorbent particle. Advection and turbulent mixing control bulk transport (see, e.g., Weber, 1981a).

The second step is transport of solute across the boundary layer to the exterior surface of the adsorbent particle. Molecular diffusion or eddy diffusion controls transport across the film or boundary layer.

For adsorbents used in water treatment, most of the active surface area occurs in pores within adsorbent particles. The third kinetic step is diffusion of solute within these pores, from the exterior of the particle to the interior surfaces of the particle. Similarly, solute may be transported along surfaces of pore walls. Molecular or eddy diffusion controls the rate of internal diffusion.

The final step is the physical or chemical binding of adsorbate to the internal surface of the adsorbent. The rate of this step is controlled by chemical kinetics at the molecular level.

The concept of the ''rate-controlling step'' is used to simplify the analysis of adsorption kinetics. The four consecutive steps may be viewed as a set of resistances in series against the mass transport of adsorbate from bulk solution to adsorbent. In most cases it can be shown that the resistance of one of the four steps is much larger than others, and for most purposes the other resistances can be neglected. When the rate-controlling step is known, the kinetics of adsorption can be simplified and characterized with respect to dependence on operating parameters such as temperature, bulk concentration, and so on.

In water treatment the steps of bulk transport and of chemical or physical bonding are generally rapid, and the overall rate of adsorption is controlled by either film diffusion or pore diffusion or both. As described in Chapter 5, the rate of mass transfer per unit area across a film by turbulent diffusion may be expressed as

$$N_A = k_f(C_{A_i} - C_A)$$

where k_f is the local mass transfer coefficient across the film, C_A is the bulk concentration of adsorbate outside the film, and C_{A_i} is the concentration of adsorbate at the inner interface of the film, assumed to be in equilibrium with the adsorbent surface. The mass transfer coefficient is related to the adsorbent and adsorbate and can be determined experimentally or, for external film diffusion, can be estimated from models (see, e.g., Chu et al., 1953; Keinath and Weber, 1968). Kinetic information can then be used to predict the ''breakthrough'' curve for a given adsorbate, which is the appearance of contaminant at increasing concentrations in the effluent from a GAC bed. Application of thermodynamic and kinetic theory to practical design and use of GAC is described in the last section of this chapter.

ADSORBENTS

In full-scale water treatment by adsorption, activated carbon is used almost exclusively. Other natural adsorbents are used in special processes. Activated alumina is widely used for drying gases and liquids, for removal of fluoride, and for neutralization of lubrication oils. Silica gel is used for drying of gases and separation of hydrocarbons. Petroleum fractions and vegetable oils and juices are refined using acid-treated clay or Fuller's earth. The synthetic aluminosilicates perform as molecular sieves, selectively adsorbing on the basis of molecular size and shape, due to their high porosity and high uniformity of pore size. The use of polymeric resins and carbonized resins specifically for improved removal of organic compounds during water treatment has been explored recently (Weber and van Vliet, 1980; Wood and DeMarco, 1980).

Properties of Activated Carbon

The physical properties of activated carbon that relate to its performance depend on whether the carbon is used in powdered form in a mixed reactor or in granular form in a bed configuration.

For PAC the two important physical properties are filterability and bulk density. If powdered carbon passes through a filter, it contaminates the product water; the ability of the carbon to be removed by filtration is controlled by particle shape and size distribution. Bulk density is the mass of carbon per unit volume. The mass is proportional to the adsorptive capacity for a given carbon, so higher bulk density gives higher removal of adsorbate per volume of adsorbent. Also, the length of a filter run is related to the volume of carbon accumulated on the filter and will be longer for higher bulk density.

For GAC the important physical properties are hardness and particle size. Much of the operating cost of GAC results from losses by attrition during handling and regeneration. Losses are smaller for harder carbons. Similarly, the friability of the carbon used in filter beds controls the rate with which particles are broken down in size, leading to short filter runs and loss during backwashing. The particle size controls bulk transport because smaller size increases availability of macropores to the bulk phase as the external surface area per unit mass increases. Particle size also influences head loss across a bed of GAC as it does through other porous media.

The adsorptive properties of activated carbon have a strong effect on the rate and capacity for adsorption and must be considered in the choice of carbon and design. These properties include specific surface area, pore size distribution, and the chemical nature of the surface.

The total specific surface area per unit mass of a carbon can be determined by the BET method in which the volume of a monolayer of nitrogen atoms covering a surface is measured (Brunauer et al., 1938). However, the effective area available to a given adsorbate is determined by the minimum pore size the molecule can pass, so the distribution of pore sizes controls adsorptive capacity for different adsorbates. A plot of pore volume against pore diameter can be measured by mercury penetration experiments (Ritter and Drake, 1945), and from this a distribution of pore diameters may be calculated (see Figure 9-9). The pore size distribution changes as adsorptive capacity is used up and pores are clogged with adsorbate. The effective surface area of GAC has an upper limit of about 1500 m^2/g because any further increase during activation yields pores smaller than 1 nm, which are inaccessible to most adsorbates.

The surface chemistry of a carbon influences the rate and capacity of adsorption because of the specific interaction between surface and adsorbates. Activated carbon is basically a set of stacked parallel graphitelike planes of condensed carbon rings. Functional groups on the carbon can strongly influence chemical affinity for sorption of classes of adsorbates; these groups include carboxylic, phenolic,

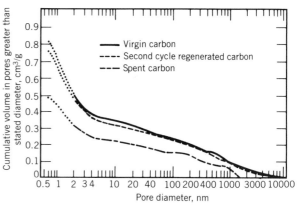

FIGURE 9-9. Pore size distribution of virgin, spent, and regenerated carbons (Weber and van Vliet, 1980).

hydroxyl, carbonyl, and peroxide groups (Ishizaki, 1973). Some of these groups demonstrate specific affinities for sorption of organics (Mattson et al., 1969).

Both the pore size distribution and chemical surface activity of carbons are strongly influenced by their origin. Among starting materials used are bituminous coal, lignite, petroleum coke (the three most common), bone char, coconut shells, and wood char (Abram, 1974). Pores are developed during activation in part by burning away carbon layers, the thickness of which depends on the structure of the starting material. For example, coconut shell–based carbons have higher density and finer pore size distribution, which make these carbons useful for adsorption of very small molecules as in gas-purifying applications rather than water treatment (Mattson and Mark, 1971).

Similarly, the parameters of the activation process influence adsorptive properties. There are three steps: starting material is dehydrated, heated in the absence of air to carbonize it, and then activated by oxidation to remove noncarbon impurities. The activation step produces the pore structure. It is accomplished either by heating to 200–1000°C in the presence of steam, carbon dioxide, or air or by wet chemical treatment at lower temperatures, with exposure to agents such as concentrated phosphoric acid, potassium hydroxide, or zinc chloride. The method and temperature of activation strongly influence pore size distribution and chemical properties. Table 9-1 and Figure 9-10 show pore size distributions and properties for a few commercially available activated carbons.

TABLE 9-1. Characteristics of Selected Activated Carbons

Activated Carbon (Mesh Size)	Manufacturer	Raw Material	Surface Area (m²/g)		Pore Volume (cm³/g)	
			BET	Macro-pore	Macro-pore	Total
Filtrasorb 400 (<14)	Calgon Corp.	Bituminous coal	1228	366	0.625	1.108
Filtrasorb 400 (20 × 30)	Calgon Corp.	Bituminous coal	1075	309	0.643	1.071
Filtrasorb 400 (40 × 60)	Calgon Corp.	Bituminous coal	1155	433	0.847	1.235
Hydrodarco 3000 (20 × 40)	ICI America	Lignite coal	575	99	0.787	0.975
Witcarb 940 (14 × 20)	Witco Chemical Corp.	Petroleum-based coke	950	106	0.208	0.599
Nuchar WV-B (20 × 35)	Westvaco	Bituminous coal	1422	778	1.290	1.865
Nuchar WV-DC (20 × 35)	Westvaco	Wood-based coal	1115	621	1.230	1.764
Nuchar WV-G (20 × 40)	Westvaco	Bituminous coal	1020	238	0.398	0.814
Nuchar WV-H (8 × 16)	Westvaco	Bituminous coal	910	133	0.251	0.610
Nuchar WV-L (20 × 30)	Westvaco	Bituminous coal	976	188	0.420	0.818
Nuchar WV-W (20 × 40)	Westvaco	Bituminous coal	861	154	0.281	0.612

SOURCE: Lee and Snoeyink (1980).

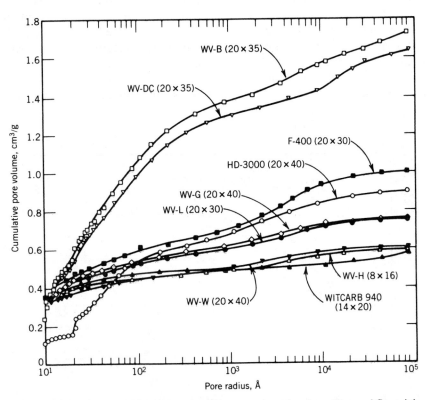

FIGURE 9-10. Pore size distributions for different activated carbons (Lee and Snoeyink, 1980).

ADSORBATES

Historically, adsorption in water treatment has been applied for removal of taste and odor, while present concerns are directed toward synthetic and natural organic compounds. This section briefly describes four classes of adsorbates, organic and inorganic compounds and chlorine and viruses, and also examines the properties of adsorbates that affect their removal by adsorption on activated carbon.

Organic Compounds

Organic contaminants can be subdivided into four categories of concern: naturally occurring taste- and odor-producing compounds, synthetic compounds of industrial or agricultural origin that cause taste and odor or have potentially adverse health effects, naturally occurring precursors of disinfection by-products, and disinfection by-products themselves.

Taste and Odor. Biologically derived earthy-musty odors in water supplies are a widespread problem. Geosmin and 2-methylisoborneol (MIB) have been identified as major causative agents of such odors (see Chapter 17). Herzing et al. (1977) observed that geosmin and MIB are strongly adsorbed by activated carbon but noted a significant reduction in adsorption of geosmin and MIB in the presence of other naturally occurring organic matter such as humic acids. Figure 9-11 shows break-

through curves for MIB over time in a GAC column test; these curves all level off before reaching C/C_0 = 1, where C/C_0 is the ratio of effluent to influent concentration. This suggests that a portion of the carbon capacity is being used very slowly and that complete saturation may not occur for a long period of time. GAC beds have been observed to produce odor-free water for several years at conventional contact times without replacing the carbon (Robeck, 1975).

Taste- and odor-producing compounds of industrial origins are also readily adsorbed by activated carbon. The chlorophenols are adsorbed strongly and will not break through conventional GAC columns for years (Snoeyink et al., 1977). Other taste and odor compounds of industrial origin, however, can saturate the capacity of a column more rapidly than naturally occurring taste and odor compounds. Granular carbon beds for the removal of such compounds at Nitro, West Virginia, could produce odor-free water for less than a month before breakthrough at influent concentrations ranging from 10 to 100 g/L.

Synthetic Organics. Concern about synthetic organic chemicals in drinking water has motivated interest in adsorption as a treatment process for removal of toxic and potentially carcinogenic compounds present in minute, but significant, concentrations. Few other processes can effect removal at such low levels. Organic compounds are discussed in detail in Chapter 16; this discussion

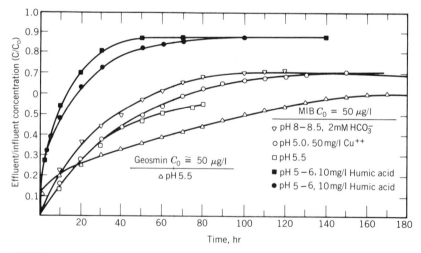

FIGURE 9-11. Column breakthrough curves of MIB under different conditions (Herzing et al., 1977).

focuses on three properties of adsorbates affecting removal by activated carbon: solubility, molecular size, and polarity.

In general, the less soluble an organic compound is in water, the better it is adsorbed from solution; this is Lundelius's rule (Lundelius, 1920; Weber, 1972). Similarly, the less polar a compound is, the better it is adsorbed from polar aqueous solution to nonpolar carbon. Both of these qualitative rules result from the need of an organic molecule to break adsorbate–solvent bonds in order to adsorb. These bonds are generally stronger when the adsorbate is more soluble in water or has dipole–dipole interactions with water. Larger organic molecules tend to be less soluble and also tend to diffuse more slowly through adsorbent pores and therefore are more poorly adsorbed than smaller molecules. They cannot penetrate the smallest pores.

Other factors affecting sorption of organics relate to specific chemical affinities between functional groups on the adsorbate and on the adsorbent. Table 9-2 shows classes of organic compounds that are easily or poorly removed by GAC in general. Most of the compounds that are poorly removed are highly soluble and have low molecular weights;

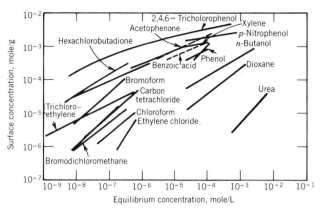

FIGURE 9-12. Freundlich isotherms for various organic compounds (Suffet, 1980).

TABLE 9-2. Classes of Organics Readily Adsorbed and Poorly Adsorbed onto Activated Carbon

Readily Adsorbed Organics

Aromatic solvents
  Benzene, toluene, nitrobenzenes, etc.
Chlorinated aromatics
  PCBs, chlorobenzenes, chloronaphthalene
Phenol and chlorophenols
Polynuclear aromatics
  Acenaphthene, benzopyrenes, etc.
Pesticides and herbicides
  DDT, aldrin, chlordane, BHCs, heptachlor, etc.
Chlorinated nonaromatics
  Carbon tetrachloride, chloroalkyl ethers,
    hexachlorobutadiene, etc.
High-MW hydrocarbons
  Dyes, gasoline, amines, humics

Poorly Adsorbed Organics

Alcohols
Low-MW ketones, acids, and aldehydes
Sugars and starches
Very-High-MW or colloidal organics
Low-MW aliphatics

SOURCE: Froelich (1978).

others have very large molecular weights. In the case of organic acids and bases, adsorption is strongly dependent on pH because of the preference for removal of neutral species from aqueous solution.

Table 9-3 summarizes adsorption data for a variety of synthetic organics of concern for health, on different carbons. Figure 9-12 shows some isotherms graphically. All these compounds are adsorbed on GAC to different degrees; substituted phenols (common odor-causing industrial contaminants) are generally more readily sorbed than halogenated methanes and ethylenes, for example. These latter volatile compounds are industrial solvents often present at higher concentrations and may be poorly removed by GAC or require frequent regeneration of beds. Polycyclic aromatic hydrocarbons and nitrosamines are readily sorbed. Because information about GAC removal of specific organic compounds is generally not available, the practicing engineer must generally rely on batch or column tests for data with which to predict removal efficiency in a full-scale bed. For general organics removal surrogate parameters such as TOC, total organic halogen, or chemical oxygen demand are used to model adsorption (Roberts and Summers, 1982).

Humic Substances as Halomethane Precursors.
Chlorination of most natural waters during treatment produces halogenated by-products such as THMs, which may have chronic health risks and carcinogenic effects (Rook, 1974, 1975; Trussell and Umphres, 1978). The natural organic precursors for these reactions are humic materials, which are polydispersed large organics, often color forming, with

TABLE 9-3. Selected Values of the Freundlich Isotherm Parameters for Adsorption of Various Organic Compounds

Priority Pollutant	k_F	$1/n$	Priority Pollutant	k_F	$1/n$
Acenaphthene	190	0.36	2,4-Dimethylphenol	78	0.44
Acenaphthylene	115	0.37	Dimethyl phthalate	97	0.41
Acrolein	1.2	0.65	4,6-Dinitro-o-cresol	237	0.32
Acrylonitrile	1.4	0.51	2,4-Dinitrophenol	160	0.37
Aldrin	651	0.92	2,4-Dinitrotoluene	146	0.31
Anthracene	376	0.70	2,6-Dinitrotoluene	145	0.32
Benzene	1.0	1.6	1,2-Diphenylhydrazine	16,000	2.0
Benzidine-dihydrochloride	110	0.35	Alpha-endosulfan	194	0.50
3,4,-Benzofluoranthene	57.0	0.37	Beta-endosulfan	615	0.83
Benzo[k]fluoroanthene	181	0.57	Endosulfan sulfate	686	0.81
Benzo[ghi]perylene	10.7	0.37	Endrin	666	0.80
Benzo[a]pyrene	33.6	0.44	Ethylbenzene	53	0.79
Alpha-BHC	303	0.43	bis-(2-Ethylhexyl)phthalate	11,300	1.5
Beta-BHC	220	0.49	Fluoranthene	664	0.61
Gamma-BHC (lindane)	256	0.49	Fluorene	330	0.28
Bromoform	19.6	0.52	Heptachlor	1220	0.95
4-Bromophenyl phenyl ether	144	0.68	Heptachlor epoxide	1038	0.70
Butylbenzyl phthalate	1520	1.26	Hexachlorobenzene	450	0.60
N-Butyl phthalate	220	0.45	Hexachlorobutadiene	258	0.45
Carbon tetrachloride	11.1	0.83	Hexachlorocyclopentadiene	370	0.17
Chlorobenzene	91	0.99	Hexachloroethane	96.5	0.38
Chlordane	245	0.38	Isophorone	32	0.39
Chloroethane	0.59	0.95	Methylene chloride	1.30	0.16
bis-(2-Chloroethoxy)methane	11	0.65	4,4'-Methylene-bis-		
bis-(2-Chloroethyl)ether	0.086	1.84	(2-chloroaniline)	190	0.64
2-Chloroethyl vinyl ether	3.9	0.80	Naphthalene	132	0.42
Chloroform	2.6	0.73	Beta-naphthylamine	150	0.30
bis-(2-Chloroisopropyl)ether	24	0.57	Nitrobenzene	68	0.43
Parachlorometa cresol	122	0.29	2-Nitrophenol	101	0.26
2-Chloronaphthalene	280	0.46	4-Nitrophenol	80.2	0.17
2-Chlorophenol	51.0	0.41	N-Nitrosodiphenylamine	220	0.37
4-Chlorophenyl phenyl ether	111	0.26	N-Nitrosodi-n-propylamine	24.4	0.26
DDE	232	0.37	PCB-1221	242	0.70
DDT	322	0.50	PCB-1232	630	0.73
Dibenzo[a,h]anthracene	69.3	0.75	PCB-1016		
1,2-Dichlorobenzene	129	0.43	PCB-1254		
1,3-Dichlorobenzene	118	0.45	Pentachlorophenol	260	0.39
1,4-Dichlorobenzene	121	0.47	Phenanthrene	215	0.44
3,3-Dichlorobenzidine	300	0.20	Phenol	21	0.54
Dichlorobromomethane	7.9	0.61	1,1,2,2-Tetrachloroethane	10.6	0.37
1,1-Dichloroethane	1.79	0.53	Tetrachloroethane	50.8	0.56
1,2-Dichloroethane	3.57	0.83	Toluene	26.1	0.44
1,2-trans-Dichloroethane	3.05	0.51	1,2,4-Trichlorobenzene	157	0.31
1,1-Dichloroethylene	4.91	0.54	1,1,1-Trichloroethane	2.48	0.34
2,4-Dichlorophenol	147	0.35	1,1,2-Trichloroethane	5.81	0.60
1,2-Dichloropropane	5.86	0.60	Trichloroethene	28.0	0.62
Dieldrin	606	0.51	Trichlorofluoromethane	5.6	0.24
Diethyl phthalate	110	0.27	2,4,6-Trichlorophenol	219	0.29
N-Dimethylnitrosamine	6.8×10^{-5}	6.6			

SOURCE: Weber (1981), and Dobbs and Cohen (1980). Refer to references for details in parameter estimation.

molecular weights ranging from a few hundreds to hundreds of thousands. Usually, the majority of organic carbon in natural waters is present as humic substances that represent a variety of molecules with different functional groups. Not surprisingly, the adsorption properties of poorly defined humic substances on GAC vary greatly and depend on the source of the material (McCreary and Snoeyink, 1981). The pore size distribution strongly influences adsorption of these large molecules by carbons; capacity decreases for molecules of larger molecular weight. Also, as pH is lowered, capacity increases because particle sizes increase and the acidic groups on the substance take on their neutral form. In general, adsorption of humic substances by GAC is poor: incomplete removal and short bed lifetimes mean limited ability to control THMs by removing their precursors.

Another important role of humic materials is to compete with other organics for adsorption sites. Competitive adsorption can result in displacement of previously sorbed compounds from a GAC bed when different compounds are introduced. In some cases the effluent concentration of the original compound will exceed influent concentration as it desorbs (see Figure 9-13). Appearance of such concentrated slugs of hazardous compounds by displacement by humic materials is not believed to

be likely in actual practice (Snoeyink et al., 1977; Love and Symons, 1978). However, humic substances can decrease capacity for removal of important trace organics and cause shorter bed times.

Disinfection By-Products. The principal by-products of concern for health in chlorination of natural waters are the THMs: chloroform ($CHCl_3$), bromodichloromethane ($CHBrCl_2$), dibromochloromethane ($CHClBr_2$), and bromoform ($CHBr_3$). Other halogenated organic by-products, such as chlorohydrins from naturally occurring fatty acids and sterols, are still poorly studied. Laboratory work indicates that GAC can effectively remove THMs down to a few micrograms per liter at equilibrium under certain conditions (Weber et al., 1977; Youssefi and Faust, 1980). The capacity for adsorption increases with increasing bromine content. However, practical application of GAC in pilot studies has shown that removal efficiency can deteriorate from near 100 to less than 10 percent within a few weeks (Wood and DeMarco, 1979; McCarty et al., 1979; Hyde, 1980; Roberts and Summers, 1982). In general, chloroform is the first THM to break through a column. There is evidence for displacement of THMs by competitive adsorption by humic materials and other chlorinated organics (Varma et al., 1980). Regenerated carbon can have much

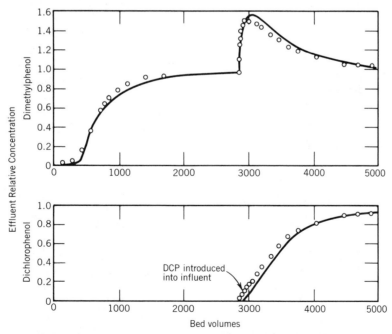

FIGURE 9-13. Displacement of adsorbed dimethylphenol from GAC by addition of dichlorophenol (Thacker, 1981).

poorer removal efficiencies for THMs than fresh carbon.

Synthetic resins have been proposed as alternative adsorbents that could specifically remove THMs. Some pilot studies have shown that carbonized polymeric resins can remove THMs and other chlorinated organics more effectively and for longer periods of time than GAC (Wood and DeMarco, 1980b; Symons et al., 1979).

Inorganic Compounds

Some inorganic compounds that have adverse effects on health can be removed by adsorption onto activated carbon. Trace metals such as mercury, arsenic, and lead can be removed onto GAC (Chow and David, 1977; Hamdy and Wheeler, 1977; Chen and Gupta, 1978). Table 9-4 shows a ranking of metals by potential for adsorption (Sigworth and Smith, 1972; Sorg and Logsdon, 1980). Most common cations and anions found in natural waters are not adsorbed, such as calcium, sodium, orthophosphate, nitrate, and halides. An exception to this is fluoride, which can be removed onto activated alumina as well as carbon.

Metals can react at oxygenated functional groups on carbon surfaces, displacing hydrogen ions or common cations in ion exchange. Carbon surfaces can also precipitate metal salts by nucleation or coagulation. Reduced metals in the original carbon structure can also react with metals in ionic solution to reduce them to a deposited form; similarly, metals can be catalytically oxidized.

Chlorine

Activated carbon can be used for dechlorination of water. Free chlorine (as HOCl) reacts with the carbon surface to form H^+, Cl^-, and a surface oxide group (Magee, 1956). This ultimately reduces the adsorption capacity of the carbon by loss of functional groups or blockage of pores, although the surface oxides may decompose to CO_2 or CO (Snoeyink and Suidan, 1975). This latter reaction also occurs during regeneration and accounts for some weight loss during that process.

Monochloramine (NH_2Cl) reacts with activated carbon but more slowly than HOCl. The carbon acts both as reducing agent and as catalyst in producing NH_3, Cl^-, and $N_2(g)$:

$$NH_2Cl + H_2O + C^* \rightarrow NH_3 + H^+ + Cl^- + C^*O$$

$$2NH_2Cl + C^*O \rightarrow N_2 + H_2O + 2Cl^- + C^*$$

where C^* is a surface carbon and C^*O is a surface oxide (Bauer and Snoeyink, 1973; Scaramelli and DiGiano, 1977). Dichloramine reacts more rapidly than HOCl with carbon to form $N_2(g)$ and Cl^- (Kim et al., 1978). The kinetics of reactions between carbon and free and combined chlorine have been extensively modeled for bed reactors (Suidan et al., 1980).

Free chlorine also reacts with organic compounds sorbed to GAC, and these reactions can be catalyzed by the presence of the carbon surface (McCreary, 1980). Products include chlorinated dihydroxybenzenes (McCreary and Snoeyink, 1981). High concentrations of free chlorine react directly with activated carbon to produce high-molecular-weight, color-forming organics as well as chloroform, trichloroethane, and chlorinated aromatics (Snoeyink et al., 1981). The degree to which these may result during water treatment practice is believed to be insignificant.

Viruses

Adsorption of viruses onto activated carbon has been studied as a process for advanced treatment

---

TABLE 9-4. Metals and Inorganic Compounds Classified by Potential for Adsorption on Activated Carbon

Metals of high adsorption potential
  Antimony, arsenic, bismuth, chromium, tin
Metals of good adsorption potential
  Silver, mercury, cobalt, zirconium
Metals of fair adsorption potential
  Lead, nickel, titanium, vanadium, iron
Metals of low adsorption potential
  Copper, cadmium, zinc, barium, selenium, molybdenum, manganese, tungsten, radium

Other Inorganic Compounds	Adsorption Potential
Nitrate	Low
Phosphate	Low
Chlorine	High
Bromine	High
Iodine	High
Chloride	Low
Bromide	Low
Iodide	Low
Fluoride	High

SOURCE: Sigworth and Smith (1972).

and possible reclamation of wastewater. Specific functional groups on carbon such as carboxyl and lactone groups are believed to act as sites for adsorption of viruses (Cookson, 1970). Degree of adsorption is strongly dependent on pH, with strongest retention at low pH, so electrostatic attraction between negatively charged groups on the carbon and positively charged groups on viruses is believed to be important (Gerba et al., 1975). Bench and pilot studies suggest that virus removal onto GAC is inconsistent and hard to control (Healy, 1975; Guy and McIver, 1977).

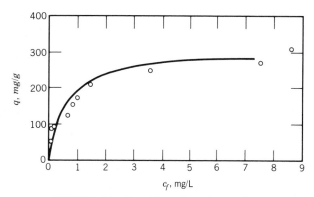

FIGURE 9-14. Benzidine adsorption isotherm.

EXAMPLE 9-1

Problem

An engineer performed the following experiment to study the adsorption of benzidine on Filtrasorb 400

granular carbon made by Calgon Corp. Several bottles containing a given concentration of benzidine were prepared; to each a different mass of carbon was added. The suspension was allowed to equilibrate for 2 hr; then the carbon was filtered out and the concentration of benzidine was measured; the results are shown in Table 9-5.

Plot the adsorption isotherm for this experiment. Using a linearized form of the Langmuir equation, replot the data and estimate the parameters of the Langmuir isotherm. Using log–log paper, replot the data and estimate the parameters of the Freundlich isotherm.

Solution

Let X = benzidine adsorbed (in mg/L) = initial concentration minus final concentration = $C_i - C_f$. The mass of benzidine adsorbed per unit mass of carbon is $q = X/M$. Table 9-6 and Figure 9-14 show the isotherm. When C_f/q is plotted against C_f, a linear fit to the data yields the parameters for the

TABLE 9-5. Adsorption of Benzidine on Filtrasorb 400

Carbon Dose, M (mg/L)	Initial Concentration, C_i (mg/L)	Final Concentration, C_f (mg/L)
3.72	9.81	8.63
8.42	9.81	7.52
24.5	9.81	3.55
39.8	9.81	1.41
1.08	1.17	0.98
2.12	1.17	0.84
4.05	1.17	0.66
10.85	1.17	0.17
11.9	1.17	0.11
21.1	1.17	0.03

TABLE 9-6. Benzidine Adsorption Isotherm Data

Final Concentration, C_f (mg/L)	Change in Concentration, X (mg/L)	Mass Adsorbed, $q = X/M$ (mg/g)	C_f/q (g/L)	$1/q$ (g/mg)	$1/C_f$ (L/mg)
8.63	1.18	317	0.0272	0.00315	0.116
7.52	2.29	272	0.0276	0.00368	0.133
3.55	6.26	256	0.0139	0.00391	0.282
1.41	8.40	212	0.0067	0.00472	0.709
0.98	0.19	176	0.0056	0.00568	1.02
0.84	0.33	156	0.0054	0.00641	1.19
0.66	0.51	126	0.0052	0.00794	1.52
0.17	1.00	92	0.0018	0.0109	5.88
0.11	1.06	89	0.0012	0.0112	9.09
0.03	1.14	54	0.0006	0.0185	333

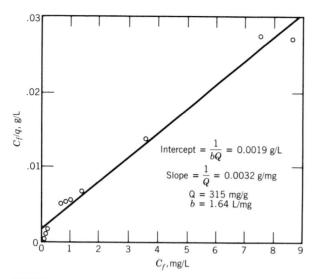

FIGURE 9-15. Langmuir isotherm in linearized form: C_f/q vs. C_f.

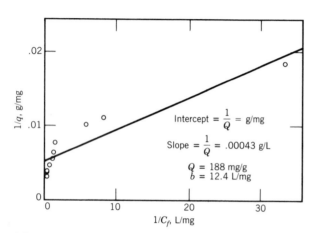

FIGURE 9-16. Langmuir isotherm in linearized form = $1/q$ vs. $1/C_f$.

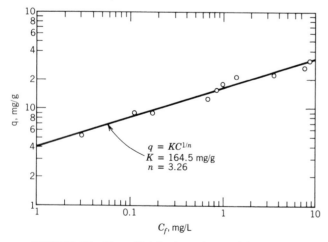

FIGURE 9-17. Freundlich isotherm for benzidine adsorption.

Langmuir equation (Figure 9-15): $b = 1.64$ L/mg, $Q = 315$ mg/g. (One may alternatively plot $1/q$ against $1/C_f$ to obtain $b = 12.4$ L/mg, $Q = 188$ mg/g; Figure 9-16 illustrates how this linearized form is more sensitive to data from low values of C_f). On a log–log plot (Figure 9-17), the fit of data to a straight line yields the parameters for the Freundlich isotherm: $K = 164.5$ mg/g, $n = 3.26$. □

CONTINUOUS CONTACTOR ADSORPTION PROCESSES

Process Description

A downflow contactor system is shown schematically in Figure 9-18. The adsorbent is a packed granular media fixed in a stationary position. Process water is applied directly to one end and forced through the bed typically by gravity, pressure, or pumping energy. For fresh adsorbent, the adsorbate is usually readily removed in the region of the bed closest to the influent. Adsorbate not removed immediately is likely to be adsorbed as it passes through successive levels of the bed. Depending on the characteristics of the adsorbate and adsorbent as well as the physical and hydraulic nature of the system, all of the adsorbate may be removed before the process water appears in the effluent. The region of the bed where adsorption and removal of the

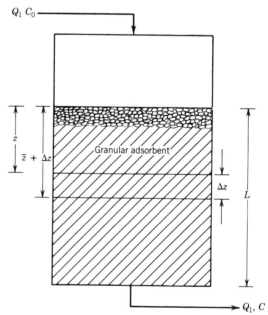

FIGURE 9-18. Detail of adsorption beds.

adsorbate takes place is referred to as the mass transfer zone or as the adsorption zone or zone of active transport. As the initially contacted adsorbent is exhausted, the mass transfer zone moves further into the bed in a wavelike manner; accordingly, the mass transfer zone is sometimes referred to as the adsorption wave. Finally, as the entire bed becomes exhausted and the mass transfer zone approaches the end of the bed, increasingly higher concentrations of the adsorbate are observed until the effluent concentration equals influent concentration and no removal occurs. This phenomenon is termed *breakthrough*. In practice, the bed would be replaced with fresh adsorbent or switched with an alternate unit once a predetermined treatment goal of the adsorbate is attained.

Process Analysis: Fixed-Bed Adsorber with Granular Media

Weber and Crittenden and co-workers (Crittenden, 1976; Crittenden and Weber, 1978a,b; Crittenden et al., 1980) have developed the homogeneous surface diffusion model (HSDM), a mathematical description of the kinetics of adsorbate removal. The following assumptions are made in the fixed-bed model:

1. There is no radial dispersion or channeling (in other words, concentration gradients exist only in the axial direction).

2. Surface diffusion flux is much greater than pore diffusion flux as an intraparticle mass transfer mechanism (see below). Therefore, pore diffusion flux is neglected. In addition, the adsorbent is assumed to be homogeneous and the surface diffusion flux can be described by Fick's law: Flux $= D_s(C/x)$.

3. The liquid phase diffusion flux can be described by the linear driving force approximation, using estimates for the film transfer coefficient K_f.

4. The adsorbent is fixed in the adsorber (backwashing is not considered).

5. Adsorption equilibria can be described by the Freundlich isotherm.

6. Plug flow exists within the bed. This is valid when the mass transfer zone is longer than 30 adsorbent particle diameters.

As shown in Figure 9-19, the adsorbate can diffuse by two mechanisms within the adsorbent, pore and surface diffusion. For pore diffusion, the adsorbate is transported within the pore fluid. For surface diffusion, the adsorbate continues to move along the surface of the adsorbent to available adsorption sites as long as it has enough energy to leave its present site. Investigations (Brecher et al., 1967; Furusawa and Smith, 1973; Komiyama and Smith, 1974; Crittenden, 1976) have demonstrated that surface diffusion is the dominant mechanism, so the contribution of pore diffusion is neglected. The assumption of adsorbent homogeneity is a simplifica-

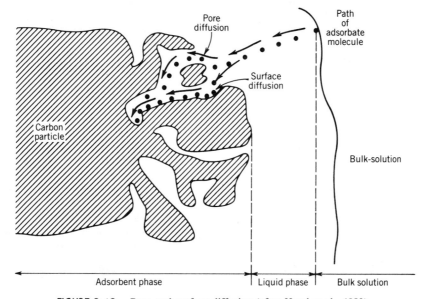

FIGURE 9-19.   Pore and surface diffusion (after Hand et al., 1982).

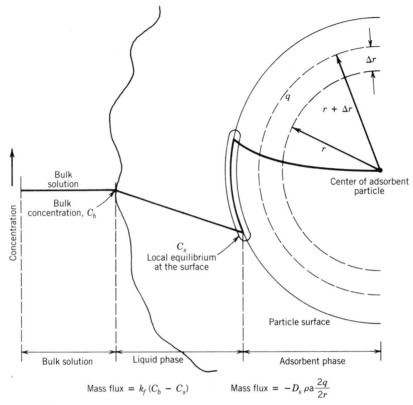

Mass flux $= k_f (C_b - C_s)$ Mass flux $= -D_s \rho a \dfrac{2q}{2r}$

FIGURE 9-20.   Mass balance for adsorbate (after Hand, et al., 1982).

tion of the complex inner pore and surface structure.

Two of the three basic differential equations of the model are mass balances for the adsorbate in the axial direction, z, shown in Figure 9-18, and in the radial direction, r, shown in Figure 9-20. The liquid phase mass balance describes the spatial and temporal variation of the adsorbate concentration in solution. The solid phase mass balance describes the removal rate of adsorbate from the liquid phase into the adsorbent at a given axial location in the bed. Dimensionless groups are defined to simplify solution of the equations and reduce the number of independent variables.

Mass throughput or dimensionless time is defined as

$$T = \frac{\text{rate of mass of adsorbate fed}}{\text{total mass of adsorbate at equilibrium}}$$

$$= \frac{QC_0 t}{Mq_e + \varepsilon V C_0}$$

where Q = influent flowrate
  C_0 = fluid phase concentration of adsorbate in influent
  t = elapsed time
  M = total mass of adsorbent in the bed
  q_e = adsorbent phase concentration at equilibrium with C_0 in fluid phase
  ε = ratio of void volume to total bed volume
  V = total bed volume.

The dimensionless solute distribution parameter D_g is defined as

$$D_g = \frac{\begin{array}{c}\text{mass of adsorbate in}\\ \text{solid phase at equilibrium}\end{array}}{\begin{array}{c}\text{mass of adsorbate in}\\ \text{liquid phase at equilibrium}\end{array}}$$

$$= \frac{\rho_a q_e (1 - \varepsilon)}{\varepsilon C_0}$$

where ρ_a = adsorbent density including pore volume.

The dimensionless Biot number, Bi, is defined as

$$\text{Bi} = \frac{\text{rate of liquid phase mass transfer}}{\substack{\text{mass of adsorbate in} \\ \text{liquid phase at equilibrium}}}$$

$$= \frac{(1 - \varepsilon)k_f R}{\varepsilon D_s D_g \phi}$$

where k_f = film transfer coefficient

D_s = surface diffusion coefficient

ϕ = sphericity (dimensionless ratio of the surface area of the equivalent volume sphere to the actual surface area of the particle)

R = particle radius.

The modified Stanton number, St, is a dimensionless measure of the bed length as compared to the length of the mass transfer zone in the case where liquid phase mass transfer resistance controls the adsorption rate:

$$St = \frac{k_f \tau (1 - \varepsilon)}{R \varepsilon \phi}$$

where τ = hydraulic residence time in the bed.

The surface diffusion modulus E_d is a dimensionless measure of bed length compared to the length of the mass transfer zone in the case where intraparticle diffusion controls adsorption rate:

$$E_d = \frac{D_s D_g \tau}{R^2} = \frac{St}{Bi}$$

Assuming the adsorbent phase, including the pore volume, is homogeneous solid, the surface diffusion flux J_s is

$$J_s = -D_s \rho_a \frac{\partial q}{\partial r}$$

Notice that D_s is a surface diffusion coefficient which is averaged over the pore tortuosity and constriction factors. In some cases D_s has been found to be dependent on adsorbate concentration (Neret-

nieks, 1976; Fritz et al., 1980). For such cases, this dependence is incorporated into an *effective* D_s.

A mass balance for the adsorbate in the solid phase system shown in Figure 9-20 is

| $\substack{\text{Rate of accu-} \\ \text{mulation of} \\ \text{absorbate} \\ \text{within the} \\ \text{spherical shell}}$ | = | $\substack{\text{rate of flow} \\ \text{of adsorbate} \\ \text{into spherical} \\ \text{shell at } r}$ | $-$ | $\substack{\text{rate of flow} \\ \text{of adsorbate} \\ \text{leaving spheri-} \\ \text{cal shell at} \\ r + \Delta r}$ |

or

$$\frac{\partial q}{\partial t} = \frac{D_s}{r^2} \frac{\partial}{\partial r} \left[r^2 \frac{\partial q}{\partial r} \right] \qquad (9\text{-}1)$$

where r = radial length of differential spherical shell

t = time.

Equation 9-1 requires one initial condition and two boundary conditions. The initial condition assumes there is no adsorbate in the adsorbent at $t = 0$:

$$q = 0 \qquad (0 \geq r \geq R, \, t = 0) \qquad (9\text{-}2)$$

By symmetry at the center of the particle, the first boundary condition is

$$\frac{\partial q}{\partial r} = 0 \qquad (r = 0, \, t \geq 0) \qquad (9\text{-}3)$$

The second boundary condition at the exterior surface of the adsorbent is derived by performing a mass balance on the spherical shell. Simply stated, the mass flux of adsorbate transferred through the liquid phase equals the mass flux of adsorbate transferred away by surface diffusion:

$$\frac{\partial q}{\partial r} = \frac{k_f}{\rho_a D_s \phi} C(t) - C_s(t) \quad (r = R, \, t) \quad (9\text{-}4)$$

Assuming the linear driving force approximation, the liquid phase mass flux J_1 can be written as $J_1 = k_f(C_b - C_s)$, where C_b = bulk fluid phase concentration of adsorbate.

A mass balance for the adsorbate in the liquid phase system defined in Figure 9-20 can be written as

$$
\begin{bmatrix}
\text{Rate of accumulation} \\
\text{of adsorbate} \\
\text{within the} \\
\text{longitudinal element}
\end{bmatrix}
=
\begin{bmatrix}
\text{rate of flow} \\
\text{of adsorbate} \\
\text{into longitudinal} \\
\text{shell by advection}
\end{bmatrix}
$$

$$
-
\begin{bmatrix}
\text{rate of flow} \\
\text{of adsorbate} \\
\text{out of longitudinal} \\
\text{shell by advection}
\end{bmatrix}
-
\begin{bmatrix}
\text{rate of removal} \\
\text{of adsorbate by} \\
\text{adsorbent}
\end{bmatrix}
$$

$$
-V \frac{\partial C}{\partial Z} = \frac{\partial C}{\partial t} + \frac{3k_f(1 - \varepsilon)}{R\phi\varepsilon}[C - C_s] \quad (9\text{-}5)
$$

where V = interstitial velocity

z = longitudinal dimension.

Note the term $3(1 - \varepsilon)/R\phi$ is the area available for liquid phase mass transfer per volume of bed.

Equation 9-5 requires one initial and one boundary condition. The initial condition assumes the adsorption wave is contained entirely within the bed at $t = 0$ to yield

$$
C = 0 \quad (t_0 < Z < L, \, t < \tau) \quad (9\text{-}6)
$$

The boundary condition assumes a constant influent adsorbate loading, stated as

$$
C = C_0(t) \quad (Z = 0, \, t \geq 0) \quad (9\text{-}7)
$$

To couple the solid and liquid phase mass balance equations, the surface concentration of adsorbate in the liquid phase $C_s(t)$ must be expressed in terms of the surface concentration of adsorbate in the solid phase $q(r = R, t)$. This equation is developed from the assumption of local adsorption equilibria adjacent to the exterior adsorbent surface, as described by the nonlinear Freundlich isotherm:

$$
q = KC_s(t)^{1/n} \quad (r = R, \, t) \quad (9\text{-}8)
$$

Process Model Solutions

Refer to Table 9-7 for a summary of the HSDM equations. The three equations contain three independent variables, mass throughput T, radial position r, and axial position z. Dependent variables are liquid phase concentration $C(z, T)$, liquid phase concentration at the exterior surface of the adsorbent $C_s(z, T)$, and solid phase concentration $q(r, z, T)$. Simultaneous solution of the system of equations results in a predictive model of fixed-

TABLE 9-7. Homogeneous Surface Diffusion Model Equations

Purpose	Equation Number
Solid phase mass balance	9-1
Initial condition	9-2
Boundary condition	9-3
Boundary condition	9-4
Liquid phase mass balance	9-5
Initial condition	9-6
Boundary condition	9-7
Freundlich isotherm coupling	9-8

bed concentration history profiles for a given set of physical and chemical properties. Unfortunately, the equations cannot be directly solved analytically. Solutions may be obtained using orthogonal collocation techniques (Crittenden et al., 1980; Thacker, 1981; Thacker et al., 1981). This numerical method reduces the system of partial differential equations to a set of ordinary differential equations which may be integrated. Recently analytical and user-oriented techniques that do not require computer software have been developed by Crittenden and co-workers (Hand et al., 1981). These solutions are available for most conditions encountered in water treatment adsorption process design.

REFERENCES

Abram, J. C., "The Characteristics of Activated Carbon," *Papers and Proceedings of the Water Research Association, Conference on Activated Carbon in Water Treatment*, Water Research Association, London (1974).

Adamson, A. W., *Physical Chemistry of Surfaces*, 2nd ed., Wiley-Interscience, New York (1967).

American Water Works Association Committee, "Measurement and Control of Organic Contaminants by Utilities," *JAWWA*, **69**, 267–271 (1977).

Bauer, R. C., and Snoeyink, V. L., "Reactions of Chloramines with Active Carbon," *JWPCF*, **45**, 2290–2301 (1973).

Baylis, J. R., "The Activated Carbons and Their Use in Removing Objectionable Tastes and Odors from Water," *JAWWA*, **21**, 787–814, (1929).

Brecher, L. E., Frantz, D. C., and Kostecki, J. A., "Combined Diffusion in Batch Adsorption Systems Displaying BET Isotherms: Part II," *AIChE Symp. Ser.*, **63**(74), 25–30 (1967).

Brunauer, S., Emmet, P. H., and Teller, E., "Adsorption of Gases in Multimolecular Layers," *J. Amer. Chem. Soc.* **60**, 309–319 (1938).

Butler, J. A. V., and Ockrent, C., "Studies in Electrocapillarity, III," *J. Phys. Chem.* **34**, 2841–2858 (1930).

Chen, K. Y., and Gupta, S. K., "Arsenic Removal by Adsorption," *JWPCF*, **50**, 493–506 (1978).

Chow, D. K., and David, M. M., "Compounds Resistant to Carbon Adsorption in Municipal Wastewater Treatment," *JAWWA*, **69**, 555–561 (1977).

Chu, J. C., Kalil, J., and Wetteroth, W. A., "Mass Transfer in a Fluidized Bed," *Chem. Eng. Prog.* **49**, 141–149 (1953).

Cookson, J. R., "Removal of Submicron Particles in Packed Beds," *Environ. Sci. Technol.*, **4**, 128–134 (1970).

Crittenden, J. C., "Mathematical Modeling of Adsorber Dynamics: Single Components and Multi-Components," Ph.D. Thesis, University of Michigan (1976).

Crittenden, J. C., and Weber, W. J., Jr., "Predictive Model for Design of Fixed-Bed Adsorbers: Parameter Estimation and Model Development," *J. Environ. Eng. Div.*, **104**, 185 (1978a).

Crittenden, J. C., and Weber, W. J., Jr., "Model for Design of Multicomponent Adsorption Systems," *J. Environ. Eng. Div.*, **104**, 1175 (1978b).

Crittenden, J. C., Wong, B. W. C., Thacker, W. E., Snoeyink, V. L., and Hinrichs, R. L., "Mathematical Modeling of Sequential Loading in Fixed-Bed Adsorbers," *JWPCF*, **52**, 2780–2795 (1980).

Croes, J. J. R., "The Filtration of Public Water Suppliers in America," *Eng. News Amer. Contr. J.* **10**, 277–281 (1883).

Dobbs, R. A., and Cohen, J. M., "Carbon Adsorption Isotherms for Toxic Organics," *U.S. EPA Report 600.880-023*, Cincinnati (1980).

Dostal, K. A., and Robeck, G., "Carbon Bed Design Criteria Study at Nitro, West Virginia," *JAWWA*, **57**, 663–674 (1965).

Eckenfelder, W. W., Jr. (ed.), *Application of Adsorption to Wastewater Treatment*, Enviro Press, Nashville, Tenn. (1981).

EPA, "National Interim Primary Drinking Water Regulations; Control of Organic Chemical Contaminants in Drinking Water," *Fed. Reg.*, **43**(28), 5755–5780 (1978).

EPA, "National Interim Primary Drinking Water Regulations; Control of Trihalomethanes in Drinking Water; Final Rule," *Fed. Reg.* **44**(231), 68624–68707 (1979).

Freundlich, H., *Colloid and Capillary Chemistry*, Methuen, London (1926).

Fritz, W., Merk, W., and Schlunder, E. U., "Competitive Adsorption of Two Dissolved Organics onto Activated Carbon," *Chem. Eng. Sci.*, **36**, 743 (1980).

Froelich, E. M., "Control of Synthetic Organic Chemicals by Granular Activated Carbon: Theory, Application and Reactivation Alternatives," Presented at the Seminar on Control of Organic Chemical Contaminants in Drinking Water, Cincinnati, October 10–11 (1978).

Furusawa, T., and Smith, J. M., "Fluid-Particle and Intra-particle Mass Transport Rates in Slurries, *Ind. Eng. Chem. Fund*, **12**(2), 197 (1973).

Gerba, C. P., Sobsey, M. D., Wallis, C., and Melnick, J. L., "Adsorption of Poliovirus onto Activated Carbon in Wastewater," *Environ. Sci. Technol.*, **9**, 727–731 (1975).

Getzen, F. W., and Ward, T. M., "Model for the Adsorption of Weak Electrolytes of Solids as a Function of pH I. Carbox-ylic Acid–Charcoal Systems," *J. Colloid Int. Sci.* **31**, 441–453 (1969).

Guy, M. D., and McIver, J. D., "The Removal of Virus by a Pilot Treatment Plant," *Water Res.* **11**, 421–428 (1977).

Hamdy, M. K., and Wheeler, S. R., "Removal of Soluble Mercury from Water by Rubber," *Bull. Environ. Contamin. Toxicol.* **17**, 150–158 (1977).

Hand, D. W., Crittenden, J. C., and Thacker, W. E., "User-Oriented Solutions to the Homogeneous Surface Diffusion Model for Adsorption Process Design Calculations, Part 1, Batch Reactor Solutions," Paper Presented at the 54th Annual Conference of the Water Pollution Control Federation, Detroit, Michigan (October 4–9, 1981).

Hassler, J. S., *Purification with Activated Carbon*, Chemical Publishing, New York (1974).

Healy, R. P., "Adsorption of Virus on Activated Carbon Under Varying Water Chemistry Conditions," University of Maryland, U.S. National Technical Information Service Report No. PB 294 530, Springfield, VA (1975).

Herzing, D. R., Snoeyink, V. L., and Wood, M. G., "Activated Carbon Adsorption of the Odorous Compounds 2-Methyl-isoborneol and Geosmin," *JAWWA*, **69**, 223–228 (1977).

Holmes, H. N., and McKelvey, J. B., "The Reversal of Traube's Rule of Adsorption," *J. Phys. Chem.* **32**, 1522–1523 (1928).

Hyde, R. A., "Removal of Halogens and Pesticides by Granular Activated Carbon," *Activated Carbon Adsorption of Organics from the Aqueous Phase*, Vol. 2, Ann Arbor Science Publishers, Ann Arbor, MI (1980).

Ishizaki, C., "Surface Chemistry of Activated Carbon: Its Influence on Adsorption from Aqueous Solution," Ph.D. Thesis, University of Maryland (1973).

James, R. O., and Healy, T. W., "Adsorption of Hydrolyzable Metal Ions at the Oxide Water Interface," *J. Coll. Int. Sci.* **40**, 42–81 (1972).

Keinath, T. M., and Weber, W. J., Jr., "A Predictive Model for the Design of Fluid Bed Adsorbers," *JWPCF*, **40**, 741–765 (1968).

Kim, B. T., Snoeyink, V. L., and Schmitz, R. A., "Removal of Dichloramine and Ammonia by Granular Carbon," *JWPCF*, **50**, 122–133 (1978).

Komiyama, H., and Smith, J. M., "Surface Diffusion on Liquid-Filled Pores," *AIChE J.*, **20**(6), 1110 (1974).

Kornegay, B. H., "Control of Synthetic Organic Chemicals by Activated Carbon: Theory, Application, and Regeneration Alternatives," Seminar on Control of Organic Chemical Contaminants in Drinking Water, Los Angeles, CA (November 14, 15, 1978).

Langmuir, I., "The Adsorption of Gases on Plane Surfaces of Glass, Mica, and Platinum," *J. Amer. Chem. Soc.* **40**, 1361–1402 (1918).

Lee, M., and Snoeyink, V. L., "Humic Substances Removal by Activated Carbon," University of Illinois at Urbana-Champaign, Water Resources Center, UILU-WRC-80-D153, Research Report 153 (November 1980).

London, F., "Properties and Applications of Molecular Forces," *Z. Phys. Chem.*, **B11**, 222–251 (1930).

Love, O. T., Jr., and Symons, J. M., "Operational Aspects of

Granular Activated Carbon Adsorption Treatment," U.S. EPA, Cincinnati (1978).

Lundelius, E. F., "Adsorption and Solubility," *Kolloid-Zeitschrift,* **26,** 145–151 (1920).

Magee, V., "The Application of Granular Activated Carbon for Dechlorination of Water Supplies," *Proc. Soc. Water Treatment Examinat.,* **5,** 17–40 (1956).

Manes, M., "The Polanyi Adsorption Theory and its Applications to Adsorption from Water Solution onto Activated Carbon," in I. H. Suffet and M. J. McGuire (eds.), *Activated Carbon Adsorption of Organics from the Aqueous Phase,* Vol. 1, Ann Arbor Science Publishers, Ann Arbor, MI (1980).

Manes, M., and Hofer, J. E., "Application of the Polanyi Adsorption Potential Theory to Adsorption from Solution on Activated Carbon," *J. Phys. Chem.* **73,** 584–90 (1969).

Mattson, J. S., and Mark, H. B., Jr., *Activated Carbon, Surface Chemistry, and Adsorption from Solution,* Marcel Dekker, New York (1971).

Mattson, J. S., et al., "Surface Chemistry of Active Carbons: Specific Adsorption of Phenols," *J. Coll. Int. Sci.,* **31,** 116–130 (1969).

McCarty, P. L., Argo, D., and Reinhard, M., "Operational Experiences with Activated Carbon Adsorbers at Water Factory 21," presented at the EPA–NATO/CCMS Conference on Practical Applications of Adsorption Techniques, Reston, VA (April 30–May 2, 1979).

McCreary, J. J., "The Reaction of Aqueous Free Chlorine with Organic Compounds Adsorbed on Activated Carbon," Ph.D. Thesis, University of Illinois at Urbana-Champaign (1980).

McCreary, J. J., and Snoeyink, V. L., "Reaction of Free Chlorine with Humic Substances Before and After Adsorption on Activated Carbon," *Environ. Sci. Technol.,* **15,** 193–197 (1981).

McGuire, M. J., and Suffet, I. H., "The Calculated Net Adsorption Energy Concept," in I. H. Suffet and M. J. McGuire (eds.), *Activated Carbon Adsorption of Organics from the Aqueous Phase,* Vol 1, Ann Arbor Science Publishers, Ann Arbor, MI (1980).

Montgomery, J. M., Consulting Engineers, *Contra Costa County Water District Water Treatment Study,* Vol 2, unpublished report, Montgomery Consulting Engineers, Pasadena, CA (1980).

Murin, C. J., and Snoeyink, V. L., "Competitive Adsorption of 2,4-Dichlorophenol in the Nanomolar to Micromolar Concentration Range," *Environ. Sci. Technol.,* **13,** 305–311 (1979).

Nemethy, G., and Scheraga, H. A., "Structure of Water and Hydrophobic Bonding in Proteins. I. A model for the Thermodynamic Properties of Liquid Water," *J. Chem. Phys.* **36,** 3382–3400 (1962).

Neretnieks, L., "Analysis of Some Adsorption Experiments with Activated Carbon," *Chem. Eng. Sci.,* **31,** 1029 (1976).

Parks, G. A., "Aqueous Surface Chemistry of Oxides and Complex Oxide Minerals, Isoelectric Point and Zero Point of Charge," in *Equilibrium Concepts in Natural Water Systems,* Advances in Chemistry Series, No. 67, American Chemical Society, Washington, DC (1967).

Parks, G. A., "Adsorption in the Marine Environment," In J. P. Riley and G. Skirrow (eds), *Chemical Oceanography,* 2d ed., Academic Press, New York (1975).

Polanyi, M., "Adsorption of Gases (Vapors) by a Solid Nonvolatile Adsorbent," *Verhandlungen, Deutsche physikalische gesellschaft,* **18,** 55–80 (1916).

Polanyi, M., "Adsorption From Solution of Substances of Limited Solubility," *Z. Phys.,* **2,** 111–116 (1920).

Puri, B. P., "Carbon Adsorption of Pure Compounds and Mixtures from Solution Phase," in I. H. Suffet and M. J. McGuire (eds.), *Activated Carbon Adsorption of Organics from the Aqueous Phase,* Vol. 1, Ann Arbor Science Publishers, Ann Arbor, MI (1980).

Ritter, H. L., and Drake, L. C., "Pore-Size Distribution in Porous Materials. Pressure Porosimeter and Determination of Complete Macropore Size Distributions," *Ind. Eng. Chem., Anal. Ed.,* **17,** 782–786 (1945).

Robeck, G. G., *Evaluation of Activated Carbon,* Water Supply Research Laboratory, National Environmental Research Center, Cincinnati, OH (1975).

Roberts, P. V., and Summers, R. S., "Granular Activated Carbon Performance for TOC Removal," *JAWWA,* **74,** 113–118 (1982).

Rook, J., "Formation of Haloforms During Chlorination of Natural Waters," *Water Treatment and Examination,* **23,** 234–243 (1974).

Rook, J., "Formation and Occurrence of Haloforms in Drinking Water," *Proc. 95th Ann. AWWA Conf.,* Minneapolis, MN (June 8–13, 1975).

Rosene, M. R., and Manes, M., "Application of the Polanyi Adsorption Potential Theory to Adsorption from Solution on Activated Carbon. VII. Competitive Adsorption of Solids from Water Solution," *J. Phys. Chem.,* **80,** 953–959 (1976).

Scaramelli, A. B., and DiGiano, F. A., "Effect of Sorbed Organics on the Efficiency of Ammonia Removal by Chloramine–Carbon Surface Reactions," *JWPCF,* **49,** 693–705 (1977).

Schay, F., "Thermodynamics of Physical Adsorption from Solutions of Nonelectrolytes at the Surface of Solid Adsorbents," *J. Coll. Int. Sci.,* **42,** 478–485 (1973).

Sigworth, E. A., and Smith, S. B., "Adsorption of Inorganic Compounds by Activated Carbon," *JAWWA,* **64,** 386–391 (1972).

Singer, P. C., and Yen, C., "Adsorption of Alkyl Phenols by Activated Carbon," in I. H. Suffet and M. J. McGuire (eds.), *Activated Carbon Adsorption of Organics From the Aqueous Phase,* Ann Arbor Science Publishers, Ann Arbor, MI (1980).

Snoeyink, V. L., McCreary, J. J., and Murin, C. J., *"Activated Carbon Adsorption of Trace Organic Compounds,"* U.S. EPA Report, EPA-600/2-77-223, Cincinnati (1977).

Snoeyink, V. L., and Suidan, M. T., "Dechlorination by Activated Carbon and Other Reducing Agents," in J. D. Johnson (ed.), *Disinfection: Water and Wastewater,* Ann Arbor Science Publishers, Ann Arbor, MI (1975).

Snoeyink, V. L., et al., "Organic Compounds Produced by the Aqueous Free Chlorine-Activated Carbon Reaction," *Environ. Sci. Technol.,* **15,** 188–192 (1981).

Sorg, T. J., and Logsdon, G. S., "Treatment Technology to Meet the Interim Primary Drinking Water Regulations for Inorganics: Part 5," *JAWWA*, **72**, 411–422 (1980).

Stumm, W., and Morgan, J. J., *Aquatic Chemistry*, 2nd ed., Wiley-Interscience, New York (1981).

Suffet, I. H., "An Evaluation of Activated Carbon for Drinking Water Treatment: A National Academy of Science Report," *JAWWA*, **72**, 41–50 (1980).

Suidan, M. R., Kim, B. R., and Snoeyink, V. L., "Reduction of Free and Combined Chlorine with Granular Activated Carbon," in I. H. Suffet and M. J. McGuire (eds.), *Activated Carbon Adsorption of Organics from the Aqueous Phase*, Vol. 1, Ann Arbor Science Publishers, Ann Arbor, MI (1980).

Summers, R. S., and Roberts, P. S., "Dynamic Behavior of Organics Removal in Full-Scale Granular Activated-Carbon Columns," presented at the 181st National Meeting, American Chemical Society, Division of Environmental Chemistry, Atlanta, GA (March 29–April 2, 1981).

Symons, J. M., et al., "Removal of Organic Contaminants from Drinking Water Using Techniques Other Than Granular Carbon Alone," from Proceedings of the USEPA/NATO-CCMS Symposium on Practical Application of Adsorption Techniques in Drinking Water Treatment, Reston, VA (April 30–May 2, 1979).

Thacker, W. E., "Modeling of Activated Carbon and Coal Gasification Char Adsorbents in Single-Solute and Bisolute Systems," Ph.D. Thesis, Department of Civil Engineering, University of Illinois (1981).

Thacker, W. E., Snoeyink, V. L., and Crittenden, J. C., "Desorption of Compounds During Operation of GAC Adsorption Systems," presented at the 101st Annual Conference of the American Water Works Association, St. Louis, MO (1981).

Traube, I., "Ueber die Capillaritats—constanten organischer Stoffe in wasserigen Losungen," *Annalen Der Chemie*, **265**, 27–55 (1891).

Trussell, R. R., and Umphres, M. D., "The Formation of Trihalomethanes," *JAWWA*, **70**, 604–612 (1978).

Varma, M. M., et al., "Adsorption of Trihalomethanes by Granular Activated Carbon. A Bench-scale Study of the Adsorption Isotherms in Binary and Multicomponent Aqueous Solutions," *Aqua*, **1980**, 157–161 (1980).

Weber, W. J., Jr., *Physicochemical Processes for Water Quality Control*, Wiley-Interscience, New York (1972).

Weber, W. J., Jr., "Concepts and Principles of Carbon Applications in Wastewater Treatment," in W. W. Eckenfelder, Jr. (ed.), *Application of Adsorption to Wastewater Treatment*, Enviro Press, Nashville (1981a).

Weber, W. J., Jr., "Pretreatment of Industrial Wastes with GAC for Removal of Priority Pollutants," in W. W. Eckenfelder, Jr. (ed.), *Application of Adsorption to Wastewater Treatment*, Enviro Press, Nashville (1981b), pp. 175–196.

Weber, W. J., Jr., and van Vliet, B. M., "Fundamental Concepts for Application of Activated Carbon in Water and Wastewater Treatment," in I. H. Suffet and M. J. McGuire (eds.), *Activated Carbon Adsorption of Organics from the Aqueous Phase*, Vol. 1, Ann Arbor Science Publishers, Ann Arbor, MI (1980).

Weber, W. J., Jr., and van Vliet, B. M., "Synthetic Adsorbents and Activated Carbons for Water Treatment: Overview and Experimental Comparisons," *JAWWA*, **73**, 420–426 (1981).

Weber, W. J., Jr., et al., "Effectiveness of Activated Carbon for Removal of Volatile Halogenated Hydrocarbons from Drinking Water," in J. A. Borchardt et al. (eds.), *Viruses and Trace Contaminants in Water and Wastewater*, Ann Arbor Science Publishers, Ann Arbor, MI (1977).

Wood, P. R., and DeMarco, J., "Treatment of Groundwater with Granular Activated Carbon," presented at USEPA/NATO-CCMS Symposium on Practical Application of Adsorption Techniques in Drinking Water Treatment, Reston, VA (April 30–May 2, 1979).

Wood, P. R., and DeMarco, J., "Removing Total Organic Carbon and Trihalomethane Precursor Substances," in I. H. Suffet and M. J. McGuire (eds.), *Activated Carbon Adsorption of Organics from the Aqueous Phase*, Vol. 2, Ann Arbor Science Publishers, Ann Arbor, MI (1980a).

Wood, P. R., and DeMarco, J., "Effectiveness of Various Adsorbents in Removing Organic Compounds from Water: Removing Purgeable Halogenated Organics," in I. H. Suffet and M. J. McGuire (eds.), *Activated Carbon Adsorption of Organics from the Aqueous Phase*, Vol. 2, Ann Arbor Science Publishers, Ann Arbor, MI (1980b).

Youssefi, M., and Faust, S. D., "Adsorption and Formation of Light Halogenated Hydrocarbons and Humic Acid in Water by Granular Activated Carbon," in I. H. Suffet and M. J. McGuire (eds.), *Activated Carbon Adsorption of Organics from the Aqueous Phase*, Vol. 1, Ann Arbor Science Publishers, Ann Arbor, MI (1980).

—10—

Ion Exchange and Demineralization

Ion exchange and membrane processes are becoming more extensively used in water and wastewater treatment. Ion exchange is primarily used for the removal of hardness ions and for water demineralization. Reverse osmosis and electrodialysis, both membrane processes, are presently used for removal of dissolved solids from brackish water and seawater. This chapter discusses the mechanisms of ion exchange, the properties of exchange media, equilibria and process kinetics, and design considerations.

Similarly, the reverse-osmosis section covers the principles and theory of reverse osmosis, membrane configurations and types, and other treatment considerations. Finally, the principles, elements, and design considerations of electrodialysis systems are presented.

ION EXCHANGE

As its name implies, ion exchange describes the physical–chemical process by which ions are transferred from a solid to a liquid phase or vice versa. Ions held by electrostatic forces to charged func-

tional groups on the surfaces of a solid are exchanged for ions of like charge in a solution in which the solid is being contacted. Because the exchange occurs at the surface of the solid and the exchanging ions undergo a phase change from a solution onto a surface or vice versa, ion exchange is typically classified as a sorption process. Chapter 9 discusses some of the principles of sorption phenomena, with specific reference to activated carbon adsorption. While adsorption is discussed in terms of thermodynamic relationships due to changes in surface tension of the solute by the adsorbate and the hydrophobic nature of certain adsorbates, ion exchange can be differentiated from adsorption by the chemical and electrical potentials that control the exchange of mobile ions between the solid and solute.

PROCESS OBJECTIVES

Currently, the two most widely used applications of ion exchange are for the removal of hardness ions (Ca^{2+} and Mg^{2+}), called softening, from domestic and industrial waters and for complete demineralization of waters for industrial purposes. The softening process replaces the calcium and magnesium in the water with sodium. This practice is being discontinued for drinking water because of the deleterious health effects of elevated sodium levels. Demineralization can be accomplished with many arrangements, but in its simplest form, it is a two-step process, with all cations being exchanged for H^+ ions in a cation exchanger and then all anions being replaced by OH^- ions in an anion exchanger.

Iron(II) and manganese(II) can also be removed from water supplies by ion exchange, although control of oxidation states is important because Fe(III) and Mn(IV) will foul the resins. Nitrates, rapidly becoming a major concern for some groundwater users, can be removed by anion exchange. In this application, sulfates will also be removed. Dealkalization can be accomplished using ion exchange.

In most applications today, synthetic ion exchange resins are used. However, some natural exchangers exhibit a high affinity for particular ions. Clinoptilolite, a sodium–calcium–aluminum silicate, has a very strong preference for ammonium ions and is used to remove them from water and wastewater. A small number of full-scale installations exist which are using activated alumina to remove fluoride. Alumina is also effective in removing phosphate and some forms of arsenic, selenium, and silica.

Recent work (Kolle, 1979) has shown that certain macroporous exchangers have the ability to remove color and TOC (total organic carbon) from water supplies. Other carbonaceous resins are effective in removing nonpolar organics like the trihalomethanes, phenols, pesticides, and certain other chlo-

rinated compounds. Ion exchange is used for the treatment of a variety of industrial wastes to recover valuable waste materials, or by-products, such as the ionic forms of gold, silver, platinum, chromium, and uranium. It is used in the production of ultrapure water for the electronic and pharmaceutical industries. Radionuclides from nuclear reactors, hospitals, and laboratories can be treated by ion exchange to remove these contaminants.

MECHANISMS OF ION EXCHANGE

Prior to discussing the various types of ion exchange media, it is important to review the basic phenomena of an ion exchange reaction. All ion exchangers, whether natural or synthetic, have fixed ionic groups that are balanced by counterions of opposite charge to maintain electroneutrality. The counterions are those ions, either cations or anions, that exchange with ions in solution. As an example, consider the schematic cation exchange resin shown in Figure 10-1.

As the resin initially containing countercation A^+ is placed in a solution containing cation B^+, a Fickian-type diffusion is established due to the concentration difference between counterions A^+ and B^+ in the resin and in solution. The equation used to described this particular exchange reaction is shown below:

$$B^+ + (R^-)A^+ \rightleftarrows A^+ + (R^-)B^+ \qquad (10\text{-}1)$$

where R^- represents the negatively charged functional group of the resin.

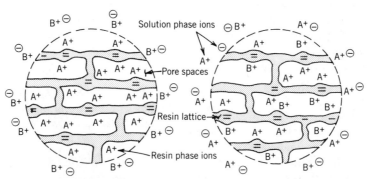

FIGURE 10-1.   Schematic diagrams of a cation exchange resin framework with fixed exchange sites prior to and following an exchange reaction. (a) Initial state prior to exchange reaction with cation, B^+. (b) Equilibrium state after exchange reaction with cation, B^+ (Weber, 1971).

In a batch reactor, exchange will continue to occur until the solution within the matrix of the resin and the bulk solution in which the resin is immersed all come to equilibrium with each other. The conditions within the solution phase contained in the resin pores will be much different than those that exist in the bulk solution due to the fixed charge sites on the resin surface and pore effects. Therefore, equilibrium will involve more than just a concentration equalization. Also, the exchange reaction will be governed by more than ion diffusion due to concentration differences. The chemical potentials of the exchanging ions in the two phases will determine, to a large part, the equilibrium that is established.

ION EXCHANGE MEDIA

The development of the first synthetic ion exchange resins in the 1940s has brought about extensive use of the ion exchange process. Synthetic resins with their high capacities and controllable properties have made large-scale ion exchange practical. However, some of the principles of ion exchange were discovered over 100 years ago. Many natural materials exhibit ion exchange capabilities, and the historical development of synthetic resins was brought about because these natural materials aroused investigators' interests.

Greensand, clay, peat, bauxite, and charred bone are natural exchangers that were tested by scientists for ion exchange properties. Thompson and Way reported in the mid-1800s that ammonia in solution was adsorbed as the liquid was passed through soil. Lime treatment of the soil afterward produced ammonia in the product water. Work in the late 1800s led to the use of natural aluminosilicates for the softening of water. These materials, called zeolites, had relatively low capacity and were subject to abrasion, but they were the first ion exchangers used in large quantities for municipal purposes. Hence the word *zeolites* is used for synthetic resins today.

In 1935, two English chemists, Adams and Holmes, observed that crushed phenolic phonograph records exhibited ion exchange properties. Advancement of this idea led to the development of the first synthetic organic ion exchange resins. In 1945, D'Alelio was granted the first patent for a synthetic cation exchange resin, and the development of new types of resins continues.

Synthetic Ion Exchange Resins

The majority of ion exchange resins are made by the copolymerization of styrene and divinylbenzene (DVB). The styrene molecules provide the basic matrix of the resin, while the DVB is used to cross-link the polymers to allow for the general insolubility and toughness of the resin. The degree of cross-linking in the resin's three-dimensional array is important because it determines the internal pore structure, which will have a large effect on the internal movement of exchanging ions. Recent developments have produced macroporous (or macroreticular) resins that have a discrete pore structure. These resins are said to be more resistant to thermal and osmotic shock as well as to the deleterious effects of oxidation. The more porous resins also are more resistant to organic fouling than the gel-type. Figure 10-2 shows electron micrographs of gel-type and macroreticular resins. There is a noticeable difference in the pore structure between these two types.

To produce the various types of cationic and anionic resins, the plastic structure is reacted with either acids or bases. A typical strong-acid ion exchange resin will be reacted with sulfuric acid, and sulfonic (HSO_3^-) groups will become the fixed ionic groups in the matrix. A weak-acid resin would be made by reacting the plastic with carbonic acid so that carboxylic ($COOH^-$) groups will make up the fixed groups. Likewise, strong- or weak-base resins are composed of functional groups of amines. Primary, secondary, or tertiary amines produce weak-base resins, while strong-base exchangers are made with quaternary amine (or ammonium) groups.

The ionizable group attached to the resin structure determines the functional capability of the exchanger. There are four general types of ion exchange resins, based on their functional groups, used in water treatment:

1. Strong-acid cation exchangers.
2. Weak-acid cation exchangers.
3. Strong-base anion exchangers.
4. Weak-base anion exchangers.

Strong-Acid Exchangers. The strong-acid cation resins can convert neutral salts into their corresponding acids if operated in the hydrogen cycle (e.g., $NaCl + R—H \rightleftarrows HCl + R—Na$). This ability is known as salt splitting, and it distinguishes

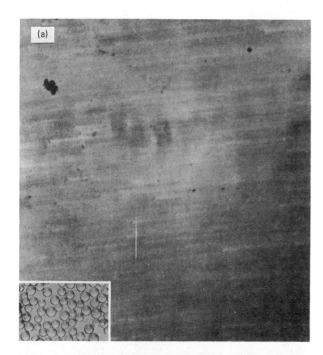

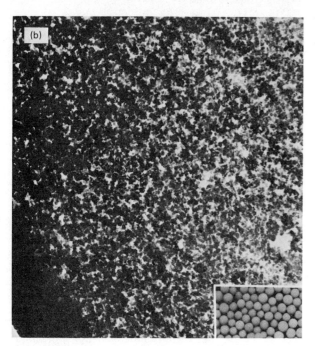

FIGURE 10-2. Electron micrograph of (a) microrecticular (gel-type) ion exchange resin and (b) macrorecticular ion exchange resin. Difference in structure is shown in these electron micrographs. Pores of the macroreticular resin (b) are clearly visible at a magnification of 50,000. However, pores of conventional gel-type resin (a), still cannot be seen on the micrograph. In the corners of each micrograph, a photograph of the resin as it appears to the human eye has been included. [Courtesy of Rohm and Haas Co.]

strong-acid resins from their weak-acid counterparts. As mentioned before, the functional groups of a strong-acid resin can be derived from sulfonic (HSO_3^-), phosphonic ($H_2PO_3^-$), or phenolic (OH^-) groups.

Strong-acid exchangers are typically operated in either a hydrogen cycle, where the resin is usually regenerated with HCl or H_2SO_4, or a sodium cycle, where the resin is regenerated with NaCl. The hydrogen cycle will remove nearly all major raw water cations and is usually the first step in demineralizing a water. It can be represented by the following reaction:

$$CaSO_4 + 2(R^- \cdot H^+) \rightleftarrows$$
$$(2R^-) \cdot Ca^{2+} + H_2SO_4 \quad (10\text{-}2)$$

The sodium cycle is used for softening waters (and also for the removal of soluble iron and manganese) and takes the following form:

$$CaSO_4 + 2(R^- \cdot Na^+) \rightleftarrows$$
$$(2R^-) \cdot Ca^{2+} + Na_2SO_4 \quad (10\text{-}3)$$

Weak-Acid Exchangers. Weak-acid exchangers differ from strong-acid resins in that weak-acid resins require the presence of some alkaline species to react with the more tightly held hydrogen ions of the resin, for example,

$$Ca(HCO_3)_2 + 2(R^- \cdot H^+) \rightleftarrows$$
$$(2R^-) \cdot Ca^{2+} + 2(H_2CO_3) \quad (10\text{-}4)$$

The exchange is, in effect, a neutralization with the alkalinity (HCO_3^-) neutralizing the H^+ of the resin. Weak-acid resins will split alkaline salts but not nonalkaline salts (e.g., $NaHCO_3$ but not NaCl or Na_2SO_4).

Weak-acid resins are very similar in properties to weak organic acids. Weak-acid cation resins are not highly dissociated and do not exchange their H^+ as readily as strong resins. But because they exhibit a higher affinity for hydrogen ions than strong-acid resins, weak-acid resins show higher regeneration efficiencies. Weak-acid resins do not require the same sort of concentration driving force required to convert strong-acid resins to the hydrogen form. The carboxylic functional groups have a high affinity for H^+ and will utilize up to 90 percent of the acid (HCl or H_2SO_4) regenerant, even with low acid concentrations. This is in contrast to strong-acid resin regeneration, where a large excess of regenerant (of which 60–75 percent goes unutilized) is required to create the concentration driving force.

Because of their high affinity for the hydrogen ion, weak-acid resins can be used only at pH's above 4 or 5. Weak-acid resins are favored when the untreated water is high in carbonate hardness (i.e., where high calcium and magnesium are associated with high alkalinity) and low in dissolved CO_2 and sodium. Weak-acid resins are used primarily for achieving simultaneous softening and dealkalization. They are sometimes used in conjunction with a strong-acid polishing resin, which allows for economic operation in terms of regenerant requirements, but also produces a treated water of quality comparable to the use of just a strong acid resin.

Strong-Base Exchangers. The strong-base anion exchange resins operate well throughout the entire pH range and will split neutral salts into their corresponding bases if operated on the hydroxide cycle (e.g., $NaCl + R\text{—}OH \rightleftarrows NaOH + R\text{—}Cl$). In this cycle, weakly ionized substances, such as silica and CO_2, can be removed. The functional sites are derived from quaternary ammonium groups. Two types of strong-base resins are available. Type 1 has three methyl groups making up the functional group as follows:

$$\begin{array}{c} CH_2 \\ | \\ (R\text{—}N\text{—}CH_2)^+ \\ | \\ CH_2 \end{array}$$

where R represents the cross-linked resin matrix. Type II resins have an ethanol group that replaces one of the methyl groups:

$$\begin{array}{c} CH_2 \\ | \\ (R\text{—}N\text{—}CH_2CH_2OH)^+ \\ | \\ CH_2 \end{array}$$

Type I has a greater chemical stability, while Type II has a slightly greater regeneration efficiency and capacity.

Strong-base exchangers are typically used after cation exchangers to remove all anions for complete demineralization. Operating on the hydroxide cycle, where the resin is regenerated with NaOH, the following reaction can be used to represent the exchange:

$$H_2SO_4 + 2(R^+ \cdot OH^-) \rightleftarrows$$
$$(2R^+) \cdot SO_4^{2-} + 2(H_2O) \quad (10\text{-}5)$$

More recently, strong-base resins have also been used for nitrate and sulfate removal from municipal water supplies. The resins are then operated in the chloride cycle, where the resin is regenerated with NaCl and removes sulfates and nitrates:

$$SO_4^{2-} + 2(R^+ \cdot Cl^-) \rightleftarrows$$
$$(2R^+) \cdot SO_4^{2-} + 2Cl^- \quad (10\text{-}6)$$
$$NO_3^- + R^+ \cdot Cl^- \rightleftarrows R^+ \cdot NO_3^- + Cl^- \quad (10\text{-}7)$$

Weak-Base Exchangers. The weak-base anion exchange resins behave much like their weak-acid counterparts. The weak-base resins remove free mineral acidity (FMA) such as HCl or H_2SO_4 but will not remove weakly ionized acids such as silicic and carbonic. For this reason, weak-base resins are often called "acid adsorbers."

Weak-base resins can be regenerated with NaOH, NH_4OH, or Na_2CO_3. Once again, the regeneration efficiencies of these resins are much greater than those for strong-base resins. Weak-base exchangers are used in conjunction with strong-base resins in demineralizing systems to reduce regenerant costs and to attract organics that might otherwise foul the strong-base resins. Where silica removal is not critical, weak-base resins may be used by themselves in conjunction with an air stripper to remove CO_2.

Physical Properties. The use of synthetic ion exchange resins gained wide acceptance because it became possible to predetermine their exchange behavior based on their synthesis. However, the exchange behavior is not the only factor that must be considered when discussing their use. The structure of the resins must have high physical and chemical resistances, thus ensuring a long, useful life. The quality of the resins must be such that oxidative, hydrolytic, or thermal influences neither produce nor accelerate a release of the resins' components during their use. A short discussion of some of the physical properties of the resins follows.

Swelling, Moisture Content, and Density. The relative insolubility of the synthetic resins is a key feature in their widespread use. However, because of certain hygroscopic properties, these resins are water-swellable structures. The swelling is produced by osmotic pressure in the interior of the exchanger against the external solution. The amount of bound water will depend mainly on two parameters: the functional group substitution and the cross-linking between matrix molecules.

The type and amount of the functional groups determine the moisture content potential of a resin, with a greater degree of substitution giving a greater tendency for moisture uptake. The behavior of strong-acid and strong-base resins is such that they have their largest swollen volumes in the hydrogen and hydroxide forms. Figure 10-3 shows this phenomena for a strong-acid cation exchanger. However, weak-acid and weak-base resins behave just the opposite. Swollen volumes increase as the weak resin is converted to salt forms.

Also, the greater the cross-linking, the greater will elastic forces resist the uptake of water. Figure 10-4 shows the effect of cross-linking on the moisture content of a strong-acid resin and a weak-acid resin.

The importance of these properties can be illustrated in the following example.

A weak-acid resin originally in the hydrogen form is to be used to dealkylize a water containing a large fraction of sodium ions. Special care must be taken in the choice of a suitable resin because the conversion of a typical weak-acid resin from the hydrogen to the sodium form will increase the relative water-swollen volume by as much as 90 percent. (That is, the resin in the sodium form will have a water-swollen volume 90 percent greater than that of the hydrogen form of the resin.) Frequent changes in the volume of the resin (shrinking and swelling) due

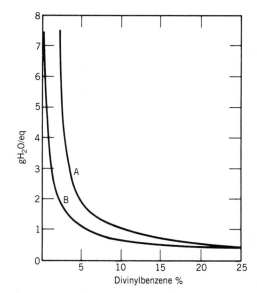

FIGURE 10-4. Moisture regain of cation exchangers of different degrees of cross-linking. A, strong-acid sulfonic acid exchanger; B, weak-acid carboxyl exchanger. (From Dorfner, 1973.)

to service followed by regeneration will greatly reduce the life of the resin by breakage and attrition unless precautions are taken. Also, increases in volume due to swelling can cause severe pressure drop problems in column operations. In this case, a highly cross-linked or macroporous resin would be recommended to minimize swelling.

Typically, moisture content is expressed as percent moisture per weight of wet or dry resin or in weight or moles of water per equivalent of exchange capacity. Manufacturers of ion exchange resins will typically list the moisture content (or water-swollen volume) of each resin as a function of ionic form. The listing is usually relative to the smallest form of the resin so that the relative degree of swelling between two forms can be easily determined. As shown in the previous example, a large volume change can take place and should be considered in process design.

Because the water content of the resin is variable, the densities of different forms of resin can also vary. This density will depend on the amount of water and the ionic form of the resin. Specific gravities of wet cationic resin will vary from 1.10 to 1.35, while wet specific gravities of anionic resins are typically 1.05–1.15. The bulk or apparent density of a resin will vary from 600 to 800 mg/L when wet.

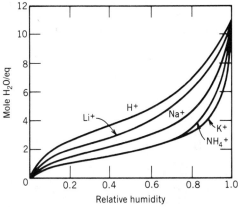

FIGURE 10-3. Moisture regain of strong-acid cation exchanger as function of relative humidity and ionic form. (From Dorfner, 1973.)

Capacity. The single most important property of an ion exchange resin is the quantity of counterions that can be taken up by the exchanger. In discussing the production of these resins, it was noted that this exchange capacity comes from substituted functional groups in the resin matrix that have exchange properties.

It is important to distinguish between the total capacity and the effective capacity of the resin. The total capacity can be determined from the total amount of counterions capable of exchange. The effective capacity is that part of the total capacity that can be utilized in a column, which is dependent on operating conditions such as a service flowrate, regeneration level, regeneration flowrate, and so on. Since the effective capacity can be quite variable, the total capacity is addressed in more detail.

The total capacity of a resin is dependent on the amount of functional groups in the copolymer bead. For strong-acid cation exchange resins, one sulfonate group, on the average, can be attached to each benzene ring in the matrix. Hence, we can determine the *dry-weight capacity* of this resin, which would be expressed in milliequivalents per gram of dry resin (meq/g). The dry-weight capacity is simply a measure of the extent of functional group substitution in the exchanger and is therefore a constant for each specific resin. For sulfonated styrene–DVB resins, the dry-weight capacity is 5.0 ± 0.1 meq/g (Anderson, 1979). For strong-base anion exchangers, more or less than one functional group can be attached to each benzene ring. Hence, the dry-weight capacity is more variable than with strong-acid resins and can range from 2 to 5 meq/g.

Because a resin will swell or shrink under different conditions, the *wet-volume capacity* will vary with moisture content. It is this capacity that is usually listed in the literature. Wet-volume capacities are expressed in equivalents per liter of resin (eq/L), although the units of kilograins of $CaCO_3$ per cubic foot (kgr/ft³) have been used in industry for quite awhile. There are 21.8 eq/L in 1 kgr/ft³. Normal strong-acid exchange capacities are 2.0 eq/L in the sodium form, while strong-base resins vary from 1.0 to 1.4 eq/L in the chloride form. The capacities of weak-acid and weak-base resins are much more variable due to their partially ionized conditions. Their capacities will be a function of pH.

The volume capacity will differ for the same resin in two different ionic forms, as it will for the same resin with different amounts of cross-linking.

The volume capacity will be inversely proportional to the swelling of the resin. The volume capacity is determined in a column under reproducible conditions, with the volume of resin usually measured after tapping in a column with excess water. This volume of resin (a bed volume) includes the volume of liquid within the interstices between the resin particles. In a backwashed and settled bed, this *void volume* is usually 35–45 percent of the total bed volume.

Particle Size. Ion exchange resins in spherical shapes are available commercially in particle diameters sizes of 0.04–1.0 mm. In the United States, the particle sizes are listed according to standard screen sizes or "mesh" values. Table 10-1 shows a comparison of mesh sizes and metric sizes. The most common size ranges used in large-scale applications are 20–50 and 50–100 mesh.

Particle size has two major influences on ion exchange applications. First, the kinetics of exchange are such that the rate of exchange is proportional to either the inverse of the particle diameter or the inverse of the square of the particle diameter. This is discussed later in the chapter. Second, particle size has a great effect on the hydraulics of column design. Smaller particle sizes increase pressure drops through the bed, requiring a higher head to push the water through the resin beads and subjecting the beads to situations that could cause breakage. In 50 percent of all ion exchange applications, the design is based on hydraulic limitations of the resin beads and the vessel rather than on ion exchange chemistry.

Manufacturers typically provide information about three parameters related to particle size: particle size range, effective size (ES), and uniformity coefficient (UC). The size range gives the maximum and minimum sizes of particles in the batch, the ES is a screen size that passes 10 percent (by weight) of

TABLE 10-1. Particle Size in U.S. Mesh and mm

U.S. Mesh	Diameter (mm)
16–20	1.2–0.85
20–50	0.85–0.30
50–100	0.30–0.15
100–200	0.15–0.08
200–400	0.08–0.04

the total quantity while 90 percent is retained, and the UC is the ratio of the mesh size in millimeters that passes 60 percent of the quantity to the effective size. Typical UCs are in the range of 1.4–1.6; however, it is possible to obtain batches with smaller uniformity coefficients if required by kinetic or hydraulic restrictions.

Stability.   The stability of an ion exchange resin under certain physical, chemical, and/or radioactive conditions will play a major role in many applications. Not only will the chemical process be affected by how a resin behaves but also the useful life; therefore, the total costs of a process will be affected. It is important to understand what physical stresses and chemical reactions can occur that will alter a resin's performance.

Typically, two types of degradation of ion exchange resins have been encountered, physical and chemical. Physical stresses may change the structure of the resin. The most common physical stress is excessive osmotic swelling and shrinking, which may break the bead. Mechanical compression or abrasion, if not controlled, will also cause excessive breakage. Excessive temperatures can rapidly degrade the physical and chemical characteristics of a resin. Manufacturer's recommendations for operating temperatures should be heeded.

Chemical degradation can occur with breaking of the polymer network, modification of the function groups, or fouling of the resin by a species in solution. The polymer networks of strong-acid resins are usually affected, while the functional groups of strong-base resins can be modified under chemical attack.

Strong-acid resins are susceptible to oxidative attack such as that from free chlorine. Oxidation will increase the moisture retention of the resin by affecting the matrix and eventually could cause excessive compression during the service cycle. In addition to network degradation, strong-acid resins can be fouled by precipitates such as calcium sulfate or ferric hydroxide. If a resin in a highly calcium form is regenerated with a strong enough concentration of sulfuric acid, calcium sulfate precipitate will form inside the resin particles, thereby reducing its exchange capacity. Excessive amounts of iron or manganese, should they become oxidized, will form precipitates that will also foul the resin. Weak-acid resins are also susceptible to calcium sulfate fouling but are not as likely to be affected by oxidants.

The major instability problem with anionic resins is the loss of exchange capacity due to modifications of the functional groups. Normal use of the Type I strong-base resins will cause a slow loss of capacity due to the loss of methyl and amine groups from the benzene rings. This loss is temperature dependent and will be greatly accelerated by elevated temperatures. The useful life could be 5–6 yr if ambient temperatures are maintained. Methanol and trimethylamine will be present in the water in very small amounts due to these losses. They present no health problems at low levels, but the odor threshold for trimethylamine is very low, and fishy odors are common to Type I resins. Type II strong-base resins lose their capacity because of splits between the nitrogen and ethanol groups. There is no loss of nitrogen, but the strong-base capacity is slowly changed to weak-base capacity. Elevated temperatures again accelerate these losses, but the useful life of Type II resins should be 3–5 yr if ambient temperatures are maintained. The by-products of the Type II resin degradation are odorless and nonvolatile.

Two major types of fouling can occur with strong-base resins, silica fouling and organic fouling. When used in the hydroxide form in a demineralization process, silicic acid is concentrated in the resin at the exchange front. Silicic acid will polymerize into an inorganic solid that will not behave as an exchangeable anion. This can accumulate in a strong-base resin to the point where silica-free water cannot be produced.

Organic complexes such as humic and fulvic acids are negatively charged and will tend to irreversibly exchange/absorb onto strong-base resins. It may take large quantities of regenerant and rinse water to bring a fouled resin back to original quality.

Weak-base resins are subject to oxidation and fouling, depending on type. Special care should be taken with these prior to their use to ensure that their stability will not be adversely affected. Manufacturers will normally provide the users with a guide list to prevent accelerated degradation.

Ion Selectivity.   The previous discussions have briefly introduced the various types of synthetic ion exchange resins. The direction, forward or reverse, of the ion exchange reactions (Eqs. 10-1 to 10-7) largely depends on the selectivity of the resin for a particular ion system. Researchers have observed preferences of resins for certain ions within classes of similar charge characteristics. These characteristics are well known for most commercial resins, and the reasons behind ion selectivity have been addressed by various investigators (Helfferich, 1962; Kunin, 1958). Without going into great detail, two main variables determine an ion's selectivity, ionic charge and ionic size. Ionic charge is the most significant influence on exchange potential.

As outlined by Kunin, the following set of empirical rules can be used to approximate selectivities:

1. At low concentrations (aqueous) and ordinary temperatures, the extent of exchange increases with increasing valence of the exchanging ion:

$$Th^{4+} > Al^{3+} > Ca^{2+} > Na^+;$$
$$PO_4^{3-} > SO_4^{2-} > Cl^-$$

2. At low concentrations (aqueous), ordinary temperatures, and constant valence, the extent of exchange increases with increasing atomic number (decreasing hydrated radius) of the exchanging ion:

$$Cs^+ > Rb^+ > K^+ > Na^+ > Li^+;$$
$$Ba^{2+} > Sr^{2+} > Ca^{2+} > Mg^{2+} > Be^{2+}$$

3. At high ionic concentrations, the differences in the exchange "potentials" of ions of different valence (Na^+ vs. Ca^{2+} or NO_3^- vs. SO_4^{2-}) diminish and, in some cases, the ion of lower valence has the higher exchange potential.

These are general, rule-of-thumb relationships; exceptions do occur. Divalent CrO_4^{2-} has a lower exchange preference than monovalent I^- and NO_3^- ions, as seen in the following series:

$$SO_4^{2-} > I^- > NO_3^- > CrO_4^{2-} > Br^-$$

Within a given series of ions, the hydrated radius is generally inversely proportional to the unhydrated ionic radius. Synthetic resins tend to prefer the ion of smallest hydrated radius, the smaller-radius ions being held more tightly by the resin.

The general rules for order of selectivity apply to ions in waters with less than approximately 1000 mg/L TDS. As the ionic strength of a solution increases, the preference for divalent ions over monovalent ions diminishes. Consider the case of a sulfonic cation exchange resin operating on the sodium cycle. In dilute solutions, calcium is much preferred over sodium; hence calcium will replace sodium on the resin structure. But at high salt concentrations (say, 100,000 mg/L TDS), the preference reverses and this enhances regeneration efficiency. Figure

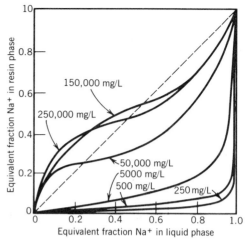

FIGURE 10-5. Na^+–Ca^{2+} equilibria for a sulfonic acid cation exchange resin. (Courtesy of Rohm and Haas.)

10-5 is a graph of Na^+–Ca^{2+} equilibria. At low concentrations, almost all of the sodium remains in the liquid phase. But as the TDS level increases, a higher proportion of the sodium can be found on the resin phase.

Another significant factor in determining selectivity is the size of organic ions or inorganic complexes. A resin will exclude some of these ions by screening or sieving. Resins that exhibit these properties are called molecular sieves. Ions too large to penetrate the matrix of the exchanger can be specifically excluded by proper selection of the resin properties. Increasing the amount of cross-linking in the resin will produce a greater screening effect.

Synthetic Resin Manufacturers. Currently, there are four major manufacturers of ion exchange resins in the United States. The companies, with their brand names in parentheses, are:

1. Duolite International (Duolite).
2. Rohm and Haas Company (Amberlite).
3. Dow Chemical Company (Dowex).
4. Sybron Chemical Division (Ionac).

Additionally, there are many foreign manufacturers of resins.

Because of continuing changes and improvements in ion exchange resins by industry, it is difficult to keep an up-to-date list of all of the major manufacturers' available resins. Rather than publish such a list in this text, it is recommended that

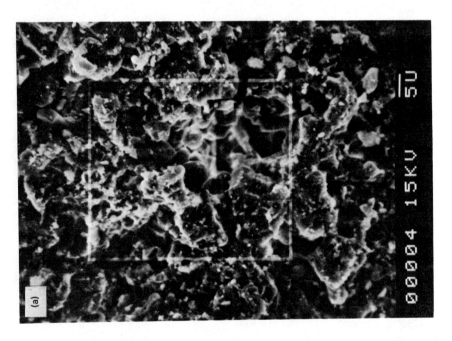

FIGURE 10-6. Scanning electronmicrographs of activated alumina: (a) 1000×; (b) 5000×.

the engineer contact these manufacturers directly regarding specific ion exchange resin applications.

Natural Ion Exchange Materials

Some inorganic ion exchangers have been used in industrial or municipal water treatment applications. The first applications, such as the use of aluminosilicates for softening, have been improved by the use of synthetic resins. Greensand (glauconite) is still used today in connection with iron and manganese removal. Activated alumina has shown promise in the removal of trace inorganics such as fluoride, arsenic, selenium, and phosphate due to its selective preference for these ions over other ions present in much greater concentrations in a water. This preferential selectivity for these ions is of major importance because the total ion exchange capacity of activated alumina is much less than that of a synthetic resin.

The majority of the natural ion exchange materials are insoluble salts of various metals, such as the phosphates and oxides of zirconium and aluminum. Zirconium phosphate, hydrous zirconium oxide, and aluminum oxide are three of the most investigated materials. Activated alumina (α-Al_2O_3) and clinoptilolite have potential for water and wastewater treatment.

While the process by which activated alumina removes particular ions from a water can be described as formal ion exchange, the term *adsorption* has been used in reference to this process. The exchange equation used to describe this reaction,

$$\begin{matrix}\equiv Al-O \\ \equiv Al-O\end{matrix} Al \cdot OH + B^- \leftrightarrows$$
$$\begin{matrix}\equiv Al-O \\ \equiv Al-O\end{matrix} > Al \cdot B + OH^- \quad (10\text{-}8)$$

is due to hydrolytic adsorption associated with aluminum and hydrogen ion exchange (Kubli, 1947). If treated with an aqueous acid solution, the alumina will be charged to a hydroxyl-bearing cation capable of binding the anions of various salts as water-soluble salts.

Figure 10-6 shows a scanning electron micrograph of the surface of activated alumina. Its porous structure provides many active sites for exchange to take place.

Activated alumina is produced by thermal treatment of hydrated alumina, the alumina being ex-

tracted from bauxite ore. Traditionally, it has been used as a dessicant for gases and liquids, since activated alumina has a great affinity for water. It is currently used in three full-scale fluoride removal facilities in the United States, and its potential for full-scale arsenic and selenium removal is being investigated. Activated alumina has also been shown to remove silica and phosphate from waters.

Studies with clinoptilolite show that it possesses properties allowing the highly selective uptake of ammonium ions (Ames, 1967). Further research has found that this material's capacity for ammonium is relatively high and that it can be used for full-scale removal of this ion (Jorgensen et al., 1979).

EQUILIBRIA

As a rule, the exchange reactions between ions in solution and ions attached to a resin surface are reversible. Because of a resin's inherent preferences for certain types of ions, these reactions will be selective and the ratios of concentrations of various ions in solution will be different from the concentration ratios in the resin phase at equilibrium. The performance of fixed-bed ion exchangers is governed by a combination of equilibrium and rate relationships. Equilibrium theory can predict ideal process performance and is much less complex mathematically than kinetic theories. Therefore, it is used quite often to predict the maximum amount of work that an ion exchange resin can do on a particular solution. Seldom are ion exchange columns run until equilibrium with the influent is reached. The limits derived from equilibrium theory should not be assumed to be operational capacities.

The two most common theories that have been developed to quantitate ion exchange equilibrium are by analogy to a chemical reaction or by analogy to a membrane exclusion phenomenon. If treated as a chemical reaction, mass action laws can be used to describe the distribution of ions in the solution and resin phases. The Donnan equilibrium model describes the behavior of ions based on the unequal distribution of ions across a membrane when an electrolyte solution on one side of the membrane contains ionic species that cannot diffuse through the membrane. The Donnan theory for ion exchange equilibria gives a thermodynamic interpretation of the phenomena which the mass action laws do not include. For this reason, it is more sophisticated and mathematically more rigorous. However, the mass action derivations will give the same equilibrium expression as the membrane theory.

Ion exchange reactions have proven to be well suited for mass action interpretation. A large amount of technical literature has been devoted to the precise description of ion exchange reactions. If ion exchange can be considered as a simple stoichiometric reaction, we can expand from Eq. 10-1 to consider a more generalized cation exchange:

$$A^{n+} + n(R^-)B^+ \leftrightarrows nB^+ + (R^-)_n A^{n+} \quad (10-9)$$

The equilibrium expression for this reaction can be written as

$$K_B^A = \frac{(a_B)^n (a_{R_nA})}{(a_A)(a_{RB})^n} \quad (10-10)$$

where a_A and a_B are the activities of ions A and B in solution; a_{R_nA} and a_{RB} are the activities of the ions in the resin; and K_B^A is typically called the equilibrium "constant," but in ion exchange it is referred to as the selectivity coefficient and is not actually a constant but depends on experimental conditions.

Most ion exchange applications are in water or wastewater treatment. In virtually all natural fresh waters and most wastewaters, the activity coefficients for the ions in solution are near unity, so a_A and a_B may be replaced in Eq. 10-10 by their respective molar concentrations $[B^+]$ and $[A^{n+}]$. However, in the resin phase, ionic concentrations can be very high and the activities cannot be assumed to be unity. Corrections for activities are usually required. Because $a = x \cdot \gamma$, the activities can be replaced by the products of the respective equivalent fractions ($\bar{X}_A = [\bar{A}^{n+}]/C$, where C is the total ionic concentration of the solution), and the activity coefficients, $\bar{\gamma}$. Equation 10-10 can be rewritten

$$K_B^A = \frac{[B^+]^n (\bar{X}_{R_nA})(\bar{\gamma}_{R_nA})}{[A^{n+}](\bar{X}_{RB})^n (\bar{\gamma}_{RB})^n} \quad (10-11)$$

or

$$K'_B{}^A = \frac{[B^+]^n (\bar{X}_{R_nA})}{[A^{n+}](\bar{X}_{RB})^n} \quad (10-12)$$

Under experimental conditions, $K'_B{}^A$ can be determined and is known as the "apparent equilibrium constant." It should be understood that this value is, in most cases, only constant for a relatively narrow concentration range. The indeterminate nature of $(\bar{\gamma}_{R_nA})/(\bar{\gamma}_{RB})^n$ makes the use of this constant meaningless unless discussed in terms of this narrow range.

Another useful parameter along the lines of the selectivity coefficient is the separation factor, defined as

$$\alpha_B^A = \frac{q_A/C_A}{q_B/C_B} \quad (10-13)$$

where q is the respective resin phase concentration and C is the solution phase concentration. For monovalent ion exchange, $\alpha_B^A = K_B^A$, but for multivalent–monovalent exchange, the two are not equal. The separation factor does not include the stoichiometric coefficients as exponents; therefore, it is considered more satisfying to some as a description of solute distribution. Its usefulness will be described in a later discussion.

In real systems, this equilibrium theory has inherent complexities that make the mathematics very rigorous. It is usually simpler to set up small-scale columns and determine resin performance for a given solution empirically rather than wade through the mathematics. Meaningful estimates can be made using simplified equilibria calculations that can predict ideal process performance, such as how much of a particular ion will exchange or how much resin will be required for removing a certain quantity of ion. Knowing this kind of information can help in determining the technical and/or economic feasibility of a proposed ion exchange application.

Typically, ion exchange equilibria is discussed first in terms of two-component systems, and then expanded into multicomponent equilibrium theory. The relative ease in discussing binary systems makes its introduction important for didactic purposes.

Two-Component Systems

The mathematical basis for solving equilibria involving two ionic species is Eq. 10-11. The notation given is for cation exchange, but it can be applied to anion exchange as well. The basic problem is determining values of the activity coefficients for ions in the resin phase. If we use equilibrium relationships to provide only a limit of the resin's capabilities, then liberties can be taken in calculating the equilibrium concentrations of the two ions. We can ignore the activity coefficients and use ion concentrations instead. Many selectivity coefficients have been estimated for various binary systems, and a list of these values is given in Tables 10-2 and 10-3. Using

TABLE 10-2. Selectivity Scale for Cations on 8 Percent Cross-Linked Strong-Acid Resin[a]

Cation	Selectivity	Cation	Selectivity
Li^+	1.0	Zn^{2+}	3.5
H^+	1.3	Co^{2+}	3.7
Na^+	2.0	Cu^{2+}	3.8
NH_4^+	2.6	Cd^{2+}	3.9
K^+	2.9	Be^{2+}	4.0
Rb^+	3.2	Mn^{2+}	4.1
Cs^+	3.3	Ni^{2+}	3.9
Ag^+	8.5	Ca^{2+}	5.2
UO^{2+}	2.5	Sr^{2+}	6.5
Mg^{2+}	3.3	Pb^{2+}	9.9
		Ba^{2+}	11.5

[a] From Bonner and Smith (1957).

these assumptions, Eq. 10-11 can be reduced to

$$K_B^A = \frac{[\bar{B}^+]^n[A^{n+}]}{[\bar{A}^{n+}][B^+]^n} \quad (10\text{-}14)$$

where the overbar above the letter refers to the species in the resin phase. The brackets indicate molar concentrations in equivalents per liter. This expression is more manageable when the terms are expressed as equivalent fractions. The equivalent ionic fractions of ions A^{n+} and B^+ in solution would be $X_{A^{n+}} = [A^{n+}]/C$ and $X_{B^+} = [B^+]/C$, where C is the total ionic concentration of the solution in equivalents per liter. In the resin phase, $\bar{X}_{A^{n+}} = [\bar{A}^{n+}]/\bar{C}$ and $\bar{X}_{B^+} = [\bar{B}^+]/\bar{C}$, $\bar{C}$ being the total exchange capacity of the resin per unit volume in equivalents per liter. Since only two exchangeable

TABLE 10-3. Approximate Selectivity Scale for Anions on Strong-Base Resins[a]

Anion	Selectivity	Anion	Selectivity
I^-	8	OH^- (Type II)	0.65
NO_3^-	4	HCO_3^-	0.4
Br^-	3	CH_3COO^-	0.2
HSO_4^-	1.6	F^-	0.1
NO_2^-	1.3	OH^- (Type I)	0.06
CN^-	1.3	SO_4^{2-}	0.15
Cl^-	1.0	CO_3^{2-}	0.03
BrO_3^-	1.0	HPO_4^{2-}	0.01

[a] From Peverson (1953).

ions are present, $X_{A^{n+}} + X_{B^+} = 1$ and $\bar{X}_{A^{n+}} + \bar{X}_{B^+} = 1$. Equation 10-14 thus becomes

$$\frac{\bar{X}_{A^{n+}}}{(1-\bar{X}_{A^{n+}})^n} = K_B^A \left(\frac{\bar{C}}{C}\right)^{n-1} \frac{X_{A^{n+}}}{(1-X_{A^{n+}})^n} \quad (10\text{-}15)$$

Note in Eq. 10-15 that when considering monovalent–monovalent exchange ($n = 1$), the terms C and $\bar{C}$ do not appear in the equation. Neither the total ionic solution concentration or the total exchange capacity of the resin will affect the equilibrium distribution of ions. However, in any other exchange reactions, such as divalent–monovalent, C and $\bar{C}$ will become factors in determining equilibrium concentrations.

Three types of limits can be defined by Eq. 10-15 (Anderson, 1979):

1. *Capacity.* The maximum exhaustion capacity will be the maximum number of equivalents of a given ion that can be removed from solution per given volume of resin.
2. *Degree of regeneration.* This will be the maximum degree to which the resin can be converted to the desired ionic form with unlimited quantities of regenerant solution of a given composition.
3. *Leakage.* Leakage is defined as the appearance of a low concentration of the undesired influent ions in the column effluent during the initial part of the exhaustion. This leakage is from residual ions in the resin at the bottom of the column due to incomplete regeneration that are displaced by other ionic species coming down the column.

Two examples are given to show the use of the equations developed (from Anderson, 1979).

EXAMPLE 10-1

Consider a monovalent exchange where we desire to remove nitrates from a chloride well water. The drinking water limit for nitrates is 0.9 meq/L or 44 mg/L as NO_3^-. Normally, other anions such as sulfate and bicarbonate would be present also, but for this case we assume that chlorides only are present. Strong-base resins have a greater affinity for sulfate than for nitrate but not for chloride. By assuming that only chloride is present, we simplify our problem. What will be the maximum volume of water that can be treated per liter of a strong-base anion exchange resin, given the water composition below? Assume a total resin capacity of 1.4 eq/L and a selectivity coefficient (K) for nitrate over chloride of 4.

Water Composition

Ca^{2+}	1.5 meq/L	Cl^-	3.5 meq/L
Mg^{2+}	1.0 meq/L	SO_4^{2-}	0.0 meq/L
Na^+	2.5 meq/L	NO_3^-	1.5 meq/L
Total cations	5.0 meq/L	Total anions	5.0 meq/L

The equivalent fraction of nitrate in solution is $1.5/5.0 = 0.30$. Using this and $K_{Cl}^{NO_3^-} = 4$, Eq. 10-11 becomes

$$\frac{\bar{X}_{NO_3^-}}{1 - \bar{X}_{NO_3^-}} = K_{Cl}^{NO_3^-} \frac{X_{NO_3}}{1 - X_{NO_3^-}} = 4\frac{0.3}{1 - 0.3} = 1.71$$

$$\bar{X}_{NO_3^-} = 0.63$$

A maximum of 63 percent of the resin sites can be loaded with nitrate ions from the given water. At that point, the resin and the water are at equilibrium with each other. The maximum useful capacity for nitrate will be 1.4 eq/L × 0.63 = 0.88 eq/L.

The maximum volume of water that can be treated per cycle will be

$$\frac{0.88 \text{ eq/L (resin)}}{1.5 \times 10^{-3} \text{ eq/L (water)}} = 588 \frac{\text{liters of water}}{\text{liters of resin}}$$

$$= 4400 \text{ gal/ft}^3 \qquad \square$$

EXAMPLE 10-2

We wish to determine the initial hardness leakage from a softener. Leakage will occur because a softening column is usually not fully regenerated due to the inefficient use of salt to completely regenerate the resin to the sodium form. The leakage will be dependent on the ionic composition of the bottom of the bed and the influent water composition. What will be the initial calcium leakage if the bottom of the softener is 80 percent in the calcium form after regeneration, the strong-acid cation exchanger has a total capacity of 2.0 eq/L and a selectivity for

calcium over sodium (K) of 3, and the influent water has the following composition?

Ca^{2+}	0.5 meq/L	SO_4^{2-}	0.5 meq/L
Mg^{2+}	0.3 meq/L	Cl^-	0.5 meq/L
Na^+	1.2 meq/L	HCO_3^-	1.0 meq/L
Total cations	2.0 meq/L	Total anions	2.0 meq/L

We know the following:

$$\bar{X}_{Ca^{2+}} = 0.80$$

$$\bar{C} = 2.0 \text{ eq/L}$$

$$C = 2 \times 10^{-3} \text{ eq/L}$$

$$K_{Na}^{Ca} = 3$$

Since we wish to know $X_{Ca^{2+}}$, we can rearrange Eq. 10-11:

$$\frac{X_{Ca}}{(1 - X_{Ca})^2} = \frac{1}{K_{Na}^{Ca}}\left[\frac{C}{\bar{C}}\frac{\bar{X}_{Ca}}{(1 - \bar{X}_{Ca})^2}\right] = 6.7 \times 10^{-3}$$

From this, $X_{Ca^{2+}} = 0.0067$ and the initial calcium leakage will be 0.0067×2 meq/L = 0.013 meq/L = 0.65 ppm as $CaCO_3$.

This shows that a low calcium leakage can be obtained when a low-TDS water is treated even with a poor regeneration of the resin in the bottom of the column. A low leakage level can be realized with a high-TDS water only if the bottom of the column is highly regenerated to the sodium form. This behavior holds for most divalent–monovalent exchanges. $\qquad \square$

These two examples and Eq. 10-15 can be used as tools for obtaining a better understanding of ion exchange resin behavior and for making preliminary assessments of process feasibility. On the other hand, they are approximate in nature and should be refined with pilot data prior to design.

Isotherms

Chapter 9 discussed equilibrium expressions for adsorption phenomena. Since ion exchange is a form of sorption from a solution, it is possible to describe the equilibrium of ions between the solid and solute phases by the Freundlich equation,

$$q_e = K_F \ln(C^{1/n}) \qquad (10\text{-}16)$$

The interpretation and application of this model can be found in Chapter 9.

The above expression was derived empirically and is used to fit the data for a given set of conditions. It is conveniently selected for its generality and simplicity of use. In simple systems, a single curve can be drawn of the equivalent concentration in the resin phase versus the equivalent concentration in the solution phase. This graphical method is based on measurements and can be useful in predicting exchange behavior. The use of the aforementioned separation factor, α, is applicable to this situation.

Equation 10-13 defined the separation factor as

$$\alpha_B^A = \frac{q_A/C_A}{q_B/C_B}$$

By normalizing the two variables and calling $X = C/C_0$, where C_0 is the total solution concentration and $y = q/Q$, where Q is the total ion exchange capacity of the resin, Eq. 10-13 can be rewritten as

$$\alpha_B^A = \frac{y_A X_B}{y_B X_A} \qquad (10\text{-}17)$$

If X versus y is plotted for a two-component system in equilibrium with a resin, a single curve can be drawn based on a constant temperature and constant solution concentration. The shape of this curve will determine the affinity of a particular ion for the resin in a binary system. Figure 10-7a shows two different curves representing favorable and unfavorable conditions for the uptake of ion A in competition with ion B. Curve 1, which is convex upward, represents a greater affinity of the resin for A, while curve 2 is concave upward and represents a greater affinity of the resin for B.

Figure 10-7b shows a method to quantitate the value of α. Since $y_A + y_B = 1$ and $X_A + X_B = 1$ for a binary system, Eq. 10-17 can be written as

$$\alpha_B^A = \frac{y_A(1 - X_A)}{(1 - y_A)X_A} = \frac{\text{area 1}}{\text{area 2}} \qquad (10\text{-}18)$$

In a narrow concentration range, α should be constant. A value of α greater than unity implies a greater affinity for ion A than B, while a value of α less than 1 indicates a greater affinity for ion B than ion A.

The usefulness of α in predicting column behavior is shown by various authors (Clifford, 1976; An-

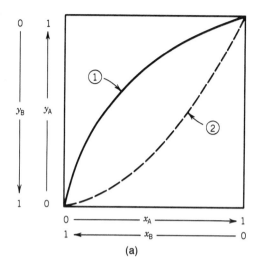

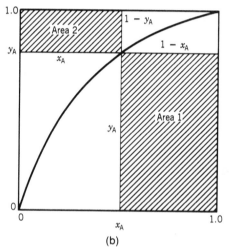

FIGURE 10-7. (a) Two types of experimental isotherms for a binary system. (b) Determination of α_B^A by the ratio of areas (Clifford, 1976).

derson, 1979). Because the separation factor experimentally takes into account the solution concentration and the total ion exchange capacity, whereas the selectivity coefficient K does not, it is preferred as a better description of binary system behavior.

The values of α and K can be related by the inclusion of the solution concentration and the total ion exchange capacity terms with the selectivity coefficient. The factor, $K_B^A(\bar{C}/C)^{n-1}$, where n is the valence of the ion to be exchanged with a monovalent ion, is similar to the separation factor, α_B^A. If $K_B^A(\bar{C}/C)^{n-1}$ is greater than unity, then α will be greater than unity. In exchanges between ions of similar charge, α will equal K. However, in exchanges between ions of dissimilar charge, α will not equal K if we assume that K remains constant. The separation factor will change as the resin converts to one form or the other.

This relationship between $K_B^A(\bar{C}/C)^{n-1}$ and α is important to note when reviewing the previously listed selectivity coefficients. The selectivity coefficients for divalent anions are less than 1; however, this does not imply that they are not preferred over monovalent anions in all cases. One must consider the factor $K_B^A(\bar{C}/C)^{n-1}$ [which equals $K(\bar{C}/C)$ in divalent–monovalent exchanges]. Thus, α may be greater than 1 while K is not. A divalent anion will be preferred on a resin at low solution concentrations giving $K(\bar{C}/C) > 1$, while higher solution concentrations making $K(\bar{C}/C) < 1$ will indicate that monovalent anions are preferred.

The separation factor can also be used to qualitatively predict column behavior. If α_B^A is much greater than unity, it means that ion B will not be retained as well as A. Initially, in a fully regenerated column, ion A will be tightly held in the top of the column, while any ions remaining after passing through the top part will be exchanged in the lower half, depending on their selectivities. As more of the influent water passes through the resin, ion A will push ion B further down the column as it preferentially replaces ion B from exchange sites. If the separation factor is high enough, ion B suddenly will appear in the effluent at concentrations as high as 2–3 times that of the influent water for a short period of time. This "chromatographic effect" is shown in Figure 10-8 for chloride exchange with a sulfate–nitrate–chloride–bicarbonate water. This effect is used advantageously in many chemical separation techniques. However, in water treatment, its occurrence might be the cause of great concern. If we are trying to remove nitrate by anion exchange with chloride, we must realize that the effluent nitrate concentration has to be monitored closely. Because of the resin's strong preference for sulfate over nitrate, the presence of increasing nitrate concentrations in the treated water will indicate that the nitrate breakthrough has occurred. Shortly, a high concentration of nitrate in the effluent will appear due to the sulfate replacing most of the nitrate on the resin. As the column approaches equilibrium, sulfate will occupy a large majority of the exchangeable sites. A slug of very high nitrate in a drinking water supply could cause serious health problems.

Multicomponent Systems

As we consider more exchangeable ions in the solution, the complexities of predicting column behavior increase. No longer can one equilibrium equation define the system. By including more variables, the mathematical solutions for generalized conditions become more complicated. However, if we acknowledge that the majority of ion exchange applications are in water treatment where the range of solution concentrations is narrow, say 100–500 ppm TDS, we can assume that the selectivity coefficients between competing ions are constant.

It has been demonstrated (Helfferich and Klein, 1970; Klein, Tondeur, and Vermeulen, 1967; Clif-

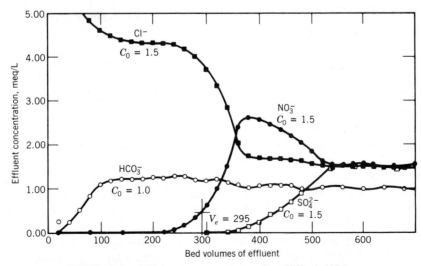

FIGURE 10-8.   Effluent concentration profile (Clifford, 1976).

ford and Weber, 1978) that multicomponent concentration profiles can be predicted from constant separation factors at a constant total solution concentration. However, the development of these predictive methods is beyond the scope of this text. For a rigorous treatment of this topic, the above-mentioned references should be read.

However, it is possible to *estimate* the equilibrium breakthrough volumes for multicomponent systems (specifically three-ion systems). The breakthrough of an ion will depend on the relative selectivities of the three ions and their concentrations. The following example will use this method to predict the breakthrough of the intermediate ion in a three-ion system (from Anderson, 1979).

EXAMPLE 10-3

As an extension of Example 10-1, consider a strong-base resin to be used to remove nitrate from a sulfate–chloride–bicarbonate–nitrate well water. What will be the total volume of water than can be treated until nitrate breakthrough given the following water composition? Assume that the resin is totally in the chloride form prior to treatment and that the total exchange capacity of the resin is 1.3 eq/L. Also assume that $K_{Cl}^{SO_4} = 0.15$, $K_{Cl}^{NO_3} = 4$, and $K_{Cl}^{HCO_3} = 0.4$.

$$
\begin{aligned}
HCO_3^- &= 25 \text{ mg/L} = 0.4 \text{ meq/L} \\
Cl^- &= 75 \text{ mg/L} = 2.1 \text{ meq/L} \\
SO_4^{2-} &= 120 \text{ mg/L} = 2.5 \text{ meq/L} \\
NO_3^- &= 80 \text{ mg/L} = 1.3 \text{ meq/L} \\
&\text{Total anions} = 6.3 \text{ meq/L}
\end{aligned}
$$

The first step is to combine the bicarbonate and chloride concentrations to simplify the situation. Since these are both less preferred on the resin at low TDS than nitrate or sulfate ($K_{Cl}^{SO_4}(\bar{C}/C) = 31$), this will not significantly alter the results. If we include the bicarbonate with the chloride, we will be making a conservative assumption since bicarbonate is less preferred than chloride. Consider the chloride concentration to be 2.5 meq/L.

In this system, sulfate will be held more tightly on the resin than nitrate and nitrate will be held more tightly than chloride. Under these conditions, we can assume that there will be an upper portion of

the resin bed that will be at equilibrium with the influent water and will contain a combination of sulfate, nitrate, and chloride depending on their selectivities. We will assume that all of the sulfate will be removed in this upper portion of the bed and that the lower part of the resin bed will contain a mixture of nitrate and chloride in equilibrium with a solution containing nitrate and chloride in approximately the same ratio that they are in the influent. Using Eq. 10-15, we can estimate the ionic compositions of the two portions of the column at nitrate breakthrough.

1. COMPOSITION OF UPPER PART OF COLUMN

$$
\frac{\bar{X}_{SO_4}}{(1 - \bar{X}_{SO_4})^2} = K_{Cl}^{SO_4}\left(\frac{\bar{C}}{C}\right)\frac{X_{SO_4}}{(1 - X_{SO_4})^2}
$$

$$
= 0.15 \frac{1.3}{0.0063}\frac{0.4}{(0.6)^2} = 34
$$

$$
\bar{X}_{SO_4} = 0.84
$$

The remaining 16 percent of the capacity in the upper portion of the column and all the lower portion of the column will contain nitrate and chloride in a ratio in equilibrium with the solution containing nitrate and chloride in the ratio 1.3 : 2.5. Therefore, $X'_{NO_3} = 1.3/(1.3 + 2.5) = 0.34$. We can use Eq. 10-15 again to predict the equilibrium distribution of nitrate and chloride.

$$
\frac{\bar{X}'_{NO_3}}{1 - \bar{X}'_{NO_3}} = K_{Cl}^{NO_3}\frac{X'_{NO_3}}{1 - X'_{NO_3}} = 4\left(\frac{0.34}{0.66}\right) = 2.1
$$

$$
\bar{X}'_{NO_3} = 0.68
$$

2. CALCULATED RESIN COMPOSITION

	Upper Portion	Lower Portion
$\bar{X}_{SO_4}$	0.84	0
$\bar{X}_{NO_3}$	$(0.68)(0.16) = 0.11$	0.68
$\bar{X}_{Cl}$	0.05	0.32

3. VOLUME OF WATER TO REACH NITRATE BREAKTHROUGH. Knowing the composition of the two parts of the column, we can calculate the volume of water until nitrate breakthrough by solving a set of simultaneous equations for the sulfate and nitrate mass balances. Let f be the fraction of column in the upper portion and v the bed volumes of feed until nitrate breakthrough.

Sulfate Balance

$$1.3 \text{ eq/L} \times 0.84f = 2.5 \times 10^{-3} \text{ eq/L} \times v$$

Nitrate Balance

$$1.3 \text{ eq/L} \times [0.11f + 0.68(1-f)]$$
$$= 1.3 \times 10^{-3} \text{ eq/L} \times v$$

Solving these two equations simultaneously yields

$$v = 290 \frac{\text{L water}}{\text{L resin}}$$
$$= 2170 \frac{\text{gal water}}{\text{ft}^3 \text{ resin}}$$

This predicts that the breakthrough of the nitrate–chloride front will start to rapidly increase at roughly 290 bed volumes of throughput. Nitrate ions will appear in the effluent shortly before this point. This is a favorable capacity.

This method can be used to predict the breakthrough of the intermediate ion for any three-ion system where the three ions are fairly well separated. Again, it should be noted that this is a rough estimate to be used for preliminary purposes only. The references previously noted have developed methods that utilize isotherm data (separation factors) and kinetic information such as mass transfer resistance to more accurately predict the breakthrough of ions in a column. This method usually requires the use of a computer for multicomponent systems. It is expected that the engineer using the equilibrium data would use it as a first cut to determine a go–no go process. □

KINETICS

Figures such as Figure 10-8 represent breakthrough curves where the effluent concentration of the species of interest is plotted versus time or volume of water passing through the column. Breaking through describes the increase in effluent concentration of a particular species due to the reduced capacity of the ion exchange resin for that species. The effects of equilibrium will determine the maximum sharpness of breakthrough, that is, how sharply separated the ion will be from other components. Kinetic effects only tend to distort the breakthrough curve from this maximum sharpness (Anderson, 1979).

Most of the fundamental work on ion exchange has been devoted to equilibrium considerations. The theory of ion exchange kinetics has been investigated far less, but its significance in large-scale column processes is of great concern. Of all the results that kinetic studies can give, the most practical information concerns the fractional attainment of equilibrium as a function of time. The general aspects of the kinetics are fairly well understood, but quantitative theories have been limited to rather simple, ideal states. A few key concepts are applied to the understanding of kinetic influences on ion exchange, for example, rate-controlling steps.

Rate-Controlling Steps

Like the adsorption phenomena discussed in Chapter 9, the rate of exchange of ions will depend on a few relative rates. These include:

1. transport of the exchanging ions to and from the bulk solution and the surface of the resin;
2. transport of the exchanging ions through the hypothetical film (or boundary layer) at the surface of the particle;
3. interstitial (pore) transport of the exchanging ions to the sites of active exchange; and
4. the actual rate of the exchange process.

Rates are generally controlled by the transport processes (1–3), with typical control being applied by film transport or pore diffusion. The actual rate of the exchange process is thought to be very rapid and is not a controlling factor. Since steps 2 and 3 occur consecutively, the slowest of these will control the rate of reaction. There could be intermediate conditions where both mechanisms affect the rate.

Film Diffusion. A convenient method for conceptualizing and modeling film diffusion is to think of a hypothetical film that surrounds each particle of exchange resin. In actuality, no film exists other than a hydrodynamic boundary layer. The most common model used to describe the rate of exchange through the "film" employs Fick's first law with a constant coefficient:

$$F_A = -D_A \text{grad } C_A \qquad (10\text{-}19)$$

where F_A is the mass transport through a unit cross section of film in unit time (flux), D_A is the constant diffusion coefficient, and grad C_A is the concentration gradient with respect to the coordinate variables.

By applying the boundary conditions $r = r_0$ (the particle radius) and $C_A = \bar{C}_A \cdot C/\bar{C}$ for $t \geq 0$ (C being the solution concentration), the following solution is obtained assuming that there is a constant boundary condition on the solution side of the film.

$$F(t) = 1 - e^{-kt} \qquad (10\text{-}20)$$

where $k = 3DC/r_0 \delta \bar{C}$, δ being the film thickness (Helfferich, 1966). The fractional attainment of equilibrium, $F(t)$, will be inversely proportional to the diffusion coefficient and the solution concentration and is proportional to the particle radius, the film thickness, and the concentration of the fixed ionic groups in the particle. It should be noted that Eq. 10-20 will also be the solution of a linear driving force function or a first-order reaction.

Electrical potential differences between the bulk solution and the particle surface can also be a driving force for film diffusion. This force will push coions into the film if the faster counterion is initially in solution and will pull coions out of the film in the opposite case.

Pore Diffusion. Intraparticle-diffusion-controlled binary ion exchange has been studied most thoroughly of all kinetic variables. A diffusion model based on Fick's second law can be used to model this phenomena. In spherical coordinates, the expression can be written

$$\frac{\partial C}{\partial t} = \frac{1}{r^2} \frac{1}{\partial r} \left(r^2 D \frac{\partial C}{\partial r} \right) - \frac{\partial q_e}{\partial t} \qquad (10\text{-}21)$$

where r is the resin particle radius, D is the diffusion coefficient, and q_e is the concentration of exchanged ions of the resin in equilibrium with a solution concentration of C (Weber, 1972). The term involving q_e represents the loss or gain of ions due to the actual ion exchange reaction on the surfaces of the solid.

When considering a binary system, the concentrations and fluxes of the exchanging counterions are rigidly coupled with one another. The fluxes of the two ions in opposite directions must be equal to magnitude, and no net transfer of electric charge may occur. The action of the electric field developed inside the particle is directed against high-concentration species, so that interdiffusion is predominantly controlled by Fickian diffusion of the low-concentration species (Helfferich, 1966). For a more complete discussion of kinetic phenomena in ion exchangers, the articles by Helferrich and Weber offer substantial theory and experimental results.

Determining the Rate-Controlling Step

A criterion has been developed for predicting whether film diffusion or particle diffusion will control the overall rate of exchange (Helfferich, 1966). The expression is dependent on ion exchange capacity, solution concentration, particle size, film thickness, diffusion coefficients, and the separation factor:

$$\frac{\bar{C}\bar{D}\delta}{CDr_0} (5 + 2\alpha_B^A) \gg 1$$

$$\text{film diffusion control} \qquad (10\text{-}22)$$

$$\frac{\bar{C}\bar{D}\delta}{CDr_0} (5 + 2\alpha_B^A) \ll 1$$

$$\text{particle diffusion control} \qquad (10\text{-}23)$$

Experimental values for the diffusion coefficients can be found in the literature (Helfferich, 1966; Weber, 1972; Dorfner, 1973). Average values for D, $\bar{D}$, and δ should be used when these are variables.

While the solution concentration will determine the rate-controlling step (low solution concentrations normally favor film diffusion control) in a great many cases, the selectivity can be an important factor. A high separation factor will tend to favor diffusion control. The above criteria predicts that in exchanges between monovalent ions on a strongly ionized resin of standard size, the intermediate solution concentration range falls between 0.1 and 1.0 N. Concentrations higher than this will allow particle diffusion to dominate. When divalent ions are being exchanged, the interdiffusion coefficients become much lower, slowing down the exchange rates within the resin.

As explained by Helfferich, interruption tests can be done to experimentally distinguish between particle and film diffusion control. If one stops a batch-ion exchange reaction by removing the particles from the solution for a short period of time and then reimmersing them, a higher rate of exchange than that at the moment of interruption will be noticed for particle diffusion control. Removal of the particles from the solution will allow internal concentration gradients to level out, thus allowing for a faster uptake upon reimmersion. Film-diffusion-controlled reactions will not exhibit this type of change.

Effects of Process Variables on Rate

A number of variables will affect the rate of an ion exchange reaction. Since the majority of ion exchange applications are done in columns as opposed to batch operations, we shall address column behavior.

1. PARTICLE SIZE. In all probability, the size of the ion exchange material will have the greatest influence on a process' kinetic behavior. Both film-diffusion- and particle-diffusion-controlled processes will have enhanced exchange rates with decreasing particle size. For film diffusion processes, the rate will vary inversely with the particle size, while particle diffusion processes will depend on the reciprocal of the particle size raised to a higher order.

This factor could come into play in situations where the engineer is restricted to vessel size requirements because of space limitations. A certain quantity of a larger-sized resin may not have the required operational capacity that a smaller resin could provide. However, the benefits of increased exchange rates with smaller particles must be weighed against the increased head loss within the columns.

2. FLOWRATE. The flowrate through the column will also play a role in determining a process' fractional attainment of equilibrium. Depending on how well an ion is preferred by the resin, lowering the flowrate (thereby increasing the contact time) can make a small or large difference in its operational capacity. Figure 10-9 shows the effect of changing service flowrates on the operating capacity of a strong-acid resin for the hardness ions, Ca^{2+} and Mg^{2+}. The resultant capacities for hardness only vary by roughly 10 percent, from 1 to 10 gpm/ft^3. Because hardness is highly preferred over sodium, the flowrate will not affect the capacity very much. This is a small difference considering that one would need a column 10 times larger with the 1-gpm/ft^3 process than the 10-gpm/ft^3 process to treat an equivalent flow. While lowering the treatment flowrate does increase the capacity in this

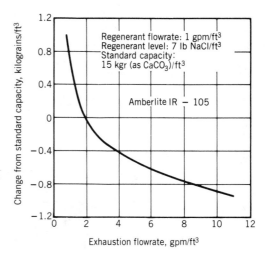

FIGURE 10-9.   Effect of exhaustion flowrate on softening capacity of a sulfonic acid cation exchange resin. (From Kunin, 1958.)

Exhaustion Flowrate (gpm/ft^3)	Capacity (lb CaCO$_3$/ft^3)
1	2.24
2	2.14
4	2.09
6	2.06
8	2.04
10	2.02

case, the advantage gained is small. The decreased capacity at higher flowrates is probably due to increased leakage of hardness ions.

Figure 10-10 shows the same resin while being regenerated. Under these conditions, particle diffusion control is being exerted. Because the sodium ions are not as preferred during regeneration as calcium and magnesium are during the service cycle, a higher contact time will allow better exchange of sodium ions to take place. Also, a much more concentrated solution is being contacted requiring more time for exchange to take place. Varying the regeneration flowrate from 5 to 0.5 gpm/ft^3 realizes a nearly 40 percent increase in operational capacity. While service flowrates less than 1 or 2 gpm/ft^3 usually are not practical because of the increased vessel size requirements, regeneration flowrates can be as low as desired depending on the turnaround time needed to put the column being regenerated back into service. In this case, regeneration flowrate makes quite a difference in the operational capacity.

3. SOLUTION CONCENTRATION. As mentioned previously, film-diffusion-controlled pro-

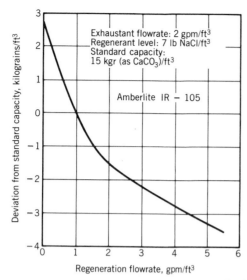

FIGURE 10-10.   Effect of regeneration flowrate on softening capacity of Amberlite IR-105, a sulfonic acid exchange resin. (From Kunin, 1958.)

Regeneration Flowrate (gpm/ft³)	Capacity (lb CaCO₃/ft³)
0.25	2.43
0.5	2.31
1	2.14
2	1.94
3	1.83
4	1.74
5	1.67

cesses are more likely with low concentrations. Pore diffusion effects are greater with increased solution concentrations. As the solution concentration increases, the flowrate must usually be lowered to achieve an equivalent operating capacity.

4. INTERSTITIAL PORE SIZE. Cross-linking is the principal factor affecting the interstitial pore size in a synthetic resin, therefore it will be a factor in particle-diffusion-controlled processes. Film diffusion processes will be affected somewhat because increased resin cross-linkage will reduce resin particle swelling, thereby reducing particle volume and surface area.

COLUMN DESIGN

Typically, most large-scale ion exchange applications are done in a column where the solution to be treated is passed through the column. Batch reac-

tions in a stirred vessel are only practiced when the exchange reactions go far to the right, as in neutralization, or when precipitation occurs. Even then, the process is cumbersome and not as convenient as column contactors. Column operations provide for an efficient use of the resin, even with a relatively low selectivity, due to the favorable driving force conditions.

Initial consideration of an ion exchange process requires definition of the problem. Depending on specific conditions, the most likely location to apply treatment should be determined. The composition of the process stream must be determined as completely as possible, including ionic and nonionic species, temperature, pH, conductivity, turbidity, and density. The presence of oxidants or reductants in the stream should be evaluated.

The actual process goals should then be outlined. These may be the required purity of treated water, maximum waste volumes allowable, and so on. Any restrictions in the availability of chemicals, space requirements, or cost limitations should be noted. An attempt should be made to estimate the process limits by evaluating the equilibrium conditions. As described previously in this chapter, the equilibrium capacity can be used to estimate the maximum capacity, the maximum degree of regeneration, or the initial leakage from a column depending on the type of resin and influent water and/or regenerant composition. If these calculated values fail to meet the specifications set for a proposed process, then the idea may be discarded because the actual operating performance will probably be poorer than the calculated limits.

Media Selection

The choice of which type of ion exchanger to use depends on the required process performance. Each major category of synthetic resin, strong- and weak-acid cation exchangers, and strong- and weak-base anion exchangers will serve a certain purpose, and it is up to the engineer to decide which type to use. The manufacturers of the different brands of resins can provide complete performance specifications of a certain resin.

Preliminary Studies

After selection of the media type (or types), it is necessary to develop criteria that will be used to design the full-scale equipment and determine the process variables. If the process is one that has been designed on a commercial basis for many years (such as softening or complete demineralization), it may be possible to use previously developed design criteria. Most ion exchange resin manufacturers have developed operating and design criteria that may be applicable to the process of interest. However, differences in water quality and a desire to minimize the capital and operating costs of treatment suggest that preliminary studies are advantageous in designing a full-scale ion exchange facility.

Laboratory or on-site small-diameter columns can develop meaningful process data if operated correctly. Because the main issues of concern are the physical–chemical diffusion and exchange of ions, 1–3-in. inside diameter columns can be scaled up directly to ultimate design if the same depth of resin is used. Hydraulic considerations such as side-wall effects do not play as important a role as in filtration, where full-scale operation cannot be modeled completely by small-scale pilot columns. However, even if full-scale depth is impossible to match in the preliminary studies, a minimum depth of 2–3 ft should be adequate to properly design an exchanger.

There are sources (Diamond Shamrock Chemical Co., 1969; Anderson, 1979) which describe the necessary hardware and components of designing small-scale testing equipment. They offer helpful hints and advice to make preliminary studies a relatively easy way to properly design a full-scale unit.

For a specific application, the main criteria that must be developed are length of removal run, service flowrate, and regenerant dose, flowrate, and concentration. Other variables such as resin stability under cyclic operation must be monitored over long periods of time and will require large-scale testing. A commercially available resin will have pressure drop versus flowrate and temperature and bed expansion (for backwash) versus flowrate and temperature data already computed which can be obtained from the manufacturer.

The two major types of data that are collected from small-scale column testing are *saturation loading curves* and *elution curves*. The loading curve is obtained by passing the process stream or a simulated stream of the same chemical composition through a fully regenerated column of resin. A sample of the actual process stream should be fed through the column for at least a couple of runs. Successive fractions of the effluent are then collected and analyzed until the effluent composition very nearly equals the influent composition. By plotting the appropriate concentration of each ion in equivalents per liter or normality versus the volume of water passed through the bed, a breakthrough or saturation loading curve can be developed. Such a curve was shown in Figure 10-8. The effluent volume is typically plotted as multiples of the actual volume of resin in the column, called "bed volumes." This will yield a normalized curve that should be the same for any size column under the same operating conditions. The breakthrough capacity for a given ion can be calculated by multiplying the bed volumes of treated water until breakthrough by the influent concentration of the ion.

There are two types of flowrates of interest in ion exchange applications. The volumetric flowrate, usually expressed in gpm/ft^3, bed volumes/hr, or L/L-hr, is inversely related to the contact time between the solution and the resin and thus the kinetics of exchange. The surface area loading rate, expressed in gpm/ft^2 or m/hr, is a measure of the linear flowrate through the resin. This must be considered in scale-up to ensure that excessive flowrates that could damage the resin do not occur.

To determine the optimum service flowrate, one must vary the rate during the saturation loading tests over a range of choices to see if any noticeable maximum in breakthrough capacity is achieved at a specific flowrate. Typically, the volumetric flowrate is the criteria used since it is directly related to the kinetics. The main goal in determining the optimum flowrate is to reduce the capital cost of equipment. The higher the acceptable flowrate, the smaller the contactor can be for a given treatment flow. Typical service flowrates range from 1 to 10 gpm/ft^3.

After completing each saturation loading curve, the resin must be eluted with an excess of regenerant to convert it fully to the desired ionic form. An elution curve is obtained, similar to a breakthrough curve, by collecting volumes of regenerant after it has passed through the bed and determining the concentrations of the ions of interest in each volume. This data can be used to choose a regeneration level that will be optimum with respect to operating capacity (resin conversion) and regenerant efficiency.

Figure 10-11 shows an elution curve for the regeneration of a strong-base resin that had been converted to the sulfate–nitrate form and was eluted within 1.0 N NaCl. Notice that all of the sulfate elutes very rapidly and is replaced by chloride ions,

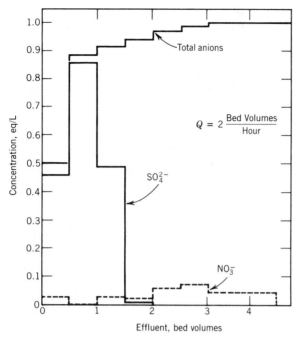

FIGURE 10-11. Elution of strong-base resin in sulfate–nitrate form with 1.0N NaCl. (Courtesy of Diamond Shamrock.)

while the nitrate takes much longer to remove. Equilibrium theory suggests that divalent ions will not be preferred in concentrated solutions; hence, they are easily replaced by chloride ions. On the other hand, the separation factor for nitrate over chloride is above unity, independent of solution concentration, and regeneration of the resin from the nitrate form to the chloride form will be inefficient.

The three variables of concern during regeneration are the concentration, the flowrate, and the dosage. A typical scheme to determine optimum conditions would be to choose a fairly slow (2–5 BV/hr or less) rate and an excess of regenerant and then vary the concentration of regenerant and develop elution curves for each concentration. An optimum concentration would elute the resin as rapidly as possible. Next, the optimum rate can be determined by keeping the optimum concentration and excess regenerant dose constant while varying the flowrate. Normally, a slower rate will allow for a more complete attainment of equilibria conditions but may not be as important a factor when the separation factor favors the ion already on the resin. This is the case in many ion exchange applications, and the only way to fully convert a resin is to use an excess of regenerant. Remember that this rate will

affect the time to complete the regeneration cycle during full-scale operations, and turnaround time may be a critical design criteria.

Using the experimentally determined values of concentration and flowrate, the optimum dose of regenerant can be determined. This dose is usually expressed in grams of regenerant per liter of resin (pounds of regenerant per cubic foot of resin). An elution curve should be developed for the optimum set of conditions. Using this curve, it is possible to determine regeneration efficiency and column utilization curves as a function of regenerant dosage.

During regeneration, an unfavorable condition ($\alpha < 1$) means that an excess of regenerant must be used to convert the resin to 100% regenerant form. Instead of converting the resin completely to this form, the amount of regenerant is chosen so that the column will be converted to a degree that will give the required quality of effluent for a reasonable run length. Figure 10-12 shows a plot of regeneration efficiency and column utilization versus regeneration level for a strong-acid exchanger being used for softening.

Regeneration efficiency is the actual hardness (or other species of interest) removed by the given amount of salt (or other regenerant) divided by the

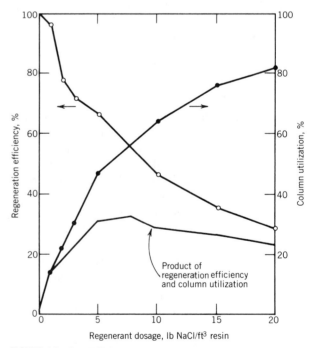

FIGURE 10-12. Efficiency and column utilization as function of regeneration level for strong-acid resin used for softening. Conditions: influent hardness, 500 mg/L as CaCO₃; service flowrate, 2 gpm/ft³; regeneration rate, 0.5 gpm/ft³, 10% NaCl.

theoretical hardness that could be removed by that amount of salt assuming 100% conversion. Column utilization represents the actual hardness removed by the regenerant divided by the total available exchange capacity of the resin in the column. The lower line in Figure 10-12 plots the product of the two percentages for a given dosage. This curve peaks roughly where the two curves intersect, and this peak usually indicates the optimum operating conditions.

Once these parameters have been established, it is necessary to operate the column to an allowable breakthrough point and leakage level for a number of cycles (3–5) to stabilize the system. A cycle can then be run, carefully analyzing the effluent versus bed volumes, to get a good indication of what can be expected in the full-scale column.

Scale-Up

Data derived from small-column experiments can be scaled up directly to any diameter column should the height of the bed remain constant. If the small-column experiments were done at a reasonable height (2–3 ft), then increasing this height in a full-scale design usually will not change the shape of the breakthrough curve. In exchanges where the separation factor is greater than 1 for the ion to be removed, the exchange zone (or front) will be relatively small with respect to the column height. Deepening the column in this case should not increase the breakthrough capacity with respect to bed volumes.

Maintaining the same volume flowrate as determined in the small-scale experiments will provide similar cycle times and effluent concentration profiles. If the height of the column is kept the same, then the surface area flowrate will also remain equal. If the column is deepened and the volume flowrate is kept the same, the surface area flowrate will be increased by the proportion that the height has increased. This should not be a problem unless a critical range of flow velocities is reached. Typical linear flowrates are in the range of 4–15 gpm/ft². Excessive linear flowrates will greatly increase the pressure drop through the column and could adversely affect the stability of the resin beads.

Once the optimum volume flowrate is known, the actual size of the full-scale contactor can be determined. The amount of resin volume needed to treat a given flow of water will be simply

Required resin volume (ft³)

$$= \frac{\text{treated water flowrate (gpm)}}{\text{volume flowrate (gpm/ft}^3)} \quad (10\text{-}24)$$

Based on this volume and the desired depth of the resin, the diameter of a single column can be determined. Should the required diameter be much larger than 12–15 ft, two or more columns should be used. Typical bed depths used in the industry are 2.5–9 ft.

Unless the treated water flow demand is intermittent, interruption of the service cycle for regeneration will require two or more columns or a treated water storage reservoir. If the exhaustion cycle is fairly long (>16–24 hr), a reservoir can provide sufficient water during regeneration time, normally 1–2 hr. Based on manufacturer's design data or laboratory studies, the regeneration requirements can be calculated. For most ion exchange applications, a typical regeneration cycle is:

1. End of service run.
2. Backwash.
3. Regeneration.
4. Slow displacement rinse.
5. Fast rinse.
6. Begin service cycle.

Backwashing is typically done to reclassify the resin so that there will be a gradual increase in particle size from top to bottom. This will help to prevent channeling. Ion exchange media will act as a good filter media, hence backwashing will remove trapped particulate matter from the resin. Bed expansion in the 50–75 percent range is normal, and proper freeboard should be allowed for in column design. This step will last 5–15 min.

Regenerant consumption based on design criteria must be determined per cycle. The rinses following regeneration are normally cocurrent; the slow rinse for one to two bed volumes at the regeneration flowrate to displace most of the regenerant from the bed, and the fast rinse at the rate of service flow for 10–30 min. An inventory of wastewater volumes must be calculated to adequately prepare for disposal. This is typically a costly part of operation and maintenance together with regenerant chemical costs. It may be the critical factor in many potential applications where disposal of concentrated brines would be a problem.

One of the major reasons for poor ion exchange performance is poor feed and outlet distribution design in the contactors. The feed must be uniformly distributed over the resin surface and uniformly collected from the bottom of the column to prevent channeling, maldistribution of flow, and density currents. If not properly designed, this will result in premature breakthrough and excessive leakage.

PROCESS CONFIGURATIONS

The vast majority of ion exchange applications are done in columns with the process flow, rinse, and regenerant all passed down or up through the resin. Batch reactions are not convenient and are usually uneconomical. The size of these columns depends on the flow requirements, while the physical design of these columns is very important in achieving one's treatment goals.

Typically, the ion exchange process is conducted in a fixed bed of resin with the water passing through the resin until a certain termination point is reached. Then the resin is taken off-stream and regenerated, while another column is used to supply continuous treatment if needed. Recently, new design applications have made it possible to operate an ion exchange system continuously with the removal run and the regeneration occurring simultaneously in different portions of a moving resin bed. These two different types of ion exchange operations will be discussed further.

Fixed-Bed Operation

The use of fixed beds has been the traditional approach to ion exchange. There are many important design features that must be addressed when considering ion exchange columns. Proper distribution and collection of flow is critical to good operation. Distribution of influent can be done with either a water dome, where the entire vessel is kept completely filled with liquid at all times, or an air dome, where the liquid level is kept a few inches above the resin level when liquid is being introduced into the column. A water dome design is much easier to instrument, operate, and control and is used most often in normal applications. Dilution is very critical when treating or feeding concentrated solutions, and while it cannot be eliminated entirely, air dome operation can reduce the dilution to a minimum.

The regeneration steps of an ion exchange resin are very important to the overall efficiency of the process. There are two methods for regenerating an ion exchange resin: cocurrent, where the regenerant is passed through the resin in the same flow direction as the solution being treated, and countercurrent, where the regenerant is passed through the resin in the opposite direction as the solution being treated. These are discussed further.

Cocurrent Operation. The most common method of operation is cocurrently, with the regeneration step being conducted in the same flow direction as the solution being treated. The direction of both flows is usually downflow. When leakage of the unwanted ion(s) is not critical and when the exchange in the regeneration step is fairly favorable, cocurrent operation is chosen. It is the simplest type of system to design, and daily operation is straightforward.

Countercurrent Operation. Countercurrent operation will result in lower leakage levels and higher chemical efficiencies than cocurrent operation. When design parameters dictate that a high-purity water is necessary, that chemical consumption be reduced to a minimum, or that the process produce as few waste volumes as possible, this method can be used with relative success.

The differences between cocurrent and countercurrent can be seen in Figure 10-13. This depicts a strong-acid resin operating in the hydrogen cycle. Under equilibrium conditions, the resin will prefer the sodium ion 1.5 times more than hydrogen; this ratio is even higher with calcium and magnesium. Typical cocurrent systems operate by using an acid regeneration level two to four times the quantity equivalent to the operating capacity. The regeneration can be said to be 200–400 percent of theory. However, countercurrent regeneration can reduce the regenerant requirement to levels of 125–150 percent of theory. The savings incurred by lower chemical requirements can be substantial. In addition, the increased efficiency means that smaller volumes of waste will have to be disposed. Disposal costs for waste regenerants can be the major operating expense of an ion exchange application. However, the capital costs for a countercurrent system will be greater than for cocurrent.

This countercurrent process will only be effective if the resin can be prevented from fluidizing during upflow operation. Either the removal cycle or regeneration can be conducted upflow. Any resin movement during the upflow cycle will destroy the ionic interface (exchange front) that ensures good

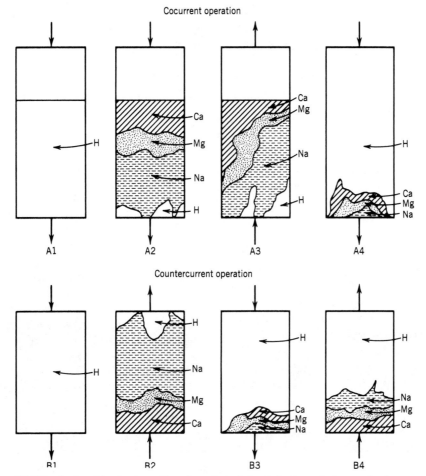

FIGURE 10-13. Distribution of ions in resin bed. In (A1), the resin is completely converted to the hydrogen form. Column (A2) depicts the approximate distribution of calcium, magnesium, and sodium ions at the end of a downflow service run. Following exhaustion, the bed is backwashed during which the resin particles expand and mix freely. As a result, the distribution of ions is scattered even more randomly than indicated in (A3). During downflow regeneration, the various cations are displaced downward by the excess of hydrogen ions, leaving a residual of calcium, magnesium, and sodium at the bottom of the column (A4). The quantity of residual cations depends mainly on the acid dosage used during regeneration. In the following service run, a few hydrogen ions are released in the upper portion of the bed and are exchanged for the residual ions at the bottom of the bed, thus causing some "leakage." This leakage is usually sodium, since this ion is most readily exchanged for hydrogen ions. If the free mineral acidity generated by ion exchange is high enough, magnesium and even calcium may also "leak" from the column. In (B1) the column is shown completely regenerated and ready for upflow service and downflow regeneration; i.e., countercurrent operation. In service the ion distribution (B2) is much the same as in (A2), but reversed. After downflow regeneration (B3), the distribution is similar to that in (A4). However, at the start of upflow service, there is virtually no leakage of cations because all of the exchange occurs at the inlet rather than the outlet end of the column (B4). (From Abrams, 1973.)

223

exchange. In addition, it is important that the resin particles remain in the same place in the bed over many cyclic operations to obtain maximum countercurrent benefits. There have been a number of methods devised to prevent resin particle movement during upflow. Some of the more commonly used methods are presented.

1. *Completely filled column.* In this operation, the ion exchange column is completely filled with resin and the service and regeneration steps are run counterflow. A reservoir tank above the column provides space for occasional backwashing.

2. *Use of inert granules to fill headspace.* Compressible inert granules are used to fill the column's headspace during the service cycle and they prevent the upward movement of the particles during upflow regeneration. A small reservoir is used to periodically withdraw the inert granules to backwash the resin.

3. *Use of air or water for blocking.* Air or water can be introduced at the top of the column during upflow regeneration to block movement of particles. Blocking water will result in increased waste volumes and has been virtually abandoned. However, only moderate air pressure is required and is used successfully in some designs.

There have been advances in the use of fixed beds which make certain processes more economical or provide better product purity. Two of these are the use of mixed beds of strong-acid and strong-base resins and the use of layered beds of a weakly ionized resin above a strong ionized resin.

Mixed Beds. For deionization purposes, a column containing strong-acid and strong-base resins intimately mixed provides better water quality than the individual resins in series. With both resins together, the effluent from the contactor will be deionized water. The reactions from salt to base to water or from salt to acid to water will occur so rapidly that there are virtually no back reactions. Hence, there is no need for completely regenerated resins to produce high-quality water. Probably the most important factor in achieving a good product is how well the resins are mixed, especially close to the exit of the bed.

Regeneration will require the separation of the two resin types into layers. This can be achieved by properly selecting the densities and particle sizes of the resins. Strong-base resins are normally lighter than strong-acid resins, hence backwashing prior to regeneration will place the basic resin above the acidic resin. During regeneration, the regenerants are introduced at the top and bottom of the respective beds and are simultaneously withdrawn at the interface. The anionic and cationic resins then must be well mixed by an air-scouring operation.

Layered Beds. The layered bed approach is taken to increase chemical efficiency and to increase the capacity of a given-sized unit without having to use another column. As mentioned before, a weak-acid resin is layered over a strong-acid resin or a weak-base resin is layered on top of a strong-base resin. Weak- and strong-acid resins will increase the regeneration efficiency if the water to be treated contains sufficient alkalinity. The weak-acid resin will remove the cations equivalent to the alkalinity, while the strong-acid resin will remove all of the other cations. Weak- and strong-base resins are used when there is a combination of strong and weak acids that must be removed.

The increased chemical efficiency arises from the fact that an excess of regenerant is needed to convert the strongly ionized resin, while the weakly ionized resin is converted rather easily. Therefore, if the strongly ionized resin is contacted first, as with upflow regeneration, the excess regenerant from the strong resin will be sufficient to convert the weak resin. This process can be refined to approach 100 percent regeneration efficiency and reduce waste acidity and basicity.

Continuous Contactors

In most ion exchange applications, the only part of the resin that is in use is at the exchange front. Resin above the exchange zone (assuming that service is downflow) has been exhausted and will not react. The resin below the exchange zone is not in use. Hence continuous contactors reduce the resin inventory by having only the amount necessary for the exchange front being contacted at one time while other portions of the moving bed are being regenerated and rinsed. The system is operated counterflow with the resin moving in one direction and the process stream, regenerant, and rinses moving in the other. The continuous design creates steady-state zones where a known quality of treated water can be produced continuously and regenerant can be supplied continuously.

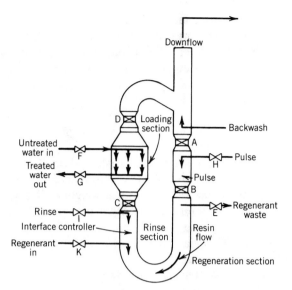

FIGURE 10-14. Schematic diagram of Chem-Seps continuous countercurrent ion exchange contactor. Valve positions during cycles: (a) Run cycle. Valves A, E, F, G, I, K open; valves B, C, D, H closed. (b) Pulse cycle. Valves B, C, D, H open; valves A, E, F, G, I, K closed. (Courtesy of Chemical Separations Corporation Oak Ridge, Tennessee.)

Figure 10-14 shows a schematic of a typical "pulsed-bed" continuous contactor. The resin is moved through the contact zones by frequent pulses of pressure. The removal or "run" operation will last from 5 to 30 min, then flow is halted for 15–30 sec while valves are opened and pressure is applied. This will move a small portion of the uppermost contaminated resin out of the treatment section and move an equal increment of regenerated resin into the lower treatment section.

With these types of systems, major problems can occur with the frequent operation of the valves. Wear on the valves can be high and the resin can suffer serious losses from abrasion.

Continuous systems may be economically attractive when either the contaminant loads or flow requirements are high. Continuous contactors are suggested when working with strong solutions where the component to be removed has a concentration roughly 0.5 N or greater. Treatability of the same solution with a fixed bed may be limited to as little as two bed volumes. For continuous ion exchange, the lower limit on the mass removal rate is estimated at 300–500 g/min.

REVERSE OSMOSIS

Reverse osmosis is a membrane process for desalting saline water by the application of hydrostatic pressure to drive the feedwater through a semipermeable membrane while a major portion of its impurity content remains behind and is discharged as waste. The pure water (known as product water or permeate) essentially emerges at near atmospheric pressure, while the waste (known as brine) practically remains at its original pressure. In this process, the permeate loses its salt content to the brine, which therefore gets highly concentrated. The operating pressure ranges from 300 to 400 psi (20.4–27.2 atm) for desalting brackish waters and 800–1000 psi (54.4–68 atm) for seawater desalination.

As illustrated schematically in Figure 10-15, the reverse-osmosis process operates with three streams:

Stream	Salt Content	Pressure
Feedwater	High	High
Brine	Higher	High
Permeate	Low	Low

Principle

If concentrated and dilute aqueous solutions are separated by a membrane, the liquid tends to flow through the membrane from the dilute to the con-

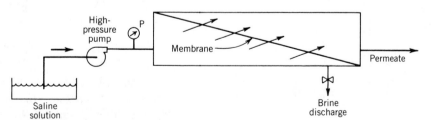

FIGURE 10-15. Elements of a reverse-osmosis system. A membrane assembly is generally symbolized as a rectangular box with a diagonal line across it representing the membrane. (Source: USAID, 1980.)

centrated side as if in an attempt to equalize concentrations on both sides of the membrane. This phenomenon is shown simplistically in Figure 10-16. Chambers A and B contain a dilute and a concentrated solution, respectively, and are separated by a membrane permeable to water but not to chemical ions. The liquid immediately starts flowing through the membrane, raising the level of liquid in the tube located at the top of chamber B. The liquid in the tube will continue to rise until the hydrostatic head of the water column in the tube is just adequate to prevent further flow through the membrane. At this point, the hydrostatic head exerted by the column of water in the tube is exactly equal to the difference in osmotic pressure of the two solutions.

Now, suppose the experiment is repeated with new dilute and concentrated solutions, and hydrostatic pressure in excess of the osmotic pressure differential is applied to the concentrated solution side. In this case, the direction of liquid flow would be reverse, that is, from the concentrated solution to the dilute solution. This phenomenon, whereby the liquid flow occurs from the concentrated solution to the dilute solution across a semipermeable membrane by the application of an external pressure or driving force, is known as reverse osmosis.

In the actual process, saline water is pumped to pressurize it against a membrane in a closed vessel, as shown in Figure 10-15. As pure water from the saline solution passes through the membrane, the remaining solution becomes more and more concentrated. This concentrated solution, brine, exits the vessel via a controlled valve and discharge piping. The pure water, which has passed through the membrane, is collected separately for use.

Terminology

Water flux is the quantity of water that can flow through the membrane with units of flow per unit membrane area per unit time (gal/ft^2/d or cm^3/cm^2/sec). *Salt flux* is the quantity of salt that can flow through the membrane in units of weight of salt passage per unit membrane area per unit time (lb/ft^2/sec or g/cm^2/sec).

Water recovery is the percent of feedwater recovered as product water.

This can be defined as follows:

$$\text{Recovery} = \frac{Q_p}{Q_f} \times 100\% \qquad (10\text{-}25)$$

where Q_p = product water flow (gpd or m^3/d)
Q_f = Feedwater flow (gpd or m^3/d)

Most reverse-osmosis plants are designed for a water recovery of about 75–80 percent in brackish water and 20–35 percent in seawater systems.

Salt rejection, which is a measure of the overall amount of salt rejected in the brine (as opposed to the rejection of different types of salts present in the feedwater), is defined as follows:

$$\text{Rejection} = \left(1 - \frac{\text{product concentration}}{\text{feedwater concentration}}\right) \times 100\% \qquad (10\text{-}26)$$

Salt rejection is determined by measuring the TDS. Sometimes, conductivity measurements are also made in lieu of TDS. No membrane is a perfect rejector of salt; some amount of salt escapes with the permeate. Furthermore, not all salts are rejected equally by any type of membrane. Salt rejections of 90 percent or above are commonly achieved in normal brackish waters.

Theory

The liquid and salt passages (or fluxes) through the membranes can be expressed mathematically by the following equations (Office of Water Research and Technology, 1979):

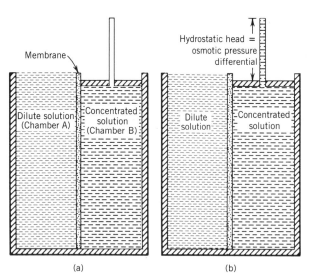

FIGURE 10-16(a). Beginning of experiment. (b) End of experiment.

$$F_w = K(\Delta P - \Delta \pi) \qquad (10\text{-}27)$$

$$F_s = k(C_b - C_p) \qquad (10\text{-}28)$$

where F_w, F_s = liquid and salt fluxes, respectively, across membrane (gal/ft^2/day or cm^3/cm^2/sec and lb/ft^2/sec or g/cm^2/sec).

ΔP = drop in total water pressure across membrane (psi or atm)

$\Delta \pi$ = difference in osmotic pressures between concentrated and dilute solutions (psi or atm)

$\Delta P - \Delta \pi$ = net driving force or liquid flow across membrane

C_b, C_p = salt concentration on two sides of membrane (lb/ft^3 or g/cm^3)

K, k = empirical constants, which depend on membrane structure, its method of manufacture, and salt type

$C_b - C_p$ = net driving force for salt or solute flow across membrane

In simple words, the liquid or solvent flow depends on the pressure gradient and the salt or solute flow depends on the concentration gradient as in a diffusion model represented by Fick's law. Thus, as the feedwater pressure increases, water flow through the membrane increases, but the solute flow is essentially constant. Therefore, the quality of permeate improves (low dissolved solids content). The reverse is true when the feedwater pressure decreases. Similarly, these equations also indicate that the quality of product water decreases as the feedwater solute concentration increases, at constant applied pressure. As more water is extracted from a given feed (higher recovery), the solute concentration becomes higher and the water flux falls off (because of higher $\Delta \pi$). The permeate quality therefore deteriorates.

More detailed theoretical concepts are available in the literature (Sourirajan, 1977; Merten, 1966; Lonsdale and Podall, 1972).

Figures 10-17 and 10-18 illustrate these points. Figure 10-17 shows the membrane flux as a function of water recovery at a constant applied pressure. The membrane flux decreases with increasing feed TDS and increasing recovery, since both result in increased osmotic pressure (thereby reducing the net effective pressure). Similarly, Figure 10-18 presents the effect of water recovery on salt rejection and hence water quality, which deteriorates

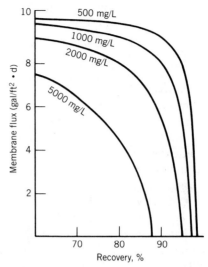

FIGURE 10-17. Membrane flux as a function of recovery.

with both the increasing feed salinity and increasing recovery. Both of these factors increase C_b in Eq. 10-28, causing an increased salt flux.

Osmotic Pressures. Theoretically, the osmotic pressure varies in the same manner as the pressure of an ideal gas:

$$\pi^\circ = \frac{\bar{n}RT}{V_m} \qquad (10\text{-}29)$$

where π° = osmotic pressure

$\bar{n}$ = number of moles of solute

V_m = molar volume of water

R = universal gas constant

T = absolute temperature

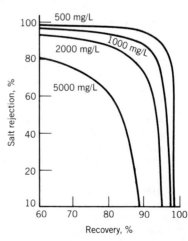

FIGURE 10-18. Salt rejection as a function of recovery.

The value determined above should be corrected by multiplying with a coefficient, Φ_c, whose value is dependent on concentration and varies from 1 for dilute solutions to <1 as the concentration increases. Thus, in all cases, the above equation provides conservative values of osmotic pressure.

As seen from the above equation, the osmotic pressure at any temperature is proportional to the molar concentration of the solute (i.e., number of moles of solute dissolved in a unit weight of water). For ionized substances, it is proportional to the molar concentration of ions. Thus, 1 mole of NaCl, common salt, dissolved in 1 L of water (58.5 g dissolved in 1 kg of water) will double the osmotic pressure of 1 mole of glycerine added in the same amount of water (92.11 g in 1 kg of water), since the former yields two ions (Na^+ and Cl^-) as opposed to only one ion produced by glycerine. Similarly, 1 mole of ferric chloride ($FeCl_3$), by yielding four ions, will double the osmotic pressure of 1 mole of NaCl. Osmotic pressures of some common substances are provided in Table 10-4 (Weber, 1971).

Membrane Configurations

Modern reverse-osmosis membranes are constructed in modular form. Two types of module configurations are most common: spiral wound and hollow fine fiber. In each case, the modules are mounted in containment pressure vessels into which the pressurized feedwater is pumped. Other configurations include the plate and frame (similar to a filter press) and tubular construction. Discus-

sion of the two commonly used configurations is provided in Chapter 21.

Membrane Types

The ideal membrane for desalination by reverse osmosis would consist of a thin imperfection-free film of a polymeric material (Office of Water Research and Technology, 1979). The transport properties of the material would be such that water could pass through with little hindrance, while presenting a virtually impermeable barrier to salts. A membrane would have to be extremely thin so that it has a large surface area for maximum flow to occur across it, possibly a few angstrom thick. Conversely, the membrane must be extremely strong in order to withstand the driving pressure of the feed stream (400–500 psi or more). A practical design must incorporate the above somewhat conflicting requirements and criteria.

The two major types of membranes manufactured are (1) cellulose acetate membranes using cellulose diacetate, cellulose triacetate, or blended cellulose diacetate–triacetate and (2) polyamide membranes.

Both the spiral-wound and hollow-fiber membrane configurations are available in one or more, but not necessarily all, of the above-listed formulations.

A brief description of these membranes follows. A detailed discussion is provided in RO Technical Manual (1979).

Cellulose Acetate Membranes. The most widely used membrane is the modified cellulose acetate film. This membrane has an asymmetric structure, which consists of a thin, dense skin on a porous support. The rejecting surface (the thin dense skin) is approximately 2000 Å (0.2 μm) thick. The porous support is a relatively spongy porous mass that contains about two-thirds water by weight, mostly in porous substrate. The original casting solution contained 15–25 percent by weight of cellulose diacetate in a solvent system of acetone, water, and magnesium percholate. Later this formulation was changed to 25 percent cellulose acetate, 30 percent formamide, and 45 percent acetone.

Cellulose triacetate membranes consist of 57 percent low-molecular-weight cellulose triacetate, 26 percent sulfalane, and 17 percent polyethylene glycol (MW < 1000).

Blended cellulose diacetate–triacetate mem-

TABLE 10-4. Typical Osmotic Pressures[a]

Compound	Concentration (mg/L)	Concentration (moles/L)	Osmotic Pressure (psi at 25°C)
NaCl	35,000	0.6	398
NaCl	1,000	0.0171	11.4
$NaHCO_3$	1,000	0.0119	12.8
Na_2SO_4	1,000	0.00705	6
$MgSO_4$	1,000	0.00831	3.6
$MgCl_2$	1,000	0.0105	9.7
$CaCl_2$	1,000	0.009	8.3
Sucrose	1,000	0.00292	1.05
Dextrose	1,000	0.00555	2.0

[a] A useful rule of thumb for estimating the osmotic pressure of a natural water is 1 psi/100 mg/L (ppm) of TDS.
SOURCE: Weber (1971).

branes, as the name suggests, are made out of a mixture of cellulose diacetate and cellulose triacetate in an attempt to take advantage of the superior strength and chemical resistance of the diacetates and the higher salt rejection of the triacetates.

Effect of Feed Temperature and pH. Both the liquid and salt fluxes (F_w and F_s in Eq. 10-27 and 10-28) are temperature dependent, since K and k are affected by it. Both K and k increase with increasing temperature, the former much more so than the latter. In the temperature range of 15–30°C, which is commonly encountered in full-scale reverse-osmosis operations, the water flux increases about 3 percent/°C increase in feedwater temperature. Salt flux does not change significantly over this temperature range. The temperature of operation is restricted, however, by potential failure of the module construction materials, particularly the plastic and elastomer components. Based on this concern, most manufacturers recommend that the feed temperatures should not exceed 38°C (100°F).

In practical applications, both the liquid and salt fluxes decrease with time, the former due to compaction of the membrane and the latter due to onset of the membrane hydrolysis. Membrane hydrolysis is a function of temperature and pH, as indicated by Figure 10-19 (Weber, 1971). The hydrolysis rate increases with increased temperature and minimizes at pH around 4–5. These considerations limit reverse-osmosis operation to a pH range of 3–7 (preferably around 5).

Effect of Pressure. As evident from Eq. 10-27, water flux is almost proportional to the effective pressure (i.e, applied pressure minus osmotic pressure differential). A rule of thumb for estimating the osmotic pressure is 1 psi/100 mg/L of TDS or 100 psi per 1 percent salt concentrations. Thus, seawater with an approximate salt content of 35,000 mg/L will have an osmotic pressure of about 350 psi, while a brackish water with a salt content of 1500 mg/L will have an osmotic pressure of about 15 psi. This explains the current practice of delivering a brackish water feed to reverse osmosis at a pressure of 400–500 psi, while the corresponding pressure for seawater is 800–1000 psi.

From the standpoint of product purity, a high feed pressure is desirable because liquid flux is proportional to the effective pressure. Thus, a high applied pressure increases liquid flow without correspondingly increasing salt passage; hence an increased product purity. Elevated pressures, however, require increased module wall thickness and stronger pipes and fittings. This is particularly important in seawater desalination, where the applied pressure is greater by a factor of 2–3. High pressures are also known to accelerate the flux decline normally experienced in reverse-osmosis applications, which should be carefully considered in design.

Additional pressure (beyond what is osmotically useful) must be applied to ensure a turbulent flow of feed through the module. The turbulence tends to minimize the stagnant layer of feed adjacent to the membrane surface. Too low a velocity will permit high local concentrations of the salts removed at the membrane surface (concentration polarization), which may not only hinder the transfer of product water through the membrane but may also initiate the precipitation of scaling substances on the membrane.

Solute Rejection. Table 10-5 lists average solute rejection data for cellulose acetate membranes over the first 12,000 hr of operation of Water Factory 21 reverse-osmosis plant (Argo, 1979). Several generalizations can be made based on these data and other published work (Weber, 1971).

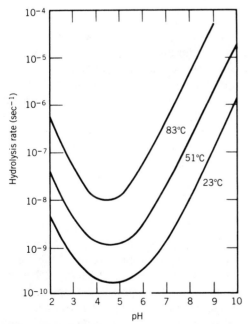

FIGURE 10-19. Cellulose acetate membrane hydrolysis rate as a function of pH and temperature.

1. Multivalent ions have higher rejection than univalent ions (calcium ion vs. sodium ion).

2. Undissociated or poorly dissociated substances have lower rejection (e.g., silica).

3. Acids and bases are rejected to a lesser extent than their salts.

4. Co-ions affect the rejection of a particular ion (e.g., sodium is better rejected as sodium sulfate than as sodium chloride).

5. Generally, low-molecular-weight organic acids are poorly rejected.

6. Undissociated low-molecular-weight organic acids are poorly rejected and their salts are well rejected.

7. Trace quantities of univalent ions are generally poorly rejected.

8. The average rejection of nitrate is significantly below that of other common monovalent ions.

Other Characteristics. Some unique characteristics of the cellulose acetate membranes are:

1. Membranes are subject to deterioration in the presence of microorganisms capable of cellulose enzyme production. Cellulose triacetate

TABLE 10-5. Removal of Contaminants by Reverse-Osmosis Demineralization

Contaminant	Unit	Influent Concentration Q-22A	Percent Removal
EC	μmho	1367	87
Na	mg/L	160	85
Cl	mg/L	195	85
SO$_4$	mg/L	223	98
TH	mg/L	223	94
Ca	mg/L	85	97
NH$_3$–N	mg/L	2.9	85
NO$_3$–N	mg/L	11.0	49
TKN	mg/L	3.6	85
COD	mg/L	13.15	89
TOC	mg/L	7.0	63
Ba	μg/L	14.8	55
Cd	μg/L	10.0	96
Cr	μg/L	7.6	65
Cu	μg/L	34	73
Fe	μg/L	23.8	74
Mn	μg/L	3.9	70
Pb	μg/L	2.4	50

SOURCE: Argo (1979).

membranes have possibly improved resistance to microorganism attack and so do the blended diacetate–triacetate membranes. A chlorine residual of up to 1 mg/L in the feedwater is often desirable.

2. Membranes are hydrolyzed at high and low pH values with the optimum value around pH 5.0. Cellulose triacetate and blended diacetate–triacetate membranes may have somewhat improved hydrolytic stability to high and low pH; this fact should not, however, be utilized in RO operations.

3. Membranes are subject to compaction and loss of productivity with time. The compaction rate increases with increasing feedwater pressure and temperature. Again, cellulose triacetate and blended diacetate–triacetate membranes show somewhat better characteristics in this regard.

4. Membranes, on an average basis (includes polyamide membranes), have a life of about 3–5 yr and thus need to be replaced at this frequency.

Polyamide Membranes. Polyamide membranes have high chemical and physical stability conducive to longer life. The polyamide materials are described as "synthetic, organic, nitrogen-linked, aromatic, substantially linear, condensation polymers."

Polyamides do not hydrolyze in water. A pH range of 3–11 is permissible in the feed. Polyamides are generally immune to bacterial attack; hence prechlorination of the feed is not required. In fact, any residual oxidant like chlorine in the feed will cause rapid deterioration of the membrane. This problem will require dechlorination of the feed if it is taken from a source that has been chlorinated. Dechlorination, if required, can be done by sodium bisulfite, sulfur dioxide, or activated carbon.

Pretreatment Requirements

The importance of adequate pretreatment of feedwater in RO systems cannot be overemphasized. Lack of this can cause rapid "fouling" of the membranes and reduced water productivity. (See Figure 10-20.) In addition, both pressure drop and salt passages are significantly affected by fouling.

The frequency of cleaning required to maintain performance may serve as a guide to the effective-

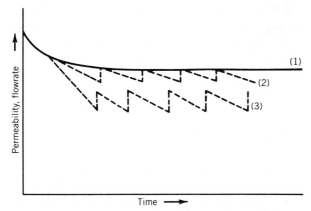

FIGURE 10-20. Productivity versus time. (1) Normal flow with nonfouling feed (adequate pretreatment). (2) Actual flow with pretreatment and periodic cleaning. (3) Actual flow with inadequate pretreatment and periodic cleaning.

ness of pretreatment. Cleaning more than once a month to maintain performance (productivity, salt rejection, pressure drop, etc.) suggests inadequate pretreatment.

Fouling involves the trapping of some type of material within the pores or on the surface of the membranes. Since the causes, symptoms, and cures are different, five types of fouling are identified (E. I. Dupont De Nemours & Company):

1. Membrane scaling.
2. Fouling by metal oxides.
3. Plugging.
4. Colloidal fouling.
5. Biological fouling.

These are discussed below.

Membrane Scaling. Membrane scaling is caused by the precipitation of some of the salts dissolved in the feedwater. Since the salts in the feedwater are usually concentrated by a factor of 2–4 (50–75% water recovery) in the reverse-osmosis process, their solubility limits can be exceeded, causing precipitation. Further, if the reject flowrates are not maintained at or above the recommended minimum required to maintain sufficiently high velocities, maldistribution within the membrane—known as polarization—can cause even higher concentration of solids in some parts of the membrane bundle. This can accentuate the problem.

The most common scales encountered in water treatment applications are calcium carbonate

($CaCO_3$) and/or calcium sulfate ($CaSO_4$). Other compounds that can cause scaling problems, although less frequently encountered, are silica, calcium phosphate, strontium sulfate, barium sulfate, and calcium fluoride.

Scaling due to calcium carbonate, which is relatively insoluble at normal pH, can be avoided by adjusting the pH of the feed stream with the addition of acid (usually sulfuric or hydrochloric) to about 5.5–6 so that the Langlier index of the brine is negative. The acid addition reduces the carbonate ion concentration by converting it to bicarbonate and/or carbon dioxide. The carbon dioxide will readily pass through the membrane and can be removed from product water by degasification.

Removal of sulfate ion from the feedwater can be accomplished by softening. If this is not practical, calcium sulfate precipitation can be avoided either by lowering the recovery (this will avoid exceeding the solubility limit) or in a borderline case by the addition of sodium hexametaphosphate (SHMP), which inhibits the precipitation of $CaSO_4$. As a general rule, the solubility limit for calcium sulfate is not exceeded at normal recovery levels (75–80 percent) unless its concentration in the feedwater is about 1500 ppm or above.

Silica can be reduced by hot or cold lime softening. The amount of silica removed is generally a strong function of the amount of $Mg(OH)_2$ precipitated. Recently, certain proprietary formulations have become available that will inhibit the formation of silica scale.

As little as 10–15 ppm of strontium (Sr^{2+}) or 0.05–0.1 ppm of barium (Ba^{2+}) may be sufficient to cause a problem. Here, again, the remedies are to reduce conversion so that the solubility product of these compounds in the brine is not exceeded beyond permissible limits; pretreat feedwater by ion exchange; or add SHMP in borderline situations.

In summary, by proper pretreatment, based necessarily on the chemical analyses of the feedwater, scaling can be minimized and/or prevented. If scaling does occur due to equipment failure or other unforeseen reasons, it is important to detect it early. When detected, carbonate scale should be dissolved with dilute hydrochloric and sulfate scale with citric acid and/or EDTA cleaning solution. Diagnostic and corrective steps should immediately be taken to prevent recurrence.

Metal Oxide Fouling. Soluble compounds in the feedwater can be oxidized in the RO system, ahead of the permeator or in the permeator itself, to form insoluble compounds that can then deposit in the permeators. Both manganese and iron can cause fouling by this mechanism; iron fouling being most prevalent. Iron fouling can occur rapidly and is caused by the oxidation of ferrous ions to ferric ions and the subsequent precipitation of the ferric hydroxides in the fiber bundle. Iron fouling may be avoided by either (1) removal of iron from feed supply or (2) prevention of oxidation from the ferrous to the ferric state.

Iron removal may be accomplished by any of several approaches. Any iron present in surface supplies is normally completely oxidized (ferric) and may be removed by clarification and/or filtration processes. Iron in groundwater supplies, before contact with air, will normally be ferrous (soluble) and may be removed by aeration followed by some type of filtration. Sodium zeolite softening is also effective for iron removal, as the resin will remove ferrous iron by the ion exchange process. Caution must be exercised, however, to avoid oxidation of the iron as that will result in serious fouling of the resin itself. If operation of the system can be controlled in a manner to prevent oxidation of the ferrous iron, the RO system will reject it as any other cation. As a general rule, iron concentrations as high as 4 ppm can be tolerated if the water contains essentially no dissolved oxygen (less than 0.1 ppm). If oxygen is present (5 ppm or above), the iron concentration should be maintained below 0.05 ppm. Citric acid, adjusted to a pH of 4 with NH_4OH, is a very effective cleaning reagent for the removal of metal oxides from permeators. With the proper pretreatment, iron fouling should not present a problem. However, due to the rapid rate of iron fouling, corrective action should be taken early if iron fouling begins to occur in the RO system.

Plugging. Plugging is caused by mechanical filtration in which particles too large to pass through the feed-brine passage (fiber bundle) are trapped by the membrane. In theory, this problem should be most severe for a hollow fiber membrane and least severe for tubular or spiral wound membranes, because of the relative size of the feed-brine flow channels. This problem is corrected by prefiltration through 5–10-μm cartridge filters, which is a minimum level of pretreatment provided in all RO systems.

Colloidal Fouling. Colloidal fouling is caused by the entrapment of colloids on the membrane surface in the RO systems. Colloids found in feedwaters usually belong to the aluminum silicate class (clays), although iron colloids from corrosion of carbon steel piping, pumps, and filters can also be present. Colloids of aluminum hydroxide can also be present if the water has been improperly clarified with alum. All of these colloids are hydrophobic and are usually in the 0.3–1.0-μm size range. These colloids are in suspension because of their electrical charge.

To determine the concentration of colloids in water, a measurement called the fouling index (also called silt density index) is used.

Several techniques can be used to lower the colloidal concentration to an acceptable level. Coagulation (using alum or iron salts as coagulants) and filtration (using sand, carbon, or other filter media) will usually reduce the colloidal concentration to acceptable levels. Small pilot columns can be used to determine the degree of colloid reduction and the best filter media. Diatomaceous earth filtration can also be used to decrease the colloidal concentration, but if the concentration is high, a rapid pressure drop will occur across the filter.

Biological Fouling. Both plugging and membrane attack can be prevented by continuous chlorine feed, in concentrations which the cellulose acetate category membranes can readily accept. Although the polyamide membranes are resistant to microbiological attack, they can accumulate growths to block salt passages. Therefore, periodic formaldehyde disinfection may be required for these membranes. Every effort should be made to disinfect the influent to the polyamide reverse-osmosis modules without exposing them to an oxidant residual. Chlorination followed by dechlorination or ultraviolet radiation is sometimes used.

Airborne algae or slime, if encouraged to multiply, may plague RO influents. Open surge tanks before RO units should be discouraged.

The types of fouling discussed cover at least 95 percent of the fouling situations that can occur. Other types of fouling, such as silica scaling, sulfur deposition from H_2S oxidation, and fouling caused by organic compounds, are possible but not common. One type of fouling can induce another. For example, if colloidal fouling occurs to a great extent, flow distribution within the permeator will be poor and this could result in scaling. Thus, it is important that all types of fouling be considered in the design of the pretreatment system.

Typical Applications

RO systems can be used in several applications:

1. Desalination of brackish water (TDS less than 10,000 mg/L) to provide potable water (TDS less than 500 mg/L). Recent work has extended its capability to treat seawater (TDS up to 45,000 mg/L).
2. Pretreatment of normal municipal water (TDS range 500–1000 mg/L) preceding ion exchange deionization to make ultrapure water for applications such as boiler feed.
3. Recovery of valuable or reusable materials from a waste via the RO reject stream. This is dictated by the concentrations of these materials in the reject stream and the market value of such materials.
4. Reduction in the volume of waste, if required.
5. Water conservation or recovery such as the cooling tower blowdown.

ELECTRODIALYSIS

Electrodialysis (ED) is another membrane process used to treat brackish water. In this process, electric energy is used to transfer ionized salts from feedwater through membranes, leaving behind a purified product water. Whereas the pure liquid moves across the membranes in RO, there is a migration of ions in ED leaving behind pure liquid.

Principle

Desalination by ED is based on the following principles (USAID, 1980).

1. Most of the salts dissolved in water are ionic in nature and thus can be removed under the influence of an electromagnetic field.
2. The ions, both positively and negatively charged, are attracted by an opposite electric charge, i.e., positive ions (cations) are attracted to a negative electrode (cathode) and negative ions (anions) are attracted to a positive electrode (anode).
3. Membranes can be devised that will be selective in the type of charged ion they will pass or reject; that is, membranes can be made that allow negative ions (anions) to pass but reject cations and vice versa.

The above principles are utilized in electrodialysis in which cation- and anion-selective membranes are placed alternately between two electrodes (cathode and anode, respectively), as shown in Figure 10-21. These cation- and anion-selective membranes are separated by cut-out plastic spacers to form what is known as a membrane stack. A typical membrane stack has several hundred cell pairs, each pair consisting of one dilute and one concentration cell.

As the electrodes are charged by a direct current source, the anions and cations move through the respective selective membranes located on either side into the concentrate (or brine) cells. Thus, concentrated and dilute/pure solutions are formed in the spaces (called cells) between alternating membranes. The concentrated and pure solutions are collected via separate piping manifolds for ultimate disposal or use.

Elements of a Typical ED System

Typically, there are three elements of an ED system: (1) supply of pressurized water, (2) membrane stack, and (3) DC power supply. Other ancillary equipment often includes the pretreatment devices for feedwater similar to RO such as filters, acid feed equipment for pH adjustment, and complexing agent (SHMP) feed equipment for preventing calcium carbonate or calcium sulfate precipitation in brine stream. (See Figure 10-22.)

Pressurized Water. Flow through an ED system varies with the pressure drop across the membrane stack. Most current designs employ a maximum gage pressure of about 60 psi; pressures higher than this are likely to cause excessive leakage through the membrane–spacer interface. This maximum pressure in turn limits the number of ED stages to 8–10 that can be connected in series without intermediate pumping.

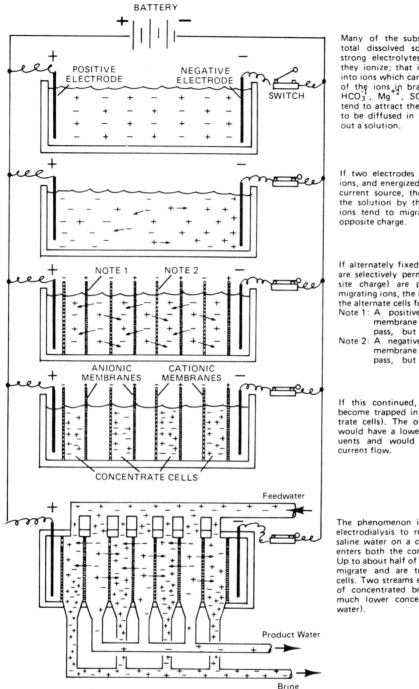

FIGURE 10-21. Movement of ions in the electrodialysis process. (Source: USAID, 1980.)

Membrane Stack. Removal per stage or rejection varies anywhere from 25–60 percent. Most applications will thus require two or more stages in series. Multiple-stage arrangements may be realized by arranging single-stage stacks in series or by using multiple stages in a single stack. For large plants, where the number of stacks required may exceed the number of stages, a single stack is usually employed as a single stage, and a sufficient number of stacks are employed in series to obtain the number of stages required to achieve the desired demineralization.

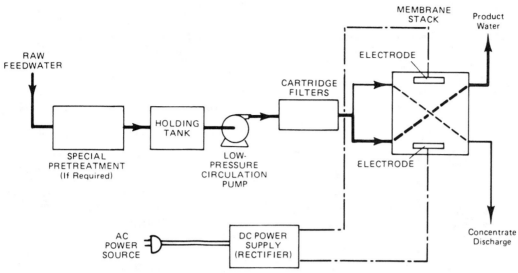

FIGURE 10-22. Basic components of an electrodialysis unit. (Source: USAID, 1980.)

DC Power Supply (Rectifier). The main element of the power supply is a rectifier, which converts alternating current to direct current. Direct current is applied to the electrodes on the membrane stack, which energizes the ions in the feed stream causing them to migrate through the selective membranes as discussed previously.

Pretreatment

As in RO systems, some pretreatment of ED feedwater is often necessary for satisfactory operation. The pretreatment processes discussed for RO are applicable for ED as well. Normally, acid addition to inhibit the precipitation of sparsely soluble salts, for example calcium carbonate and calcium sulfate, and filtration to remove any suspended impurities constitutes minimum level of pretreatment. A small dosage of SHMP added to the feedwater helps in inhibiting the precipitation of some salts, particularly in borderline cases.

A recent innovation by a U.S. manufacturer of ED systems has significantly minimized the need for pretreatment. This innovation, known as electrodialysis reversal (EDR), operates on the same basic principle as the standard electrodialysis unit, except that both the product and the brine cells are identical in construction. At a frequency of 3–4 times/hr, the polarity of the electrodes is reversed and the flows are simultaneously switched by automatic valves, so that the product cell becomes the brine cell and the brine cell becomes the product

cell. The salts are thus transferred in opposite directions across the membranes. Following the reversal of polarity and flow, the product water is discharged to waste until the cells and lines are flushed out and the desired water quality is restored. This process takes about 1–2 min. The reversal process aids in breaking up and flushing out scale, slimes, and other deposits in the cells. Scaling in the electrode compartments is minimized due to the continuous alternation of the environment from basic to acidic in the cells. Reversal of polarity, therefore, eliminates the need to continuously add acid and/or SHMP. Some cleaning of the membrane stacks is still required, although with significantly reduced frequency than would be otherwise necessary.

REFERENCES

Abrams, I. M., "Countercurrent Ion Exchange with Fixed Beds," *Ind. Water Eng.*, **10**(1), 18–26 (Jan./Feb. 1973).

Ames, L. L., "The Application of Clinoptilolite for Ammonium Removal," *Proceedings of the 13th Pacific Northwest Industrial Waste Conference*, University of Washington, Seattle (1967).

Anderson, R. A., "Ion Exchange Separations," in P. A. Schweitzer (ed), *Handbook of Separation Techniques for Chemical Engineers*, McGraw-Hill, New York (1979).

Applebaum, S. B., *Demineralization by Ion Exchange in Water Treatment and Chemical Processing of Other Liquids*, Academic Press, New York (1968).

Argo, D. G., *Evaluation of Membrane Processes and Their Role in Wastewater Reclamation*, U.S. Department of the Interior, Office of Water Research and Technology, Washington, DC (November 1979).

AWWA, *Water Treatment Plant Design Manual*, American Water Works Association, New York (1969).

Betz Chemical Co., *Handbook of Industrial Water Conditioning*, 7th ed., Trevose, PA (1976).

Bonner, O. D., and Smith, L. L., *J. Phys. Chem.*, **61**, 326 (1957).

Bresler, S. A., and Miller, E. F., "Economics of Ion Exchange Techniques for Municipal Water Quality Improvement," *JAWWA*, **64**(11), 764–772 (Nov. 1972).

Calmon, C., "Development of Anion Exchangers for Silica Removal," *Ind. Water Eng.*, **16**(2), 10–16 (Mar./April 1979).

Calmon, C., and Gold, H., "New Directions in Ion Exchange," *Env. Sci. and Tech.*, **10**(10), 980–984 (Oct. 1976).

Clifford, D. A., and Weber, W. J., "Multi-Component Ion Exchange: Nitrate Removal Process With Land Disposal Regenerant," *Ind. Water. Eng.*, **15**(2), 18–26 (March 1978).

Clifford, D., "Nitrate Removal from Water Supplies by Ion Exchange," Ph.D. thesis, University Microfilms International, Ann Arbor, MI (1976).

Diamond Shamrock Chemical Co., *Duolite Ion Exchange Manual*, Redwood City, CA (1969).

Dorfner, K., *Ion Exchange, Properties and Applications*, Ann Arbor Science, Ann Arbor, MI (1973).

Dow Chemical Co., *Dowex Ion Exchange Resins*, Vols. 1 and 2, Midland, MI (1972).

EI DuPont De Nemours & Company, *Engineering Design Manual, Permasep Permeator*, Wilmington, DE (1978).

Helfferich, F., "Ion Exchange Kinetics," in J. Marinsky (ed.), *Ion Exchange: A Series of Advances*, Vol. 1, Marcel Dekker, New York (1966).

Helfferich, F., *Ion Exchange*, McGraw-Hill, New York (1962).

Helfferich, F., and Klein, G., *Multicomponent Chromatography*, Marcel Dekker, New York (1970).

USAID, *The USAID Desalination Manual*, International Desalination and Environmental Association (August 1980).

Johnson, M. S., and Musterman, J. L., "Removal of Sulfate by Ion Exchange," *JAWWA*, **71**(6), 343–348 (June 1979).

Jorgensen, S. E., et al., "Equilibrium and Capacity Data of Clinoptilolite," *Water Res.*, **13**(2), 159–165 (Feb. 1979).

Kemmer, F. A. (ed.), *NALCO Water Handbook*, McGraw-Hill, New York (1979).

Klein, G., Tondeur, D., and Vermeulen, T., "Multicomponent Ion Exchange in Fixed Beds," *I.E.C. Fund.*, **6**(3), 351–361 (August 1967).

Kolle, W., "Resin Treatment Improves High Color Groundwater," *Water and Sewage Works*, **126**(1), 68–69 (Jan. 1979).

Kubli, H., "On the Separation of Anions by Adsorption on Alumina," *Helv. Chem. Acta*, **30**(2), 453–463 (1947).

Kunin, R., *Ion Exchange Resins*, 2nd ed., Wiley, New York (1958).

Lonsdale, H. K., and Podall, H. E. (eds.), *Reverse Osmosis Membrane Research*, Plenum Press, New York (1972).

Merten, U. (ed.), *Desalination by Reverse Osmosis*, MIT Press, Cambridge, MA (1966).

Office of Water Research and Technology, *Reverse Osmosis Technical Manual*, Washington, DC (July 1979).

Peverson, S., *Ann., N.Y. Acad. Sci.*, **57**, 144 (1953).

Rohm and Haas Co., *Amberlite Ion Exchange Resin*, Vols. 1 and 2, Philadelphia (1973).

Slater, M., "Continuous Ion Exchange: A Survey of Recent Applications," *Effluent and Water Treatment Journal* (Sept. 1981) Vol. 21, No. 9, p 416–422.

Sorg, T., and Logsdon, G., "Treatment Technologics to Meet the Interim Primary Drinking Water Regulations for Inorganics," *JAWWA*, Part 1, **70**(2), 105–112 (Feb. 1978); Part 2, **70**(7), 379–393 (July 1978); Part 3, **70**(12), 680–691 (Dec. 1978).

Sourirajan, S. (ed.), *Reverse Osmosis and Synthetic Membranes*, National Research Council, Canada (1977).

Sourirajan, S., "Reverse Osmosis: A New Field of Applied Chemistry and Chemical Engineering," *ACS Symposium on Synthetic Membranes, and Their Applications*, San Francisco (1980).

Vermeulen, T., Klein, G., and Hiester, N., "Adsorption and Ion Exchange," in R. Perry and C. Chilton (eds.), *Chemical Engineer's Handbook*, 5th ed., McGraw-Hill, New York (1973).

Weber, W. J., Jr., *Physiochemical Processes for Water Quality Control*, Wiley-Interscience, New York (1972).

—11—

Gas Transfer

PROCESS DESCRIPTION

The transfer of material from the gas phase to water or from water to gas plays an important role in water treatment. Oxidation of reduced inorganic compounds and destruction of pathogenic bacteria can be achieved by dissolution of chlorine or ozone gas in water. Conversely, chemical conditioning of certain low-pH groundwaters having excess carbon dioxide (CO_2) depends on removal of the gas from water. The addition of gases to water is known as *gas absorption,* while the opposite process is designated *desorption* or, more commonly, as *stripping*.

In both processes, material is transferred from one bulk phase to another across a gas–liquid interface. As was discussed in Chapter 5, mass transfer occurs in the direction of decreasing concentration. For any given gas transfer problem, equipment must be selected and sized to provide the maximum rate of mass transfer at a minimum cost. Analytical design of this equipment requires knowledge of gas–water equilibria, equipment hydraulics, and the effects of system variables on mass transfer coefficients. This chapter covers these topics and their application to process design of absorption and stripping equipment. Mechanical equipment details and commonly used design criteria for gas–liquid contacting systems will be discussed in Chapter 21.

Applications in Water Treatment

A wide range of water treatment objectives can be achieved through gas transfer, as summarized in Table 11-1. Absorption of ozone (O_3), chlorine (Cl_2), or chlorine dioxide (ClO_2) in water produces chemical species that inactivate various microorganisms and oxidize selected organic and inorganic contaminants. Oxygen (O_2), either as a pure gas or in air, is often added to water to accelerate oxidation of various inorganic ions. Conversely, this slightly soluble gas (9 g/m^3 at 20°C) must be stripped from boiler feedwater to minimize corrosion. The process is often applied in an attempt to strip taste- and odor-producing compounds from the water and into the air by contacting the two phases in appropriate equipment.

Air stripping has also been found effective for removal of ammonia gas from wastewaters and removal of a variety of volatile organic compounds that have been found in ground and surface waters with alarming frequency in the United States.

Gas–Liquid Contacting System

Equipment for gas–liquid contacting is designed to create and continuously renew the interfacial area between the two phases, across which mass trans-

237

TABLE 11-1. Applications of Gas Transfer in Water Treatment

Examples	Water Treatment Objectives
Gas Absorption	
O_2	Oxidation of Fe^{2+}, Mn^{2+}, S^{2-}; lake destratification
O_3	Disinfection, color removal, oxidation of selected organic compounds
Cl_2	Disinfection; oxidation of Fe^{2+}, Mn^{2+}, H_2S
ClO_2	Disinfection
CO_2	pH control
SO_2	Dechlorination
NH_3	Chloramine formation for disinfection
Stripping	
CO_2	Corrosion control
O_2	Corrosion control
H_2S	Odor control
NH_3	Nutrient removal
Volatile organics (e.g., CH_4, $CHCl_3$, C_2HCl_3)	Taste and odor control, removal of potential carcinogens

fer occurs. Increasing the interfacial area produced per unit of mechanical energy expended will increase the efficiency of mass transfer. The two principal categories of contacting systems are gas dispersed in water and water dispersed in gas. Examples of the first category include injection of gas bubbles into a quiescent or mechanical mixed column or tank of water and intense mechanical agitation of water in a tank, creating contact between the bulk liquid and the air–water interface at the tank surface.

Contacting equipment in the second category are designed to produce thin films or small droplets of water that will promote rapid mass transfer. Droplets are produced by spray devices designed to provide the desired droplet size and contact time with the gas phase (usually air). Thin films of water are created in cascade trays and packed towers or columns. Water flows by gravity over the surfaces of assorted materials placed in the column or tower, which continuously disrupts the liquid films produced.

Each type of equipment offers process and cost advantages for a specific gas transfer problem. Se-

lection of the appropriate equipment is based on relative removal efficiencies, available hydraulic head, ease of maintenance, and, of course, operating costs.

Flow Arrangements

In addition to selecting the appropriate equipment configuration for gas–liquid contacting, the engineer must also choose the proper flow arrangement. Several possible alternatives are shown in Figure 11-1. In packed columns, the gas and water flow in opposite directions (countercurrent) or at right angles to each other (cross-flow). Gas injection systems are often operated in the cross-flow arrangement, with bubbles rising vertically through the water flowing horizontally (e.g., activated sludge), or in the counterflow arrangement with the water entering the top of the columns and the gas entering the bottom (e.g., ozonation). The relative advantages of alternative flow configurations will be discussed subsequently.

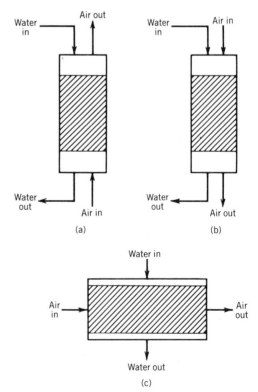

FIGURE 11-1. Several configurations of gas transfer equipment: (a) countercurrent flow, (b) cocurrent flow; and (c) cross-flow.

GAS–LIQUID EQUILIBRIUM

When gas-free water is exposed to air, oxygen and nitrogen will pass from the gas phase into the water until the concentration of these gases in the water reaches equilibrium. Conversely, if water in deep wells is brought to the ground surface, dissolved gases such as methane or carbon dioxide will be released to the air because gas concentrations in the deep well water exceed equilibrium conditions at the lower pressure and higher temperature. The eruption of a carbonated beverage after it is opened is a more familiar example. In each case, the driving force for mass transfer is the difference between the existing and equilibrium concentrations in the two phases as discussed in Chapter 5.

Henry's Law

The equilibrium concentrations of gases or volatile liquids in water depends on the temperature of the phases, the total pressure, and the molecular interactions occurring between the dissolved substance (solute) and water (solvent). At equilibrium, the concentration or partial pressure of a substance in the gas phase is proportional to its concentration in the liquid phase (Henry's law). When the gas phase is ideal, the total pressure of a gaseous mixture is the sum of the partial pressure of the individual components (Dalton's law). Combining these two laws, we obtain

$$Y_i = \frac{X_i H_i}{P_T} \qquad (11\text{-}1)$$

Where Y_i is the mole fraction of i in the gas phase, X_i is the mole fraction of i in the liquid, H_i is Henry's constant, and P_T is the total pressure.

Knowing Henry's constant and the total pressure permits computation of the equilibrium concentration in a gas–liquid or vapor–liquid system. Qualitatively, the greater the Henry constant, the more easily a compound can be removed from solution. Conversely, a low value of the Henry constant indicates high solubility of the compound in water.

Effect of Temperature and Water Chemistry

In general, increasing the system temperature will increase the partial pressure of a component in the gas phase in equilibrium with a specified solution concentration. From a thermodynamic analysis, the temperature dependence of the Henry constant can be modeled by a van't Hoff-type relation, given in integrated form by

$$\log H = \frac{-\Delta H}{RT} + K \qquad (11\text{-}2)$$

where ΔH is the heat absorbed in the evaporation of 1 mole of component A from solution at constant temperature and pressure, R is the gas constant (1.987 kcal/K kmole), and K is an empirical constant. Tables 11-2 and 11-3 contain values of ΔH and K for several gases and volatile organics obtained by linear regression of literature data. Use of this model is illustrated in Example 11-1.

EXAMPLE 11-1

Problem

Treated water in an enclosed storage reservoir contains a chloroform concentration of 100 μg/L. Assuming equilibrium, what is the concentration of

TABLE 11-2. Henry's Constants and Temperature Dependence for Gases in Water[a]

Gas	Formula	Henry's Constant at 20°C (atm)[b]	Temperature Dependence ΔH (kcal/kmole, $\times 10^{-3}$)	K
Oxygen	O_2	4.3×10^4	1.45	7.11
Nitrogen	N_2	8.6×10^4	1.12	6.85
Methane	CH_4	3.8×10^4	1.54	7.22
Ozone	O_3	5.0×10^3	2.52	8.05
Carbon dioxide	CO_2	1.51×10^2	2.07	6.73
Hydrogen sulfide	H_2S	5.15×10^2	1.85	5.88
Chlorine	Cl_2	5.85×10^2	1.74	5.75
Chlorine dioxide	ClO_2	54	2.93	6.76
Sulfur dioxide	SO_2	38	2.40	5.68
Ammonia	NH_3	0.76	3.75	6.31

[a] From Landolt-Boernstein (1976) and International Critical Tables.
[b] 1 atm = 101.3 kPa.

TABLE 11-3. Henry's Constants and Temperature Dependence for Selected Volatile Organics in Water[a]

Gas	Formula	Henry's Constant at 20°C (atm)[b]	Temperature Dependence ΔH (kcal/kmole, $\times 10^{-3}$)	K
Vinyl chloride	CH_2CHCl	3.55×10^5	—	—
Carbon tetrachloride	CCl_4	1.29×10^3	4.05	10.06
Tetrachloroethylene	C_2Cl_4	1.10×10^3	4.29	10.38
Trichloroethylene	$CCHCl_3$	550	3.41	8.59
Benzene	C_6H_6	240	3.68	8.68
Chloroform	$CHCl_3$	170	4.00	9.10
Bromoform	$CHBr_3$	35	—	—
Pentachlorophenol	C_6OHCl_5	0.12	—	—

[a] From Landolt-Boerstein (1976).
[b] Computed from water solubility data and partial pressure of pure liquid at specified temperature.

chloroform (MW = 119) in the air space in the reservoir at 10°C? □

Solution

Compute Henry's constant for $CHCl_3$ at 10°C (283 K).

$$\log H = \frac{-\Delta H}{RT} + K$$

$$\log H = \frac{-4 \times 10^3 \text{ kcal/kmole}}{(1.987 \text{ kcal/kmole K})(283 \text{ K})} + 9.10$$

$$H = 97 \text{ atm} (9.82 \times 10^3 \text{ kPa})$$

From Eq. 11-1,

$$y_i = \frac{H}{P_T} x_i$$

$$x_i = \frac{C_i}{C} = \frac{(100 \ \mu g/L \div 119 \text{ g/mole})}{55.6 \text{ kmole/m}^3}$$

$$\times (10^{-6} \text{ g/}\mu g) = 1.51 \times 10^{-8}$$

$$y_i = \frac{97 \text{ atm}}{1 \text{ atm}} (1.51 \times 10^{-8}) = 1.46 \times 10^{-6}$$

The partial pressure is

$$p_i = y_i P_T = 1.46 \times 10^{-6} \text{ atm}$$

Assume ideal gas and apply the ideal gas law:

$$p_i V = n_i RT$$

$$\frac{n_i}{V} = \frac{p_i}{RT} = \frac{1.46 \times 10^{-6} \text{ atm}}{(.08285 \text{ atm-L/mole K}) \times (283 \text{ K})}$$

$$= 6.31 \times 10^{-8} \text{ mole/L} = 7.51 \times 10^{-6} \text{ gm/L}$$

Therefore,

$$C_i = 7.51 \text{ mg/m}^3 \ (6 \times 10^{-6} \text{ kg/kg dry air}) \text{ or 6 ppm}$$

NIOSH Occupational Exposure Standard (1974) =
10 ppm □

Gas solubility data are reported in a variety of formats. In lieu of, or in addition to, Henry's constants, some texts and handbooks use the Bunsen, or absorption, coefficient b_B, defined as the volume of gas (in milliliters) at standard temperature and pressure (0°C and 1 atm) dissolved in 1 mL of water at the partial pressure of the gas (in atmospheres). It can be shown that b_B and Henry's constant are related by

$$H = \frac{1.246 \times 10^3}{b_B} \qquad (11-3)$$

for dilute solutions ($X_i < 10^{-2}$).

The Henry's constants listed in Tables 11-2 and 11-3 apply to solubilities of gases and volatile organics in pure water. As the salt content of the water increases, solubilities will gradually decrease. For example, the solubility of oxygen in seawater at 20° is 7.7 mg/L, about 13 percent lower than the solubility in distilled water. Empirical corrections to the change in Henry's constants with changing ionic composition are summarized by Danckwerts (1970).

RATE OF GAS TRANSFER

The rate at which a gas or volatile liquid is transferred across the air–water interface depends on the extent of deviation from equilibrium and the intensity of mixing at the interface. In order to pass from one phase to another, a dissolved gas must proceed through consecutive steps:

1. transfer from the bulk fluid to the interface;
2. transfer across the interface; and
3. transfer away from the interface into the bulk of the new phase.

We wish to determine the flux, N_A, defined as the quantity of dissolved gas, A, transferred per unit time per unit area of interface, for the design of gas–liquid contacting equipment for both absorption and stripping problems.

Gas Transfer without Chemical Reaction

Assuming that no chemical reactions are occurring and when flow of the fluids is laminar, the rate of mass transfer can be estimated analytically for a variety of physical systems (Sherwood et al., 1975). For most gas–liquid contacting systems of practical interest, however, the flow of one or both of the fluids is turbulent, and few useful analytical solutions are available. Recourse must be made to empirical approaches as developed in the section on mass transfer between phases in Chapter 5.

Controlling Resistances. The overall mass transfer coefficients incorporate the mass transfer resistances in both fluid phases. For gases and volatile liquids obeying Henry's law, $p^* = H'C$ where $H' = H/C_{H_2O}$ with C_{H_2O} = molar density of water (55.6 kmole/m³). Combining this equilibrium condition with Eqs. 5-19, it can be shown that the various mass transfer coefficients are related by

$$\frac{1}{K_G} = \frac{1}{k_G} + \frac{H'}{k_L} \qquad (11\text{-}4)$$

$$\frac{1}{K_L} = \frac{1}{k_L} + \frac{1}{H'k_G} \qquad (11\text{-}5)$$

Depending on the magnitude of the local mass transfer coefficient and the magnitude of Henry's constant, the rate of mass transfer may be controlled either by resistances in the liquid or the gas phase. Rearranging Eq. 11-5, the relative contribution of the two resistances is shown by

$$\frac{k_L}{K_L} = 1 + \frac{k_L}{k_G H'} \qquad (11\text{-}6)$$

Literature values for k_G are scarce; however, some data suggest (Liss, 1974; Makay, 1979) that the ratio for k_L/k_G may be between 0.1 and 0.5 m³/kmole in water, the value depending on the type of mass transfer device. Figure 11-2 illustrates the dependence of K_L on the two local mass transfer coefficients and the Henry constant.

As can be seen, for $k_L/k_G = 0.1$ and a Henry constant of 8.0 atm., half of the total resistance to mass transfer occurs in the liquid side (film) of the interface.

As seen in Tables 11-2 and 11-3, many organic compounds found in surface or groundwaters have Henry constants greater than 50 atm. Thus, the overall rate of mass transfer should generally be controlled by liquid phase resistance to mass transfer. However, additional data on local gas phase transfer coefficients are needed to confirm this.

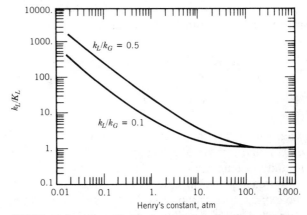

FIGURE 11-2. Dependence of overall mass transfer coefficient, K_L on Henry's constant and ratio of local mass transfer coefficients: curve A, $k_L/k_G = 0.1$; curve B, $k_L/k_G = 0.5$.

Mass Transfer Coefficients

Theoretical models have not been successful in predicting the liquid phase mass transfer coefficient, k_L, except under laminar flow conditions. Thus, in turbulent flow, experimental determination of k_L is required. A similar situation exists for prediction of the gas phase mass transfer coefficient, k_G. When experimental data cannot be obtained, the engineer can use mass transfer correlations developed for most of the gas–liquid contacting devices commonly used in water treatment. These correlations are summarized in several handbooks and texts (Perry, 1973; Sherwood et al., 1975; Treybal, 1980) and will be discussed where appropriate.

For most gas–liquid problems in water treatment, resistances in the liquid phase control the rate of mass transfer. Typically, in the absence of any chemical reactions, k_L values range from 10^{-4} to 10^{-1} m/sec in common gas–liquid contacting devices such as packed columns, diffused air, and mechanically agitated diffused air systems. This contrasts with k_L values reported in natural water systems (lakes, oceans, and rivers) ranging from 10^{-7} m/sec for mass transfer of low-molecular-weight organic compounds in quiescent water up to 7×10^{-5} m/sec for reaeration in highly turbulent rivers (e.g., Fortescue and Pearson, 1967; Liss and Slater, 1974; Emerson, 1975; Schwarzenbach et al., 1979).

Interfacial Area. The *volumetric rate of mass transfer* is defined as the quantity of material transferred per unit time per unit volume. Computation of this rate requires estimation of the total effective interfacial area A_m in the system across which mass transfer is occurring. Using Eq. 11-7, the volumetric rate of mass transfer M_A is seen to be

$$M_A = \frac{A_m}{V} N_A = K_L a (C_A^* - C_A)$$

$$= K_G a (p_A - p_A^*) \qquad (11\text{-}7)$$

where A_m is the interfacial area in the system volume V and $a = A_m/V$ is the specific interfacial area in units of length^{-1}.

Increasing the specific interfacial area requires the expenditure of work on the system. Thus, optimum design of gas–liquid contacting devices requires a balance between increased mass transfer and increased energy costs. In turbulent fluid flow, the specific interfacial area is difficult to determine. Common practice is to measure the product of the mass transfer coefficient and a, $K_L a$, or $K_G a$, known as the volumetric mass transfer coefficient. Many mass transfer correlations provide data on the volumetric coefficients when the interfacial area is poorly defined due to irregular geometries or highly turbulent fluid systems.

GAS TRANSFER WITH CHEMICAL REACTION

If a chemical reaction occurs after a gas passes through the gas–water interface, the rate of absorption will generally be increased relative to pure physical gas absorption. Several gases commonly used in water treatment applications, notably chlorine, sulfur dioxide, carbon dioxide, and ozone, undergo hydrolysis reactions or rapid chemical reactions with other solutes in water. The increase in rate of mass transfer must be accounted for in the process design of absorption equipment. The magnitude of this increase depends in complex ways on the type of chemical reaction (reversible, irreversible, series), reaction order, rate constant, concentration of reactants, and the solute molecular diffusion constants (Danckwerts, 1970) as discussed in Chapter 5.

While innumerable reaction combinations are possible, chemical reactions for gases in water used in water treatment fall mainly in two categories, first-order irreversible reactions and rapid or instantaneous reversible reactions. Consider the first case, a first-order irreversible reaction in the water phase, with rate constant k_1, as illustrated by the nearly irreversible reaction of SO_2 in water:

$$SO_2(g) + H_2O \xrightarrow{k_1} HSO_3^- + H^+$$

Using the two-resistance, or two-film, model, as it is often called, Danckwerts (1970) has shown that the volumetric rate of gas transfer is given by

$$M_A = k_L a \left[C_A^* - \frac{C_A}{\cosh M^{1/2}} \right] \frac{M^{1/2}}{\tanh M^{1/2}} \quad (11\text{-}8)$$

where $M = D_A k_1 / k_L^2$ and D_A is the molecular diffusion coefficient of the solution in water. This assumes that mass transfer resistances in the gas phase are negligible. When a single gas is absorbed, this condition is always satisfied. If $M^{1/2} \gg 1$, the reaction occurs entirely in the water near the interface, and no gas reaches the bulk fluid, so that $C_A = 0$. Thus, Eq. 11-8 becomes

$$M_A = aC_A^*(D_A k_1)^{1/2} \qquad (11-9)$$

For the case of SO_2, k_1 at 20°C is 3.4×10^6 sec^{-1}. Using $D_A = 10^{-9}$ m^2/sec and $k_L = 10^{-4}$ m/sec, a typical value of the mass transfer coefficient in a diffused gas system, $M = 580 \gg 1$, indicating that the SO_2 will disappear shortly after diffusing across the gas–liquid interface.

Comparing Eq. 11-8 with the rate of absorption without reaction Eq. 11-7, the absorption rate due to physical absorption alone has been increased by a factor $(D_A k_1)^{1/2}/k_L$. In general,

$$M_A = k_L a C_A^* E \qquad (11-10)$$

where E is the so-called enhancement factor, which predicts the increase in absorption rates due to chemical reaction and is equal to $(D_A k_1/k_L^2)^{1/2}$ for first-order irreversible reactions. Enhancement factors for many other chemical reaction types are summarized by several authors (Danckwerts, 1970; Sherwood et al., 1975). Example 11-3 illustrates this concept further.

EXAMPLE 11-2

Many gases added to water during water treatment undergo rapid hydrolysis reactions that accelerate the rate of absorption. Compare the initial relative rates of absorption of pure gases chlorine (Cl_2), carbon dioxide (CO_2), and sulfur dioxide (SO_2) with the rate of oxygen absorption in oxygen-free distilled water, pH at 20°C.

Solution

Oxygen is absorbed by physical absorption only. With the initial bulk concentration $C_A = 0$, the rate of absorption (Eq. 11-13) becomes

$$M_A = K_L a C_A^*$$

Chlorine reacts with water as follows:

(a) $Cl_2(g) + H_2O \rightleftharpoons HOCl + H^+ + Cl^-$

(b) $HOCl \rightleftharpoons OCl^- + H^+$

At pH 8, most of the HOCl dissociates instantaneously to OCl^-. Thus, reaction (a) can be considered a pseudo-first-order irreversible reaction,

$$CL_2(g) \xrightarrow{k_1} \text{products}$$

with rate constant $k_1 = 11$ sec^{-1} at 20°C. For $C_A = 0$, the enhanced rate of absorption becomes

$$E = \frac{M^{1/2}}{\tanh M^{1/2}}$$

$$(M)^{1/2} = \frac{(D_A k_1)^{1/2}}{k_L}$$

$$= \frac{(10^{-9}\ \text{m}^2/\text{sec}\ 11\ \text{sec}^{-1})^{1/2}}{10^{-4}\ \text{m/sec}} = 1.05$$

Therefore,

$$\tanh(M)^{1/2} = 0.78$$

and

$$E = 1.34$$

Thus, the initial rate of absorption is approximately one-third faster than pure physical absorption.

For carbon dioxide, the reactions are

(c) $CO_2(g) + H_2O \rightleftharpoons H_2CO_3$

(d) $H_2CO_3 \rightleftharpoons HCO_3^- + H^+$

(e) $HCO_3^- \rightleftharpoons CO_3^- + H^+$

Again, the absorption of this gas is added by the reaction

$$CO_2(g) \xrightarrow{k_1} \text{products}$$

which approximates a pseudo-first-order irreversible reaction with $k_1 = 0.02$ sec^{-1}. In this case,

$$M^{1/2} \cong 4.5 \times 10^{-2}$$

and

$$E = 1.01$$

The reaction is too slow to enhance the rate of absorption.

For SO$_2$, the reactions are

(f) $SO_2(g) + H_2O \rightarrow H_2SO_3$

(g) $H_2SO_3 \rightleftharpoons HSO_3^- + H^+$

(h) $HSO_3^- \rightleftharpoons SO_3^= + H^+$

At pH 8, H$_2$SO$_3$ rapidly dissociates, and reaction (f) becomes pseudo first order, given as

$$SO_2(g) \xrightarrow{k_1} \text{products}$$

with $k_1 = 3.4 \times 10^6$ sec^{-1} at 20°C

Thus,

$$M^{1/2} = 583$$

and

$$E = M^{1/2} = 583$$

For SO$_2$, the rapid hydrolysis reaction dramatically enhances the rate of absorption.   □

GAS–LIQUID CONTACTING SYSTEMS

A variety of gas transfer processes are used in water treatment. In general, they can be classified as either gas in water or water in air as illustrated in Table 11-4. For many applications of gas transfer in water treatment, such as stripping of volatile contaminants or addition of a reactant gas, one or more of the processes may be used. Capital and operating costs generally are the deciding factors. Most gas transfer equipment is relatively simple. Complexity or heavy maintenance is typically not a major consideration. The capital cost of any one of these processes is closely related to the mass transfer efficiency. The poorer the transfer efficiency, the larger the facility required to achieve a certain removal. Operating cost is primarily a function of the hydraulics of the process and the method of gas dispersion.

The exception to these generalizations is the addition of reactive gases such as chlorine, ammonia, or ozone. Here, the principal cost is the chemical itself and/or generation of the compound and to a

TABLE 11-4. Characteristics of Gas–Water Contacting Systems

Type of System	Flow Patterns	Mass Transfer Characteristics		Gas–Water Pressure Drop	Water Treatment Applications
		Interfacial Area/Volume Reactor (m^2/m^3)	Water Volume Fraction		
Gas in water					
Gas diffusers, Spargers	Cross-flow or cocurrent	20–200	0.9–0.98	Gas compression required	Mixing, O$_2$ absorption, oxidation
Diffuser with mechanical mixing	Cross-flow	30–500	0.8–0.9	Gas compression required	O$_2$, O$_3$ absorption, oxidation
Mechanical mixing	Complete mixing	50–600	<0.9	Mixer power	O$_2$ absorption, stripping of organics
Ejectors (venturi)	Cocurrent	—	<0.9	Gas compression	Absorption of CO$_2$, Cl$_2$
Water in air					
Spray systems	Cocurrent, countercurrent	20–60	0.005	High water pressure drop	Stripping of CO$_2$, NH$_3$, organics
Cascade systems	Cocurrent	—	0.2–0.5	Water pumping	Stripping of gases
Packed towers	Cross-flow countercurrent	50–100	0.05–0.1	Gas, 50–400 N/m^2/m; H$_2$O pumping	Cooling towers, stripping of NH$_3$, CO$_2$, organics
Packed columns	Countercurrent	100–150	—	Gas, 50–400 N/m^2/m; H$_2$O pumping	Stripping of CO$_2$, O$_2$

lesser degree the feeding equipment and gas transfer contact tank (if any).

Another important consideration in selecting a process for a specific application is the upper limit of transfer efficiency that can be economically achieved by the process. Many of the processes shown in Table 11-1 have limited transfer efficiency. Commercial cascade aerators cannot achieve greater than 50 or 60 percent removal of chloroform, a relatively volatile organic contaminant. Two types of gas transfer applications in water treatment illustrate process selection principals: (1) absorption of oxygen and CO_2 and reactive gases and (2) desorption or stripping of volatile compounds.

Absorption

With the exception of air–water contact systems, gas adsorption facilities are generally designed for diffused gas application. The highly soluble gases used in water treatment typically require only moderate and not very efficient gas contact facilities. In many cases, the gas would be applied by simple diffuser in the existing channel or contact basin. The less soluble gases require more efficiently designed gas transfer facilities, often with the capability to recycle off gases. The absorption of gases from air such as O_2 and CO_2 present a special case of gas transfer in water treatment. These systems rely on the ambient partial pressure. Consequently, there is little cost in providing the gas and therefore no need for recycling of off gases.

Because of their rapid hydrolysis reactions, ammonia, sulfur dioxide, and chlorine are highly soluble in water. Sulfur dioxide presents a special case as the hydrolysis may enhance absorption directly by increasing the mass transfer coefficient. The overall absorption of these gases in the diffused gas system is efficient because of the low aqueous concentration of the diffusing gas. Near neutral pH, less than 5 percent of the aqueous concentration is NH_3 (aqueous), Cl_2 (aqueous), or SO_2 (aqueous). With proper mixing and adequate gas dispersion, complete absorption of the gas can be achieved. Because of the efficient transfer, specially designed gas transfer facilities are generally not necessary. Simple systems for these gases may consist of application to an existing transmission pipe, channel, or process basin.

In contrast, pure oxygen and ozone are relatively insoluble in water and may require special, highly efficient gas transfer facilities. Pure oxygen is included in this category because of the high production costs and necessity for recycling off gases. Unlike the highly soluble gases, ozone and oxygen do not hydrolyze upon addition to water. Furthermore, the Henry constants are quite low. Even at high partial pressures of gases, the equilibrium concentration in water is low. Transfer of these less soluble gases is limited by the equilibrium concentration. The overall efficiency of the process can be increased by the reactivity of these gases in a water having a high oxygen demand. Rapid reduction by oxidizable inorganics and organics will maintain a low apparent equilibrium concentration. Consequently, the driving force for mass transfer is substantially greater than would be in the pure water system.

Absorption of CO_2 and O_2 represents a special case of gas absorption since they can be applied as either a pure gas or as a mixture. The absorption of CO_2 or O_2 from air can either be designed as a gas-in-liquid or liquid-in-gas system. Liquid-in-gas systems such as packed towers or slat, countercurrent towers can be designed for much higher transfer efficiencies than can most air-in-water systems. This is principally due to the difference in interfacial area provided by packed systems, as indicated in Table 11-4, and the more efficient countercurrent plug flow contacting in both phases in the water-in-air systems.

Desorption

Often, stripping of volatile organics may require a more efficient system such as a slat tray or packed tower. Gas-in-liquid systems can be used for volatile organic stripping where the volatility is quite high and the required removals are fairly low. The two principal factors controlling the process selection for desorption are the desired degree of removal and the Henry constant. This relation is summarized in Figure 11-3. When required removals are less than 90 percent, both spray and diffused aeration systems including mechanical aeration may be economically attractive. In general, the usefulness of a diffused aeration system is the modification of an existing contact basin or reservoir with diffusers or surface aerators. Design of a diffused aeration basin specifically for stripping is impractical because of the relatively high cost of the facility.

The required percent removal of volatile organics such as chloroform will be determined by federal

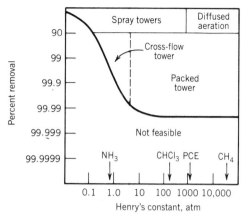

FIGURE 11-3. Schematic diagram for selection of feasible aeration process for control of volatile compounds (Kavanaugh and Trussell, 1981).

or state drinking water regulations. Under such restrictions on finished water quality, either the design procedure for the stripping process must be exact and/or the process must be pilot tested to ensure that the process will provide the requisite removal. Typical gas-in-liquid systems in water treatment are not easily scaled up from pilot tests, nor are design procedures as well established compared to packed towers.

Liquid-in-gas systems generally require pumping the water an additional 6–20 ft depending on the interfacial surface area required. Methane, CO_2, and hydrogen sulfide occasionally encountered in well water are quite easily removed. An aeration tower height of 4–10 ft is generally sufficient. This is easily provided by a multiple-tray or cascade-type aerator. It is often possible to modify a well pump to lift water into an aerator or spray nozzle, which can then spray into an existing reservoir. Using a spray nozzle in such applications avoids elevating the cascade or slat tower so that the water will then flow by gravity into the reservoir. Tower aerators mounted on the ground otherwise require repumping into the reservoir or water distribution system. Cascade or coke tray towers and spray nozzles are by far the simplest design for the liquid-in-gas systems. The following sections will discuss some of the important factors in the design and operation of the most common gas transfer processes listed in Table 11-4.

PACKED TOWERS

Efficient mass transfer can be achieved using packed towers consisting of a cylindrical column or a rectangular tower containing packing that disrupts the flow of liquid, thus producing and renewing the air and water interface. Various types of packing shapes and materials are available commercially. They are placed either randomly or in a fixed configuration inside the column or tower.

Water is pumped to the top of these units and allowed to flow by gravity countercurrent to the air flow produced either by natural or forced draft. This system is characterized by high liquid interfacial area compared to the total volume of liquid in the column. Either countercurrent or cross-flow process configurations can be used.

Generally, packed towers have void volumes greater than 90 percent, which minimizes air pressure drop through the tower. The principal design parameters include the volumetric air–water ratio, which ranges from 10 to 5000, the packing type, size and depth, column diameter, water and air loading rates (superficial velocities), and the gas pressure drop. The process is widely used in the chemical process industries for gas–liquid contacting and for cooling.

Design of Packed Towers

In a countercurrent packed tower, the liquid feed flows downward countercurrent to the rising airflow. A dissolved contaminant will be stripped from the water into the air depending on the relative volatility (Henry's constant) of the contaminant and the column design. The water concentration of the contaminant decreases as it passes through the column. Correspondingly, the air concentration or partial pressure of the contaminant will increase from the bottom to the top of the column. For any given substance, the required dimensions of the column, the size of the packing, and the necessary air flowrate must be known to achieve a specified degree of removal.

The height of the column packing, Z, required to achieve the desired removal of solute is the product of two quantities, the height of a transfer unit (HTU) and the number of transfer units (NTU). That is,

$$Z = (HTU)(NTU) \qquad (11\text{-}11)$$

As shown in Chapter 5, HTU is inversely proportional to the product of the overall liquid phase mass transfer coefficient and the interfacial area, $K_L a$, and characterizes the efficiency of mass transfer from water to air. The integral term NTU characterizes the difficulty of removing solute A from the liquid phase. When the difference between the equilibrium and actual concentration of solute A is large, the mass transfer driving force and the rate of removal will be large. As the actual concentration approaches the equilibrium value, however, the integral term approaches infinity and separation becomes exceedingly difficult, requiring a large number of transfer units and an unreasonably tall packed column.

For dilute solutions ($X_A < 10^{-2}$) and solutes obeying Henry's law, the integral expression for NTU can be solved analytically. At any point in the packed column, the mole fractions of A in the solute, X_A, and the gas, Y_A, are related by a material balance around either the upper or lower section of the column.

In addition, the mole fraction of A in the air is related to the equilibrium mole fraction of A in the water through Henry's law.

Substituting these relationships in the integral expression for NTU with $y_1 = 0$, the number of transfer units becomes

$$NTU = \frac{R}{R - 1} \ln \frac{x_1/x_2(R - 1) + 1}{R} \quad (11\text{-}12)$$

where R is the stripping factor defined as $R = H_A G/L$; x_1 is the influent mole fraction; x_2 is the effluent mole fraction; G is the superficial molar air flowrate (kmole sec · m^2); and L the superficial molar water flowrate (kmole/sec · m^2). Thus, the NTU depends on the desired removal efficiency, the air–water ratio, and the Henry constant.

Equation 11-12 has been plotted in Figure 11-4. For a desired removal efficiency and a specific Henry constant, the number of transfer units in a packed column can be computed for any given stripping factor or ratio of the air to water flowrates. As shown in Figure 11-5, for a stripping factor greater than 3, little improvement in the NTU is observed.

Computation of the height of the transfer unit requires data on mass transfer coefficients for the system under consideration. Often, packing manufacturers supply mass transfer data as a function of temperature and flowrates. Values of $K_L a$ also may be determined from pilot studies on the contaminated water. In the absence of this data, however, mass transfer correlations from the literature can be used for feasibility analyses. A typical empirical

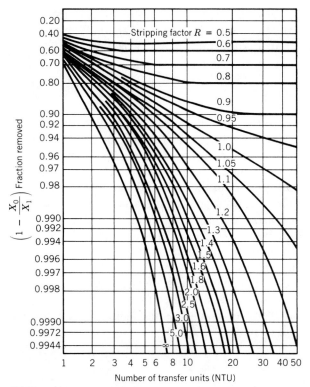

FIGURE 11-4. Dependence of NTU on removal efficiency and stripping factor (Treybal, 1980).

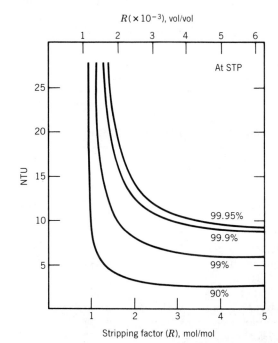

FIGURE 11-5. Relationship between stripping factor and NTU. Shows influence of stripping factor (R) and percent removal on NTUs required. *Note:* R (vol/vol) must be greater than 1 to get much removal, but increasing R ceases to yield improved results above values of 4–5 (Kavanaugh and Trussell, 1981).

correlation used for liquid phase mass transfer coefficients in towers containing randomly packed materials is shown in Eq. 11-13 (Holloway, 1940).

$$\frac{K_L a}{D_A} = m \left(\frac{L'}{\mu_L}\right)^{(1-n)} \frac{\mu_L}{\rho_L D_A} 0.5 \qquad (11\text{-}13)$$

where D_A is the molecular diffusion coefficient of A in water (ft²/hr), μ_L and ρ_L are the liquid viscosity and density, L' is the liquid mass superficial velocity (lb/ft² hr), and m and n are empirical constants that depend on the type and size of packing materials. The value of m ranges between 20 and 200, and n ranges between 0.2 and 0.5. The molecular diffusion coefficient of most nonelectrolytes in water at low concentrations ranges from 0.5 to 2 × 10^{-9} m²/sec.

Where data is absent, the molecular diffusion coefficient can be estimated from a number of empirical correlations such as the Wilke–Chang correlation (Wilke, 1955).

$$D_A = K_A \frac{T}{\mu_L} \qquad (11\text{-}14)$$

where T is the absolute temperature (K) and K_A depends on the molar volume of contaminant A in water. Values of K_A for a number of compounds are shown in Table 11-5. Generally this correlation is accurate within ±10 percent.

TABLE 11-5. Computation of Molecular Diffusion Coefficients using Wilke–Chang Relation[a]

Compounds	Formula	K_A (×10⁸)	D_{AB} at 20°C × 10¹⁰ (m²/sec)
Vinyl chloride	C_2H_3Cl	3.85	11.3
Methane	CH_4	6.18	18.1
Carbon tetrachloride	CCl_4	2.76	8.08
Tetra chloroethylene	C_2Cl_4	2.57	7.52
Trichlorethylene	$CCHCl_3$	2.86	8.37
Chloromethane	CH_3Cl	4.49	13.1
1,1,1-Trichloroethane	$C_2H_3Cl_3$	2.75	8.04
Benzene	C_6H_6	3.04	8.91
Chloroform	$CHCl_3$	3.12	9.15
1,2-Dichloroethane	$C_2H_4Cl_2$	3.10	9.08
Bromoform	$CHBr_3$	2.99	8.75

[a] $D_{AB} = K_A T / \mu_B$; $T(°K)$; $\mu_B(cP)$; D_{AB} (ft²/hr) = $3.875 \times 10^4 D_{AB}$ (m²/sec).

Gas Phase Pressure Drop

The pressure drop of gas rising countercurrent to liquid flowing through a packed power typically follows the pattern illustrated in Figure 11-6. For a given liquid loading rate, gas pressure drop increases approximately in proportion to the square of the gas velocity. At very high gas rates, entrainment of the liquid by the rising gas may occur, characterized by a sudden rapid increase in the gas pressure drop.

Packed towers are usually designed to operate with a gas pressure drop well below flooding conditions. Most stripping towers are designed for gas pressure drops of 200–400 N/m² per meter of packing depth (0.25–0.5 in. H₂O/ft). For removal of volatile compounds with $H_A > 100$ atm, however, a lower design pressure drop may be more cost-effective. For any given gas loading rate and a selected design gas pressure drop, the allowable liquid loading can be obtained from data similar to that shown in Figure 11-6.

A generalized pressure drop correlation commonly used for design of towers with random packing is shown in Figure 11-7 (Treybal, 1980). At a

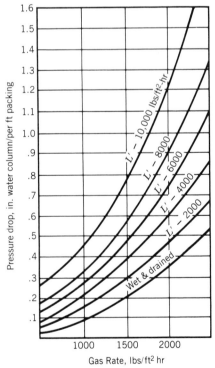

FIGURE 11-6. Influence of liquid and gas rates on pressure drop (L' = liquid rate, lb/ft² hr). Packing 1 in. Tellerettes. (Ceilcote Co., 140 Sheldon Road, Berea, Ohio 44017, Brochure 1200.)

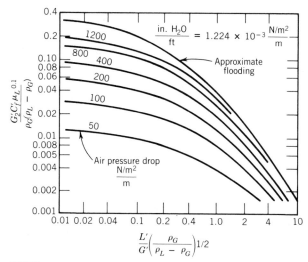

FIGURE 11-7. Generalized air pressure drop correlation in packed towers. (Adapted from Eckert, 1970.)

specified gas pressure drop, the allowable gas loading rate is specified by the gas and liquid density, the liquid viscosity, packing factor C_f, and the characteristics of the type and size of the packing selected. Packing factors for several commonly used packings are shown in Table 11-6. Once stripping factor R and the design gas pressure drop are selected, the gas loading rate is obtained from the ordinate axis in Figure 11-7 (Treybal, 1980).

When the value of the abscissa rises above 4, very high liquid loading rates produce back-cycling of the gas and unstable hydrodynamic behavior. This tends to reduce the stripping efficiency of the tower. Similarly, high gas flowrates in the region of abscissa values of less than 0.02 cause liquid entrainment, which also reduces the overall efficiency of the tower. For values of the abscissa outside

these boundaries, pilot studies are recommended to determine the effect of high or low gas flowrates on removal efficiencies. Towers operating in these regions may require additional packing depth to compensate for the efficiency reductions.

For the air–water system of 20°C and 1 atm, the abscissa in Figure 11-8 is approximately 0.0347 (L'/G'). Substituting the stripping factor R, this becomes 0.0217 (H/R). Thus, for values of $R = 5$, use of Figure 11-8 for stripping tower design is restricted to dissolved compounds with Henry constants ranging from about 4.6 to 920 atm. Pilot studies are strongly recommended for tower design when the contaminant exhibits Henry constants outside this range.

Design Procedures for Packed Towers

The packed tower design methodology can be illustrated by evaluating the impact of various design parameters on the size of packed towers. Given an influent concentration of an organic contaminant, an effluent water quality goal, flowrate and design temperature, the above methodology is applied as follows:

1. Select an efficient packing material that is expected to give good mass transfer at low gas head loss. For this packing, determine head loss and mass transfer characteristics from commercially available data.

2. Select a stripping factor.

3. Select a desirable gas phase pressure drop.

TABLE 11-6. Packing Factor C_f for Common Plastic Packings and Raschig Rings[a]

	C_f For Nominal Packing Size Shown				
Packing Type	1 in.	1½ in.	2 in.	3 in.	3½ in.
Super intalox	33	—	21	16	—
Pall rings	52	32	25	—	16
Tellerettes	40	—	20	—	—
Maspack	—	—	32	20	—
Heil-Pack	45	—	18	15	—
Raschig rings	155	95	65	37	—
Berl saddles	110	65	45	—	—

[a] From commercial literature.

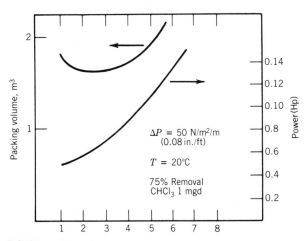

FIGURE 11-8. Influence of stripping factor on volume of packing and blower size for removal of chloroform (Kavanaugh and Trussell, 1981).

4. Determine the number of transfer units from Eq. 11-12.

5. Using the head loss correlation (Figure 11-7), determine the allowable superficial gas rate.

6. Based on the stripping factor and this gas rate, determine the liquid loading rate and the diameter of the column.

7. Compare the liquid loading rate to allowable liquid loading rates in commercially available equipment. If the liquid loading rate exceeds recommended values, reduce the air head loss and repeat the computation.

8. Repeat for various values of the stripping factor and the air head loss and determine the optimum, or least-cost, design.

Figure 11-8 illustrates the influence of the stripping factor on equipment size for removal of $CHCl_3$ in a 1-MGD plant, assuming a gas head loss of 50 $N/m^2/m$ and a design temperature of 20°C. Shown on the plot is the dependence of the total volume of packing and air blower horsepower as a function of the stripping factor. Depending on the relative costs of these two design parameters, an optimum cost can be determined. A minimum packing volume occurs at a stripping factor of approximately 2, which would be the likely economic optimum in this case.

Figure 11-9 illustrates the effect of the allowable air pressure drop on the volume of packing and the blower horsepower for chloroform removal with a

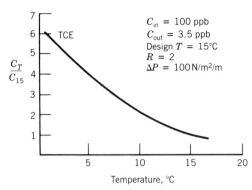

FIGURE 11-10. Influent of temperature on removal of TCE for a packed tower, design temperature 15°C (Kavanaugh and Trussell, 1981).

stripping factor of 1.5. As the air pressure drop increases, the required blower size increases proportionally. The higher pressure drop permits a smaller column diameter, however, and consequently a decrease in the overall volume of packing.

Temperature influences both the rate of mass transfer and the Henry constant and thus impacts equipment size as well as the removal efficiency in an existing packed tower. Figure 11-10 illustrates the effect of temperature changes on the removal efficiency of an existing column designed to remove trichloroethylene from an influent concentration of 100 $\mu g/L$ to an effluent concentration of 3.5 $\mu g/L$, with a design temperature of 15°C, $R = 2$, and a pressure drop of 100 $N/m^2/m$. As the temperature decreases, the removal efficiency, as expected, decreases. A reduction in temperature from 15 to 5°C will result in a fourfold increase of the predicted effluent concentration.

Often, a contaminated groundwater may contain excessive quantities of several organic contaminants, each with a different target effluent concentration. Under these conditions, the design procedure must be modified to account for differences in the removal efficiency of each compound and varying Henry constants (Kavanaugh and Trussell, 1981).

These examples illustrate the utility of the design methodology discussed. It must be emphasized, however, that the methodology is based on empirical mass transfer correlations. Pilot studies should be considered prior to equipment design. The methodology discussed above is a useful tool in designing and evaluating such pilot studies and for feasibility studies of packed columns for control of organic contaminants in groundwater.

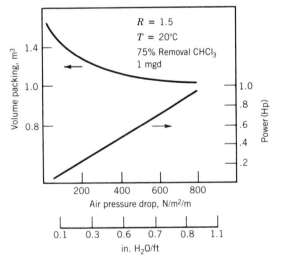

FIGURE 11-9. Effect of air pressure drop on volume of packing and blower size for removal of chloroform (Kavanaugh and Trussell, 1981).

CROSS-FLOW TOWERS

The general approach to model a single-stage cross-flow tower is to partition the tower into n incremental volumes that are geometrically similar to the shape of the cross section (Thibodeaux, 1977). The mass transfer driving force for each of the n elements is summed to give an equation analogous to Eq. 11-11 for countercurrent packed towers.

$$V = VTU_0(NTU)^{1/2} \qquad (11\text{-}14)$$

where V is the volume of the cross-flow tower and VTU_0 is the volume of a transfer unit (cubic feet) given by Eq. 11-15.

$$VTU_0 = \frac{(HG_T L_T)^{1/2}}{K_L a} \qquad (11\text{-}15)$$

where $K_L a$ is the overall liquid mass transfer coefficient (in units of lb moles/ft^3 hr), G_T and L_T are the total gas and liquid rates, respectively (in lb/mole hr), and H is the unitless Henry constant. Figure 11-11 gives the value of NTU as a function of a unitless separation factor S_a (Eq. 11-16) and A is the adsorption factor.

$$S_a = \frac{Y_1 - HX_2}{Y_1 - HX_1} \qquad (11\text{-}16)$$

Where X and Y are the mole fractions in influent liquid and gas, respectively, and X_2 is the mole fraction in the effluent liquid.

Figure 11-11 also can be used for stripping of a volatile gas in which $1/A$ is the stripping factor. For the same packing volume, the transfer efficiency of a cross-flow device is less than a countercurrent device. One of the principal attributes of a cross-flow configuration is that the gas flow area and the liquid flow area are independent, thus providing the designer with considerable flexibility. This feature allows much higher loading rates and lower pressure drops.

SPRAY TOWERS

There are a variety of configurations in which spraying can be used. In addition to those configurations that are analogous to packed tower designs, there are a number of more complex designs typically used for air pollution control, such as cyclone scrubbers and venturi scrubbers. Historically, spray systems have been used in water treatment for aeration, degasification of well water, and odor removal.

Only a few studies (Pigford, 1951; Davies, 1981; Ip, 1978) have been conducted on mass transfer in clean water systems. The research done to date has suggested that spray systems are limited with respect to the removals that can be achieved. Typically, one to three transfer units are reported as a maximum limit that can be achieved in spraying systems. This apparent limitation in percent removal is the result of back-mixing of air, and spray disturbance due to wall or adjacent spray contact.

Significant removal in a spray system may occur at the nozzle. Figure 11-12 shows the number of transfer units (NTU) as a function of the height of the spray tower. The residual NTU at zero height, in this case between 0.1 and 0.2 transfer units, is the result of the removal occurring at the nozzle. Some process designs may take advantage of the removal occurring at the nozzles by recycling flow or by incorporating several banks of nozzles.

The aforementioned nonideal effects have hindered development of a general empirical design model. With the data presently available, one cannot design a spray tower for a precise removal.

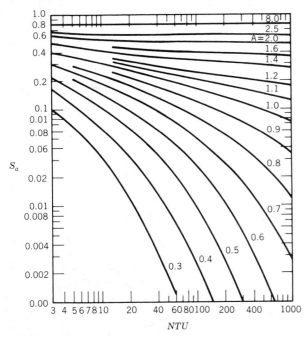

FIGURE 11-11. Number of transfer units in a cross-flow packed tower as a function of absorption factor A and S_a. (Courtesy of Thibodeaux, 1977.)

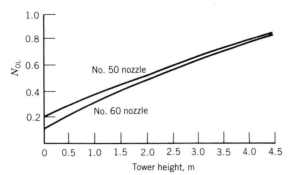

FIGURE 11-12. Number of transfer units in a pilot spray tower as a function of nozzle height. (From Davies, 1981.)

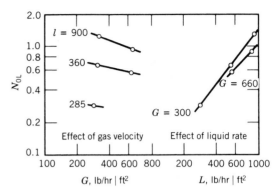

FIGURE 11-13. Number of transfer units in a pilot spray tower as a function of air and water rate. (From Pigford, 1951.)

Rather, the design approach, as described below, serves merely as a basis with which the approximate efficiency of a spray system can be estimated.

Design Basis

The transfer unit design approach, as derived in the preceding discussion of packed towers, has been suggested for spray chambers (Downing, 1957; Calvert, 1972). The basis is to consider that the bulk of the mass transport occurs in the region after the spray is fully developed and before it contacts the chamber wall or base of the chamber.

Spray systems designed for mass transfer require small droplets, on the order of 500–1000 μm. For this size range, penetration theory is used to determined the mass transfer coefficient for a single drop. The mass transfer coefficient calculated in this manner is a function of time and is therefore dependent on the drop's relative velocity and distance of fall. The transfer unit approach as used in packed towers, however, is not strictly valid. Since the mass transfer coefficient varies with height, the overall mass transfer coefficient then is actually an average value for a given tower height and air flow.

Effect of Gas and Liquid Rates

The work of Pigford (Figure 11-13) shows the effect of gas and liquid rate on the overall mass transfer coefficient, K_La, for spray nozzles. As computed from Figure 11-13, K_L suggests a proportionality to L.

Gas rate apparently has little effect on the overall liquid transfer coefficient (Figure 11-13) and a slight effect on the overall gas phase mass transfer coefficient as estimated from the data for ammonia stripping. The term K_La is a function of the difference

between the drop terminal velocity and the superficial gas velocity. In general, the gas velocity is a small fraction of the drop terminal velocity, and the difference is not expected to vary significantly with gas rate. At higher gas rates, the sensitivity should be more pronounced.

When mass transfer data is not available for a specific nozzle and spray chamber configuration, K_La may be estimated using the procedure outlined below.

Estimation of K_La

The overall mass transfer coefficient is computed from either Eq. 11-17 or 11-18, depending on the value of the dimensionless quantity $(D_A\theta^{1/2}/r_d)$ (Higbie, 1935; Jury, 1967; Calvert, 1972).

$$\text{For} \quad \frac{D_A\theta^{1/2}}{r_d} < 0.22 \quad K_L = 2\left(\frac{D_A}{\pi\theta}\right)^{1/2} \quad (11\text{-}17)$$

$$\text{For} \quad \frac{D_A\theta^{1/2}}{r_d} > 0.22 \quad K_L = \frac{10D_A}{d_p} \quad (11\text{-}18)$$

where D_A is the diffusivity (in cm²/sec), θ is the diffusion time or time (sec) of exposure of a surface element of a drop, and r_d and d_p are the drop radius and diameter (in.), respectively.

The diffusion time of a droplet is simply the drop terminal axial velocity times the falling distance. In most gas transfer processes, K_L is dependent on the physical properties of the liquid and the compound diffusivities. But this approach suggests a dependence on gas rate and unit size. In a countercurrent spray chamber, the diffusion time is determined from

$$\theta = Z(U_t - U_g) \quad (11\text{-}19)$$

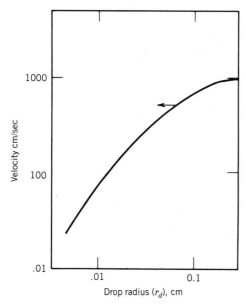

FIGURE 11-14. Terminal fall velocities of droplets of water. (From Fuchs, 1964.)

where the quantity $U_t - U_g$ is the relative velocity of the drop. A falling drop does not behave strictly as a solid sphere. The actual velocity deviates from Stokes's law because of the resistance due to circulation. At very small drop diameters, the terminal velocity approaches the value estimated from Stokes' law. Fuchs (1964) has measured the terminal velocity of drops as a function of droplet size (Figure 11-14). The volumetric surface area, a, is calculated from Eq. 11-20.

$$a = \frac{3U_L}{r_d(U_t - U_g)} \qquad (11\text{-}20)$$

where U_L is the liquid superficial velocity (cm/sec).

Drop Size

For a given nozzle, the drop diameter is inversely proportional to the cubed root of the pressure drop across the nozzle, sauter mean diameter (SMD), where SMD $\propto P^{-1/3}$ (Perry, 1973). Empirical relationships may be used to calculate the diameter from the nozzle size and pressure drop. However, in practice, reliable droplet size distribution data are typically provided by the manufacturer. The mean or median diameter may be reported in a number of ways: number mean diameter, median volume diameter, and SMD. Manufacturers typically report

median volume diameter (MVD). Mean number diameter is less widely used because of the inaccuracy of the measurement (Pigford, 1951).

The SMD is related to the mean surface area and is therefore appropriate for Eq. 11-27. SMD is obtained by dividing the total volume of a given spray by the total surface area. It is the diameter of a drop having the same volume–surface area ratio and is found to be approximately 70–90 percent of the MVD depending on the nozzle design and operating pressures.

Nozzles

The numerous commercially available nozzles are generally classified as pressure, two-fluid, or centrifugal nozzles. Pressure nozzles are most common to the water treatment industry. Of these, there are three generic types: hollow cone, in which the nozzle imparts a tangential velocity to the fluid; full cone; and fan spray (Figure 11-15). Although hollow cones do not distribute droplets as a full cone nozzle would, they are generally preferred over full cone for mass transfer because of the smaller drop diameter and larger nozzle orifice.

Flow capacity of various types of nozzles is provided by manufacturers over a broad range of pressure drops and orifice size. Flowrate is proportional to the square root of the pressure. The flow change with pressure for a given nozzle is proportional to the ratio of the square of the change in pressure drop:

$$\frac{Q_2}{Q_1} = \left(\frac{\Delta P_2}{\Delta P_1}\right)^2 \qquad (11\text{-}21)$$

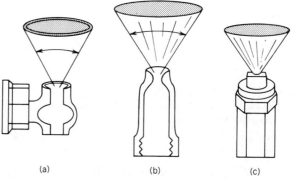

FIGURE 11-15. Common spray nozzles: (a) Hollow cone, (b) full cone, and (c) fan spray.

Nozzles are also available in wide or narrow spray angles; 60–80° for a full spray angle is common. For the range of drop diameters of interest in mass transfer systems, the orifice of the spray nozzles is generally 1–20 mm. To achieve a mean surface area diameter of 500 μm, a hollow cone nozzle diameter would be roughly 5 mm. This range of nozzle diameters may be prone to clogging. In-line strainers in the spray nozzle manifold are usually recommended.

Air

Air dispersion greatly affects process efficiency and must coexist with good liquid distribution. In designing a spray chamber, the orientation and inlets of air should be designed to avoid high inlet velocities and turbulence, which promote back-mixing. Air velocities through the unit are restricted by the drop terminal velocity. Fine sprays may be entrained and create a substantial loss of water if air velocity is too high. Generally 6400–12,800 kg/hr/ m^2 (400–800 lb/hr/ft^3) is recommended for maximum air velocities. Mist eliminators to avoid excess liquid loss are recommended. The air pressure drop through a countercurrent spray chamber is due to the pressure required for liquid hold-up, momentum transfer near the nozzle, and frictional drag from the chamber sides. If the latter two effects are neglected, the pressure drop is obtained from

$$P(\text{bars}) = \frac{U_L \gamma_L Z \times 10^{-2}}{U_t - U_g} \qquad (11\text{-}22)$$

where γ_L is the specific weight of the liquid (N/m^3) and Z is the height of the chamber (m).

Spray chambers are typically designed for a pressure drop of approximately 1 in. H$_2$O.

EXAMPLE 11-3

A well water contains 2 mg/L of methane. The well pumps 0.08 m^3/sec (1268 gpm) into a 12 m (39.4 ft) diameter reservoir. The pump has the capacity to deliver the same flow under an additional 14 m (20 psi) of head. The reservoir is to be equipped with a blower and spray nozzles to remove methane. How many nozzles are required and what removal is expected? The average height of water is 2 m (6.56 ft) from the top of the reservoir.   □

Solution

1. Determine the droplet size and number of nozzles. A median volume diameter (MVD) of 1030 μm is assumed. Droplet size data provided by the manufacturer is for 40 psi. The droplet diameter at 20 psi is determined from the diameter at 40 psi times the cubed root of the pressure ratios:

$$(1030) \sqrt[3]{\frac{40 \text{ psi}}{20 \text{ psi}}} = 1300 \ \mu\text{m}$$

The nozzle providing this MVD has a capacity of 16 L/min (4.6 gpm). Thus, a total of 300 nozzles are required.

2. Determine the air velocity. Using a Henry constant of 38,000 atm for methane and assuming a stripping factor of 5, the air–water ratio is

$$\text{A/W} = 1334 \left(\frac{R}{H_A}\right) = \frac{(1334)(5)}{38,000} = 0.176$$

The air flow is

$$(0.08 \text{ m}^3/\text{sec})(0.176) = 0.014 \text{ m}^3/\text{sec (30 cfm)}$$

The air velocity is

$$\frac{0.014 \text{ m}^3\text{sec}}{(12 \text{ m})^2 \pi/4} = 1.24 \times 10^{-4} \text{ m/sec}$$
$$= 0.124 \text{ cm/sec (4.07} \times 10^{-3} \text{ ft/sec)}$$

3. Determine the droplet fall time. From Figure 11-14, a droplet with a SMD of 950 μm has a terminal velocity of 350 cm/sec. SMD is equal to 73 percent of the MVD. The drop fall time from Eq. 11-19 is

$$\theta = \frac{(2 \text{ m})(100 \text{ cm/m})}{320 - 0.124} = 0.63 \text{ sec}$$

The diffusivity of methane in water is 1.7×100^{-5} cm^2/sec.

$$\left(\frac{D_L \theta}{r_d}\right)^{1/2} = \left(\frac{(1.7 \times 10^{-5} \text{ cm}^2/\text{sec})(0.63 \text{ sec})}{0.048 \text{ cm}}\right)^{1/2}$$
$$= 0.005$$

Since $0.005 \ll 0.22$, Eq. 11-14 is used to calculate k_L.

4. Estimate the overall liquid mass transfer co-

efficient. From Eq. 11-14, the liquid mass transfer coefficient is

$$k_L = 2 \left(\frac{(1.7 \times 10^{-5} \text{ cm}^2/\text{sec})}{(3.141)(0.63 \text{ sec})} \right)^{1/2}$$

$$= 5.86 \times 10^{-3} \text{ cm/sec} \ (1.9 \times 10^{-4} \text{ ft/sec})$$

From Eq. 11-20, the interfacial volumetric surface area is

$$a = \frac{(3)(0.071 \text{ cm/sec})}{(0.048 \text{ cm})(320 - 0.124)}$$

$$= 0.014 \text{ cm}^2/\text{cm}^3 \ (0.43 \text{ ft}^2/\text{ft}^3)$$

The volumetric mass transfer coefficient is

$$k_L a = (0.014 \text{ cm}^2/\text{cm}^3)(5.86 \times 10^{-3} \text{ cm/sec})$$
$$= 8.2 \times 10^{-5} \text{ sec}^{-1}$$

5.   Determine the NTU. Assuming liquid phase controls mass transfer, the height of a transfer unit is

$$\text{HTU} = \frac{L}{k_L a} = \frac{0.071 \text{ cm/sec}}{8.2 \times 10^{-5} \text{ sec}^{-1}}$$

$$= 866 \text{ cm} = 8.6 \text{ m} \ (28 \text{ ft})$$

The number of transfer units is

$$\text{NTU} = \frac{Z}{\text{HTU}} = \frac{2 \text{ m}}{8.6} = 0.23$$

Another 0.1 transfer units is added for removal at the nozzle giving a total of 0.33 NTU for the system.

7.   Estimate the effluent methane concentration. Rearranging Eq. 11-18

$$\frac{X_{\text{in}}}{X_{\text{out}}} = \frac{R \exp[(\text{NTU})R/R - 1] - 1}{R - 1}$$

$$= \frac{5 \exp[(0.33)(5)/(4)] - 1}{4} = 1.64$$

The expected effluent concentration is 1.2 mg/L. Forty percent is removed.

In practice, the removal at the nozzle may be as high as 1 NTU. This would be affected by the water flow per nozzle and the pressure.   □

DIFFUSED AERATION

Diffused aeration is the most generally applicable of the gas transfer processes. The principles of diffused aeration apply to, for example, countercurrent bubble columns such as those used for ozonation, simple pipe or orifice diffusers often used for ammonia addition, or large aeration basins or agitated vessels and surface aerated basins used extensively in wastewater treatment.

Diffused gas systems are generally referred to as single-bubble, bubble swarm, and agitated diffused gas systems. Mass transfer in a single rising bubble is most easily described. This system is characteristic of orifice diffusers and bubble column systems with low superficial gas velocities. Bubble swarms are characteristic of large-scale gas-diffusing processes, particularly at high superficial gas velocities. Most mass transfer data has been developed for bubble swarms in small-scale countercurrent or cocurrent bubble columns. The mass transfer efficiency is determined from separate correlations of mass transfer coefficient and interfacial area, derived from gas holdup. In general, the design of large-scale agitated vessels must use the mass transfer data provided by aeration equipment manufacturers. Mass transfer coefficients in these systems are typically correlated with mixing intensity. These correlations are less general because of the dependence on basin geometry and flow characteristics.

Process Design

A general model for gas-in-liquid systems has been proposed by Matter-Müller (1981) for estimating mass transfer efficiency. The model is based on the mass transfer flux (Eq. 11-23) derived from the assumption of bubbles rising in a completely mixed vessel.

$$F = Q_G H \left(\frac{C_G}{H} - C_L \right) \left[1 - \exp \left(\frac{K_L a V_L}{H Q_L} \right) \right] \quad (11\text{-}23)$$

where Q_G is the gas flowrate (L/min); C_G is the inlet gas concentration (moles/L); H is the dimensionless Henry constant expressed as the ratio of liquid and gas phase concentration, Q_L, equals the liquid flow (L/min); $K_L a$ is the overall mass transfer coefficient, and V_L is the total liquid volume.

The mass flux equation (11-23) can be used with a general mass balance equation (11-24) for a complete-mix reactor. For steady-state conditions, this leads to the general mass balance relation shown in Eq. 11-25.

$$V_L \frac{dC_L}{dt} = Q_L(C_{L_i} - C_L) - F_y \quad (11\text{-}24)$$

$$\frac{C_G/H - C_L}{C_G/H - C_{L_i}}$$

$$= \frac{1}{(1 + HQ_G/Q_L)[1 - \exp(-K_L aV/HQ_L)]} \quad (11\text{-}25)$$

where C_{Li} is the concentration of the substance in the reactor influent. Although this model has been derived for a single-bubble diffused gas process, Matter-Müller has demonstrated its use for mechanically aerated surface systems as well. Equation 11-25 can be further reduced for two extreme cases:

$$\theta = \frac{K_L aV_L}{Q_G H} \gg 1 \qquad \frac{C_L}{C_G/H} = \frac{1}{1 + Q_G H/Q_L} \quad (11\text{-}26)$$

$$\theta = \frac{K_L aV_L}{Q_G H} \ll 1 \qquad \frac{C_L}{C_G/H}$$

$$= \frac{1}{1 + K_L aV_L/Q_L} \quad (11\text{-}27)$$

These limiting cases provide valuable insight with respect to the controlling variables in a given process and how the process may be designed or modified to enhance desorption or absorption. For the diffused air system, either mechanically mixed or stagnant, Eq. 11-26 describes the transfer of a compound with very low Henry constants, such as ammonia. Bubbles exiting from the top of the liquid surface would be saturated with ammonia in the stripping process. Removal could be further enhanced by increasing the air flow.

Equation 11-27 describes the situation in which the existing gases are undersaturated. In this case, the efficiency can be improved by increasing the overall mass transfer coefficient by either using a finer diffuser or by increasing the mixing intensity in the aeration vessel. This is generally the case for oxygenation in which the typical value for the term $\theta = K_L aV_L/Q_G H$ is 0.1. Surface-aerated systems present a special case where the gas flow, Q_G, is unknown but is generally considered to be quite large. Equation 11-27 is then appropriate. The removal or absorption in surface-aerated systems would therefore be independent of air flowrate.

It should be noted that the choice of either Eq. 11-26 or 11-27 is not merely a matter of Henry's constant but of the entire unitless quantity, $K_a V_L/HQ_G$. This requires knowledge of the overall mass transfer coefficient. It is quite important in the intermediate range in which Henry's constant, $K_L a$, and gas flowrate are all significant. Matter-Müller has summarized the typical ranges of θ for various aeration processes (Figure 11-16). Note that typical diffused aeration systems fall within the intermediate range and may require the use of Eq. 11-26.

Given this approach to process design, the first task is to understand the factors affecting and methods for estimating the overall mass transfer coefficient for the various diffused aeration processes.

Mass Transfer Correlations

Mass transfer data for gas-in-liquid systems for both mechanically mixed and unmixed bubble dispersion systems is provided in the literature for many variations of mixing and diffuser equipment. The data, however, is applicable only within the range of the experimental conditions. Several general correlations presented in the following discussions may, however, provide a reasonable estimate. Sideman (1966) provides a comprehensive review and reference for mass transfer in gas-in-liquid systems.

The liquid phase overall volumetric mass transfer coefficient, $K_L a$, may be estimated by one of three methods: (1) general correlations of K_L with a separate correlation for the volumetric interfacial surface area a, (2) direct correlations of $K_L a$ for a specific system typically as a function of the holdup ratio ε, and (3) estimate of $K_L a$ from the oxygen transfer efficiency, which is often provided in the form of tabulated or graphical data provided by the manufacturer. Estimating $K_L a$ for other volatile components from the $K_L a$ from oxygen is valid if the Henry constants are sufficiently high. Otherwise, the gas phase mass transfer coefficient $k_G a$ must be known. These three different approaches for estimating the mass transfer coefficient are discussed below.

Mass Transfer Coefficients

For all diffused gas systems, the interfacial surface area is related to the holdup and bubble diameter by Eq. 11-28.

$$a = \frac{6\varepsilon}{d_b} \quad (11\text{-}28)$$

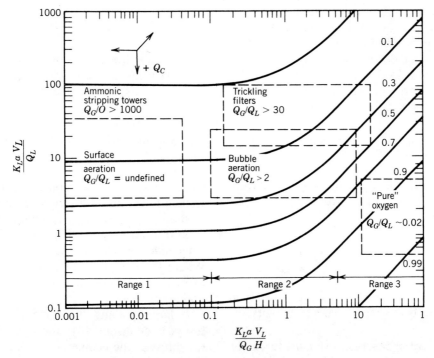

FIGURE 11-16. Common aeration processes in relation to the unitless parameter θ. (Courtesy of Matter-Müller et al., 1981.)

where d_b is the bubble diameter and ε is the holdup ratio.

Agitated Diffused Gas Vessels. Most correlations for mass transfer coefficients are generally applicable over a limited range of conditions and provide only a rough estimate. Apparently, K_L does not vary appreciably with variations of bubble diameter, depth, or unit power input. Rather, the dependence of the overall mass transfer coefficient $K_L a$ is principally dependent on interfacial area a. For well-agitated vessels with small bubble diameters, less than 0.25 cm, Calderbank suggests the correlation in Eq. 11-29.

$$K_L = 0.31 \left(\frac{\Delta\rho \ \mu_L g}{\rho_L}\right)^{1/3} \left(\frac{\mu_L}{\rho_L D_A}\right) \quad (11\text{-}29)$$

where μ_L and ρ_L are the viscosity and density of water, $\Delta\rho$ is the density difference essentially equal to the density of water, and D_A is the diffusivity in water. The volumetric interfacial area, a, for pure liquids is given by Eq. 11-30 where surface aeration is negligible.

$$a = 215 \left(\frac{P}{V}\right)^{0.4} \frac{\gamma}{\sigma^{0.6}} \left(\frac{U_g}{U_r}\right)^{0.5} \quad (11\text{-}30)$$

where a = specific interfacial area, cm²/cm³
  P/V = agitator horsepower per vessel volume, hp/ft³
  γ = water specific gravity
  σ = surface tension, dyne/cm
  U_g = superficial gas velocity, ft/sec
  U_r = terminal bubble rise velocity

Generally, when surface aeration is substantial, interfacial surface area is estimated by correlations that include terms for the impeller speed and characteristics (Perry, 1973).

Alternatively, for a system equipped with a turbine impeller with subsurface aeration, Eq. 11-31 can be used (Westerterp, 1963).

$$\frac{aZ}{1-\varepsilon} = 0.79\mu_L(n - n_0)D_i \sqrt{\frac{\rho_i T_i}{\sigma}} \quad (11\text{-}31)$$

where μ_L is the viscosity, ρ_i is liquid density (kg/m³), σ is the surface tension (kg/sec²), Z is the depth of the liquid, D_i is the diameter of the impeller, and T_i is the vessel diameter. The constant n_0 is given by Eq. 11-32.

$$n_0 = \left(\frac{1.22}{D_i} + 125\frac{T_i}{D_i^2}\right)\left(\frac{\sigma G}{\rho_L}\right)^{1/4} \quad (11\text{-}32)$$

Bubble Columns. Mass transfer data for large un-agitated diffused gas systems is not available in the literature. There is considerable data for bubble columns with diameters in the range of 6–24 in. Akita and Yoshida (1974) have reported extensive studies of gas absorption using an orifice sparger. The correlation for $K_L < 0.14$ is given by

$$K_L = 0.5g^{5/8}D_A\rho^{3/8}\sigma^{-3/8}d_b^{1/2} \qquad (11\text{-}33)$$

The interfacial surface area a is then estimated from Eq. 11-28 and ε is given by 11-34.

$$\frac{\varepsilon}{(1-\varepsilon)^4} = 0.0268\gamma_L^{-1/6}U_G\left(\frac{\sigma}{\rho_L}\right)^{-1/8} \qquad (11\text{-}34)$$

where ρ_L is the kinematic viscosity of the liquid (cm^2/sec) and U_G is the superficial gas velocity.

Large Diffused Gas Processes. Manufacturers of large diffuser systems and mechanical aeration systems typically specify performance as either oxygen transfer efficiency, E_0, or oxygenation efficiency, E_P, as a function of the tank depth. These can be used to compute the oxygen mass transfer coefficient, which as previously noted can be used to estimate mass transfer rates for other volatile constituents assuming that the Henry constants are within the same range. Oxygen transfer efficiency can be expressed as

$$E_0 = \frac{R_0(100)}{1.015Q_G} \qquad (11\text{-}35)$$

where R_0 is the transfer rate (kg/hr) and Q_G is the air rate (L/sec). The transfer rate R_0 is given by

$$R_0 = \frac{V_LK_La(C_G/H - C_L)}{1000} \qquad (11\text{-}36)$$

where C_G is the oxygen concentration in gas (mg/L) at 20°C and 760 mm of mercury pressure, K_La is the overall mass transfer coefficient and V_L is the volume of the vessel, and K_La is estimated from Eq. 11-37.

$$K_La = \frac{E_0(1.015)Q_G}{V_LC_G/H} \qquad (11\text{-}37)$$

or from the oxygenation efficiency

$$K_La = \frac{E_pB_{hp}}{V_LC_G/H} \qquad (11\text{-}38)$$

where B_{hp} is the brake horsepower of the mixer.

Mass transfer coefficients for constituents other than oxygen are computed from the ratio of the diffusivity of oxygen to the compound of interest. This ratio is usually expressed as β, where β (Table 11-7) is the ratio of the diffusivities to some power ν. Roberts (1980) has indicated that ν typically falls within a range of $\frac{1}{2}$–$\frac{2}{3}$ for surface-aerated systems. Other workers have reported similar ranges for packed tower and diffused air systems. Use of this method for estimating K_La requires that the liquid mass transfer is controlling; otherwise, there will be a dependence on other factors, such as the gas rate. Where the gas phase resistance is controlling, the mass transfer coefficient must be provided by experimental data.

In large systems, the number and spacing of diffusers is determined by the individual capacity of the diffuser, the geometry of the basin, and the desired mixing.

TABLE 11-7. Reported Values of β

Substance	β
1,1-Dichloroethylene[a]	0.62
Tetrachloroethylene[a]	0.55
Benzene[a]	0.53
Toluene[a]	0.53
1,2,4-Trimethylbenzene[a]	0.39
1,4-Dichlorobenzene[a]	0.39
Freon[b]	0.66
Chloroform[b]	0.56
1,1,1-Trichloroethane[b]	0.61
Carbon Tetrachloride[b]	0.617
Trichloroethylene[b]	0.615
Tetrachloroethylene[b]	0.608
Krypton[c]	0.82
Carbon dioxide[c]	0.89
Nitrogen[c]	0.91
Ethylene[d]	0.89
Propane[d]	0.72
Chloroform[e]	0.66
1,1-Dichloroethane[e]	0.71
Hydrogen[f]	1.3
Helium[f]	1.3

[a] From Matter-Müller (1980).
[b] From Roberts.
[c] From Tsivogloci (1967).
[d] From Rathbun (1973).
[e] From Smith and Bonnberger (1980).
[f] From Hutchinson (1932).

EXAMPLE 11-4

An ozonation facility is designed to apply 2.5 mg/L of ozone to a raw surface water. The average flow through the facility is 0.2 m³/sec (3170 gpm). Air is used for the ozone generation source. The generator converts 1.5 percent of the oxygen to ozone. The contact basin is a covered concrete tank 4.5 m (14.8 ft) deep with 10 min contact time. A gridwork of porous diffusers at the bottom of the tank produces 0.2-cm bubbles at the design air rate. Estimate the overall mass transfer coefficient assuming even distribution of air and no enhancement of $K_L a$ due to chemical reactions and the ozone transfer efficiency.

Solution

1. Determine the superficial air velocity. The area of the basin is

$$A = \frac{Q\theta}{Z} = \frac{(0.2 \text{ m}^3/\text{sec})(10 \text{ min} \times 60 \text{ sec/min})}{4.5}$$

$$= 26.7 \text{ m}^2$$

The air rate Q_A is

$$\frac{(Q_W)(\text{dosage})}{(\rho_A)(\%O_3)} = \frac{(2 \text{ g/m}^3)(0.2 \text{ m}^3/\text{sec})}{(1208 \text{ g/m}^3)(0.015)(0.2)}$$

$$= 0.107 \text{ m}^3/\text{sec}$$

The superficial air velocity, U_G, is then 0.4 cm/sec.

$$\frac{0.107 \text{ m}^3/\text{sec}}{26.7 \text{ m}^2} \times 100 \text{ cm/m} = 0.4 \text{ cm/sec}$$

2. Estimate the holdup, ε, and the volumetric interfacial area.

$$\frac{\varepsilon}{(1 - \varepsilon)^4} = 0.0268\nu_L - \frac{1}{6} U_G \left(\frac{\sigma}{\rho_L}\right)^{-1/8}$$

$$= (0.0268)(0.01 \text{ cm}^2/\text{sec})^{-1/6}(0.4 \text{ cm/sec})$$

$$\left(\frac{72 \text{ gm/sec}}{1 \text{ gm/cm}^3}\right)^{-1/8}$$

$$= 0.0135$$

By trial and error, $\varepsilon = 0.014$.
The volumetric interfacial area from Eq. 11-35 is

$$a = \frac{6\varepsilon}{d_b} = \frac{(6)(0.014)}{(0.2 \text{ cm})} = 0.42 \text{ cm}^2/\text{cm}^3$$

3. Estimate the liquid mass transfer coefficient given a diffusivity of $2.3 \times 10^{-5} \text{ cm}^2/\text{sec}$. From Eq. 11-33,

$$k_L = 0.5(980 \text{ cm}^2/\text{sec})^{5/8}(2.3 \times 10^{-5} \text{ cm}^2/\text{sec})^{1/2}$$
$$(1 \text{ gm/cm}^3)^{3/8} \times (72 \text{ gm/sec})^{-3/8}(0.2 \text{ cm})^{1/2}$$
$$= 0.016 \text{ cm/sec}$$

4. Calculate the overall mass transfer coefficient. Assume the liquid phase is controlling and there is no effect of chemical reaction.

$$K_L a = k_L a = (0.016 \text{ cm/sec})(0.42 \text{ cm}^2/\text{cm}^3)$$
$$= 0.0067 \text{ sec}^{-1}$$

5. Estimate the percent of ozone transferred. The decay rate of ozone in water has a substantial effect on the transfer efficiency. Assume a rate of 0.4 min⁻¹ with first-order decay. The mass balance for the reactor assuming complete mix is

$$V_L \frac{dC_L}{dt} = F - Q_L C_L - 0.4 C_L V_L$$

where

$$F = Q_G H(C_{L,\text{sat}} - C_L) \left[1 - \exp\left(\frac{K_L a V_L}{Q_G H}\right)\right]$$

$$\frac{K_L a V_L}{Q_G H} = \frac{(0.0067 \text{ sec}^{-1})(120 \text{ m}^3)}{(0.107 \text{ m}^3/5)(2.97)} = 2.53$$

At steady state,

$$0 = Q_G H(C_{L,\text{sat}} - C_L)0.92 - Q_L C_L - 0.006 C_L V_L$$

and

$$\frac{C_L}{C_{L,\text{sat}}} = \frac{1}{1 + (Q_L/0.92Q_G H)/(0.006V_L/0.92Q_G H)}$$

$$= 0.24$$

$$C_{L,\text{sat}} = \frac{C_G}{H}$$

$$= \frac{(0.015)(0.2)(41.6 \text{ m}M/\text{L})948 \text{ mg/m}M)}{2.97(\text{mg/L/mg/L})}$$

$$= 2.0 \text{ mg/L}$$

$$C_L = (0.24)(2.0 \text{ mg/L}) = 0.48 \text{ mg/L}$$

and

$$\text{Percent transferred} = \frac{100(Q_L C_L + 0.006V_L C_L)}{Q_G C_G}$$

$$= 69\% \qquad \square$$

MECHANICAL AERATORS

Mechanical aerators commonly used in water and wastewater treatment are surface or submerged turbine type. Surface aerators can be used in water treatment as an alternative to diffused aeration systems for stripping of volatile contaminants. Two types of surface aerators are prevalent; large-diameter, slow-speed turbines that operate just below the free surface of the turbine and small-diameter, high-speed propellers operating in draft tubes. The submerged turbine is equivalent to an agitated vessel as discussed in the preceding section.

Mass Transfer

The most widespread application of surface aeration is in oxygenation. Consequently, manufacturers provide mass transfer data expressed in terms of oxygenation efficiency. In addition, generalized correlations for $K_L a$ as a function of power input are available. These may be used to estimate the $K_L a$ for other volatile constituents.

In the turbulent mixing regime, the overall mass transfer coefficient is nearly proportional to the specific power input (Roberts, 1982).

$$(K_L a)_{O_2} = 3.0 \times 10^{-5} \left(\frac{P}{V}\right)^{0.95} \quad (11\text{-}39)$$

Where $K_L a$ is expressed as $\sec^{-1}$ and P/V is the specific power (W/m³). The lower limit of turbulent mixing corresponds to a specific power input of 10 W/m³. The $K_L a$ of volatile constituents were found to be related to $(K_L a)_{O_2}$ by Eq. 11-40 (Roberts, 1982).

$$\frac{(K_L a)_i}{(K_L a)_{O_2}} = \left(\frac{D_i}{D_{O_2}}\right)^{0.66} \quad (11\text{-}40)$$

Where D_i and D_{O_2} are diffusivities of the volatile constituent i and oxygen.

Mass transfer in surface-aerated systems is difficult to model because of the two different processes occurring: mass transfer to and from a spray and to and from entrained air bubbles. Eckenfelder has suggested that 60 percent of the transfer of oxygen results from the spray and the remaining from entrainment. Roberts (1981) points out that the driving force is different in these cases, and care must be taken in selecting the proper correlation. In estimat-ing the percent removals or absorption, one should consider the mass driving force as being in contact with air rather than a bubble with changing concentration of gas. The overall oxygen uptake in these systems is typically measured in this manner.

MULTIPLE-TRAY AERATORS

Gas transfer by multiple-tray aerators is analogous to the countercurrent packed tower. A variety of commercially available aerators are typically used for iron, manganese, methane, CO_2, and hydrogen sulfide removal in water treatment. The three most common types of multiple-tray aerators are the wood slat or plastic slat, coke tray, and cascade-type aerator as described in Chapter 21.

Mass Transfer

Detailed mass transfer data are not available for multiple-tray-type aerators. Mass transfer in cascade and coke tray aerators is nearly impossible to predict because of the uncertainty of the air-to-water contact. One approach has been to estimate the contact time of the free-flowing water. If one assumes an approximate mass transfer coefficient and contact time with atmospheric concentrations of the gases, then one can estimate the degree of absorption or desorption of the gas.

Although there is very little mass transfer data on slat-type tray towers, reasonable estimates can be made if one uses the transfer unit model (Eq. 11-11). Slat tray towers are often designed as a cross-flow configuration, particularly for cooling towers.

For closely spaced slats providing a large specific area, the $K_L a$ is proportional to the surface area of the wooden slats. For widely spaced slats, the surface area of the wood has only a small affect on $K_L a$ since the interfacial areas are mostly in the form of water droplets or entrained air.

Commercially available slat towers range in height from 10 to 18 ft in general. Manufacturers apparently design these towers for height of a transfer unit for oxygen in the range of 6 ft, roughly three to four times that for a comparable packed tower. The commercial limit should therefore provide one to three transfer units for components with very high Henry constants.

REFERENCES

Akita, K., and Yoshida, F., "Bubble Size Interfacial Area and Liquid Phase Mass Transfer Coefficient in Bubble Columns," *Ind. Eng. Chem. Process Des. Develop.*, **13**, 1 (1974).

Danckwerts, P. V., "Significance of Liquid-Film Coefficients in Gas Absorption," *Ind. Eng. Chem.*, **43**(6), 1460–1467 (1951).

Danckwerts, P. V., *Gas-Liquid Reactions*, McGraw-Hill, New York (1970).

Davies, T. H., and Ip, S. Y., "The Droplet Size and Height Effects in Ammonia Removal in a Spray Tower," *Water Res.*, **15**(5), 525 (1981).

Downing, A. L., "Aeration in Relation to Water Treatment," *Society of Water Treatment Examiner Proceedings*. (1957)

Eckert, J. S., "How Tower Packings Behave," *Chem. Eng.*, **82**, 70 (April 14, 1975).

Emerson, S., *Limnol. Oceanogr.*, **62**, 1–13 (1975).

Fortescue, G., and Pearson, J., "On Gas Absorption into a Turbulent Liquid," *Amer. Eng. Sci.*, **22**, 1163–1176 (1967).

Fuchs, N. A., *The Mechanics of Aerosols*, Pergamon Press, New York (1964).

Higbie, R., "The Rate of Adsorption of a Pure Gas into a Still Liquid During Short Periods of Exposure," *Trans. Am. Inst. Chem. Engrs.*, **31**, 365 (1935).

Ip, S. Y., and Raper, W., "Ammonia Stripping with Spray Towers," *Prog. Wat. Tech.*, **10**, 587–605 (1978).

Jury, S. H., *A.I.Ch.E. J.*, **13**, 1924 (1967).

Kavanaugh, M. C., and Trussell, R. R., "Air Stripping as a Treatment Process," Paper Presented at the AWWA Annual Conference, St. Louis, MO (1981).

Liss, P. S., and Slater, P. G., "Flux of Gases Across the Air–Sea Interface," *Nature*, **247**, 181 (January 25, 1974).

Mackay, D., Shiu, W. Y., and Sutherland, R. P., "Determination of Air–Water Henry's Law Constants for Hydrophobic Pollutants," *Environ. Sci Technol.*, **13**, 3 (1979).

Matter-Müller, C., Gujer, W., and Giger, W., "Transfer of Volatile Substances from Water to the Atmosphere," *Water Res.*, **5**, 1271–1279 (1981).

Perry, R. H., and Chilton, C. H. (eds.), *Chemical Engineers'*

Handbook, 5th ed., Sections 14 and 18, McGraw-Hill, New York (1973).

Pigford, R. L., and Pyle, C., "Performance Characteristics of Spray-Type Absorption Equipment," *Ind. Eng. Chem.*, **43**(7), 1649 (1951).

Rathbun, R., Stephens, D., Shultz, D., and Tai, D., "Laboratory Studies of Gas Tracers for Reaeration," *J. Envir. Eng. Div.*, **104**, 215–229 (1973).

Roberts, P., Dändliker, P., Matter, C., and Muniz, C., *Volatilization of trace organic contaminants during surface aeration: Model Studies*, Tech. Report No. 257, Dept. of Civil Eng., Stanford University (1981).

Roberts, P. V., Dändliker, P., "Mass Transfer of Volatile Organic Contaminants During Surface Aeration," Proceedings of the 1982 Annual AWWA Conf., Miami (1982).

Sherwood, T. K., and Holloway, F. A. L., "Performance of Packed Towers—Liquid Film Data For Several Packings," *Trans. Am. Inst. Chem. Eng.*, **36**, 39 (1940).

Sherwood, T. J., Shipley, G. H., and Holloway, F. A. I., "Flooding Velocities in Packed Columns," *Ind. Eng. Chem.*, **30**(7), 765–796 (1938).

Sherwood, T. K., Pigford, R. L., and Wilke, C. R., *Mass Transfer*, McGraw-Hill, New York (1975).

Sideman, S., Hortaesu, and Wells, N., "Mass Transfer in Gas–Liquid Contacting Systems," *Ind. Eng. Chem.*, **58**, 7 (1966).

Schwarzenbach, R., Molnar-Kubica, E., Giger, W., and Wakeham, S., "Distribution, Residence Time and Fluxes of Tetrachloroethylene and 1,4-Dichlorobenzene 1-Labre," Zurich, Switzerland, *Environ. Sci. Technol.*, **13**, 11 (1979).

Smith, J., and Bomberger, D., "Prediction of Volatilization Rates of Chemicals in Water," in B. Afghan and D. Mackay (eds.). *Hydrocarbons and Halogenated Hydrocarbons in the Aquatic Environment*, Plenum Press, New York (1980).

Thibodeaux, L. J., et al., "Mass Transfer Units in Single and Multiple-Stage Packed Towers," *Ind. Eng. Chem. Process Des. Dev.*, **16**(3), 325–330 (1977).

Treybal, R. E., *Mass Transfer Operations*, McGraw-Hill, New York (1980).

Westerterp, K. R., Van Dierendonck, L. L., and De Kraa, J. A., "Interfacial Areas in Agitated Gas–Liquid Contactors," *Chem. Eng. Sci.*, **18** (1963).

Wilke, C. R., Chang, P., "Correlation of Diffusion Coefficients in Dilute Solutions." *A.J.Ch.E.J.*, (June, 1955).

Disinfection

Water disinfection, the process by which pathogenic microorganisms are destroyed, provides essential public health protection. This chapter touches on the development of modern disinfection practices and on the objectives and mechanisms of disinfection. It also covers the factors that govern effectiveness, the kinetic models that aid in describing the disinfection process, and the chemistry of specific agents. Finally, it develops design relationships for disinfection facilities from dispersion and hydraulic aspects.

HISTORICAL PERSPECTIVE

Dr. John Snow established that water could be a mode of communication of dread diseases after a careful epidemiological study of the 1854 London cholera epidemic (Rosenau, 1940). At first, slow sand filtration was the only means employed for purifying public water supplies. Then, when Louis Pasteur and Robert Koch developed the germ theory of disease in the 1870s, things began to move more rapidly. In 1881, Koch demonstrated in the laboratory that chlorine could kill bacteria. Following an outbreak of typhoid fever in London, continuous chlorination of a public water supply was used for the first time in 1905. The regular use of disinfec-

tion in the United States was initiated by G. Johnson at the Bubbley Creek Filtration Plant in Chicago in 1908. This is about the same time that Dr. Hariette Chick first advanced her famous theory of disinfection.

The practice of chlorination increased rapidly after the development of liquid chlorine application to large-scale facilities at Niagara Falls in 1912. Previously, relatively unstable hypochlorites were used for disinfection. Although chlorination has been the dominant method employed for disinfection, ozonation has been widely used particularly in France, Germany, Canada, and the USSR. Recently, chlorine dioxide is gaining in popularity in European disinfection practice.

PROCESS ALTERNATIVES

Disinfection is a unit process whose objective is the destruction or otherwise inactivation of pathogenic microorganisms, including bacteria, amoebic cysts, algae, spores, and viruses.

Alternative disinfection systems available for disinfection are numerous. Generally, they can be divided into two groups: (1) chemical agents and (2) nonchemical agents. Chemical agents include an array of compounds with oxidation potential including chlorine, chlorine dioxide, bromine, iodine, bromine chloride, and ozone. Nonchemical, or energy-related, means of disinfection include ultraviolet (UV) radiation and gamma radiation.

DISINFECTION MECHANISMS

A general classification (Stanier, 1963) of disinfectant actions identifies three major modes of action: (1) destruction or impairment of cellular structural organization, (2) interference with energy-yielding metabolism, and (3) interference with biosynthesis and growth. The first group of agents that adversely affect the structural integrity of cells do so by destroying major cell constituents, destroying the cell wall, impairing functions of semipermeable membranes, or combining with nucleic acids within the nucleus or cytoplasm. Interference with metabolism can occur through enzyme inactivation by denaturing protein, combination with prosthetic groups of enzymes, competition with enzyme substrates, or prevention of oxidative phosphorylation. Interference with biosynthesis and growth can be caused by preventing synthesis of normal proteins, nucleic acids, coenzymes, or the cell wall. Various combinations of these mechanisms can be responsible for disinfection, depending on the disinfection agent and the type of microorganism.

In water treatment, two types of mechanisms are postulated to be the predominant factors controlling the effectiveness of disinfection: (1) oxidation or rupture of the cell wall with consequent cellular disintegration and (2) diffusion into the cell and interference with cellular activity. Thus, the ability to oxidize biological molecules and the ability to diffuse through the cell walls are requirements of any effective disinfectant.

FACTORS GOVERNING EFFECTIVENESS

The effectiveness of disinfection is a complex function of several variables including type and dose of disinfectant, type and concentration of microorganisms, contact time, and water quality characteristics. The effect of each of these factors on disinfection efficiency (i.e., ability to destroy microorganisms) is presented briefly below for illustrative purposes. Of the variables that contribute to disinfection effectiveness, contact time and disinfectant characteristics (type, dose, application technique) are subject to technical control.

Disinfectant Characteristics

Selection of proper disinfectant type and dose is a critical step in providing a process to meet water quality objectives. One measure of a disinfectant's ability to oxidize organic material is the standard reduction potential, an electrochemical characteristic that varies depending on the oxidant, as listed in Table 12-1.

The higher the oxidation potential, the easier that compound is able to oxidize organic materials. If oxidation was the only mechanism responsible for disinfection, the order of disinfection action would be ozone > chlorine dioxide > chlorine > bromine > iodine. In many instances, oxidation is the governing factor which would lead to selecting ozone, chlorine dioxide, or chlorine (depending on costs) as the most effective disinfectant; however, the selection is more complex because of other factors. For example, diffusion into the cell, cell permeability, and germicidal properties are dependent on molecular weight, charge on the chemical species, and other factors. Within the halogen series, the diffusion order is iodine > bromine > chlorine, just op-

TABLE 12-1. Standard Potentials of Selected Chemical Disinfectants

Compound	Formula	Potential (V)	Assumed Reactions
Chlorine	Cl_2	1.36	$Cl_2 + 2e^- \rightleftarrows 2Cl^-$
Bromine	Br_2	1.09	$Br_2 + 2e^- \rightleftarrows 2Br^-$
Iodine	I_2	0.54	$I_2 + 2e^- \rightleftarrows 2I^-$
Ozone	O_3	2.07	$O_3 + 2e^- + 2H^+ \rightleftarrows O_2 + H_2O$
Chlorine dioxide	ClO_2	1.91[a]	$ClO_2 + 5e^- + 2H_2O \rightleftarrows Cl^- + 4OH^-$
		0.95[b]	$ClO_2 + e^- \rightleftarrows ClO_2^-$

[a] Complete reaction, traditionally used to describe chlorine dioxide.
[b] Reaction that often occurs in water.

posite to the oxidation potentials. Thus, in most cases empirical evidence, operating experience, pilot plant studies, and other practical considerations should guide the selection process. This is particularly the case for selection of the required dose of disinfectant, which is a function of water quality. The dose can only be determined experimentally and must be controlled to respond to fluctuating water quality conditions. Because dose is also intrinsically linked to contact time in establishing disinfectant efficiency, further discussion of dose is deferred to that section.

Microorganism Characteristics

Pathogens may be divided into four groups, listed in decreasing order of resistance: (1) bacterial spores, (2) protozoan spores, (3) viruses, and (4) vegetative bacteria. Their relative resistance to disinfection is basically attributed to differences in cytostructure. The resistance of the spore wall, cytochemical changes, such as loss of cations and storage of basic ions, and the partially dehydrated state of the spore protoplasm may be reasons for the increased resistance of spores (Chang, 1971). Similarly, the resistance of the cyst wall is a major factor in determining the cysticidal activity of disinfectants because the diffusiveness of the disinfectant plays an important part. The high resistance of enteric viruses is associated with their lack of enzymes and other sensitive systems. Inactivation of viruses principally involves denaturing of their protein capsid. Destruction of the metabolic systems occurs very rapidly with vegetative bacteria because respiration takes place on the surface of the cell and highly active systems are present very close to the cell wall.

Microorganisms in wastewater differ in their initial concentrations, die-off rates, and susceptibility to various disinfectants. In most supplies, bacteria outnumber viruses and other pathogenic microorganisms by several orders of magnitude. For this reason, among others, bacteria are typically included in water quality standards as the sole indicator of microbiological quality.

The relative susceptibility of different microorganisms has been determined experimentally under various conditions. Many models have been proposed to predict the rate of microorganism die-off when disinfectants are used. Some of these are discussed in the section on disinfection models.

Contact Time

Disinfection effectiveness depends on contact time. For this reason, contact time is a critical yet manageable factor in disinfection and should be carefully considered in formulating design criteria.

Empirical relationships have been developed that indicate that a given percentage destruction of a certain organism, based on a water quality objective, can be obtained by balancing dose against contact time. The equations describing concentration versus time for disinfection are discussed in more detail in the section on kinetics of disinfection.

Water Quality Characteristics

Influent water quality characteristics other than microbiological components influence disinfectant efficiency. Among these are turbidity, organics, pH, and temperature. Turbidity has been demonstrated to interfere with disinfection (Symons and Hoff, 1975; Hoff, 1978; Hijkal et al., 1979; Foster et al., 1980; Boyce et al., 1981; Emerson et al., 1982) because particulates responsible for turbidity can also surround and shield microorganisms from disinfectant action. Organic materials can decrease disinfection efficiency by adhering to cell surfaces and hindering attack by the disinfectant, reacting with the disinfectant to form compounds with weaker germicidal properties, or reacting irreversibly with the disinfectant to produce products with no disinfection capabilities. Likewise, compounds such as iron, manganese, hydrogen sulfide, cyanides, and nitrates can decrease disinfection efficiency as they are rapidly oxidized by the disinfectant. This reaction of inorganic compounds with the disinfectant, such as chlorine, creates a demand that must be met before the disinfectant can act on the microorganisms.

The pH of the water, by affecting the chemical form of the disinfectant in aqueous solution, can influence microbial destruction. For example, the most active chlorine species for disinfection is hypochlorous acid (HOCl), which predominates in water if the pH is less than 7. Temperature affects the reaction rate of certain steps in the disinfection process, such as diffusion of the disinfectant through cell walls or the reaction rate with key enzymes, and can thus influence the rate of disinfection.

OTHER CONSIDERATIONS

When selecting a disinfectant, the first criteria has traditionally been the ability of the compound or process to destroy microorganisms. However, other properties of the disinfectant should be considered as they affect treated water quality, plant performance, and economics.

In the area of effluent quality, the same compound that inactivates microorganisms can also affect water quality parameters, some beneficially and others adversely. Certain disinfectants are also used for taste and odor control. The action of some disinfectants like chlorine and ozone can decrease color by oxidizing organic color-causing compounds. However, the reaction of certain disinfectants with organics is not without its drawbacks. Chlorine, for example, reacts with humic material and certain low molecular weight natural organic compounds to produce chlorinated organics, some of which are believed to be carcinogenic. Ozone forms aldehydes. Chlorine dioxide forms chlorate and chlorite. The by-products of each disinfectant must be assessed for possible deleterious impact on the environment or adverse health effects.

Other considerations include the impact of the selected disinfectant on other portions of the treatment plant. For example, the maintenance of a chlorine residual through the filters has been shown to be advantageous, in that the filters operate in an inert state and do not become affected by slime growths. Prechlorination also prevents growths in sedimentation basins. Both ozone and chlorine pretreatment can lead to decreased filtered water turbidity.

In addition, the handling characteristics and general safety will affect plant operations because disinfectants can be toxic to people as well as microorganisms. Other operation and maintenance considerations include the level of skill required of personnel, the degree of control and supervision necessary to achieve satisfactory performance, and the equipment maintenance requirements. Another important factor in an evaluation concerns the disinfectant's ability to maintain a residual through the distribution system and provide a final safeguard for public health protection. Finally, implementation capacity and public acceptance must be evaluated.

KINETICS OF DISINFECTION

In most disinfection operations, inactivation or destruction of microorganisms is a gradual process that involves a series of physical–chemical and biochemical steps. In an effort to predict the outcome of disinfection, various disinfection models have been proposed based on laboratory data and refined with field data. The following section describes the historical development of such models, how they relate to actual disinfection data, and their utility in comparing the relative effectiveness of different disinfectants.

Chick–Watson

Although a good deal of work has recently been done on modeling disinfection, the principal disinfection theory used today is still the Chick or Chick–Watson model. Chick's law (1908) expresses the rate of destruction of microorganisms by the relationship of a first-order chemical reaction:

$$\ln \frac{N}{N_0} = -kt \qquad (12\text{-}1)$$

where N = number of organisms present at time t
N_0 = number of organisms present at time 0
k = rate constant characteristic of type of disinfectant, microorganism, and water quality aspects of system
t = time

Watson (1908), using Chick's data, refined the equation to produce an empirical relation that includes changes in the disinfectant concentration:

$$\ln \left(\frac{N}{N_0}\right) = -\Lambda C^n t \qquad (12\text{-}2)$$

where C = concentration of disinfectant
Λ = coefficient of specific lethality
n = coefficient of dilution

The value of n in the Chick–Watson equation depends on the disinfectant and pH value. It is often found to be near unity.

Figure 12-1 illustrates data from disinfection of poliovirus by bromine, which are well represented by the Chick–Watson relationship, where the loga-

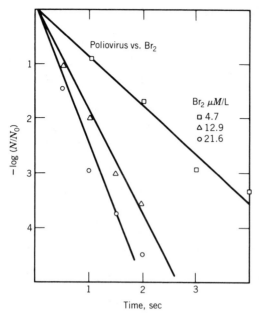

FIGURE 12-1. Sample "linear" disinfection data (Floyd et al., 1978).

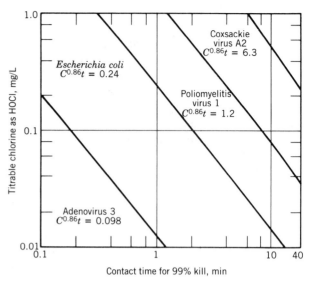

FIGURE 12-2. Time–concentration relationship for chlorine. (Source: Berg, 1964.)

rithm of the survival rate (N/N_0) plots as a straight line versus time.

The coefficient Λ is a specific lethality coefficient representing the relative potencies of disinfectants at a unit concentration for a unit time. The simplification that $n = 1.0$ and the assumption that Λ is a constant for a given reagent and organism at a specified temperature are used in establishing coefficient values. Table 12-2 shows the specific lethality coefficient for several disinfectants with respect to specific species of bacteria, viruses, and cysts for a 99 percent inactivation at or near pH 7.0 and 20°C (National Academy of Science, 1980). The concept of

specific lethality is illustrated in Figure 12-2, which shows the concentration versus time relationship for chlorine (as HOCl) and several types of organisms. This figure shows the differential capability of chlorine for destroying different types of organisms. These values serve only as crude indications of the relative lethality of these agents because there are so many variables in disinfection experiments that make it difficult to compare results.

Table 12-2 indicates that the most effective disinfectant is ozone, followed by hypochlorous acid, chlorine dioxide, hypochlorite ion, and the chloramines. This generalization follows for most microorganism types. Also, the specific lethalities show that enteric bacteria are easier to kill than viruses and that cysts are particularly resistant.

Another method frequently employed to compare disinfectants uses plots of residual versus time for a specified level of kill, usually 99 percent. From survival curves, the 99 percent inactivation points are extrapolated to give the time necessary for 99 percent destruction of the organisms. Figure 12-3 is such a plot of data by Scarpino et al. (1977) for poliovirus 1 inactivation. The relationships among the disinfectants shown are similar to those in Table 12-2. That is, HOCl is the most powerful disinfectant, followed by ClO_2, OCl^-, NH_2Cl, and $NHCl_2$. Both methods of comparison, tabulated values of specific lethality coefficients and graphs of 99 percent inactivation, assume linear relationships based on the Chick–Watson law.

TABLE 12-2. Specific Lethality of Alternative Disinfectants[a]

Disinfectant	E. coli Bacteria	Poliovirus 1	Entamoeba histolytica Cysts
O_3	2300	920	3.1
HOCl	120	4.6	0.23
ClO_2	16	2.4	—
OCl^-	5.0	0.44	—
$NHCl_2$	0.84	0.00092	—
NH_2Cl	0.12	0.014	—

[a] Adapted from NAS (1980). Based on 99 percent inactivation of microorganisms and conditions closest to pH 7.0 and 20°C.

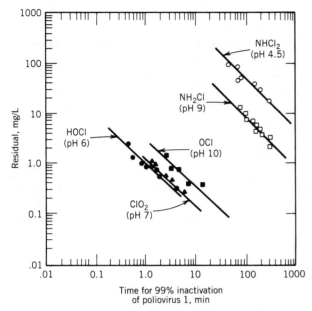

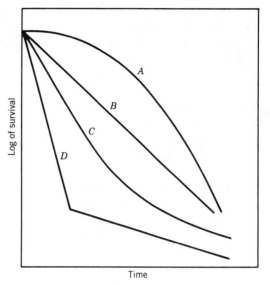

FIGURE 12-4. Types of survival curves.

FIGURE 12-3. Plots of residual versus time for fixed kill using classical kinetics. (Source: Scarpino et al., 1977.)

Anomalies

Unfortunately, the Chick–Watson theory is of limited usefulness in most practical disinfection operations. Generally, the rate of kill does not remain constant. Rather, it increases or decreases with time depending on the type of organism, the changing concentration or form of disinfectant, and other operating conditions.

The shapes of the disinfection curves that typically occur are shown in Figure 12-4. Curve *A* represents an increase in the rate of kill over most of the contact time, which Rahn (1973) cited as typical of the survival curve for multicellular organisms. Curves of this type are also common in disinfection of coliform with chloramines and chlorine dioxide. Curve *B*, described by Chick's law, indicates a constant rate of kill. Curve *C*, with a decreasing rate of kill with time, is also common in chlorination. Curve *D* has two steps, each following a different rate of kill.

In water or wastewater chlorination, an initial lag time normally exists before the disinfection process begins. Figure 12-5, from the data used to construct the 99 percent plot in Figure 12-3, shows this effect with the disinfection of *Escherichia coli* with chlorine dioxide (Scarpino et al., 1977). Possible explanations have been offered for this initial lag:

1. It may represent the time required for disinfectant diffusion to and transport across the cell membrane and for the reaction with vital constituents (Collins and Selleck, 1972).

2. Random collisions between molecules of a disinfectant and microorganisms may be ex-

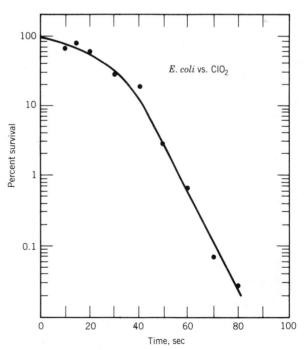

FIGURE 12-5. Common aberrations: initial lag. (Source: Scarpino et al., 1977.)

pressed as a Poisson probability. Because organisms may exist in clumps of various sizes, the probability of collision would be based on a multi-Poisson distribution (Wei and Chang, 1975).

3. The survival curve of a population of organisms, each with a number of sensitive units all of which must be inactivated before the organism will lose its viability, may be modeled with a multihit curve (Kimball, 1953).

Another common aberration observed in disinfection data is a decreasing rate of disinfection with time. Figure 12-6 illustrates this principle using the data of Floyd et al. (1978). Several hypotheses have been formulated to explain the decrease in the rate of kill with time:

1. There is a decrease in the germicidal properties of the disinfecting agent with time (Gard, 1957).

2. The microorganisms have a natural heterogeneity in size and resistance to the disinfectant. The observed decreasing rate would then be the sum of several first-order rate constants (Hess et al., 1953).

3. Exposure of the organisms to the disinfecting agent induces an increased resistance to the agent, that is, induced heterogeneity (Gard, 1957).

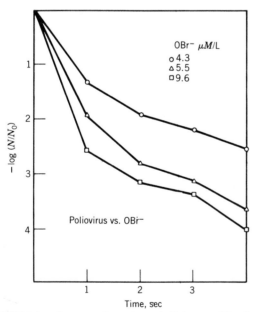

FIGURE 12-6. Common aberrations: declining rate (Floyd et al., 1978).

Other Models: Efforts to Fit Anomalies

In the last 40 years, there have been a number of disinfection models developed in the sanitary engineering and medical literature. Each has tried to deal with some of the common aberrations in disinfection data. In sanitary engineering, these include (1) a relationship between disinfection and chlorine species (Fair et al., 1947), (2) a reciprocal relation between coliform MPN and residual chlorine (Eliassen and Krieger, 1956), (3) a multihit model (Kimball, 1953), (4) declining rate model (Gard, 1957), (5) coefficients on concentration and time (Hom, 1972), (6) a modified declining rate model (Collins and Selleck, 1972), (7) an intermediate organism–disinfectant complex model (Haas, 1980), and (8) a predator–prey model (Hass, 1981).

Gard. In the medical literature, the importance of the declining rate of inactivation was recognized by Gard. In his investigation of the kinetics of chemical virus inactivation, Gard (1957) presented evidence that the rate of inactivation was not first order but decreased over time, as shown in Figure 12-7. Gard proposed the following equation to express the declining rate:

$$- \frac{dN}{dt} = \frac{kN}{1 + a(Ct)} \qquad (12\text{-}3)$$

or in the integrated form

$$\frac{N}{N_0} = [1 + a(Ct)]^{-k/a} \qquad (12\text{-}4)$$

where N = concentration of viable organisms at time t

C = disinfection concentration held constant over time

k = first-order rate of deactivation effected at time 0

a = rate coefficient

Collins and Selleck. Collins and Selleck (1972) observed that actual disinfection data often produce an initial lag as well as a declining rate of inactivation, as represented in Figure 12-8. If the log of survival is plotted versus time, as in many other disinfection plots, the form of Figure 12-9a is produced. After the initial lag, the declining portion of the curve is evident. Their proposed model curve, shown schematically in Figure 12-9b, has survival

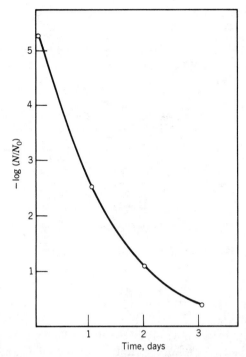

FIGURE 12-7.   Inactivation of poliovirus by formaldehyde, means of 12 experiments. (Source: Gard, 1957.)

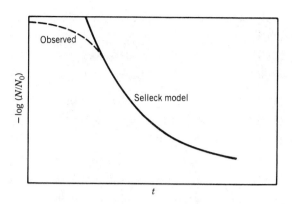

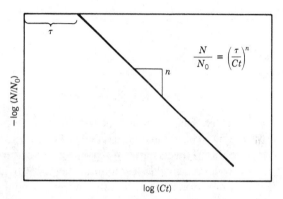

$$\frac{N}{N_0} = \left(\frac{\tau}{Ct}\right)^n$$

FIGURE 12-9.   Form of plot for Selleck model: (a) Versus time; (b) versus product of concentration and time.

(N/N_0) plotted against the product of concentration and time (Ct), not time alone.

Applying Scarpino et al.'s (1977) data to Selleck's model generates a good fit, as shown in Figure 12-10. Though previously representing the common lag aberration in Figure 12-5, the data illustrate

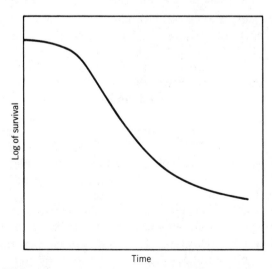

FIGURE 12-8.   Model disinfection curve. (Source: Collins and Selleck, 1972.)

the usefulness of Selleck's model to incorporate both the initial lag and the declining rate.

It can be seen that the Collins–Selleck curve can be used to explain each of Rahn's typical disinfection curves, shown in Figure 12-4. First, a curve of type A is produced if the initial lag is long and a short experiment is conducted. Curve B, which is fit by Chick's law, is the result if the initial lag is short and the experiment is short. Curve C, the declining rate, is obtained in long experiments. Finally, Curve D may just be another way of representing declining rate data.

The rate equation for the Collins–Selleck model is as follows:

$$\frac{dN}{dt} = -kN \qquad (12\text{-}5)$$

where $k = 0$ for $Ct < \tau$

$k = k'$ for $Ct = \tau$

$k = \dfrac{k'}{b(Ct)}$ for $Ct > \tau$

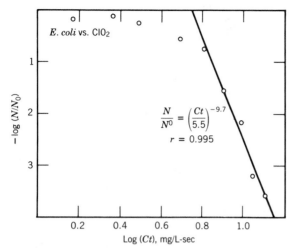

FIGURE 12-10. Application of model chlorine dioxide disinfection (Scarpino et al., 1977).

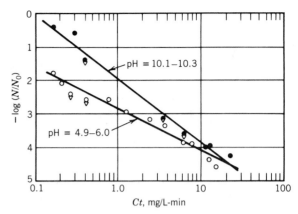

FIGURE 12-11. Free chlorine versus coliform (HOCl and OCl$^-$). (Source: Selleck et al., 1978.)

In this case, τ is a product of the concentration and time (Ct), which must be exceeded before inactivation begins. This assumes no disinfection occurs until the time (t) exceeds a minimum lag time (t').

After integration and application of the boundary conditions, the rate expression becomes

$$\frac{N}{N_0} = 1 \quad \text{for } Ct < \tau \tag{12-6}$$

$$\frac{N}{N_0} = \left(\frac{Ct}{\tau}\right)^{-n} \quad \text{for } Ct > \tau \tag{12-7}$$

where $n = \dfrac{k'}{b}$

Equation 12-6 applies to the lag period and Equation 12-7 models the declining rate. Implicit in the integration was the assumption that concentration remains constant. When chlorine and chlorine dioxide are used in wastewater disinfection, however, C decreases with time. Selleck et al. (1978) mathematically extrapolated an estimate of the active chlorine residual concentration after the immediate wastewater demand had been satisfied. He then assumed the bactericidal properties of the residual to remain constant.

The model curve (Figure 12-8) developed by Collins and Selleck (1972) can be used to explain the data observed for several disinfectants under the same conditions. Selleck et al. (1978) conducted pilot plant experiments to evaluate the various forms of chlorine on tap water seeded with coliform bacte-

ria. In Figure 12-11, chlorine was directly added to the seeded tap water and the pH altered between the two studies. (Each point plotted represents the average of four to five samples.) At the low pH, chlorine was 98 percent undissociated HOCl; at the high pH, it was predominantly OCl$^-$. Regression of the coliform data with Eq. 12-7 was a good fit with $r^2 = 0.97$ for the lower pH and $r^2 = 0.98$ for the higher pH. The HOCl residual showed a much smaller lag coefficient (τ) with the bacteria than the predominantly OCl$^-$ solution produced. Within the limit of the Ct products investigated ($Ct < 20$ mg/L-min), the high- and low-pH experiments approached each other in bacterial inactivation.

Selleck also ran a series of batch studies on chlorinated primary effluent (combined chlorine) with indigenous bacteria, as shown in Figure 12-12. The regression analysis of the coliform bacteria with Eq. 12-7 again showed a significant correlation with $r^2 = 0.96$. Roberts et al. (1980), using Selleck's model, plotted the inactivation of coliform bacteria by chlorine dioxide, as illustrated in Figure 12-13. These researchers also obtained similar results with free and combined chlorine.

Data from studies done by various researchers with different disinfectants yield further justification for the use of Selleck's model. Representative values for coefficients in Selleck's disinfection model are tabulated in Table 12-3. The high regression coefficients achieved with different disinfectants suggests the applicability of Selleck's model.

Because of its consistently good correlation with the experimental results, Selleck's model provides a common basis of comparison. When the various disinfectants are plotted on one graph as in Figure

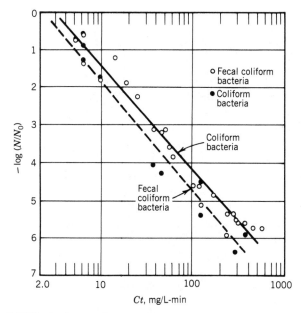

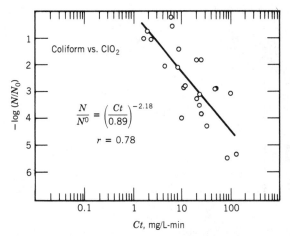

FIGURE 12-13. Application of model chlorine dioxide disinfection. (Source: Roberts et al., 1980.)

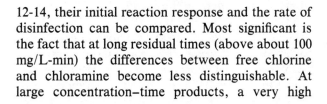

FIGURE 12-12. Application of model combined chlorine disinfection. (Source: Selleck et al., 1978.)

12-14, their initial reaction response and the rate of disinfection can be compared. Most significant is the fact that at long residual times (above about 100 mg/L-min) the differences between free chlorine and chloramine become less distinguishable. At large concentration–time products, a very high level of kill is achieved with all the disinfectants on the plot. This suggests that at the long residual times commonly found in practice, alternative disinfectants such as chloramines may be as satisfactory in disinfection as free chlorine.

Although the Selleck model has no rational mechanistic basis in describing chemical disinfection, it empirically approximates the behavior of real systems. As a predictor of disinfection phenomena, it therefore becomes a useful design tool (Roberts et al., 1980).

TABLE 12-3. Applications of Selleck's Model

	τ	n^a	r^b	Organism	Source
Chlorine					
Secondary effluent (combined residual)	4.06	−2.82	0.86	Coliform	Roberts et al. (1980)
Nitrified effluent (free residual)	1.33	−2.40	0.87	Coliform	Roberts et al. (1980)
Nitrified effluent (combined residual)	6.74	−3.17	0.91	Coliform	Roberts et al. (1980)
Nitrified, filtered (free residual)	0.59	−2.12	0.69	Coliform	Roberts et al. (1980)
Chlorine dioxide					
Secondary effluent	0.89	−2.18	0.78	Coliform	Roberts et al. (1980)
Nitrified effluent	0.78	−2.06	0.75	Coliform	Roberts et al. (1980)
Nitrified, filtered	0.003	−1.13	0.81	Coliform	Roberts et al. (1980)
Ozone					
Secondary, filtered[c]	8.65	−4.46	0.96	Coliform	Venosa et al. (1979)
Ozone-demand free water[d]	0.002	−2.51	0.76	*E. coli*	Katzenelson et al. (1974)
UV					
Secondary effluent	8.01	−1.99	0.70	Fecal coliform	Wolf et al. (1979)

[a] For definitions of τ and n, see Eqs. 12-5 and 12-7.

[b] Correlation coefficient.

[c] Results obtained using bubble diffuser ozone contactor and naturally occurring coliforms.

[d] Results obtained using batch reactor and laboratory-grown *E. coli* at 1°C.

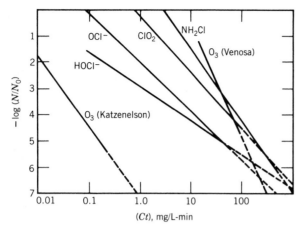

FIGURE 12-14. Application of model comparison of disinfectants.

CHEMISTRY OF DISINFECTANTS

Although chlorine has been used extensively throughout the United States, other disinfectants have been gaining acceptance. The reaction chemistry of the more common disinfectants will be discussed. Of the chemical agents, the disinfectants include chlorine, chloramines, chlorine dioxide, and ozone. The nonchemical disinfectant, ultraviolet radiation, is also included. Because of the very limited practical application of bromine, iodine, and bromine chloride in water treatment, they will not be discussed.

Chlorine

Disinfectant capabilities of chlorine depend on its chemical form in water, which in turn is dependent on pH, temperature, organic content of the water, and other water quality factors. Chlorine chemistry in water can be briefly described as follows: gaseous chlorine, when added to water, rapidly hydrolyzes to hypochlorous acid (HOCl) and hydrochloric acid (HCl).

$$Cl_2 + H_2O \rightleftarrows HOCl + H^+ + Cl^- \quad (12\text{-}8)$$

In dilute solution and at pH levels above 4, the reaction proceeds essentially to completion.

The hypochlorous acid then is subject to additional reaction which can include disinfection, reaction with various organic and inorganic compounds, or dissociation to hydrogen and hypochlorite ions

(OCl$^-$), as below:

$$HOCl \rightleftarrows H^+ + OCl^- \quad (12\text{-}9)$$

Hypochlorous acid is weakly acidic and its acid dissociation constant (K_a) at 20°C is 2.611×10^{-8} moles/L (Morris, 1966). Thus, the pH of water affects the relative amounts of HOCl and OCl$^-$. Figure 12-15 is a distribution diagram for the various chlorine species (Cl$_2$, HOCl, and OCl$^-$) over a broad pH range. Between pH 6 and 9, the relative fraction of HOCl decreases, while the corresponding fraction of OCl$^-$ increases.

The dissociation of hypochlorous acid is also temperature dependent. The effect of temperature is such that at a given pH, the fraction of HOCl will be lower at higher temperatures. The best-fit formula, developed by Morris (1966), is shown in Eq. 12-10:

$$pK_a = \frac{3000.00}{T} - 10.0686 + 0.0253T \quad (12\text{-}10)$$

where T = temperature in °K (°C + 273).

Generally, the disinfection capabilities of hypochlorous acid are greater than that of hypochlorite, especially at short contact times. However, as shown in Figure 12-13, the disinfection model suggests the equivalency of bacterial inactivation by these free chlorine products at long contact times.

Chloramines

When chlorine (Cl$_2$) and ammonia (NH$_3$) are both present in the water, they react to form products collectively known as chloramines. As opposed to the free chlorine described above, the chloramines are referred to as "combined chlorine."

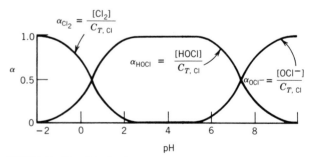

FIGURE 12-15. Distribution diagram for chlorine species: 25°C, [Cl$^-$] = 10^{-3} M, $C_{T, Cl}$ = [Cl$_2$] + [HOCl] + [OCl$^-$]. (Source: Snoeyink and Jenkins, 1980.)

The inorganic chloramines consist of three species: monochloramine (NH_2Cl), dichloramine ($NHCl_2$), and trichloramine, or nitrogen trichloride (NCl_3). The species of chloramines formed as a result of the combination of chlorine and ammonia depend on a number of factors, including the ratio of chlorine to ammonia–nitrogen, chlorine dose, temperature, pH, and alkalinity. As higher chlorine to ammonia–nitrogen ratios are reached, the ammonia is eventually oxidized to nitrogen gas (N_2), a small amount of nitrate (NO_3^-), or a variety of nitrogen-containing inorganic oxidation products.

The principal reactions for chloramine formation are shown in Eqs. 12-11 through 12-13. The product in Eq. 12-11 is monochloramine, followed by dichloramine in Eq. 12-12 and trichloramine in Eq. 12-13.

$$NH_3(aq) + HOCl \rightleftarrows NH_2Cl + H_2O \quad (12\text{-}11)$$

$$NH_2Cl + HOCl \rightleftarrows NHCl_2 + H_2O \quad (12\text{-}12)$$

$$NHCl_2 + HOCl \rightleftarrows NCl_3 + H_2O \quad (12\text{-}13)$$

At low pH, other reactions in the combined region are fairly significant, as shown in Eqs. 12-14 and 12-15:

$$NH_2Cl + H^+ \rightleftarrows NH_3Cl^+ \quad (12\text{-}14)$$

$$NH_3Cl^+ + NH_2Cl \rightleftarrows NHCl_2 + NH_4^+ \quad (12\text{-}15)$$

At high pH, these reactions for forming dichloramine from monochloramine would not be favored.

In addition to chlorinating the ammonia, as shown above, chlorine reacts to oxidize ammonia to species that are chlorine-free products. The two most common end products of ammonia oxidation by chlorine are nitrogen gas and nitrate, as shown in oxidation–reduction (redox) Eqs. 12-16 and 12-17, respectively:

$$3Cl_2 + 2NH_3 \rightleftarrows N_2(g) + 6HCl \quad (12\text{-}16)$$

$$4Cl_2 + NH_3 + 3H_2O \rightleftarrows \\ 8Cl^- + NO_3^- + 9H^+ \quad (12\text{-}17)$$

If ammonia is present, either as a natural constituent of the raw water or as a chemical deliberately added to produce combined chlorine rather than free chlorine, a hump-shaped breakpoint curve similar to Figure 12-16 is produced. If inorganic chlorine demand (such as from iron or manganese) is present, the initial chlorine dose produces no resid-

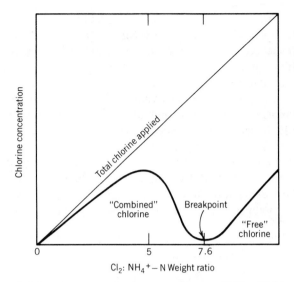

FIGURE 12-16. Breakpoint curve. (Source: White, 1978.)

ual and the residual versus dose curve would be flat until the demand is satisfied. Figure 12-16 illustrates the breakpoint curve as a function of chlorine to ammonia–nitrogen dose on a weight basis. As the chlorine dose increases (or the chlorine to ammonia–nitrogen ratio increases), the chlorine residual first rises to a maximum and then declines to a minimum.

Up to a chlorine to ammonia weight ratio of 5, the predominant product formed is monochloramine. On the declining side of the hump, the monochloramine disappears by forming nitrogen gas or a trace of dichloramine. Prior to the breakpoint, the chlorine residual is a combined residual. After the breakpoint, which occurs at a chlorine to ammonia–nitrogen weight ratio of approximately 7.6, there is no ammonia left to react with the chlorine. Therefore, the residual shown in the second rising portion of the curve is free chlorine. There may also be traces of dichloramine and trichloramine as the equilibrium reactions continue.

In addition to the chlorine to ammonia ratio, reaction conditions are also important in determining the final end product of the chlorine reaction. Of these conditions, pH is the most important. Snoeyink and Jenkins (1980) describe in detail the effect of pH on the formation of monochloramine, since both reactants (ammonia and hypochlorous acid) are affected by pH. The optimum pH for forming monochloramine is around 8.4. In general, monochloramine is formed above pH 7.

The effect of pH on the formation of the various

chloramine species is illustrated in Figure 12-17, after Palin (1975). Using a given initial ammonia–nitrogen concentration of 0.5 mg/L and a 1-day reaction time, Palin showed that at pH 6, both monochloramine and dichloramine were formed before the breakpoint and that trichloramine and free chlorine existed together after the breakpoint. At pH 7, the principal species before the breakpoint was monochloramine, with only a small amount of dichloramine; trichloramine continued to exist in combination with free chlorine after the breakpoint. However, at pH 8, only a trace of dichloramine was formed prior to the breakpoint, while the remainder was monochloramine; furthermore, there was no trichloramine after the breakpoint.

Chlorine Dioxide

Chlorine dioxide has not been widely used for water or wastewater disinfection in the United States. Interest in chlorine dioxide as a disinfectant has been heightened by evidence indicating that chlorine dioxide does not produce significant amounts of THMs as by-products from reactions with organics (Chow and Roberts, 1981; EPA, 1981).

The chemistry of chlorine dioxide (ClO_2) in water is relatively complex. In acid solution, reduction to chloride predominates:

$$ClO_2 + 5e^- + 4H^+ \rightleftharpoons Cl^- + 2H_2O \quad (12\text{-}18)$$

If Eq. 12-18 proceeded in water, Table 12-1 would indicate that chlorine dioxide has about 1.4 times the oxidizing power of chlorine. However, at the relatively neutral pH found in most natural waters, it is generally accepted that the following reduction to chlorite predominates:

$$ClO_2 + e^- \rightleftharpoons ClO_2^- \quad (12\text{-}19)$$

With a reduction potential of about 0.95 volts, chlorine dioxide as depicted in Eq. 12-19 only has about 70 percent of the oxidizing power of chlorine. Thus, the total oxidizing capacity of chlorine dioxide is not typically used in water treatment practices. Other aqueous reactions of chlorine dioxide and its oxidized forms are discussed by White (1972).

Chlorine dioxide, when used in water treatment applications, is almost always generated on-site directly prior to application. As a gas, chlorine dioxide is explosive at elevated temperatures, on exposure to light, or in the presence of organic substances, so it is usually never shipped in the gaseous state. Therefore, the method of choice in the United States for obtaining chlorine dioxide has been the chlorine–chlorite ($NaClO_2$) process:

$$2NaClO_2 + Cl_2 \rightarrow 2ClO_2 + NaCl \quad (12\text{-}20)$$

The major drawback of this process is that the ClO_2 gas is only 60–70 percent pure and contains a considerable amount of chlorine. This chlorine is then available to produce the undesirable by-products

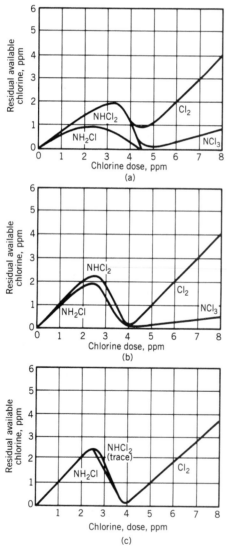

FIGURE 12-17. Effect of pH on chloramine speciation (after 1 day, initial NH_3–N = 0.5 ppm): (a) pH 6.0, (b) pH 7.0, (c) pH 8.0. (Source: Palin, 1975.)

that selection of the chlorine dioxide process was originally trying to avoid.

Other processes have been developed to produce chlorine dioxide (White, 1978). A recent European system (CIFEC, France) is able to produce 95–98 percent pure chlorine dioxide solution through the use of an enrichment loop for the chlorine utilized (CIFEC, 1976).

Several factors are noteworthy in the chemistry of chlorine dioxide. In contrast to chlorine, chlorine dioxide remains in molecular form as ClO_2 in the pH range typically found in natural waters, does not react with ammonia or nitrogenous compounds, and does not react with precursors to form chloroform (Roberts et al., 1980). However, chlorine dioxide in water produces inorganic breakdown products, chlorite (ClO_2^-) and chlorate (ClO_3^-), for which the health effects are not well understood. Initial work with a variety of laboratory animals demonstrated some oxidative changes in blood cells exposed to high concentrations of chlorine dioxide or its breakdown products. More recent work with humans at concentrations more typical of drinking water disinfection have shown little or no change in blood chemistry relative to controls (Bull, 1980).

Ozone

Ozone is one of the most powerful oxidizing agents that has practical applications for water and wastewater treatment, as shown in Table 12-1. Ozone (O_3), an allotrope of oxygen (O_2), is a highly reactive gas formed by electrical discharges in the presence of oxygen as follows:

$$3O_2 + energy \rightleftarrows 2O_3 \qquad (12\text{-}21)$$

Substantial amounts of energy are required to split the stable oxygen–oxygen covalent bond to form ozone.

Ozone's high level of chemical energy is also the driving force for its decomposition. The ozone molecule readily reverts to elemental oxygen during the oxidation–reduction reaction. Hoigne and Bader (1975a; 1975b; 1976) demonstrated that the rate of O_3 decomposition is a complex function of temperature, pH, and concentration of organic solutes and inorganic constituents.

Figure 12-18 shows reaction pathways of ozone as they have been described by these authors. Once ozone enters solution, it follows two basic modes of reaction: direct oxidation, which is rather slow and

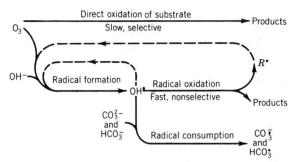

FIGURE 12-18. Reaction pathways of ozone. (Source: Hoigne and Bader, 1976.)

extremely selective, and autodecomposition to the hydroxyl radical. Autodecomposition to the hydroxyl radical (OH·) is catalyzed by the presence of hydroxyl radicals, by organic radicals, or by high concentrations of hydroxide ion. The hydroxyl radical is extremely fast and nonselective in its oxidation of organic compounds. But at the same time, the hydroxyl radical is scavenged by carbonate and bicarbonate ions to form carbonate (CO_3^-) and bicarbonate ($HCO_3·$) radicals. These radicals are of no consequence in organic reactions. Further, the hydroxyl radicals and organic radicals produced by the autodecomposition become chain carriers and enter back into the autodecomposition reaction to accelerate it. Thus, conditions of low pH favor the slow direct oxidation reactions involving O_3, and high pH conditions or high concentrations of organic matter favor the autodecomposition route. In general, better disinfection would be expected at lower pH's.

Ultraviolet Radiation

Destruction of microorganisms by ultraviolet (UV) radiation occurs when the UV energy is absorbed by the genetic material of the cells. Maximum destructive activity is assumed to occur at 265 nm, which corresponds with the maximum absorption of nucleic acids (Stanier et al., 1963). When the genetic material in the cells absorbs the UV energy, pyrimidine dimers are formed. These dimers, by causing distortions in the DNA, prevent the proper replication of the DNA strands. Under certain conditions, however, the genetic damage may be reversed. When the injured organism is exposed to visible light energy (310–500 nm), photoreactivation occurs, wherein the dimerization is reversed (Harm, 1976; Witkin, 1976; Kelner, 1949).

TABLE 12-4. Summary of Disinfectant Characteristics

Characteristics	Free Chlorine	Chloramines	Chlorine Dioxide	Ozone	Ultraviolet Radiation
Disinfection					
Bacteria	Excellent (as HOCl)	Moderate	Excellent	Excellent	Good
Viruses	Excellent (as HOCl)	Low (good at long contact times)	Excellent	Excellent	Good
pH influence	Efficiency decreases with increase in pH	Dichloramine predominates pH 5 and below; monochloramine predominates pH 7 and above. Overall relatively independent of pH.	Slightly more efficient at higher pH's	Residuals last longer at low pH	Insensitive
Residual in distribution system	Yes	Yes	Yes	No	No
By-products					
THM formation	Yes	Unlikely	Unlikely	Unlikely	Unlikely
Other	Uncharacterized chlorinated and oxidized intermediates; chloramines; chlorophenols	Unknown	Chlorinated aromatic compounds; chlorate chlorite	Aldehydes; aromatic carboxylic acids; phthalates	Unknown
Experience	Widespread use in the U.S.	Widespread use in the U.S.	Widespread use in Europe; limited in the U.S.	Widespread use in Europe and Canada; limited in the U.S.	Use limited to small systems
Typical applied dose, mg/L	2–20	0.5–3.0	[a]	1–5	[a]
$/lb[b]	0.07	0.16[c]	1.44[d]	0.48[e]	—
Pound equivalent weight[f]	35.5	25.8[g]	13.4[h]	24	—
Cost per pound equivalent weight ($/lb)	2.49	4.13	19.3	11.5	—

[a] Insufficient operating knowledge

[b] Effective January 1982. Does not include shipping.

[c] Assumes 1.4 lb Cl_2 + .46 lb $NH_3 \rightarrow$ 1 lb NH_2Cl

[d] Assumes 0.5 lb Cl_2 + 1.4 lb $NaClO_2 \rightarrow$ 1 lb ClO_2 + NaCl

[e] Assumes 12 kw-hr per lb O_3; energy cost = \$0.04 per kw-hr

[f] Weight of compound per 1 electron change in oxidation-reduction reaction, based on equations in Table 12-1.

[g] Assumes NH_2Cl + 2e^- + $H_2O \rightarrow NH_3$ + Cl^- + OH^-

[h] Assumes complete reaction: ClO_2 + 5e^- + 2$H_2O \rightarrow Cl^-$ + 4OH^-

SOURCES: National Academy of Science (1980), EPA (1981), Lawrence et al. (1980).

Although all microorganisms are susceptible to UV radiation, the sensitivity of the organisms varies, depending on their resistance to penetration of UV energy. The chemical composition of the cell wall and its thickness determines the relative resistance of an organism. An organism's resistance is measured by the time needed to kill a certain percentage of organisms by a specific UV dose (Scheible and Bassell, 1981).

Table 12-4 presents significant characteristics of the five disinfectants considered. As mentioned in the section on general concepts, the selection of an appropriate disinfectant requires the weighted consideration of many factors such as germicidal efficiency, process and design characteristics, and environmental consequences.

DESIGN OF DISINFECTANT CONTACT FACILITIES

Many wastewater treatment facilities are designed and evaluated on the dispersion characteristics of their disinfectant contact basins. Unfortunately, this has not been the practice in the water treatment field. The design of facilities for the primary disinfection of drinking water should take advantage of some of the design principles developed in the field of wastewater treatment.

Disinfection process performance is substantially affected by hydraulics during the initial mixing of the disinfectant and the water to be treated and in the subsequent contact chamber. However, the mechanisms of these processes are not well understood.

Rapid dispersion of chemicals, particularly with halogens, through turbulent mixing enhances the disinfection of microorganisms by providing the transportation means to allow the added material to reach all parts of the water. Devices to provide varying degrees of mixing conditions for chemical dispersion include mixing chambers with pumps, baffled reactors and hydraulic jumps. These are discussed further in Chapter 21.

Evaluating the geometry of design for disinfectant contact facilities is important because the effectiveness of the reaction between the organism and the disinfectant, particularly for relatively slow reactions, is affected by the geometry. Often, design of contact chambers is evaluated on the basis of crude rules of thumb with true performance being determined only after construction. A completely rational basis for contact chamber design requires a relation between the kinetics of the disinfection reaction and the distribution of residence times as well as a means of estimating the distribution of the residual times for a proposed contact chamber design. This section presents the form and nature of these relations and develops them to the degree possible at the present time based on the earlier work of Trussell and Chao (1977). The general configuration of the contact facilities for chlorine, chlorine dioxide, and ozone is addressed assuming proper management of the effects of other important design aspects such as initial mixing, inlet and outlet design, and so on.

Disinfection Relation

As presented in the section on the kinetics of disinfection, Collins and Selleck (1972) developed a general kinetic expression for the effect of combined chlorine residuals on both total and fecal coliform. Because of the applicability and reliability of this model to various disinfectants as expressed in Eqs. 12-6 and 12-7, it will be used in the remaining discussion.

Dispersion

Plug Flow. A number of parameters have been used to evaluate the hydraulic efficiency of chlorine contact basins. Marske and Boyle (1973) found that the most reproducible of them is the dispersion number, d. Figure 12-19 shows the shape of the dye curves that might emanate from a vessel to which a pulse of dye is added. The curves are shown for various values of the dispersion number to illustrate its significance. As the dispersion number in a reactor decreases, the reactor approaches closer to "perfect plug flow." A perfect plug-flow, or batch, reactor has a dispersion number of zero. As indicated by Marske and Boyle (1973), however, the dispersion number for a given reactor under given conditions is reproducible because it is derived from the entire dye curve rather than just one or two points.

The dispersion number is also more valuable from the standpoint of chemical kinetics. For a given type of reactor, mathematical relations can be developed to describe the completion of a reaction given the dispersion number and the basic kinetic equation of the reaction.

It is only possible to obtain an accurate analytical expression for the dye curve (frequency distribution

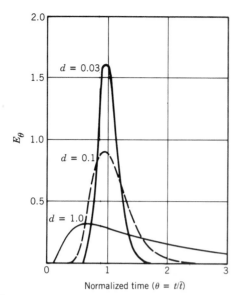

FIGURE 12-19. Frequency distribution of ages versus dispersion number. (Source: Trussell and Chao, 1977.)

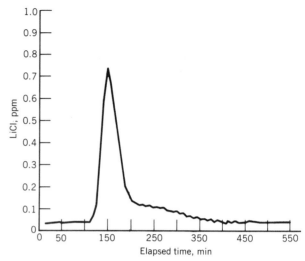

FIGURE 12-20. Lithium chloride tracer test (dispersion number: 0.04).

of ages) for the "open vessel" case where flow is presumed to be undisturbed at both the entrance and exit of the contact vessel. In the strict sense, a chlorine contact tank does not meet these criteria; however, since chlorine contact tanks are designed to minimize influent and effluent disturbance, the open vessel formula may prove satisfactory. As developed by Levenspiel and Smith (1957), this formula is as follows:

$$E_\theta = \frac{1}{\sqrt{4\pi\theta d}} \exp\left[\frac{-(1 - \theta)^2}{4\theta d}\right] \quad (12\text{-}22)$$

where d = dispersion number
θ = normalized contact time $(t/\bar{t})$
$\bar{t}$ = hydraulic residence time, min (V/Q)

Thus, an expression is now available relating the dispersion number and the dye curve in the basin. This equation describes the dye curves shown in Figure 12-19.

To generate a dispersion curve, a tracer study is typically conducted. A tracer is injected into the inlet basin and monitored at the outlet over time. The tracer selected should present no adverse health effects, be a chemical which is not already present in the water, and be convenient to analyze. Lithium salts, particularly lithium chloride, are tracers that meet those requirements. The tracer test in Figure 12-20 was generated from adding lithium chloride as a pulse to the inlet of a serpentine

chlorine contact basin. Samples from the basin outlet were collected at approximately 7-min intervals and analyzed for lithium by atomic absorption. From the information provided by the tracer test, the dispersion number was calculated from Eq. 12-22.

As indicated before, it is also possible to develop a relation between the dye curve or frequency distribution of the ages and the completion of a given chemical reaction. In developing this relation, the degree of segregation among reacting components must be considered. Bacterial disinfection is an excellent example of a system that may be treated with a segregated model where aggregates of molecules would travel together throughout their stay in the reactor. Using the concept of a segregated system, the fraction of bacteria remaining after reaction will be the sum of the batch reactions of all the small aggregates, or:

$$\frac{N}{N_0} = \frac{1}{\bar{t}} \int_0^\infty \left(\frac{N}{N_0}\right)_{\text{batch}} E_\theta \, d\theta \quad (12\text{-}23)$$

Combining Eqs. 12-7, 12-22, and 12-23,

$$\frac{N}{N_0} = \frac{1}{\bar{t}} \int^{\tau/\bar{t}} \frac{1}{\sqrt{4\pi\theta d}} \exp\left[\frac{-(1 - \theta)^2}{4\theta d}\right] d\theta$$
$$+ \frac{1}{\bar{t}} \int_{\tau/\bar{t}}^\infty \left(\frac{\tau}{C\bar{t}\theta}\right)^n \frac{1}{\sqrt{4\pi\theta d}}$$
$$\exp\left[\frac{-(1 - \theta)^2}{4\theta d}\right] d\theta \quad (12\text{-}24)$$

Thus, given the environmental coefficient or induction time, τ, the contact time, and the initial chlorine residual, it should be possible to show the relation between the survival ratio of total or fecal coliform and dispersion. Figure 12-21 shows such a relationship for various chlorine residuals and a contact time of 1 hr with total coliform from the secondary effluent. It should be noted that little benefit is derived from attempting to bring the dispersion number below 0.01. About a 60 percent increase in kill is accomplished by bringing the dispersion from 0.1 to 0.01, whereas the increase in kill between 0.01 and 0.001 is only 5 or 6 percent. For dispersion numbers below 0.1, it appears that the influence of dispersion on contact chamber performance is relatively minor compared to the influence of chlorine residual.

Completely Stirred Tank Reactor. The extremely rapid reaction of ozone disinfection creates difficulties in attempting to translate experimental information directly into design practice. Unlike chlorine, ozone does not form a longlasting residual. As a result, prediction of disinfection performance requires a vigorous prediction of the changes in the magnitude of the ozone residual as a function of time. It would be relatively difficult to develop such a prediction for water free of organic contamination, and for most natural waters the task is nearly

impossible. Another difficulty is that of adequately monitoring the ozone residual. Because ozone is so reactive, its residual is always in a state of change and this makes the residual monitoring task difficult. The final difficulty is the complexity of the distribution of residence times in the three-phase countercurrent reactor configuration most often encountered (Trussell and Russell, 1978).

As shown earlier, Eq. 12-23 predicts the performance of a two-phase countercurrent reactor if the distribution of residence times is known. However, because ozone is a gas, a countercurrent reactor configuration is generally employed and this makes the application of Eq. 12-24 very difficult.

To develop a model of the typical diffused air contact chamber, two assumptions are necessary: (1) the chamber behaves as a completely stirred tank reactor (CSTR) and (2) the ozone residual present during the process can be approximated as one-half the effluent residual.

If these assumptions are made, then the Selleck formula (Eq. 12-7) can be used to represent the batch disinfection reaction, and the CSTR dispersion equation can be used as the normalized frequency distribution of ages (E_θ). For a cascade of m stirred tanks, the CSTR dispersion equation (Kramer and Westerterp, 1963) is

$$E_\theta = \frac{m(m\theta)^{m-1}}{(m-1)!}\exp(-m\theta) \qquad m \neq 1 \quad (12\text{-}25)$$

Equation 12-25 may be combined with Eq. 12-23 to develop a relation for disinfection of coliforms in cascades of stirred tanks:

$$\frac{N}{N_0} = \frac{1}{\bar{t}}\int_0^{\tau/\bar{t}} \frac{m(m\theta)^{m-1}}{(m-1)!}\exp(-m\theta)\,d\theta$$
$$+ \frac{1}{\bar{t}}\int_{\tau/\bar{t}}^{\infty}\left(\frac{\tau}{C\bar{t}\theta}\right)^n \frac{m(m\theta)^{m-1}}{(m-1)!}$$
$$\exp(-m\theta)\,d\theta \qquad (12\text{-}26)$$

The plot in Figure 12-22 compares the results obtained by assuming a batch reaction and by using Eq. 12-26 to evaluate the impact of the distribution of residence times. It can be shown that a CSTR can be made to approach the performance of a batch reactor more closely if it is subdivided into individual compartments. Figure 12-22 compares the performance of a batch reactor with the CSTR having the same total volume divided into 1–20 compartments in series. It can be observed that batch per-

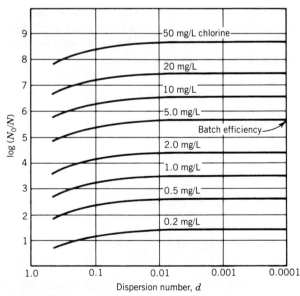

FIGURE 12-21. Dispersion and reduction of total coliform by combined chlorine (60 min contact time). (Source: Trussell and Chao, 1977.)

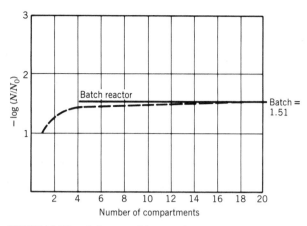

FIGURE 12-22. Influence of degree of compartmentalization on performance of CSTRs: $C_0 = 0.5$ mg/L; $t = 600$ sec; $\tau = 600$ sec; $n = 4$.

formance is achieved after only 3–5 compartments are used. As a result compartmentalization is probably both economical and desirable in the design of ozone contact chambers.

CHAMBER DESIGN

Several forms of contact chamber design are in use today. Among these are pipes, baffle basins, settling tanks, circular tanks, and annular rings around secondary clarifiers. Dispersion numbers may be estimated for long pipes and can be developed with some degree of reliability. A crude method can be used for estimating the dispersion in long channels.

Long Pipes

Axial dispersion in pipeline flow is the most straightforward case that will be considered. Taylor (1954) shows that the longitudinal dispersion coefficient (D_L) can be described by the following formula:

$$D_L = 5.05 D U_* \qquad (12\text{-}27)$$

In the above formula, U_* is the shear velocity or friction velocity in a pipe of diameter D and may be defined in terms of the velocity of flow, U, and the Darcy–Weisbach friction factor, f.

$$U_* = \sqrt{\frac{fU^2}{8}} \qquad (12\text{-}28)$$

The dispersion number is defined in terms of a longitudinal dispersion coefficient, the velocity of flow, and a characteristic length, in this case, the length of the pipe:

$$d = \frac{D_L}{UL} \qquad (12\text{-}29)$$

Combining Eqs. 12-27 through 12-29 results in a formula describing the dispersion of flow in a pipe:

$$d = 5.50 \left(\frac{D}{L}\right) \sqrt{\frac{f}{8}} \qquad (12\text{-}30)$$

Sjenitzer (1958) gathered a great number of measurements, both in the laboratory and in the field, and correlated them to produce the following empirical expression:

$$d = 89500 f^{3.6} \left(\frac{D}{L}\right)^{0.859} \qquad (12\text{-}31)$$

This relation, however, is only accurate for a long pipeline without bends, restrictions, or other disturbances to flow.

Long Channels

Although no work has been done on longitudinal dispersion of open-channel flow in narrow conveyances, extensive studies have been recorded concerning dispersion in open-channel flow. Most of the models developed for the dispersion coefficient and this type of flow have been developed from the following form of the Taylor equation:

$$D_L = CR_h U_* \qquad (12\text{-}32)$$

where R_h = hydraulic radius
U_* = shear velocity
C = a coefficient

Combined with Eq. 12-29, this relation leads to the formula for the dispersion number given below:

$$d = \frac{CR_h U_*}{UL} \qquad (12\text{-}33)$$

As noted by Graber (1972), coefficient C is a function of channel geometry and the Reynolds number. For narrow conveyances, channel geometry becomes very significant. In the work cited above, open channels were considered two-dimensional owing to their large width–depth ratio.

Elder (1959) developed a model that applied Taylor's concept of longitudinal dispersion to a logarithmic velocity profile. The formula for C that corresponds to this model is shown below:

$$C = \frac{0.404}{k^3} + \frac{k}{6} \qquad (12\text{-}34)$$

where k is the von Karman turbulence coefficient and is generally assumed to have a value of about 0.4. Under these circumstances, the corresponding value of C would be 5.9.

For uniform flow in an open channel, the friction velocity, U_*, might be defined as follows:

$$U_* = \frac{3.82nU}{R_h^{1/6}}$$

$$(n = \text{Manning coefficient}) \qquad (12\text{-}35)$$

Combining Eqs. 12-33 through 12-35, the following crude formula for dispersion number in a long open channel is obtained:

$$d = \frac{22.7nR_h^{5/6}}{L} \qquad (12\text{-}36)$$

Given certain coefficients of channel geometry, Eq. 12-36 may be rewritten to describe the dispersion using the channel volume, given

$$\alpha = H/W$$

$$\beta = L/W$$

$$\nu = \text{volume of channel, ft}^3$$

Then

$$d = 22.7 \frac{n}{\beta} \left(\frac{\alpha}{2\alpha + 1}\right)^{5/6} \left(\frac{\alpha\beta}{\nu}\right)^{1/18} \qquad (12\text{-}37)$$

Equation 12-37 is not very sensitive to either α or ν, and for most concrete basins, the following abbreviated form is satisfactory:

$$d = \frac{0.14}{\beta} \qquad (12\text{-}38)$$

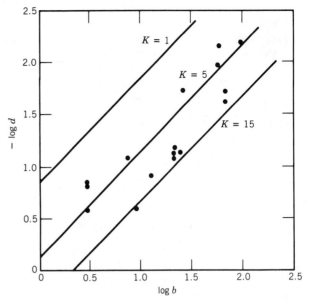

FIGURE 12-23. Log of length-to-width ratio versus log of dispersion number. (Source: Trussell and Chao, 1977.)

In addition to the shortcomings of the theoretical development that led to Eq. 12-37, a number of other considerations, such as baffling, lead to some nonideality. A coefficient of nonideality K may be used to describe these deviations in the following manner:

$$d = \frac{0.14K}{\beta} \qquad (12\text{-}39)$$

Figure 12-23 shows a plot of Eq. 12-39 in relation to full-scale field data with K values of 1, 5, and 15. It is suggested that conservative designs should assume a nonideality coefficient of 15. However, it appears that nonideality can be reduced to as low as 3–5 by good hydraulic design.

REFERENCES

Berg, G., "The Virus Hazard in Water Supplies," *J. New England Water Works Assoc.*, **78**, 79 (1964).

Boyce, D. S., Sproul, O. J., and Buck, C. E., "The Effect of Bentonite Clay on Ozone Disinfection of Bacteria and Viruses in Water," *Water Research*, **15**, 759–767 (1981).

Bull, R. J., "Health Effects of Alternative Disinfectants and Their Reaction Products," *JAWWA*, **72**(5), 299–303 (1980).

Chang, S. L., "Modern Concept of Disinfection," *J. Div. Eng., Proc. Am. Soc. Civ. Eng.*, **97**(SA5), 689–707 (October 1971).

Chick, H., "An Investigation of the Laws of Disinfection," *J. Hygiene, 8*, 92 (1908).

CIFEC, "Chlorine Dioxide Generator: The French Method with Enrichment Loop," Cat. 167 CIFEC, Paris, France (1976).

Chow, B. M., and Roberts, P. V., "Halogenated Byproduct Formation by ClO_2 and Cl_2," *J. Div. Eng., Proc. Am. Soc. Civ. Eng., 107*(EE4), 609–618 (August 1981).

Collins, H., and Selleck, R., "Process Kinetics of Wastewater Chlorination," *SERL Report No. 72-5,* University of California, Berkeley (1972).

Elder, J. W., "The Dispersion of Marked Fluid in Turbulent Shear Flow," *J. Fluid Mech., 5*, 544 (1959).

Eliassen, R., and Krieger, H. L., "Control of Bacterial Numbers in Chlorinated Sewage Effluents," *Sewage Ind. Wastes, 22*(1), 47–54 (January 1956).

Emerson, M., Sproul, O., and Buck, C., "Ozone Inactivation of Cell-Associated Viruses," *Appl. Env. Microb., 43*(3), 603–608 (March 1982).

EPA, "Treatment Techniques for Controlling Trihalomethanes in Drinking Water," EPA-600/2-81-156 (September 1981).

Fair, G. M., Morris, J. C., and Chang, S. L., "The Dynamics of Water Chlorination," *J. NEWWA, LXI, 4*, 285 (1947).

Floyd, R., Sharp, D. G., and Johnson, J. D., "Inactivation of Single Poliovirus Particulates in Water by Hypobromite Ion, Molecular Bromine, Dibromamine, and Tribromamine," *ES&T, 12*, 1031 (1978).

Foster, D. M., et al., "Ozone Inactivation of Cell- and Fecal-Associated Viruses and Bacteria," *J. Water Pollut. Control Fed., 52*(9), 2174–2184 (1980).

Gard, S., "Chemical Inactivation of Viruses," in *CIBA Foundation Symposium on the Nature of Viruses,* Little Brown & Co., Boston (1957), p. 123.

Graber, D. S., "Discussion/Communication," *J. Water Pollut. Control Fed., 44*, 2029 (1972).

Haas, C., "A Mechanistic Kinetic Model for Chlorine Disinfection," *ES&T, 14*, 339 (1980).

Haas, C., "Application of Predator-Prey Models to Disinfection," *J. Water Pollut. Control Fed., 53*, 378 (1981).

Harm, H., "Repair of Ultraviolet Irradiated Biological Systems: Photoreactivation," in *Photochemistry and Photobiology of Nucleic Acids,* Vol. 2, Academic Press, New York (1976).

Hess, S., Diachishin, A., and De Falco, Jr., P., "Bactericidal Effects of Sewage Chlorination, Theoretical Aspects," *Sewage Ind. Wastes, 25*, 909 (1953).

Hijkal, T. W., et al., "Survival of Poliovirus Within Organic Solids During Chlorination," *Appl. Environ. Microbiol., 38*(11), 114 (July 1979).

Hoff, J. E., "The Relationship of Turbidity to Disinfection of Potable Water," in C. H. Hendricks (ed.) *Evaluation of the Microbiology Standards for Drinking Water,* EPA-570/9-78-00C (1978).

Hoigne, J., and Bader, H., "Ozonation of Water: Role of Hydroxyl Radicals as Oxidizing Intermediates," *Science, 190*, 782 (1975a).

Hoigne, J., and Bader, H., "Identification and Kinetic Properties of the Oxidizing Decomposition Products of Ozone in Water and its Impacts on Water Purification," Presented at the Second International Ozone Symposium, International Ozone Institute, Montreal, Canada (May 1975b).

Hoigne, J., and Bader, H., "Role of Hydroxyl Radical Reactions in Ozonation Processes in Aqueous Solutions," *Water Res., 10*, 377 (1976).

Hom, L. W., "Kinetics of Chlorine Disinfection in an Ecosystem," *J. Sanitary Eng. Div.,* ASCE, SA1, 183–194 (February 1972).

Katzenelson, E., Kletter, B., Schechter, H., and Shuval, H., "Inactivation of Viruses and Bacteria by Ozone," in *Chemistry of Water Supply, Treatment, and Distribution,* A. Rubin (ed.), Ann Arbor Science, Ann Arbor, MI (1974).

Kelner, A., "Effects of Visible Light on the Recovery of *Streptomyces Griseus Conidia* from Ultraviolet Radiation Injury," *Proc. Natl. Acad. Sci. 35*, 73–79 (1949).

Kimball, A. W., "The Fitting of Multi-Hit Survival Curves," *Biometrics,* 201–211 (June 1953).

Kramers, H., Westertrep, K. R., *Elements of Chemical Reactor Design and Operation,* Academic Press, New York (1963).

Lawrence, J., Tosine, H., Onuska, F. E., and Comba, M. E., "The Ozonation of Natural Waters: Product Identification," *Ozone: Sci. Eng., 2*, 55–64 (1980).

Levenspiel, O., and Smith, W. K., "Notes on the Diffusion—Type Model for the Longitudinal Mixing of Fluids in Flow," *Chem. Engr. Sci., 6*, 227 (1957).

Marske, D. M., and Boyle, J. D., "Chlorine Contact Chamber Design—A Field Design Evaluation," *Water Sewage Works, 70* (January 1973).

Morris, J. C., "The Acid Ionization Constant of HOCl from 5° to 35° C," *J. Phys. Chem., 70*, 3798 (December 1966).

National Academy of Science, *Drinking Water and Health,* Vol. 2, National Academy Press, Washington, DC (1980).

Palin, A. T., "Water Disinfection—Chemical Aspects and Analytical Control," in J. D. Johnson, (ed.), *Disinfection—Water and Wastewater,* Ann Arbor Science, Ann Arbor, MI (1975).

Rahn, O., *Physiology of Bacteria,* Blankston's and Son, Philadelphia (1973).

Roberts, P. V., Aieta, E. M., Berg, J. D., and Chow, B. M., "Chlorine Dioxide for Wastewater Disinfection: A Feasibility Evaluation," Stanford University Technical Report 251 (October 1980).

Rosenau, M. J., *Preventative Medicine and Hygiene,* 6th ed., Appleton-Century-Crofts, New York (1940).

Scarpino, P. V., Cronier, S., Zink, M. L., and Brigano, F. A. O., "Effect of Particulates on Disinfection of Enteroviruses and Coliform Bacteria in Water by Chlorine Dioxide," in Proceedings of the AWWA Water Quality Technology Conference, Kansas City, Missouri (December 1977).

Scheible, O. K., and Bassell, C. D., "Ultraviolet Disinfection of a Secondary Wastewater Treatment Plant Effluent," Municipal Environmental Research Laboratory, Cincinnati, EPA-600/2-81-152 (August 1981).

Selleck, R. E., Saunier, B. M., and Collins, H. F., "Kinetics of Bacterial Deactivation with Chlorine," *J. Div. Env. Eng., Proc. Am. Soc. Civ. Eng., 104*(EE6), 1197–1212 (December 1978).

Sjenitzer, F., "How Much do Products Mix in a Pipeline?", *The Pipeline Engineer,* D-31 (December 1958).

Snoeyink, V. L., and Jenkins, D., *Water Chemistry,* Wiley, New York (1980).

Stanier, R. Y., Doudoroff, M., and Adelberg, E. A., *The Microbial World,* Prentice-Hall, Englewood Cliffs, NJ (1963).

Symons, J. M., and Hoff, J. C., "Rationale for Turbidity Maximum Contaminant Level," in Proceedings of Third Annual Water Quality Technology Conference, AWWA, Atlanta (December 1975).

Taylor, G. I., "The Dispersion of Matter in Turbulent Flow Through a Pipe," *Proc. Roy. Soc.,* **A223,** 446 (1954).

Trussell, R. R., and Chao, J.-L., "Rational Design of Chlorine Contact Facilities," *J. Water Pollut. Control Fed.,* **49**(4), 659–667 (April 1977).

Trussell, R. R., and Russell, L. L., "Application of Ozonation to Water Treatment," James M. Montgomery, Consulting Engineers, Inc. (January 1978).

Venosa, A. D., Meckes, M. C., Optaken, E. J., and Evans, J. W., "Comparative Efficiencies of Ozone Utilization and Microorganism Reduction in Different Ozone Contactors," in Proceedings of the National Symposium, Progress in Wastewater Disinfection Technology, Cincinnati, Ohio, September 1978, EPA-600/9-79-018 (June 1979).

Watson, H. E., "A Note on the Variation of the Rate of Disinfection with Change in the Concentration of the Disinfectant," *J. Hygiene,* **8,** 536 (1908).

Wei, J. H., and Chang, S. L., "A Multi-Poisson Distribution Model for Treating Disinfection Data," in J. D. Johnson (ed.), *Disinfection Water and Wastewater,* Ann Arbor Science, Ann Arbor, MI (1975).

White, G. C., *Handbook of Chlorination,* Litton Educational Publishing, New York (1972).

White, G. D., *Disinfection of Wastewater and Water for Reuse,* Van Nostrand Reinhold Company, New York (1978).

Witkin, E. M., "Ultraviolet Repair in *Escherichia coli,*" *Bacteri. Rev.,* **40,** 869 (December 1976).

Wolf, H. W., Petrasek, Jr., A. C., and Esmond, S. E., "Utility of UV Disinfection of Secondary Effluent," in Proceedings of the National Symposium, Progress in Wastewater Disinfection Technology, Cincinnati, Ohio, September 1978, EPA-600/9-79-018 (June 1979).

—13—

Residuals Management

In most water treatment processes, the objective is to remove certain materials from the water, to purify it. Water treatment residuals are those materials removed during the treatment process, along with any transport water that is removed with them. These residuals include the turbidity-causing materials in any transport raw water, organic and inorganic solids, algae, bacteria, viruses, colloids, precipitated chemicals from the raw water and those added in treatment, and dissolved salts. They are referred to as "residuals" rather than "wastes" to encourage the engineer to consider them as by-products of the treatment process, and potentially useful, rather than simply as waste materials to be discarded or disposed of as conveniently as possible.

Residuals management is a term used to describe the planning, designing, and operating of facilities to reuse or dispose of water treatment residuals. From a technical standpoint, the objective in residuals management is usually to minimize the amount of material that must ultimately be disposed of by recovering recycleable materials and reducing the water content of the residuals. In most cases, the cost of transporting and ultimately disposing of the residuals makes up the major fraction of residuals management costs, and the most economical solution is to reduce the quantity of material for ultimate

disposal. Other considerations include minimizing environmental impacts and meeting discharge requirements established by regulatory agencies.

The processes that have shown the most successful and significant capabilities for dewatering sludges from water treatment plants are:

Drying beds
Vacuum filtration
Pressure filter press
Belt filter press
Centrifugation
Alum and lime recovery

A generalized process diagram showing the various unit processes that may be used in residuals management and the sequence in which they may be assembled to form complete systems is shown in Figure 13-1. A complete residuals management system would be made up of one unit process from one or more of the process steps shown (thickening, conditioning, etc.) and must include one of the unit processes from the final disposal step. Some typical residuals management processes are:

For alum sludge disposal: gravity thickening, chemical conditioning, centrifugation, and final disposal to sanitary landfill.

For alum sludge disposal: sludge lagoons, backwash recovery, and final disposal to sanitary landfill or sanitary sewerage system.

For lime sludge disposal: gravity thickening, filter press dewatering, heat drying, lime calcining, and reuse.

For lime sludge disposal: sludge lagooning, drying beds, cropland disposal, or landfill.

For brines: final disposal directly to the ocean.

The selection of process steps and unit processes for a specific installation depends on factors such as space available at the site, local weather conditions, type of residual, cost of reuse versus disposal, sophistication of the plant operations personnel, distance to an ultimate disposal site, and types of ultimate disposal available. In most cases, it will be possible to rule out most of the options by inspection (e.g., a small site surrounded by residential development would not have space available for sludge lagoons or drying beds, an inland site could not consider ocean disposal), leaving a limited number of options that can be evaluated in more depth to determine the least-cost alternative. This chapter discusses the types of residues produced during water treatment, the various methods practiced for residuals management, and the considerations in process selection.

DEFINING THE PROBLEM

The problem of residuals management can be quantified in two ways: with respect to the quantities and costs of handling residuals and with respect to the environmental and regulatory constraints any engineered solution must meet.

Scale of Problem

Solid and liquid residuals may constitute as much as 3–5 percent of the volume of the raw water entering a typical water treatment plant. The bulk of that will be the filter backwash water, which typically contains less than 10 percent of the removed solids in a conventional treatment plant. Sedimentation basins sludge should be on the order of 0.1 to 0.3 percent of the plant flow but will contain most of the solids removed. In a direct or in-line filtration plant, however, all solids removal is accomplished by the filters. Table 13-1 summarizes the quantities of residuals produced by various treatment processes.

The costs of handling these residuals are dependent on the type of handling provided and the nature of the materials. In most cases, treatment of the residuals is minimal, and the major portion of the cost is in transport and ultimate disposal.

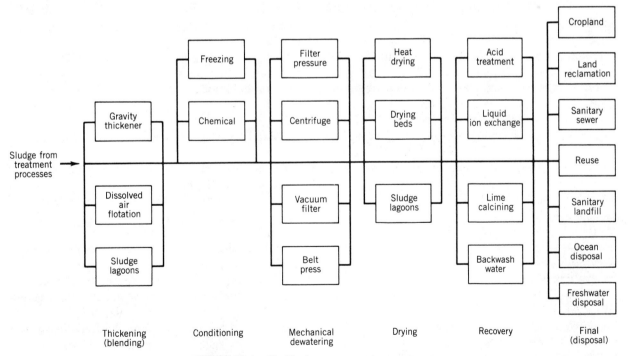

FIGURE 13-1. Residuals management processes.

TABLE 13-1. Typical Residue Production in Water Treatment Facilities

Type of Residue	Quantity Produced	Reference
Alum sludge	8–210 kg/1000 m³, average 48 kg/ 1000 m³	AWWA, 1971
	67–1800 lb/mil gal[a], average 400 lb/ mil gal	AWWA, 1971
	Over 2×10^6 tons/yr in U.S.	
Iron sludge	80 kg/1000 m³	AWWA, 1971
	650 lb/mil gal[a]	
	Over 300,000 tons/yr in U.S.	
Filter backwash water	1–5% of treated water	Linsley and
	Average 2% of treated water	Franzini, 1979
Softening sludge	0.3–6% of treated water	AWWA, 1969
	2 mg/L per mg/L hardness removed	
	Over 600,000 tons/yr in U.S.	
Softening brines	1.5–10% of treated water	AWWA, 1971
	Average 3.5% of treated water	AWWA, 1969
Microstrainer washwater	1–5% of treated water	AWWA, 1969
	Average 2.2% of treated water	AWWA, 1969

[a] Dry solids basis.

Environmental Constraints

Regulatory constraints on residuals disposal have become increasingly severe in recent years. Prior to the late 1960s there was little concern for disposal of water treatment residuals. In most cases, they were simply returned to the nearest receiving water, usually the source of water supply. In the late 1960s some states began considering these residuals as pollutants and began establishing treatment or discharge standards for them. The 1972 Federal Water Pollution Control Act classified water treatment plant wastes as pollutants and placed them in the industrial waste category. As such they are now required to meet standards for Best Practicable Control Technology Currently Available (BPT), and to meet Best Available Technology Economically Achievable (BAT) by 1983. Specific standards for either of those categories have not yet been established; however, the next logical step is zero discharge of pollutants. There has also been legislation, both federal and state, to control toxic and hazardous substances. Such regulations add to the cost of disposal.

The impacts of water treatment residuals on the environment can be characterized as aesthetic or impacts on biota. Aesthetic impacts include discoloration or increased turbidity in receiving waters, buildup of sludge deposits and occupying large land areas with lagoons. Impacts on the biota are, for sludges and backwash waters, primarily the impact on fish of increased water turbidity, pH, and hardness. Redissolved iron and aluminum may also pose a problem. In the cases of brine, there may be toxic effects of high salt concentrations, especially in localized areas around the discharge. Most sludges, if spread on land to any depth, will prevent or inhibit plant growth; however, if adequately mixed into the soil, there may be little or no impact. Lime sludges may have beneficial impacts on the soil if used in appropriate amounts.

QUANTITIES AND PROPERTIES OF RESIDUES

The quantities of residues produced by treatment processes, and the properties of those residues, are primarily a function of the type of residue and the quality of water being treated. An understanding of the quantities and properties of residues is fundamental to determining appropriate management techniques and to designing facilities to implement those techniques. In this section, residues have been grouped into four general categories: coagulation sludges, lime sludges, diatomaceous earth sludges, and brines. The sources, quantities, and properties of each of these classes of residues are discussed below.

Coagulation Sludges

Coagulation sludges are produced by the coagulation and settling of natural turbidity by added coagulant chemicals. In a treatment plant, they are collected in the sedimentation basins and on the filters. The proportions of the sludge collected in the basins and the filters depend on the water quality, type and dose of coagulant used, efficiency of operation, plant design, and other factors. For typical plants using alum as the coagulant, between 60 and 90 percent of the total sludge will be collected in the basins with the remainder in the filters. The sludge collected on the filters is removed during backwashing and, if the backwash water is recovered, is removed from the backwash water by settling and eventually added to the basin sludge for further handling or disposal.

Sludge from the sedimentation basins may be removed continuously or, more commonly, on an intermittent basis. Removal may be by various mechanical devices (see Chapter 21) or by draining and manually washing down the basin. If basins are manually cleaned, the frequency may be once every 3 months or more. Mechanical cleaning equipment is usually designed to operate between once a week and once every few hours or continuously.

Filters are typically backwashed every 24–72 hr, with a relatively large volume of backwash water produced in a short time. However, some proprietary filter types (such as automatic backwash filters) may backwash as frequently as every 2–6 hr. Backwash water recovery facilities must be designed to accept the high, intermittant flows. Coagulation sludges can be grouped according to the type of primary coagulant employed: alum, iron salts (e.g., ferric chloride), or polymers are the three major alternatives. Of these three, alum is most commonly used, iron salts and polymers are used more rarely as primary coagulants. Polymers are commonly employed as coagulant aids with alum or iron to reduce the amount of primary coagulant required and improve its efficiency.

Quantities. The quantities of sludges produced were summarized in Table 13-1. For design purposes, the amount of sludge anticipated at a plant can be estimated on the basis of coagulant, dose, and raw water quality. The suspended solids fraction of the sludge may be safely assumed to be equal to the suspended solids of the raw water, or, if suspended solids data is not available, it may be estimated from turbidity data. For turbidities less than 100 turbidity units (TU) the suspended solids in mg/L has been shown to be approximately equal to the turbidity in turbidity units (Nielsen et al., 1973).

For alum or iron sludges, the precipitated coagulant is largely aluminum or iron hydroxide, respectively. Both of these will include some bound water that will not be released in a standard total solids test. That bound water must be considered in estimating solids loadings that will later be measured by standard total solids testing of the sludge produced. In the case of alum sludge, an empirical formula for the aluminum hydroxide plus bound water fraction of the sludge (Nielsen et al., 1973) has been determined to be $Al(OH)_3 \cdot 1.25\ H_2O$. Using that formula, it can be calculated that a total of 0.33 lb of sludge on a dry-solids basis will be produced for each pound of alum $[Al_2(SO_4)_3 \cdot 14\ H_2O]$ added. If a similar proportion of bound water is assumed for iron hydroxide sludges, the value for use of ferric sulfate $Fe_2(SO_4)_3$ would be about 0.59 lb of sludge per lb of coagulant added and 0.48 for ferric chloride.

The total sludge produced can be estimated by adding the suspended solids removed to the coagulant added, then multiplying by an appropriate factor. For polymer sludges or sludges with polymer used as coagulant aid, the amount of polymer added should also be included in the calculation. If other coagulant aids such as bentonite or actived silica are used, they should also be considered in the calculation, as well as any other chemicals that may be collected in the basins or filters, such as activated carbon. For example, for a 500 L/sec (11.4 mgd) plant with an average raw water turbidity of 25 TU and an average alum dose of 30 mg/L, the sludge produced would be calculated as follows:

Total sludge
  = sludge from turbidity + sludge from alum

kg/d dry solids

$$= \frac{86{,}400\ \text{sec/d}[(0.5\ \text{m}^3/\text{sec} \times 25\ \text{g/m}^3) + (0.5\ \text{m}^3/\text{sec} \times 30\ \text{g/m}^3 \times 0.33)]}{1000\ \text{g/kg}}$$

= 1508 kg/d

lb/d dry solids
  = (25 ppm × 8.34 lb/gal × 11.4 mgd)
    + (30 ppm × 8.34 lb/gal × 11.4 mgd × 0.33)
  = 3318 lb/d

Assuming the specific gravity of 5 percent solids sludge is 1.01, volume of 5 percent solids sludge can be estimated as:

$$1508 \text{ kg/d} \div (0.05 \times 1.01 \times 1000 \text{ kg/m}^3)$$
$$= 30 \text{ m}^3/\text{d}$$

or

$$3318 \text{ lb/d} \div (0.05 \times 1.01 \times 62.4 \text{ lb/ft}^3)$$
$$= 1053 \text{ ft}^3/\text{d}$$

Physical–Chemical Properties. The physical–chemical properties of various residuals are summarized in Tables 13-2 and 13-3. The solids concentrations and physical properties are the most important for sizing and design of residuals management facilities.

Solids concentrations depend on the design and operation of the sedimentation basins in addition to the type of sludge and its composition. For example, alum sludge from an upflow clarifier would typically be drawn off at a concentration of 0.1–0.3 percent solids, that from a horizontal flow basin at 0.2–1.0 percent or more. If it is allowed to accumulate for a month or longer in a horizontal flow clarifier, sludge may thicken to 4–6 percent solids. Sludges that have relatively high proportions of alum or iron coagulant, as would result from treating low-turbidity water, will have lower solids concentrations than will those with relatively higher proportions of turbidity or silt. Coagulation of waters having substantial algae concentrations will also result in light, low solids concentration sludges. The addition of polymers generally tends to produce higher solids concentrations.

The other chemical characteristics of sludges are directly related to the chemical content of the raw water and the coagulant chemicals. The BOD, COD, and related organic content are representative of the dissolved and suspended organic materials and algae removed from the water. The inorganic solids are derived from the coagulant chemicals and the clay and sediments removed from the raw water. The pH and dissolved solids of the sludge are about the same as those in the water being treated. Complete chemical analyses of an alum sludge were reported by Schmitt and Hall (1975). The major elements found were (in order of decreasing predominance) silicon, aluminum, iron, titanium, calcium, potassium, magnesium, and manganese. A total of 72 elements were detected.

Coagulation sludges will also contain bacteria removed in the sedimentation process. Bacterial counts in the range of 2300/100 mL for 1 percent solids sludge have been reported (Nielsen et al., 1973), but this number should be highly variable depending on the quality of the raw water and treatment process employed (prechlorination, etc.). Presumably, viruses are also present in coagulation sludges, although no analyses have been reported.

Sludge density is dependent on the moisture content, varying from the density of water for sludges below about 1 percent to 1200–1520 kg/m^3 (75 to 95 lb/ft^3) when dry. A reasonable estimate of the density of wet inorganic sludges, typical of alum or iron salts, can be made by assuming the density of the dry solids is about 2300 kg/m^3 (145 lb/ft^3) and calculating the density of the solids plus water using the following:

$$\text{Density} = \cfrac{100}{\cfrac{\text{percent solids}}{\text{density solids}} + \cfrac{100 - \text{percent solids}}{\text{density water}}}$$

This equation only applies to sludges with a solids content of approximately 50 percent or less; at solids contents of greater than 50 percent, cracking and increased voids tend to lower the overall average density in sludge lagoons.

Sludge viscosity, settling velocity, and other physical properties are dependent on solids concentrations and the relative proportions of coagulant and other materials in the sludge. Specific resistance increases at solids concentrations below about 2 percent, but is relatively constant above that concentration as shown in Figure 13-2 (Hawkins et al., 1974). Shear strength and viscosity increase as solids concentrations increase, as shown in Figures 13-3 and 13-4.

As coagulation sludges are dewatered and dried, there is a gradual transformation from a liquid to a solid. For the purposes of designing residuals management facilities, it is important to know when that transformation occurs as it will determine the type of equipment required to handle the sludge. As a liquid, the sludge can be pumped, piped, and transported in tank trucks while as a solid it must be shoveled and transported on a conveyor or in open trucks. Unfortunately, the transition is not sharply defined nor is the transition point the same for all sludges. Coagulation sludges that have high proportions of gelatinous aluminum or iron hydroxides will act as liquids at higher solids concentrations than

TABLE 13-2. Typical Chemical Properties of Residuals

Type of Residue	Chemical Properties	Reference
Alum sludge	BOD—30–300 mg/L	AWWA, 1971
	COD—30–5000 mg/L	AWWA, 1971
	pH—6 to 8	AWWA, 1971
	Total solids—0.1–4%	King et al., 1975
	Solids:	
	15–40% $Al_2O_3 \cdot 5.5\ H_2O$	Nielsen et al., 1973
	35–70% silicates and inert materials	
	15–25% organics	King et al., 1975
Iron sludge	pH—7.4–8.5	Pigeon et al., 1978
	Total solids—0.25–3.5%	King et al., 1975
	Solids:	
	4.6–20.6% Fe	Pigeon et al., 1978
	5.1–14.1% volatiles	
Filter backwash water	BOD—2–10 mg/L	AWWA, 1971
	COD—28–160 mg/L	
	pH—7.2–7.8	
	Total solids—0.01 to 0.1%	
	Solids:	
	25–50% Al_2O_3	
	34–35% SiO_2	
	15–22% carbon and organics	
Softening sludges (lime sludges)	BOD and COD—low to zero	AWWA, 1971
	Total solids—2–15%	
	Solids from low-magnesium water:	
	75% $CaCO_3$ = 42% calcium as CaO	Watt and Angelbeck, 1977
	6% silica as SiO_2	
	7% total carbon	
	3% aluminum as Al_2O_3	
	2% magnesium as MgO	
Softening brines	Total dissolved solids 15,000–35,000 mg/L	AWWA, 1969
	Chloride 9000–22,000	
	Sodium 2000–5000	
	Calcium 3000–6000	
	Magnesium 1000–2000	
	Sulfate 328	
Diatomaceous earth filter washwater	BOD—105 mg/L	AWWA, 1969
	COD—340 mg/L	
	pH 7.6	
	Total solids 7500 mg/L	
	Volatile solids 275 mg/L	
	Total suspended solids 7600 mg/L	
	Volatile suspended solids 260 mg/L	

TABLE 13-3. Physical Properties of Residuals

Type of Residue	Physical Properties	Reference
Alum sludge	Specific resistance	
	$1.65–5.4 \times 10^{11}$ sec^2/g	King et al., 1975
	$1.0–4.4 \times 10^9$ sec^2/g	AWWA, 1969
	Viscosity 0.03 g/cm-sec	
	(non-Newtonian)	AWWA, 1971
	Initial settling velocity 0.06–0.15 cm/sec	King et al., 1975
	Dry density 1200–1520 kg/m^3	
	(75–95 lb/ft^3)	AWWA, 1969
Iron sludge	Specific resistance	
	1.95×10^{12} sec^2/g	King et al., 1975
	$4.1–14.8 \times 10^8$ sec^2/g	Calkins and Novak, 1973
	Initial settling velocity 0.007 cm/sec	King et al., 1975
Softening sludges	Specific resistance	
	$5.45–25.1 \times 10^7$ sec^2/g	Calkins and Novak, 1973
	$0.20–25.6 \times 10^7$ sec^2/g	O'Connor and Novak, 1978
	Plastic viscosity 0.06 g/cm-sec	Weber, 1972
	Settling velocity 0.001–0.1 cm/sec	Weber, 1972
	Dry solids specific gravity 2.75	Weber, 1972
	Wet density of 25% solids 1920 kg/m^3 (120 lb/ft^3)	AWWA, 1969
Diatomaceous earth (DE)	Dry density of DE 160 kg/m^3 (10 lb/ft^3)	AWWA, 1969
	Specific gravity 2.0	

will those that contain more clay and sediments. These sludges are also thixotropic, that is, on standing they will seem to solidify, but when disturbed they revert to a liquid. The minimum concentration at which sludge can be considered a solid is about 16 percent solids; however, for design purposes a value in the range of 40–50 percent is recommended.

Structural Properties. The variation in physical properties among sludges of various compositions is due to the physical structure of the sludge. A coagulation sludge is made up of the suspended material in the raw water, metal hydroxides, and a large amount of bound and entrapped water in a loose structure. The suspended materials are clay and sediment particles, color-causing colloids, algae, and other similar materials. Clays and sediments are structurally solid and have a specific gravity of around 2.6; the other materials are agglomerations of individual metal hydroxide molecules with various ions and water molecules, all loosely held together by electrostatic bonds. Metal hydroxides become attached to the suspended materials by electrostatic bonds, and also physically entrap suspended materials as well as water molecules. In the coagulation–flocculation process, the suspended particles and metal hydroxides are brought together to form the flocs that then settle and make sludges. When the individual flocs come in contact with each other in the sludge, they become loosely bound by the same electrostatic forces that hold the individual flocs together. The extent of the bonding depends on the extent of the contact between the flocs, which is limited by the entrapped water that separates the flocs. As more and more water is removed by draining, pressure, or other means, the particles contact more and the sludge becomes increasingly solid. Because the metal hydroxides have water molecules in their structure, direct particle contact is more difficult than for other suspended materials that do not have water in

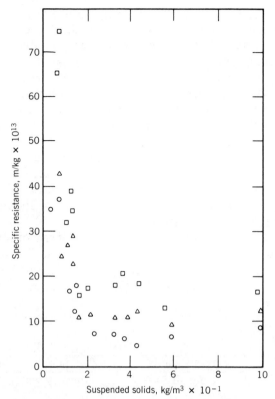

FIGURE 13-2. Relationship between suspended solids and specific resistance of alum sludges (○), vacuum, 5 in. Hg; (△), vacuum, 10 in. Hg; (□), vacuum, 15 in. Hg (Hawkins et al., 1974).

their structure. Therefore, sludges with high proportions of metal hydroxides are not as easily dewatered as are sludges that have high proportions of other materials, unless the sludge is conditioned with large amounts of lime or an appropriate polymer dose.

If polymers are added to coagulation sludges, either as sludge conditioners or as a part of the coagulation process, the sludge produced will have a more solid structure. This is due to the polymer molecules forming bridges between floc particles and improving the bond between particles.

Lime Sludges

Lime sludges are produced from the precipitation of calcium carbonate and magnesium hydroxide in the lime–soda softening process. They may be essentially pure chemical sludges, or they may include suspended materials from the raw water if turbidity removal is combined with softening. Similar sludges are produced in the recently developed magnesium carbonate coagulation process (EPA, 1971).

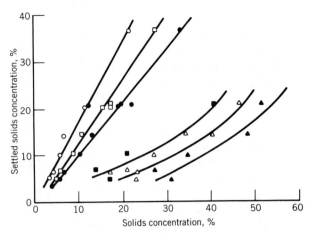

FIGURE 13-3. Variation in sludge physical properties with solids concentration. Viscosity: (○), 1 P; (□), 3 P; (●), 8 P. Shear: (■), 0.01 tons/ft²; (△), 0.04 tons/ft²; (▲), 0.1 tons/ft² (Novak and Calkins, 1975). Reprinted with permission.

Quantities. Typical quantities of lime sludge produced from water softening were summarized in Table 13-1. For design purposes, the amount of sludge anticipated at a plant can be estimated based on the chemical treatment and raw water quality. The sludge is essentially composed of precipitated calcium carbonate and magnesium hydroxide, any turbidity or suspended solids present in the raw water that are not removed prior to softening, and any insoluble impurities present in the treatment chemicals, such as lime grit. The suspended contribution can be estimated from turbidity data, as discussed above for coagulation sludges. The amounts of pre-

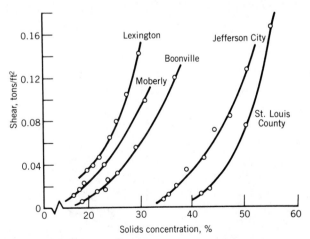

FIGURE 13-4. Effect of concentration on the shear of five different sludges as they are dried (Novak and Calkins, 1975). Reprinted with permission.

cipitated calcium carbonate and magnesium hydroxide can be directly calculated from the anticipated calcium and magnesium removals. It is not necessary to include any water of hydration in the formula of the precipitates as in the case of coagulation sludges, so removal of 1.0 mg of calcium (expressed as Ca) results in 2.5 mg of $CaCO_3$ in the sludge. Similarly, removal of 1.0 mg of magnesium (expressed as Mg) results in 2.4 mg of $Mg(OH)_2$ in the sludge. Figure 13-5 provides a graphical means of estimating sludge production.

Physical–Chemical Properties. The physical–chemical properties of lime sludges are most important. In larger plants, it is often economical to recover lime from the sludge, in which case the chemical content of the sludge also becomes important.

Solids concentrations are dependent on the treatment process and on the proportions of the various chemical precipitates in the sludge. Sludges that are high in $CaCO_3$ typically have higher solids concentrations than sludges with more $Mg(OH)_2$, because $CaCO_3$ is a fine grained, dense precipitate while $Mg(OH)_2$ is a more gelatinous material. Figure 15-6 demonstrates this relationship. The typical treatment process utilizing either upflow or horizontal flow sedimentation basins following chemical addi-

tion and reaction produce fine-grained precipitates similar in nature to mud or silt deposits. An alternative process, the spiractor process, produces a granular precipitate that is more like sand.

Lime sludges typically have a high pH (10.5–11.5) and are white unless colored by turbidity or iron and manganese. Generally they are odorless, with little or no BOD or COD. Because of the high pH, lime sludges do not contain significant numbers of viable bacteria. Typical chemical characteristics are summarized in Table 13-2. The specific chemical content of sludge from any given plant can be determined from the raw water quality and the chemical treatment used.

As with coagulation sludges, the density of lime sludges is dependent on the water content. The dry $CaCO_3$ solids have a specific gravity of about 2.71. The density of wet sludge can, therefore, be calculated as

Density
$$= \frac{100}{\dfrac{\text{percent solids}}{2.71 \times \text{density water}} + \dfrac{100 - \text{percent solids}}{\text{density water}}}$$

or, using 62.4 lb/ft³ for density of water:

$$\text{Density} = \frac{16{,}910}{271 - 1.71(\text{percent solids})}$$

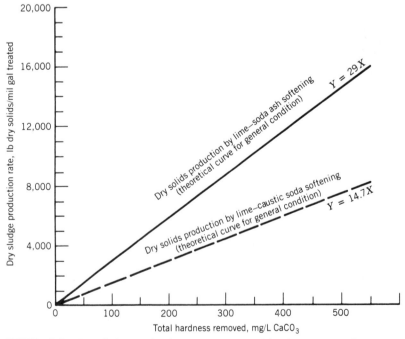

FIGURE 13-5. Dry sludge production rate versus total hardness removed.

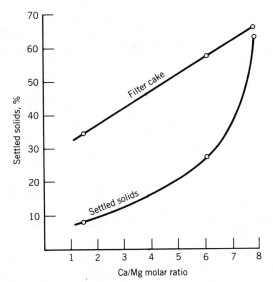

FIGURE 13-6. Effect of Ca/Mg ratio on settled solids (lime sludges) (Calkins and Novak, 1973).

Diatomaceous Earth Sludges

Sludges are produced during backwash of diatomaceous earth filters. Generally, the diatomaceous earth filter process is operated such that the filter cake contains about two parts of diatomaceous earth for each part of impurities removed from the water (AWWA, 1969). As a result, the sludge characteristics are predominantly those of the diatomaceous earth. Some reported characteristics are presented in Tables 13-2 and 13-3. The sludge is equivalent to the diatomaceous earth added in the process, plus the suspended materials removed in the filter. There are no chemical reactions involved so quantity estimates are straightforward.

Brines

The primary source of waste brines from water treatment processes is ion exchange softener regeneration. Other sources include reverse osmosis brines, ion exchange demineralizers, and blowdown from cooling towers and boilers. All of these processes produce wastes that are high in dissolved solids but low in suspended solids. The following discussion will emphasize softening brines as these are the major area of concern in municipal water supply.

Quantities. Typical quantities of brine produced in ion exchange softening were shown in Table 13-1.

Specific quantities and characteristics of the brine from a particular softening plant will depend on the resin selected, the raw water characteristics, and the operation of the regeneration process. Quantities of brine required for regeneration are determined during design of an ion exchange softening facility, as is discussed in Chapter 10. The brine to be disposed of is equal in volume to that used for regeneration, but instead of being entirely composed of water and sodium chloride, it would be a mixture of the excess sodium required to drive the regeneration process and the calcium, magnesium and other ions removed by the ion exchange. In addition, there will be rinse water used to flush the brine out of the resin between the regeneration cycle and operation, and there may be wastewater from an initial backwash cycle to remove any suspended materials collected by the bed. Depending on the design of the facilities, the various waste streams may be separated or combined. Ideally, the brine and the freshwater streams would be segregated, as would be less contaminated portions of the regenerant brine, so as to allow some brine recovery.

Physical–Chemical Properties. The chemical characteristics of typical ion exchange softener brines were summarized in Table 13-2. Physically, these waste brines are clear, with the specific gravity being dependent on the salt concentration but typically in the range of 1.02–1.11. If backwash water is included, any suspended material present in the raw water would also be included in the waste brine. The detailed chemical content of a waste brine can be determined from a mass balance on the process and, as discussed above, depends on the raw water quality and the detailed design and operation of the system.

RESIDUAL MANAGEMENT PROCESSES

After an understanding of the characteristics of residues, development of a complete management process train requires an understanding of the various process steps. In this section, all of the major unit processes that are employed in residuals management will be reviewed. Design of these processes is covered in Chapter 21. Some of the processes shown in Figure 13-1 are omitted because they are either seldom used or there was insufficient data available on their application.

Gravity Thickening

Gravity thickening is commonly the first step in the residuals management process. It can be described generally as the process of removal of excess water by decanting, and concentration of the solids by settling. The decanted water is usually recovered unless the water contains objectionable taste, odor, or large numbers of algae or other microorganisms while the solids are processed further or disposed. The most common method of gravity thickening of sludge is by lagooning with a decantation operation.

Process Principles. With most water treatment residuals, the general treatment objective is to reduce the volume as much as possible in order to reduce the costs of subsequent recovery or disposal steps. Because most of the residuals from the treatment process are fairly dilute, the simplest and cheapest first step in reducing volume is gravity thickening. The theory of settling and thickening has been covered in Chapter 7 and will not be repeated here. For coagulation or softening sludges, the primary process involved is compaction of the sludge, while for filter backwash water the processes of settling and hindered settling are most important. A detailed study was conducted (Kos, 1977) in which it was determined that thickening could be modeled based on two parameters of the sludge: the modulus of linear compressibility and the intrinsic conductivity. The reader is referred to that article for further details.

Thickener Design. The theoretical basis for thickener design is presented in Chapter 7, and general considerations on facilities design are presented in Chapter 21. For design of a mechanical thickener it would generally be desirable to conduct bench scale or pilot testing to obtain data on subsidence rates, underflow concentrations, and supernatant quality. In designing such a study, consideration should be given to annual or seasonal variation in sludge characteristics, testing of polymers or other conditioning methods, and reasonably accurate simulation of such process variables as depth of the sludge layer, mixing, and wall effects.

In the absence of pilot testing, design criteria for thickeners being used on similar sludges may be adopted. Table 13-4 lists design criteria for several typical mechanical thickeners.

Thickening can also be achieved in lagoons, which are commonly used where land is readily available at the treatment plant site. In many cases it is mistakenly assumed that lagoons are an ultimate disposal method requiring little attention. As a result, their performance is often found unsatisfactory. However, if designed and operated as a thickening process, lagoons are quite satisfactory and will provide economical sludge thickening. In the absence of freezing or polyelectrolyte conditioning, alum or iron coagulation sludges can be expected to thicken to 4–6 percent solids after a month or more in a lagoon with continuous decantation. Concentrations up to 10–15 percent can be achieved at the bottom of a lagoon after half a year or more of lagooning; however, the upper layers will have lower concentrations. For design purposes, 5 percent solids after two or three months of thickening can be assumed. Lime sludges thicken much more readily in a lagoon, reaching 30 percent solids or more in a few months.

TABLE 13-4. Typical Mechanical Thickeners

| Plant Type | Thickener | | Conditioning | Solids Concentration | | Ultimate Disposal | Reference |
	Capacity	Size ($D \times W \times L$)		Inlet (%)	Outlet (%)		
Alum coagulation	500 gpm 31.5 L/sec	$20 \times 40 \times 40$ ft $6 \times 12 \times 12$ m	Yes	0.2–1.0	5	Dewatering and land filled	Nielsen et al., 1973
Lime–magnesium carbonate	50 ton/d 45.3×10^3 kg/d	75 ft diameter 23 m diameter	No	—	—	Dewatering–recovery	Thompson and Mooney, 1978
Lime softening	7.5 ton/d 6.8×10^3 kg/d	25 ft diameter 7.6 m diameter	No	1–4	28–32	Dewatering–road base	AWWA, 1969

Sludge Conditioning

Sludge conditioning may be employed to aid in gravity thickening but is more frequently used in mechanical dewatering processes. Conditioning most commonly involves the addition of polymers to aid in settling and dewatering. Other processes such as lime addition (for alum sludge), heat treatment, freezing, and thawing have also been used.

Successful dewatering often depends on proper conditioning of the sludge in advance. The objectives of conditioning are to improve the physical properties of the sludge so that water will be released easily from the sludge, improve the structural properties of the sludge to allow free draining of the released water, improve the solids recovery of the process (i.e., to reduce the fraction of solids lost in the removed water), and minimize dewatering process cycle times.

Polymers are the most commonly used conditioners for dewatering water treatment sludges. A number of studies (Bugg et al., 1970; Novak and O'Brien, 1975) on polymer conditioning of sludges have found that most types of polymers will improve the dewatering characteristics of sludges. The selection of a polymer for a given application should be based on bench tests, or, preferably, pilot or full-scale tests. Also, in general, higher-molecular-weight polymers are more effective except that at very high molecular weights the viscosity causes handling problems.

Successful use of polymers is dependent on good dispersion of the polymer into the sludge to be conditioned. As with any chemical addition, provision must be made for good initial mixing.

Sizing of polymer feeding equipment should be based on bench-scale determination of dosage requirements. If that is not practical (such as in design of a new plant), facilities can be sized based on estimated sludge solids concentrations and quantities. Polymer doses required are typically in the range of $10-100 \times 10^{-4}$ g polymer per gram of sludge solids for metal hydroxide sludges.

Freezing is very effective for metal hydroxide sludges such as alum and iron sludges. The effect is to completely destroy the gelatinous structure, leaving the sludge (after thawing) in the form of a fairly coarse granular material like sand or coffee grounds. The process is irreversible. Unfortunately, the mechanical efficiencies of equipment for freezing and thawing sludge are low, so this process is usually applied only where natural freezing will

occur in a lagoon. With natural freezing of alum sludge, a 2 percent solids sludge can be converted to a 20 percent solids granular slurry that will readily drain to over 30 percent solids and can be easily handled (AWWA, 1969).

Heat treatment has been investigated (AWWA, 1969) as a sludge conditioning process, but the results are not as dramatic as with freezing. It is not being employed on a full scale. With rapidly rising energy costs, heat treatment would be considered as an undesirable alternative for sludge conditioning.

Another conditioning step often applied in pressure filtration of alum sludges is the addition of lime or inert granular materials like fly ash or diatomaceous earth. Relatively high proportions of these materials are required. Conditioning requirements of specific dewatering processes are summarized in Table 13-4.

Gravity Dewatering

Gravity dewatering involves filtration, collecting the sludge solids on sand, and drainage of water from the sludge through the filter media. The process will produce a relatively dry, solid sludge for further treatment or disposal. It may be combined with drying to produce a sludge of any desired dryness. Gravity dewatering is applicable to dewatering of sludge discharged directly from sedimentation basins, or following thickening. It has also been used for dewatering of diatomaceous earth from backwashing filter plants.

Bed Area. The size of the filter bed required is usually the factor that determines the feasibility of gravity dewatering at any given plant site. If land is readily available, gravity dewatering is the method of choice, otherwise a more sophisticated mechanical system will be required. Multiple filter beds must be sized to allow spreading of a relatively thin layer of sludge (6 in. to 1 ft), and allowing sufficient time between spreading cycles to permit drainage, drying, and removal of the sludge. At least three and preferably four or more beds should be provided to allow discharge of sludge to one while the others are draining, drying, and being cleaned.

A common approach used at many water treatment plants in the United States is to use lagoons not only as thickeners (with continuous decanting) but also as drying beds after a predetermined filling period. Three months of filling and an average drying cycle of 3 months are most commonly used to design (see Chapter 21). The required area of lagoons can be determined using a sludge loading rate of 40–80 kg dry solids per square meter of lagoon area (8.2–16.4 lb/ft^2); 40 should be used for wet regions and 80 for relatively dry areas.

For example, the effective area of lagoons required to handle alum sludge from a conventional 1 m^3/sec average flow rate (22.8 mgd) water treatment plant, assuming the average turbidity of the raw water is 15 TU and an alum dosage of 12 mg/L, can be approximated as follows:

Dry sludge production rate =

$$\frac{86,400 \text{ sec/d}[(1 \text{ m}^3/\text{sec} \times 15 \text{ g/m}^3) + (1 \text{ m}^3/\text{sec} \times 12 \text{ g/m}^3 \times 0.33)]}{1000 \text{ g/kg}} = 1640 \text{ kg/d}$$

If ft-lb units are to be used:

$$[(15 \times 8.34) + (12 \times 8.34 \times 0.33)] \times 22.8 = 3605 \text{ lb/d}$$

Assuming a total of four lagoons and an average of 100 days filling cycle (detention) per lagoon in a dry region, an effective lagoon area for each lagoon is

$$\text{Effective lagoon areas} = \frac{1640 \times 100}{80}$$
$$= 2050 \text{ m}^2 \quad \text{or} \quad 0.5 \text{ acres}$$

Actual area required for a lagoon would be at least 1.5 times the area computed because of the additional area required for berms and access roads.

Underdrains. An underdrain system must be provided to collect the water drained from the sludge if sludge beds are constructed in wet regions. Underdrains are not required in most dry regions. This water can then be either recycled, if the quality is good, to the plant inlet or discharged. Underdrains typically consist of gravel and perforated clay or PVC pipes.

Cycle Time, Weather, and Conditioning. As indicated above under bed area, cycle time includes time for filling the bed, sludge draining and drying, and cleaning the bed. The major portion of the cycle time is the drainage and drying time. Ideally, these should be determined by bench or pilot testing, with due consideration given to variations in climate and

sludge characteristics. In the absence of actual testing, the engineer must estimate the extent to which the sludge will drain. The use of polymers to condition alum sludge can reduce draining time, but will probably not substantially increase the drained solids concentration.

After draining is completed, further dewatering occurs by evaporation. The time required to reach the desired dryness can be calculated from evaporation/precipitation data. Once the sludge is drained, rainfall will drain through or be decanted from the surface of the bed, rather than rewetting the sludge. For conservative design, the net evaporation rate (evaporation minus precipitation) should be used as a reference for sizing lagoons.

Vacuum Filtration

Vacuum filters have been used to dewater sludge from wastewater treatment plants for many years, but their application to water treatment sludge is a recent development and has seen limited use. The unit employed is almost exclusively a rotary drum vacuum filter. There are two basic types: (1) traveling media and (2) precoat media filters. The precoat filter is used mainly for dewatering coagulated sludges such as alum sludge. However, successful operation of vacuum filters requires the use of polymer or lime as sludge conditioning chemicals for alum and ferric hydroxide sludges. Lime sludges, however, generally do not require conditioning prior to vacuum filtration.

The variables to be considered in designing a vacuum filter system are the size and type of filter, the cake discharge mechanism, the filter media, vacuum level, cycle time, and sludge conditioning. Typical results for alum and lime sludge dewatering with vacuum filtration are shown in Tables 13-5 and 13-6.

Pressure Filter Sludge Presses

Filter presses have been used for dewatering alum sludges in Great Britain, Japan, and, to some degree, in the United States. Conditioning of the sludge prior to filtration is required, and the degree of conditioning dictates the performance. In general, a filter cake of about 30–40 percent solids concentration is expected after pressure filtration with lime and polymers as sludge conditioning chemi-

TABLE 13-5. Typical Precoat Rotary Vacuum
Filter Performance Data on Alum Sludge
Dewatering

Parameter	Data Range
Feed concentration (%)	2–6
Feed rate	
(L/m²/hr)	0.7–2.1
(gal/ft²/hr)	2–6
Dry solids yield	
(kg/m²/hr)	0.2–0.3
(lb/ft²/hr)	1.0–1.5
Cake concentration (%)	30–35
Filtrate suspended solids (mg/L)	10–20
Solids recovery (%)	99+
Precoat recovery (%)	30–35
Precoat rate	
(kg/m²/hr)	0.02–0.04
(lb/ft²/hr)	0.1–0.2
Precoat thickness	
(mm)	38.1–63.5
(in.)	1.5–2.5
Drum speed	
(rpm)	0.2–0.3
Operating vacuum	
(mm Hg)	127–508
(in. Hg)	5–20

SOURCE: Westerhoff and Daly (1974). Reprinted with permission.

cals. The filtrate may contain less than 10 mg/L of suspended solids if the sludge is properly conditioned. Both capital and operating and maintenance costs for this process are relatively high.

TABLE 13-6. Average Performance and
Operating Data by Traveling Belt
Vacuum Filter on Lime Softening Sludge

Parameter	Data Range
Free solids (%)	5–30
Cake concentration (%)	40–70
Cake yield	
(kg/m²/hr)	0.8–4.0
(lb/ft²/hr)	4–20
Filtrate solids (mg/L)	950–1500
Solids recovery (%)	95–99
Filter speed (rpm)	0.2–0.5
Operating vacuum	
(mm Hg)	381–635
(in. Hg)	15–25

SOURCE: Ingersoll Rand (1970–1977).

Belt Filter Sludge Presses

The application of belt filters for water sludge dewatering is rather recent in the United States. Some recent data indicate that a filter cake of 30–40 percent solids is feasible with proper conditioning. The capital cost and space requirements for this process are generally significantly lower than pressure filter sludge systems.

Centrifugation

Dewatering of alum sludge by centrifugation requires conditioning of the sludge with polymers and lime. Twenty to 25 percent solids may be obtained from 3–4 percent solids alum sludge. Both capital and operating and maintenance costs for this process are relatively high. There are two basic types of centrifuges: solid-bowl and basket centrifuges.

Recovery of Coagulant

Aluminum and iron recovery can be accomplished by adding acid (normally sulfuric acid) to sludges to solubilize the metal ion salts. Lime recovery from lime sludge has also been practiced by the recalcination process.

Alum recovery from alum sludges was very popular in Japan in the 1960s when about 15 plants were in operation. Normally, over 80 percent of alum recovery is achieved at a pH of about 2.5 by acidification. However, an accumulation of heavy metals, manganese, and other organic compounds was experienced in recovered alum. This fact, as well as rising costs, has halted alum recovery operations in recent years. In order to solve the problem of degradation of the quality of recovered alum, the liquid ion exchange technique has been evaluated (Cornwell et al., 1980).

Lime recovery by recalcination has been practiced in a number of locations in the United States, and the primary consideration in selecting this process is the overall economy because this process is very energy intensive.

Magnesium Bicarbonate Recovery. The magnesium bicarbonate recovery process was developed as a part of an overall treatment process that would use magnesium bicarbonate (precipitating magnesium hydroxide and calcium carbonate) as the primary coagulant. In a water having the appropriate characteristics, the process would produce more magnesium carbonate than is required for treatment, thereby producing a salable by-product.

During the late 1960s and the early 1970s, recycling magnesium carbonate was a developing and promising technology. Numerous pilot studies were conducted in such places as Dayton, Ohio, Montgomery, Alabama, and Kansas City, Missouri, to prove the usefulness of the recovery process. Although these pilot studies did indicate that the magnesium bicarbonate recovery process is technically feasible, the economics could not justify scale-up to a full-size facility. Presently, there are no known existing full-scale facilities that employ the magnesium bicarbonate recovery process. Recovery of magnesium bicarbonate is being practiced in the paper and pulp industry but not in the water treatment industry.

ULTIMATE DISPOSAL

Several alternatives are available for the disposal of water treatment plant residuals. In practice, the options available for ultimate disposal of water treatment plant residuals frequently dictate the type of in-plant handling system necessary. Selection of an alternative should be based on economic as well as regulatory considerations. The type of sludge and sludge characteristics are also important criteria to be used in developing disposal alternatives. It is critical that the ultimate solids disposal program be a reliable, environmentally sound practice to ensure that it does not affect the primary goal of the treatment plant, production of potable water.

Disposal Alternatives

Alternatives available for disposal of residuals include disposal to surface waters, disposal on land, disposal to the sanitary sewer, and deep-well injection. Surface water alternatives include both freshwater disposal and saltwater disposal. Land spreading, lagooning, and landfilling are typical land disposal programs. Residuals disposed of in a sanitary sewer system end up in the wastewater treatment plant where they are removed and disposed of with sewage sludge. Deep-well injection of ion exchange residuals to saline aquifers simply returns the brine to an aquifer with similar characteristics. Each of these alternatives is discussed in detail in the following sections.

Disposal to Surface Waters. Disposal of water treatment plant residuals to surface waters falls under the jurisdiction of the Federal Water Pollution Control Act Amendments of 1972 (PL 92-500) and the Clean Water Act of 1977. Basically, these laws consider water treatment and supply to be an industry, and, therefore, consider the discharge of residuals to surface waters to be an industrial discharge. To dispose of residuals to a surface water, a National Pollutant Discharge Elimination System (NPDES) permit would be required. The requirements of the NPDES permit will vary depending on the type of watercourse to which the residual is being discharged, and the type of residual being discharged. The advantage of a surface water disposal system is the relatively low capital and operational and maintenance costs. Facilities needed for a surface water discharge system are usually simple to design, construct, operate, and maintain, keeping costs lower than for other disposal systems.

Disadvantages include the uncertainty of continued allowance of this practice in the future and the potential for creating a water pollution problem. It may be possible, however, to discharge brines to the ocean with minimal environmental impact. Under the NPDES system, extensive monitoring of the residual and the discharge water body would be required. It is possible that some type of treatment of the residual would also be required prior to discharge.

Considerations for the design of a surface water disposal system include location of the outfall, pumping requirements, flow equalization, and outfall design. Outfall location is an extremely important concern. The outfall should be located such that it discharges to a point of maximum dispersion. Similarly, the outfall should be designed to disperse the wastes across the well-mixed zone of the water body. The location and design of the outfall significantly impacts the pumping requirements. Consideration should be given to equalizing the residual flow to minimize pump and motor sizing, and limit discharge of large residual slugs to the receiving water.

Land Disposal. Disposal of water treatment plant residuals on land is governed by the Solid Waste Disposal Act (PL 91-512) and the Resource Conservation and Recovery Act of 1976 (PL 94-580). Under these regulations, disposal of liquid or semiliquid residuals is limited to specific types of landfills designated for this type of waste. Dewatered residuals would, however, be acceptable to most sanitary landfills. Land spreading is also covered under these laws.

Landfilling can be a relatively high capital cost system. The dewatered residuals are usually transported to a municipal landfill. The cost of transportation may or may not be significant, depending on the haul distance. It may be possible to contract out sludge hauling to a private business, decreasing the capital investment required for a truck. Other capital costs, however, include sludge dewatering and storage facilities which may offset any savings seen in handling equipment. Land spreading usually requires the purchase of land dedicated solely to the disposal of water treatment residuals. Capital costs also include spreading trucks and sludge storage facilities. Operation and maintenance costs are low because no residuals treatment facilities are required.

Design considerations for a landfilling system include landfill availability, dewatering requirements, and transportation requirements. Municipal landfills should be available and fairly close to the treatment plant. Moisture content requirements of the landfill and acceptability of residuals during the rainy season and winter months will dictate the type of dewatering and the amount of on-site storage required. The land location will determine transportation requirements and may indicate a need for odor or noise control if land is close to residential areas. Soil and geological characteristics are needed to determine water percolation rates, proximity to groundwater, and useful life of the site.

Sanitary Sewer Disposal. The same laws that govern surface water discharge apply to sewer disposal. In addition, local pretreatment guidelines would cover the discharge from a water treatment plant to the sewer. The amounts and type of liquids and solids that may be added to the sewer may be limited by the capacity of the sewer or the wastewater treatment plant and the types of processes and operations at the wastewater facility.

Direct discharge to the sewer system has a low capital cost and may also have a low operation and maintenance cost depending on monitoring requirements and sewer use fees. A condition of discharge may be continuous monitoring of the organic strength and solid content of the residual flow. An attempt should be made to assess the impact of the residuals on the wastewater treatment facility prior to the selection of this alternative. It is possible that disposal of alum sludge may enhance phosphate removal at the wastewater treatment plant if any of the alum activity remains. It also may enhance primary sedimentation.

Design of a sewer disposal system must provide for controlled discharges to eliminate the possibility of large slugs of residuals upsetting the wastewater treatment facility. Discharge should be coordinated with the wastewater treatment plant operators so that they may optimize the performance of their process units.

Deep-Well Injection of Brines. Discharge of ion exchange brines to a saline aquifer may be controlled by local environmental regulations, and is certainly dependent on the geology and groundwater hydrology of the area. This system is generally limited to clear brine solutions, as suspended solids tend to clog the injection well screen. Such a system tends to be fairly expensive due to well drilling cost. Depending on the local groundwater hydrology, there may be a significant potential for groundwater contamination. In addition, the high pressure at the bottom of the injection well and saline solution tend to enhance the corrosion potential of the well screen and casing. Selection of materials resistant to corrosion under those conditions may prolong the operating life of an injection facility.

PROCESS SELECTION

Selection of a particular dewatering, disposal, or recovery option should be based on both economic and operational factors. Economic factors that must be considered are capital, operation and maintenance, and solids disposal costs. Capital costs should include such items as construction costs, trucks, and special equipment needed for the process. Operation and maintenance costs should include power, chemicals, labor, parts replacement, and equipment repair costs. Disposal costs are typically such items as fees at the landfill or sewer discharge fees.

Operational factors that must be evaluated are the plant location, size, and reliability. An extremely complicated system would probably not be either economical or effective for a small, simple plant. Similarly, the plant reliability history must be evaluated to determine if a complex system would be well operated and maintained at the particular facility.

REFERENCES

American Water Works Association, *Water Quality and Treatment*, McGraw-Hill, New York (1971).

AWWA Research Foundation Report, "Disposal of Wastes from Water Treatment Plants—Part 3," *JAWWA*, **61**(12) 681 (December 1969).

Bugg, H. M., King, P. H., and Randall, C. W., "Polyelectrolyte Conditioning of Alum Sludges," *JAWWA*, **62,** 792, (December 1970).

Calkins, R. J., and Novak, J. T., "Characterization of Chemical Sludges," *JAWWA*, 423, (June 1973).

Cornwell, D. A., and Lemunyon, R. M., "Feasibility Studies on Liquid Ion Exchange for Alum Recovery from Water Treatment Plant Sludges," *JAWWA*, **72**(1), 64 (January 1980).

Environmental Protection Agency, "Magnesium Carbonate, A Recycled Coagulant for Water Treatment," EP 1.16:12120 ASW (June 1971).

Hawkins, F. C., Judkins, Jr., J. F., and Morgan, J. M., "Water Treatment Sludge Filtration Studies," *JAWWA* (November 1974) 66:653–658

Ingersoll Rand, "Operating and Pilot Data," Nasau, NH (1970–77).

King, P. H., Chen, B. H., and Weeks, R. K., "Recovery & Reuse of Coagulants from Treatment of Water and Wastewater," *Virginia Water Resources Research Center Bulletin,* **77** (1975).

Kos, P., "Gravity Thickening of Water-Treatment Plant Sludges," *JAWWA,* 272 (May 1977).

Linsley, R., and Franzini, J., *Water Resources Engineering,* McGraw-Hill, New York (1979).

Nielsen, H. L., Carns, K. E., and DeBoice, J. N., "Alum Sludge Thickening & Disposal," *JAWWA*, 65:385–394, 385 (June 1973).

Novak, J. T., and Calkins, D. C., "Sludge Dewatering and its Physical Properties," *JAWWA*, 67:42–45 42 (January 1975).

Novak, J. T., and O'Brien, J. M., "The Influence of Sludge Characteristics on Polymer Conditioning of Chemical Sludges," *JWPCF*, **47**(10), 2397 (October 1975).

O'Connor, J. T., and Novak, J. T., "Management of Water Treatment Plant Residues," AWWA Seminar Proceedings, AWWA Conference (June 25, 1978).

Pigeon, P. E., Linstedt, K. D., and Bennett, E. R., "Recovery and Reuse of Iron Coagulants in Water Treatment," *JAWWA*, 70:397 397, (July 1978).

Schmitt, C. R., and Hall, J. E., "Analytical Characterization of Water-Treatment-Plant Sludge," *JAWWA*, 40–42 (January 1975).

Thompson, C. G., and Mooney, G. A., "Case History—Recovery of Lime and Magnesium Compounds from Water Plant Sludge," AWWA Seminar Proceedings, AWWA Conference (June 25, 1978).

Watt, R. D., and Angelbeck, D. I., "Incorporation of a Water Softening Sludge into Pozzolanic Paving Material," *JAWWA*, 69:175 175, (March 1977).

Weber, W., "Physicochemical Processes for Water Quality Control," Wiley-Interscience, New York (1972).

Westerhoff, G. P., and Daly, M. P., "Water Treatment Plant Wastes Disposal—Part II," *JAWWA*, **66**(6), 378 (June 1974).

— 14 —

Water Reuse

WASTEWATER RECLAMATION AND REUSE

Unplanned, indirect wastewater reuse, through effluent discharge to streams and groundwater basins for subsequent downstream use, has been a long-accepted practice throughout the world. Many communities located at the end of major waterways, such as New Orleans and London, ingest water that has already been used as many as five times via repeated river withdrawal and discharge. Similarly, rivers or percolation basins may recharge underlying groundwater aquifers with reclaimed wastewater, which in turn is withdrawn by subsequent communities. For example, the effluent from over 140 wastewater treatment plants partially replenishes the groundwater tapped by the water supply system for London (Eden et al., 1977). This means of effluent disposal, known as unplanned, indirect reuse, has become a generally accepted practice for a variety of agricultural, industrial, and domestic interests.

Planned, direct reuse is practiced on a smaller scale for a limited number of purposes, primarily industrial and agricultural. The key distinction between the terms *direct* and *indirect* relates to whether the discharge has achieved "a loss of identity" via some intermediate step. When accountability for the discharged effluent is lost, then subse-

quent downstream use is termed *indirect*. The terms *planned* and *unplanned* refer to whether the subsequent reuse was designed as a conscious act following the effluent discharge or whether subsequent reuse was an unintentional by-product of effluent discharge. A planned reuse scheme incorporates a wastewater reclamation system designed to meet not only effluent discharge standards but also reuse standards promulgated by health authorities. This chapter will deal primarily with planned direct and indirect reuse.

A further distinction is made between the concept of planned reuse and recycling. Recycling, which already also occurs on a widespread basis, is defined as the internal reuse of water by the original user prior to discharge to a treatment system or other point of disposal. In the United States, for example, approximately 526 billion liters per day (GL/d) or 139 billion gallons per day (bgd) of freshwater were recycled in the stream-generated power, manufacturing, and mineral industries in 1975 (WRC, 1978). This quantity of recycled water is projected to increase to 3274 GL/d (865 bgd) by the year 2000. In contrast, the term *reuse* is applied to discharged wastewaters that are subsequently withdrawn by users other than the discharger. In the United States, planned reuse, which has only begun to recognize its potential, was estimated at 2.6

TABLE 14-1. Historical (1975) Planned Wastewater Reuse in the United States

| | Historical Wastewater Reuse | |
	(GL/d)	(bgd)
Irrigation		
Agricultural	0.76	0.20
Landscape	0.11	0.30
Not specified	0.72	0.19
Subtotal	1.59	0.42
Industrial		
Process	0.26	0.07
Cooling	0.53	0.14
Boiler feed	0.04	0.01
Subtotal	0.83	0.22
Groundwater recharge	0.11	0.03
Other (recreation, fish, and		
wildlife, etc.)	0.04	0.01
Total	2.57	0.68

Source: WRC (January 1978).

GL/d (0.7 bgd) in 1975 (0.4 percent of available wastewater) and is projected to increase to only 18.2 GL/d (4.8 bgd) by the year 2000 (4.0 percent of available wastewater). Of the planned wastewater reuse, approximately one-third can be attributed to industrial needs either in cooling, process, or boiler feed applications and two-thirds to agricultural uptake (Table 14-1).

This chapter presents the water qualities and quantities required to meet planned reuse purposes, that is, those subject to the rules, regulations, and policies of regulatory agencies. Also presented will be those planning considerations that need to be addressed in order to achieve a viable reuse project. Several existing wastewater reclamation and reuse projects are presented as examples of current activities, and the potential for additional uses over the next two decades will be summarized. Water quality criteria are established for the major beneficial uses of reclaimed wastewater, along with the appropriate treatment technologies required to remove contaminants prior to subsequent reuse.

WASTEWATER REUSE POTENTIAL

Two factors will play major roles in the development of wastewater reuse during the next 20 years:

the inadequacy of existing freshwater resources in selected areas of the world and the overall upgrading of wastewater treatment plant effluents through at least the secondary level of treatment. Because of identified shortfalls between available freshwater resources and total needs, both water purveyors and low-quality users are turning to reclaimed wastewater as a reliable water source, thereby freeing available freshwater resources to higher quality users. In the United States, the passage of the Clean Water Act of 1977 (PL 95-217) has placed the burden of producing secondary effluent onto the wastewater discharger, thereby providing the potential reuser with a valuable resource for subsequent reclamation activities.

Reuse Applications

Potentially large quantities of reclaimed wastewater can be used in four sectors of activities. On the basis of quantity of use, irrigation will undoubtedly rank highest. Overall irrigation use, consisting of both agriculture and landscape applications, is projected to account for 54 percent of total United States freshwater withdrawals by the year 2000. Because of the general acceptability of disinfected secondary effluent for irrigation use and the advantage of using existing wastewater nutrients as fertilizer, the potential for direct planned reuse in irrigation is large.

The second major potential user of reclaimed wastewater is industry, primarily for cooling and process needs. In general, industrial reuse offers two significant advantages over agricultural reuse: proximity to the source of domestic wastewaters and continuous needs on a 24-hr per day, 365-day per year basis. Agricultural reuse, on the other hand, is distributed both temporally (season of the year) and spacially (outlying regions). A further advantage is that many industrial water quality requirements can be met by secondary effluent, or secondary effluent upgraded with some physical–chemical treatment step.

The third major potential reuse application is groundwater recharge, either via percolation following landspreading, or direct injection. Groundwater recharge takes advantage of the subsoil's natural ability for biodegradation and filtration, thereby providing the necessary polishing of secondary effluent. Furthermore, groundwater recharge provides the psychologically important ''loss of identity'' that makes the reclaimed wastewater acceptable for a wider variety of uses, both potable and subpotable. Planned recharge of groundwater basins with reclaimed wastewater has the greatest potential in semiarid areas, such as Israel and the southwestern United States, which are characterized by inadequate surface supplies and groundwater overdrafts.

A fourth sector of activity is characterized as miscellaneous subpotable uses, such as for recreational lakes, aquaculture, and industrial utility needs. These subpotable uses are projected to be minor wastewater reuse applications, accounting for less than 5 percent of total wastewater reuse. However, in selected instances, they may provide an ideal means of effluent disposal or a supply adequate for low-quality needs.

In general, the needs for subpotable quality water as outlined above already greatly exceed the supply of available wastewater. Some communities, however, are now implementing plans for direct planned potable reuse of sewage effluent where no other possibilities exist for supplemental freshwater supplies. Direct planned reuse for potable purposes could also be called recycling, since wastewater basically is used again by the entity producing it in a closed system. While the quantities involved in these projects are small, these projects will be discussed because of their technological and public health interest.

Wastewater discharges available for reuse are divided into treated and untreated categories. Treated wastewater includes (1) secondary effluent from municipal wastewater treatment plants, (2) industrial discharges treated by best available technology, (3) discharges from steam-generated power plants, and (4) discharges from fish hatcheries. Untreated wastewaters are irrigation return flows that could be collected and reused. Most irrigation return flows reach streams and underground basins through underground or overland flow and are unavailable for planned reuse. Emphasis is primarily on treated wastewater flows which are more readily available for reuse.

Reuse Potential

In order to place the various reuse alternatives in perspective, this section briefly outlines existing wastewater reuse and estimates the potential for future wastewater reuse in the United States. While wastewater reuse plays an important role in water supply in several semiarid regions of the world, Is-

rael and South Africa for example, the United States experience was selected because of the diversity of its climate, water resources, and water needs. The estimates of potential reuse are based on data from the U.S. Water Resources Council (WRC) which prepares periodic assessments of the water situation in the United States.

A great deal of publicity attends those projects that utilize reclaimed wastewater; however, perspective is needed to evaluate the potential for reclaimed wastewater use and the magnitude of this reuse in comparison with overall wastewater discharge and wastewater recycling. Existing wastewater reuse in the United States, including irrigation, industrial, groundwater recharge, and miscellaneous subpotable uses, is approximately 2.65 GL/d (0.7 bgd). This quantity pales in comparison with total freshwater withdrawals of 1373 GL/d (362.7 bgd) and wastewater discharges of 925.5 GL/d (244.5 bgd) (OWRT, 1979). As shown in Table 14-2, total U.S. freshwater withdrawals and wastewater discharges are expected to decrease over the next 20 years, especially in the areas of steam-generated power and manufacturing. Thus, the requirements for freshwater are decreasing in two areas that have the greatest potential for reclaimed wastewater usage.

In the same vein, the Safe Drinking Water (SDWA:PL93-523) and the Clean Water Act (CWA:PL95-217) impact the quality of wastewater discharges and the extent of wastewater recycling. The SDWA indirectly affects the quality of wastewater because many wastewaters are discharged into streams that are also used for public water supplies. The SDWA requires upstream industries and municipalities to pretreat their discharge so as to protect the drinking water quality of downstream users. However, once treated, the former wastewater may have new value for internal recycling. The CWA similarly improves the general quality of wastewater through more stringent control of industrial waste discharge. Clearly, major changes are anticipated in the withdrawal of freshwater by major users and the extent to which that water is then discharged back to the environment.

On a positive note, of the projected 1253 GL/d (330.9 bgd) of total freshwater withdrawal (2000), approximately 1098 GL/d (290 bgd) could be replaced with reclaimed wastewater, at least on a quality basis. However, as shown in Table 14-3, only 458 GL/d (121 bgd) of wastewater discharge will be readily available for subsequent treatment

TABLE 14-2.   Historical (1975) and Projected (2000) U.S. Freshwater Withdrawals and Wastewater Discharges

	Freshwater Withdrawal					Wastewater Discharge				
	Quantity				Percent Change 1975– 2000	Quantity				Percent Change 1975– 2000
	1975		2000			1975		2000		
Area	GL/d	bgd	GL/d	bgd		GL/d	bgd	GL/d	bgd	
Agriculture	699	184.7	683	180.3	−2	321	84.7	273	72.0	−15
Steam-generated power	337	89.1	303	80.1	−10	332	87.7	263	69.5	−21
Manufacturing	194	51.2	75	19.7	−62	171	45.2	19	5.0	−89
Municipal	109	28.8	140	37.1	+29	81	21.4	104	27.6	+29
Minerals	27	7.1	43	11.3	+59	18.5	4.9	29	7.7	+57
Public lands	4.5	1.2	6.4	1.7	+42	0.01	0.003	0.02	0.006	+100
Fish hatcheries	2.3	0.6	2.6	0.7	+17	2.3	0.6	2.6	0.7	+17
Total	1372.8	362.2	1253.0	330.9	−9	925.8	244.5	690.6	182.5	−25

SOURCE: WRC (January 1978).

and reuse. If the agricultural irrigation freshwater withdrawal and the untreated agricultural discharges are ignored (both are unlikely to participate in major reclamation projects), the projected freshwater withdrawals capable of replacement are approximately equal to the treated wastewater discharges available for reuse. Thus, if wastewater discharges from the steam-generated power, industrial, and municipal sectors [amounting to approximately 412 GL/d (109 bgd)] were treated to acceptable quality, they could be paired directly with projected freshwater withdrawals for steam-generated power and industrial users (amounting to 420 GL/d or 111 bgd). For a variety of reasons, this potential for growth in wastewater reuse will not be achieved. Little more than 18 GL/d (4.8 bgd) of wastewater reuse is projected for the year 2000, or less than 3 percent of the treated wastewater discharge available for reuse. Projected wastewater recycling will be approximately 1575 GL/d (865 bgd), or 180 times greater than wastewater reuse.

The total freshwater withdrawal from ground and surface water sources is expected to decrease from about 1373 GL/d (362 bgd) in 1975 to 1253 GL/d (331 bgd) by the year 2000. This is achieved principally by increased recycling of water in the manufacturing and steam-generated power industries in order to meet water quality requirements prior to discharge. As shown in Figure 14-1, by the year 2000 the steam-generated power, manufacturing, and mineral sectors will account for only 34 percent

of the total U.S. freshwater withdrawal as opposed to the current 41 percent. However, because of the close proximity of these users to wastewater reclamation plants, and their generally constant water quality and quantity requirements, these industries remain prime candidates for municipal wastewater reuse.

Agriculture will remain the single largest user of freshwater withdrawals, accounting for roughly 54 percent of total withdrawals in the year 2000, but unfortunately large-scale agricultural water demands are not located near large wastewater discharges. The potential for wastewater reuse in the other sectors will be small, partly because these other sectors will account for less than 10 percent of total U.S. freshwater withdrawals in the year 2000. The following sections present water quantity needs by the major use categories outlined above.

Irrigation.   Water for irrigation purposes can be used for rural agriculture and livestock watering, and for urban landscape watering. Irrigation represents the single largest use of water and, therefore, represents opportunities for significant water conservation and reuse. Irrigation consumption includes both transpiration by crops and incidental losses. Most of the demand for irrigation water occurs in the Pacific Northwest, California, and the north central states, which together account for more than 63 percent of all water withdrawn for agricultural irrigation. Although the extent of U.S.

TABLE 14-3. Projected (2000) Wastewater Discharges Are Much Less Than Projected Subpotable Needs

		Quantity	
		GL/d	bgd
I.	Freshwater withdrawals capable of replacement with wastewater		
	Agricultural irrigation	673	177.8
	Landscape irrigation	5.3	1.4
	Steam-generated power	303	80.1
	Industrial cooling	64	16.9
	Industrial, other	53	14.1
	Subtotal	1098.3	290.3
II.	Treated wastewater available for reuse		
	Steam-generated power	263	69.5
	Industry	48	12.7
	Municipal	101	26.7
	Fish hatcheries	2.6	0.7
	Subtotal	414.6	109.6
	Untreated agricultural discharges for reuse	43.5	11.5
III.	Projected wastewater reuse (2000)	18.2	4.8
	Projected wastewater recycle (2000)	1574.6	865.5

SOURCE: WRC (January 1978).

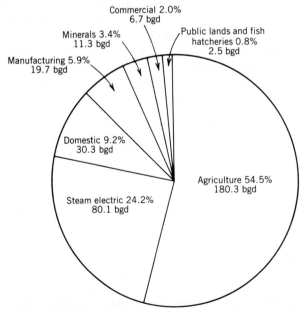

FIGURE 14-1. Total projected vs freshwater withdrawals. Potential industrial wastewater reuse can occur in the steam electric, manufacturing, and minerals sectors. These sectors will account for III.I bgd (33.6 percent) of total national freshwater withdrawals in the year 2000 (OWRT, 1979).

irrigated acreage is estimated to increase over the next 20 years, total annual irrigation withdrawals are projected to decrease from 693 GL/d (183 bgd) to 674 GL/d (178 bgd) as a result of increased irrigation efficiency.

At the present time, reclaimed water is used occasionally for agricultural and landscape irrigation in the southwestern United States. In general, the reclaimed water irrigates fruit trees and row crops, thereby reducing the need for freshwater and some fertilizers. In urban areas, small quantities of reclaimed wastewater are applied for landscape irrigation (city parks, golf courses, median strips, and industrial plantings). The WRC has estimated the total current irrigation reuse of reclaimed wastewater at about 1.6 GL/d (0.42 bgd), with by far the largest reuse in the southwestern United States.

Industrial. Three industrial groupings were defined for estimation of wastewater reuse potential: steam-generated power, manufacturing industries, and mineral industries. Cooling water for steam generating plants represents the second largest requirement for freshwater in the United States (337 GL/d (89 bgd) in 1975 and an estimated 303 GL/d (80 bgd) in 2000 (WRC, January, 1978). Most of this demand is in the eastern United States with the Great Lakes and Ohio regions accounting for over 50 percent of the total national requirement (OWRT, 1979). Both increased recycling and wastewater reuse are expected to supply a significant part of the makeup water. Since saltwater is used for steam electric cooling at coastal plants, obviously low-quality water is acceptable for once-through cooling purposes. With adequate pretreatment, reclaimed wastewater offers the potential for several cycles of concentration through steam electric cooling condensers. As less once-through cooling is utilized, more cooling towers will be constructed and the actual consumptive use of water by steam electric plants will increase over seven times between 1975 and 2000 (WRC, January 1978).

The manufacturing industries consist of primary metals, chemicals and allied products, paper and allied products, petroleum and coal products, food and kindred products, transportation equipment, textile mill products, and other manufacturing. The first three categories accounted for about 77 percent of the total freshwater requirements for manufacturing industries in 1975. Water use in manufacturing industries is primarily for three purposes: cooling, boiler feed, and processing. About 60 percent of all industrial use is for cooling, and this represents the greatest potential for wastewater reuse and recycling. Boiler feed water and most process waters require high quality; the potential for reuse of wastewater in these areas is limited.

In 1975, total water use in U.S. manufacturing industries was about 515 GL/d (136 bgd). However, more than one-half of this demand was met by recycling. Consequently, as shown in Table 14-4, only 194 GL/d (51.2 bgd) was actually freshwater withdrawal to meet manufacturing needs. Because of further recycling within plants, freshwater withdrawals are projected to decrease to approximately 76 GL/d (20 bgd) by the year 2000. As shown in Figure 14-2, three major product manufacturers—paper, chemical, and primary metals—will account for roughly 70 percent of the freshwater withdrawals in the year 2000. Since these industries utilize a large percentage of their water for cooling (Table 14-4), or in the case of paper and allied products, for process water which could be replaced with reclaimed wastewater, reuse potential is large. The other industries in the manufacturing group have much smaller freshwater withdrawals, or have requirements for high-quality water, such as the food and kindred products industries. These have limited potential for reclaimed wastewater usage. Although water quality requirements have been established for process uses and for boiler makeup waters, in general the cost of reclaiming wastewater to meet these quality constraints is too high to justify planning reclaimed wastewater usage.

The mineral industries include metal mining, fuels, and nonmetal minerals. As shown in Table 14-5,

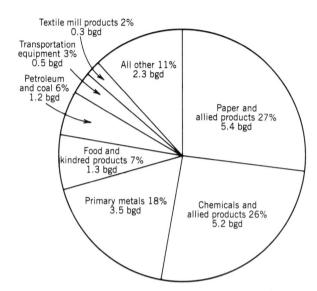

FIGURE 14-2. Projected freshwater withdrawals in the manufacturing industries will fall from 51.2 bgd (1975) to 19.7 bgd (2000). Three major product groups, paper, chemicals, and primary metals, will make 70 percent of the withdrawals (OWRT, 1979).

growth of freshwater withdrawals by the mineral industries through the year 2000 will bring them to within 60 percent of the water withdrawals by the manufacturing industries. Within the mineral industries, most water is utilized for processes such as coal washing, sand and gravel washing, subsurface injection, and iron and copper ore extraction. The water quality requirements for these process applications are minimal, and thus the potential for reclaimed wastewater usage in these sectors is large.

TABLE 14-4. Freshwater Withdrawals and Point of Use in U.S. Manufacturing Industries

| | Freshwater Withdrawals | | | | Percent Usage at Point of Application | | |
| | 1975 | | 2000 | | | | |
Manufacturing Industry	GL/d	bgd	GL/d	bgd	Cooling	Process	Other
Primary metals	66.6	17.6	13.2	3.5	70	25	5
Chemicals and allied products	51.1	13.5	19.7	5.2	80	15	5
Paper and allied products	31.8	8.4	20.4	5.4	35	60	5
Food and kindred products	9.8	2.6	4.9	1.3	50	40	10
Petroleum and coal	9.5	2.5	4.5	1.2	65	5	30
Transportation equipment	4.9	1.3	1.9	0.5	35	50	15
Textile mill products	2.3	0.6	1.1	0.3	—	—	—
All other	17.8	4.7	8.7	2.3	—	—	—
Totals	193.8	51.2	74.4	19.7	60	25	15

SOURCE: WRC (April 1978). Reproduced with permission.

TABLE 14-5. Freshwater Withdrawals and Point of Use in U.S. Mineral Industries

| Mineral Industry | Freshwater Withdrawals | | | | Percent Usage at Point of Application | | |
| | 1975 | | 2000 | | | | |
	GL/d	bgd	GL/d	bgd	Cooling	Process	Other
Nonmetals	14.0	3.7	23.1	6.1	5	90	5
Fuels	9.8	2.6	14.4	3.8	45	45	10
Metals	3.0	0.8	5.3	1.4	—	95	5
Totals	26.8	7.1	42.8	11.3	20	75	5

SOURCE: OWRT (1979).

Unfortunately, most mineral deposits are not located near large sources of reclaimed wastewater, and the cost of bringing wastewater to the deposits may be prohibitive.

Groundwater Recharge. Water reuse through groundwater recharge is accomplished by percolation or injection through wells. The simplest and most widely used recharge method is percolation. Treated effluent is conveyed to basins where the water is allowed to percolate through the soil zone to the groundwater table. Percolation is primarily practiced in the southwestern United States with the majority of projects in California. Injection through wells is the other method of artificial recharge currently in use. The injected water can be used for two purposes: first to replenish local groundwater supplies and second to prevent seawater intrusion into an existing groundwater supply. A recent review of groundwater recharge with reclaimed wastewater reports six existing percolation projects and one injection project with two others under construction (Schmidt and Clements, 1977). The amount of reclaimed wastewater actually recharged by percolation and injection is small, amounting to less than 0.4 GL/d (0.1 bgd) in 1980 in the United States.

Miscellaneous Subpotable. Miscellaneous subpotable uses of reclaimed wastewater do occur in very small quantities or unusual situations. The two major subpotable uses are recreation and aquaculture. Recreational reuse, defined as any direct use of reclaimed water to provide recreational enjoyment, includes recreational lakes and streams, wetlands enhancement, swimming pools, and others. Recreational reuse is generally limited in size to small treatment plants whose effluent is used to aug-

ment local water sources. At the present time, recreational reuse accounts for less than 0.08 GL/d (0.02 bgd). High costs and adverse public attitudes rather than a lack of technology will probably limit future growth of recreational reuse programs. Recreational facilities in which body contact is anticipated will require high degrees of treatment, yet only limited monetary return can be provided.

The second subpotable use of domestic wastewater is for aquaculture, primarily the production of fish. While present aquaculture reuse systems include both water plants and fish, most plant systems are used to provide additional treatment rather than a useful crop. Some systems have been designed to maximize plant growth with the concept of using the plant either as a supplemental feed crop or as a potential new energy source. Currently, less than 0.04 GL/d (0.01 bgd) of effluent is used in aquaculture operations, and the possibilities for further growth are limited by water quality problems and the expense of building and operating a full-scale aquaculture reuse system.

Potable. The first documented case of direct potable reuse of wastewater occurred in Chanute, Kansas. A severe drought from 1952–1957 caused the city's normal water source, Neosho River, to cease flowing in the summer of 1956. During the ensuing 5 months, secondary chlorinated effluent was collected behind a dam on the river and used as intake water for the city's water treatment plant. After coagulation, settling, filtration, and chlorination at the water treatment plant, this water served as the city's sole potable water supply. Although this tap water always met USPHS drinking water standards, public acceptance was poor. The water had a pale yellow color, an unpleasant taste, and foamed when drawn from the tap. Aesthetically, it was very unpleasant. At the end of the drought, local and federal health authorities concluded that no illnesses could be directly attributed to the use of the reclaimed wastewater.

Currently, the only full-scale potable reuse facilities in the world are at Windhoek and Stander, South Africa. Both the Windhoek plant, commissioned in 1968, and the Stander plant, commissioned in 1970, are rated at 52.6 L/sec (1.2 mgd). Effluent from both these plants is blended with existing freshwater resources so that the reclaimed wastewater constitutes less than 15 percent of the total water supply. A somewhat parallel situation occurs along the Octoquan River, Virginia. The Octoquan watershed drains into the Octoquan Reservoir, a water supply serving approximately 600,000 people. This reservoir is also the recipient of reclaimed wastewater representing 12 percent of the total annual inflow and about 46 percent of the July low flow. In order to ensure a consistently high water quality, a 15-mgd wastewater reclamation plant was constructed to provide high-quality effluent on a fail-safe basis. Operation of the reclamation plant began in 1978, and the quality of water discharged to the Octoquan Reservoir has shown substantial improvement over the subsequent years. During the 1977 drought, as much as 80 percent of the flow to the reservoir was from treated wastewater discharges.

While the development of direct potable use of reclaimed wastewater is obviously limited to extreme situations where no other water supply is available, planning is currently underway for additional projects incorporating direct reuse. For example, Denver, Colorado, is completing the construction of a 43.8-L/sec (1-mgd) pilot plant to evaluate the reuse of secondary effluent for potable water demands. The pilot plant, to be operated and evaluated for the next 5 yr, will provide the technical data needed to construct and operate a 100-mgd reuse plant for direct potable use. Denver's plan is to comingle this water with the existing supplies to increase the total available potable water resources. Because of the extensive analytical and health effects program, Denver will not begin to rely on wastewater reclamation for potable use as an increment to the existing supply until the turn of the century.

The major element in considering the direct reuse of reclaimed wastewater is the shortage of available freshwater supplies. The WRC has concluded that U.S. freshwater supplies are sufficient to meet the requirements for all beneficial uses, especially in light of the anticipated reduction in water withdrawals over the next 20 yr. In spite of this overall optimistic assessment, major water supply problems exist in most water use regions and severe local problems exist in some subregions located mainly in the Southwest and Midwest. For example, in the lower Colorado region only 3.4 L (0.9 gal) of freshwater are available for each gallon required, in the Rio Grande region only 4.2 L (1.1 gal) are available, and in California only 9.5 L (2.5 gal). As a general guideline, 18.9 L (5 gal) or more should be available for withdrawal for each gallon actually needed, so as to avoid shortages during low flows or droughts. In each of these cases, the situation is expected to deteriorate even further over the remainder of this century. Unfortunately, the water shortages cannot be offset directly by wastewater available for reuse. The extent of shortages only provides an indication of the potential for reuse.

Many factors will affect the degree of potential reuse actually achieved: (1) the specific geographic relationship between dischargers and users will influence whether economic transfer of water can be made between them; (2) the timing of water needs and discharges will determine storage requirements and costs; and (3) the cost of treating and delivering available wastewater will be compared to available freshwater supplies. In spite of these obstacles, the use of reclaimed wastewater by agriculture and industry can help to alleviate existing water imbalances and will enable freshwater resources to be devoted to higher quality needs.

WATER QUALITY CONSIDERATIONS

Irrigation

On a quality basis, the use of reclaimed water for irrigation is governed by chemical parameters of the water that affect plant growth and survival, soil conditions, and the underlying groundwater. The chemical parameters of major importance are salinity, sodium adsorption ratio (SAR), and the concentrations of specific ions known to have toxic effects on plants. These parameters interact with each other and with local conditions such as soil permeability, plant types, and irrigation practices so that no absolute standard can be stated for their combined affect. A useful and widely accepted guide for evaluating the suitability of irrigation water and

TABLE 14-6. UC Cooperative Extension Guidelines for Irrigation Water Quality (UCCES)

Problems and Related Constituent	Water Quality Guidelines[a]		
	No Problems	Increasing Problems	Severe
Salinity[b]			
EC of irrigation water (mho/cm)	<750	750–3000	>3000
Permeability			
EC of irrigation water (mho/cm)	<500	500–2000	>2000
SAR[c]	<6.0	6.0–9.0	>9.0
Specific ion toxicity[d]			
From root adsorption			
Sodium (meq/L)	<3	3–9	>9
Chloride			
(meq/L)	<4	4–10	>10
(mg/L)	<142	142–355	>355
Boron (mg/L)	<0.5	0.5–2.0	>2.0–10.0
From foliar adsorption[e] (Sprinklers)			
Sodium			
(meq/L)	<3.0	>3.0	—
(mg/L)	<70	>70	—
Chloride			
(meq/L)	<3.0	>3.0	—
(mg/L)	<106	>106	—
Miscellaneous[f]			
NH$_4$ as N			
(mg/L) for sensitive crops	<5	5–30	>30
NO$_3$ (only with overhead sprinklers)			
(meq/L)	<1.5	1.5–8.5	>8.5
(mg/L)	<90	90–520	>520
pH (units)	Normal range	6.5–8.4	—

[a] Interpretations are based on possible effects of constituents on crops and/or soils. Guidelines are flexible and should be modified when warranted by local experience or special conditions of crop, soil, and method of irrigation.

[b] Assumes water for crop plus water for leaching requirement (LR) will be applied. Crops vary in tolerance to salinity. Refer to tables for crop tolerance and LR. Mho/cm × 0.64 = total dissolved solids (TDS) in mg/L or ppm.

[c] SAR (sodium adsorption ratio) is calculated from a modified equation developed by U.S. Salinity Laboratory to include added effects of precipitation and dissolution of calcium and is related to CO_3 + HCO_3 concentrations.

[d] Most tree crops and woody ornamentals are sensitive to sodium and chloride (use values shown). Most annual crops are not sensitive (use salinity tolerance tables). For boron sensitivity, refer to boron tolerance tables.

[e] Leaf areas wet by sprinklers (rotating heads) may show a leaf burn caused by sodium or chloride adsorption under low-humidity/high-evaporation conditions. (Evaporation increases ion concentration in water films on leaves between rotations of sprinkler heads).

[f] Excess N may affect production or quality of certain crops; e.g., sugar beets, citrus, avocados, apricots, etc. Water containing HCO_3 distributed by overhead sprinklers may cause a white carbonate deposit to form on fruit and leaves.

identifying potential areas of concern is presented in Table 14-6, developed by the University of California Cooperative Extension Service (UCCES). The following sections discuss problems associated with the four major properties associated with plant growth: salinity, sodium adsorption ratio, specific ion toxicity, and trace constituents.

Salinity. The salinity of irrigation waters is commonly reported as electrical conductivity or total dissolved solids. If the soil salinity exceeds the level of tolerance for a specific plant, osmotic regulation by the root membrane inhibits the adsorption of water. The plant will then dehydrate and wilt even though the soil shows an apparent excess of moisture. Table 14-6 presents the range of salinities usually encountered and the resulting effects on plant productivity.

Sodium Adsorption Ratio (SAR). SAR evaluates the sodium hazard in water by rationing the sodium ion concentration to the divalent magnesium and calcium ion concentrations as discussed in Chapter 2.

A diagram for classifying irrigation waters by interrelating the conductivity and the sodium adsorption ratio is shown in Figure 14-3. The lines dividing SAR classes are empirical and based on results from field and greenhouse studies. The conductivity classes from C_1 to C_4 represent water with increasing hazards from total salt concentration, while the sodium adsorption ratios, from S_1 to S_4, represent waters of increasing hazards from exchangeable sodium accumulation in irrigated soils. Using the foregoing figures and tables, one can readily assess the suitability of a reclaimed water supply for irrigation.

Specific Ion Toxicity. Boron toxicity in many areas can be traced to the use of irrigation waters with a boron concentration in excess of 1 mg/L. At the present time, no economically feasible method of removing boron from irrigation water is available. Similarly, no chemical or soil amendment can be added economically to the soil to render the boron nontoxic. However, marginal boron concentrations, like salinity concentrations, can be alleviated by irrigating slightly in excess of plant consumptive use. Also, boron toxicity is plant specific, so alterations in cropping practices may alleviate the toxicity problem.

Chloride ions are known to have toxic affects on plants, particularly on woody, perennial shrubs and

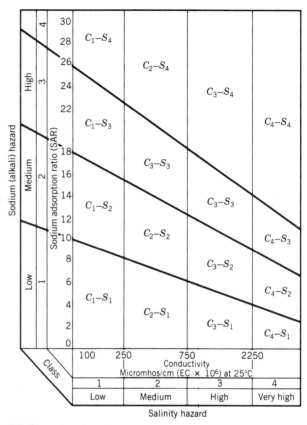

FIGURE 14-3. Classification of irrigation waters by conductivity and SAR (UCCES).

fruit trees. Foliar deterioration and leaf scorching occur for chloride concentrations between 140 and 350 mg/L if the ions are adsorbed by the roots, and for concentrations above 100 mg/L if spray irrigation is used and the ions adsorb to the leaves. With adequate care, chloride toxicity should not be a problem except possibly for sensitive species being spray irrigated. While sodium may be beneficial to plants in small quantities, similar leaf scorch observations have been made for irrigation waters high in sodium. Again, adequate care should prevent sodium toxicity.

Trace Elements. In addition to the major ion requirements, numerous trace elements in reclaimed water may affect plant growth. Recommended maximum concentrations of various trace elements in irrigation water are presented in Table 14-7 (DHS).

Other Concerns. An additional concern when reclaimed wastewater is used for irrigation is protec-

TABLE 14-7. Recommended Maximum Concentrations of Trace Elements in Irrigation Waters (DHS)

Element	Maximum Concentration (mg/L)[a]	
	For Waters Used Continuously on All Types of Soil	For Use up to 20 yr on Fine Textured Soils With pH 6.0–8.5
Aluminum (Al)	5.0	20.0
Arsenic (As)	0.10	2.0
Beryllium (Be)	0.10	0.50
Boron (B)	0.75	2.0
Cadmium (Cd)	0.010	0.05
Chromium (Cr)	0.10	1.0
Cobalt (Co)	0.050	5.0
Copper (Cu)	0.20	5.0
Fluoride (F)	1.0	15.0
Iron (Fe)	5.0	20.0
Lead (Pb)	5.0	10.0
Lithium (Li)[b]	2.5	2.5
Manganese (Mn)	0.20	10.0
Molybdenum (Mo)	0.01	0.5[c]
Nickel (Ni)	0.20	2.0
Selenium (Se)	0.02	0.02
Vanadium (V)	0.10	1.0
Zinc (Zn)	2.0	10.0

[a] These levels will normally not adversely affect plants or soils. No data are available for mercury, silver, tin, titanium, and tungsten.
[b] Recommended maximum concentration for irrigating citrus is 0.075 mg/L.
[c] Only for acidic, fine textured soils or acidic soils with relatively high iron oxide contents.

tion of the public health from contamination of edible crops, pasture lands, and feed crops by direct contact with disease agents carried in reclaimed water or aerosols from spray irrigation. Criteria established to protect public health during agricultural or landscape irrigation depend not only on the type of crop irrigated but also on the method of irrigation and degree of public contact. Requirements developed by the California Department of Health Services (DHS) are shown in Tables 14-8 and 14-9. Similar requirements have been established by other states and countries.

Finally, irrigation and drainage can reach groundwater supplies by percolation or surface waters by runoff. Buildup of minerals in the ground or surface water supplies must be controlled to protect other beneficial uses of freshwater supplies. For these reasons, irrigation with reclaimed wastewater is often judged on a case-by-case basis to protect against buildup of contaminants in freshwater supplies.

Industrial

This section describes water quality requirements for the following major industrial users of reclaimed wastewater: (1) steam-generated power cooling, (2) industrial cooling, and (3) other industrial users. Generally, industries are willing to accept water that meets drinking water standards. Occasionally, industrial water quality requirements are more stringent than drinking water criteria, such as water used in high-pressure boilers or in selected industrial processes. In these instances, the industry provides additional treatment as required. From an industrial viewpoint, the primary criterion is that the water supply be of consistent quality so that pretreatment costs can be minimized. Once a consistent water quality is achieved, then pretreatment steps are maintained routinely. Inadequate water quality can cause three major categories of difficulties, as listed in Table 14-10. Virtually all these difficulties can be eliminated via good operational control of a typical municipal wastewater reclamation plant.

Steam-Generated Power and Industrial Cooling. The basic quality considerations for cooling water are that (1) scale is not deposited, (2) the cooling system is not corroded, (3) nutrients are not present for slime-forming organisms, and (4) delignification does not occur in the cooling towers. Water quality requirements meeting these basic criteria are presented in Table 14-11. Requirements for a recirculation makeup system are more stringent than for once-through cooling. Particular problems are associated with the presence of hardness ions such as calcium and magnesium, which can combine with available phosphate or sulfate to form a hard heat-transfer-limiting scale. Delignification of cooling towers has been studied in considerable depth, and the primary cause found to be sodium carbonate in the cooling water. By maintaining the pH between 6.5 and 7.5 a favorable bicarbonate equilibrium can be achieved, thereby minimizing this problem.

Treated wastewater for cooling represents the single largest reuse application by industry and may require special considerations in regard to effluent quality control and water treatment. Fortunately,

TABLE 14-8. California State Department of Health Services Reuse Criteria for Agricultural Irrigation

Reuse	Required Treatment[a]	Allowable Coliforms[b] (MPN/ 100 mL)
Food crops Spray irrigation	Biooxidation, coagulation/ clarification, filtration, disinfection	2.2[c]
Surface irrigation General	Biooxidation, disinfection	2.2
Orchards and vineyards with no fruit contact with water or ground	Primary sedimentation	NS[d]
Exceptions	Considered on individual basis if crop undergoes pathogen destroying processing	NS[d]
Fodder, Fiber, and Seed Crops Fodder, fiber and seed crops	Primary sedimentation	NS
Pasture for milking animals	Biooxidation, disinfection	23

[a] Primary effluent settleable solids must be less than 0.5 mL/L per hour. Filtration must provide an average effluent turbidity less than 2 TU and a maximum less than 5 TU. Alternative methods of treatment may be accepted if the applicant demonstrates to the satisfaction of DHS an equal degree of treatment and reliability.
[b] Median as determined from results of last 7 days for which analyses have been completed.
[c] Shall not exceed 23 MPN/100 mL in more than one sample within any 30-day period.
[d] Requirement not specified (NS).

TABLE 14-9. California State Department of Health Services Reuse Criteria for Landscape Irrigation

Reuse	Required Treatment[a]	Allowable Coliforms[b] (MPN/ 100 mL)
Landscape irrigation Golf courses, cemeteries, freeway landscaping	Biooxidation, disinfection	23[c]
Parks, playgrounds, and schoolyards	Biooxidation, coagulation/ clarification, filtration, disinfection	2.2[d]

[a] Primary effluent settleable solids less than 0.5 mL/L per hour. Filtration must provide an average effluent turbidity less than 2 TU and a maximum less than 5 TU. Alternative methods of treatment may be accepted if the applicant demonstrates to the satisfaction of DHS an equal degree of treatment and reliability.
[b] Median as determined from results of last 7 days for which analyses have been completed.
[c] Shall not exceed 240 MPN/100 mL in any sample.
[d] Shall not exceed 23 MPN/100 mL in more than one sample within any 30-day period.

TABLE 14-10. Three Major Difficulties Encountered by Industry when Using Improper Water Quality

Product degradation
   Contamination via biological activity
   Staining
   Corrosion
   Chemical reaction and contamination
Equipment deterioration
   Corrosion
   Erosion
   Scale deposition
Reduction of efficiency or capacity
   Tuberculation
   Sludge formation
   Scale deposition
   Foaming
   Organic growths

SOURCE: McKee and Wolf (1963).

most reclaimed water is used for once-through cooling, and no additional treatment is necessary before use. Where recirculating systems are used, industrial users of reclaimed wastewater must provide additional treatment since a much better quality is required. In general, these users treat the municipal effluent by lime clarification in order to reduce phosphates, organics, silica, and suspended solids. Following lime treatment, the effluent is filtered, neutralized by acid addition, and chlorinated for disinfection. Table 14-12 summarizes typical water quality problems associated with cooling water makeup and possible treatments to minimize each problem.

High-pressure boilers (pressure greater than 10,340 Pa or 1500 psig) have the most stringent water quality requirements of any major industrial use; however, as boiler pressures decrease, the quality requirements become less demanding, as indicated in Table 14-11. Pretreatment of existing potable water supplies is expected for boiler feed water. Silica and aluminum are especially critical materials in that they tend to form a hard scale on the boiler heat exchange tubes. Excessive sodium or potassium can cause foaming of the boiler water.

Other Industrial Users. The quality requirements for manufacturing and mineral industries, clustered under "other industrial users," are totally general in nature since the broad industrial categories cover a wide range of individual industrial plant types and

processes. For example, the chemical and allied products category includes manufacturing operations for a variety of inorganic and organic chemicals, plastics, and fertilizers. Similarly, the paper and allied products category includes a variety of paper, paper products, and pulp operations. Both manufacturing and mineral industry operating methods vary from one company to another and often from one location to another in the same company. Definite quality requirements for each process water or the character of each wastewater discharge cannot be specified for the broad industrial categories but must be investigated on a case-by-case basis. However, several water quality parameters are provided in Table 14-11 to indicate whether the reclaimed wastewater meets the general requirements of various industrial users.

Groundwater Recharge

Reclaimed water quality standards for groundwater recharge are based on nondegradation of existing groundwater supplies, maintenance of public health standards, protection of recharge capability, and prevention of nuisance conditions. Quality requirements will vary depending on basin geology, recharge method employed, and extent of recharge operations.

The concept of nondegradation, when applied to water supplies, requires maintenance of existing water qualities while recharge operations are underway. This ensures that existing water quality is maintained for the benefit of all concerned users. Similarly, reclaimed wastewater that percolates or is injected into the groundwater basin must not adversely affect the public health. This consideration is particularly important for direct injection systems since injected wastewater does not receive additional filtration through the aeration zone. Parameters of specific concern are pathogenic organisms, viruses, toxic materials, and carcinogenic or mutagenic materials.

Studies linking wastewater contamination of water supplies to outbreaks of waterborne diseases underline the importance of eliminating bacterial contamination of potable water supplies. For this reason, wastewaters used for groundwater recharge of potable aquifers generally receive secondary treatment followed by disinfection or an equivalent method of treatment. Treatment beyond secondary is not required if surface spreading is used and suf-

TABLE 14-11. Water Quality Requirements (mg/L) Are Most Stringent for High-Pressure Boiler Feedwater: Requirements for Many Other Uses Can Be Achieved by a Well-Run Secondary Treatment Plant

Characteristics	BF 0–150	BF 150–700	BF 700–1500	BF 1500–5000	Once-Through Fresh	Once-Through Brackish	Makeup Fresh	Makeup Brackish	Textile	Lumber	Pulp & Paper	Chem.	Petroleum & Coal Products	Prim. Metals	Food Canning	Bottled and Canned Soft Drinks	Tanning
Silica (SiO₂)	30	10	1.0	0.01	50	25	50	25	—	—	50	50	60	—	50	—	50
Aluminum (Al)	5	0.1	0.01	0.01	—	—	0.1	—	—	—	—	—	—	—	—	—	—
Iron (Fe)	1	0.3	0.05	0.01	—	—	0.5	—	0.1	—	0.3	0.1	1.0	—	0.2	0.3	0.2
Manganese (Mn)	0.3	0.01	0.01	—	—	—	0.5	—	0.01	—	0.1	0.1	—	—	0.2	0.05	—
Copper (Cu)	0.5	0.05	0.05	0.01	—	—	—	—	0.05	—	—	—	—	—	—	—	—
Calcium (Ca)	—	0	0	a	200	420	50	420	—	—	20	70	75	—	100	—	60
Magnesium (Mg)	—	0	0	—	—	—	—	—	—	—	12	20	30	—	—	—	—
Sodium and potassium	—	—	—	—	—	—	—	—	—	—	—	—	230	—	—	—	—
Ammonia (NH₃)	0.1	0.1	0.1	0.7	—	—	—	—	—	—	—	—	40	—	—	—	—
Bicarbonate (HCO₃)	170	120	50	a	600	—	25	—	—	—	—	130	480	—	—	—	—
Sulfate (SO₄)	—	—	—	—	680	2,700	200	2,700	—	—	—	100	600	—	250	500	250
Chloride (Cl)	—	—	—	—	600	19,000	500	19,000	—	—	200	500	300	500	250	500	250
Fluoride (F)	—	—	—	—	600	—	500	—	—	—	—	5	1.2	—	1	1.7	—
Nitrate (NO₃)	—	—	—	—	—	—	—	—	—	—	—	—	10	—	10	—	—
Phosphate (PO₄)	—	—	—	—	—	—	—	—	—	—	—	—	—	—	—	—	—
Dissolved solids	700	500	200	0.5	1,000	35,000	500	35,000	100	—	100	1,000	1,000	1,500	500	—	—
Suspended solids	10	5	0	0	5,000	2,500	100	100	5	<3 mm in dia.	10	5	10	3,000	10	—	—
Hardness (CaCO₃)	20	1.0	0.1	0.07	850	6,250	130	6,250	25	—	475	250	350	1,000	250	—	150
Alkalinity (CaCO₃)	140	100	40	0	500	115	20	115	—	—	—	125	500	200	250	85	—
Acidity (CaCO₃)	—	—	—	—	—	—	—	—	—	—	—	—	—	75	—	—	—
pH (units)	8.0–10.0	8.0–10.0	8.2–9.2	8.8–9.2	5.0–8.3	—	—	—	6.0–8.0	5.0–9.0	4.6–9.4	5.5–9.0	6.0–9.0	5.0–9.0	6.5–8.5	—	6.0–8.0
Color (units)	—	—	—	—	100	—	—	—	5	—	10	20	25	—	5	10	5
Organics MBAS	—	—	—	—	—	—	1	—	—	—	—	—	—	—	—	—	—
CCl₄	—	—	—	—	—	—	1	—	—	—	—	—	—	—	—	—	—
COD	5	5	0.5	0	75	75	75	75	—	—	—	—	—	30	—	—	—
Dissolved oxygen	<0.03	<0.03	<0.03	<0.005	—	—	—	—	—	—	—	—	—	—	—	—	—
Temperature, °F	120	120	120	120	100	120	100	120	—	—	100	—	—	100	—	—	—
Turbidity (JTU)	10	5	0.5	0.05	5,000	100	—	—	—	—	—	—	—	—	—	—	0

Column groupings: Boiler Feedwater (psig): 0–150, 150–700, 700–1500, 1500–5000. Cooling Water — Once-Through: Fresh, Brackish; Makeup for Recirculation: Fresh, Brackish. Process Water by Industry: Textile, Lumber, Pulp & Paper, Chem., Petroleum & Coal Products, Prim. Metals, Food Canning, Bottled and Canned Soft Drinks, Tanning.

ᵃ Determined by treatment of other constituents.

TABLE 14-12. Water Quality Problems Associated with Cooling Water Makeup and Possible Treatment to Minimize the Problem

Problem	Factors Contributing to Problem	Treatment or Solution
Corrosion	Oxygen and other dissolved gases, dissolved or suspended solids, pH, velocity, temperature, and microbial growth	Deaeration, pH adjustment, chlorination, addition of chemical inhibitors that suppress reaction or protect metal by covering surface with protective film
Scale formation	Concentration of calcium, magnesium, silica, aluminum, total dissolved solids, and alkalinity	Softening processes such as hot or cold lime–soda softening. For silica removal distillation or highly basic anion exchangers will also work
Sediments or foulants	Concentration of dirt, silt, microbial masses, or general debris in influent. Corrosion products and precipitates also add to the problem	Sediments can be removed by normal settling processes. Foulants can be removed by addition of organic reagents that disperse, dissolve or suspend the foulant for removal in blowdown stream
Biological growth	Concentration of nutrients and existing microbial growth in influent	Oxidizing chemicals such as chlorine and bromine and/or biocides such as chlorophenates, copper salts, and quaternary amines.
Foaming	MBAS levels above 0.5 mg/L and/or concentrations of phosphates and organic reagents. High pH and high alkalinity	Lower foaming agents to safe level or add antifoaming agent. Adjust pH and alkalinity by addition of acid
Dilignification	Presence of sodium carbonate in influent	Adjust pH to between 6.5–7.5 for more favorable carbonate–bicarbonate balance

ficient travel distance exists through the aeration zone before reaching the groundwater.

Trace concentrations of toxic materials, such as heavy metals, are generally adsorbed to the soil matrix and are of little concern to the public health. Not all species, however, are removed in the soil mantle and water quality testing must be undertaken before discharge takes place. Similarly, strict water quality requirements may be imposed on the discharger to ensure that organic carcinogenic or mutagenic compounds are not present in the effluent.

Finally, discharge quality requirements are instituted to maintain groundwater recharge capacity and prevent nuisance conditions. Inadequately treated wastewater can result in decreased soil permeability, production of odors, and creation of breeding grounds for mosquitoes and flies.

Recharged water quality requirements will vary depending on aeration and purification capacity of the underlying soil and/or the effect recharge has on groundwater quality. The factors affecting purification capacity and groundwater quality include aeration zone lithology and geochemistry, depth to groundwater table, existing groundwater quality, assimilative capacity of aquifer, distance in travel time to extraction point, and method of recharge. These factors must be considered in establishing both effluent quality and treatment requirements.

In the United States, the EPA has proposed regulations for direct injection of reclaimed wastewater into groundwater aquifers. These proposed regulations generally allow the states a maximum amount of flexibility in setting water quality requirements to prevent contamination of potable water supplies. Federal requirements for groundwater recharge by surface spreading, other than those covered in their requirements for land disposal of wastewater, have not been established. In general, requirements for surface spreading are less stringent than for direct injection, thereby taking advantage of additional purification provided as the recharge water flows through the aeration zone.

Miscellaneous Subpotable

In general, water requirements for miscellaneous subpotable uses are determined on a case-by-case basis. California does have specific recommendations for recreational impoundments and waterways that contain reclaimed water. These requirements, presented in Table 14-13, specify the required treatment as well as the effluent quality. These requirements for recreational impoundments reflect the need to protect the public from contamination caused by contact or ingestion of inadequately treated wastewater or fish contaminated by that wastewater. Constituents of concern include pathogenic bacteria, viruses, and toxic materials. The extent of removal is generally tailored to the extent of contact of the water with the public.

With respect to aquaculture reuse, the two parameters of greatest concern to the preservation of a healthy fish environment are the dissolved oxygen and free ammonia concentrations. A dissolved oxygen concentration of 5 mg/L or greater must be maintained at all times. The dissolved oxygen level can be depressed by the presence of BOD or reduced nitrogen compounds, both of which oxidize into CO_2 and NO_3, thereby consuming oxygen. Free ammonia is toxic to fish even in small concentrations. The toxicity of ammonia and ammonia salts is directly related to the amount of undissociated ammonia, itself inversely related to the pH of the wastewater.

Other parameters of concern are pesticides, herbicides, heavy metals, and viruses, especially if the fish are grown for human consumption. While these substances may not be present in sufficient concentrations in domestic wastewaters to be lethal to fish life, they can be accumulated in the fish and thereby transmitted to humans. Aquaculture reuse will generally require some type of tertiary wastewater treatment, both to achieve water quality objectives and to achieve high fish productivity. While many species are able to survive in primary effluent, production of high-quality fish, such as brown and rainbow trout, will require tertiary effluent if the fish are to survive.

Potable

Health officials, regulatory agencies, and the general public have expressed two major concerns associated with the quality of reclaimed water used for potable purposes. First, conventional drinking water standards were originally developed for water withdrawn from groundwater sources or from protected surface sources which have received little contamination. These standards were not developed to apply to direct reuse situations. Consequently, more stringent standards will have to be developed to define acceptable potable reuse quality. The second concern relates to the variability in effluent quality, especially that produced by secondary treatment facilities. Inherent variability in the quality and quantity of raw wastewater, and our ability to treat this wastewater, may require special provisions for equalization, storage, and process reliability. The cost of providing directly potable water on a consistent basis will be substantially greater than the cost to achieve current discharge standards. Process reliability can be obtained through

TABLE 14-13. Water Quality Requirements for Recreational Reuse of Wastewater

Reuse	Required Treatment[a]	Allowable Coliform[b] (MPN/100 mL)
Recreational impoundments		
Nonrestricted access	Biooxidation, coagulation/clarification, filtration, disinfection	2.2[c]
Restricted to nonbody contact	Biooxidation, disinfection	2.2
Landscape noncontact	Biooxidation, disinfection	23

[a] Primary effluent must not contain more than 0.5 mL/L per hour of settleable solids. Filtration must provide an effluent with a turbidity that does not exceed an average of 2 turbidity units and does not exceed a maximum of 5 turbidity units. Alternative methods of treatment may be accepted if the applicant demonstrates to the satisfaction of DHS that they will assure an equal degree of treatment and reliability.
[b] Median as determined from results of last 7 days for which analyses have been completed.
[c] Shall not exceed 23 MPN/100 mL in more than one sample within any 30-day period.

duplication of equipment, careful process control, and a well-qualified and skilled operating staff.

With the lack of water quality standards for potable reuse water, two alternatives are available for addressing the question of the suitability of renovated water for potable reuse: (1) in-depth toxicological and epidemiological studies for a period of years to identify any health problems and provide background criteria for the development of meaningful standards and (2) provision of renovation capabilities adequate to remove all the contaminants added between the water supply and the wastewater treatment facility, that is, the use increment of contaminants. Most of the work to date has centered on this latter technique, that is, the removal of the use increment of contaminants.

To provide an acceptable water source, the treatment plants in South Africa and the planned facility in Denver have adopted roughly parallel treatment processes. The first process, high-pH lime precipitation, is included to provide phosphorus and heavy metal removal, suspended solids precipitation, and virus and bacteria kills at pH levels of 11.2 or greater. This is followed by ammonia removal; either ammonia stripping or selective ion exchange. Following recarbonation, filtration is included for additional suspended solids removal. Two-stage activated carbon adsorption along with final organics oxidation via either chlorine or ozone is provided for removal of residual organic material from the process stream, and for virus and pathogenic organism kill. In addition, the Denver plant provides for demineralization by reverse osmosis (RO) to reduce the dissolved solids and also to provide a final physical barrier against viruses and bacteria, and most organic materials. The RO unit thereby provides additional redundancy to the other processes in the treatment scheme.

Microbiological and epidemilogical studies related to the reuse of reclaimed wastewater have been conducted at the Windhoek site in South Africa and are planned for the Denver reuse study.

PLANNING CONSIDERATIONS

This section will briefly outline the planning considerations associated with treating and delivering wastewater to the potential user as they apply to wastewater reuse in general. Specific planning elements are design, economics, public health, institutional arrangements, environment, and energy.

Design

Wastewater reuse may require treatment, storage, and conveyance facilities whose design is very similar to general water and wastewater treatment, storage, and conveyance units. These general design procedures and parameters are covered in detail elsewhere in this text as well as in numerous other sources (Metcalf and Eddy, 1972; Weber, 1972; Culp et al., 1978; USEPA). This section will outline some of the design concepts specific to wastewater reuse facilities.

Treatment. Extensive treatment is required to transform raw sewage into an acceptable water for reuse. The level of treatment provided at a wastewater reuse facility is dependent on both the influent wastewater quality and the intended reuse application. Figure 14-4 illustrates the various levels of treatment for specific categories of reuse and some typical treatment processes utilized to achieve the desired effluent quality. Representative effluent qualities after each step of the treatment train are presented in Table 14-14.

Many communities within the United States already provide a level of treatment equal to or better than secondary treatment. Consequently, much of the wastewater being generated can be utilized for agricultural irrigation, lower-quality industrial uses, and groundwater percolation without additional treatment. The total cost to a water purveyor to meet higher effluent qualities is substantially reduced by the presence of secondary-treated wastewater produced by the discharger.

Of particular importance to treatment facility design for wastewater reuse is the acceptable variation of effluent quality. Effluent quality from a typical wastewater treatment plant varies widely throughout the day and year because of varying influent quality and quantity, varying climatic conditions, and operation and maintenance problems. NPDES permits for typical wastewater treatment facilities make allowances for these variations by stating the water quality requirements in terms of daily, weekly, or monthly averages. This may be acceptable for some reuse categories utilizing lower-quality effluent but is generally unacceptable for those reuse categories requiring higher-quality effluent. Hence, treatment facilities designed for certain wastewater reuse categories must achieve smaller variations in effluent quality, usually through (1) judicious choice of treatment processes

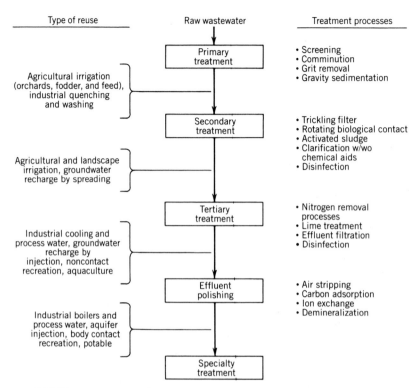

FIGURE 14-4. Levels of treatment required for selected reuse applications.

that produce less variation in effluent quality, (2) equalizing influent flow and quality, or (3) redundancy in the key treatment processes. Additionally, overdesign of a treatment process can achieve an average effluent quality substantially better than re-

quired, so that variations in effluent quality will not exceed the acceptable level (Rope et al., 1979).

Storage. Storage facilities may be required in a wastewater reuse scheme. Design of these facilities

TABLE 14-14. Typical Quality Characteristics Following Various Treatment Levels

	Quality Parameters							
Treatment Level	BOD (mg/L)	COD (mg/L)	SS (mg/L)	Turbidity (TU)	Phosphorus (mg/L)	Ammonia–N (mg/L)	Color (units)	Coliforms (MPN/100 mL)
Raw wastewater	300	500	250	—	12	25	—	10^8
Primary (1)	200	250	100	—	9	25	—	10^7
Secondary (2)	30	60	30	—	6	25^a	—	10^{6b}
1 + 2 + filtration (3)	5	40	10	5	6	25	30	$<2.2^c$
1 + 2 + coagulation and sedimentation (4) + 3	2	30	<1	<1	<1	25	30	$<2.2^c$
1 + 2 + 4 + activated carbon (5) + 3	<1	10	<1	<1	<1	25	10	$<2.2^c$

[a] Ammonia–N may be reduced at this point if nitrification is provided.
[b] Coliform may be reduced at this point if disinfection is provided.
[c] Actual coliform level dependent on mode of operation and degree of disinfection provided.

is essentially identical to design of general water storage facilities, as discussed in several other sources (Whiting, 1976; USBR, 1973; Linsley and Franzini, 1972). Factors specific to storage of wastewater include public access, water quality, and combined-use systems.

Depending on the degree supply matches demand, whether alternative methods of wastewater disposal are available, whether alternative sources of freshwater are available, and the control required by local or state ordinances, storage of reclaimed wastewater may be required. The normal diurnal variation in wastewater flow generally creates a need for some storage or flow equalization. If demand fluctuates widely and the supply of reclaimed wastewater cannot otherwise be disposed of, storage capacity will be required. A typical example is agricultural irrigation, in which the reclaimed wastewater does not meet discharge requirements to a water body. Demand for reclaimed water by agriculture is usually dependent on irrigation rate and climatic conditions. Both of these factors are highly variable, whereas the supply of wastewater is, for most communities, relatively constant from day to day. This can result in a significant amount of storage.

Storage requirements also depend on whether reclaimed wastewater can be supplemented from other sources and on the percent of the total demand supplied by reclaimed wastewater. Storage requirements can be greatly reduced or even eliminated if another source of water can be utilized to balance the peaks between supply and demand. Likewise, if reclaimed wastewater constitutes only a relatively small proportion of the total demand, the normal diurnal variations will not produce a need for storage facilities.

State or local ordinances may directly or indirectly mandate storage facilities. In California, for instance, landscape irrigation of parks, golf courses, and other areas resulting in direct exposure of the public to reclaimed wastewater is prohibited. To meet this requirement, irrigation is generally conducted only at night when human exposure is minimized. Consequently, flows generated during the day must be stored until they can be utilized at night.

In most reuse projects requiring storage facilities, reclaimed wastewater has not been treated to a level sufficient for body contact recreation. Hence, public access to the reservoir must be limited. The relatively low level of treatment provided may lead to water-quality-related problems such as odors and algal blooms. Odors can be alleviated by addition of chemicals, aerators, or judicious siting of storage facilities. Algal blooms can be alleviated by use of algicides or control of detention time in the facility. If the storage reservoir overlies a groundwater aquifer utilized for potable supplies, the reservoir may have to be lined with an impermeable barrier to prevent contamination of the aquifer.

A cost-effective system involving large aerated or nonaerated lagoons for both treatment and storage of wastewater prior to reuse is used successfully at several sites in the Midwest and southwestern United States. These combined-use facilities consist of a number of wastewater lagoons in series with the latter lagoons designed for storage, as well as treatment, of the wastewater. A particularly well-documented case is Muskegon, Michigan (Malhotra and Myers, 1975).

Conveyance. Distribution of reclaimed wastewater to users requires pump stations, pipes, and appurtenant facilities. Normal design procedures for potable water systems are generally applicable to reclaimed wastewater systems and, hence, can be utilized for design purposes (Vennard and Street, 1975; Symons, 1967; ASCE, 1975). An important distinction between potable and reclaimed systems, however, is the difference in water quality.

Reclaimed wastewater conveyance systems must be designed to convey water with a higher suspended solids concentration. The higher solids content leads to increased clogging of pipelines, hose bibs, and irrigation systems unless adequate preventive measures are taken such as adequate pipe velocities, strainers, and oversized holes in nozzles for irrigation systems.

The inherently poorer water quality in a reclaimed wastewater system means careful design of wastewater conveyance facilities to avoid the possibility of cross connections with potable water systems. The most important criterion in both cases is minimum separation distance between the two systems as specified in local and state ordinances. An example ordinance is the Los Angeles County Department of Health Services' (LADOHS) "Environmental Management Guideline for Construction and Installation of Reclaimed Wastewater Pipelines" (LADOHS, 1969). This document details minimum separation distances and other requirements to help assure protection of the public health.

Economics

The cost elements considered in the economic feasibility of reuse include reclaimed wastewater supply, storage and conveyance system, on-site repiping, and additional operations and maintenance requirements. All relevant cost elements are taken into account in evaluating the economics of a reuse project including those imposed on the user as a result of the generally poorer water quality. However, in areas of water shortage, the economic evaluation may not have to show a favorable ratio of reclaimed wastewater cost to potable water cost. Rather, the economics of reclaimed wastewater usage need only appear reasonable in terms of the users' ability to pay versus the consequences of not having enough water.

Reclaimed Wastewater Supply. The base cost to the potential user of reclaimed wastewater usually covers the supplier's capital and/or operating and maintenance costs for treatment provided above that required by law (NPDES permit). This cost is paid through a monthly charge, through capital outlays for improvements, or through a combination of both. For example, Bethlehem Steel in Maryland required cooling water quality better than that provided by the local treatment plant. The industry undertook a program to locate and control the sources of high salinity in the wastewater (seawater intrusion into the sewers), provided funding for upgrading the trickling filter system to activated sludge system, and participated in the ongoing operating and maintenance costs.

Storage and Conveyance System. The cost of storage facilities, if required, should be included in the total economic evaluations. Historically, storage facilities have been constructed, owned, and maintained by the supplier since they are generally located at the same site as the treatment facility and are tied into operation of the treatment plant.

Conveyance facilities for transporting reclaimed wastewater to the reuse site are usually constructed, owned, and maintained by the user. Conveyance facilities often represent the major cost to the potential user of reclaimed wastewater. These costs vary widely depending on the distance between the user and supplier and the number of users. In those cases where the majority of potential users are not located near reclaimed water supplies,

the cost of delivery can be consolidated on a regional basis.

On-Site Repiping. Once the reclaimed wastewater is delivered to the reuse site, it must be distributed to the actual points of use. To minimize the potential for cross connections and provide backup capabilities from a separate source of freshwater, some on-site repiping will generally be required. The amount of repiping will largely depend on the number of actual points of reuse and the actual use of reclaimed wastewater. If reclaimed wastewater is to be utilized as a source of irrigation water for individual homeowners, then a complete dual water system may be required. Use of reclaimed wastewater for irrigation of a golf course might only require installation of a suitable backflow prevention device and changing over sprinkler heads to those with larger-diameter orifices.

Operations and Maintenance. The cost of operating and maintaining the wastewater reuse facilities is often a significant cost and must be included in the overall economic evaluation. The operating and maintenance costs may include costs for normal treatment facilities as well as those facilities specific to the reuse capabilities of the system, such as the internal chemical feed system required at cooling towers to inhibit scale formation, corrosion, and biofouling.

Specific Requirements. Numerous other cost elements, specific to a particular reuse project, should be included in the total economic evaluation. These elements are mentioned here primarily to illustrate their specificity. Cost elements found in industrial reuse include pretreatment, internal treatment, and sludge and brine disposal above that associated with potable water use. Cost elements specific to irrigation reuse include (1) a generally greater volume of water required because of the higher total dissolved solids and (2) periodic cleaning of sprinkler heads because of a higher level of suspended solids. An offsetting cost element for irrigation reuse is the fertilizer value of the wastewater.

Intangibles. Numerous intangible elements have an effect upon the economic evaluation of reuse, but are difficult to quantify. These include the uncertainty of future potable supplies, the reliability of locally developed reclaimed water supplies, and a public policy that generally favors reuse. These elements may combine to make a questionable or unfavorable economic alternative acceptable.

Public Health

The level of public health risk associated with use of reclaimed wastewater is related to the extent of human contact and the overall level and reliability of treatment provided. For example, microbial and chemical contaminants in reclaimed wastewater utilized for surface irrigation of forage crops are of minimal concern when compared with direct potable reuse of wastewater. This section briefly describes the types of contaminants of concern to public health, health considerations involved in reuse projects, and the general level of treatment required to protect the public health for each type of reuse. A more detailed discussion of these topics can be found in cited references (Shuval et al., 1977; Kasperson and Kasperson, 1977; Crook, 1978).

Contaminants of concern to public health found in reclaimed wastewater are broadly classified as either pathogens or organic compounds. The transmission of waterborne diseases via pathogens in wastewater has been well documented (Crook, 1978). Contamination of water supplies by viruses and other pathogens is generally manifested in outbreaks of diseases such as bacillary and amoebic dysentary, salmonellosis, cholera, vibriosis, and infectious hepatitis. The protection of public health dictates that pathogenic agents be eliminated prior to reuse of reclaimed wastewater or that use of reclaimed wastewater be limited to areas that will not result in direct or indirect human contact.

An increase in the use of hundreds of new chemical compounds by industry and agriculture has resulted in appearance of many of these compounds in municipal wastewater. Many of these compounds, such as pesticides, are toxic and can cause serious health problems if ingested. Some have a

TABLE 14-15. Health Considerations by Type of Reuse

Type of Reuse	Health Considerations	Mitigating Measures
Agricultural	Contamination of crops with pathogens	Provide adequate treatment, eliminate direct contact of wastewater with food crop
	Direct contamination of workers	Educate workers about proper personal hygiene, provide appropriate sanitary facilities
	Contamination of general public via aerosols	Provide adequate treatment, buffer zones, and barriers as needed
	Contamination of animals that graze on pasture irrigated with reclaimed wastewater	Provide adequate treatment
	Contamination of fish or fish ponds	Provide adequate treatment and control of intermediate host
Industrial	Cross connections of potable and reclaimed wastewater pipelines within plant site and with main potable water line entering plant site	Provide adequate treatment, color-coded pipes, air gap, or acceptable back flow prevention device on potable water line entering plant site
	If used as process water, potential contamination of food products and workers	Provide adequate treatment
Recreational	Waterborne enteric infections and ear, eye, and nose infections	Provide adequate treatment, restrict direct body contact
	Ingestion of toxic chemicals or skin or eye irritations due to chemical industrial wastes	Reduce or eliminate industrial discharges, provide adequate treatment, restrict direct body contact
Groundwater recharge	Contamination of groundwater aquifer utilized as source of potable supply of water	Provide adequate treatment prior to spreading, provide adequate travel distances through soil, provide adequate dilution within aquifer, provide minimum hydraulic detention time in aquifer
Municipal nonpotable	Cross connections of potable and reclaimed water lines	Provide adequate treatment, color-coded pipes, provide minimum separation distance between pipes

TABLE 14-16. Suggested Treatment Processes to Meet the Given Health Criteria for Wastewater Reuse[a]

	Irrigation			Recreation		Industrial Reuse	Municipal Reuse	
	Crops Not for Direct Human Consumption	Crops Eaten Cooked; Fish Culture	Crops Eaten Raw	No Contact	Contact		Non-potable	Potable
Health criteria (see below for explanation of symbols)	A + F	B + F or D + F	D + F	B	D + G	C or D	C	E
Primary treatment	●●●	●●●	●●●	●●●	●●●	●●●	●●●	●●●
Secondary treatment		●●●	●●●	●●●	●●●	●●●	●●●	●●●
Sand filtration or equivalent polishing methods		●	●	●	●●●	●	●●●	●●●
Nitrification						●		●●●
Denitrification								●●●
Chemical clarification						●		●●●
Carbon adsorption								●●●
Ion exchange or other means of removing ions						●		●●●
Disinfection		●	●●●	●	●●●	●	●●●	●●●[a]

[a] Health criteria:

A Freedom from gross solids; significant removal of parasite eggs.
B As A, plus significant removal of bacteria.
C As A, plus more effective removal of bacteria, plus some removal of viruses.
D Not more than 100 coliform organisms per 100 ml in 80% of samples.
E No fecal coliform organisms in 100 ml, plus no virus particles in 1000 ml, plus no toxic effects on man, and other drinking-water criteria.
F No chemicals that lead to undesirable residues in crops or fish.
G No chemicals that lead to irritation of mucous membranes and skin.

In order to meet the given health criteria, processes marked ●●● will be essential. In addition, one or more processes marked ●● will also be essential, and further processes marked ● may sometimes be required.

SOURCE: WHO (1973).

322

chronic effect that can only be detected after a long period of time. Bioaccumulation of pesticides in animal and plant life has been documented and may pose a serious threat to public health. Experimental studies have also shown that other man-made organic compounds are carcinogenic.

Table 14-15 contains a brief description of the health considerations associated with the various types of wastewater reuse. This table also includes a listing of mitigation measures that might be followed to address these health considerations. One of the more obvious mitigation measures is to assure that the reclaimed wastewater is adequately treated for its intended reuse. Table 14-16 contains a brief listing of the level of treatment required by type of reuse to assure an adequate level of protection to the public health.

Institutional

Institutional considerations play an important role in determining feasibility of a wastewater reuse program. A reuse project that is economically and technically feasible may be impossible to implement because of institutional constraints. By their very nature, institutional considerations vary from locale to locale and must be dealt with on an individual basis. Some prominent institutional considerations include (1) water rights of present downstream users of treated wastewater, (2) establishment of an agency to sell wastewater, (3) identification and coordination with present water purveyor serving same user, and (4) gaining public and regulatory agency acceptance. Implementation of a viable wastewater reuse program may, in some cases, require as much effort in solving institutional problems as solving the engineering problems.

Environmental Considerations

In general, wastewater reuse has an overall beneficial effect upon the environment by minimizing discharge of pollutants to a water body and by eliminating adverse environmental impacts associated with the new facilities otherwise needed to develop new water supplies. However, adverse environmental impacts associated with building and operating the additional treatment, storage, and conveyance facilities to recover wastewater must be considered.

Energy Considerations

The cost of transporting the treated wastewater will parallel increases in energy costs. This is particularly true in those areas where water is conveyed over a long distance to serve the community. A case-by-case evaluation is needed to compare the energy cost of reclaimed wastewater with the energy cost of alternative supplies. In many instances, the energy cost associated with providing additional treatment necessary to reclaim wastewater will be significantly less than that necessary to transport and treat an alternative freshwater supply.

Examples of energy calculations are found in the literature (Roberts and Hagen, 1976; SCS Engineers, 1975). A case study of Orange County Water District, Orange County, California, compared the energy cost of treating and distributing wastewater with the energy cost of importing freshwater from the California State Water Project and the Colorado River Aqueduct (Cline and Argo, 1978). Reclaimed water, comparable in quality to State Project Water, required about 10 percent more energy than State Project Water, but slightly less than Colorado River water. Since reclaimed wastewater need not be treated to a high degree for many reuse applications, it can often show favorable energy economies when compared with freshwater importation.

REFERENCES

American Society of Civil Engineers (ASCE), *Pipeline Design for Water and Wastewater,* Report to the Task Committee on Engineering Practice in the Design of Pipelines, New York (1975).

Cline, N. M., and Argo, D. G., "Energy Trade offs in Advanced Wastewater Treatment in Southern California" (1978).

County of Los Angeles, Department of Health Services (LA-DOHS) "Environmental Management, Reclaimed Wastewater—A Guide to Pipeline Construction and Installation" (1969).

Crook, J., "Health Aspects of Water Reuse in California," *J. Div. Environ. Eng., EE4* (1978).

Culp, R. L., Wesner, G. M., and Culp, G. L., *Handbook of Advanced Wastewater Treatment,* 2nd ed., Van Nostrand Reinhold, San Francisco (1978).

Eden, G., Bailey, D., and Jones, K., "Water Reuse in the United Kingdom," in H. Shuval (ed.), *Water Renovation and Reuse,* Academic Press, New York pp. 398–428 (1977).

James M. Montgomery, Consulting Engineers, Inc., "Industrial Wastewater Reuse: Cost Analysis and Pricing Strategies," 1–12 (October 1979).

Kasperson, R. E., and Kasperson, J. K. (eds.), *Water Re-Use and the Cities,* Clark University Press, Hanover, England (1977).

Linsley, R. K., and Franzini, J. B., *Water Resources and Environmental Engineering,* 2nd ed., McGraw-Hill, New York (1972).

Malhotra, K., and Myers, E. A., "Design, Operation, and Monitoring of Municipal Irrigation Systems," *J. Water Pollution Control Fed.,* **47**(11) (1975).

McKee, J. E., and Wolf, H. W., *Water Quality Criteria,* Pub. No. 3-A, California State Water Quality Control Board, p. 92 (1963).

Metcalf and Eddy, Inc., *Wastewater Engineering: Treatment, Disposal, Reuse,* 2nd ed., McGraw-Hill Book, New York (1972).

Office of Water Research and Technology (OWRT), *Water Reuse and Recycling Volume 1: Evaluation of Needs and Potential,* OWRT/RU-79 1 (April 1979).

Roberts, E. B., and Hagen, R. M., "Explore the Energy Economics of Wastewater Treatment and Reuse," *Bulletin,* **13**(1) (1976).

Roper, R. E., Dickey, R. O., Marman, S., Kim, S. W., and Yandt, R. W., "Design Effluent Quality," *J. Div. Environ. Eng.* EE2 (1979).

Schmidt, C. J., and Clements, E. V., "Reuse of Municipal Wastewater for Groundwater Recharge," Office of Research and Development, United States Environmental Protection Agency, Cincinnati, Ohio EPA-600/2-77-183 (1977).

SCS Engineers, "Demonstrated Technology and Research Needs for Reuse of Municipal Wastewater," National Environmental Research Center (1975).

Shuval, H. I., (ed.) *Water Renovation and Reuse,* Academic Press, San Francisco (1977).

State of California Department of Health Service (DHS), "Guidelines for Use of Reclaimed Water for Spray Irrigation of Crops," "Guidelines for Use of Reclaimed Water for Surface Irrigation of Crops," "Guidelines for Use of Reclaimed Water for Landscape Irrigation," "Guidelines for Use of Reclaimed Water for Impoundments," "Guidelines for Worker Protection at Water Reclamation Areas," (undated).

Symons, G. E., *Water Systems Pipes and Piping,* Water and Waste Engineering Manual of Practice No. 2, Donnelley Publishing, New York (1967).

United States Department of Interior, Bureau of Reclamation (USBR), "Design of Small Dams," 2nd ed., U.S. Govt. Printing Office, Washington, D.C. (1973).

United States Environmental Protection Agency, *Process Design Manual for Land Treatment of Municipal Wastewater.* EPA 625 1-81-013 (October 1981).

U.S. Water Resources Council (WRC), "The Nation's Water Resources, The Second National Assessment," Draft (January 1978).

U.S. Water Resources Council, "The Nation's Water Resources, The Second National Assessment," Appendix B, "Methodologies and Assumptions for Socio-Economic Characteristics and Patterns of Change and Water Use and Water Supply Data," Preliminary Review Draft (April 1978).

University of California, Cooperative Extension Service (UC-CES), "Guidelines for Interpretation of Quality of Water for Irrigation." (undated)

Vennard, J. K., and Street, R. L., *Elementary Fluid Mechanics,* 5th ed., Wiley, New York (1975).

Weber, W. J., *Physicochemical Processes for Water Quality Control,* Wiley-Interscience, New York (1972).

Whiting, D. H., "Use of Climatic Data in Estimating Storage Days for Soil Treatment Systems," EPA, Office of Research and Development, EPA-IAG-D5-F694 (1976).

Williams, R. B., "Water Reuse—An Assessment of the Potential and Technology," in E. Middlebrooks (ed.), *Water Reuse,* Ann Arbor Science, Ann Arbor, MI, pp. 87–136 (1982).

World Health Organization, "Reuse of Effluents: Methods of Wastewater Treatment and Health Safeguards" WHO Technical Report Series No-517, Geneva (1973).

— 15 —

Inorganics

The presence of certain inorganic solutes can cause significant concern with respect to drinking water quality, aesthetics, and industrial use. The concerns over inorganic solutes range from their toxicity to their impact on process operation, product operation, and product quality in industrial processes. To control and limit the overall impact of inorganic solutes, the Environmental Protection Agency (EPA) has defined primary and secondary drinking water standards (see Chapter 4). While the majority of these standards are related to health impacts, wherein toxicity of the compounds has been identified and concentrations are limited to control impact on the user, there are also secondary standards (such as iron and manganese) that cover a number of aesthetic and product uses. In addition, there are some compounds not regulated that are cause for concern for certain types of water use.

The purpose of this chapter is to address the significant inorganic contaminants and the methods of removal and control of these materials. A thorough discussion is presented for the typical range of concentrations, the chemistry, and the methods of removal for each major element or compound. The chapter is organized to discuss the major solutes that frequently occur and are subject to the primary drinking water standards. Then other major solutes that make up those regulated by the secondary standards are discussed.

IDENTIFICATION OF PROBLEM INORGANICS

Drinking Water Contaminants

Of all the water supplies in the United States that have violations of primary drinking water standards, nitrate and fluoride are the most common offenders. A recent survey (AWWA, 1985) discovered that 369 out of 1583 violations of primary drinking water standards in the United States were for excessive nitrate. The fluoride limit was exceeded in 907 of the cases. Because the major sources of nitrate are domestic wastewaters, refuse dumps, fertilized agricultural land, and septic tanks, the presence of nitrate generally indicates that the water supply has been exposed to pollution. There is a tendency for all nitrogenous materials in natural waters to be converted to nitrate, the thermodynamically stable form of combined nitrogen. Therefore, all sources of combined nitrogen must be regarded as potential sources of nitrate. Many areas throughout the country, mainly past and present agricultural centers, have experienced nitrate concentrations in excess of the 10 mg/L maximum contaminant limit (MCL) for years. A few of these locations include southern and central California, Oklahoma, Iowa, Texas, Wisconsin, and New York. The normal range of nitrate-laden waters is 10–25 mg/L as nitrogen. There have been some extreme cases with nitrate levels rising as high as 100 mg/L as nitrogen.

Naturally occurring fluoride is present in many groundwater supplies, predominantly in South Carolina, Texas, New Mexico, and other parts of the southwestern United States. The presence of fluoride is limited mainly to groundwater, where fluoride-containing soils contact the water. Some high fluoride concentrations in surface water exist due to mining operations. Of all the fluoride violations exceeding the 1.4–2.4 mg/L (temperature dependent) MCL, the typical range of concentrations found was 1.5–6.0 mg/L. Fluoride concentrations in excess of 10 mg/L are unusual in natural drinking waters.

There are two other primary drinking water contaminants that occur with enough frequency for their treatment to be discussed: arsenic and selenium. Both of these are naturally occurring in some groundwaters. In addition, mining and manufacturing industries could be sources of these in water supplies. Arsenic has been found in elevated levels of California, Nevada, Arizona, Texas, and in the southeast. Elevated concentrations of arsenic have been found up to 0.5 mg/L. Selenium-contaminated waters have been found in California, Colorado, Nebraska, New Mexico, Oklahoma, and South Dakota. Selenium has been found in some wells up to 0.15 mg/L.

There are four heavy metals that are included in the secondary regulations: iron, manganese, copper, and zinc. The presence of iron in the ferrous state and manganese in the manganous state occurs under conditions normally found in groundwaters and in lakes when water is drawn from below the thermocline. The exposure of these reduced forms of iron and manganese to highly oxidative environments will cause immediate precipitation, intense staining in a red rusty color from the iron or black color from the manganese, and substantial taste impacts when the EPA secondary drinking standards limits are exceeded.

The presence of iron and manganese, especially in groundwater systems generally assumed to be homogeneous, is in reality highly variable due to the spatial variation in the deposition of the minerals, which are being dissolved to provide the reduced iron and manganese. As a result, concentration ranges varying within an order of magnitude are not unusual in groundwater systems (JMM, 1977).

Another cause of iron in groundwater systems is the growth of *Crenothrix* and other iron bacteria that depend on iron as an electron donor. These organisms thrive in areas such as well screens and other related water supply and distribution system equipment. These bacteria can clog well screens and cause considerable consumer complaint with respect to their presence in the water as turbidity and the presence of oxidized iron precipitates. However, these organisms can be controlled with intermittent down-well chlorination.

The normal range of iron and manganese in water supplies is up to 10 mg/L and 2 mg/L, respectively. Typically, iron concentrations will be less than 2 mg/L and manganese less than 0.5 mg/L.

The other two heavy metals, copper and zinc, will be present at concentrations that exceed the standards only in raw water supplies that are contaminated from mining or manufacturing activities. For the consumer, internal corrosion of piping may create high concentrations of copper and zinc at the tap. In general, the preferred approach for these constituents is source control and anticorrosion techniques, which are discussed in Chapter 18.

Other Contaminants

In addition to the regulated and recommended standards, there are at least two other inorganic parameters that are frequently considered for treatment. These are sulfide and hardness. Sulfide, typically present as hydrogen sulfide, is a foul-smelling gas present in noxious quantities in some groundwaters. Its removal prior to reaching the consumer should definitely be considered. Hardness, which is defined by analytical techniques as the sum of all polyvalent cations, is mainly composed of calcium and magnesium. Hard waters have traditionally been softened to reduce scaling and precipitation in hot water pipes and to lower the usage of soap and detergent for cleaning. It is a major treatment step for many industries requiring soft water, especially those industries with boilers and turbines.

In summary, the treatment of the following inorganic constituents will be discussed in this chapter on an individual basis: nitrate, fluoride, arsenic, selenium, iron and manganese, sulfide, and hardness. It is important to consider that treatment is typically the last resort for obtaining acceptable water quality. If alternatives to treatment such as developing a new water supply, blending with a better quality source, or connecting to another utility's supply do not appear viable, then treatment to remove an inorganic contaminant should be considered.

NITRATE REMOVAL

Several treatment techniques have been studied for removal of nitrate from drinking water: chemical coagulation, lime softening, chemical reduction, biological denitrification, ion exchange, and reverse osmosis (Sorg, 1980). Because of the high solubility and the lack of coprecipitation and adsorption of nitrate, neither chemical coagulation nor lime softening is effective. Chemical reduction involves reducing nitrate to nitrogen gas. Ferrous iron appears economically attractive as a reductant, however, certain drawbacks limit its practicality: copper is required as a catalyst, only about 70 percent of the nitrate can be removed, and large amounts of ferrous iron (resulting in large amounts of sludge) are required (O'Brien, 1966). Biological denitrification, reverse osmosis, and ion exchange are discussed further.

Biological Denitrification

Biological denitrification requires the use of special organisms to reduce the nitrate to nitrogen gas. These bacteria require an organic energy source, hence groundwaters having a low TOC would require the addition of "food-grade" ethanol or methanol. While this process may be feasible for wastewater treatment, the idea of maintaining bacteria in the flow stream and adding organic material to the water has not been accepted by the water utility industry in the United States. However, it is gaining acceptance in Europe and will be discussed further.

Figure 15-1 shows one of two types of biological denitrification processes being studied: fluidized media beds. The other, a fixed media bed, operates similarly except for the retainment of the media in an enclosed column. The fluidized system requires a fluidized reactor, a sand separator, a clarifier, and a filter. The clarifier and filter remove the biological organisms that are grown for nitrate reduction. The denitrified effluent is then put into storage or into the distribution system. Certain drawbacks make its acceptability limited. After clarification, aeration is required to remove the nitrogen gas. In addition to adding food-grade TOC, increased disinfection would be required to control any organism that might leave the clarifier. If free chlorine is used as the disinfectant, THM production could be a problem.

Reverse Osmosis

Reverse osmosis (RO) will reduce nitrate levels in drinking water, but this process is used primarily for treating high-TDS and saltwater. Its cost solely for nitrate removal from a low-TDS water would be much higher than ion exchange. Table 15-1 shows typical data for RO performance. These data show a 65 percent removal of nitrate. This is relatively low compared with the removal of other anions by RO. It has been demonstrated that RO could be used before ion exchange to successfully pretreat a water for nitrate removal because RO removes a very large percentage of sulfate. With the sulfate gone, treatment by ion exchange is much more efficient.

Ion Exchange

It seems that the only advanced treatment process for nitrate removal that can operate economically is ion exchange with strong-base anionic resins. It is

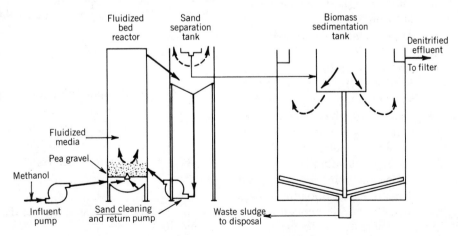

FIGURE 15-1. Biological denitrification using fluidized coarse media beds.

TABLE 15-1. Typical Performance of RO for Nitrate Removal[a]

Constituents	Pretreated Feed (mg/L)	Product (mg/L)	Brine (mg/L)
Calcium (Ca)	154	7.0	590
Magnesium (Mg)	3.8	0.17	15
Sodium (Na)	92	11	345
Potassium (K)	3.6	0.5	12.8
Carbonate (CO_3)	0.0	0.0	0.0
Bicarbonate (HCO_3)	7.8	5.2	45.9
Chloride (Cl)	92.8	6.0	346.9
Sulfate (SO_4)	380	5	1,500
Nitrate (NO_3)	93.0	31.9	270.2
Fluoride (F)	0.06	0.03	0.13
Iron (Fe)	<0.05	<0.05	0.08
Manganese (Mn)	<0.01	<0.01	0.01
Arsenic (As)	<0.01	<0.01	<0.01
Copper (Cu)	<0.01	<0.01	<0.02
Zinc (Zn)	0.01	0.007	0.02
MBAS	<0.1	<0.1	0.2
Hardness as $CaCO_3$	401.0	18.2	1,538
Total solids at 105°C	823	64	3,120
pH	5.2	5.6	5.9
Electrical conductivity mhos/cm ($K \times 10^6$) at 25°C	890	120	3,800

[a] Product rate = 13.0 gpm, feed pressure = 390 psig, water recovery = 74.3 percent.
SOURCE: Guter (1981).

the only method presently being used to remove nitrate on a full scale. Because the full-scale nitrate removal facility operated by the Garden City Park Water District, Nassau County, New York, has not been operated on a full-time basis, there has been a lack of data on nitrate removal by ion exchange. However, bench-scale and pilot-scale studies (Clifford, 1976; Anderson, 1978; Guter, 1981) have developed enough design and operating data to successfully discuss this topic. The site studied by Guter now has a full-scale facility.

As explained in Chapter 10, the nitrate ion, in competition with the other major anions in a water supply, such as sulfate, bicarbonate, and chloride, will be removed by strongly basic anion exchange resins. The amount of nitrate removed per given amount of resin will depend mainly upon the sulfate concentration, the TDS, the nitrate concentration, and the total exchange capacity of the resin. Because the resin has only a limited exchange capacity

and, in low-TDS waters, sulfate is preferred over nitrate on the resin, these factors will determine to a large degree the cost of ion exchange.

Chapter 10 introduced the use of equilibrium expressions to predict exchange capacities for particular ions based on a given water quality and resin properties. In the specific case of nitrate removal in a water with TDS less than 1000 mg/L, nitrate will be the second most preferred ion by the resin in competition with sulfate, chloride, and bicarbonate, sulfate being most preferred. As shown in Example 10-3, it is possible to predict nitrate removal for a given water quality based on equilibrium expressions. Work done by Anderson (1978) and Guter (1981) have shown that these equilibrium estimates are comparable to actual column results.

The theory behind the development of a predictive equilibrium model for nitrate removal is based on the following assumption: at nitrate breakthrough (the time when nitrate ions begin to appear in the effluent from the ion exchange contactor), the column of resin will contain two zones. The upper portion will have a composition in equilibrium with the influent water. The lower portion will have a composition in equilibrium with the water coming from the upper zone. Because sulfate is so highly preferred on the resin over nitrate and the other monovalent anions, there will be no sulfate in the lower zone. Therefore, the volume of influent necessary per volume of resin (bed volumes) to reach this equilibrium condition can be calculated as done in Example 10-3.

Based on these assumptions, Anderson (1979) derived the following equation predicting nitrate breakthrough:

$$V = \frac{\bar{C}}{a + b + b/n} \text{ bed volumes} \quad (15\text{-}1)$$

where $n = K_{Cl}^{NO_3} \dfrac{0.8a + b}{C - 0.8a - b}$

$\bar{C}$ = total capacity of resin, eq/L
C = total solution concentration, eq/L
a = sulfate concentration, eq/L
b = nitrate concentration, eq/L
$K_{Cl}^{NO_3}$ = selectivity for nitrate over chloride, typically 4

This was derived using a generalized case of Example 10-3.

This equation states that the capacity for nitrate

will be directly proportional to the total exchange capacity of the resin and will be roughly inversely proportional to the sum of nitrate and sulfate concentration. The selectivity for nitrate over chloride and the total solution concentration, C, will effect the value of n, but the volume of water treated until nitrate breakthrough will be relatively insensitive to n. It should be noted that if n is much greater than b, then b/n becomes insignificant. If n is much smaller than b, then b/n will become controlling. This will only happen when chloride represents a major fraction of the anions.

Note that even if the selectivity coefficient for nitrate over chloride could be increased, the increase would have little effect on nitrate capacity. An increase of $K_{Cl}^{NO_3}$ would also decrease the regeneration efficiency with sodium chloride, hence there would probably be an overall increase in treatment costs. The parameter $(0.8a + b)$ in n is a gross approximation of the nitrate concentration entering the lower portion of the resin bed, which should be somewhat less than $a + b$. If the particular case in Example 10-3 is reviewed, it would be appropriate to replace the 0.8 factor with 0.68 (this being $\bar{X}'_{NO_3}$), however, because n plays a minor role in the nitrate capacity, the approximation $(0.8a + b)$ will be sufficient for most cases. These estimates should be done only to determine feasibility and preliminary costs. Pilot studies should be done to obtain actual long-term operating data.

The selectivity for sulfate over nitrate does not appear in Eq. 15-1. Clifford (1976) determined that the sulfate–nitrate selectivity is nearly irrelevant in determining nitrate capacity in column studies. He discovered, in fact, that slight increases in nitrate retention are possible if the sulfate–nitrate selectivity is *increased*. This seems to go against logic, however, Clifford offered this explanation, which supports the assumptions Anderson used in developing his predictive model: "(1) all of the sulfate will be removed from the feed water regardless of its actual selectivity because it is the most preferred species, and, (2) high sulfate selectivity promotes a short, sulfate-rich zone near the column entrance in which almost no nitrate is removed, leaving essentially all of that species to compete with the lesser-preferred chloride in the second equilibrium zone of the column. This second zone is where nearly all of the nitrate is concentrated."

It should be noted that Eq. 15-1 and the discussion up to now has considered a strong-base resin regenerated completely to the chloride form prior to

nitrate removal. During regeneration with NaCl, it would take excessive amounts of regenerant to completely convert the resin to the chloride form because of the resin's preference for nitrate over chloride (see Figure 10-8). Therefore, some economical level of regeneration is chosen that will provide an acceptable nitrate breakthrough capacity. Chapter 10 describes this selection process. Because of the incomplete regeneration, leakage of nitrate will appear in the effluent before actual nitrate breakthrough if co-current regeneration is applied. Countercurrent regeneration will minimize this leakage by pushing the residual nitrate on the resin away from the exit of the column.

Anderson (1978) developed estimates of regeneration requirements based on influent water quality. Figure 15-2 shows these estimates, while Table 15-2 lists the various water qualities and also includes an estimate of the nitrate breakthrough capacity based on Eq. 15-1. The regeneration estimates were based, in part, upon preliminary column studies. For waters B, C, and D, the nitrate capacity is relatively insensitive to the amount of salt used above 5 lb/ft^3 (175 percent of theoretical capacity). This is because the major anionic load on the resin is sulfate and sulfate can be eluted almost stoichiometrically with NaCl. Water A will require a higher salt dosage to remove a greater amount of nitrate. However, this dosage would be acceptable because the nitrate capacity is so much higher than B, C, or D. These

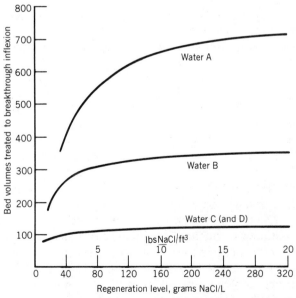

FIGURE 15-2. Nitrate removal versus salt dosage for waters listed in Table 15-2.

TABLE 15-2. Estimated Resin Performance on Four Typical Waters (Fully Regenerated Duolite A-101D; Capacity of 1.3 eq/L)

Water	A	B	C	D
Anionic constituents, meq/L				
Cl^-	1.04	4.03	2.8	2.0
HCO_3^-	0.25	0.98	1.6	1.6
$SO_4^=$	0.31	2.15	8.3	8.3
NO_3^-	1.21	1.21	1.2	2.0
Total anions	2.81	7.47	13.9	13.9
Estimated volume to break-through inflexion,				
bed volumes	706	345	122	115
gal/ft^3	5300	2580	910	860
NO_3^- capacity, eq/L	0.85	0.42	0.15	0.23

SOURCE: Anderson (1978).

estimates can be used to develop preliminary cost estimates and backwash inventories for regeneration.

Design Considerations. Based on the above-mentioned knowledge and data from small column or pilot studies, it is possible to design a full-scale nitrate removal facility. The initial design steps would include choosing the proper volume of resin and deciding what will be done during regeneration of a column (will there be a stand-by column or will storage be available?). Because nitrate removal is sensitive to flowrate, it has been recommended that volume loading rates be kept under 2.5 gpm/ft^3 where possible. The kinetics of exchange are such that chloride is favored under high flowrate conditions. Excess nitrate leakage and reduced capacity could occur under high loading rates. Additionally, the depth of the resin in the column should be kept above 2 ft. A decision must be made as to what percentage of the flow will have to be treated. A portion of the flow could go untreated and blended with treated water to achieve a nitrate level under the standard. To do this properly, the engineer must determine the acceptable nitrate leakage of the treated water based on a certain regenerant dosage.

An obvious need for a nitrate removal facility is a continuous and accurate method for determining the effluent nitrate concentration. This is necessary because once nitrate breakthrough occurs, the effluent concentration could far exceed the influent concentration for some period of time (see Figure 10-9). There are ion-specific nitrate probes on the market, but they have not been tested on a continuous basis.

Presently, they require maintenance and calibration fairly often.

The design salt dosage and concentration (determined in pilot tests) will allow for proper sizing of the brine tank. The concentration should be kept as high as possible to minimize waste volumes. Five percent to 10 percent NaCl is a typical range. Once the regeneration requirements are known, the regeneration time can be determined. This will include backwash, regeneration, slow rinse, and fast rinse. Criteria for these are listed in Chapter 10. From this, backwash volumes can be determined. This could be a high operating and maintenance cost if cheap disposal is not available. A facility near the coast may be able to cheaply dispose of the brine waste into the ocean, while other areas will have to consider lined evaporation ponds, if possible, or hauling.

There are many ion exchange equipment manufacturers that can provide the necessary equipment for a specific design. Complete skid-mounted packages including vessels, piping, valves, regeneration pumps, brine tanks, controls, and other appurtenances are available. It may be possible to adapt a commercial water softener by replacing the resin and installing adequate controls if the treated water requirement is not too large.

FLUORIDE REMOVAL

Various treatment techniques for fluoride removal have been studied and there appears to be only one method, ion exchange with activated alumina, that can economically reduce fluoride levels to below drinking water standards. However, both lime softening and alum coagulation are capable of reducing fluoride levels to some degree.

Lime Softening

Lime softening will remove fluoride from water both by forming an insoluble precipitate and by coprecipitation with magnesium hydroxide [$Mg(OH_2)$]. The relatively high solubility of calcium fluoride (roughly 10 mg/L at pH 10) limits the reduction of fluoride by softening alone; but it can reduce extremely high fluoride levels (such as may be found in mining discharges) to more acceptable levels. Waters high in magnesium that would be softened anyway can have their fluoride levels reduced appreciably by coprecipitation. A theoretical equation has been developed to predict the amount of magnesium required to reduce the fluoride level from a given value (Culp and Stoltenberg, 1958; Boruff, 1934):

$$F_{residual} = F_{initial} - (0.07\ F_{initial} \times \sqrt{Mg})\quad (15\text{-}2)$$

with all concentrations in milligrams per liter. This equation suggests that to reduce the fluoride level from 5.0 to 1.5 mg/L would require 100 mg/L of magnesium. This is a high level of magnesium to occur naturally in a water. The other option, which is not feasible at all, is to add magnesium for softening. The other drawback to this process is the large amounts of sludge produced.

Alum Coagulation

Alum coagulation will reduce fluoride levels to acceptable drinking water standards, but requires very large amounts of alum to do so. Studies conducted by Culp and Stoltenberg (1958) showed that fluoride was reduced from 3.6 to 1.4 mg/L using 250 mg/L of alum during conventional treatment. The pH for fluoride removal by coagulation seemed to follow optimal pH requirements for typical alum flocculation, that is, pH 5.5–7.0. While alum coagulation may not be the sole answer to reducing fluoride levels, this method could be used as a trimming function. The limit to this method, as it is with softening, is the large amounts of sludge that would be produced.

Activated Alumina

Since the 1930s, it has been known that contact of fluoride-containing water with activated alumina would remove fluoride. Continued research into this phenomena has led to increased knowledge about the removal of fluoride by activated alumina. In fact, there are a few full-scale fluoride removal treatment plants in operation in the United States. Activated alumina is used in much the same way as ion exchange resins. Strong-base anion exchange resins have a very low affinity for fluoride ($K_{Cl}^{F} <$ 0.1). The operating experience from the full-scale plants plus many other bench-scale studies have developed criteria for fluoride removal by activated alumina.

As explained in Chapter 10, it is thought that activated alumina removes certain species from water due to hydrolytic adsorption. It is an amphoteric substance and its isoelectric point is approximately pH 9.5. It will remove anions below this pH and

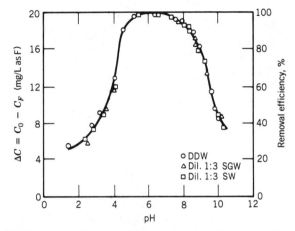

FIGURE 15-3. Effects of pH on the removal efficiencies of fluoride in deionized, distilled water (DDW) and in the fourfold dilutions of seawater (dil. 1:3 SW) and simulated geothermal water (dil. 1:3 SGW) by activated alumina: $C_0 = 20$ mg/L of F; adsorbent dosage = 25 g/L (Choi and Chen, 1979). Reprinted with permission.

cations above. The affinity of alumina for an anion seems to be inversely related to the solubility of its aluminum salt. Therefore, when treated with an acid solution, alumina behaves like an anion exchanger and fluoride is very high on the selectivity list.

Activated alumina can be regenerated with NaOH. Hydroxide ions are the most preferred anions by alumina and will readily replace the fluoride from the alumina given enough time.

The following characteristics of fluoride removal by activated alumina have been determined:

1. pH Effects. From Figures 15-3 and 15-4, it appears that optimum removal of fluoride occurs in the range of pH 5 to pH 8. Figure 15-3 shows the

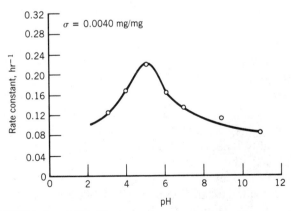

FIGURE 15-4. Fluoride removal rate constant as a function of pH. (From Wu, 1978.) Reprinted with permission.

effects of varying pH on the equilibrium capacity for fluoride, while Figure 15-4 shows that the most rapid uptake of fluoride occurs at pH 5.5. Full-scale work (Rubel, 1978) confirmed that pH adjustment down to 5.5 increases the fluoride capacity of activated alumina.

2. CAPACITY. The capacity for fluoride in a continuous-flow column will depend on three important factors: the initial fluoride concentration, the influent pH, and the particle size of the activated alumina. Each of these will be briefly discussed.

A. *Initial fluoride concentration.* Studies (Choi, 1979) show that the capacity of activated alumina for fluoride increases with increasing concentration. In side-by-side tests, water with 10 and 20 mg/L F^- were each reduced to less than 1 mg/L F^- for roughly the same volume of water. It appears that activated alumina, being a kinetically slow material, will allow a greater penetration of ions into unused sites as the concentration gradient is increased.

B. *pH.* As shown before, the pH of the influent water will influence the volume of water treated until breakthrough. A pH of 5.5 appears to be optimum.

C. *Activated Alumina Size.* Commercially available alumina is available in four typical size ranges, 8–10, 14–28, 28–48, and 48–100 mesh, from largest to smallest. The kinetics of removal are such that the smallest particle sizes provide the most rapid uptake of fluoride. Some trade-off will have to be made because the smaller particles will have a greater tendency to be washed out of the bed during backwash, and they are more susceptible to rapid dissolution by NaOH. Material of a 28–48 mesh is most often used.

Full-scale tests (Rubel, 1978) indicate that up to 6.7 g/L (3000 grains/ft³) of fluoride were removed at a pH of 5.5 and an initial fluoride concentration of 5 mg/L.

3. KINETICS. The minimum recommended empty bed contact time (EBCT) for fluoride removal is 5 min (Rubel, 1978). For example, a 0.028-m³ (1-ft³) bed of alumina (28.3 L; 7.48 gal) should have a maximum flow rate of 0.095 L/sec (1.5 gpm) to provide at least 5 min of EBCT between the water and the alumina.

4. REGENERATION. The steps involved in eluting the fluoride from the activated alumina are very important in the overall process.

A. *Backwash.* Upon termination of the treatment run (there are reliable commercially available fluoride-specific probes that can monitor the effluent concentration), the bed should be backwashed up to 50 percent bed expansion for 10–15 min to remove trapped suspended material, break up the packed bed, and redistribute the media.

B. *Regeneration.* Bench-scale or pilot-scale tests should be done to determine the specific design criteria for NaOH elution of the fluoride. As explained in Chapter 10, it will be necessary to choose the NaOH concentration, dosage, and flowrate to economically prepare the bed for another fluoride removal cycle. Typical values for NaOH concentration range from 0.5 to 2 percent, while flowrates of 0.25–3 gpm/ft³ (0.56–6.7 L/sec/m³) for 1–1.5 hr are typical. One item to note is that activated alumina is dissolved to some degree by NaOH.

C. *Neutralization.* Once the alumina has been eluted with NaOH, the bed must be returned to an acidic condition to allow removal of fluoride and other anions and to prevent prolonged high pHs in the effluent. First, excess caustic soda must be rinsed from the bed (raw water can be used for one to two bed volumes at the NaOH flowrate). Then the bed must be rinsed with an acid solution, either H_2SO_4 or HCl. The simplest method for doing this is to pass raw water with its pH adjusted to 2.0 or 2.5 through the bed at the treatment flowrate while monitoring the effluent pH. This effluent should be disposed with the regenerant. When the pH of the effluent drops to the desired level, the water can then be fed into the system again while the influent pH of the water is changed to the desired level.

The disposal of the regenerant waste is a problem inherent with all ion exchange processes. Because most operating fluoride removal facilities are located inland, the typical method for treatment of this brine is to discharge it into a lined evaporation pond, then haul the dried material to an appropriate dump site. If evaporation is not feasible, then the waste must be disposed in some other manner. The waste disposal costs will likely be a large part of the overall cost for operating and maintaining a fluoride removal facility.

Due to the small number of existing full-scale fluoride removal facilities, standard operating procedures and design criteria for activated alumina have not been determined as have those with ion exchange resins. Rather, it is recommended that adequate pilot testing be done prior to any full-scale design to determine the design and operating criteria. Once the full-scale facility is built, it is recommended that adequate monitoring be conducted to ensure that the facility is operating as planned.

ARSENIC REMOVAL

Chemistry

Prior to discussing the treatment methods available for removing arsenic from water, it is essential to discuss some of the chemistry behind aqueous arsenic solutions. Arsenic can occur in four oxidation states in water (+5, +3, 0, −3), but is generally found in only the trivalent and pentavalent states. Figure 15-5 shows distribution diagrams for arsenite, As(III), and arsenate, As(V), as a function of

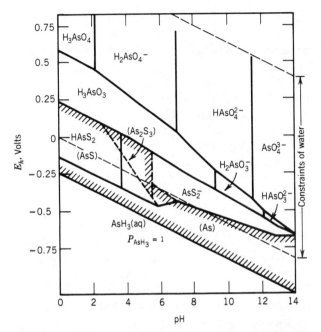

FIGURE 15-6. The E_h–pH diagram for As at 25°C and 1 atm with total arsenic 10^{-5} mol L^{-1} and total sulfur 10^{-3} mol L^{-1}. Solid species are enclosed in parentheses in cross-hatched area, which indicates solubility less than $10^{-5.3}$ mol L^{-1}. (Ferguson and Gavis, 1972). Reprinted with permission.

pH. H_3AsO_3, an undissociated weak acid, is predominant in the pH range of 2–9. Therefore, any As(III) present in a typical water supply would occur as H_3AsO_3. As(V), on the other hand, will occur as a strong acid and dissociates according to the diagram in Figure 15-5. Note that $HAsO_4^{2-}$ predominates from pH 7 to pH 11.5, indicating that this would most likely occur in normal water supplies. At a pH less than 7, $H_2AsO_4^-$ dominates.

Figure 15-6 is an E_h–pH diagram for arsenic in a system including oxygen, H$_2$O, and sulfur. This diagram represents the equilibrium conditions of arsenic under various redox potentials. Well-aerated surface waters would tend to induce high E_h values, therefore, any arsenic present should be in the arsenate [As(V)] form. Mildly reducing conditions, such as can be found in well water, should produce arsenite [As(III)]. Note the two dashed lines representing the limits that constrain water.

Conventional Coagulation

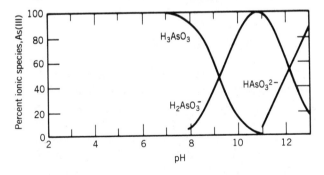

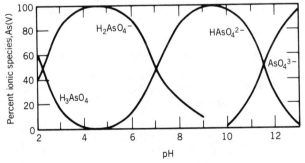

FIGURE 15-5. Predominance diagram for As(III) and As(V) as a function of pH. (Gupta and Chen, 1978).

In light of this chemistry, some of the treatment methods for removing arsenic can be reviewed. Conventional coagulation with iron or aluminum salts is effective for removing greater than 90 percent of As(V) (with initial concentrations of roughly 0.1 mg/L) at pH of 7 or below. Above pH 7, iron coagulation will still remove As(V) effectively, but alum coagulation efficiency drops off. Figures 15-7 and 15-8 show this. Lime softening is very efficient for As(V) (>95 percent removal) at pH above 10.5. These processes are not as successful with As(III). Iron coagulation will remove roughly 50 percent, while alum coagulation will not reduce As(III) levels more than 20 percent. Lime softening above pH 11 can remove up to 80 percent of As(III) (Sorg and Logsdon, 1978). Note that at pH 11, almost 100 percent of the As(III) is present as the anion $H_2AsO_3^-$ instead of H_3AsO_3. Increased removal of As(III) can be accomplished by oxidation of the As(III) to As(V) prior to the treatment. Chlorine or potassium permanganate will easily do this.

The most common type of water supply that is contaminated with arsenic is well water. The above-mentioned treatment methods are typically practiced for turbid surface waters or hard waters; the expense of this treatment for a well supply may not be warranted. Ion exchange treatment using activated alumina, or possibly a strong base anion exchange resin, can also remove arsenic from water.

Activated Alumina

As with fluoride, As(V) is strongly adsorbed/exchanged by activated alumina. In fact, column studies indicate that arsenic is retained better than fluoride (Bellack, 1971; Rubel and Williams, 1980). On one hand, this will lead to higher capacities, however, it will require higher doses and higher concentrations of NaOH to elute an acceptable amount of As(V) from the alumina. Rubel and Williams found that a 4 percent solution of NaOH was required to

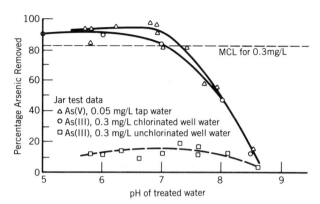

FIGURE 15-8. Effect of pH on arsenic removal by alum (30 mg/L) coagulation (Sorg and Logsdon, 1978). Reprinted with permission.

elute the arsenic from the alumina; a 1 percent solution was used to elute the fluoride. Activated alumina has an equilibrium capacity for As(III) up to 10 times less than that for As(V) (Gupta and Chen, 1978). The uncharged nature of the As(III) molecule at pH below 9 is probably the cause.

An activated alumina process for removing arsenic from a drinking water supply would be operated and designed very similarly to a fluoride removal facility. The process would include removal, backwash, regeneration, neutralization, and rinse steps. Pilot studies would have to be conducted to develop design and operating criteria such as optimum pH, operating capacity, removal flowrate, and regeneration and neutralization steps.

Gupta and Chen (1978) showed that equilibrium capacities for As(V) were maximized at pH less than 7, while As(III) was removed best at pH less than 9. As with fluoride, this equilibrium data might not take into account kinetic effects that would dominate in a column operation. Because of activated alumina's poor (slow) kinetics, an optimum pH for column operations should be determined. The speciation of arsenic must be determined prior to removal studies. While the atomic adsorption spectrophotometer (AAS) can detect total arsenic, it cannot speciate. Therefore, care must be taken to accurately determine the speciation using another test in addition to the AAS. If As(III) is found to predominate, then preoxidation should be considered to convert As(III) to As(V).

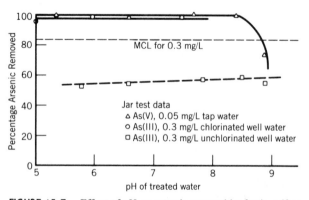

FIGURE 15-7. Effect of pH on arsenic removal by ferric sulfate (30 mg/L) coagulation (Sorg and Logsdon, 1978). Reprinted with permission.

Ion Exchange with Strong-Base Resins

Limited testing has been done with strong-base anion exchange resin. As(V) present as the divalent anion $HAsO_4^{2-}$ appears to be preferred on the resin over monovalent anions in the water supply. Although equilibrium calculations predict a low capacity for trace constituents because of their low equivalent fraction, actual kinetic studies may show an acceptable capacity for $HAsO_4^{2-}$. The benefit in using a strong-base resin would be during regeneration, where sodium chloride could be used instead of caustic soda followed by an acid neutralization. The initial cost of the resin will probably be higher than activated alumina, but the lower cost of NaCl and its easier handling may make it cheaper in the long run. This process is being studied further.

SELENIUM REMOVAL

Chemistry

Similar to arsenic, the chemistry of aqueous selenium species is important in discussing removal techniques. Two oxidation states, selenium (IV) and selenium (VI), are predominant in water. Figure 15-9 shows the dependence of pH on the forms of the two oxidation states of selenium. Se(IV), as the weak acid H_2SeO_3, will be present as $HSeO_3^-$ in the normal pH range of most waters. Se(VI), as the strong acid H_2SeO_4, will be SeO_4^{2-} above pH 1.5. Note the similarity between Se(VI) and sulfur. In fact, the solubility of most selenate salts is similar to the solubility of sulfate salts. Most selenate salts are more soluble than selenite salts.

Figure 15-10 shows the E_h–pH diagram for selenium. The shaded region represents the conditions likely to occur in normal waters. Well-aerated surface waters containing selenium would be expected to have SeO_4^{2-} in predominance, while mildly reduced waters would tend to have more $HSeO_3^-$ present.

Conventional Coagulation

Conventional treatment techniques using alum or ferric sulfate coagulation and lime softening have been investigated for selenium removal. Se(IV) removals of only 30 percent or less from an initial concentration of 0.1 mg/L were achieved using ferric sulfate coagulation in the pH range of 6–7. Increasing pH above this reduced removals appreciably. Alum coagulation was more effective for Se(IV) removal, with up to 85 percent being removed at pH 5.5. Higher pHs reduced alum's effectiveness (see Figure 15-11). Lime softening up to a pH of 11.5 removed only 50 percent of Se(IV). Neither coagulation or softening techniques could

achieve greater than 10 percent removal of Se(VI) (Sorg and Logsdon, 1978).

Activated Alumina

Activated alumina was investigated for its potential to remove Se(IV) and Se(VI) (Trussell et al., 1980). Se(IV) is removed well from water by activated alumina. The following capacities for Se(IV) were estimated at an influent concentration of 0.10 mg/L:

pH 5: 1200 bed volumes

$$\left(\frac{120 \text{ mg Se(IV)}}{\text{liter of activated alumina}} \right)$$

pH 6: 900 bed volumes (90 mg/L)

pH 7: 500 bed volumes (50 mg/L)

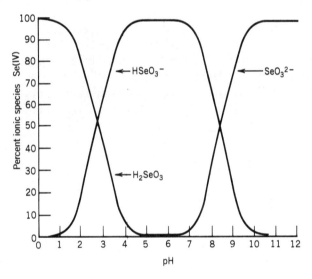

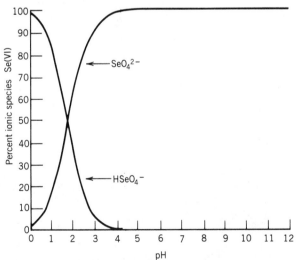

FIGURE 15-9. Predominance diagram for Se(IV) and Se(VI) as a function of pH.

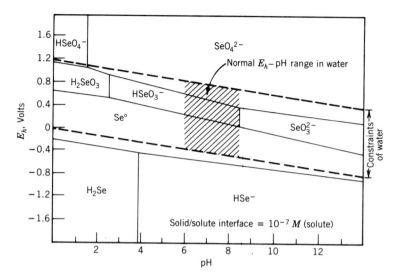

FIGURE 15-10. E_h–pH diagram for selenium.

These capacities were based on regenerating the alumina with 5 bed volumes of 0.5 percent NaOH at 0.34 L/sec/m² (0.5 gpm/ft²) followed by appropriate rinses and neutralization.

Se(VI) removal by activated alumina was poorer than for Se(IV). The following capacities for Se(VI) were estimated at an influent concentration of 0.10 mg/L:

pH 5: 100 bed volumes

$$\left(\frac{10 \text{ mg Se(VI)}}{\text{liter of activated alumina}} \right)$$

pH 6: 70 bed volumes (7 mg/L)

pH 7: 35 bed volumes (3.5 mg/L)

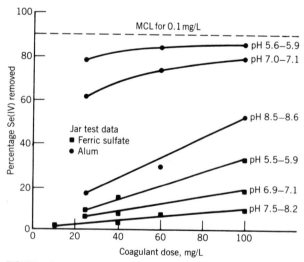

FIGURE 15-11. Effect of coagulant dose on selenium(IV) (0.1 mg/L) removal from well water (Sorg and Logsdon, 1978). Reprinted with permission.

Because Se(VI) is not retained as well as Se(IV) on alumina, NaOH requirements are less than those described for Se(IV).

The above capacities for Se(IV) and Se(VI) were based on a water quality with 100 mg/L sulfate and the alkalinity ranging from 20 to 200 mg/L as $CaCO_3$ depending on pH. It was noted that the Se(VI) capacity on activated alumina was greatly increased by reducing the sulfate concentration and marginally increased by reducing the alkalinity. The following Se(VI) capacities were estimated for waters having varied sulfate concentrations at a pH of 6:

Sulfate (mg/L)	Se(VI) Capacity (Bed Volumes)
500	5
100	70
50	150
5	450

From this work, it was noted that neutralization of an NaOH-regenerated bed would have to be done with HCl since H_2SO_4 would put sulfate ions on the bed and therefore interfere with Se(VI) removal.

The speciation between Se(IV) and Se(VI) must be determined before removal tests are conducted. Activated alumina appears to be feasible only for Se(IV) removal unless the sulfate concentration in the water is very low, thereby allowing for Se(VI) removal also.

Ion Exchange with Strong-Base Resins

Although strong-base anion exchange resins have not been thoroughly investigated for selenium removal, one study (Maneval, et al 1985) determined that ion exchange could be a practical method for selenium removal. On strong-base resins, $SeO_4^=$ is preferred slightly more than $SO_4^=$. Therefore, the removal of Se(VI) (as $SeO_4^=$) will continue as long as sulfate is being removed by ion exchange. Estimates of capacity for sulfates can be made using expressions given in Chapter 10. Se(IV) is less preferred on strong-base resins than nitrate (as $HSeO_4^-$), but more than chloride or bicarbonate hence the amount of Se(IV) which can be removed is dependent on sulfate and nitrate concentrations in the water supply. Se(IV) could be oxidized to Se(IV) with chlorine prior to ion exchange.

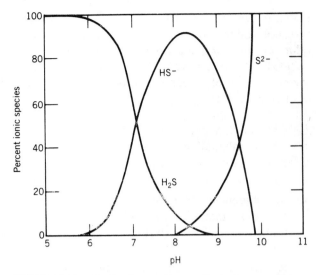

FIGURE 15-12. Predominance diagram for hydrogen sulfide—sulfide as a function of pH. (32 mg H_2S/L) (Sawyer and McCarty, 1978). Reprinted with permission.

SULFIDE REMOVAL

Chemistry

Hydrogen sulfide (H_2S), though frequently associated with the handling and treatment of wastewater, is occasionally present in groundwaters. Its readily identifiable rotten egg odor is the major reason it is not desirable in drinking water. However, it also is a cause for marked corrosiveness of waters to metal and concrete. Sulfides also promote the growth of various filamentous sulfur bacteria, leading to a general degradation of water quality. Sulfides are formed by the anaerobic reduction of sulfates and organic matter into sulfides, and bisulfides by the following reactions:

$$SO_4^{2-} + \begin{array}{c} \text{organic} \\ \text{matter} \end{array} + \begin{array}{c} \text{anaerobic} \\ \text{bacteria} \end{array} \rightarrow$$

$$S^{2-} + CO_2 + H_2O \quad (15\text{-}3)$$

$$S^{2-} + 2H^+ \rightleftharpoons HS^- + H^+ \rightleftharpoons H_2S \quad (15\text{-}4)$$

Equation 15-4 is pH dependent and Figure 15-12 shows the effect of pH on the sulfide equilibria. As can be seen, hydrogen sulfide predominates at pH less than 7, bisulfide ion dominates between pH 7 and 9.5 and sulfide ion is predominant above pH 9.5. Hydrogen sulfide is the species that causes the odors. Concentrations of up to 40 mg/L have been noted in literature.

Oxidation

Odor problems from sulfide-bearing waters have typically been treated by aeration or oxidation by chlorination. The assumed reactions are as follows:

$$2S^{2-} + 2O_2 \rightleftharpoons SO_4^{2-} + S^0 \quad \text{(solid)} \quad (15\text{-}5)$$

$$H_2S + Cl_2 \rightleftharpoons 2HCl + S^0 \quad \text{(solid)} \quad (15\text{-}6)$$

It has been assumed that if enough chlorine was added to satisfy the natural chlorine demand of the water plus the chlorine demand of the sulfide, free sulfur would be produced. This free sulfur is colloidal in nature and imparts to the water a milky blue turbidity. Unfortunately, it has been observed that waters treated for sulfide removal by oxidation followed by filtration to remove the free sulfur turbidity still exhibit a certain amount of sulfurous odor, not necessarily hydrogen sulfide-like, in hot water. The reasons for this not being uncovered earlier can probably be attributed to two things: (1) threshold odor tests are rarely made on treated waters after chlorine addition and (2) samples taken from the distribution system that exhibit an odor are typically considered to be caused by degradation within the system.

Observations made in Santa Barbara, California (Monscvitz and Ainsworth, 1969) suggested that the presence of these odors was due to the formation of polysulfides (HS_n^-) during the oxidation reaction. Although the quantity of the polysulfides produced is typically very small, they have very low threshold odor levels and will produce noticeable odors even when it appears that all of the hydrogen sulfide has been removed. These observations led to the suggestion of a four-step process to ensure an odor-free water:

1. Chlorination and/or aeration to convert the sulfide to colloidal sulfur and polysulfides.
2. Filtration to remove a majority of the colloidal sulfur.
3. Subsequent dechlorination with sulfur dioxide or sodium metabisulfite, which will convert the remaining colloidal sulfur and polysulfides to thiosulfate.
4. Rechlorination to convert the thiosulfate to sulfate.

The filtration step may not be necessary if there is not too much sulfide present in the water and the water to be treated would not be filtered otherwise.

The following equations express steps 3 and 4 above:

$$S^0 + SO_3^{2-} \rightleftharpoons S_2O_3^{2-} \qquad (15\text{-}7)$$
$$\updownarrow$$
$$HS_n^-$$

$$S_2O_3^{2-} + HOCl \rightleftharpoons S_4O_6^{2-} + SO_4^{2-} \qquad (15\text{-}8)$$

Equation 15-7 is very rapid and should require virtually no more than good mixing for up to 10 min after the addition of the sulfite. Rechlorination (Eq. 15-8) will form tetrathionate ($S_4O_6^{2-}$) in addition to sulfate, but this will convert rapidly to sulfate and will not cause an odor problem.

Chen (1973) has shown that polysulfides do not appear to form above pH 9. This may be one reason why many hard groundwaters that probably contain sulfides do not exhibit threshold odors after chlorination. These waters have been treated typically by lime softening and will not promote formation of polysulfides.

IRON AND MANGANESE REMOVAL

Chemistry of Iron

The basic principles of iron and manganese removal are similar. However, the widely varying chemistry of these compounds makes them difficult to discuss

in parallel. For that reason the chemistry will be reviewed, first for iron and then for manganese.

The chemistry of iron in natural water systems involves a number of factors ranging from the E_h to the impact of organic complexing agents. The highly complex nature of the water chemistry of these species serves to complicate the literature on the subject. As a result, the typical phenomenological approach utilized in the water treatment studies, wherein chemicals are added and percent removals are observed, frequently does more to confuse the state of the art than to advance it. A number of factors such as pH, which has been indicated by Stumm and Lee (1961) to increase the rate of ferrous iron oxidation by 100-fold per pH unit, bicarbonate, sulfate, and dissolved silica, which were indicated by Schenk and Weber (1968), have significant impacts on the rate of oxidation.

For instance, the fate of iron in oxidative environments is not clearly understood. It is normally assumed that a hydroxide precipitate is formed following oxidation of iron. However, depending on the amount of carbonate alkalinity (>250 mg/L as $CaCO_3$; O'Connor, 1969) the chemistry may be such that ferric carbonate is formed rather than the hydroxide (Cleasby, 1975). Based on Cleasby's analysis, it appears that the more rapidly the iron is oxidized, that is, through the use of strong oxidants such as permanganate, chlorine, or ozone, the more likely the end product will be the hydroxide. However, when the oxidation is conducted more slowly through the use of aeration, it then appears that the most likely end product would be carbonate in a water having a high alkalinity.

Further complicating the reaction system for iron is the impact of organic complexing agents. Various humics and similar materials can act to complex the iron and to slow down the kinetics of the oxidation. As a result, frequent reports in the literature will indicate that a particular approach, which has been effective in numerous locations, was not successful at that time. This is generally due to the inability to break the complex at a sufficiently rapid rate to allow the oxidation of the iron within the treatment system. The solution to problems of this nature is to more rapidly and completely oxidize the system to ensure that the system is kept at a very high state of E_h. In a highly oxidative state, the complex can be broken and the iron can be oxidized to ferric readily. In general, the use of aeration has been shown to be totally ineffective in complex waters of this nature (Oldham and Gloyna, 1968).

To ensure the complete and effective removal of iron, it is essential that the right combination of oxidizing agents be utilized in the right pH environment to produce total oxidation to the ferric state. In terms most commonly associated with chlorination, it is essential that a condition analogous to the "breakpoint" be reached during oxidation. That is, the concentration of the oxidant must exceed the demand, with a slight excess of oxidant remaining. The mechanisms for oxidizing iron will be discussed in the subsequent section.

Chemistry of Manganese

The chemistry of manganese is substantially more complex than that of iron, and only a limited understanding of manganese oxidation exists. As indicated by Kessick and Morado (1975), the most common precipitate that forms from manganous oxidation is $MnOOH$ rather than MnO_2, which has typically been reported as the result of manganous oxidation.

The oxidation and control of manganese is complicated by factors that range from misunderstanding of the reaction chemistry to the relatively slow kinetics and the numerous oxidation states that result from this oxidation. Manganese precipitation is also affected by the formation of organic complexes. Prior to oxidation of the manganese, these complexes must be broken down in the same manner as discussed for iron. In general, the removal of manganese is enhanced by increased pH and by increasing the E_h. Fortunately, manganese removal, although substantially more complex than that of iron, is less of a problem because the frequency of occurrence of manganese in water systems is less than that for iron.

Further adding to the complexity of manganese chemistry is the fact that there are difficulties with analytical techniques, as indicated by Morgan and Stumm (1965). The gross analytical techniques do not allow for differentiation of the speciation of the manganese. Therefore, presumptions are made with respect to the oxidative states of the manganese without thorough knowledge of its speciation. As a result, an incorrect approach may be utilized in the oxidative system design. A better understanding of the speciation of the manganese would help engineers approach the design of manganese removal systems more efficiently.

Oxidation and Removal

The basic approach utilized to remove iron and manganese involves oxidation and removal of the suspended material by either sedimentation or filtration. The single most important factor in the control of iron and manganese oxidation is that sufficient oxidation be conducted with sufficient detention time to allow for complete and efficient removal. Without a doubt, this is the most significant problem in most iron and manganese removal system designs. In general, the designer has a vague understanding of the reaction chemistry of iron and manganese and, therefore, is subject to making errors with respect to judgment on the detention time required for the oxidative reaction of iron and manganese.

Table 15-3 shows the relative oxidative requirements to achieve precipitation of iron and manganese. From our experience, the most successful approach for removal of iron and manganese involves pH adjustment, chlorination, and direct filtration on a monomedia anthracite filter. Figure 15-13 shows a typical iron and manganese removal system. This approach ensures that the reaction kinetics are taken into account and that proper filtration techniques are applied to remove the oxidized iron and manganese. Observing the discharge following the oxidation step indicates that even after 5 min of detention time, the water appears to be clear and still contains a substantial amount of microscopic

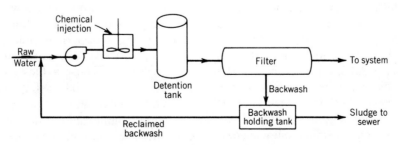

FIGURE 15-13. Typical iron and manganese removal system.

TABLE 15-3.   Iron and Manganese Chemistry

Element	Precipitate	Conditions
Fe^{2+}	$Fe(OH)_2$	In absence of O_2, CO_3^{2-} (pH $\geq$ 10)
Fe^{2+}	$FeCO_3$	In absence of O_2, S^{2-} (pH $\geq$ 8, alk > 10^{-2} eq/L)
$Fe^{2+} \rightarrow Fe^{3+}$	$Fe(OH)_3$	1. $4Fe^{2+} + 2H^+ + O_2 \rightarrow 4Fe^{3+} + 2OH^-$ or 7 mg Fe/mg O_2 2. $2Fe^{2+} + Cl_2 \rightarrow 2Fe^{3+} + 2Cl^-$ or 1.6 mg Fe/mg Cl_2 3. $3Fe^{2+} + MnO_4^- + 4H^+ \rightarrow 3Fe^{3+} + MnO_2 + 2H_2O$ or 1.06 mg Fe/mg MnO_4
$Mn^{2+} \rightarrow Mn^{4+}$	$MnOOH$	1. $2H^+ + Mn^{2+} + \frac{1}{2}O_2 \rightarrow Mn^{4+} + H_2O$ or 3.5 mg Mn/mg O_2 2. $Mn^{2+} + Cl_2 \rightarrow Mn^{4+}\ 2Cl^-$ or 1.3 mg Mn/mg Cl_2 3. $3Mn^{2+}\ 2Mn^{7+} \rightarrow 5\ Mn^{4+}$ or 0.52 mg Mn/mg MnO_4

colloidal (i.e., uncoalesced) iron and manganese. Frequently, no observed precipitation will occur until the solution reaches the filter bed, wherein previously precipitated iron and manganese act as a catalyst for precipitating the compounds on the media. During backwash, substantial quantities of dark red and black precipitates are observed, indicating that an efficient removal of iron and manganese has occurred. It is interesting that even though the reacting bed is required to complete this reaction, it does not appear to have excessive buildup of material on the media, which would eventually cause clogging and failure of the filter due to high head loss. Operation studies (JMM, 1977) indicate that the filter can run at high rates in excess of 5 gpm/ft^2 for extended periods without excessive head loss buildup when iron concentrations are less than 0.5 mg/L. A study conducted by JMM (1982), indicated that a water containing 2.4 mg/L of iron and 0.5 mg/L of manganese did cause excessive head loss in the filter due to suspended solids accumulation. It should be noted that each milligram per liter of iron results in 2 mg/L of suspended solids following oxidation. Thus, a water such as this one with 2.4

mg/L of iron would have nearly 5 mg/L of suspended solids following oxidation. These levels of suspended solids are approaching the upper limit for direct filtration. As a result, it is clear that this is a viable approach and should be considered in all iron and manganese removal plants.

Design Considerations.   The selection of media for the filtration unit is important. The media should have a large effective size (>1.5 mm) to reduce head loss and should have a low uniformity coefficient. Coal works well for this application because lower backwashing velocities can be utilized to clean the bed due to its low specific gravity. In general, the use of greensand, a natural zeolite which allows for more rapid oxidation and removal of iron and manganese, is not recommended because of the high head loss due to its small effective size. Sand or coal will generally acclimate within 3 weeks and produce high-quality water with low iron and manganese.

The greensand media is generally not justified for large water treatment systems. In small systems, the use of the greensand has some significant advantages in that it can be precharged with permanganate on an intermittent basis. In this manner, it can be utilized without continuous chemical feed for periods of time. Thus, a small unit can be conveniently charged and left without attendance for a period of several weeks. The permanganate which is adsorbed on the media acts as an adsorbent of iron and manganese.

When permanganate is used as the oxidizing agent, it is essential that a media such as greensand be utilized which will adsorb any excess permanganate during treatment. Otherwise, excess permanganate will bleed through as a pink color and will cause substantial consumer complaints in the water supply system. In the past, the use of permanganate was not justified because chlorine is a superior oxidizing agent and is more easily handled by the water treatment personnel. However, the formation of trihalomethanes with chlorine and the resulting use of alternative disinfectants may encourage the use of permanganate. Table 15-4 shows the design criteria recommended for iron and manganese control. These values are considered to be conservative and can be safely used in the design of iron and manganese removal systems.

Another significant consideration with respect to conventional iron and manganese removal systems is the recovery of the backwash water and the dis-

TABLE 15-4. Design Criteria for Iron and Manganese Control

Element	Units
1. Detention time	5–30 min at average flow after chemical feed
2. Filter media	1.5 mm effective size or greater anthracite
Uniformity coefficient	~1.2–1.4
Q/A	5 gpm/ft^2 to 10 gpm/ft^2
Backwash Q/A	15–25 gpm/ft^2
3. Backwash tank size	20 min flow at 5 times normal Q/A

posal of the sludge. Following backwash of the filtration system, the high-quality water, which has been utilized for backwashing, can be collected and settled to remove the precipitated iron and manganese.

Based on field experiences, approximately three-quarters of the backwash water can be reclaimed with a very simple control system that consists of two pumps. One pump is located at approximately three-quarters of the depth of the holding tank for the backwash water and the other is located in a sump to pump the sludge. After approximately 2 hr of settling, the first pump can be initiated and the backwash water can be returned to the system for subsequent treatment. Following this period, the sludge is removed and pumped to an appropriate holding basin or to the domestic sewer for discharge. The discharge of these materials into the sewer is a viable alternative and will act to control sulfides to some extent if they are produced in the wastewater system. Backwash reclamation reduces the volume of wastewater that must be disposed and should be considered for all systems.

Ion Exchange

The use of water softeners for iron and manganese removal is fairly common. Due to the divalent nature of these compounds, they are readily removed on ion exchange materials. One of the major difficulties involved in the control of iron and manganese with this method is that, if these materials are oxidized by the dissolved oxygen in the water, the media can become coated and fouled. As a result, it is very important to control the levels of dissolved oxygen in ion exchange beds which are used on waters containing iron and manganese levels above

the drinking water standards. It is possible to restore the original ion exchange capacity by dissolving the precipitated iron and manganese from the ion exchange materials with acid. However, this will reduce the useful life of the resin.

In-Situ Oxidation

In-situ oxidation for iron and manganese control involves a patented process called VYREDOX® (Hallberg and Martinell, 1976). In this process, oxygen, through the use of oxygenated water, is injected around the water supply well. The concept behind this treatment method is that the E_h is raised in the vicinity of the discharge of the well. Theoretically, a more oxidative environment will change the oxidation state of these materials and stop the dissolution of iron and manganese in the area of the well. Concerns with respect to the operation of this type of process involve the potential for contamination of the aquifer system, excessive bacterial growth in the area of the well, and subsequent clogging of the aquifer. Additional concerns include the impact of corrosion on the well screen and pumping system due to the introduction of oxygenated water into the aquifer.

The basic operation of the process involves the intermittent injection of oxygenated water approximately every 2 weeks for a period of 20 hr as indicated by the vendor. It is difficult to assess the true feasibility of this process at this time. It has reportedly been used extensively in Europe; however, the number of installations in the United States is limited.

Polyphosphates

Another iron and manganese control alternative, considered to be adequate only in special cases, involves the use of polyphosphates. Polyphosphates can be instituted quickly and economically to control the precipitated iron and manganese. The polyphosphate reacts to complex the iron and manganese and hold it in solution so that the consumer is not aware of its presence. Polyphosphates are dosed at approximately two times the concentration of iron and manganese (Cox, 1969). However, O'Connor (1969) indicated that a dose of sodium hexametaphosphate of 5 mg/mg of iron plus manganese is required. O'Connor also indicated that polyphosphates will not be suitable when the total iron and manganese concentration exceeds 1 mg/L.

TABLE 15-5. Chemical Costs for
Iron and Manganese Control

Oxidant	$/mil gal[a]
Cl_2[b]	2.10
$KMnO_4$[c]	5.75
Aeration	0.50
Chelation	
Polyphosphate[d]	32.00

[a] Concentration of Fe = 1 mg/L; Mn = 0.5 mg/L.
[b] $0.11/lb.
[c] $0.55/lb.
[d] $0.50/lb.

Table 15-5 shows the relative cost of iron and manganese removal utilizing various treatment methods.

HARDNESS REMOVAL

In the past the hardness of water was expressed as "temporary" or "permanent." Today these are referred to by the terms *carbonate* and *noncarbonate* hardness. The total hardness is the sum of the concentrations of the polyvalent cations, all expressed as the equivalent concentrations of calcium carbonate. Of the total, that part that is associated with the carbonate alkalinity is called carbonate and the rest as noncarbonate; thus, the total hardness is also the sum of the carbonate and noncarbonate hardnesses. Since the major polyvalent ions in potable water are calcium and magnesium, the sum of these two (expressed as calcium carbonate) is usually taken as the total hardness.

The distribution of hard waters in the United States is shown in Figure 15-14. Although waters above 150 mg/L hardness (as $CaCO_3$) are considered hard, many utilities do not soften the water. Other utilities soften when the total hardness exceeds 150 mg/L. The finished water hardness produced by a utility softening plant may vary from a low of 50 mg/L to a high of 150 mg/L depending upon economics and public acceptance. The AWWA Water Quality Goals (AWWA, 1968) recommended a range of 80–100 mg/L as desirable, both because of aesthetics and corrosion control.

Removal of the hardness, that is, softening, may be accomplished either by precipitation as insoluble compounds or by ion exchange. Softening may also be accomplished in conjunction with demineralization by reverse osmosis. Low-pressure reverse osmosis may be used specifically for softening of relatively low-TDS waters. Further discussion of reverse osmosis can be found in Chapter 10.

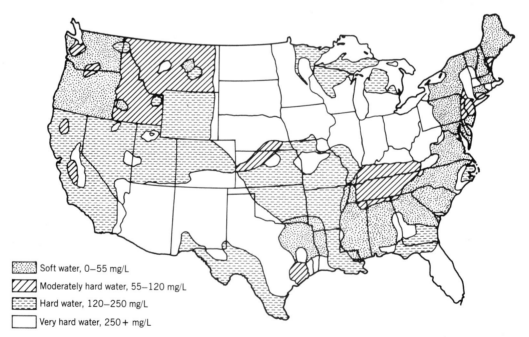

Soft water, 0–55 mg/L

Moderately hard water, 55–120 mg/L

Hard water, 120–250 mg/L

Very hard water, 250+ mg/L

FIGURE 15-14. Distribution of hard water in the United States. The areas shown define approximate hardness values for municipal water supplies. Reprinted with permission from Ciaccio, L. (ed.), *Water and Water Pollution Handbook*, Marcel Dekker, Inc. N.Y. (1971).

Precipitative Softening

In precipitative softening, use is made of the relative insolubilities of calcium carbonate and magnesium hydroxide. The choice of precipitating agents depends on the raw water quality, which complicates the selection of the optimum treatment process, although the chemistry itself is simple.

The equilibrium solubilities of calcium carbonate and magnesium hydroxide are shown in Figures 15-15 and 15-16. From these it can be seen that it is necessary to increase the pH for either calcium or magnesium precipitation; therefore, the compounds most commonly used are lime and caustic soda. The choice between the two depends on economics and water quality. When the carbonate hardness is adequate, the softening required can be accomplished with pH adjustment alone and both calcium carbonate and magnesium hydroxide can be precipitated. When the carbonate hardness is too low, the carbonate content must be supplemented by soda ash (sodium carbonate). The minimum hardness that can be achieved depends on the solubilities of the calcium hydroxide and magnesium hydroxide at the pH of softening as well as the subsequent processes.

Reactions. The relevant chemical reactions are shown below for the major precipitative softening alternatives.

Lime softening

$$CO_2 + Ca(OH)_2 \rightarrow \underline{CaCO_3} + H_2O \quad (15\text{-}9)$$

$$Ca(HCO_3)_2 + Ca(OH)_2 \rightarrow$$
$$2\underline{CaCO_3} + H_2O \quad (15\text{-}10)$$

$$Mg(HCO_3)_2 + 2Ca(OH)_2 \rightarrow$$
$$\underline{Mg(OH)_2} + 2\underline{CaCO_3} + 2H_2O \quad (15\text{-}11)$$

$$MgSO_4 + Ca(OH)_2 \rightarrow$$
$$\underline{Mg(OH)_2} + CaSO_4 \quad (15\text{-}12)$$

It should be noted that the removal of CO_2, as in Eq. 15-9, does not reduce the hardness, but must precede the subsequent reactions since CO_2 is a stronger acid than HCO_3^-. The conversion of bicarbonate to carbonate as a function of pH is shown in Figure 15-17. This would indicate that complete conversion would require a pH value greater than 12 to attain complete utilization of the bicarbonate alkalinity for calcium precipitation. In practice, the

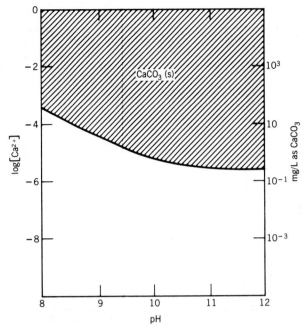

FIGURE 15-15. Solubility of $CaCO_3$ as function of pH ($K_{sp} = 4 \times 10^{-9}$). $C_T = 2 \times 10^{-3}$ Mole/L.

optimum pH for maximum calcium carbonate precipitation may be as low as 9.3 because of the shift in equilibrium as calcium carbonate precipitates. The actual optimum pH is a function of the concentrations of both calcium and bicarbonate ions. Eq. 15-11 shows that the removal of magnesium, even

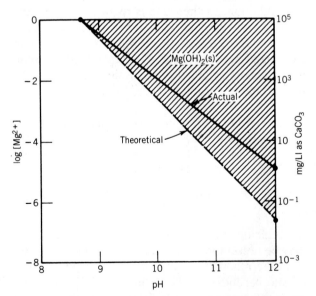

FIGURE 15-16. Solubility of $Mg(OH)_2$ as a function of pH ($K_{sp} = 2.5 \times 10^{-11}$). Solid line is for solubility in 23 natural waters as determined by Thompson et al. (1972).

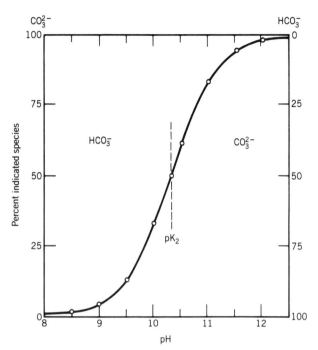

FIGURE 15-17. Distribution of carbonate and bicarbonate as a function of pH (25°C).

when it is present as the bicarbonate, (i.e., magnesium carbonate hardness), requires extra lime. Equation 15-12 further illustrates the extra lime required to remove magnesium noncarbonate hardness since no reduction of hardness results because of the solubility of $CaSO_4$.

In addition, further lime is required to raise the pH to a value greater than 10.5 in order to precipitate $Mg(OH)_2$. Lime, then, will remove CO_2 and carbonate hardness and will replace magnesium with calcium when the magnesium is present as noncarbonate hardness.

Lime–Soda Softening. When inadequate carbonate alkalinity is present to react with lime it is necessary to provide an external source, usually soda ash, Na_2CO_3. The relevant reaction is

$$CaSO_4 + Na_2CO_3 \rightarrow \underline{CaCO_3} + Na_2SO_4 \quad (15\text{-}13)$$

The calcium noncarbonate hardness in Eq. 15-13 may be present in the untreated water or may result from the reaction of lime with magnesium noncarbonate hardness as shown in Eq. 15-12. In either case, the amount of soda ash required depends on the amount of noncarbonate hardness to be removed.

Caustic Soda Softening. Another alternative, caustic soda (NaOH), has been utilized when the untreated water has inadequate carbonate hardness to react with lime. The choice between the two (soda ash or caustic soda) depends not only on economics but also on such factors as ease of handling (NaOH is purchased as a 50 percent solution) and magnesium content. The use of a sodium-containing compound may be cause for concern by some, but noncarbonate hardness removal by either of the two above alternatives results in replacement of the divalent hardness ions with sodium.

The relevant reactions of caustic soda softening are

$$CO_2 + 2NaOH \rightarrow Na_2CO_3 + H_2O \quad (15\text{-}14)$$

$$Ca(HCO_3)_2 + 2NaOH \rightarrow \\ \underline{CaCO_3} + Na_2CO_3 + 2H_2O \quad (15\text{-}15)$$

$$Mg(HCO_3)_2 + 2NaOH \rightarrow \\ \underline{Mg(OH)_2} + Na_2CO_3 + H_2O \quad (15\text{-}16)$$

$$MgSO_4 + 2NaOH \rightarrow \\ \underline{Mg(OH)_2} + Na_2SO_4 \quad (15\text{-}17)$$

$$CaSO_4 + Na_2CO_3 \rightarrow \\ \underline{CaCO_3} + Na_2SO_4 \quad (15\text{-}18)$$

The sodium carbonate shown in Eq. 15-18 may be that produced in the reactions represented in Eqs. 15-14–15-17. In the event that more sodium carbonate is produced than there is calcium noncarbonate hardness to remove, the excess must remain in solution. This presents a problem of very high carbonate alkalinity which can be reduced by acidification, but none of the sodium will precipitate. In contrast, when lime is used, the excess calcium is precipitated as the carbonate. This is particularly important in those cases where high pH (and the concomitant high caustic alkalinity) must be attained in order to precipitate magnesium as $Mg(OH)_2$. This type of water can be treated by use of a mixture of lime and caustic soda with the ratio determined by the water quality, both raw and finished.

Recarbonation. When the water after softening has a pH value greater than the saturation pH, pH_s, it is necessary to reduce that value. The most common and economical method is the use of carbon dioxide, CO_2. When excess caustic (noncarbonate) alkalinity is present, the reaction that occurs is

$$2OH^- + CO_2 \rightleftharpoons CO_3^{2-} + H_2O \quad (15\text{-}19)$$

The CO_3^{2-} that results can precipitate any calcium ion that is present above the saturation of calcium carbonate. The pH may be lowered to the point of saturation equilibrium by use of additional CO_2:

$$CO_3^{2-} + CO_2 + H_2O \rightleftharpoons 2HCO_3^- \quad (15\text{-}20)$$

Until recently, CO_2 has been produced by the combustion of diesel oil or natural gas in either underwater or external burners. The availability of bulk CO_2 has led to its utilization because of convenience and reduced operation and maintenance problems.

The recarbonation stage in a plant may be either single or double. The double stage, according to Eqs. 15-19 and 15-20 has been called "split recarbonation" but is practiced in few plants since it requires two-stage setting, thus twice the settling basin capacity. On the other hand, the water quality produced by this process is softer and lower in alkalinity than single-stage softening when magnesium reduction is desired.

When the softened water contains little or no excess caustic alkalinity, only single-stage recarbonation is possible. It is used to reduce the pH to the saturation pH.

Process Modifications. The quality of the raw water available and the treated water desired dictate the optimum treatment process(es) to be used. Several modifications to or combinations of the above softening processes have been designed for specific purposes:

1. SPLIT RECARBONATION (DISCUSSED ABOVE). This is used when magnesium removal is required from water with relatively high noncarbonate hardness.

2. UNDERSOFTENING. This term is applied to the use of lime softening of waters with little or no noncarbonate or magnesium hardness. Softening to a final hardness of 70–100 mg/L as $CaCO_3$ can be accomplished at a pH of 8.7–9.0 producing a water very close to the pH_s. Post-chlorination is frequently sufficient to reduce the pH to the pH_s.

3. SPLIT TREATMENT. There are several different processes that fit this description, wherein two or more streams are treated differently and then combined. Among them are:

   A. Split Treatment with Excess Lime. This process is used where magnesium must be reduced and the raw water contains relatively little noncarbonate hardness. One portion of the influent flow is softened with excess lime to remove the magnesium and then blended with the rest of the raw influent to precipitate the excess lime. The alkalinity in the by-passed stream is used to neutralize the excess caustic alkalinity required to reduce the magnesium in the treated stream. The ratio of by-pass to treated flow is determined by a mass balance on the magnesium. The desired magnesium hardness of the finished water is predetermined and the ratio thus fixed. The magnesium in the treated portion is usually reduced to a very low concentration (<2 mg/L) at high pH (>11.0), resulting in excess caustic alkalinities of about 70 mg/L as $CaCO_3$. The blended waters are allowed to settle again before filtration.

   B. Parallel Softening and Coagulation. This process has been used on waters with high magnesium and turbidity and/or color. A third by-passed stream has been used in at least one case to provide additional hardness for neutralization of the excess caustic alkalinity in the softened stream. The coagulated stream has low turbidity and color but also low alkalinity. The blend has a balance of good color, turbidity and hardness.

   C. Blend Lime-Softened Stream with Another Stream, Either Ion Exchange or Reverse Osmosis Treated Water.

   a. Ion-exchange-softened stream. This alternative can be used when there is high noncarbonate hardness in the raw water but the use of soda ash or caustic soda would not be economical or in the case of an expansion of an existing ion exchange plant.

   b. Reverse-osmosis-softened or demineralized stream. This alternative can be used to provide additional reduction of dissolved solids.

Dose Calculation Methods. An estimate of the dosages of chemicals in lime–soda softening can be made by any one of three methods: stoichiometry, the solution of simultaneous equilibria equations, and laboratory studies. All have applications in estimating dosages with experimental methods providing results that most closely approximate plant performance. The first two methods are described in more detail below.

Stoichiometry. The stoichiometric method is based on the assumption that all of the relevant reactions go to completion. Corrections can be made for the solubility of calcium carbonate and for the excess hydroxide alkalinity required for magnesium hydroxide precipitation. As an example, for lime–soda softening an approximation of the lime and soda ash requirements can be made based on Eq. 15-9–15-13. The lime, as 100 percent CaO, is calculated by noting that it serves the function of CO_2 removal, bicarbonate conversion to carbonate, and magnesium reduction (to approximately 40 mg/L), or

$$\text{mg/L 100\% CaO} = (CO_2 + \text{carbonate hardness} + Mg_r* + 30*)\frac{56}{100}$$

where all terms within the parentheses are expressed as mg/L $CaCO_3$. Mg_r* is the magnesium removed (i.e., initial magnesium concentration $-$ 40 mg/L), the 30* represents an approximation of the excess alkalinity required to precipitate magnesium as the hydroxide, and $\frac{56}{100}$ is the gravimetric factor for converting calcium carbonate to calcium oxide.

The soda ash requirement can be estimated from Eq. 15-13 by noting that soda ash is used only for noncarbonate hardness reduction. Since it requires about 1.9 times as much 100 percent Na_2CO_3 as 100 percent CaO to remove the same amount of hardness and soda ash is more expensive per pound than CaO, it is desirable to minimize the amount of soda ash used (i.e., the amount of noncarbonate hardness removed). This can be accomplished by reducing the residual carbonate hardness to a minimum or allowing the finished total hardness to be higher, or both.

This is true since the total hardness in the finished water is the sum of the carbonate (CH) and noncarbonate hardness (NCH) or

$$TH_f = CH_f + NCH_f \qquad (15\text{-}21)$$

Since the minimum practical CH is about 30 mg/L, the equation could be modified to

$$TH_f = 30 + NCH_f \qquad (15\text{-}22)$$

and the noncarbonate hardness to be removed (NCH_r) is that present in the raw water (NCH_i) minus that in the finished water

$$NCH_r = NCH_i - NCH_f \qquad (15\text{-}23)$$

or substituting Eq. 15-22 into 15-23

$$NCH_r = NCH_i - TH_f + 30 \qquad (15\text{-}24)$$

From this it can be seen that the choice of the finished total hardness to be achieved directly impacts the amount of soda ash required, or

$$\text{mg/L 100\% } Na_2CO_3 = (NCH_r)\frac{106}{100}$$

where $\frac{106}{100}$ is the gravimetric factor for converting calcium carbonate to soda ash (sodium carbonate).

Equilibrium. Completely rigorous solution of the equilibria involves in softening requires consideration of a number of ion complexes and several interrelated equilibria, so many that a computer is required. A number of programs have been written that enable this sort of calculation, notably the RIDEQL and MINEQL series (Morel, 1983) and the WATEQ series developed by USGS (Ball et al., 1979). For serious research concerning softening kinetics or equilibria, these programs and their progeny should be considered.

Fortunately for the practitioner, a series of diagrams have been developed that enable the conduct of equilibrium calculations with less special training. Research on diagrams of this type began with Langelier's stability diagram (Langelier, 1946) and culminated with the Caldwell–Lawrence diagram (Caldwell and Lawrence, 1953). Since that time, the diagrams themselves have been updated (Lowenthal and Marais, 1976; Merrill and Sanks, 1977). The discussion by Merrill and Sanks is a good summary of the application of these diagrams to water conditioning problems. Updated versions of the diagrams for several different temperatures and ionic strengths may be obtained from the American Water Works Association. The principles of the diagrams and their application will be only briefly discussed here.

Each diagram is drawn with a set of specified conditions (temperature and ionic strength). The major axes of the diagram are acidity and alkalinity–calcium. On the diagram are isopleths for pH, alkalinity, and calcium hardness. A direction diagram is provided that shows the direction to move for the addition of various chemicals. A diagram is also provided for the solubility relationship between magnesium and the pH. All values on the diagram, except for pH, are expressed as milligrams of $CaCO_3$ per liter. It is important to remember that the diagram is based upon equilibrium with calcium carbonate. The following are some problems that illustrate the use of the diagram:

EXAMPLE 15-1
Determining $CaCO_3$ Saturation/Equilibrium

For the two water qualities shown below: (i) Determine if they are undersaturated, supersaturated, or at equilibrium with $CaCO_3$ and (ii) Determine the condition they would reach if they were equilibrated with solid $CaCO_3$:

Water A

Alkalinity (Alk) = 120 mg/L

Calcium = 140 mg/L

pH = 8.2

Water B

Alk = 120 mg/L

Calcium = 200 mg/L

pH = 7.2

STEP 1:   Locate the lines corresponding to the two water qualities on the water conditioning diagram (see Figure 15-18).
STEP 2:   Draw a horizontal line through the intersection of the pH and alkalinity lines and a vertical line through the intersection of the alkalinity and calcium lines.
STEP 3:   The intersection of the vertical and horizontal lines described above represents the condition of the water at equilibrium with $CaCO_3$:

Water A

Equilibrium Quality

Alk = 110 mg/L

Calcium = 130 mg/L

pH = 7.7

The equilibrium calcium is lower than that for water A, so the original water is supersaturated with $CaCO_3$.

Water B

Equilibrium Quality

Alk = 130 mg/L

Calcium = 210 mg/L

pH = 7.4

The equilibrium calcium is greater than that for water B, so the original water is undersaturated with $CaCO_3$.   □

EXAMPLE 15-2
Lime and Soda Requirements for Softening Calcium Hardness

A.   For the water quality given in water B above, determine the lime dose to achieve a calcium hardness of 100 mg/L as $CaCO_3$.

STEP 1:   Locate equilibrium point on the diagram.
STEP 2:   Note the direction for lime on the direction diagram (vertical) and draw a line to the appropriate calcium hardness (point Q)
STEP 3:   Take difference in conditioning chemical lines from the vertical axes of the diagram to get the lime dose.

Answer: lime dose = 150 − 20 = 130 mg $CaCO_3$/L. (Note that the lowest hardness that can be obtained from this water with lime alone is slightly less than 90 mg $CaCO_3$/L.)

B.   For the above example, find the soda ash dose required to further reduce the calcium hardness to 60 mg $CaCO_3$/L.

STEP 4:   Note the direction for soda on the direction diagram (to the right) and draw a line to the appropriate calcium hardness (point R)
STEP 5:   Take the difference in conditioning chemical lines to get the soda dose.

Answer: soda ash dose = 80 − 40 = 40 mg $CaCO_3$/L. (Note that the hardness can be further reduced by manipulating both the lime and soda doses.)   □

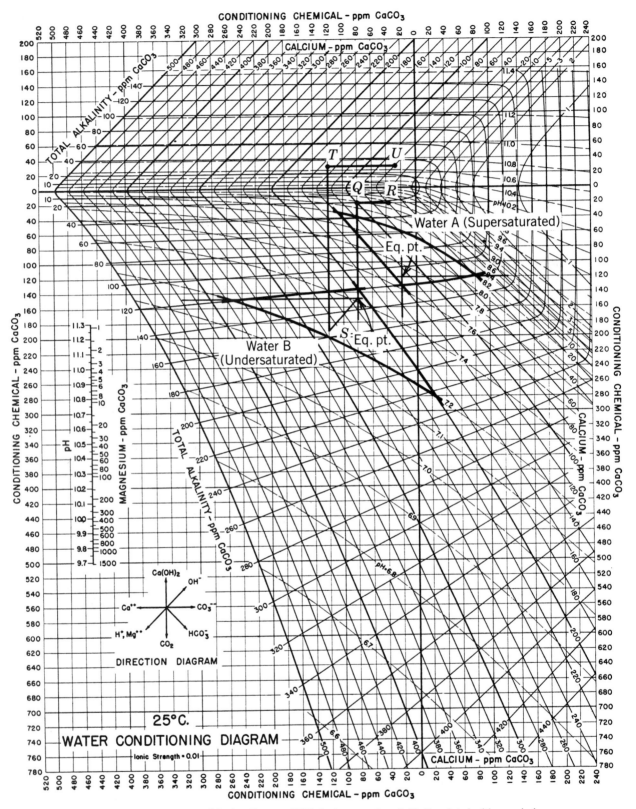

FIGURE 15-18. Water conditioning diagram, 25°C, ionic strength = 0.01. Reprinted with permission.

348

EXAMPLE 15-3
Softening with Magnesium Removal

A. Assume the water in the example above also has magnesium at a level of 60 mg $CaCO_3$/L and we wish to reduce the magnesium to 20 mg $CaCO_3$/L. What lime dose is required and what will the final conditions be?

STEP 1: From the Mg–pH diagram, we determine that a magnesium level of 20 mg/L as $CaCO_3$ corresponds to an equilibrium pH of about 10.63.

STEP 2: Beginning at the equilibrium point and using the direction diagram, mark off the distance corresponding to the magnesium that will precipitate (point S). Then, to determine the lime dose, mark off the distance required to obtain a pH of 10.63 (point T).

STEP 3: Determine the lime dose by reading from the vertical axis.

Answer: lime required = 190 + 30 = 220 mg $CaCO_3$/L.

B. If a calcium hardness of 60 mg $CaCO_3$/L is also required, what soda dose is required?

STEP 4: Draw a horizontal line to the appropriate calcium hardness.

STEP 5: Determine the soda dose by reading from the horizontal axis (point U):

Answer: soda required = 120 − 30 = 90 mg $CaCO_3$/L.   □

Design Considerations. The removal of hardness by lime–soda softening requires that a chemical feed and reaction system be installed along with sedimentation and filtration. The chemical feed equipment is utilized to add lime, soda ash, and CO_2 if required. The amounts of material utilized will vary substantially with the water quality of the incoming water. In general, large-scale lime-handling facilities are required for this process along with substantial sludge dewatering and disposal facilities. A typical lime–soda treatment scheme is shown in Figure 15-19.

As noted previously, substantial quantities of sludge are produced by this process. For this reason, such handling must be given adequate review during the design. Dewatering facilities and disposal facilities will be required. It is possible to reclaim the lime from the sludge by heating it or recalcining it. However, with current energy considerations, it is frequently not justified to use this approach because of large quantities of water that must be evaporated from the sludge. However, in the event that a large unit were installed, solids handling and recalcining would be major elements of the design consideration. Such installations exist at the Metro-Dade Water and Sewer Authority facility in Dade County, Florida and at Dayton, Ohio.

Ion Exchange

The other major option for water softening involves sodium-based ion exchange. This process is discussed in depth in Chapter 10 and will not be discussed in detail in this section. However, it should be noted that on an equivalent basis, the exchange of sodium for calcium and magnesium results in an increase in total dissolved solids due to the difference in equivalent weights between calcium and magnesium and sodium.

One of the major drawbacks to this approach is the large amount of salt required to ensure complete regeneration of the bed. The amount of excess salt varies with the efficiency of the design of the unit and can range from 120 to 200 percent of theoretical requirements. The discharge of substantial amounts

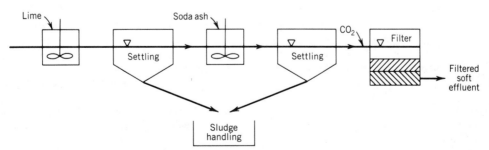

FIGURE 15-19. Lime–soda softening system.

of salt with the regenerant is expensive and involves the disposal of large amounts of very salty water. With increasing concerns over the impacts of waste discharges on groundwater basins, the control of these materials becomes increasingly more stringent.

Another disadvantage of sodium-based exchangers is the reported health concerns regarding the ratio of monovalent to divalent ions in a water supply. Additionally, sodium will replace calcium and magnesium, and the health effects of excess sodium in one's diet are cause for concern. The ion exchange unit has the capability of producing a "dead soft" water that has a hardness of less than 10 mg/L as $CaCO_3$. The lime–soda process can generally produce a water with the hardness in the range of 50 to 80 mg/L as $CaCO_3$.

REFERENCES

AWWA, "An AWWA Survey of Inorganic Contaminants in Water Supplies," *JAWWA*, **77**(5), 67–72 (May 1985).

AWWA, "Quality Control for Potable Water," JAWWA **60**(12), 1317–1322 (December 1968).

Anderson, R. E., "Removal of Nitrate from Well Water," in-house communications, Diamond Shamrock Corporation, Functional Polymers Division, Redwood City, CA (August and October 1978).

Anderson, R. E., "Removal of Nitrate from Well Water; Estimation of Nitrate Breakthrough," in-house communication, Diamond Shamrock Corporation, Functional Polymers Division, Redwood City, CA (March 1979).

Ball, J. W., Jeanne, E. A., and Nordstrom, D. K., "WATEQ2—a Computerized Chemical Model for Trace and Major Element Speciation and Mineral Equilibria of Natural Waters," in Jeanne, E. A. (ed.), *Chemical Modeling in Aqueous Systems*, Washington, DC, ACS Symposium Series 93, 815–835 (1979).

Baumann, E. R., "Nitrate Removal for Production of Drinking Water," Prepared for EPA Region VIII Water Treatment Methods Short Course Presented by the AWWA Research Foundation, Denver, Colorado (March 1982).

Bellack, E., "Arsenic Removal from Potable Water," *JAWWA* (July 1971).

Black, A. P., and Goodson, J. B., "The Oxidation of Sulfides by Chlorine in Aqueous Solutions," *JAWWA* (April 1952).

Boruff, C. S., "Removal of Fluorides from Drinking Water," *Ind. Engr. Chem.*, **26**, 1 (January 1934).

Caldwell, D., and Lawrence, W. B., "Diagrams for Water Treatment Calculations," *Ind. Engr. Chem.*, **45**, 535 (March 1953a).

Caldwell, D. H., and Lawrence, W. B., "Water Softening and Conditioning Problems," *Ind. Engr. Chem.*, **45**, 535 (1953b).

Chen, K. Y., "Chemistry of Sulfur Species and Their Removal from Water Supply," A. J. Rubin (ed.), *Chemistry of Water Supply, Treatment and Distribution*, Ann Arbor Science, Ann Arbor, MI (1973).

Choi, W. W., and Chen, K. Y., "The Removal of Fluoride from Waters by Adsorption," *JAWWA* (October 1979).

Cleasby, J., "Iron and Manganese Removal—A Case Study," *JAWWA*, 147, (March 1975).

Clifford, D. A., "Nitrate Removal from Water Supplies by Ion Exchange: Resin Selectivity and Multicomponent, Chromatographic Column Behavior of Sulfate, Nitrate, Chloride and Bicarbonate," Ph.D. dissertation, Water Resources Engineering, University of Michigan (1976).

Clifford, D. A., and Weber, W. J., "Multicomponent Ion Exchange: Nitrate Removal Process with Land-Disposable Regenerant," *Ind. Water Engr.* (March 1978).

Cox, C. R., *Operating and Control of Water Treatment Processes*, WHO (1969).

Culp, R. L., and Stoltenberg, H. A., "Fluoride Reduction at LaCrosse, Kansas," *JAWWA*, **50**, 3 (March 1958).

DeBoice, J. N., "Chemical Treatment for Phosphate Control," Ph.D. dissertation, University of California at Berkeley (1974).

Evans, S., "Nitrate Removal by Ion Exchange," *JWPCF*, **45**(4) (April 1973).

Ferguson, J. F., and Gavis, J., "A Review of the Arsenic Cycle in Natural Waters," *Water Res.*, **6**, 1259–1274 (June 1972).

Gupta, S. K., and Chen, K. Y., "Arsenic Removal by Adsorption," *JWPCF* (March 1978).

Guter, G. A., "Removal of Nitrate from Contaminated Water Supplies for Public Use—Interim Report," EPA Publication 600/2-81-029, Cincinnati (February 1981).

Hallberg, R. O., and Martinell, R., "Vyredox-In Situ Purification of Groundwater," *Groundwater*, **14**(2), 88–93 (Mar./Apr. 1976).

James M. Montgomery, Consulting Engineers, Inc., "Water Quality Enhancement Study," Report to the County of Sacramento, California (January 1977).

James M. Montgomery, Consulting Engineers, Inc., "Pilot Study for Fairbanks, Alaska" (September 1982).

Kessick, M. A., and Morado, J. J., "Mechanism of Autodation of Manganese in Aqueous Solution," *ES&T*, **9**(2) (February 1975).

Langelier, W. F., "Chemical Equilibria in Water Treatment," *JAWWA*, **38**, 169–178 (1946).

Lowenthal, R. E., and Mards, G. V. R., *"Carbonate Chemistry of Aquatic Systems: Theory and Application,"* Ann Arbor Science Publishers, Ann Arbor, MI (1976).

Maneval, J., Klein, G., and Sinkovic, J. "Selenium Removal from Drinking Water by Ion Exchange" A Study Conducted for the U.S.E.P.A., Municipal Environmental Research Laboratory. EPA Contract No. 81-02-5401 (March 1985).

Merrill, D. T., and Sanks, R. L., "Corrosion Control by Deposition of Calcium Carbonate Films; Part I, A Practical Approach for Plant Operators," *JAWWA*, **69**, 592–599 (Nov. 1977).

Monel, F., and Morgan, J. J., "A Numerical Method for Computing Equilibria in Aqueous Systems," *ES&T*, **6**, 58 (1972).

Monscvitz, J. T., and Ainsworth, L. D., "Hydrogen Polysulfide in Water Systems," American Chemical Society Meeting, Division of Water, Air and Waste Chemistry, Minneapolis (April 1969).

Morel, F. M. M., *Principles of Aquatic Chemistry,* Wiley, New York (1983).

Morgan, J. J., and Stumm, W., "Analytical Chemistry of Aqueous Manganese," *JAWWA,* **57,** 107–119 (January 1965).

Musterman, J., "Removal of Inorganics from Drinking Water Supplies," Paper Presented at 20th Annual ASCE Water Resources Design Conference, Ames, Iowa (1982).

O'Brien, W. J., "Chemical Removal of Nitrate from Potable Water Supplies," Kansas W.R.R.I., Manhattan, Kansas (1966).

O'Connor, J. T., "Iron and Manganese Removal: Methods and Practice," *Water Sewage Works,* **116,** R-68 (1969).

Oldham, W. K., and Gloyna, E. F., "Effect of Colored Organics on Iron Removal," *JAWWA,* 610 (1968).

Ricci, J. E., *Hydrogen Ion Concentration—New Concepts in a Systematic Treatment,* Princeton University Press, Princeton, NJ (1952).

Rubel, F., and Williams, F. S., "Pilot Study of Fluoride and Arsenic Removal from Potable Water," EPA Publication 600/2-80-100, Cincinnati, (August 1980).

Rubel, F. R., and Woosley, R. D., "Removal of Excess Fluoride from Drinking Water," EPA Publication 570/9-78-001, Washington, DC (January 1978).

Sawyer, C. N., and McCarty, P. L., *Chemistry for Environmental Engineering,* McGraw-Hill, New York (1978).

Schenk, J. E., and Weber, Jr., W. J., "Chemical Interactions of Dissolved Silica with Iron (II) and (III)," *JAWWA,* **60,** 199–212 (February 1968).

Scott, R. D., Kimberly, A. E., Van Horn, A. L., Ery, L. F., and Waring, F. H. "Fluoride in Ohio Water Supplies—Its Effects, Occurrence and Reduction," JAWWA, **29**(1), 9–25 (January, 1937).

Shen, Y. S., "Study of Arsenic Removal from Drinking Water," *JAWWA,* (August 1973).

Sorg, T. J., "Compare Nitrate Removal Methods," *Water Wastes Engr.,* (December 1980).

Sorg, T. J., and Logsdon, G. S., "Treatment Technology to Meet the Interim Primary Drinking Water Regulations for Inorganics: Part 1 and Part 2," *JAWWA* (April and July 1978).

St. Ammant, P. P., and McCarty, P. L., "Treatment of High Nitrate Waters," *JAWWA,* **61**(12), 659 (1969).

Stumm, W., and Lee, T., "Oxygenation of Ferrous Iron," *Ind. Engr., Chem.,* **53** (1961).

Thompson, C. G., Singley, J. E., and Black, A. P., "Magnesium Carbonate: A Recycled Coagulant—II, *JAWWA,* **64,** 93 (February 1972).

Trussell, R. R., "Systematic Aqueous Chemistry for Engineers," Ph.D. dissertation, University of California at Berkeley (1972).

Trussell, R. R., Trussell, A. R., and Kreft, P., "Selenium Removal from Groundwater Using Activated Alumina," EPA Publication 600/2-80-153, Cincinnati (August 1980).

White, G. C., "Hydrogen Sulfide Control and Removal," *Handbook of Chlorination,* Van Nostrand Reinhold, New York (1972) pp. 326–331.

Wu, Y. C., "Activated Alumina Removes Fluoride Ions from Water," *Water Sewage Works* (June 1978).

Zipf, K. A., and Luthy, R. G., "Chemical Equilibria in Split-Treatment Softening of Water," *JAWWA,* **73,** 304–3111 (June 1981).

— 16 —

Organics

The presence of organic compounds in water is an important factor effecting water quality. Since the inception of the water treatment industry, several types of organic chemicals have been found to be the cause of disagreeable tastes and odors in drinking water (Ettinger and Ruchoft, 1951; Burttschell et al., 1959; Lillard and Powers, 1975; Verschueren, 1977). In today's industrialized society, water purveyors must be aware of the fact that toxic organic chemicals can occur in water supplies. Vinyl chloride, benzene, and other organic contaminants that have been found in drinking water are known to be carcinogenic (Searle, 1976). Other frequently detected substances such as chloroform are classified as cancer-suspect agents. These anthropogenic pollutants are generally present at extremely low concentration. Under these circumstances, these compounds do not appear to pose an immediate hazard. However, numerous scientists are currently attempting to determine "at which level do trace organic contaminants exert an impact on human health?" It seems likely that this question will remain unanswered for some time. In order to further understand this issue, it would be prudent to review the fundamental aspects of the chemical composition and properties of organic compounds, the potential sources for their introduction to the environment, the techniques employed for their analysis,

and the unit processes that can be applied to expedite their removal from aquatic systems.

DEFINITION AND PROPERTIES

The term *organics* refers to the general class of chemicals composed of carbon (C) and one or more of the following elements: hydrogen (H), nitrogen (N), and oxygen (O) (Hendrickson et al., 1970). This nomenclature dates to the very early studies of chemistry when substances were categorized as inorganic when they were obtained from mineral sources and as organic when derived from living organisms. Today, many organic compounds are procured from sources other than biological activity. A wide variety of materials are synthesized by the chemicals industry. The molecular structure of these compounds may also contain atoms of sulfur (S), phosphorus (P), and/or one or more of the halogens, that is, fluorine (F), chlorine (Cl), bromine (Br), and iodine (I). There are many chemical species that are commonly considered to be inorganic, in spite of having C, H, O, and N within their structure. Examples of such compounds include carbon monoxide (CO), carbon dioxide (CO_2), carbonate (CO_3^{2-}), bicarbonate (HCO_3^-), and cyanide (CN^-). The principal structural feature that distinguishes

352

organic compounds from inorganic substances is the existence of strong carbon–carbon bonds.

From an environmental standpoint, it is especially convenient to classify organic substances into groups according to their chemical or physical properties. A knowledge of these properties facilitates the selection of appropriate methods for the analysis and treatment of these materials in water. One important property of organic compounds is molecular size. The dimension of organic molecules varies from less than 1 nm for simple compounds such as chloroform ($CHCl_3$) to approximately 1 μm for complex humic acid polymers. The data in Figure 16-1 depict the relative size of organic molecules with respect to microorganisms and other material commonly found in aquatic systems. Another parameter closely correlated with the size of a molecule is molecular weight (MW). The molecular weight of organic compounds ranges from 16 g/mole for methane (CH_4) to values approaching 1 million (10^6) for polymeric materials. The polarity of an organic substance describes the degree to which one segment of a molecule is either positively or negatively charged with respect to another part of the molecular structure (Paul et al., 1967). A frequently used measure of the polarity of a compound is given by the dipole moment. The dipole moment of organic substances can vary from a value of 0 debye (D) for molecules such as carbon tetrachloride (CCl_4), which have a highly symmetric spatial distribution of electron density about their bonding structures, to approximately 1.74 D for substances such as acetic acid, which possess a considerable degree of ionic character (Weast, 1973). The volatility of an organic substance is generally reflected by its boiling point or vapor pressure. At ambient atmospheric pressure (760 mm

Hg), the boiling points of organic contaminants may vary from as low as $-13.4°C$ for highly volatile compounds such as vinyl chloride to temperatures in excess of 400°C for nonvolatile polycyclic aromatic hydrocarbons such as chrysene (b.p. = 448°C) (Weast, 1973).

A striking variety of organic compounds have been identified in the aquatic environment. Figure 16-2 illustrates a scheme for the categorization of the organics that are known to exist in drinking water, based on their volatility, molecular weight, and polarity (Trussell and Umphres, 1978). This diagram presents a highly simplified picture of a complex mixture of substances. Therefore, there are many overlapping areas, omissions, and exceptions among these classifications. Shown in Figure 16-3 are representative compounds from each region of the chart in Figure 16-2 (Trussell and Umphres, 1978a). The size and complexity of the organic molecules increase dramatically as a function of molecular weight until the structures become polymeric in nature. The highly polar compounds listed at the top of each column contain a large number of oxygenated functional groups relative to the nonpolar substances. While a great many structural variations are possible within each chemical class, for example, substitution by halogen atoms, the general trends depicted by this scheme are valid.

SOURCES AND CONCENTRATIONS

There are three major sources from which organics may be introduced to water:

1. Dissolution of naturally occurring organic materials.

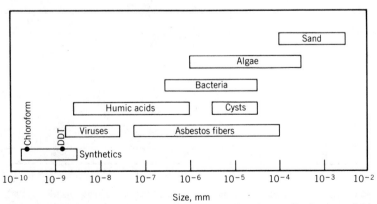

FIGURE 16-1. Relative size of organics to microorganisms and other material in water.

Volatility

	Volatile	Semivolatile	Nonvolatile
Polar	Alcohols Ketones Carboxylic acids	Alcohols Ketones Carboxylic acids Phenols	Polyelectrolytes Carbohydrates Fulvic acids
Semipolar	Ethers Esters Aldehydes	Ethers Esters Aldehydes Epoxides Heterocyclics	Proteins Carbohydrates Humic acids
Nonpolar	Aliphatic hydrocarbons Aromatic hydrocarbons	Aliphatics Aromatics Alicyclics Arenes	Nonionic polymers Lignins Hymatomelanic acid
	Low	Medium	High

Polarity (left axis label)

Molecular weight

FIGURE 16-2. Schematic classification of organic compounds found in water. (Reprinted from *JAWWA*, Vol. 70, No. 11 (November 1978) by permission. Copyright 1978, The American Water Works Association.)

2. Compounds formed through chemical reactions that occur during disinfection/treatment and transmission of water.
3. Contaminants originating directly from commercial activity.

In general, the attention of the water treatment industry has focused primarily on reducing the in-put of compounds from the latter two categories of sources. However, substances such as pesticides and chlorinated hydrocarbons which result from man-made sources comprise only a very small fraction of the total mass of organic material present in most water systems. Typically, anthropogenic organic pollutants occur in the aquatic environment at low parts-per-billion (μg/L or ppb) concentrations,

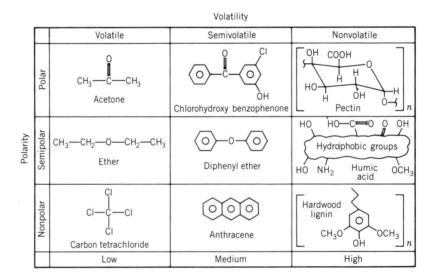

FIGURE 16-3. Example compounds in classification schematic. (Reprinted from *JAWWA*, Vol. 70, No. 11 (November 1978) by permission. Copyright 1978, The American Water Works Association.)

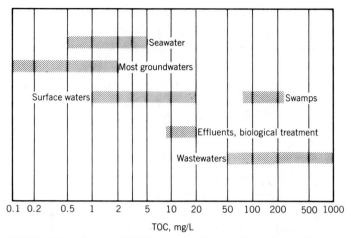

FIGURE 16-4. Ranges of TOC reported for a variety of natural waters. (Reprinted from *JAWWA*, Vol. 70, No. 11 (November 1978) by permission. Copyright 1978, The American Water Works Association.)

whereas, the total organic carbon (TOC) content of water may approach levels of several parts per million (mg/L or ppm) (Kavanaugh, 1978). The data summarized in Figure 16-4 show the TOC concentrations that have been reported for a variety of natural waters. The organic carbon concentrations of ground and surface waters often fall in the ranges of 0.1–2 and 1–20 mg/L, respectively. By contrast, the highly colored waters of most swamps may possess considerably higher TOC levels which approach several hundred mg/L.

Examination of Figure 16-5 helps to clarify the point made in the preceding paragraph concerning the distribution of organic material in water. The diagram depicts the size distribution of organic substances detected in the Mississippi River according

to molecular weight (Anderson and Maier, 1977). The majority of the contaminants introduced by man have MWs of less than 400 g/mole. The shaded region of the drawing shows that a large percentage of the chemicals in this molecular weight range are amenable to measurement by gas chromatography (GC). The remaining compounds are either too polar or too thermally unstable for analysis by GC. The high-molecular-weight fraction of the organic material, which constitutes the remainder of the diagram, is comprised of polymers derived from the natural chemical and microbial degradation of vegetation. These substances are termed *aquatic humus*.

Naturally Occurring Organics

The vast majority of organic material present in water, as mentioned above, is comprised of naturally occurring humic material. A general mechanism for the synthesis of these organic substances is outlined in Figure 16-6 (Gjessing, 1976). According to this pathway, carbohydrates and proteins are decomposed through chemical and microbiological oxidation into smaller chemical units such as dihydroxybenzoic acids and amino acids. In a separate series of reactions, phenolic polymers are formed by intramolecular condensation of ligins and tannins derived from plants. These reactants then combine with the metabolites of living and deceased microorganisms to yield an amorphous polyheterocondensate product.

Based on their solubility in acid and alkali,

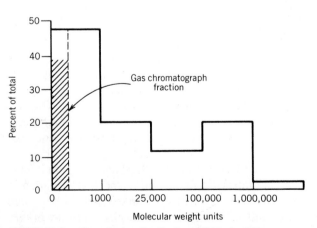

FIGURE 16-5. Organic size distribution, Mississippi River water. (After Anderson and Maier, 1977.)

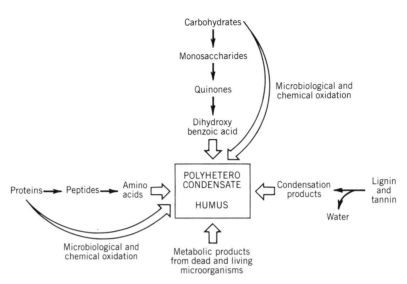

FIGURE 16-6. Natural synthesis of humus. (After Gjessing, 1976.)

aquatic humates are usually divided into two principle components: humic acid (HA), which is soluble in dilute alkaline media but is precipitated upon acidification and fulvic acid (FA), which remains in solution at low pH (Oden, 1919; Black and Christman, 1963). The structural features of HA and FA are similar, but the two fractions differ considerably in molecular weight and functional group composition (Steelink, 1973, 1977). Laboratory analysis of waterborne humic substances by gel filtration indicates that the molecular weight of fulvic acids varies from 200 to 1000 (Ghassemi and Christman, 1968) whereas that of HA ranges up to 200,000 (Gjessing and Lee, 1967). The FA fraction also possesses a higher content of oxygen-containing substituents per unit weight than humic acid. However, due to the arbitrary nature of this classification scheme, the term *humic acid* is frequently used in reference to an aggregate of FA and HA.

Although the origin and primary components of aquatic humus are well understood, very little definitive information is available concerning the details of its chemical structure. The application of modern instrumental techniques such as infrared (IR) and nuclear magnetic resonance (NMR) spectroscopy has met with only limited success due to the structural complexity of humic acid molecules. An abundant amount of data has been obtained from experiments in which samples of the humic acid polymer were treated with a variety of different chemical reagents. The monomeric decomposition products of these reactions are then identified using chro-

matographic and spectroscopic methods. Numerous investigators have utilized the data from this type of experiment in order to construct structural models of aquatic humus. Two examples of these models are depicted in Figure 16-7. Christman and Ghassemi (1966) in Figure 16-7a suggested that the basic structural unit of HA is composed of a "chicken-wire" pattern of aromatic carbon rings. Perdue (1978) in Figure 16-7b recently proposed a simplified formulation of the humate polymer in which the polar hydrophilic functionalities are clustered about the periphery of the macromolecular structure and the nonpolar substituents congregate to form a hydrophobic "pocket" at the center of the molecule.

There are two additional sources of naturally occurring organics in water: microorganisms and petroleum residues. Algae, bacteria, and actinomycetes are ubiquitous contaminants in surface water supplies. Under conditions in which the populations of these organisms flourish, as in the case of an algal bloom, their cellular mass can contribute significantly to the total organic carbon content of water. In addition to cellular matter, a variety of plants and microorganisms release metabolites to aqueous solution through active excretion processes. For example, methylisoborneol (MIB) and geosmin impart the characteristic musty odor and taste to aquatic systems that are densely inhabited by blue-green algae and actinomycetes. Murphy and co-workers (1976) have recently demonstrated that blue-green algae exert dominance over other algal species in

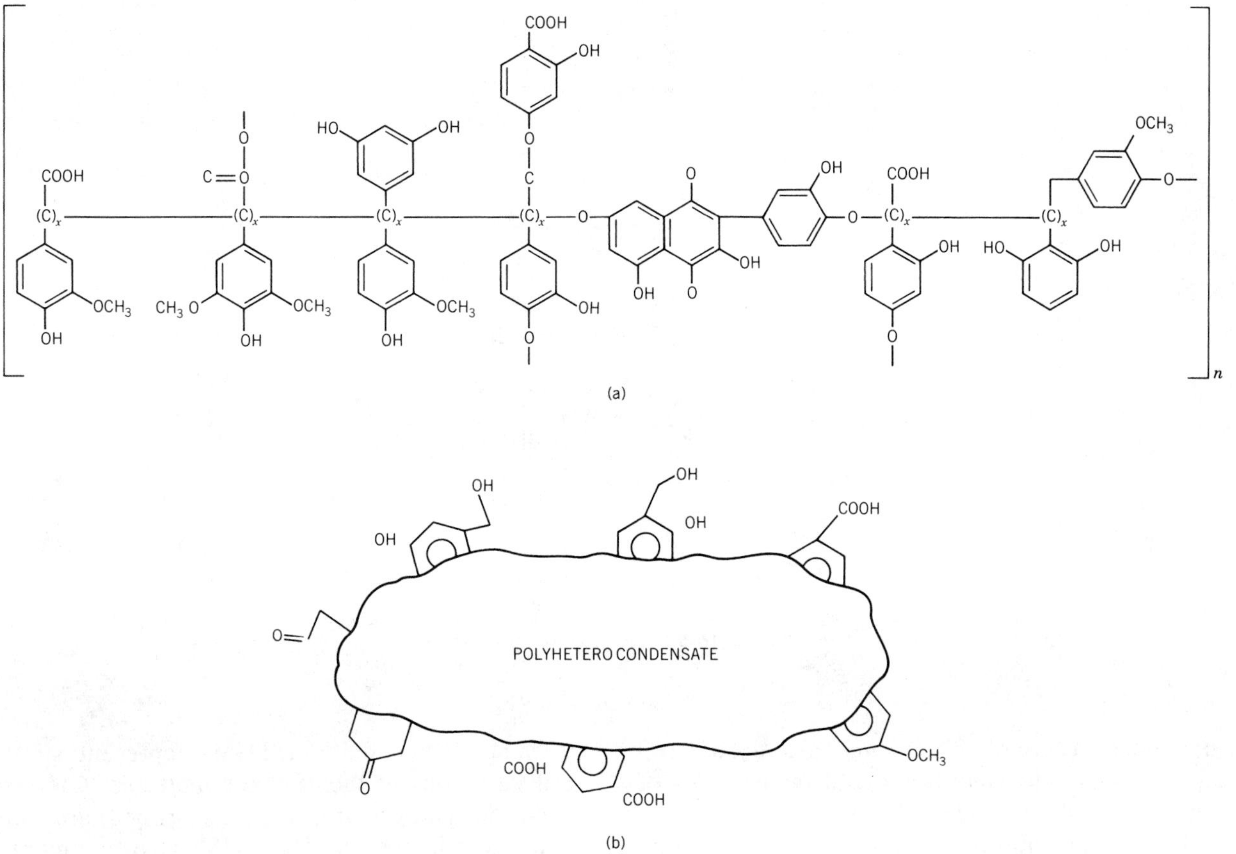

FIGURE 16-7. Proposed models for humic acid structure: (a) from Christman and Ghassemi (1966); (b) from Perdue (1978).

eutrophic waters through excretion of a phytotoxic hydroxamine acid derivative that readily binds dissolved iron. In contrast to the substances of biological origin, petroleum-based chemicals are seldom encountered in surface waters. Instead, methane and higher-molecular-weight aliphatic and aromatic hydrocarbon pollutants often contaminate groundwaters in regions where deposits of natural gas and oil come in contact with subterranean aquifers.

Organics Introduced during Water Disinfection Treatment and Transmission

The processing of water for commercial applications and human consumption introduces a variety of organic compounds. Numerous contaminants are formed through chemical transformations of naturally occurring organic matter during water disinfection. For example, chlorine can efficiently convert humic substances to trihalomethanes and other organohalogen oxidation products under the reaction conditions encountered in water treatment systems

(Rook, 1976, 1977; Babcock and Singer, 1979; Trussell and Umphres, 1978b; Stevens et al., 1977). Other treatment chemicals may also pass through the purification process. This inventory of pollutants is augmented further by leachates from coal tar enamel coatings and plastic materials used in the construction of water distribution systems.

Chlorine (Cl_2) is the most commonly used chemical for the disinfection of water and wastewaters. The impacts of chlorination on the chemical composition of organic matter in aqueous systems has become the subject of intensive study since Rook (1974) initially reported the formation of chloroform in drinking water. Tabulated in Figure 16-8 are some of the "new" compounds that are produced during water chlorination. Of the compounds listed in Figure 16-8, carbon dioxide represents the principal carbonaceous reaction product from the complete oxidation of organic precursors with chlorine. This fact has often been neglected by researchers who have attempted to inventory the fate of organics in the treatment process. Under conditions of

Humics + Br⁻ + NH₃ + Cl₂ ⟶ CO₂ + *New Organics:* + Smaller humics + N₂ + Cl⁻

Trihalomethanes
+
Dihaloacetonitriles
+
Halogenated carboxylic acids
+
Halogenated amines
+
Halogenated phenols
+
Halogenated ketones
+
Halogenated aromatics
+
Halogenated humics
+
Aldehydes
+
Aromatics
+
Phthlatates

FIGURE 16-8.   Chlorination by-products.

incomplete reaction, chlorine may also decompose aquatic humus to yield fulvic and humic acids of lower molecular weight.

The composition of the organics formed during the disinfection of water is now known to be far more complex than originally discovered in 1974. In addition to the trihalomethanes, 11 classes of compounds have been detected in drinking water supplies (see Figure 16-8). For example, analytical surveys indicate that dihaloacetonitriles are present in a high percentage of treatment plant effluents (Trehy and Bieber, 1981). Other types of organic chemicals such as halogenated carboxylic acids and halogenated amines have been reported to occur in only a limited number of water supplies. However, infrequent detection of these substances may simply reflect the limitations in available analytical methods. The total organic halogen (TOX) content of finished waters is typically three to four times higher than values which can be attributed to trihalomethanes. Future improvements in techniques for trace organics analysis, therefore, should provide a wealth of new information about the chemical reactions that occur as the result of water disinfection.

The trihalomethanes (THMs) are the most common by-products of water chlorination that are detected using currently available analytical techniques. These pollutants generally occur at elevated concentrations relative to other organohalide con-

taminants. By definition, THMs represent structural variations of the methane molecule (CH_4) in which four halogen atoms (F, Cl, Br, or I) are substituted for hydrogen. The THMs that commonly occur in drinking water supplies include chloroform ($CHCl_3$), bromodichloromethane ($CHBrCl_2$), dibromochloromethane ($CHBr_2Cl$), bromoform ($CHBr_3$), dichloroidomethane ($CHCl_2I$), and bromochloroidomethane ($CHIBrCl$). Several of the factors that influence the formation of THMs are shown in Figure 16-9. The first graph demonstrates that THM production increases in proportion to increases in the pH of the aqueous system. This results from a key step in the pathway of trihalomethane formation which becomes favored under alkaline reaction conditions (Trussell and Umphres, 1978b). Second, increases in the concentration of bromide ion (Br^-) enhance the yield of THMs and rate of reaction. Ammonia can be employed to arrest the reactions responsible for THM production as shown by the final drawing in Figure 16-9. Addition of ammonia readily converts residual free chlorine to chloramines which cannot undergo rapid reaction with humic substances to give THMs (Stevens et al., 1977).

Utilization of ozone for water disinfection causes the formation of numerous types of organic compounds, but not the production of THMs. The products of ozonation of naturally occurring organic

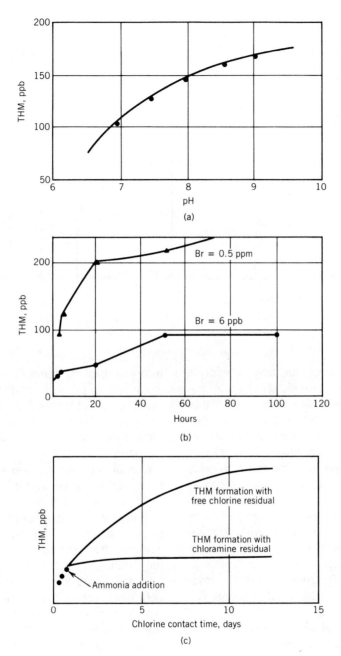

FIGURE 16-9. Factors influencing formation of THMs. (a) pH effect; (b) bromide effect; (c) ammonia effect.

substrates include aldehydes, ketones, carboxylic acids, phthalates, and, of course, carbon dioxide. The gas chromatogram depicted in Figure 16-10 shows the volatile organic by-products that resulted from treatment of a sample of water from the Colorado River with ozone. The most notable treatment products were a series of aliphatic aldehydes which contained from 4 to 10 carbon atoms. In this experi-

ment, the ozone was applied at elevated levels (20 mg/L) in order to ensure the formation of products at sufficient concentrations (1–10 ppb) for identification by gas chromatography/mass spectrometry. At dosages of ozone typically used for water treatment (3 mg/L), aldehyde concentrations may range from 0.01 to 1 ppb.

Similar by-products are formed from the reac-

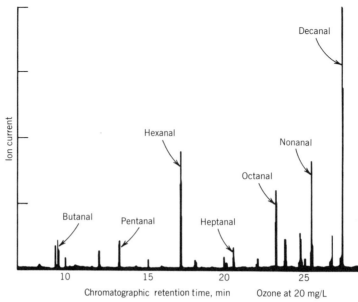

FIGURE 16-10.   Volatile by-products of ozonation of Colorado River water.

tions of less popular disinfectants such as chlorine dioxide (ClO$_2$), hydrogen peroxide (H$_2$O$_2$), and potassium permanganate (KMnO$_4$) with organic substrates dissolved in raw water. It appears that the use of these disinfectants does not lead to the production of halogenated contaminants. Formation of THMs may occur through application of chloride dioxide due to the presence of molecular chlorine as an impurity from the manufacture of chlorine dioxide. Evidence compiled by Singer and co-workers (1980) demonstrated that disinfection with permanganate in conjunction with chlorine significantly reduced the trihalomethane formation potential of surface water. In the results from the U.S. Environmental Protection Agency's National Organics Reconnaisance Survey, Symons et al. (1975) reported only trace concentrations of trihalomethanes in municipal drinking water supplies that were treated with ozone as an alternative or supplement to chlorine. Presumably, initial addition of KMnO$_4$ and O$_3$ acts to oxidize some of the primary trihalomethane precursors.

Bulk chemicals used in water treatment plants can also contribute to the organic content of finished water supplies. Under certain conditions, organic polymers added to improve the efficiency of coagulation and filtration steps can break through the treatment process. In addition, low concentrations of the monomeric substances that make up the structure of coagulant polymers have also been

found in drinking water. Contaminants in treatment chemicals represent another source of pollution that may potentially affect an aquatic system. For example, solvents such as carbon tetrachloride can be routinely employed in cleaning cylinders used for storage and delivery of chlorine. Therefore, traces of carbon tetrachloride may be dissolved in water upon disinfection with chlorine.

Materials used in the construction of water distribution systems can influence water quality. The use of coal tar enamel coatings has been discontinued because these substances can release carcinogenic polynuclear aromatic compounds to aqueous solution. Increased use of plastic pipes has raised concern over three other potential sources of organics contamination: (1) dissolution of monomeric impurities from polymer plastics, (2) leaching of solvents from pipe-joint adhesives (Dressman and McFarren, 1978; Boettner et al., 1978) and (3) leaching of organic compounds into plastic pipe systems from contaminated soils. Improvements in manufacturing processes have eliminated many problems associated with leaching of vinyl chloride from polyvinylchloride (PVC) formulations. Pollution of water systems by chemicals such as methylethylketone (MEK), trichloroethene (TCE), and tetrachloroethene (PCE) derived from solvent welds may exert a wide-ranging impact on human health. Experiments conducted in the Montgomery Laboratories showed a rapid decline in the diffusion of

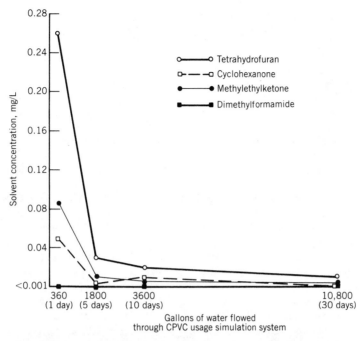

FIGURE 16-11. Solvent diffusion from chlorinated polyvinyl chloride pipe during usage simulation study.

tetrahydrofuran, dimethylformamide, cyclohexanone, and MEK during a 30-day usage simulation test, as shown in Figure 16-11. However, high concentrations of these solvents accumulated in dead ends of distribution networks with low flowrates (California Department of Health Services, 1980; Reich et al., 1981).

Organic Pollutants

Organic chemicals from industry, agriculture, and municipal effluents contribute extensively to the pollution of many water supplies. Surface waters are especially vulnerable to these types of pollution, but groundwater systems can also become contaminated. Contaminants that originate from a specific site are defined as "point-source pollutants," whereas substances that enter the aquatic environment over a broad area are referred to as "nonpoint-source pollutants." Groundwaters are more commonly subjected to point-source contamination. By contrast, large-scale surface water basins may contain organic chemicals which cannot be traced to a single site.

Industries that utilize large quantities of chemicals in manufacturing processes can be major sources of organic pollutants. Commercial facilities are often located in close proximity to major bodies of surface water. For example, the Mississippi and Ohio Rivers provide a plentiful supply of water for cooling to a major fraction of industries in the United States. Consequently, effluents from these activities can introduce a broad range of chemical contaminants to these river systems depending on the nature of the materials being processed and the degree of treatment at each facility.

Municipal sewage treatment plants constitute a second major point-source of organic contamination. Listed in Table 16-1 are 50 organic chemicals that were most often detected in municipal effluents (Shackleford, 1981). The substances have been ranked in order of decreasing frequency of occurrence. Chlorinated aliphatic and aromatic solvents dominate the top portion of this tabulation.

The quantity of agricultural pesticides used annually is staggering. In California alone, over 30,000 tons of chemicals have been applied annually. The vast majority of these substances are organic chemicals. The data compiled in Table 16-2 shows the 50 leading pesticides employed in California during 1976 (Li et al., 1979). Pesticide treatments are evenly distributed over a large acreage. Modern agricultural practice has been directed toward the use of nonrefractory pesticides, such as organophos-

TABLE 16-1.   Compounds Most Frequently Reported in Municipal Effluents

Reported Frequency Ranking	Compound	Reported Frequency Ranking	Compound
1	Tetrachloroethene	26	1,1-Dichloroethene
2	Dichloromethane	27	o-Ethyltoluene
3	Trichloroethene	28	Benzoic acid
4	2-(2-Butoxyethoxy)ethanol	29	2-N-Butoxyethanol
5	Benzene	30	Dimethyldisulfide
6	Toluene	31	Diethyl-D-phthalate
7	Chloroform	32	Lauric acid
8	Ethylbenzene		(N-dodecanoic acid)
9	1,1,1-Trichloroethane	33	1,8-Dimethyldisulfide
10	Phenol	34	2-Propanone (acetone)
11	p-Cresol	35	Tetradecanoic acid
12	Caffeine	36	Decanoic acid
13	m-Cresol	37	Methylisobutylketone
14	Cycloheptatriene	38	2,7-Dimethylnaphthalene
15	Octadecane	39	n-Pentadecane
16	Phenylacetic acid	40	Dibutylphthalate
17	Dioctylphthalate	41	1-Hydroxy-2-phenylbenzene
18	1,4-Dimethylbenzene	42	1,2,4-Trimethylbenzene
	(p-xylene)	43	Indole
19	1-Methylnaphthlene	44	n-Hexane
20	m-Xylene	45	n-Eicosane
21	Hexadecane (practical)	46	Dioctylphthalate
22	2-Methylnaphthalene	47	1,3-Dimethylnaphthalene
23	o-Cresol	48	p-Ethyltoluene
24	α-Terpineol	49	2,4-Dimethylphenol
25	Naphthalene	50	1,3,5-Trimethylbenzene

SOURCE: EPA (1980).

phate, that degrade rapidly in the environment following application. This development has helped to minimize the risk of water contamination. Nevertheless, the use of such magnitudes of pesticide chemicals demands that programs be developed to monitor water supplies that are subject to agricultural runoff.

ORGANICS ANALYSIS

Fundamentally, the analysis of trace organics involves the determination of the type and amount of individual carbon-containing molecules present in the aqueous environment. To accomplish this goal, numerous methods are available for the isolation, resolution, identification, and quantitation of the complex mixtures of organic compounds dissolved in water.

In the past decade, dramatic advancements have been achieved in the development of efficient techniques for the analysis of organic compounds at trace levels. However, available technology only permits the characterization of a limited number of substances that are likely to exist in aquatic systems. Approximately 500 of the almost 2 million known organic compounds have been detected in samples of drinking water (Donaldson, 1977). It may be assumed that an enhancement in the sensitivity of analytical techniques will bring about a proportional increase in the number of compounds that are accessible to measurement. The diagram in Figure 16-12 provides a graphical representation of this relationship. Under the conditions outlined in Figure 16-12, each order of magnitude (10 times) increase in analytical sensitivity will permit detection of a 10-fold higher number of substances. If the number of known organic chemicals should grow to 5 million, then all of the compounds might be found in water

TABLE 16-2. Leading 50 Pesticide Chemicals Applied in California in 1976 Based on Acres Treated

Rank	Chemical	Applied		Area Treated	
		(kg × 10⁴)	(lb × 10⁴)	(Ha × 10⁴)	(acres × 10⁴)
1	Methomyl	37	81	51	127
2	Paraquat (cation)	20	43	50	123
3	Aromatic petroleum solvents	68	150	46	113
4	Xylene	43	94	45	111
5	Sulfur	953	2100	39	97
6	Cacodylic acid	15	32	36	90
7	Parathion	34	74	36	89
8	Methyl paration	21	46	34	84
9	2,4-D, alkyl (Cl2) amine salts	53	117	28	69
10	Zinc sulfate	23	50	25	63
11	Petroleum distillate, aromatic	16	36	24	59
12	Sodium chlorate	91	201	23	57
13	Phosdrin-R	10	22	22	54
14	Phosdrin-R, other related	7	15	22	54
15	Di-syston-R	23	50	21	51
16	Azodrin-R	15	32	21	51
17	Dimethoate	25	56	20	50
18	Diazinon	16	36	20	49
19	Kelthane-R	37	81	19	48
20	MCPA acid	19	42	18	44
21	Carbaryl	35	78	17	43
22	Omite-R	34	76	17	43
23	Endosulfan	18	40	17	43
24	Petroleum hydrocarbons	426	939	17	42
25	Fundal-R	8	18	16	39
26	Phorate	15	33	14	35
27	Guthion-R	15	34	14	34
28	Petroleum oil, unclassified	574	1265	14	34
29	DEF	23	51	13	32
30	BTB	0.3	0.6	10.9	27
31	Ordram-R	34	76	10.5	26
32	Difolatan-R	20	45	9.3	23
33	Toxaphene	39	86	9.3	23
34	Strychnine	0.4	0.9	8.5	21
35	Folex-R	15	32	8.5	21
36	Folex-R, other related	0.9	2	8.5	21
37	Chlorothalonil	17	38	8.5	21
38	Benomyl	5	12	8.5	21
39	DNBP	59	130	8.5	21
40	Xylene range, aromatic solvent	15	33	8.1	20
41	Petroleum distillates	10	23	8.1	20
42	Meta-Systox	4	9	7.7	19
43	Captan	21	47	7.3	18
44	Carbofuran	5	11	7.3	18
45	Aldicarb	10	22	6.9	17
46	Monitor-R	5	12	6.9	17
47	Naled	7	16	6.9	17
48	Trifluralin	6	13	6.5	16
49	Dithane	10	23	5.7	14
50	Maneb	12	27	5.7	14

at the time a minimum analytical detection limit of 1 pg (10^{-12} g) per liter is attained. This point is designated by a ? in Figure 16-12. As analytical methodology evolves beyond this stage, the frequency of detection of organic substances in water could increase to 100 percent.

As the efficiency of analytical methodology improves, scientific debate shall no longer concern the potential existence of organic contamination in a given water supply, but shall focus on whether or not a measured level of pollution represents a significant health hazard. This question is left for toxicologists to resolve.

At this point it is clear that only the tip of the iceberg has been exposed with respect to organic contamination of water. However, abundant information has been gathered concerning the characteristics of many types of organic compounds that commonly occur in the aquatic environment. In this context, it would be useful to briefly discuss existing methods for the isolation of organic chemicals from water and the analytical instrumentation employed for their measurement and identification.

Isolation of Organics

Several methods for the isolation of organics from water are listed in Figure 16-13 (Trussell and Umphres, 1978a). Together, these techniques provide the analyst access to the determination of a broad range of chemicals. Highly polar, water-soluble compounds are generally analyzed directly without prior separation from aqueous solution. For example, direct aqueous injection gas chromatography is often used for the analysis of volatile carboxylic acids. Volatile organics of lower polarity can be removed from water samples by allowing the substances to partition into a small volume of gas phase (headspace) above the solution. Alternatively, the partition equilibrium may be enhanced through the use of a dynamic headspace technique which involves passing an inert gas, for example, nitrogen or helium, into the sample and collecting the organic solutes on a solid adsorbent as they enter the vapor phase. Compounds of moderate polarity that are not efficiently separated from water by direct utilization of dynamic headspace analysis can frequently be isolated following application of a preliminary concentration procedure such as distillation.

Four procedures routinely used to isolate the organic contaminants from water are dynamic headspace partitioning (purge-and-trap analysis), liquid–liquid solvent extraction, resin adsorption–elution, and closed-loop gas stripping (Bellar and Lichtenberg, 1974; Keith, 1976, 1981; Grob, 1973). Figure 16-14 gives the boiling points and aqueous phase concentrations for organic compounds that are amenable to each form of analysis.

Analysis for volatile organic solutes at microgram-per-liter concentrations is generally accomplished by dynamic stripping of the compounds from water with an inert gas. The contaminants are isolated from the vapor phase using an adsorbent trap. This procedure is commonly referred to as the purge-and-trap method or volatile organics analysis (VOA). Many volatile substances that are highly polar and hence water soluble cannot be efficiently removed from aqueous solution using the dynamic headspace technique.

Liquid–liquid extraction (LLE) techniques involve mixing an aqueous sample with a nonpolar organic solvent that is insoluble in water. Upon contact with the organic solvent, contaminants in the water will partition favorably into the solvent phase. In order to achieve minimum limits of detection, the excess solvent is generally removed by evaporation (boiling) prior to analysis. Examples of extractable compounds include pesticides, polychlorinated biphenyls (PCBs), organic acids, and neutral species.

In general, LLE methods are not utilized for the

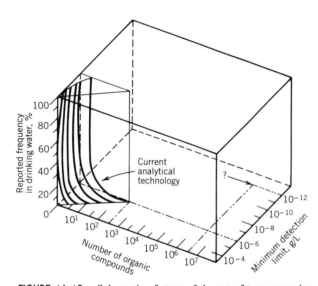

FIGURE 16-12. Schematic of state of the art of trace organics analysis in water. (Reprinted from *JAWWA*, Vol. 70, No. 11 (November 1978) by permission. Copyright 1978, The American Water Works Association.)

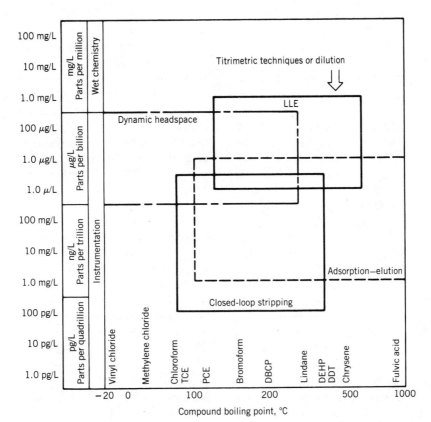

FIGURE 16-13. Isolation of organic compounds found in water volatility. (Reprinted from *JAWWA*, Vol. 70, No. 11 (November 1978) by permission. Copyright 1978, The American Water Works Association.)

FIGURE 16-14. Analytical approaches to measuring organic contaminants in water.

quantitative analysis of highly volatile organic solutes due to losses of compounds that can be incurred during concentration. However, the applicability of LLE without sample concentration has recently become popular for the routine measurement of trihalomethanes in drinking water (Keith, 1976; Richard and Junk, 1977; Mieure, 1977). Addition of salts to aqueous solution and selection of appropriate pH conditions allows a variety of polar substances to be readily extracted from water.

The adsorption–elution method is particularly well suited to the separation of trace organic pollutants from large quantities of water. In this technique, organic solutes are concentrated from aqueous solution by passing water through a column containing an adsorbent such as polymer resins or activated carbon. Subsequently, the compounds of interest are desorbed from the solid phase with an organic solvent. The use of different types of adsorbents enables optimal recovery of organics of different polarity, just as a wide range of extraction conditions can be selectively employed in LLE procedures. Resin adsorption is more efficient than liquid–liquid extraction for the analysis of many types of high-molecular-weight organics because the rate of adsorption of most solutes to a solid surface generally exceeds the rate at which solvent–water partition equilibria are attained.

Closed-loop stripping represents a variation of the purge-and-trap procedure in which a stream of air is repeatedly passed through a sample of water. Organic contaminants that partition into the gas phase are adsorbed in a small column packed with activated carbon. A schematic of the apparatus used in this analysis is pictured in Figure 16-15. Following removal of the volatilized organics, the air is recirculated to the sample through a pump that is specially constructed from noncontaminating materials. After treating the sample for 1–2 hr, the column is removed and the organic substances are extracted from the activated carbon with a small volume of organic solvent.

Shown in Figure 16-14 are several regions in which each of several analytical techniques can be applied to the analysis of organic contaminants in water. Under different concentration conditions, a specific procedure may be required for detection of a specific type of compound. Titrimetric methods can be utilized successfully for the analysis of many substances at high concentration. Alternatively, the water sample can be diluted to a concentration range that is appropriate for measurement by other

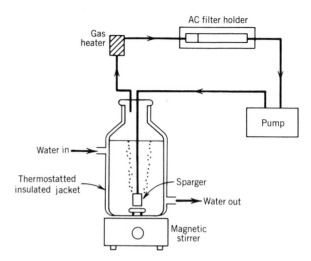

FIGURE 16-15.   Closed-loop stripping apparatus.

techniques. The purge-and-trap technique is the only technique that permits the recovery of solutes that have low boiling points. Closed-loop stripping provides the lowest minimum detection limits of any of the analytical procedures described in this chapter. However, nonvolatile substances such as DDT and chrysene cannot be readily separated from water using this technique. These compounds are amenable to analysis by liquid–liquid extraction or resin adsorption. The latter process is especially appropriate for the isolation of trace quantities of high-molecular-weight organics from large samples.

Several highly specialized techniques are currently being developed for the isolation of organic compounds of high molecular weight. The ultimate objective of this research is to provide methods for obtaining samples of naturally occurring organic polymers and other high-molecular-weight substances from water such that the chemical structure of the material remains unchanged during the separation process. Vacuum distillation and lyophilization (freeze-drying) selectively remove water from an aqueous sample and thereby leave behind an organic residue. Organic materials of high polarity can be isolated successfully using these techniques. Nonpolar substances of high molecular weight are amenable to separation by reverse osmosis and ultrafiltration. In reverse osmosis, a pressure-induced reverse gradient is established in order to concentrate solutes on the low-pressure side of a membrane (see Chapter 10). Ultrafiltration is based on a similar process; however, fractionation of organic solutes is achieved according to molecular size rather than chemical composition.

Identification and Quantification of Organics

Samples obtained by the isolation procedures described above will generally contain a myriad of organic compounds. Choice of appropriate procedural conditions can simplify the composition of the isolate to some degree. However, efficient separation of the complex mixture into distinct components is required for unambiguous identification and quantitation of individual chemicals. The most effective technique for the resolution of this mixture is chromatography. Two types of chromatographic procedures, gas chromatography (GC) and high-pressure liquid chromatography (HPLC) are routinely applied to the analysis of organic pollutants in the environment. In gas chromatography, individual substances are separated through either a partition or adsorption mechanism in which the volatilized compounds are reversibly transferred from an inert carrier gas, for example, nitrogen or helium, to a solid surface (adsorption) or a high boiling liquid bonded to a solid support (partition). In HPLC, the carrier stream is composed of a solvent or mixture of solvents maintained under high pressure. The efficiency of the separation process is dependent on the chemical and physical properties of the sample mixture, the carrier, and the solid or liquid stationary phase. For a detailed discussion of these factors, the reader should consult review compilations on this subject (Keith, 1976; Keith, 1981 and references therein). However, it should be noted that careful selection of these parameters will permit the selective separation of a wide variety of organic substances.

The identity and concentration of individual chemical species must be accurately determined once the organic material has been successfully removed from aqueous solution. Tabulated in Figure 16-16 are various detection devices that can be used to monitor the chemical and physical properties of the components eluted from chromatographic columns (Trussell and Umphres, 1978a). Although the properties of many types of organic compounds are quite similar, several of these detectors allow differentiation between subtle differences in chemical structure. For the analysis of volatile substances, the instruments are commonly employed in conjunction with gas chromatography.

The flame ionization detector (FID) is a moderately sensitive, nonspecific detector for most categories of organic compounds. By contrast, the electron capture detector (ECD) system is highly specific for molecular structures that contain halogen, oxygen, and nitrogen atoms as well as a few other types of substituents. Detection limits of less than 1 pg can be achieved with the use of the ECD for the analysis of halogenated organics. Thermionic or alkali flame ionization detectors also provide efficient detection of specific classes of chemicals: phosphorus-substituted compounds and

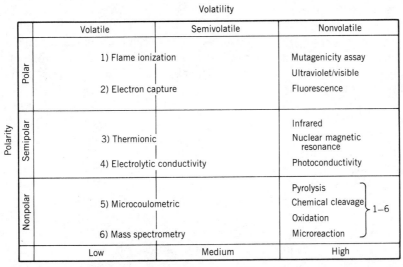

FIGURE 16-16.   Identification and quantification of organics in water. (Reprinted from *JAWWA*, Vol. 70, No. 11 (November 1978) by permission. Copyright 1978, The American Water Works Association.)

nitrogeneous derivatives. The microcoulometric and electrolytic conductivity detectors are sensitive to compounds bearing halogen, sulfur, and nitrogen substituents. Each of these detectors offers a linear dynamic range in excess of 10^3 and, hence, is suitable for the identification and quantitation of organic compounds over a broad range of concentrations.

Combined gas chromatography/mass spectrometry (GC/MS) has been utilized extensively for the analysis of environmental contaminants. In this instrument, molecules of a substance are ionized in the gas phase, generally through bombardment with a beam of high-energy electrons. At low sample pressures (10^{-7}–10^{-6} mm Hg), the excess energy of collision is transmitted to the chemical bonds between atoms, thereby initiating a series of secondary intramolecular decomposition reactions. The primary and secondary ionic products are focused to a collector through systematic variation of applied magnetic and/or electric fields. The mass-to-charge ratio (m/e) of these ions provides a unique structural "fingerprint" (mass spectrum) for each compound. While mass spectrometric technique can often permit identification of an organic compound, it has a more limited linear range than other detectors used for quantitative GC measurements.

Nonvolatile organics of high molecular weight present a more difficult problem with respect to identification and quantitation. For example, molecules of naturally occurring humic and fulvic acids generally contain many different organic functional groups within the various parts of their chemical structure. Therefore, the physical and chemical properties of these substances represent a composite picture of these substituents. Instruments that can be used in conjunction with high-performance liquid chromatography include detectors that record ultraviolet, visible and infrared absorption of radiation, fluorescence intensity, nuclear magnetic resonance, and photoconductivity. Another method that has recently been introduced involves the use of microbiological assay systems. Application of organic isolates to a mutagenicity assay, for example, the Ames test, can sometimes identify a potentially carcinogenic substance.

A great deal of research is needed to elucidate further the fundamental characteristics of the high-molecular-weight organics present in water supplies. Attainment of this objective will require the use of existing analytical technology as well as the development of new fractionation procedures. Per-

haps, a systematic approach to the analysis of organic polymers could be designed using available isolation methods such as reverse osmosis in conjunction with chemical reactions that would selectively divide the substrate into identifiable fractions. Techniques, such as oxidation, reduction, hydrolysis, and high-temperature pyrolysis, have been used with moderate success for the analysis of soil-derived humic substances. Following chemical degradation, modern chromatographic methods would facilitate the separation and analysis of the monomeric decomposition products. The identity of these products would provide important information about the molecular structure of the original polymer.

Surrogate Measurements

While the details of the chemical composition of organic matter present in drinking water are of interest, it is neither technologically nor economically feasible to perform comprehensive quantitative analyses for each type of organic compound on a routine basis. Therefore, it is important to consider the use of surrogate measurements that provide a representative picture of the organics in a water supply without requiring the application of sophisticated instrumentation. Three parameters that fulfill these criteria are total organic carbon (TOC), total organic halogen (TOX) and standard trihalomethane potential (STP).

Total organic carbon (TOC) is the parameter most commonly used to define the organic content of an aquatic system. In the mid-1970s, instrumentation became available for the reliable determination of TOC at concentrations of less than 1 mg/L (Takahashi, 1976). Since that time, measurements of organic carbon data have been reported for a full spectrum of drinking water supplies (see Figure 16-4). TOC has replaced the use of other measurements of water quality such as chemical oxygen demand (COD) and carbon–chloroform extracts (CCE) for the assessments of potential organic contamination.

Instrumentation used for the measurement of TOC rely on the conversion of dissolved organic compounds to a single chemical form while excluding inorganic carbon compounds from the analysis. Generally, analyses are conducted through the quantitative oxidation of organic carbon to carbon dioxide. In some processes, the CO_2 oxidation product is reduced to methane prior to detection. A wide variety of techniques exist for the transformation of organic matter to CO_2. Two of the most popular methods are pyrolysis and wet chemical oxidation. Pyrolysis, which gives the highest yield of CO_2, is conducted at 800°C in the presence of molecular oxygen (O_2) or oxides of manganese (MnO_2) and copper (CuO). Inorganic carbon is excluded from the analysis by acidification of the sample to convert dissolved carbonates to CO_2. Aeration of the acidified solution removes the excess carbon dioxide before pyrolysis. Unfortunately, the pyrolysis process can only accommodate small volumes (microliters) of water and as a result lacks the sensitivity of techniques based on chemical oxidation. Wet chemical systems utilize acid persulfate in the presence of either heat or ultraviolet light to effect oxidation. Lower limits of detection are attained with this process through the use of relatively large sample volumes (milliliters). Compared to the pyrolysis technique, the major limitation of the process lies in the fact that quantitative recoveries of refractory organics may not be achieved due to less vigorous oxidation conditions.

Total organic halogen (TOX) refers to the total mass concentration of organically bound halogen atoms (X = Cl, Br, of I) present in water. From the standpoint of water quality, this parameter is especially significant because it accounts not only for volatile-halogen-containing compounds such as the trihalomethanes, trichloroethene, tetrachloroethene, and so on, but it also includes the contribution of halogenated substances of high molecular weight that are also suspected health hazards.

The most popular method for TOX analysis involves the adsorption of organohalide solutes onto activated carbon (Dressman et al., 1977). The particles of carbon are then washed in order to displace inorganic halides (predominantly Cl^-). After treatment with nitrate, the carbon adsorbent is subjected to pyrohydrolysis which converts the organically bound halogen to hydrogen halides (HX) and hypohalous acids (HOX). The aqueous effluent from the pyrohydrolysis step (pyrohydrozylate) can be analyzed for halide ion using a specific ion probe or by direct injection of the sample into a microcoulometric titration cell.

The standard trihalomethane potential (STP) is employed to assess the maximum tendency of the organic compounds in a given water supply to form THMs upon disinfection. Aqueous systems that exhibit low STP values are considered to be superior when it becomes necessary to choose between alternative sources of water. The measurement of STP is performed by first adding bromide to the water to produce an aqueous phase concentration of 10 mg/L. After adjusting the pH of the water to 11, the sample is treated with chlorine at a level of 100 mg/L and is then incubated at 35°C for 24 hr. Chromatographic analysis of the water sample for trihalomethane reaction products is performed following dechlorination and pH neutralization. Once the STP data have been utilized to select potential sources of potable water, an analysis of total potential trihalomethane (TPTHM) formation is generally performed.

The measurement of TPTHM is preceded by a short-term test for chlorine demand in order to determine the dose of chlorine necessary to attain a 1 mg/L free chlorine concentration after 24 hr of contact time. This test ensures that an adequate Cl_2 dosage is selected to maintain a chlorine residual through the end of the TPTHM analysis. The water sample is chlorinated at the appropriate level and measurements of pH, free chlorine, and THM concentrations are conducted after 0, 1, 10, 100, 1000, and 10,000 min. The time profile constructed from this data (TPTHM curve) combined with measurements of TOC and bromide ion concentration allows the evaluation of the potential for THM production under reaction conditions that closely simulate those encountered in the actual treatment and distribution system.

MONITORING

Fate and Transport of Organic Chemicals in the Environment

In designing a program to monitor water systems for organics pollution, it is important to consider the fundamental physical and chemical processes that govern the fate of organic chemicals in the aquatic environment. Figure 16-17 shows the principal mechanisms for the introduction, transportation, and degradation of organic substances in water, soil, and atmospheric systems.

Organic contaminants of anthropogenic origin can enter the aquatic environment in numerous ways. The pathways of introduction include disposal of chemical wastes in landfills, accidental spillage during the storage and transfer of chemicals, waste effluents from manufacturing facilities and other commercial activity, and the discharge of

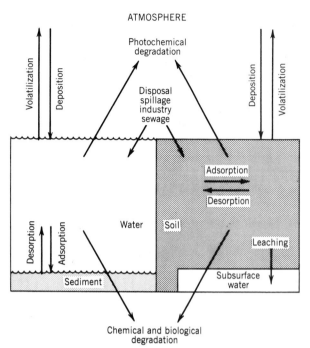

FIGURE 16-17. Fate and transport of organics in the environment.

sewage directly from municipal treatment plants or indirectly through groundwater injection. Naturally occurring substances are present in water as the result of the chemical and microbial degradation of vegetation and soil-derived materials. Wet and dry deposition of organic compounds from the atmosphere may also take place. Substances are released to air through combustion of organic matter and volatilization from land and water surfaces. In aqueous media, organic solutes can also be adsorbed by bottom sediments as well as soil particles suspended in water. Adsorption–desorption equilibria determine the distribution of contaminants between the aqueous and solid phases. In soils, these processes also affect the quantity of material that is leached into groundwater.

Monitoring Programs

In order to develop a comprehensive plan for the analysis of trace organics, the following steps are recommended (Trussell and Trussell, 1980, and references therein):

1. Preliminary review of data generated in previous or related studies of similar water systems.

2. Conduct a survey of contaminant sources to delineate the type of pollutants and locations from which chemicals may be discharged.

3. Select analytical procedures to detect the compounds of interest (at the very least, chemicals revealed in the preliminary contaminant survey as well as substances governed by pertinent regulations should be included in the analysis).

4. Determine the sampling sites that are to be used in the monitoring program.

5. Establish an appropriate sampling frequency based on available data, statistical effects, and financial resources.

Considering the potential impact of decisions based on the results of such a program, it is important to ensure that the sampling data accurately reflect the water quality of the aqueous system. Composite sampling techniques should be used as an alternative to the collection of grab samples. For example, a composite sample can be prepared for the analysis of volatile organic compounds using a graduated cylinder fitted with a movable Teflon plunger seal. Water is introduced to the apparatus through an automatic solenoid valve at preset time intervals. If the collection of field composites is not feasible, then grab samples can be obtained at scheduled times which can later be composited in the laboratory.

Samples should be collected using bottles that have been vigorously cleaned to remove potential contaminants. Chemical preservatives are added to the sample in order to prevent chemical or microbiological degradation of the organic solutes prior to analysis. Samples of uncontaminated water obtained from the laboratory should also be transported to the sampling site and exposed to the environment in a manner similar to that used for the other sample containers. The samples are then returned to the laboratory for analysis of chemical artifacts that may have been introduced during the sampling and shipment procedures. The so-called travel blank has become an essential part of monitoring programs that are designed to measure organic contaminants in water at parts-per-trillion (ng/L) concentrations.

The components of a generalized monitoring program are listed in Table 16-3. This hypothetical scheme incorporates each of the analytical procedures discussed in previous sections of this chapter. Displayed in the table is a suggested sampling

TABLE 16-3. Example Organics Monitoring Program[a]

Analysis	Blue River	Creek Junction	Canal Outlet	Plant Influent	Plant Effluent	Distribution System
Purgeables	M	M	M	M	M	M
Taste and odor purgeables	M*	M*	M*	M*	M*	M*
Base/neutral/acid extractable	Q	Q	Q	Q	Q	Q
Pesticide/herbicide extractable	Q	Q	Q	Q	Q	Q
Closed-loop stripping	Q	Q	Q	Q	Q	Q
Total organic carbon	BW	BW	BW	BW	BW	BW
Total organic halogen	BW	BW	BW	BW	BW	BW
Total potential trihalomethanes	Q	Q	Q	Q	—	—

[a] Q, Quarterly; M, monthly; BW, biweekly; M*, monthly, but omit sample at quarters.

schedule based on typical budgetary constraints. Less frequent analyses are recommended for more expensive procedures such as closed-loop stripping. Analyses of water for volatile compounds such as trihalomethanes may be condensed on a monthly basis, since these organic contaminants are frequently already analyzed in the purgeable fraction.

REMOVAL

The elimination of organic compounds from potable water can be accomplished through the use of unit processes such as reverse osmosis (RO), ion exchange, air stripping, adsorption, oxidation, and co-agulation (Trussell and Trussell, 1980). Figure 16-18 presents an overview of the types of organic compounds to which each unit process may be applied. The boundaries between each region of the diagram represent a generalization based on the aqueous solubility and molecular weight of the organic substances. Therefore, there exist many overlapping areas and omissions and exceptions to these examples. Nonetheless, the trends shown in this drawing are valid and are intended to be instructive rather than an absolute reference. Reverse osmosis has been omitted from the figure because current advances in membrane technology are rapidly increasing the types of organics to which RO techniques can be applied. Although organics removal by reverse osmosis is a rapidly growing area of research,

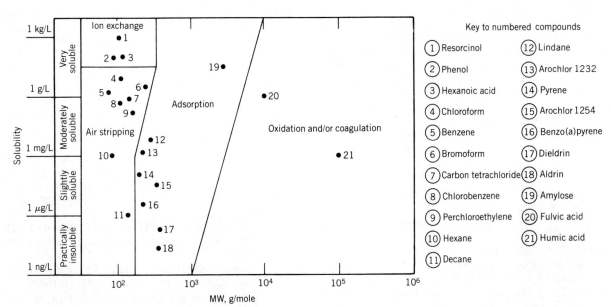

FIGURE 16-18. Treatment processes for organics removal.

large-scale application of this process remains prohibitively expensive.

Ion exchange removes from aqueous solution polar and ionized organic compounds such as phenols and amines that are highly water soluble. Air stripping effectively eliminates volatile organics, for example, carbon tetrachloride, that are of low solubility and molecular weight. Adsorption is recommended for use in the removal of refractory substances such as pesticides and polynuclear aromatic hydrocarbons. In general, oxidation and/or coagulation are used to degrade and precipitate high-molecular-weight humic and fulvic acids. The latter two techniques could be applied as a first-stage process prior to ion exchange, air stripping, and adsorption for the comprehensive removal of a wide range of organic pollutants from water. The reader is referred to appropriate chapters within this volume for a detailed discussion of the theory and practice of each unit process.

REFERENCES

Anderson, C., and Maier, W. J., "The Removal of Organic Matter from Water Supplies by Ion Exchange," Report for Office of Water Research and Technology, Washington, DC (1977).

Babcock, D. B., and Singer, P. C., "Chlorination and Coagulation of Humic and Fulvic Acids," *JAWWA*, **71**, 149–152 (1979).

Bellar, T. A., and Lichtenberg, J. J., "Determining Volatile Organics at Microgram-per-Litre Levels by Gas Chromatography," *JAWWA*, **66**, 739–744 (1974).

Black, A., and Christman, R., "Chemical Characteristics of Fulvic Acids," *JAWWA*, **55**, 897–912 (1963).

Boettner, E. A., Ball, G. L., Hollingsworth, Z., and Aquino, R., "Organic and Organotin Compounds Leached from PVC and CPVC Pipe," U.S. EPA Project Summary, Health Effects Research Laboratory, Cincinnati (1982).

Burttschell, R. H., Rosen, A. A., Middleton, F. M., and Ettinger, M. B., "Chlorine Derivatives of Phenol Causing Taste and Odor," *JAWWA*, **51**, 205–214 (1959).

California Department of Health Services, "Solvent Leaching from Potable Water Plastic Pipes," J. M. Montgomery Consulting Engineers, Final Report (1980).

Christman, R. F., and Ghassermi, M., "Chemical Nature of Organic Color in Water," *JAWWA*, **58**, 723–741 (1966).

Donaldson, W. T., "Trace Organics in Water," *Environ. Sci. Tech.*, **11**, 348 (1977).

Dressman, R. C., and McFarren, E. F., "Determination of Vinyl Chloride Migration from Polyvinyl Chloride Pipe into Water," *JAWWA*, **70**, 29–30 (1978).

Dressman, R. C., McFarren, E. F., and Symons, J. H., "An Evaluation of the Determination of Total Organic Chlorine

in Water by Adsorption onto Ground Granular Activated Carbon, Pyrohydrolysis, and Chloride Ion Measurement," Proceedings AWWA WQTC, Kansas City, MO (1977).

Ettinger, M. B., and Ruchhoft, C. C., "Effect of Stepwise Chlorination on Taste- and Odor-Producing Intensity of Source Phenolic Compounds," *JAWWA*, **43**, 561–567 (1951).

Ghassemi, M., and Christman, R. F., "Properties of Yellow Organic Acids of Natural Waters," *Limnol. Oceanog.*, **13**, 583–597 (1968).

Gjessing, E. T., *Physical and Chemical Characteristics of Aquatic Humus*, Ann Arbor Science, Ann Arbor, MI (1976).

Gjessing, E., and Lee, G. F., "Fractionation of Organic Matter in Natural Waters on Sephadex Columns," *Environ. Sci. Technol.*, **1**, 631–638 (1967).

Grob, K., "Organic Substances in Potable Water and in Its Precursor. Part I. Methods for Their Determination by Gas-Liquid Chromatography," *J. Chromatog.*, **84**, 255–273 (1973).

Hendrickson, J. B., Cram, D. J., and Hammond, G. S., *Organic Chemistry*, McGraw-Hill, New York (1970).

Kavanaugh, M. C., "Modified Coagulation for Improved Removal of Trihalomethane Precursors," *JAWWA*, **70**, 613–620 (1978).

Keith, L. H. (ed.), *Identification and Analysis of Organic Pollutants in Water*, Ann Arbor Science, Ann Arbor, MI (1976).

Keith, L. H. (ed.), *Advances in the Identification and Analysis of Organic Pollutants in Water*, Ann Arbor Science, Ann Arbor, MI (1981).

Li, M-V., Kilgore, W. W., and Aref, K. E., "Pesticides and Toxic Chemicals in Irrigated Agriculture," Report to California State Water Resources Control Board (1979).

Lillard, D. A., and Powers, J. J., "Aqueous Odor Thresholds of Organic Pollutants," EPA-660/4-75-002 (1975).

Mieure, J. P., "A Rapid and Sensitive Method for Determining Volatile Organohalides in Water," *JAWWA*, **69**, 60–62 (1977).

Murphy, T. R., Lean, D. R. S., and Nalewajko, C., "Blue-Green Algae, Their Excretion of Ion-Selective Chelators Enables Them to Dominate Other Algae," *Science*, **192**, 900–902 (1976).

Oden, S., "Humic Acids," *Kolloidchem Beineftc.*, **11**, 75 (1919).

Paul, M. A., and King, E. J., and Farinholt, L. A., *General Chemistry*, Harcourt, Brace and World, New York (1967).

Perdue, E. M., "Solution Thermochemistry of Humic Substances. I. Acid-Base Equilibria of Humic Acid," in Proceedings ACS 175th National Meeting, pp. 375–378 (1978).

Reich, K. D., Trussell, A. R., Lieu, F. Y., Leong, L. Y. C., and Trussell, R. R., "Diffusion of Organics from Solvent-Bonded Plastic Pipes Used for Potable Water Plumbing," in Proceedings AWWA Annual Conference, Part 2, pp. 1249–1260 (1981).

Richard, J. J., and Junk, G. A., "Liquid Extraction for the Rapid Determination of Halomethanes in Water," *JAWWA*, **69**, 62–64 (1977).

Rook, J., "Formation of Haloforms During Chlorination of Natural Waters," *Water Treatment Exam.*, **23**, 234 (1974).

Rook, J. J., "Haloforms in Drinking Water," *JAWWA*, **68**, 168–172 (1976).

Rook, J. J., "Chlorination Reactions of Fulvic Acids in Natural Waters," *Environ. Sci. Technol.*, **11**, 478–482 (1977).

Searle, C. E. (ed.), *Chemical Carcinogens*, ACS Monograph 173, American Chemical Society, Washington, DC (1976).

Shackleford, W., "A Computer Survey of Gas Chromatography/Mass Spectrometry Data Acquired in The U.S. Environmental Protection Agency Priority Pollutant Screening Analysis: System and Results," in L. H. Keith (ed.), *Advances in the Identification and Analysis of Organic Pollutants in Water*, Vol. 2, Ann Arbor Science, Ann Arbor, MI pp. 555–593 (1981).

Singer, P. C., Borchardt, J. H., and Calthurst, J. M., "The Effects of Permanganate Pretreatment on Trihalomethane Formation in Drinking Water," *JAWWA*, **72**, 573–578 (1980).

Steelink, C., "What is Humic Acid?" *J. Chem. Educ.*, **40**, 379–384 (1973).

Steelink, C., "Humates and Other Natural Organic Substances in the Environment," *J. Chem. Educ.*, **54**, 599–603 (1977).

Stevens, A. A., and Symons, J. M., "Measurement of Trihalomethane and Precursor Concentration Changes," *JAWWA*, **64**, 546–554 (1977).

Symons, J., Bellar, T. A., Carswell, J. K., DeMarco, J., Kropp, K. L., Robeck, G. C., Seeger, D. R., Slocum, C. J., Smith, B. L., and Stevens, A. A., "National Organics Reconnaissance Survey for Halogenated Organics," *JAWWA*, **67**, 634–647 (1975).

Takahashi, Y., "Ultra Low Level TOC Analysis of Potable Waters," Proceedings AWWA WQTC, San Diego, CA (1976).

Trehy, M. L., and Bieber, T. I., "Detection, Identification and Quantitative Analysis of Dihaloacentonitules in Chlorinated Natural Waters," in L. H. Keith (ed.), *Advances in the Identification and Analysis of Organic Pollutants in Water*, Vol. 2. Ann Arbor Science, Ann Arbor, MI pp. 941–975 (1981).

Trussell, A. R., and Umphres, M. D., "An Overview of the Analysis of Trace Organics in Water," *JAWWA*, **70**, 595–603 (1978a).

Trussell, R. R., and Trussell, A. R., "Evaluation and Treatment of Synthetic Organics in Drinking Water Supplies," *JAWWA*, **72**, 458–470 (1980).

Trussell, R. R., and Umphres, M. D., "The Formation of Trihalomethanes," *JAWWA*, **70**, 604–612 (1978b).

Verschueren, K., *Handbook of Environmental Data on Organic Chemicals*, Van Nostrand Reinhold, New York (1977).

Weast, R. C. (ed.), *Handbook of Chemistry and Physics*, 54th ed. CRC Press Cleveland (1973).

— 17 —

Taste and Odor

The human senses of taste and smell are stimulated by myriad chemical compounds, both inorganic and organic. Certain of these compounds are found occasionally in domestic water supplies and, more than any other factor, they influence the palatability of the product. Virtually all water treatment plants contain provisions for the feeding of taste and odor control chemicals, adsorbents, or both. Additionally, many agencies practice a variety of preventative and control measures in raw water reservoirs, lakes, and rivers. The purveyor of drinking water who does not experience taste and odor problems, either periodically or sporadically, is indeed the exception rather than the rule. It is impossible to accurately estimate the annual expenditure, nationally, on taste and odor control measures, but this can be expected to be a significant portion of total water treatment costs and may amount to many millions of dollars. Recommended limits on odors are set by the Environmental Protection Agency (EPA) and are to be considered by the individual states in their secondary drinking water standards.

This discussion will describe the mechanisms of taste and smell, discuss some of the possible sources of tastes and odor compounds in drinking water, and outline means to prevent their development or to remove them once they have appeared.

PHYSIOLOGY OF ODOR AND TASTE PERCEPTION

Both taste and odor are aspects of human perception of aesthetic quality. Taste is an accumulation of several sensory responses from olfactory receptors in the upper part of the nasal cleft, gustatory receptors primarily on the tongue, and other receptors on the skin throughout the nasal cavity. The sense of smell, however, is a much more sensitive mechanism.

The olfactory receptors are located in the olfactory epithelium at the roof of the nasal cavity (Figure 17-1a). Embedded in the epithelium tissue are nerve fibers whose endings receive odorous molecules and send signals to the olfactory bulb in the brain (Amoore et al., 1964).

There are two principal theories presently used to describe the process of odor detection: the stereochemical theory and the chromatographic theory. The stereochemical theory asserts that there are only seven primary odors, and that molecules of compounds having similar odors are also similar in shape. Receptor sites in the nose have corresponding depressions or slots where these molecules may attach and cause a sensory response. The seven

TABLE 17-1. The Primary Odor Qualities

Primary Odor	Familiar Substance
Camphoraceous	Moth repellent
Pepperminty	Mint candy
Floral	Roses
Ethereal	Dry cleaning fluid
Pungent	Vinegar
Putrid	Rotten egg

SOURCE: Amoore et al. (1964).

primary odors are shown in Table 17-1 with familiar examples.

The second theory argues that the tremendous surface area of the olfactory epithelium (600 cm^2) allows it to function as a gas chromatograph, separating and distinguishing between various compounds (Roderick, 1966).

The taste receptors consist of four types of papillae or gustatory cells with hairlike fibers attached. These receptors allow the perception of four basic taste qualities: salt, sour, sweet, and bitter (Figure 17-1b). The structure of some typical compounds that produce these flavors are presented in Table 17-2. As shown, relatively simple ionic compounds produce salty and sour tastes, whereas more complex organic species cause sweet and bitter tastes. Electron micrographs have shown that papillae that respond to salty and sour tastes have similar shapes and blood supplies; correspondingly, a second set of papillae, which respond to bitter and sweet tastes, can also be characterized by common shapes and blood supplies. Recent gustatory theory suggests that the salt–sour papillae detect taste with an ion exchange phenomenon, while the bitter–sweet papillae detect taste with an enzyme inhibition mechanism.

While there are only four different types of taste, there is evidence that olfactory stimulants are highly varied. Table 17-3 shows different odorous substances along with their odor threshold in air. Table 17-4 lists several compounds and their odor thresholds in water. Though certainly the human olfactory function is less sensitive than many animals and insects, it is extremely sensitive to certain compounds. Methylisoborneal (MIB) and geosmin are examples of this extreme sensitivity with threshold odor concentrations in the range of 5 parts per trillion based on research conducted by the Metropolitan Water District of Southern California.

The flavor of a substance results from a simulta-

TABLE 17-2. Examples of the Four Primary Taste Qualities

Taste	Compound	Formula	Structure
Sour	Hydrochloric acid	HCl	$H^+ + Cl^-$
Salty	Sodium chloride	NaCl	$Na^+ + Cl^-$
Sweet	Sucrose	$C_{12}H_{22}O_{11}$	
Bitter	Caffeine	$C_8H_{10}N_4O_2$	

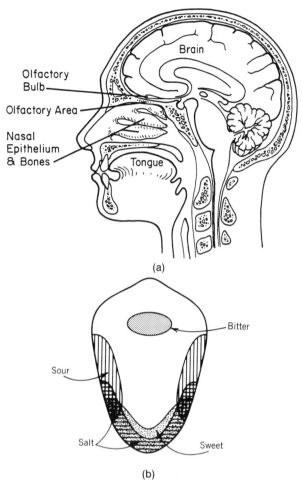

(a)

(b)

FIGURE 17-1. (a) Location of olfactory epithelium in the head. (Source: Amoore et al, 1964.) (b) Specific regions of the tongue responsible for four tastes: salt, sweet, bitter, sour.

TABLE 17-4. Odor Thresholds of Various Substances in Water

Compound	Threshold Odor Concentration (mg/L)
2-Octanol	0.13
Styrene	0.05
Ethylbenzene	0.1
Nathalene	0.007
p-Dichlorobenzene	0.15
Chloroform	20.0
Nonanal	0.001
Methyl sulfide	0.003
Geosmin	0.000005
Methylisoborneol (MIB)	0.000005
Trichloroethene	2.6
Tetrachlorethene	2.8
Dichloromethane	24.0
Toluene	0.14

neous sensation of taste on the tongue and odor in the olfactory center in the nose. When a substance is taken into the mouth, a flow of air containing aromatic chemicals travels up a passageway to the nose, eliciting simultaneous stimulation of various specialized taste and odor receptors both in the nose and on the tongue. As a result, flavors are often composed of a number of complex, interrelated sensations. For example, 11 different compounds are responsible for the flavor of a blueberry, but no single compound has a blueberry flavor by itself.

TABLE 17-3. Odor Thresholds of Various Substances in Air

Substance	Description	Concentration Causing Faint Odor (mg/L)
Allyl sulfide	Garlic odor	0.00005
Amyl acetate (*iso*)	Banana odor	0.0006
Benzaldehyde	Odor of bitter almonds	0.003
Chlorine	Pungent and irritating odor	0.010
Coumarine	Vanilla odor, pleasant	0.00034
Crotyl mercaptan	Skunk odor	0.000029
Diphenylchlorarsine	Shoe polish odor	0.0003
Hydrogen sulfide	Odor of rotten eggs, nauseating	0.0011
Ozone	Slightly pungent, irritating odor	0.001
Phenyl isothiocyanate	Cinnamon odor, pleasant	0.0024
Phosgene	Odor of ensilage or fresh-cut hay	0.0044

SOURCE: Adapted from Olishifski (1971).

TASTE AND ODOR MEASUREMENT

Tastes and odors in potable water supplies are typically assessed for both strength and character. Testing procedures for these properties are described in *Standard Methods* (American Public Health Association, 1980) and should be adhered to whenever taste or odor measurements are attempted.

Levels of taste- or odor-causing contaminants are quantified in a gross manner through threshold odor or taste tests. The water to be examined is diluted with taste- and odor-free water in a series of dilutions and presented in order of highest to lowest dilution to a panel of selected persons. The sample average at which odors or tastes are just detected determines the threshold number according to the following formula:

$$\text{Threshold number} = \frac{A + B}{A}$$

where A is the volume of the sample tested and B is the volume of taste- and odor-free dilution water used. Thus, if a 100-mL sample is diluted with 100 mL of odor-free water and odor is just detectable in the product water, the threshold odor number (TON) is calculated to be 2.

A precise, accurate technique for detecting several taste- and odor-causing organic compounds in water, sediments, and cultures is closed-loop stripping analysis (CLSA). This method, which strips semivolatile organic compounds from water for identification on a gas chromatograph–mass spectrometer (GC–MS), has equal or greater sensitivity than the human nose (McGuire et al., 1981). Table 17-5, which lists five odorous compounds, their CLSA detection limits in nanograms per liter (ng/L), and their lowest reported threshold odor concentrations, shows that CLSA is capable of detecting at or below the lowest reported threshold odor concentrations.

Qualitative characterization of tastes and odors in drinking water is much more subjective than the quantitative measures just described. Typically, panelists are asked to define the nature of the taste or odor they are experiencing in terms of standard descriptions or in their own words. An overall trend is sought and recorded in the test records. This information is primarily useful in determining odor or taste origins and in defining corrective measures. It must be noted that not all tastes or odors are undesirable. Distilled water and other tasteless, odorless

TABLE 17-5. Odor Thresholds and CLSA Detection Limits

Compounds	CLSA Detection Limit (ng/L)	Lowest Reported Threshold Odor Concentration (ng/L)
Geosmin	2	10
2-Methylisoborneol	2	29
2-Isopropyl-3-methoxy pyrazine	2	2
2-Isobutyl-3-methoxy pyrazine	2	2
2,3,6-Trichloranisole	5	7

SOURCE: McGuire et al. (1981). Reported with permission.

waters are often described as flat and unattractive. Among the many qualitative descriptors frequently applied to drinking water tastes and odors are swampy, grassy, musty, medicinal, phenolic, septic, sweet, and fishy.

SOURCES OF TASTES AND ODORS IN WATER SUPPLIES

Tastes and odors in water supplies can generally be attributed to two different causative elements: the actions of human beings upon the aquatic environment and natural forces within the environment. But these sources of tastes and odors are not unrelated: for example, odors due to biological degradation of algae and their waste products may sometimes be traced to an upstream nutrient input of human origin. This section will examine the sources directly responsible for taste and odor production. The examination of these sources is essential when attempting to identify particular tastes or odors.

Surface water supplies are more commonly linked to tastes and odors than groundwater supplies among large water utilities, both proportionally and quantitatively. A 1974 taste and odor survey (AWWA, 1976) of 120 large water purveyors in the United states and Canada showed that while only 40 percent of the utilities with only groundwater supplies had taste and odor problems, approximately 85 percent of the utilities using surface water had to deal with taste and odor problems. In addition, utilities with partial or total surface water

supplies comprised five-sixths of all utilities surveyed. This combined impact of surface water supplies makes taste and odor a prevalent problem for
water treatment plants in North America.

Groundwater

Most odors and tastes in groundwater supplies are
natural in origin. They are caused by bacterial
actions within the groundwater aquifers or the dissolution of salts and minerals as groundwater percolates and flows through geologic deposits. Intrusion
of salt or mineral-bearing waters (such as seawater)
may also result in taste or odor problems. Recently,
tastes and odors in some groundwaters have been
attributed to human sources, such as landfill
leachate.

One of the most common odor problems in
groundwater supplies is hydrogen sulfide (H_2S). It is
frequently characterized as a rotten egg odor, but at
low concentrations it may also impart a swampy,
musty odor. The odor threshold concentration of
H_2S in water is less than 100 ng/L (0.0001 mg/L)
and odors from waters containing 0.1–0.5 mg/L or
greater are offensive (Pomeroy and Cruze, 1969;
Lochrane, 1979). Sulfides in groundwater result
from anaerobic bacterial action on organic sulfur,
elemental sulfur, sulfates, and sulfites. A detailed
description of these processes and reactions may be
found in a number of references (e.g. Lochrane,
1979; Chen, 1975).

Reduced iron and manganese may also pose taste
problems in groundwater. Although tastes due to
dissolved iron or manganese are not particularly
noxious, they can render a water unpalatable and
cause problems in pipelines, water services, laundry facilities, and so on.

High salt content, as characterized by total dissolved solids (TDS) or conductivity, can result in
taste problems but does not usually result in objectionable odors. Consumers generally prefer waters
with lower TDS content, although this preference is
not strong (Bruvold, 1967). The current revival of
bottled mineral-bearing waters, however, may indicate that other psychophysical effects may affect
taste preference.

Human-induced tastes and odors in groundwater
occur as a result of chemical dumping, landfill disposal, mining and agricultural activities, or industrial waste disposal. A variety of synthetic organic
chemicals have been identified in groundwater supplies. Examples include trichloroethylene (TCE),

which has been at highly objectionable concentrations in wells throughout the country and numerous
toxic substances found at the Love Canal dumping
ground in New York. In 1976, the Resource Conservation and Recovery Act was legislated in the
United States to control, regulate, and monitor disposal of potentially hazardous wastes, but because
of the extremely slow movement of groundwater,
current taste and odor problems may take many
years to correct.

Surface Waters

Taste and odor problems are proportionally more
common in surface waters than in groundwaters
largely because of the presence of algae. In addition, direct organic inputs such as autumnal leaf
fall, runoff, and agricultural drainage provide ample
nutrients for microorganisms that can often generate taste- and odor-producing compounds.

Decaying vegetation from leaf fall and other
sources may result in brown-colored, sweet-smell-
ing water. These effects are due to suspended and
dissolved glucosides, such as tannin, which originate in vegetative matter. Other suspended particulates, such as colloidal silts and clays, may render a
water unpalatable if not removed in treatment.

Effects of Microorganisms. The most significant
natural odor problems in domestic water supplies
derive from the growth and decay of algae, actinomycetes, and other microorganisms. Table 17-6
shows a qualitative description of the tastes and
odors associated with the presence of a number of
common algae. A more extensive list is given in
Palmer (1962). Blue-green algae, green algae,
diatoms, and flagellates are the algae groups identified as the cause of various taste and odor problems. Within these groups, important genera are
Anabaena and *Oscillatoria,* blue-green algae frequently responsible for odor, and *Ulothrix* and
Scenedesmus, green algae that may produce sweet,
grassy odors. *Asterionella,* a diatom responsible for
taste problems at a number of locations, is shown in
Figure 17-2.

Odors from algae are a complex set of phenomena, resulting from the bacterial degradation of algae, algal waste products, or the algae themselves.
Geosmin is the most well-known odor-producing
metabolite. It is frequently responsible for musty,
earthy odors in water supplies and gives soil its
characteristic earthy odor. Geosmin was first identi-

TABLE 17-6. Odors, Tastes, and Tongue Sensations Association with Various Algae

Algal Genus	Algal Group	Odor When Algae Are: Moderate	Odor When Algae Are: Abundant	Taste	Tongue Sensation
Anabaena	Blue-green	Grassy, nasturtium, musty	Septic	—	—
Anacystis	Blue-green	Grassy	Septic	Sweet	—
Aphanizomenon	Blue-green	Grassy, nasturtium, musty	Septic	Sweet	Dry
Asterionella	Diatom	Geranium, spicy	Fishy	—	—
Ceratium	Flagellate	Fishy	Septic	Bitter	—
Dinobryon	Flagellate	Violet	Fishy	—	Slick
Oscillatoria	Blue-green	Grassy	Musty, spicy	—	—
Scenedesmus	Green	—	Grassy	—	—
Spirogyra	Green	—	Grassy	—	—
Synura	Flagellate	Cucumber, muskmelon, spicy	Fishy	Bitter	Dry, metallic slick
Tabellaria	Diatom	Geranium	Fishy	—	—
Ulothrix	Green	—	Grassy	—	—
Volvox	Flagellate	Fishy	Fishy	—	—

SOURCE: Adapted from Palmer (1962).

fied as a product of actinomycete cultures (Gerber and Lechevalier, 1965) but has since been noted as a product of various blue-green algae species. MIB is another common taste and odor compound produced by actinomycetes and blue-green algae, particularly *Oscillatoria*. The aquatic conditions necessary to stimulate algal production of taste- and odor-causing metabolites are not clearly understood. Countless metabolites produced by many algal species may give rise to taste and odor problems. Thus, it is impossible to describe the specific cause-and-effect relationship in algal taste and odor incidents without familiarity with the particular species and its prevalent conditions.

Actinomycetes are described as moldlike bacteria due to their method of sporulation and mycelial (filamentous) development. They are primary decomposers of microbiota and have been observed to multiply greatly following blooms of various blue-green algae (Silvey and Roach, 1964). The most common genus encountered in water supplies is *Streptomyces,* but *Micromonospora* and *Nocardia* may also be found. In a surface water, it is difficult to attribute tastes and odors to algal growth or biological degradation of that growth since a symbiotic relationship often exists between various organisms, such as algae and actinomycetes.

In recent years, the chemical characteristics of the odorous compounds produced by algae and actinomycetes have been studied in depth (Silvey et al., 1950; Romano and Safferman, 1963; Dougherty et al., 1966; Safferman et al., 1967; Jenkins et al., 1967; Dougherty and Morris, 1967; Silvey et al., 1968; Medsker et al., 1968; Izaguirre et al., 1982).

FIGURE 17-2. *Asterionella,* a diatom frequently implicated in taste and odor problems.

Table 17-7 summarizes some of the information developed. As might be expected, some of these odor-producing compounds are quite intricately structured while others, such as the mercaptans, are relatively simple.

Table 17-7 shows that odors can be generated by the algae themselves as well as by actinomycetes during algal decay. In the future, new techniques, such as closed-loop stripping analysis and high-pressure liquid chromatography, will aid in answering questions about identified compounds (e.g., whether mucidone is a unique compound) and in discovering new ones.

Effects of Inorganics and Synthetic Organics. Certain naturally occurring inorganic compounds can also be responsible for tastes or odors in surface waters. For example, ferrous and manganous ions

(Fe^{2+} and Mn^{2+}) and sulfides appear in surface waters for several reasons: thermal stratification and oxygen depletion in the benthic layer of lakes and reservoirs may trigger the reduction of insoluble iron and manganese deposits, and these substances are released when the lake or reservoir overturns and mixes. The specific compounds are determined by conditions such as pH and dissolved oxygen (DO) levels. Stumm and Morgan (1981) have presented a thorough discussion of iron and manganese speciation and transformation in surface waters.

Other metal ions may also produce tastes when present in sufficient concentrations. Table 17-8 presents the taste thresholds of several metals and some inorganic and synthetic organic substances (Cohen et al., 1960).

Probably the most common cause of consumer complaints about human-origin tastes and odors is

TABLE 17-7. Structure of Various Compounds Isolated from Odor-Causing Aquatic Organisms

Compound	Structure	Associated Organisms
Methylisoborneol (MIB)		*Actinomycetes* *Oscillatoria curviceps* *Oscillatoria tenuis*
Geosmin		*Actinomycetes* *Sympioca muscoum* *Oscillatoria tenuis* *Oscillatoria simplicissima* *Anabaena scheremetievi*
Mucidone		*Actinomycetes*
Isobutyl mercaptan	$CH_3{>}CHCH_2{-}SH$ (from CH_3)	*Microcystis flos-aquae*
N-Butyl mercaptan	$CH_3(CH_2)_3{-}SH$	*Microcystis flos-aquae* *Oscillatoria chalybea*
Isopropyl mercaptan	$CH_3{>}CHSH$ (from CH_3)	*Microcystis flos-aquae*
Dimethyl disulfide	$CH_3{-}S{-}S{-}CH_3$	*Microcystis flos-aquae* *Oscillatoria chalybea*
Dimethyl sulfide	$CH_3{-}S{-}CH_3$	*Oscillatoria chalybea* *Anabaena*
Methyl mercaptan	CH_3SH	*Microcystis flos-aquae* *Oscillatoria chalybea*

TABLE 17-8. Taste Thresholds for Selected Materials

Material	Taste Threshold (mg/L)
Zn^{2+}	4–9[b]
Cu^{2+}	2–5[b]
Fe^{2+}	0.04–0.1[b]
Mn^{2+}	4–30[b]
2-Chlorophenol[a]	0.004
2,4-Chlorophenol[a]	0.008
2,6-Chlorophenol[a]	0.002
Phenol	>1.0
Fluoride	10[b]

[a] Created by the action of chlorine on phenol.
[b] Threshold detected by 5 percent of panel.
SOURCE: Cohen (1960).

chlorination in the water treatment plant. Studies by Bryan et al. (1973) have shown that the taste threshold for chlorine in water at a neutral pH is about 0.2 mg/L. The threshold increases to about 0.5 mg/L at pH 9.0. The odor threshold for certain reaction products of chlorine (such as nitrogen trichloride, NCl_3) are much lower. In studies of the taste of fluoridated water, Cox and Nathans (1952) found that the taste threshold of sodium fluoride is about 2.4 mg/L fluoride. Unless a particular taste is recognizable as an undesirable compound, however, the consumer may not necessarily reject the water supply even with concentrations of compounds above those taste thresholds (see, e.g., Bruvold et al., 1967).

Phenolic odors are frequently encountered in waters from industrial rivers. These reported medicinal odors are often not really phenol but actually chlorinated compounds formed by chlorinating raw water, or other synthetic organics (Lin, 1976). In addition, pesticides and organic solvents from industrial installations may cause highly objectionable tastes and odors. These compounds have caused severe problems, but with increased governmental control, they may be effectively prevented from entering water supplies.

PREVENTION AND CONTROL OF TASTES AND ODORS AT SOURCE

Taste and odor prevention and control may be accomplished at the source, in the treatment plant, and, to a certain extent, in the distribution system.

Undoubtedly the most satisfactory site for control in surface supplies is at the source. This generally means controlling the growth of algae and related organisms. For groundwater supplies, source control must be accomplished through watershed management, a difficult task. For surface reservoirs, algicides, destratification/aeration, and watershed management are used as control methods. Purveyors using continuous draft intakes with negligible raw water storage or detention most often address taste and odor problems in-plant rather than at the source.

Algicides

Although a number of other compounds can be used as algicides, copper sulfate almost exclusively dominates the field of potable water treatment. Presently, copper sulfate and chlorine are the only effective algicides that satisfy all necessary legal requirements. Most other synthetic algicides involve the use of mercury or tertiary and quaternary amines. Both copper sulfate and chlorine have been used as algicides for a number of decades. For example, Hale's summary of copper sulfate and chlorine doses required for a wide variety of biota was published in 1930 (see Table 17-9). As most information available in the literature is based on batch-scale laboratory work or on unique field experience, a given purveyor's own field experience is probably the best guide as to the dose necessary.

Where copper is concerned, this experience generally indicates that maintaining a residual between 0.1 and 0.2 mg/L for several hours is necessary. In order to maintain this residual, ordinarily a copper dose of about 0.5 mg/L is required. One of the principal problems in copper sulfate treatment is very little of the copper added is present as free copper ion; the majority settles to the bottom as a precipitate. An example of this is shown in Figure 17-3, which is a solubility diagram of copper (II) in water from the Owens River, California (JMM, 1977). In the normal pH range, the solubility of Cu(II) decreases with increasing pH. The insoluble species, based on equilibrium chemistry, is tenorite (CuO) and the principal soluble species is not cupric ion but the $CuCO_3$ ion pair complex. After the addition of 0.5 mg/L of copper to Owens River water, 99.5 percent was tenorite particulates and only 0.014 percent was cupric ion.

Microbiologists (Stanier et al., 1963) agree that copper and other heavy metals kill algae and bacteria by combining with the sulfhydryl (SH) groups of

TABLE 17-9.   Concentration of Copper and Chlorine Required to Kill Troublesome Growths of Organisms

	Organism	Trouble	Copper Sulfate (mg/L)	Chlorine (mg/L)
Algae				
Diatoms	*Asterionella, Synedra, Tabellaria*	Odor: aromatic to fishy	0.1–0.5	0.5–1.0
	Fragillaria, Navicula	Turbidity	0.1–0.3	—
	Melosira	Turbidity	0.2	2.0
Grass-green	*Eudorina,[a] Pandorina[a]*	Odor: fishy	2–10	—
	Volvox[a]	Odor: fishy	0.25	0.3–1.0
	Chara, Cladophora	Turbidity, scum	0.1–0.5	—
	Coelastrum, Spirogyra	Turbidity, scum	0.1–0.3	1.0–1.5
Blue-green	*Anabaena, Aphanizomenon*	Odor: moldy, grassy, vile	0.1–0.5	0.5–1.0
	Clathrocystis, Coelosphaerium	Odor: grassy, vile	0.1–0.3	0.5–1.0
	Oscillatoria	Turbidity	0.2–0.5	1.1
Golden or	*Cryptomonas[b]*	Odor: aromatic	0.2–0.5	—
yellow-brown	*Dinobryon*	Odor: aromatic to fishy	0.2	0.3–1.0
	Mallomonas	Odor: aromatic	0.2–0.5	—
	Synura	Taste: cucumber	0.1–0.3	0.3–1.0
	Uroglenopsis	Odor: fishy. Taste: oily	0.1–0.2	0.3–1.0
Dinoflagellates	*Ceratium*	Odor: fishy, vile	0.2–0.3	0.3–1.0
	Glenodinium	Odor: fishy	0.2–0.5	—
	Peridinium	Odor: fishy	0.5–2.0	—
Filamentous bacteria	*Beggiatoa* (sulfur)	Odor: decayed. Pipe growths	5.0	—
	Crenothrix (iron)	Odor: decayed. Pipe growths	0.3–0.5	0.5
Crustacea	*Cyclops*	[c]	—	1.0–3.0
	Daphnia	[c]	2.0	1.0–3.0
Miscellaneous	*Chironomus* (bloodworm)	[c]	—	15–50
	Craspedacusta (jellyfish)	[c]	0.3	—

[a] These organisms are classified also as flagellate protozoa.
[b] Classification uncertain.
[c] These organisms are individually visible and cause consumer complaints.
SOURCE: Hale (1930).

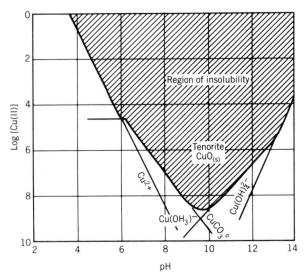

FIGURE 17-3.   Solubility of copper(11) in Owens River water constants (Schindler, 1967).

many enzymes that play an important role in catalysis and energy-yielding metabolism. In order to accomplish this, copper must diffuse through the cell wall. It is most likely that only the soluble Cu(II) species directly participate in this activity. Thus, the efficiency of algal control through copper sulfate addition is limited since most of the copper added is lost through precipitation and sedimentation (Button et al., 1977).

The fact that most of the copper added is soon converted to particulate form in most water supplies has been recognized for some time, and various methods have been employed in attempts to circumvent the problem. One of the most common is to complex the copper added with citric acid. Normally, when citric acid is used to chelate copper ion for reservoir treatment, a mixture of 2:1 copper sulfate to citric acid by weight is used (Barnett, 1976). Using recent national market figures for both compounds (*Chemical Marketing Reporter*, 1984) that is, $1.01/kg ($0.46/lb) $CuSO_4$ and $1.78/kg ($0.81/lb) citrate, it is calculated that the resultant mix costs $1.89/kg ($0.86/lb) of copper sulfate. Thus "citrated" copper must be approximately twice as effective as copper sulfate to be economical. This seems likely, however, because as the copper-citrate complex is all in the soluble form, it should be effective at a lower residual and should not be subject to sedimentation losses. In the final analysis, only long-term full-scale tests will determine when citric acid treatment is effective. The scope of any test program should set out to (1) refine the ratio of copper sulfate to citric acid that is most effective; (2) determine the dose of the copper-citrate complex necessary to establish a given residual in the full-scale system; and (3) determine the minimum residual that can be used to maintain satisfactory plankton counts.

One important shortcoming with any copper treatment, however, is the association that has been postulated between the corrosion of new iron and galvanized pipe and the presence of significant amounts of copper in the water supply (Treweek et al., 1978). In this regard, the use of citric acid probably has a detrimental effect. When copper sulfate is used alone, most of the copper is insoluble, and filtration will usually bring the level of Cu(II) to below 0.005 mg/L (JMM, 1977). When citric acid is used, either the ligand complex must be destroyed or the bulk of the copper must be removed by adsorption prior to the water's entrance to the distribution system and home plumbing unit. If this is not accomplished, the complexed copper will move through the water treatment plant, and accelerated pitting corrosion may occur in new galvanized surfaces. Because of this corrosion problem, copper algicides should be used with caution. Perhaps the ultimate solution is either deleting galvanized pipe from the plumbing code (as has been done in one California county plagued with corrosion problems) or finding alternative algicides.

Chlorine is also used for the control of algal growths in water storage reservoirs, particularly treated-water reservoirs. Its usefulness, however, is impaired by difficulty in maintaining adequate residuals. Residuals are difficult to maintain because the chlorine reacts rapidly with natural aquatic humus and spontaneously decomposes in the presence of light. Nevertheless, some agencies successfully employ chlorine as the principal means of algal control in large, open reservoirs (JMM, 1980).

Reservoir Destratification

Thermal stratification is a phenomenon of lakes and reservoirs that may lead to taste and odor problems. Thermal stratification results when a warm water zone overlays a colder zone; temperature and density gradients form, and mixing is prevented. In the bottom layers, anaerobic conditions may prevail, producing reduced iron, manganese, and hydrogen sulfide at troublesome concentrations. At the surface, a stressed, warm water layer may also form, where taste- and odor-producing blue-green algae flourish. Destratification neutralizes these effects: it increases the dissolved oxygen throughout the water column so that anaerobic conditions do not prevail at the lower levels, and it eliminates or reduces the temperature gradient from surface to bottom so that the warm layer of surface water does not form. In temperate zones, natural destratification occurs in the fall and spring.

If lake or reservoir intakes are located in the middle depths, and if the top and bottom layers are never mixed, tastes and odors may never pose a problem. However, layers frequently mix in the spring and fall, and tastes and odors often result from reduced metal compounds, algal blooms, or other sources. For example, Figure 17-4a is a plot of dissolved oxygen (DO) and MIB concentrations versus depth from the Lake Casitas, California, drinking water reservoir. These samples, taken prior to turnover, show high DO and MIB concentrations in the upper mixing zone of the reservoir. As this zone dropped during turnover, the MIB concentration increased at lower depths. Samples taken a week later (Figure 17-4b) indicated that the MIB levels were high at the level of the water intake. A taste and odor episode in the drinking water distribution system followed shortly thereafter.

Destratification may be accomplished with mechanical mixing devices, such as pumps, or forced aeration through injection of compressed air. Although mechanical mixing and forced aeration are both effective (Symons et al., 1970), most U.S. water supplies who artificially destratify their reservoirs employ forced aeration systems.

As a result of aeration or mixing, dissolved iron, manganese, and hydrogen sulfide in the raw water are eliminated. The oxidized iron and manganese materials are insoluble and settle out. However, the major reason to install destratification facilities is for the prevention of odors from algal growths. Indications are that mixing not only reduces the total

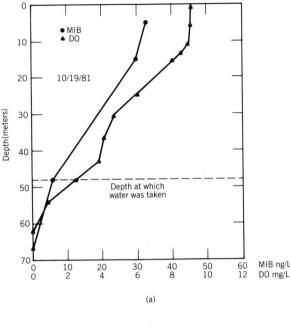

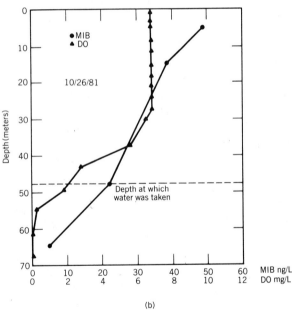

FIGURE 17-4. MIB and DO concentrations versus depth (a) before and (b) during reservoir turnover.

production of algal biomass, but reduces blue-green algae more effectively than green algae or diatoms (Symons, 1969). This selectivity is probably due to the destruction of the very warm layer in the top few centimeters of the lake surface where most blue-greens flourish. Lorenzen and Mitchell (1975) proposed a systematic model for evaluating the effectiveness of a proposed reservoir destratification project; however, their proposed method does not currently include the ability to describe selectivity for the types of algae that would be present. Using the model as outlined would probably cost more for most utilities than the alternative of actually conducting field destratification tests. If reservoir destratification proves effective in a given installation, it would reduce but may not eliminate the need for an algicide treatment program.

Biological Control

Silvey and others at North Texas State University are primarily responsible for the development of proposed biological control methods. In Silvey and Roach's studies (1964) of microbiotic cycles in southwestern reservoirs, they determined that certain *Bacillus* bacteria species were apparently able to oxidize metabolites of actinomycetes. Actinomycete growth was in turn stimulated by the presence of large numbers of blue-green algae. The order of appearance in cycles or blooms was observed to be blue-green algae, actinomycetes, and *Bacillus*. This suggested that taste and odor problems in reservoirs or lakes that have experienced blue-green algal blooms may be circumvented by seeding with *Bacillus* cultures. This approach was field tested at Lake Hefner in Oklahoma City in 1963 and 1964. *Bacillus cereus* was applied to the surface after an algal bloom, with appurtenant taste and odor problems, had occurred. When concentrations reached a level of 10^5 bacteria per milliliter, it was observed that odor-causing agents were greatly diminished within several days (Hoehn, 1965). It was subsequently found that several species (notably *Bacillus subtilis* and *Bacillus cereus*) are capable of and efficient at oxidizing the taste and odor compound geosmin (Narayan and Nunez, 1974).

Despite these findings, biological control of tastes and odors is not widely accepted. *Bacillus* species form spores resistant to disinfection that may pass conventional water treatment measures. Once restored to their vegetative form, they can produce toxins in food stuff and stimulate the production of nitrites and nitrosamines. For these reasons, further research is necessary prior to the adoption of biological control methods in water supplies.

Watershed Management

Control and management of watershed areas is an obvious yet often overlooked method of preventing the development of taste and odor problems. If possible, nutrient inputs should be limited to prevent abnormal algal growth. Municipal and industrial discharges should be regulated and monitored; drainage from septic tanks in the watershed should be minimized. It is impossible to accomplish all of these goals in urban areas or large river basins, but the potentially adverse effects of point and nonpoint pollution sources in a water supply basin should be minimized.

REMOVAL OF TASTES AND ODORS IN THE WATER TREATMENT PLANT

Because of the various combinations of inorganic and organic compounds that cause tastes and odors in water supplies, a wide variety of treatment processes are employed. However, because most known taste and odor compounds are in reduced form, some form of oxidation is usually effective. Generally, no simple treatment process is cost effective for all taste and odors that may develop, and a case-by-case analysis is recommended.

Oxidation

The compounds most commonly employed for removal of tastes and odors via oxidation are chlorine, permanganate, ozone, and chlorine dioxide.

Chlorine.   Studies of taste and odor removal with chlorination have generally shown this oxidant to be very effective with low-level inorganic odors, such as hydrogen sulfide. However, when the hydrogen sulfide levels are too high, this technique becomes very complex. Chlorination must then be preceded by air stripping, or heavy chlorination must be used followed by complete dechlorination and subsequent rechlorination.

Chlorination often increases odor problems when used for odors of industrial or certain algal origin (Burttschell, 1959). In this case, superchlorination to a free residual is necessary (Dougherty and Morris, 1976), usually followed by partial dechlorination. An example of this is the reaction of phenolic compounds from industrial sources with chlorine. When low dosages of chlorine are added to water that contains phenols, chlorophenol compounds are formed and impart an objectional medicinal taste to the water. The taste-producing intensity of the water increases up to a maximum, after which increasing chlorine doses reduce and finally eliminate chlorophenolic tastes (Ettinger and Ruchhoft, 1951;

Riddick, 1951). The regulations on chlorinated organics recently promulgated by EPA and the cost of the process make this alternative less attractive today than it has been in the past. Chlorine is most effective with the organic sulfides, disulfides, and mercaptans, whereas low-level chlorination may only accentuate the odor of certain more complex algal products such as MIB or geosmin. This limitation may be the reason why chlorination is often unsuccessful for treating odors from actinomycete blooms. In any case, laboratory tests should be evaluated prior to full-scale chlorination for odor removal.

Permanganate.   Unlike chlorination, potassium permanganate ($KMnO_4$) treatment has been especially effective for certain industrial and algal odors as used by several utilities (see, e.g., Swanger, 1969). Permanganate is more effective at alkaline rather than neutral or acid pHs and generally requires doses of 1–3 mg/L and a contact time of at least 1–2 hr. As exceptions, doses as high as 10 mg/L are used without adverse effects, and sometimes contact times of 10–15 min are adequate. To obtain longer contact times than would be available in the treatment plant basins, many utilities add $KMnO_4$ to raw water pipelines at their sources.

If excessive permanganate is used in the oxidation process, it will pass through the downstream filters and enter the distribution system. The soluble manganese will form manganese dioxide (MnO_2) and blacken the water. In addition, manganese concentration in the final treated water will increase and may exceed the levels prescribed in the secondary regulations.

Generally, overdosing is controlled by settling out the excess in the sedimentation basin noted by the disappearance of permanganate's characteristic pink color. However, for plants without flocculation and sedimentation steps, such as in-line or direct filtration plants, special monitoring equipment must be used to prevent permanganate from passing through the filters.

Excess permanganate residual can sometimes be removed through powdered activated carbon (PAC) addition. The PAC is added after the oxidation step to avoid reducing the permanganate's efficiency. PAC may, however, pass through the filters and move into the distribution system.

Ozone.   Ozone is widely known for its ability to oxidize tastes and odors. An allotrope of oxygen, ozone is the strongest oxidizing agent available for water treatment and the only effective oxidant that does not increase the TDS. Ozone must be generated on-site to meet demand and when introduced into solution it shortly dissipates (Trussell, 1977). Generally, ozone doses of 0.5–5 mg/L have been required for effective odor control.

McLaughlin (1943), evaluating ozone for Philadelphia's water supply, observed that 1.7 mg/L decreased odor thresholds of 49 to 19. Gullerd and Volin (1961) indicated that threshold odors of 10 are regularly eliminated by 0.7–1 mg/L ozone. Lepage (1975) observed that algal odors as high as 8 were eliminated by 1 mg/L of ozone. On the other hand, ozone has not been universally successful at such low levels. Sommerville (1972) found levels of 2–5 mg/L were necessary to reduce Shoal Lake water from a threshold odor of 4.7 to 2.5. Generally, experiences have been positive with natural aquatic odors, but high dosages have been required for some odors of industrial origin. For example, at Whiting, Indiana, a plant treating a water heavily laden with industrial wastes, ozonation was only able to reduce odor levels by about 50 percent.

Chlorine Dioxide. Although not as widely used as chlorine, permanganate, or ozone, chlorine dioxide is also a strong oxidizing agent. Like ozone, it must be generated on-site. To accomplish this, most manufacturers take advantage of the reaction between chlorine gas and sodium chlorite to produce chlorine dioxide (see Chapter 12).

A stable, aqueous form of chlorine dioxide is also available, which eliminates the need for on-site generation but, at present, it is too costly for regular use.

In a survey of several plants along the Niagara border, Aston (1947) found that chlorine dioxide was being successfully used to control both algae and phenolic odors. Chlorine dioxide is especially effective for odors resulting from phenols and chlorinated phenols (Ingolls and Ridenour, 1948). Although chlorine dioxide can be used for control of algal odors, its cost is usually prohibitive.

Activated Carbon Adsorption

Activated carbon has long been viewed as a most reliable last recourse for removal of tastes and odors (Cox, 1964; AWWA, 1971; Fair et al., 1968; Babbitt et al., 1963). Activated carbon can be used in either of two forms, powdered activated carbon (PAC) or granular activated carbon (GAC). In the past, powdered activated carbon was the most widely used in normal taste and odor applications, but tastes and odors from industrial sources are increasingly requiring the use of GAC (see, e.g., Hansen, 1975).

As currently used in conventional treatment plants, PAC works inefficiently in a co-current mode, with the most saturated carbon treating the cleanest water. PAC's low capital cost can offset its low efficiency, and with infrequent use can bring the overall cost far below that of equivalent GAC treatment.

Powered activated carbon, in slurry form, is usually added before coagulation or immediately before the filters. When relatively low doses are required (say 10 mg/L or less) and filter runs are long enough, PAC is best added immediately ahead of the filters. Experience has shown that a given amount of PAC is more effective when deposited in the filter (Cox, 1964). Care must be exercised to prevent PAC from passing the filter in this case. When relatively high doses of PAC are required, or when filter runs are already short, PAC is best added ahead of the coagulation–flocculation zone so that as much may be removed by sedimentation (if conventional treatment is employed) as possible. In normal practice, the doses of PAC required range from 3 to 15 mg/L; however, doses as high as 240 mg/L have been reported. For direct or in-line filtration plants, the PAC dose that can be effectively used is probably limited to between 10 and 15 mg/L.

Table 17-10 shows a partial listing of U.S. treatment plants presently using activated carbon treatment for taste and odor control. In general, the capacity of GAC to remove humic substances is rather limited (Snoeyink et al., 1975; Demarco and Wood, 1978).

Research at the University of Illinois indicates that GAC is highly selective for some of the most troublesome odor compounds, MIB and geosmin (Herzing et al., 1977), even in the presence of substantial amounts of humic substances. As a result, GAC could be expected to have a very long life (approximately 2 yr) where removal of these specific compounds is concerned. Practical experience confirms that GAC can be used for 2–3 yr for odor removal in certain cases.

Aeration

Aeration as applied in water treatment practices today is not useful for taste and odor control, except

TABLE 17-10. Summary of Field Experience of GAC for Taste and Odor Removal

	Capacity			Carbon Depth		Filter Rate		Turbidity (FTU)		Threshold Odor	
Plant	(ML/d)	(mgd)	Raw Water Source	(cm)	(in.)	(m/hr)	(gpm/ft²)	In	Out	In	Out
1. Ashtabula, OH	15	4	Lake Erie	61	24	4.8	2	22	0.10	1,000	1
2. Davenport, IN	76	20	Mississippi R.	46	18	4.8	2	124	0.23	4,000	1
3. E. St. Louis, IL	106	28	Mississippi R.	46	18	4.8	2	147	0.40	10	1
4. Granite City, IL	38	10	Mississippi R.	46	18	4.8	2	97	0.15	5	1
5. Kokomo, IN	38	10	Wildcat Crk.	51	20	4.8	2	54	0.56	4,000	1
6. Lexington, KY	61	16	Private Res.	30	12	4.8	2	15	0.57	4,000	1
7. Marion, OH	23	6	Wells & Scioto R.	46	18	4.8	2	38	0.11	1,000	1
8. Muncie, IN	19	5	White R.	61	24	4.8	2	44	0.50	2	1
9. Peoria, IL	38	10	Illinois R.	107	42	7.2	3	107	0.16	26	1
10. Richmond, IN	30	8	Private Res.	61	24	4.8	2	11	0.26	3,000	1.5
11. Terre Haute, IN	19	5	Wells & Wabash R.	76	30	7.2	3	91	0.16	30,000	1.3
12. Mount Uemens	30	8	Lake St. Claire	61	24	4.8	2	30	0.21		1

SOURCE: Blanck (1976).

for the most easily removed substances such as sulfides or volatile organics. Hydrogen sulfide may be effectively removed from a raw water through aeration, and both surface and groundwater supplies are frequently aerated for this purpose. Ferrous ion (Fe^{2+}) at low concentrations may be oxidized to a ferric precipitate through aeration, but at high concentrations ferrous ion requires chemical oxidant addition or long detention times. Aeration does not effectively remove manganous ion (Mn^{2+}) in normal detention times encountered in traditional water treatment plants (Cleasby, 1975). In a 1973 study (Anderson et al., 1973), it was reported that of those treatment plants in Nebraska using aeration and conventional treatment only 45 percent met the USPHS drinking water standard of 0.3 mg/L iron and 0.05 mg/L manganese. In summation, aeration is not an effective method of oxidizing materials in a raw water but may be effective at removing dissolved, odor-causing gases and various volatile substances.

WATER SUPPLY SYSTEM DESIGN

Taste and odor problems in a water supply are difficult to foresee. Their origins and nature are so varied and diverse that it is virtually impossible to predict and to entirely avoid the occurrence of tastes and odors either in the plant or at the consumer's tap. However, through the prudent design and selection of raw water sources, treatment processes,

and the distribution system, many tastes and odors may be avoided and their impacts minimized.

Controlled upstream watersheds provide the most taste- and odor-free water in most cases. Industrial and urban influences on upstream sources may often be avoided, and monitoring and control of nonpoint pollutional sources such as nutrients due to fertilizer usage is easier than in downstream watersheds. This, however, does not guarantee taste- and odor-free water. Mining and agricultural activities as well as natural algal and actinomycete blooms may produce odors in upstream sources. In selection of a new raw water source, it is doubtful that the benefits of improved taste and odor control in an upstream source would outweigh cost advantages with a downstream sources. However, the benefits are real and should be considered. Records of taste and odor episodes should be examined when evaluating the relative merits of potential sources; obviously, it is to a purveyor's advantage to have more than one source available, as taste and odor problems may be localized (see, e.g., McGuire et al., 1981).

If a source of raw water is to be used in which taste and odor problems are severe or frequent, conventional treatment (consisting of flocculation, sedimentation, and filtration units) probably provides the most flexibility in meeting the demands of taste and odor removal. Any of the various oxidants and adsorbents may be used in conventional treatment, and taste and odor removal may be maximized. Figure 17-5 shows a flow schematic for taste

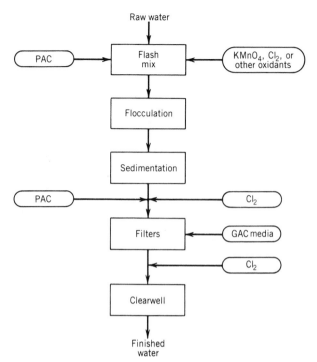

FIGURE 17-5. Potential application points of taste and odor control additives for a conventional water treatment plant.

and odor control in a conventional water treatment plant. Typically, potassium permanganate ($KMnO_4$) is applied during flash mixing. If organic substances with the potential for forming odorous chlorocompounds are present, chlorine should not be applied until the permanganate has been given adequate time to oxidize and alter these compounds. Similarly, PAC and $KMnO_4$ should not be added concurrently; the addition of PAC should be delayed until the permanganate is consumed. When odor problems are extreme, PAC may be applied at the mixing basin. The use of chlorine and PAC at the same time should be avoided.

Direct filtration plants (in which the sedimentation basin is omitted) may prove adequate if influent turbidity is relatively low and extended oxidation is not necessary. Within the specific plant design, chemical feed facilities should provide for PAC and oxidant feed at a variety of locations including preflocculation, presedimentation, and just prior to the filters. In direct filtration plants, it may be desirable to provide feed lines for both oxidant and PAC slurry upstream of the flocculation basin influent, perhaps in the raw water transmission mains. This allows for a longer contact time than would be provided in the plant alone. The use of

ozonation or a GAC media should also be considered in direct filtration designs.

In distribution systems, dead ends and low-flow zones, where odors due to reduced inorganic and bacterial substances occur, should be avoided. The low velocities and extended detention time allow chlorine residuals to be depleted, bacterial growth to occur, and iron tubercles to form. Flow disturbances may then result in taste and odor episodes. Water mains should be looped when possible, flushed regularly, and perhaps constructed in a staged manner so that velocities in the first years after construction are not extremely low. If substantial tuberculation occurs, reduced capacity and persistent taste and odor complaints may result. This can be corrected by rehabilitating (cleaning and cement-mortar lining) older mains.

TROUBLESHOOTING A TASTE AND ODOR PROBLEM

Identifying and eliminating a particular taste and odor problem can prove to be extremely difficult. Many problems simply go away and never are identified, even though they may have been successfully dealt with through the use of PAC or an oxidant. A number of interim measures can be addressed at taste and odor problems and often prove useful. Proper and timely response with algicides, oxidants, and adsorbents while attempting to isolate the causative element can successfully resolve most taste and odor problems.

Interim Measures

Interim treatment include oxidants (usually chlorine or permanganate), adsorbents (usually PAC), algicides (usually copper sulfate), or reservoir aeration. If an alternative odor-free water source is available, it can either completely or partially replace the taste- or odor-bearing source.

Oxidants such as chlorine or potassium permanganate are frequently the first response to a taste or odor problem. Potassium permanganate will often alter or reduce musty odors resulting from actinomycetes or blue-green algae, but frequently it will only make tastes and odors less objectionable. In a conventional treatment scheme, detention in the flocculation and sedimentation basins usually provides sufficient contact time for oxidation of taste- and odor-causing substances.

Powdered activated carbon is perhaps most responsive to the broadest spectrum of taste and odor problems. It may be necessary to add very high doses in some cases, and costs may be substantial. The point of application may vary from the head of the plant to just prior to the filters, and some care should be taken to ensure that PAC does not clog or pass the filters.

Ozone and chlorine dioxide are typically not available for interim measures as on-site generation is necessary. However, if the facilities are available, they may prove useful at increased dosage. Chlorine dioxide capability can be added at modest cost; GAC requires substantial expenditures and does not represent a short-term solution except under special circumstances.

Algicides may be (and usually are) applied to raw water sources to combat tastes and odors in short-term situations. The methods of application are many and are well documented in the literature (AWWA, 1979). Copper sulfate will usually result in improved tastes and odors, but some of the most persistent, such as actinomycete metabolites, do not always respond to treatment.

Short-term aeration of reservoirs may significantly improve tastes and odors. However, due to equipment and labor costs, persistent taste and odor problems are more efficiently dealt with by permanent installations.

IDENTIFYING THE CAUSE

Along with treating the effects of taste and odor problems the attempt comes to identify their causes. The first step is to isolate the problem to its physical origin (e.g., the plant, the distribution system, reservoirs, or lakes).

Once the problem is located, sampling, and analysis techniques may be used to isolate the chemical or biological cause. For example, microscopic examination can indicate the presence or predominance of various taste- and odor-causing microorganisms. Standard methods ASTM (American Public Health Association, 1980) contains many of the necessary techniques and illustrative plates to identify common taste- and odor-causing microorganisms. Gas chromatographic (GC) analysis is also useful in determining the metabolites responsible for the offensive water quantities (e.g., geosmin, MIB). Temperature, DO, and threshold odor profiles in reservoirs may indicate the widespread na-

ture of a problem, the presence of anaerobic benthic conditions, thermal stratification, or surface layer blooms. Manganese, iron, and sulfide measurements may also prove useful. If industrial pollution is suspected, gas chromatography/mass spectroscopy (GC/MS) may determine precisely which organic compound is responsible.

These techniques are not absolute identifiers. Concentrations are usually too low for detection and synergistic effects may mask the true cause of many tastes and odors. This synergy was previously cited concerning the 11 compounds that produce the blueberry flavor.

A CASE STUDY: METROPOLITAN WATER DISTRICT OF SOUTHERN CALIFORNIA

In 1974 the Metropolitan Water District of Southern California (MWD) began to experience recurrent, seasonal odor problems in their Colorado River supply. The district's problem and its response are fully documented by Pearson, Sundberg, and Beard (1976) but are briefly recounted here as one approach to controlling and preventing tastes and odors in a public water supply.

In October of 1974, customers began to complain of earthy-musty odors in water from a number of MWD's raw water reservoirs, including Lake Mathews. Until this time, mild odors from this Colorado River supply had been easily removed in MWD's treatment plants. The threshold odor numbers (TONs) ranged up to 70 during this episode, but averaged between 12 and 17 throughout the raw water system. An identification and an interim treatment program was undertaken to handle the odors.

Bacterial samples eventually isolated actinomycetes as the causative organism. *Streptomyces* were isolated from Lake Mathews depth samples and GC analyses of absorbed organics from lake water filtered through activated carbon compared favorably to GC analysis of purified geosmin. Due to the lack of significant numbers of blue-green algae in Lake Mathews samples, it was assumed that actinomycetes were the odor-causing agent.

PAC was applied and found to be completely effective at high dosages (12 mg/L). Lesser doses, however, were administered (4–7 mg/L) to somewhat reduce the odor since insufficient carbon supplies were available to maintain higher doses. Permanganate, at doses of 0.6–0.8 mg/L, was also added to alter the nature of the odor to a less objectionable "bland" characteristic odor. This combined usage of PAC and permanganate resulted in acceptable finished-water quality throughout the period of malodourous raw water.

Other methods were also tested. Chlorination was not effective at destroying the odor-causing substance (except with superchlorination followed by rapid dechlorination), but bench tests demonstrated that ozonation at normal doses entirely destroyed the musty-earthy odor. Continuing studies show that GAC may reduce the odor but not eliminate it.

After the odor problem was associated with fall overturn in reservoirs along MWD's Colorado River system, a variety of physical and chemical parameters were monitored weekly including pH, DO, temperature, sulfides, manganese, alkalinity, and turbidity. A marked thermal stratification was discerned in July through October with the highest TONs occurring below the thermocline because of actinomycete activities on the bottom. Therefore, destratification was tested to solve the recurrent odor problems. A single 15-m (50 ft) diameter aeration ring was installed in a small 54 million m³ (44,000 acre-ft) reservoir, Lake Skinner, and operated at various air inputs. Results indicated that artificial destratification could be easily maintained, DO controlled to any desirable level, manganese concentrations lowered, and odors substantially reduced so that normal treatment procedures produced acceptable water of good quality.

A cost comparison of destratification by aeration, ozonation, PAC and potassium permanganate treatment, and GAC filtration favored the aeration alternative by a substantial margin. After this study, permanent aeration facilities were installed in Lakes Skinner and Mathews. It must be noted that while the aeration system in Lake Skinner has successfully dealt with stratification and related odor problems, the system installed in the larger Lake Mathews (183,000 acre-ft) has not been adequate to reduce odors and eliminate stratification. However, through blending with other, odor-free, sources MWD has been able to maintain the aesthetic quality of treated water distributed to the public.

REFERENCES

American Public Health Association, American Water Works Association, and Water Pollution Control Federation, *Standard Methods for the Examination of Water and Wastewater*, 15th ed., American Public Health Association, New York (1980).

American Water Works Association, *Water Quality and Treatment*, 3rd ed., McGraw-Hill, New York (1971).

American Water Works Association, "Treating the Reservoir for Algae Control," *Opflow*, **5**(4), 1–8 (1979).

American Water Works Research Foundation, *Handbook of Taste and Odor Control Experiences in the U.S. and Canada*, American Water Works Association, Denver (1976).

Amoore, J., Johnston, J., and Rubin, M., "The Stereochemical Theory of Odor," *Scientific American*, 42 (February 1964).

Anderson, D. R., Row, D. D., and Sindelar, G. E., "Iron and Manganese Studies of Nebraska Water Supplies," *JAWWA*, **65**(10), 635–641 (1973).

Aston, R., "Chlorine Dioxide Use in Plants on the Niagra Border," *JAWWA*, **39**, 687 (1947).

Babbitt, H., Duland, J., and Cleasby, J., *Water Supply Engineering*, McGraw-Hill, New York (1963).

Barnett, R., "The Laboratory's Role in Quality Control in Reservoirs," Presented at AWWA Water Technology Conference, San Diego (December 1976).

Blanck, C., "Taste and Odor Control Utilizing GAC in the Plains Region of the American Water Works System," Am. Wtr. Wrks. Serv. Co., Plains Region (1976).

Bruvold, W., Ongerth, H., and Dillehay, R., "Consumer Attitudes Toward Mineral Taste in Domestic Water," *JAWWA*, **59**, 547 (1967).

Bryan, P., Kuzcinski, L., Sawyer, M., and Feng, R., "Taste Threshold of Halogens in Water," *JAWWA*, **65**, 363 (1973).

Burttschell, R., Rosen, A., Middleton, F., and Ettinger, M., "Chlorine Deviations of Phenol Causing Taste and Odor," *JAWWA*, **51**, 205 (1959).

Button, K., Hostetter, H., and Mair, D., "Copper Dispersal in a Water Supply Reservoir," *Water Res.*, **11**, 539 (1977).

Chemical Marketing Reporter (December 3, 1984).

Chen, K., "Chemistry of Sulfur Species and Their Removal from Water Supply," in A. Rubin (ed.), *Chemistry of Water Supply, Treatment and Distribution*, Ann Arbor Science, Ann Arbor, MI (1975).

Cleasby, J. L., "Iron and Manganese Removal, Case Study," *JAWWA*, **67**(3), 147–149 (1975).

Cohen, J., Kamphaki, L., Harris, E., and Woodward, R., "Taste Threshold Concentrations of Metals in Drinking Water," *JAWWA*, **52**, 660 (1960).

Cox, C., "Operation and Control of Water Treatment Processes," *WHO Geneva*, 46 (1964).

Cox, G., and Nathans, J., "A Study of the Taste of Fluoridated Water," *JAWWA*, **44**, 941 (1952).

DeMarco, J., and Wood, P., "Design Data for Organics Removal by Carbon Beds," Presented at 98th AWWA Conference, Atlantic City (June 1978).

Dougherty, J. D., and Morris, R. L., "Studies on the Removal of Actinomycete Musty Tastes and Odors in Water Supplies," *JAWWA*, **54**, 1520 (1967).

Dougherty, J. D., Campbell, R. D., and Morris, R. L., "Actinomycete: Isolation and Identification of Agent Responsible for Musty Odors," *Science*, **152**, 1372 (1966).

Ettinger, M., and Ruchhoft, C., "Stepwise Chlorination on

Taste and Odor Producing Intensity of Same Phenolic Compounds," *JAWWA*, **43**, 561 (1951).

Fair, G., Geyer, J., and Okun, D., *Water and Wastewater Engineering*, Wiley, New York (1968).

Gerber, N., and Lechevalier, H., "Geosmin, and Earthy Smelling Substances Isolated from Actinomycetes," *Applied Microbiology*, **13**, 935 (1965).

Gullerd, J., and Volin, C., "Treatment by Ozone," *L'Eaux* (May 1961).

Hale, F. E., *The Use of Copper Sulfate in Control of Microscopic Organisms*, Nichols Copper, New York (1930).

Hansen, R., "Problems Solved During 92 Months of Operation of Activated Granular Carbon Filters," in *Proceedings of 95th Annual AWWA Conference*, paper 4B-6 (1975).

Herzing, D., Snoeyink, V., and Wood, N., "Activated Carbon Adsorption of the Odorous Compounds 2-Methylisoborneol and Geosmin," *JAWWA*, **69**(4), 223 (April 1977).

Hohen, R. C., "Biological Methods for the Control of Taste and Odors," *Southwest Water Works J.*, **47**, 3, 26, 28, 30 (1965).

Ingolls, R., and Ridenour, G., "Elimination of Phenolic Tastes by Chloro-Oxidation," *W & SW*, **95**, 197 (1948).

Izaquirre, G., Hwang, C., Krasner, S. W., and McGuire, M. J., "Geosmin and 2-Methylisoborneol from Cyanobacteria in Three Water Supply Systems," *Appl. Environ. Microbio.*, **43**, 708 (1982).

James M. Montgomery, Consulting Engineers, Inc. and Daniel, Mann, Johnson & Mendenhall, "Water Quality Studies—Alternative Treatment Processes" Vol. III, Los Angeles Department of Water and Power (June 1977).

James M. Montgomery, Consulting Engineers, Inc., "Metropolitan Water District of Southern California, San Joaquin Reservoir Water Quality Study," (April 1980).

Jenkins, D., Mesker, L., and Thomas, J., "Odorous Compounds in Natural Waters. Some Sulfur Compounds Associated with Blue-Green Algae," *ES & T*, **1**, 731 (1967).

Kessell, R. G., and Kardon, R. H., *Tissues and Organs: A Text-Atlas of Scanning Electron Microscopy*, W. H. Freeman and Company, San Francisco (1979).

Lepage, W., "Ozone Studies at Monroe," *Proceedings of 2nd International Ozone Conference*, Montreal (1975).

Lin, S. D., "Sources of Tastes and Odors in Water," *Water Sewage Works*, **123**(7), 64–67 (1976).

Lochrane, T. G., "Ridding Groundwater of Hydrogen Sulfide," *Water Sewage Works*, **126**(2), 48 (Part 1); **126**(4), 66 (Part 2) (1979).

Lorenzen, M., and Mitchell, R., "An Evaluation of Artificial Destratification for Control of Algae Blooms," *JAWWA*, **67**, 373 (1975).

McGuire, M. J., Krasner, S., Hwang, C., and Izaquiree, G., "Closed-Loop Stripping Analysis as a Tool for Solving Taste and Odor Problems," *JAWWA*, **73**(10), 530–537 (October 1981).

McLaughlin, M. J., "Ozonation Tests on Philadelphia Water Supply," *Water Works Eng.*, **96**, 1131–1134, 1158–1159 (1943).

Medsker, L., Jenkins, D., and Thomas, J., "An Earthy-Smelling Compound Associated with Blue-Green Algae and Actinomycetes," *ES & T*, **2**, 461 (1968).

Narayan, L. V., and Nunez, III, W. J., "Biological Control: Isolation and Bacterial Oxidation of the Taste and Odor Compound Geosmin," *JAWWA*, **66**(9), 532–536 (1974).

Olishifski, J., "Toxicology," in Olishifski and McElroy (eds.), *Fundamentals of Industrial Hygiene*, National Safety Council, Chicago (1971).

Palmer, C. M., *Algae in Water Supplies*, U.S. Dept. HEW, Water Supply (1962).

Pearson, H. E., Sundberg, H. W., and Beard, J. D., "Facing Consumers' Expectations on Taste and Odor Problems," presented at Annual Conference of the California Section, AWWA at Stateline, Nevada (November 3, 1976).

Pomeroy, R., and Cruze, H., "Hydrogen Sulfide Odor Threshold," *JAWWA*, **61**, 677 (1969).

Riddick, T. M., "Controlling Taste, Odor, and Color with Free Residual Chlorination," *JAWWA*, **43**, 545–552 (1951).

Roderick, W., "Current Ideas on the Chemical Basis of Olfaction," *J. Chem. Ed.*, **43**, 510 (1966).

Romano, A. H., and Safferman, R. S., "Studies on Actinomycetes and Their Odors," *JAWWA*, **55**(2) 169–176 (1963).

Safferman, R., Rosen, R. S., Mashni, C. I., and Morris, M. E., "Earthy-Smelling Substances from a Blue-Green Alga," *ES & T*, **1**, 429 (1967).

Schindler, P., "Heterogeneous Equilibria Involving Oxides, Hydroxides, Carbonates, and Hydroxide Carbonates," in Adv. in Chem. Series No. 67 Chmn. W. Strumm (1967).

Silvey, J., Russell, J. C., Redden, D. R., and McCormick, W. C., "Actinomycetes and Common Tastes and Odors," *JAWWA*, **42**, 1018 (1950).

Silvey, J., Glaze, W., Hendricks, A., Henley, D., and Matlock, J., "Gas Chromatographic Studies on Taste and Odor in Water," *AWWA*, **60**, 440 (1968).

Silvey, J. K. G., and Roach, A. W., "Studies on Microbiotic Cycles in Surface Waters," *JAWWA*, **56**(1), 60–72 (1964).

Snoeyink, V., Wood, N., Herzing, D., and Murin, C., "Laboratory Studies of Trace Organics Removal by Activated Carbon," *Proceedings 95th Annual AWWA Conference*, paper No. 5B (1975).

Sommerville, C., and Rempel, G., "Ozone for Water Supply Treatment," *AWWA*, **64**, 376 (1972).

Stanier, R., Doudoroff, M., and Adelberg, E., "The Microbial World," 2nd ed., Prentice-Hall, Englewood Cliffs, NJ pp. 347–350 (1963).

Stumm, W., and Morgan, J., *Aquatic Chemistry*, 2nd ed., Wiley-Interscience, New York (1981).

Swanger, G., "Use of Potassium Permanganate at Bowling Green, Ohio," *W & W Engr.*, 46 (October 1969).

Symons, J., *Water Quality Behavior in Reservoirs*, Off. of Int. Dept. of H & W, Govt. Printing Off., Cincinnati (1969).

Symons, J. M., Carswell, J. K., and Robeck, G. G., "Mixing of Water Supply Reservoirs for Quality Control," *JAWWA*, **62**(5), 322–334 (1970).

Treweek, G., Trussell, R., and Pomeroy, R., "Rapid Pitting of New Galvanized Piping Associated with the Use of Copper Sulfate as an Algacide," presented at AWWA WQTC, Louisville (December 1978).

Trussell, R., "Application of Ozone to Water Treatment" (unpublished) (1977).

— 18 —

Corrosion

This chapter describes the internal corrosion of water conduits, with emphasis on the gradual decay of pipe materials and their release into the aqueous environment. Topics covered in this discussion include: (1) the electrochemistry of corrosion reactions, including basic principles of thermodynamic stability and electrode kinetics; (2) properties of water that affect corrosion rate; (3) the problems and behavior of particular conduit materials; (4) the use of corrosion indices to infer likely behavior of water and materials; (5) the role of microbiological organisms in corrosion; and (6) practical applications of the body of knowledge to the measurement of corrosion and alternatives for treatment to control the process of corrosion.

Corrosion is of concern because it can reduce the life of a pipe by reducing wall thickness until there are leaks because it can result in encrustations that reduce effective carrying capacity and because it is often responsible for undesirable corrosion by-products in the consumer's tap. Each of these is a separate, distinct phenomena and, although they are related, we do not have sufficient understanding of the chemistry involved to predict one knowing the other. In this discussion, the emphasis will be on characterizing the rate of corrosion (weight loss from the metal). However, the issues of encrusta-

tion and the release of corrosion by-products are addressed briefly.

A wide variety of materials have historically been used in the construction of water conduits. At one time, stone, terra-cotta, wood, and lead were common materials. Presently, the most widely used materials include cast iron, "ductile iron," asbestos-cement, and steel in large distribution conduits and copper, galvanized iron, and plastics in consumer plumbing.

Among ferrous materials, cast iron pipelines have been in continuous service for over a century in some older cities. In recent years, "ductile iron" (cast iron containing small amounts of alloying elements, such as magnesium or cerium, to reduce brittleness) has replaced cast iron in new installations. Both materials are typically lined with a bitumastic seal coat, cement mortar, or both to reduce corrosion, although in the past thin coal tar coatings or no lining at all have been used. Steel has also been employed as an alternative to cast iron, since it is stronger and more malleable. Although it is equally susceptible to corrosion, steel is relatively long-lived in most systems when a cement mortar lining is used. Galvanized iron pipe (iron with a zinc coating) has been in general use for small distribution pipes, service lines, or consumer plumbing.

Lead was once used extensively in service lines and interior plumbing because of its ductility and its relatively long life. Its long life is due to a generally low corrosion rate and to its resistance to encrustation when compared to ferrous pipe. A few utilities still use lead "pigtails" between distribution mains and service connections. However, this practice is rapidly disappearing because of widespread concern about the health significance of lead corrosion by-products in the drinking water.

Since 1950, copper tubing has gradually displaced galvanized pipe in consumer plumbing applications. Plastic materials are also reaching increased use in interior plumbing. Reinforced concrete pressure pipe is used for large-diameter transmission pipelines, while asbestos–cement (A/C) pipe is applied for both transmission and distribution lines. Neither of these latter materials is typically lined, although A/C pipe occasionally is used with a lining. The Portland cement material associated with both types is subject to corrosive attack by soft, low-pH waters.

All of these materials, when used in water systems, are subject to chemical deterioration. The basic thermodynamic principles that affect corrosion of materials include solubility, described by chemical equilibria among materials and constituents in the water, and the rate of dissolution, which is described by chemical and electrochemical kinetics. For metallic conduits, a change of oxidation state takes place in the corrosion and dissolution reactions, so the electrochemistry of this oxidation is significant. The next section describes the basic electrochemical theory of corrosion of metals.

THERMODYNAMIC AND KINETIC THEORY OF CORROSION

Most metals used in water conduits are thermodynamically unstable in the presence of water, particularly when the water contains dissolved oxygen and/or chlorine species, as is typically the case in the waterworks industry. The thermodynamics of metals exposed to water describe the underlying relationships of free energy that drive the corrosion reaction. On the other hand, the fact that a metal is thermodynamically unstable is a necessary, but not sufficient, condition for actual corrosion. The rate at which corrosion takes place is a question of electrode kinetics, which are determined by a very complex function of surface conditions, electrical behavior, and solution chemistry. Under most conditions, the theory of electrode kinetics cannot be used to make quantitative predictions of corrosion rates, although the first principles of kinetics give important insight into the nature of observed corrosion phenomena.

Electrochemical Thermodynamics

Considerations of electrochemical or oxidation–reduction equilibria determine the oxidation state of a material that will be stable under a given set of conditions. Other chemical equilibria determine whether the stable form of a material is soluble or insoluble; if soluble species are stable, it is possible for a metal to rapidly corrode into these aqueous forms. If insoluble species are stable, the metal surface may be protected by a natural scale of these materials, inhibiting corrosion.

For the oxidation of a given metal, a simple half-reaction describing the change in oxidation state may be written:

$$Fe = Fe^{2+} + 2e^- \qquad (18\text{-}1)$$

In this case, metallic iron oxidizes to the dissolved ferrous ion. This ion has an oxidation state of $+2$, whereas the metallic form has an oxidation state of zero, so the process of oxidation releases two electrons.

As for any chemical reaction, the oxidation of a metal has a standard free energy of reaction, $\Delta G°$. This describes the change of free energy associated with the reaction under standard conditions (1 atm pressure, 25°C, unit activities of reactants and products), in units of energy/gram-equivalent. For electrochemical reactions, the standard free energy is usually translated into a "standard potential" of the reaction, $E°$, which is expressed as a voltage rather than as an energy change. Faraday's law expresses the relationship between $E°$ and $\Delta G°$:

$$\Delta G° = -nFE° \qquad (18\text{-}2)$$

where n is the number of electrons transferred in the reaction and F is Faraday's constant (96,501 Coulombs/g-eq).

Thus, if $\Delta G°$ is expressed in joules/gram-equivalent, $E°$ is in units of volts (a joule is equivalent to a volt-coulomb). The $E°$ has a sign opposite to that of $\Delta G°$.

When a metal is oxidized, it generates electrons.

Free electrons cannot accumulate because of the repulsive nature of like electrostatic charges. Therefore, oxidative half-reactions must always be coupled to reductive half-reactions, which consume electrons. The reaction couple for metallic iron corroding in an acid solution is shown below and illustrated in Figure 18-1:

Reaction	$E°$ (V)	
$Fe = Fe^{2+} + 2e^-$	$-(-0.44)$	Oxidation
$2H^+ + 2e^- = H_2$	0.00	Reduction
$Fe + 2H^+ = Fe^{2+} + H_2$	$+0.44$	

$$(18-3)$$

Note that the oxidative half-reaction and the reductive half-reaction do not need to take place at the same physical location, since free electrons can migrate through conducting metals at the speed of light. The location of the metal-oxidizing reaction is called the "anode" and the location where reduction occurs is the "cathode." This convention is the basis for the terms *cation* (ions that migrate from solution toward the cathode, where they are reduced, in this case H^+) and *anion*. The coupled half-reactions from an overall reaction sometimes referred to as a galvanic cell when the reaction proceeds spontaneously.

Table 18-1 shows standard reduction potentials of several reductive half-reactions.[†] When two metallic elements and their associated ions are available under standard conditions, the half-reaction with the more positive standard potential will proceed as a reduction, and the half-reaction with the more negative standard potential will proceed as an oxidation (in reverse from the direction shown in Table 18-1). In this manner, the net standard potential of the coupled half-reactions is positive, and the free energy of the coupled system decreases as the reaction proceeds ($\Delta G°$ is negative). The more positive standard potential is referred to as more "noble," (the metals with the highest positive standard

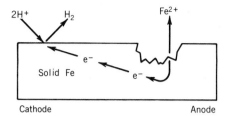

FIGURE 18-1. Reaction couple for iron corroding in acid solution. Iron metal loses two electrons and goes into solution as ferrous iron. Hydrogen ions near metal surface (probably absorbed to it) accept electrons generated by corroding iron, become reduced to nascent hydrogen, and eventually combine to form hydrogen molecules and leave the surface. Under these conditions, ideal potential of corroding iron specimen can be determined from the Nernst equation and is a function of pH and concentration of Fe^{2+} near anode.

potentials, such as gold, are noble metals which tend not to oxidize) while more negative standard potential is referred to as more "active" (metals with negative standard potentials, such as aluminum, tend to be oxidized when exposed to aqueous systems). For the iron corrosion couple shown above, the standard potential of the overall reaction is $-(-0.44) - 0.00 = +0.44$ V. Note that all standard potentials are measured as voltages relative to the "standard hydrogen electrode," the half-reaction associated with the oxidation of hydrogen ions.

In the pH range of most natural waters, the elec-

[†] The convention used in this chapter for reduction potentials is the IUPAC convention, which expresses standard potentials for reductive half-reactions; standard potentials have a sign opposite to the free-energy change. The sign of the net potential for the coupled reaction is the same as the sign of the logarithm of the equilibrium constant; therefore, coupled reactions with positive net standard potential are thermodynamically favorable and driven to the "right." Latimer (1952) established a convention commonly used by corrosion engineers that expresses oxidation potentials with the opposite sign, for which signs in Eqs. 18-2, 18-8, 18-9, 18-10, and so on, must be reversed.

TABLE 18-1. Selected Standard Oxidation–Reduction (Redox) Potentials[a]

$Au = Au^{3+} + 3e^-$	$+1.498$
$O_2 + 4H^+ + 4e^- = 2H_2O$	$+1.229$
$Pt = Pt^{2+} + 2e^-$	$+1.2$
$Fe^{3+} + e^- = Fe^{2+}$	$+0.771$
$4OH = O_2 + 2H_2O + 4e^-$	$+0.401$
$Cu = Cu^{2+} + 2e^-$	$+0.337$
$Sn^{4+} + 2e^- = Sn^{2+}$	$+0.15$
$2H^+ + 2e^- = H_2$	0.000
$Pb = Pb^{2+} + 2e^-$	-0.126
$Sn = Sn^{2+} + 2e^-$	-0.136
$Ni = Ni^{2+} + 2e^-$	-0.250
$Co = Co^{2+} + 2e^-$	-0.277
$Cd = Cd^{2+} + 2e^-$	-0.403
$Fe = Fe^{2+} + 2e^-$	-0.440
$Cr = Cr^{3+} + 3e^-$	-0.744
$Zn = Zn^{2+} + 2e^-$	-0.763
$Al = Al^{3+} + 3e^-$	-1.662

[a] 25°C, volts versus normal hydrogen electrode.

SOURCE: Bethune and Loud (1964).

trons produced by corrosion of iron are typically consumed by the reduction of dissolved oxygen

Reaction		E° (V)	
	$Fe = Fe^{2+} + 2e^-$	$-(-0.44)$	(oxidation)
$\frac{1}{2}O_2 + H_2O + 2e^- = 2OH^-$		$(0.401)/2$	(reduction)
$\frac{1}{2}O_2 + H_2O + Fe = Fe^{2+} + 2(OH^-)$		$+0.64$	

$$(18\text{-}4)$$

Note that the standard potential for the reduction of ferrous iron must be corrected by a factor of -1 because the direction of the reaction is reversed to show an oxidation of iron metal; the standard potential for the reduction of oxygen must be corrected by a factor of $\frac{1}{2}$ to balance the number of electrons transferred from the oxidation reaction. Whether or not the coupled reaction can proceed spontaneously under *actual* conditions (not *standard* conditions with unit molar concentrations of dissolved species) is determined by the overall equilibrium potential of the reaction. The potential of the reaction is proportional to the change in free energy of the reaction as shown by Faraday's law (recall that thermodynamic principles drive a reaction when the free energy of the system decreases). The potential of a reaction taking place under equilibrium conditions may be determined by means of the Nernst equation. This may be obtained from the thermodynamic relation describing the molar change in free energy of a reacting solution:

$$\Delta G = \Delta G^\circ + RT \ln \left(\frac{[\text{products}]}{[\text{reactants}]} \right) \quad (18\text{-}5)$$

where ΔG° = standard free energy of the reaction, J/mole
ΔG = molar change in free energy, J/mole
R = the universal gas constant, about 8.314 J/degree-mole
T = absolute temperature, °K

If a half-reaction is written in standard form as a reduction:

$$\text{Oxidized species} + ne^- = \text{reduced species} \quad (18\text{-}6)$$

then Eq. 18-5 can be used to relate the free energy of the reduction to the relative predominance of the oxidized and reduced forms:

$$\Delta G = \Delta G^\circ = 2.3RT \log \left(\frac{[\text{reduced}]}{[\text{oxidized}]} \right)$$
$$-2.3nRT \log(e^-) \quad (18\text{-}7)$$

rather than of hydrogen ions. Under these conditions, the overall reaction may be described by:

In a system at equilibrium, ΔG is equal to zero. The concept of equilibrium potential, E, may now be introduced as an intrinsic variable related to the chemical activity of the electrons of the system:

$$E = \frac{-2.3RT \log[e^-]}{F} \quad (18\text{-}8)$$

In this definition, the value of E (in volts) becomes more positive when the chemical activity of electrons becomes smaller, that is, when the system becomes more oxidizing. Similarly, the value of E becomes more negative when the activity of electrons increases and the electrons in the system tend to reduce chemical species. The concept of electrons having a chemical activity is useful for understanding the meaning of equilibrium potential as an intrinsic system variable governing oxidation–reduction behavior.

Alternatively, the equilibrium potential can be viewed as a driving force proportional to an effective nonzero free energy for the system at equilibrium, G, if the electrons are not considered to have an activity:

$$E = \frac{-\Delta G}{nF} \quad (18\text{-}9)$$

Combined with Faraday's law, this yields the Nernst equation,

$$E = E^\circ + \frac{2.303RT}{nF} \log \left(\frac{[\text{oxidized}]}{[\text{reduced}]} \right) \quad (18\text{-}10)$$

At 25°C, the quantity $2.303RT/F$ is approximately 0.059 V-eq. The Nernst equation may be used to describe equilibrium potential in systems of corroding metal; for example, a corroding iron system would be described by

$$E = -0.44 + 0.03 \log \left(\frac{[Fe^{2+}]}{Fe} \right)$$
$$= -0.44 + 0.03 \log([Fe^{2+}]) \quad (18\text{-}11)$$

since the activity of a pure solid is unity. The higher the concentration of the oxidized ferrous form, the higher the equilibrium potential of the system.

If the Fe/Fe^{2+} half-reaction is coupled to the H_2/H^+ half-reaction under acid conditions, one may also describe the relative concentrations of these species by the Nernst equation:

$$E = 0.00 + 0.03 \log\left(\frac{[H^+]^2}{[H_2]}\right) \quad (18\text{-}12)$$

If the oxidation of iron is coupled to the reduction of oxygen under alkaline conditions, a different relationship is obtained:

$$E = 0.401 + 0.015 \log\left(\frac{[O_2][H_2O]^2}{[OH^-]^4}\right) \quad (18\text{-}13)$$

Depending on which cathodic reaction is coupled to the anodic oxidation of iron, the potential E must be equal for both anodic and cathodic half-reactions at equilibrium, giving an equilibrium condition relating the various concentration terms. For example, combining Eqs. 18-11 and 18-12 gives the equilibrium condition

$$0 = -0.44 + 0.03(\log[H^2][Fe^{2+}] + 2\ pH) \quad (18\text{-}14)$$

Similarly, the expressions for the potential associated with the oxidation of iron and reduction of oxygen yield

$$0 = -1.68 + 0.03(\log[Fe^{2+}] - \tfrac{1}{2}\log[O_2] + 2\ pH) \quad (18\text{-}15)$$

Thus, the equilibrium potential of corroding iron depends on the concentration of ferrous ion ($[Fe^{2+}]$) in solution at the anode under both acid and alkaline conditions and on the concentration of dissolved hydrogen ($[H_2]$) and pH at the cathode for acid conditions or on the concentration of dissolved oxygen ($[O_2]$) and pH at the cathode under alkaline conditions. The magnitudes of the concentrations determine the thermodynamic driving force toward oxidation of metallic iron.

Ferrous ion is not the only form of iron produced during corrosion reactions. In an alkaline, oxygenated solution, ferrous ion is thermodynamically unstable and will itself by oxidized to ferric ion (Fe^{3+}), which, in turn, will be hydrolyzed to ferric hydroxide (represented by $Fe(OH)_3$, although this is not an

actual molecular structure) and eventually to Fe_2O_3:

$$2Fe^{2+} + \tfrac{1}{2}O_2 + H_2O = 2Fe^{3+} + 2OH^- \quad \text{(oxidation)}$$
$$Fe^{3+} + 3H_2O = Fe(OH)_3 + 3H^+ \quad \text{(hydrolysis)} \quad (18\text{-}16)$$
$$Fe(OH)_3 = Fe_2O_3 + 3H_2O \quad \text{(dehydration)}$$

In summary, known thermodynamic relationships based on standard free energies, standard potentials, and other equilibrium constants can be used to describe which species are energetically stable under given conditions. The environmental factors in aqueous solution that determine which species are stable include concentrations of dissolved reactants and products, pH, and the potential of the system under consideration. These dependencies have significant consequences for practical design of conduits and water treatment.

One way to envision the concept of equilibrium potential E is to consider it to be an intrinsic property of an equilibrated solid–solution system, an intrinsic variable that describes the overall oxidation state of the system. For example, in Eq. 18-11, the potential of an equilibrated iron–ferrous ion system is more positive when the concentration of ferrous ion, $[Fe^{2+}]$, is larger. When the reaction associated with Eq. 18-11 is shifted toward the right, the iron in the system is shifted toward more oxidized forms, the free energy or chemical activity associated with electrons in the system is higher, and the system is more "oxidizing." If oxidants such as chlorine or more oxygen are added to such a system, the equilibrium between iron and ferrous ion will be driven further toward the oxidized ferrous form, and the potential as calculated by Eq. 18-11 will increase (become more "noble"). If reductants such as sulfide were added to the system, the equilibrium would shift toward the left and the potential would decrease (become more "active").

Using the concept of equilibrium potential as a measure of the oxidation state of a solid–solution system, it is possible to describe the possible range of equilibrium potential (as E_h) versus pH. Given sional diagram. Pourbaix (1949) developed this approach to describe equilibrated systems with a plot of equilibrium potential (as Eh) versus pH. Given the potential and the pH of a system and assuming certain standard conditions (25°C, a total inorganic carbon content of 1 mM), Pourbaix constructed diagrams showing which species of a metal would be thermodynamically stable under those values of E

and pH. Figure 18-2 shows these diagrams for four metals commonly used as water conduits. For example, metallic iron is stable only for values of potential below about -0.75 V, representing a highly reducing environment; Fe_2O_3 is the thermodynamically favored form in strongly oxidizing environments (more positive values of potential), while ferrous ion is favored in an intermediate range of potential, but only when pH is low.

This graphical representation of thermodynamically stable species was also used by Sillen (1959) as "pE–pH" diagrams. The dashed lines in Figure 18-2 represent thermodynamic bounds within which water is stable; above the upper line, water is oxidized to oxygen, while below the lower dashed line, water is reduced to hydrogen gas. Actual environmental conditions are constrained between these lines.

In theory, the E_h of any system can be measured as the potential between an electrode composed of the metal of concern and a hydrogen electrode, so that Pourbaix's diagram could be directly used to estimate the driving force toward the corrosion of the metal. In practice, however, extremely low exchange currents, mixed potentials, and other interferences make such measurements possible only under carefully controlled conditions (Stumm and Morgan, 1980) and only qualitative conclusions can

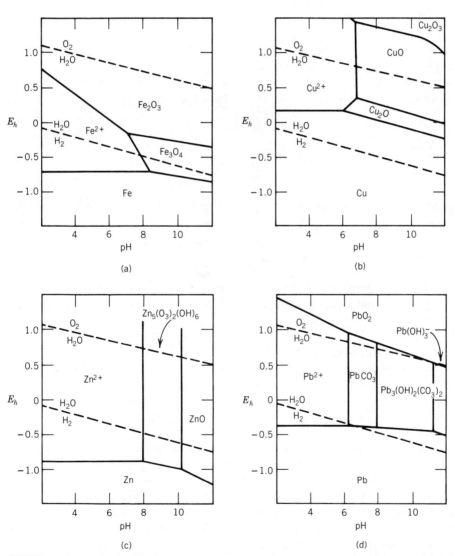

FIGURE 18-2. Pourbaix diagram for four metals used in domestic water piping: (a) iron, (b) copper, (c) zinc, (d) lead.

be drawn from E_h measurements in the field. As a result E_h–pH diagrams are principally useful as a didactic tool illustrating the relationship between pH, potential, and other solution properties and the stable species of the material of concern. Even when E_h can be accurately determined, concepts of equilibrium thermodynamics can only be used to assess whether corrosion is feasible or not. The rate of corrosion is determined by kinetic mechanisms, as described in the next section.

Kinetics of Corrosion

In order to understand the complex factors that govern the rate of corrosion of conduit materials, it is necessary to first examine the mechanism by which a material oxidizes. Three steps are involved: (1) transport of dissolved reactants to the metal surface; (2) electron transfer at the surface; and (3) transport of dissolved products from the reaction site. When either or both of the transport steps are the slowest, "rate-limiting step," the corrosion reaction is said to be under "transport control." When the transfer of electrons at the metal surface is rate limiting, the reaction is said to be under "activation control." The formation of solid natural protective scales that inhibit transport may become an important factor.

The following three sections describe key concepts about corrosion kinetics, including (1) activation control and the dependence of corrosion rates on potential ("activation polarization"); (2) transport control and the role of "concentration polarization" in limiting the rate of corrosion; and (3) the key phenomenon of "passivation" of some metal surfaces.

Activation Polarization: Corrosion Rate and Electrode Potential.

The previous sections have described the equilibrium relationships that may be employed to understand conditions under which corrosion is possible and the impact that system changes may have on the potential for corrosion. Although some comfort can be taken in understanding these relationships, the actual rate of corrosion is the primary issue of concern for the water industry. Considerations of electrochemical equilibrium can give only limited insight into the rate of corrosion.

For corrosion in water-conveying conduits to occur, the system must be in disequilibrium; a thermodynamic force drives a spontaneous change away from the original metallic form toward more energetically favored species. Typically, there is a net flow of electrons from anodic sites to cathodic sites as metal spontaneously oxidizes and goes into solution. The system is out of thermodynamic equilibrium; the current acts to bring the metal and solution closer to an equilibrium state.

Corrosion is often described in terms of small galvanic microcells on the surface of the corroding metal. In actuality, such localized anodes and cathodes are not fixed on the surface, but are statistically distributed on the exposed metal over space and time. The potential of the surface, which could be determined by measurement of voltage relative to a reference electrode, is determined by the mixed contributions to potential of both the cathodic and anodic reactions, averaged over time and over the surface area. Indeed, both the individual anodic and cathodic half-reactions are reversible and occur in both directions at the same time. For example, on an iron surface corroding under acid conditions, both the reactions

$$2H^+ + 2e^- = H_2 \quad \text{(reducing)} \quad (18\text{-}17)$$

and

$$H_2 = 2H^+ + 2e^- \quad \text{(oxidizing)} \quad (18\text{-}18)$$

are taking place at the cathode. The net direction of the H^+ to H_2 half-reaction summed over the surface of the iron depends on the *actual* electrode potential. This half-reaction proceeds at an equal rate in both oxidative and reductive directions (with no *net* formation or consumption of electrons) when the electrode is at its particular *equilibrium* potential (in this case, 0.00 V). Under this condition, there are electrons being consumed by the half-reaction in the direction of Eq. 18-17 and electrons being produced by the half-reaction in the direction of Eq. 18-18. The rates of the two directions of the half-reaction are equal and are associated with electric currents through the metal. If one could actually measure the rate of H_2 being produced and the rate at which H_2 was being consumed, the two equal chemical rates (in moles per square centimeter per second) could be translated into an "exchange current density," i_0. This term, i_0, is formally an electric current density rather than a chemical reaction rate and represents the equilibrium rate of the forward and backward directions of the half-reaction, in units of amperes per square centimeter. The exchange current density is equal to the equilibrium rate of oxidation (or reduction) in moles per square centimeter per second times a conversion factor nF (in coulombs of electrons per mole of H_2 produced or consumed).

The exchange current density, therefore, is a measure of the rate of oxidation and reduction of species for a half-reaction when those rates are *at equilibrium* (no net reaction) at the appropriate equilibrium electrode potential for the half-reaction in question. The exchange current density depends on the particular half-reaction and on the nature of the metal surface.

If the actual electrode potential differs from the equilibrium electrode potential (given by the Nernst equation), then the rates of the forward and backward directions of the half-reaction will also differ. The system will be out of equilibrium; there *will* be net production or consumption of hydrogen gas by this particular half-reaction over the surface (also net consumption or production of hydrogen ions). If the potential is higher than the standard potential of the half-reaction (more noble, more oxidizing), then the rate for reaction 18-18, oxidizing H_2 to H^+, will tend to exceed the rate of back reaction 18-17, reducing H^+ to H_2. With higher potential, a more oxidizing environment, the net rate of the half-reaction will be toward the oxidized form H^+. In this case, the system is said to have a positive "overvoltage," which is the excess of actual potential over equilibrium potential for the half-reaction. This could be viewed as an electrolytic cell in which an applied voltage difference causes disequilibrium and induces a current.

If the electrode potential is *less* than the equilibrium potential (more "active," more reducing environment), then the opposite is true. The reducing rate will exceed the oxidizing rate, so that the net difference favors the production of the reduced form, H_2. The overpotential in this case is negative in sign.

The relationship between the actual electrode potential and the current density is determined by statistical mechanics; this relates to the fact that the sites for the oxidizing and reducing directions are statistically distributed over the surface of the metal. This relationship is described by the Tafel equation:

$$i = i_0 \exp \left[\pm \frac{\eta}{\beta} \right] \qquad (18\text{-}19)$$

where i = actual current density, $\mu A/cm^2$
i_0 = exchange current density, $\mu A/cm^2$
η = overvoltage (electrode potential − equilibrium potential), mV
β = "Tafel slope," mV

Here the actual current density is a representation of either the forward or backward direction of the half-reaction translated into electric current densities on the surface of the metal. The sign within the exponential term depends on whether the current density is for the oxidizing or reducing direction of the half-reaction; the sign is positive for the oxidizing direction, leading to higher current densities for the oxidation of H_2 to H^+ when the overvoltage is positive (more noble, more oxidizing).

The Tafel slope is a basic thermodynamic concept that can be derived from the energy of activation associated with the transfer of electrons at the surface. It typically ranges from 50 to 100 mV depending on the surface and the half-reaction. The consequences of the relationship in the Tafel equation are that (1) the exchange current density, i_0, and the Tafel slope, β, are the principal constants that characterize the rate of the electrode reaction as a function of the potential; and (2) the overall rate of the reaction increases roughly exponentially with the absolute value of the overvoltage. The latter is true because when the overvoltage is sufficiently large; the rate of one direction of the half-reaction totally dominates the rate of the other direction. Values for exchange current density and the Tafel slope are given in Table 18-2 for some reactions of interest.

Given these results, it is possible to construct a diagram that relates the forward and backward rates

TABLE 18-2.   Selected Reaction Data

Reaction	Electrode Surface	Exchange Current, i_0 (A/cm²)	Tafel Slope (mV)
$2H^+ + 2e^- \rightleftharpoons H_2$	Fe	10^{-6}	120
$2H^+ + 2e^- \rightleftharpoons H_2$	Cu	10^{-7}	120
$2H^+ + 2e^- \rightleftharpoons H_2$	Zn	10^{-8}	120
$2H^+ + 2e^- \rightleftharpoons H_2$	Al	10^{-10}	100
$2H^+ + 2e^- \rightleftharpoons H_2$	Pb	10^{-11}	130
$Fe \rightleftharpoons Fe^{2+} + 2e^-$	Fe	10^{-8}	40
$Cu \rightleftharpoons Cu^{2+} + 2e^-$	Cu	10^{-5}	40
$Zn \rightleftharpoons Zn^{2+} + 2e^-$	Zn	10^{-5}	40

SOURCE: Donahue (1969).

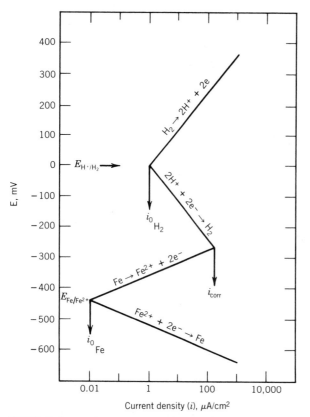

FIGURE 18-3. Activation–polarization plot for iron corroding in acid solution.

of half-reactions to the electrode potential of the surface. Such a diagram is shown in Figure 18-3 for the Fe/Fe^{2+} and H_2/H^+ half-reactions (constructed assuming unit molar concentrations). The exponential dependence of current density on overvoltage translates into straight lines when current density is plotted on a logarithmic scale. This allows a figure to be constructed, starting from the two equilibrium points given by the equilibrium potentials (Eq. 18-10) and exchange current densities of the two half-reactions (0 mV and 1 $\mu A/cm^2$ for the H_2/H^+ half-reaction, -440 mV and 0.01 $\mu A/cm^2$ for the Fe/Fe^{2+} half-reaction). As potential increases away from an equilibrium point, the positive overvoltage leads to increasing rate of the oxidizing direction of a half-reaction (slope of the line for voltage vs. current are given by the Tafel slope). Similarly, as potential decreases from an equilibrium point, the negative overvoltage leads to an increasing rate of the reducing direction for a half-reaction. Thus, on the portion of the figure below -440 mV, the line shows the current density for the reduction of Fe^{2+} to Fe; between -440 and -270 mV, the line shows the

rate of oxidation of Fe to Fe^{2+}; between -270 and 0 mV, the line shows the rate of reduction of H^+ to H_2; and above 0 mV, the line shows the rate of oxidation of H_2 to H^+.

In a hypothetical system where the Fe/Fe^{2+} and H_2/H^+ half-reactions are the only ones allowed to proceed, this diagram allows the prediction of the actual corrosion rate and electrode potential. If only these two half-reactions were possible, then the charge balance requirement that no *net* electrons be generated or consumed would require that both oxidative and reductive reactions take place at an *equal* rate. This is given in the diagram by the intersection of the two lines describing oxidation of Fe and reduction of H^+. The point of intersection determines the hypothetical corrosion rate, i_{corr}, or "corrosion current density," and the "mixed corrosion potential." In this example, i_{corr} is approximately 200 $\mu A/cm^2$, and the potential of the system is about -270 mV.

Figure 18-3 is often called an "activation polarization" diagram. The concept of "polarization" describes a system in which a potential difference is caused by the existence of a net current. In this case, both half-reactions are shifted away from equilibrium by the net current between anodic and cathodic sites. "Activation" describes the fact that the relationship that controls the potential difference is determined by the activation energy of the rate-limiting electron transfer step at the surface. The figure is also known as an "Evans diagram," after Evans (1948), who popularized the concept. If properly understood, it gives considerable insight into the electrochemical factors that determine corrosion rates in pipes.

For example, exchange currents are important in determining corrosion rate. Higher values of i_0 lead to higher values of i_{corr}. The fact that lead has a rather low value of i_0 explains the resistance of lead plumbing to corrosion by acid. Similarly, the value of the equilibrium potential for either half-reaction is important in determining i_{corr}. In Figure 18-3, if the pH is raised, the equilibrium potential for the H_2/H^+ half-reaction will be decreased, and the corrosion current will also be reduced.

Concentration Polarization: Corrosion Rate and Diffusive Transport. In most systems of interest in water conveyance, the rate of corrosion is subject to limitation by a transport step for motion of dissolved species to or from the surface sites. The rate of the cathodic reaction in the Fe/H^+ couple, for example, may be limited by the rate with which H^+ ions can diffuse to the iron surface. Since this is a familiar mass transport phenomenon describing diffusion through a boundary layer, one may characterize the limiting diffusion current density, i_L, by the mass transfer equation:

$$i_L = nFk_fC_B \qquad (18\text{-}20)$$

where k_f = mass transfer constant for reacting ions, cm/sec

C_B = concentration of reacting ions in bulk solution

For a clean surface of simple shape and relatively uniform flow conditions, a number of techniques are available for estimating k_f (Levich, 1962; Donahue, 1972; Donahue, 1979). By inspection, i_L increases linearly with increasing bulk concentration of ions, C_B. Also, increasing agitation in the aqueous phase will increase k_f and i_L. In the laminar flow region, k_f increases in proportion to the square root of bulk velocity; in the turbulent flow region, k_f increases linearly with increasing velocity.

The phenomenon of rate limitation by transport is referred to in corrosion literature as "concentration polarization." This suggests that the polarization, or difference in potential, in a corroding metal system is related to concentration differences between the surface and the bulk solution.

Where surface scales are present, the rate of transport of reacting ions through the scale may be rate limiting. Under these conditions, the limiting current density becomes:

$$i_L = \frac{nFDC_B}{d} \qquad (8\text{-}21)$$

where D = coefficient of diffusivity for reacting ions diffusing through the scale, cm²/sec

d = thickness of scale, cm

It can be seen that the limiting corrosion rate will decrease as the scale becomes thicker.

Given the various factors that can affect the rate of corrosion, Wagner and Traud (1938) presented a framework to describe the combined effect of activation polarization and concentration polarization. Figure 18-4a illustrates the effect of transport limitation on a cathodic reaction such as the H^+/H_2 half-reaction. At more positive, noble potentials, the

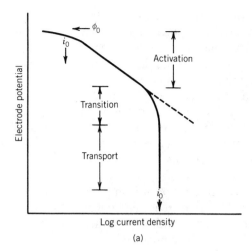

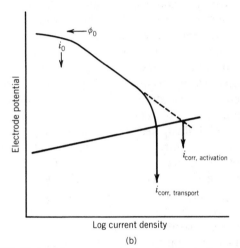

FIGURE 18-4. Mass transport control of electrode process. Influence of limiting current density on (a) cathode processes and (b) corrosion rate.

half-reaction is controlled by the activation energy of electron transport, and the relationship between potential and the logarithm of current density is linear. As the potential is decreased (made more active), the current flow increases, until the limiting current density is approached. Beyond that point, the current flow cannot increase further because of transport limitation. Figure 18-4b illustrates how transport limitation affects the corrosion current of the system, which couples the same cathodic reaction to an anodic reaction; the corrosion current is also decreased by the effect of transport limitation.

The system described in the activation polarization diagram is one of the simplest hypothetical situations possible, with only one metal being oxidized and one reducing half-reaction. Under actual conditions on corroding surfaces, the environment is much more complicated and a number of cathodic half-reactions may be involved. For example, in many water systems, the corrosion of iron may involve the reduction of oxygen, chlorine, and ferric ions. The reduction of Cu^{2+} and Zn^{2+} may also occasionally occur. Mixed potential theory suggests that all these reactions are additive in their effect. Therefore, many factors can influence the corrosion rate in real systems.

Passivity. If a piece of iron is immersed in dilute nitric acid, it will react immediately to evolve hydrogen gas and oxidize. On the other hand, if a piece of iron is immersed in *concentrated* nitric acid, it will *not* react and will still not react even if the solution is subsequently diluted with distilled water. However, once the solution is dilute, the iron specimen will begin to react violently if it is scratched. These phenomena may be explained by the mechanism of passivation. The passivity of certain metals to oxidation is a key concept affecting corrosion control in conduits.

Figure 18-5 illustrates how passivation of a metal can take place. For "ordinary" metals, as the potential of the surface and the oxidizing power of the environment is increased, a metal can be expected to corrode more rapidly, as shown in Figure 18-5a. This corresponds to the anodic reaction line in the activation polarization diagram.

With certain transition metals, however, a critical potential is reached at which the corrosion rate suddenly decreases. As potential is further increased, the corrosion rate remains low over a considerable range. In this zone of potential, the metal is described as passivated. This behavior is illustrated in Figure 18-5b. The region of increasing corrosion rate is the "active zone," the region of potential where corrosion rate is reduced is the "passive zone," and the region of higher potential where corrosion rate again increases is the "transpassive zone." In the transition from the active to the passive zone, a reduction in corrosion rate by a factor of 10^3 to 10^6 can be observed.

Passivation behavior is observed with metals such as iron, nickel, silicon, chromium, titanium, and their alloys. The potential above which passivation takes place is the "passivation potential," E_{pp}, and the maximum current density achieved before the onset of passivation is the "critical passivation current density," i_c.

The actual process by which the passivation of

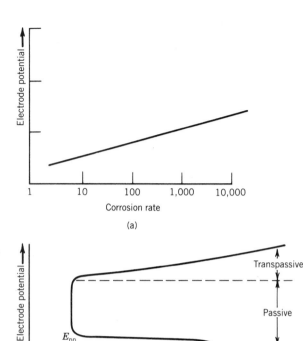

FIGURE 18-5. Corrosion rate versus solution oxidizing power or electrode potential for (a) ordinary and (b) active–passive metals. Several metals, notably iron and its alloys, are passivated by high oxidizing potentials. The passivation potential, E_{pp} is the potential below which passivation is lost, and i_c is the critical passivation in current density, the maximum current density achieved before the onset of passivation.

transition metals takes place is poorly understood. It appears to be due to the formation of a surface film that is stable over a considerable range of potentials. The nature of the film is not understood. This mechanism helps to explain the behavior of iron in dilute and in concentrated nitric acid; in concentrated solution, the iron surface is passivated and remains passivated when the acid is diluted but will corrode rapidly if the surface film is scratched.

Certain corrosion control strategies exploit the passivity of transition metals for protection of conduits. In "anodic protection," an external electromotive force is applied to bring the potential of the corroding surface above the passivation potential. This can sharply reduce rate of corrosion. The addition to solution of certain oxidizing compounds, such as the chromates, is also used to raise the potential into the passive zone in systems such as recirculating cooling towers. It should be noted that copper, zinc, and lead are not transition metals and will not be favorably influenced by practices that bring ferrous materials into the passive region; instead, these other metals are generally subject to increased corrosion when potential is increased.

PROPERTIES OF WATER AFFECTING CORROSION RATE

Given the thermodynamic basis for corrosion described above and the body of knowledge about kinetic factors that affect the rate of corrosion of metals, it is possible to identify several properties of the water passing through a conduit that influence the rate of corrosion. These properties have been empirically observed to affect the nature of corrosive attack of metals and the lifetime of pipes; therefore, knowledge of the properties of a given water supply can enable a practicing engineer to anticipate potential corrosion problems or address strategies for corrosion control. The effects of these factors can be explained in terms of the electrochemical theory. Some of the water-related properties include (1) concentration of dissolved oxygen, (2) pH, (3) temperature, (4) velocity of the water, (5) chlorine residual, and (6) concentration of chloride ions.

Dissolved Oxygen

Dissolved oxygen (DO) is one of the most important factors influencing the rate of corrosion of all the metals of concern. At ordinary temperatures, ferrous metals corrode very slowly in the absence of DO. Oxygen is a direct participant in the corrosion reaction, accepting electrons at the cathode:

$$4e^- + O_2 + 4H^+ = 2H_2O \qquad (18\text{-}22)$$

At most levels of DO, this cathodic half-reaction is transport limited. Therefore, increasing the level of DO in the bulk solution increases the rate at which oxygen is transported to the corroding metal surface. As a result, the rate of corrosion for most metals increases with any increase in the level of DO. Under certain conditions, however, some transition metals and transition metal alloys are passi-

vated by DO. The reversible potential (E_{O_2/H_2O}) for reduction of oxygen at the cathode is above the passivation potential (E_{pp}) for many alloys (See Figure 18-5b). These alloys can be passivated in an oxygenated solution provided that the limiting current density for the cathode reaction (i_L) is above the particular alloy's critical current density (i_c); otherwise, the passivation potential cannot be attained.

Figure 18-6 illustrates the principles involved. In Figure 18-6a, an activation polarization plot is shown for a typical transition metal with a family of transport-limited cathode polarization plots given for various bulk concentrations of DO ranging from 5 to 30 mg O_2/L. On the same figure, the anodic polarization plot for oxidation of iron is shown, with its passivation behavior characterized by critical current density i_c occurring at a potential just below the passivation potential E_{pp}. For a DO level of 5 mg/L, i_L is far below i_c. As a result, the corrosion rate reaches i_L before the polarization potential is reached and passivation will not occur. As the DO is increased, i_L increases in a linear fashion (see Eq. 18-9) until, at a DO level of slightly above 20 mg/L, i_L exceeds i_c. From this point on, the metal can surmount the passivation potential and the corrosion current drops sharply in the passive zone. Figure 18-6b illustrates the behavior of the corrosion rate plotted against DO level which might result. The dashed line is an idealized plot based on the mechanism described, and the points are experimental data for the corrosion rate of mild steel after 48 hr of exposure to slowly moving distilled water at 25°C and various levels of DO (Uhlig et al., 1955). The experimental data seem to verify the principle of activation polarization plot.

These same relationships can be used to illustrate the reasons for the relative corrosion resistance of certain steel alloys. Figure 18-7 shows the anodic polarization curves of iron, 18Cr stainless steel (Type 430), 18Cr–8Ni stainless steel (Type 316), all in unagitated 1 N sulfuric acid at room temperature. Each additional element alloyed with iron lowers the critical current density of the metal (i_c). Alloying with chromium significantly reduces the passivation potential (E_{pp}). In air-saturated solutions of 1 N sulfuric acid, with no agitation, the limiting current density i_L for cathodic reduction of oxygen is about 100 μA/cm^2 and the reversible potential for this reaction is about −0.06 V. As a result, both iron and 430SS are vulnerable in such a solution because of their very high critical current densities. Type

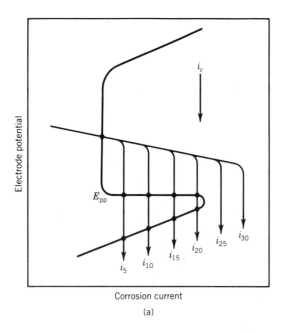

Corrosion current

(a)

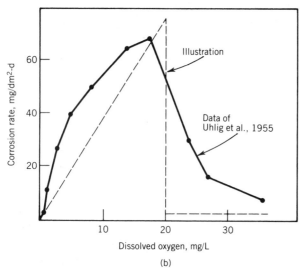

Dissolved oxygen, mg/L

(b)

FIGURE 18-6. Influence of DO on iron. At low levels of dissolved oxygen, corrosion current is diffusion limited. As DO level is increased, limiting current density ($i_5 = i_L$ for DO = 5 mg/L) increase also. Eventually, i_L passivation occurs, and further increases in DO do not affect the corrosion rate. (Data from Uhlig, et al., 1955.)

304L may be stable under some conditions and Type 316SS should be fairly resistant at all DO levels because its critical current density is substantially less than 100 μA/cm^2. Greene and Wilde (1969) have shown that the critical passivation current density for 18Cr–8Ni stainless steel can be estimated by the following formula:

$$i_c = 15.47 + 0.37\text{Mn} - 1.2\text{Cu} - 0.75\text{Mo} + 76\text{S} - 6.5\text{C} - 0.94\text{Cr} \qquad (18\text{-}23)$$

where i_c is in milliamps per square centimeter and the solute elements are in weight percent.

Generally, oxygen passivation is not a practical measure in domestic water systems with low flow velocities. In fact, overall corrosion rates are often lower in waters from deep wells where DO is absent. The Pourbaix diagram in Figure 18-2 shows that iron is nearly immune to low-DO water for pH above 8.5. It is common practice in industrial systems to deaerate or remove the oxygen from the water to be used in certain processes. Steam deaerators, cobalt-catalyzed bisulfite, and hydrazine are often used for this purpose.

The Effect of pH

At very low pH's, the cathodic reaction is the reduction of hydrogen ion to produce hydrogen gas. At a pH of 4 or more, the principal cathodic reaction is the reduction of oxygen to water. For this cathodic reaction, the reversible potential of the reaction is related to the pH in the following manner:

$$E_{\text{rev}} = 1.229 + 0.0148 \log O_2 - 0.059 \text{ pH} \qquad (18\text{-}24)$$

Over the range of pH experienced in water, the reversible potential for this reaction changes by about 200 mV. If the reaction were always activation controlled, the corrosion rate would drop by two orders of magnitude between pH 6 and 10. Figure 18-8a is a polarization resistance diagram that illustrates some of the principles involved in the oxygen-dominated region. At low pH, the reaction is transport controlled and the corrosion current density does not change with a change in pH. Once the pH increases beyond a certain point, the curve for the cathodic reaction intersects the anodic polarization curve below i_L. From this point on, an increase in pH will reduce the corrosion rate. In the early part of this century, Whitman et al. (1924) performed a simple experiment that developed a curve much like that shown in this figure. Whitman's data are shown in Figure 18-8b, illustrating the process in Figure 18-8a.

Previous sections have discussed the effect of pH on short-term corrosion rates. These results suggest little impact of pH on corrosion rate over the range of pH experienced in the domestic water

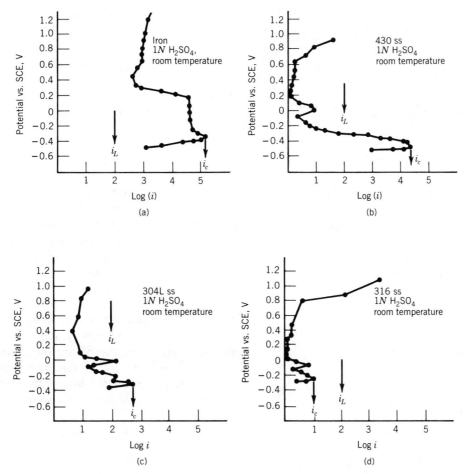

FIGURE 18-7. Potentiostatic anodic polarization curves of (a) iron and type (b) 430, (c) 304L, and (d) 316 stainless steels. Also shown is approximate limiting diffusion current for oxygen (i_L). With iron and 430SS, critical current density (i) is so far above i_L that oxygen passivation is not likely. The value of i_c for 304LSS is comparable to i_L, suggesting passivation under some conditions. For 316SS, i_c is far below i_L, suggesting that passivation will occur even at low DO levels. (Data from Fontana and Greene, 1978).

industry. On the other hand, pH is the master variable controlling the solubility, the rates of reaction, and, to some extent, the surface chemistry characteristics of most of the metal species in the corrosion reactions. The reaction chemistry of domestic waters is extremely complex; attempting to characterize the effect in the immediate vicinity of a corroding surface is even more difficult. Long-term corrosion tests conducted in dead soft waters show an increasing rate of corrosion with increasing pH (Skold and Larson, 1957; Larson and Skold, 1958a; Larson and Skold, 1958b). Stumm (1960) made similar observations concerning the corrosion of iron in waters of various hardnesses.

Both Stumm and Larson and Skold illustrated the effect of pH on the character of a corroding cast iron surface, noting uniform corrosion at lower pH, and increased unevenness and tuberculation at higher pH (Stumm, 1960; Larson and Skold, 1958a). It is likely that this effect is due in part to the influence of pH on the charge of ferric hydroxide micelles. The isoelectric point of ferric hydroxide is near a neutral pH. Thus, at low pH, the micelles have a positive charge and will migrate toward the cathode. This reduces the rate of mass transport of other ions to the cathode surface and alters the distribution of anodic and cathodic areas. At more alkaline pH, the $Fe(OH)_3$ micelles take on a more negative charge, causing them to remain at the anode; this increases the potential difference between the cathode and the anode, the rate of corrosion, and the heterogeneity of the corroding surface.

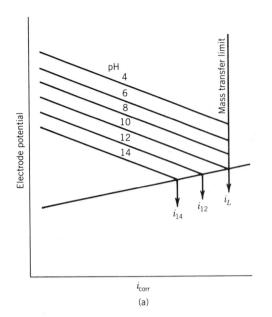

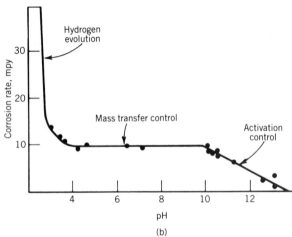

FIGURE 18-8. Influence of pH on corrosion reaction of iron: (a) Passivation–polarization plot with varying pH; (b) pH and corrosion rate. At high pH's, reaction is slow and apparently activation controlled. At pH 3–10, reaction rate limited by rate of mass transfer of oxygen to cathode surface. At pH's less than 3, hydrogen evolution occurs and reaction is once again activation controlled. Data in (b) (from Whitman et al., 1924) conducted with iron in aerated solution at room temperature.

Other effects of pH will be further discussed in the section on corrosion indices.

Temperature

Corrosion simply represents a particular group of chemical reactions. As a result, the rate of any particular corrosion reaction will increase with increas-

ing temperature according to the Arrhenius equation:

$$K = Ae^{-E_a/RT} \qquad (18\text{-}25)$$

where k = reaction rate constant (or corrosion rate)
A = preexponential factor
E_a = activation energy
R = universal gas constant
T = temperature, K

Figure 18-9 is an Arrhenius plot of corrosion data for iron in water after 8 hr of exposure (Butler and Ison, 1966). The plot shows a fairly good Arrhenius dependence with a corresponding activation energy (E_a) of 2.3 kcal/mole. This corresponds to a 75 percent increase in the corrosion rate for every 10°C rise in temperature.

Changes in temperature can influence the chemical composition and physical properties of the water, the character of any scales formed on the metals surface, and the nature of the metal itself. Shifts in equilibria with temperature may have a significant impact on the long-term protection afforded. Temperature affects the solubility of certain

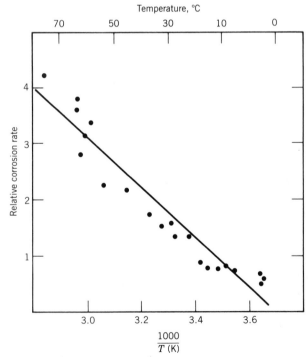

FIGURE 18-9. Arrhenius plot on relative corrosion rate of iron in water after 8 h exposure; corresponds to an activation energy of 2300 cal/mole or a change of 75 percent 10°C rise in temperature. (Data from Butler and Ison, 1966)

important dissolved gases, such as oxygen. All things considered, an increase in corrosion activity should generally be expected at higher temperatures.

Velocity

The effects of velocity on corrosion rate are also highly complex and are dependent on a number of properties of solutions and metals. Generally, if a reaction is activation controlled, changes in velocity will have little effect. In the case of a system with a normal metal and a corrosion rate that is diffusion controlled, increases in velocity will improve mass transfer and increase corrosion rates until the reaction is activation controlled (see Figures 18-10a,b). The rate of diffusion is controlled by Eq. 18-9, with linear increases in the turbulent region and increases in proportion to the square root of the velocity in the laminar region. In the case of an active–passive metal, passivation may occur at high velocities if i_L increases above i_c (see Figures 18-10c and d). Thus, in either system, the corrosion rate should become independent of velocity above some level. Moderately high velocities are desirable for most corrosion inhibitor systems.

The above is true for substantial rates of flow, but at extremely high velocities, the rate of corrosion may dramatically increase and erosion–corrosion may occur. The velocity at which erosion–corrosion begins to occur is called the "critical velocity." For copper in hot, soft water, the critical

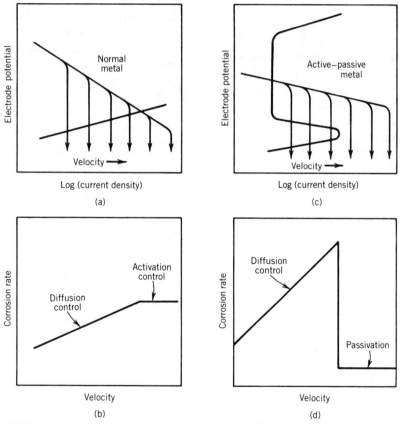

FIGURE 18-10. Influence of velocity on corrosion. (a) Velocity and activation–polarization plot for normal metal under diffusion control. (b) Velocity versus corrosion rate (normal metal). (c) Velocity and activation–polarization plot for active–passive metal under diffusion control at low velocities. (d) Velocity versus corrosion rate (active–passive metal). For most metals under diffusion control, increasing velocities will result in a corresponding increase in corrosion rate [(a) and (b)] until the limiting current density exceeds the level corresponding to activation control. For active–passive metals, if i_L is anywhere near i_0, it is sometimes possible to increase i_L with higher velocities until the surface is passified [(c) and (d)].

velocity is about 4 fps (1.3 m/sec). Erosion–corrosion involves the removal of dissolved ions, and it may include the removal of solid corrosion products mechanically swept from the metal surface. It is typically characterized by grooves, gulleys, waves, and so on, on the inside of the pipe, especially near points of turbulence. Tees and elbows are often the first to fail when excessive velocities occur.

Chlorine Residual

Chlorine is an effective disinfectant because it is a strong oxidizing agent, considerably stronger than DO. Like oxygen or any other oxidizing agent, chlorine can participate directly in cathode depolarization:

$$Cl_2 + 2e^- = 2Cl^- \qquad (18\text{-}26)$$

Figure 18-11, taken from the work of Larson (1975), shows the influence of a free chlorine residual on the corrosion rate of steel. Larson also conducted tests with combined chlorine and found that, with this disinfectant, residuals as high as 3.6 mg/L showed no effect in accelerating corrosion. This suggests that use of a combined residual in a treated water system may not affect corrosion rates to the extent that free chlorine would.

An 18-month study conducted in Portland, Oregon, compared corrosion rates of copper, iron, galvanized iron, and lead pipe in the presence of free and combined chlorine. The relative corrosion rates are shown in Table 18-3.

TABLE 18-3. Comparison of Corrosion Rates (Mils per Year) with Two Disinfectants

Metal Type	Free Chlorine	Chloramine
Black iron	2.8	3.0
Galvanized steel	0.5	0.41
Copper	0.28	0.27
Lead-coated copper	0.04	0.02
Lead	0.24	0.16

Chloride

In corrosion discussions, it is quite common to divide principal anions between those that are aggressive (accelerate corrosion) and those that are passivating (decelerate corrosion). Increasing evidence indicates that the chloride ion is the major aggressive ion where the corrosion of most metals used to transport water is concerned. In the absence of chloride ion, the most widely accepted reaction scheme for iron at the anode is the following (Cohen, 1979):

$$Fe + H_2O = Fe(H_2O)_{ads}$$
$$Fe(H_2O)_{ads} = Fe(OH^-)_{ads} + H^+$$
$$Fe(OH^-)_{ads} = Fe(OH)_{ads} + e^-$$
$$Fe(OH)_{ads} = Fe(OH)^+ + e^-$$
$$\text{(rate-determining step)}$$
$$Fe(OH)^+ + H^+ = Fe^{2+} + H_2O$$

Lorenz postulated a mechanism that suggests how a halide (X) such as chloride might act to alter the reaction and introduce a new rate-determining step (Lorenz, 1965). The reaction takes the following form:

$$Fe + H_2O = Fe(H_2O)_{ads}$$
$$Fe(H_2O)_{ads} + X^- = (FeX^-)_{ads} + H_2O$$
$$(FeX^-)_{ads} + OH^- = FeOH^+X^- + 2e^-$$
$$\text{(rate-determining step)}$$
$$Fe(OH)^+ + H^+ = Fe^{2+} + H_2O$$

In both reactions, the last step of the reaction is strongly pH dependent. The velocity of the rate-determining step in the Lorenz mechanism is dependent on the chloride concentration. As a result,

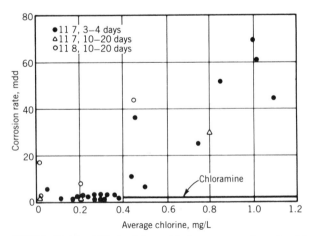

FIGURE 18-11. Effect of free chlorine on corrosion of mild steel. (Source: Larson, 1975.)

increasing chloride levels will result in increasing corrosion rates. In the presence of oxygen, Fe^{2+} will be rapidly oxidized to Fe^{3+}. The oxidation of Fe^{2+} is dependent on both DO and the pH, as shown below (Stumm and Morgan, 1980):

$$\frac{d \ln[\text{Fe(II)}]}{dt} = \frac{-k_H[O_2(\text{aq})]}{[H^+]^2} \qquad (18\text{-}27)$$

At 20°C, $k_H = 10^{-12}$ min^{-1} mole L^{-1}. For a given pH and DO, level the rate increases about 10-fold for each 15°C temperature increase. This reaction occurs for pH above 4.0. At lower pH, the oxidation rate is not affected by change in pH. The hydrolysis of Fe^{3+} appears to be generally faster than the oxidation of Fe^{2+} to Fe^{3+}.

When iron is exposed to dry air, it will form a protective oxide film of a cubic oxide of the Fe_3O_4–γ-Fe_2O_3 type, which will grow rapidly to a thickness of 15–20 Å (Cohen, 1979). A similar film formed in solution will approach a thickness of 40 Å. Its composition probably varies with depth, having more reduced forms near the metal–oxide interface and γFe_2O_3 near the oxide–water interface. These anhydrous oxide films are usually quite difficult to dissolve. It appears that the major cause of deterioration of oxide films on iron is the reductive dissolution of the oxide. The ferric oxide is reduced to Fe^{2+}, a much more soluble ion. This reaction can occur at the metal–oxide interface, loosening the oxide seals above it. Chloride can also play a role here because such dissolution reactions occur in the anodic pores. Chloride may also form ligand complexes with iron species, further aggravating the problem.

Corrosion Behavior of Conduit Materials

This section describes special topics in the corrosion of conduit materials. The phenomenon of pitting in the decomposition of pipe surfaces is described; this behavior occurs in various metals and can play a key role in pipe failure. Other aspects of the corrosion behavior of ferrous materials are discussed in separate sections, including passivation and the role of iron bacteria. Specific behavior of copper, lead, and galvanized pipe during corrosion is discussed in this section. The role of natural scale formation as a mechanism for protecting particular materials from corrosion is dealt with. Finally, the chemistry of concrete and its decomposition is described.

Pitting. Pitting describes a common phenomenon in corrosive attack of ferrous materials and galvanized and copper pipes. Formation of a pit in the interior surface of a pipe is accompanied by conditions that further the corrosion of the material at this site. Pitting may lead to pipe failure. This section specifically describes pitting of ferrous materials, although this behavior can be generalized to other materials; pitting of copper and galvanized surfaces are described separately.

There have been several good reviews written on the subject of pitting (Greene and Fontana, 1959; Robinson, 1960; Foley, 1970; Szklarska-Smialowska, 1971). Greene and Fontana discussed several theories of pitting corrosion; of these, only the acid theory and the adsorption theory presently have wide support. The acid theory (Hoar, 1947) postulates that chloride ions migrate to the positively charged anodic areas, increasing hydrogen ion concentration at these locations through hydrolysis. As the pH drops locally, the corroding metal will dissolve as a soluble salt.

The absorption theory postulates that chloride ions are preferentially adsorbed, displacing oxygen on the metal surface and creating a site for potential pitting attack. A corresponding theory of passivity hypothesizes that adsorbed oxygen protects the metal surface from attack. The preferential adsorption of the chloride destroys the passivating oxygen layer. Note that the acid theory and the adsorption theory are not mutually exclusive but describe mechanisms that could operate in parallel.

The pitting process is autocatalytic in nature. Figure 18-12 shows corrosion processes taking place in a pit (Fontana and Greene, 1978). Once a pit is started, rapid dissolution of metal at the site produces an excess of positive charge within the pit. This attracts negatively charged chloride ions, causing hydrolysis and a further increase in acidity, which increases the corrosion rate. Pits normally grow downward in the direction of gravity because of the important role the dense, concentrated chloride solution plays in the pitting process. Oxygen reduction takes place on surfaces adjacent to the pit; however, the high salinity of the solution in the pit precludes significant reduction of oxygen there. High levels of chloride accelerate pit growth and seem to be associated with more frequent pit formation.

A great deal of controversy exists regarding the mechanism by which aggressive ions influence pit initiation and growth. However, it is generally

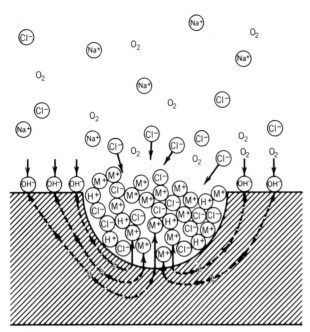

FIGURE 18-12. Autocatalytic processes in a corrosion pit. (Source: Fontana and Greene, 1978.) Reprinted with permission.

agreed that, for a given metal, as the concentration of aggressive ions increases, the pitting potential (the potential above which pitting occurs) decreases. Thus, as the concentration of the aggressive ion increases, the concentration of passivating ions necessary to prevent pitting also increases. The relationship generally takes the following form (Matsuda and Uhlig, 1964):

$$\log M_A = a + b \log M_p \qquad (18\text{-}28)$$

where M_A = molarity of aggressive ion necessary to start pitting

M_p = molarity of passivating ion

a, b = empirical constants

Typical values of a and b for stainless steel are summarized in Table 18-4. Figure 18-13 shows the effect of sodium chloride concentration on corrosion of iron in air-saturated sodium nitrite solution (Matsude and Uhlig, 1964). This illustrates the relationship between the critical concentrations of chloride and the concentration of the nitrite passivator at which no pitting would be expected.

In addition to influencing the frequency and rate of growth of pits, aggressive ions, particularly chloride, also affect the induction time that precedes the initiation of the first pit. It has been shown (Engell

TABLE 18-4. Values of Constants a and b^a

Solution	a	b
Na_2CrO_4 + NaCl	−1.19	0.85
Na_2CrO_4 + Na_2SO_4	−1.09	0.79
$NaNO_2$ + NaCl	1.10	1.13
$NaNO_2$ + Na_2SO_4	0.30	1.31

^{a} $\log M_s = a + b \log M_p$, where M_s = critical molarity of chloride or sulfate, above which increased corrosion and pitting of iron occurs, given M_p = molarity of passivator solution (chromate or nitrite).

and Stolica, 1959; Szkarska-Smialowska, 1971) that, all else being equal, the induction time (T_i) can be predicted by the following relation:

$$T_i = \frac{1}{K[Cl^-]} \qquad (18\text{-}29)$$

The equations shown for passivating pitting and for estimating pitting induction time are given here to demonstrate the importance of aggressive ions when corrosive attack involves pitting. Unfortunately for the waterworks industry, no constants are currently available for either equation for the combination of metals, aggressive ions, and passive ions of importance in water supply. The equations do suggest that sufficient levels of passivating ions

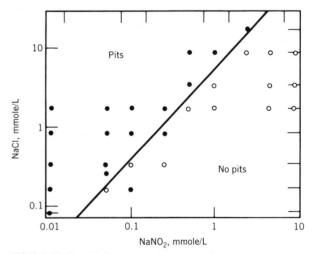

FIGURE 18-13. Pitting passivation plot: Effect of NaCl addition on corrosion of iron in air-saturated $NaNO_2$ solution at 25°C (by weight loss for 5 day immersion). Open Circle, no pitting, <1 mdd; half-dark circle, some pitting, <10 mdd; dark circle, extensive pitting or general corrosion >10 mdd.

can prevent pitting and that the induction time preceding the pitting process is decreased in inverse proportion to increasing chloride ion concentration. Because pitting corrosion can result in rapid failure, the induction time is usually an important factor in pipe life when pitting failure may occur.

Corrosion of Copper.

Copper tubing in water systems is subject to three forms of corrosion: (1) general corrosion; (2) impingement attack; and (3) pitting attack (Hatch, 1961; Cohen and Lyman, 1972). General corrosive attack is rare and generally occurs at low rates so that the service life of the tubing is not significantly reduced. Impingement and pitting attack can be both rapid and severe.

General Corrosion. General corrosive attack of copper ("green water") is most often associated with soft, acid waters. It usually proceeds at a low rate and is characterized by a gradual buildup of corrosion products, generally basic cupric salts.

The most important factors influencing general corrosion are (1) pH, (2) hardness, (3) temperature, (4) age of pipe, and (5) oxygen. As with most metal corrosion, soft, low-pH waters attack the protective oxide layer. Waters with pH lower than 6.5 and softness less than 60 mg/L as $CaCO_3$ are aggressive to copper. Corrosive attack is more severe in hot water systems. Obrecht and Quill (1960, 1961) report that temperatures exceeding 60°C (140°F) cause an increased rate of corrosive attack. It has been frequently found that the most serious cases of corrosion occur in new pipe installations where protective oxide layer had not yet formed.

The impacts of general corrosion are mostly of a nuisance nature. Green water is caused by dispersion of copper corrosion products into the water. A related problem is blue or green staining of plumbing fixtures. Such water often exhibits a rather unpleasant taste due to high concentrations of dissolved copper. General corrosion can be controlled by raising the pH and increasing the hardness of the water.

Impingement Attack. Impingement attack is the result of excessive flow velocities (greater than 4 fps) and was at one time thought to be purely mechanical in nature (Hatch, 1961). It is now believed that high velocities disrupt formation of protective films, allowing electrochemical attack to proceed more rapidly. Besides velocity, factors that aggravate impingement attack are soft water, high

temperature, and low pH. For hard, cold water, velocities of 8 fps (2.4 mg/sec) can be tolerated. Soft waters, temperatures above 140°F (60°C), and pH below 6.5 all contribute to destruction of the protective film.

Impingement attack is characterized by a rough surface, often accompanied by horseshoe, or U-shaped pits. In severe cases, perforation of the tube wall occurs in as little as 6 months. It is most severe at points of turbulence, such as sites downstream of fittings, and is most prevalent in recirculating systems.

Pitting Attack. Pitting of copper tubing is most commonly associated with hard well waters. Pitting most often occurs in cold water piping. Usually, dissolved carbon dioxide exceeds 5 mg/L and dissolved oxygen is high (Rambow and Holmgren, 1966; Cohen and Lyman, 1972). The water quality parameters typical for water systems having copper pitting problems are not the typical soft, low-pH waters normally associated with corrosion of copper. Surface waters containing organic or humic substances are not associated with pitting attack (Campbell, 1954; Lucey, 1967). In England, the presence of a carbon film due to residues of drawing lubricants from manufacturing processes was found to be important (Campbell, 1950).

Pitting occurs most often in horizontal runs of piping with the deepest pits concentrated in the bottom of the pipe. This suggests that gravity holds dense solutions of copper salts in the pit, sustaining the corrosion reaction (Cruse and Pomeroy, 1974). Horizontal surfaces are vulnerable to attack (Lucey, 1967), suggesting that corrosion products stream from the vertical surface and concentrate gravitationally.

Pitting attack is most common in new installations, with 80–90 percent of the reported failures occurring within the first 2–3 years. In some extreme cases, failure ocurs in as little as 3 months (Cruse and Pomeroy, 1974). Pitting occurs in all three standard types of copper tubing (K, L, and M). According to Cohen and Lyman, the relative frequency of failure does not appear to vary among types of tubing, even though type M has a thinner wall. If unfavorable water quality conditions occur before the protective coat has formed, then serious pitting attack may occur. After 3 or 4 years, the incidence of failure drops significantly even in systems having serious incidence of pitting.

Lucey (1967) has proposed a mechanism for copper pitting that assumes the formation of a membrane cell, as shown in Figure 18-14. The cuprous oxide membrane covers the pit cavity, which contains cuprous chloride and cuprous oxide crystals. The cuprous oxide membrane acts as a bipolar electrode with oxidation taking place on the inside face and reduction taking place on the outside face. There are reactions simultaneously with the calcium bicarbonate in the waters, which result in the precipitation of calcium carbonate and basic cupric salts such as malachite ($CuCO_3 \cdot Cu(OH)_2$) in a mound above the membrane.

Corrosion of Lead.   Pipes made of lead (plumbum) were first used for plumbing by the Romans; this material is ductile and can be bent to desired shapes. Use of lead in plumbing declined beginning in the nineteenth century as iron became competitive as a pipe material. Later, health concerns about lead poisoning from corrosion by-products in drinking water further reduced the use of lead in water conduits, although household lead plumbing is still in place in some older cities. Lead ''pigtails'' are also in use as an easily deformable connection between distribution mains and household connections.

Figure 18-2d shows the forms of lead that are thermodynamically stable under various conditions. Under typical conditions in water, the favored form of lead depends strongly on pH. At lower pH, the dissolved plumbic ion (Pb^{2+}) is the most stable form, while lead carbonate, or cerussite ($PbCO_3$), is favored at neutral pH, and the hydroxycarbonate $Pb_3(OH)_2(CO_3)_2$ or the hydroxide $Pb(OH)_2$ are favored at higher pH. Since the equilibrium between Pb^{2+} and $PbCO_3$ tends to govern the

distribution of oxidized forms of lead, the solubility of lead increases dramatically as pH decreases below pH 8 (for fixed alkalinity), as shown Figure 18-15, since the equilibrium concentration of carbonate decreases by about 2 orders of magnitude for each decrease of one pH unit when H_2CO_3 is the predominating carbonate species. Also, since Pb^{2+} is a dissolved ion, while $PbCO_3$ is a solid scale, the rate of dissolution of lead tends to be rapidly accelerated by low pH.

The equilibrium thermodynamics of lead species also suggest that waters with low alkalinity will tend to lose lead more rapidly because of the lower concentration of carbonate and less favorable conditions for cerussite. Increasing alkalinity will reduce equilibrium lead concentration when cerussite is the stable oxidized form. However, when the hydroxy-carbonate $Pb_3(OH)_2(CO_3)_2$ is the stable form, increasing alkalinity will increase the equilibrium lead concentration (Sheiham and Jackson, 1981). Another complication in the chemistry of dissolved lead species is that at moderate-to-high alkalinities the formation of soluble ion pairs such as $PbCO_3^0$ dominates the concentration of Pb^{2+} in solution.

Although equilibrium calculations may be used to predict stable forms of lead for given pH and alkalinity, actual observations show that both the carbonate and the hydroxycarbonate solid scales may form when only the carbonate is predicted. This indicates that kinetic mechanisms may limit the approach to equilibrium. However, the equilibrium chemistry does yield the useful observation that low-pH, low-alkalinity waters promote the dissolution of lead. Empirical surveys of dissolved lead in tap water show correlations of ''first-draw'' concentrations with these factors (Pocock, 1980);

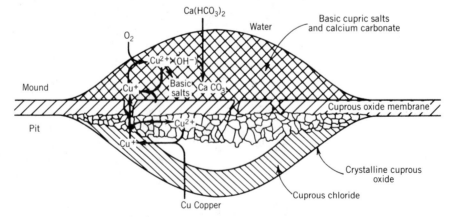

FIGURE 18-14.   Diagram of structure of copper pit. (After Lucey, 1967.)

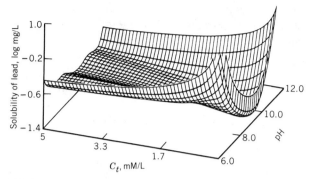

FIGURE 18-15. Three-dimensional plot of relationship of lead solubility as log mg/L (log(Pb)) to pH and total inorganic carbonate (TIC) concentration as mg CaCO₃/L. (Source: Schock 1984.) Reprinted with permission.

the concentration of lead decreases after the first flushing of standing water in pipes. The relationship between lead intake and public health problems such as retardation, hypertension, and renal failure raises concern for use of lead plumbing with soft, acidic waters. It should be emphasized that the rate of weight loss of lead from plumbing is not necessarily correlated with the concentration of dissolved lead at the tap because of the complexity of the reactions among metallic, oxidized solid, dissolved ionic, and dissolved ion pair lead. The issue here is not pipe failure but public health protection.

In many systems, it appears that elevated levels of lead at the tap are not due to the presence of lead piping but rather to the use of lead-soldered joints used to connect copper pipes (Benson and Klein, 1983). This raises concern about protection of health from lead corrosion even in water distribution systems that lack lead plumbing short of the consumer system. One major alternative for control of lead in systems as these is the use alternate solders, such as 95 : 5 tin–antimony solder, rather than the traditional 50 : 50 lead–tin solder to reduce the presence of lead in plumbing systems. The tin–antimony solder is somewhat more difficult to apply.

As suggested by the equilibrium chemistry of lead in water, one strategy for the control of lead corrosion by-products in tap water is to adjust pH upward into the range of 8.0–8.5 in order to lower the solubility of Pb^{2+}. This and other generalized corrosion control techniques are discussed in a later section.

A treatment alternative that may have specific effects on lead corrosion is the choice of disinfectant. Studies at Portland, Oregon (JMM, 1983), observed that the use of chloramines for disinfection had the effect of decreasing the level of lead corrosion by-products lost to solution compared to the use of free chlorine, although the final pH of the two alternative treated waters were approximately equal. This could be attributed to lower potential in the system with combined chlorine. However, the addition of ammonia to chlorinated water to generate chloramines tended to increase the concentration of lead by-products in solution. The role of ammonia in possible complexation of metal ions and its effect on corrosion rates are poorly understood.

Corrosion of Galvanized Pipe by Pitting. Pitting corrosion of galvanized pipe has been the subject of many studies (Kenworthy, 1943; Cruse, 1971; Hoxeng and Prutton, 1949; Hoxeng, 1950; Treweek et al., 1978; Trussell and Wagner, 1985). In southern California, the problem has been associated most recently with the rapid pitting failure of galvanized steel pipes in tract homes constructed within the past few years. The resulting litigation, brought about by homeowner groups against developers, pipe suppliers, plumbers, and water purveyors, has produced expensive settlements in favor of the homeowners. Possible factors in pitting include pipe quality, presence of dissolved copper, and potential reversal.

Pipe Quality. Quality of galvanized pipe under most U.S. plumbing codes is governed by ASTM A-120 pipe specifications for Schedule 40 "standard weight" pipe (ASTM, 1983). This pipe is intended for ordinary uses in steam, water, gas, and air lines. No mechanical tests are specified for this pipe except for the hydrostatic pressure test performed at the mill. An average weight of zinc coating is specified, along with minimum tolerances for the weight and diameter. The major shortcoming of the ASTM specification is the description of the pipe and zinc finish. In some pipe (Treweek et al., 1978), the weld seam provides a pronounced irregularity in the otherwise smooth interior surface. Treweek observed that pitting corrosion initiated at a point to the right of the seam, but as the pit developed, it reached the seam itself and quickly deteriorated through the wall thickness, resulting in pipe failure. Many of the pipes in the housing development investigated (Treweek et al., 1978) failed in this manner, leading the homeowners to conclude that the pipe seam itself was improperly constructed. Trussell and Wagner

(1985) describe problems of galvanized pipe quality in much greater detail.

Effect of Copper on Pitting of Galvanized Pipe. As previously mentioned, the presence of low concentrations of copper has sometimes been found to be an important factor in rapid pitting of galvanized iron. Copper may be introduced into the water supply either as an algacide or as a result of corrosion of copper in the system upstream of the galvanized iron pipe. Concentrations of copper as low as 0.1 mg/L have been shown to trigger pitting attack. Copper-induced pitting occurs only in new pipe. It occurs in both hot and cold pipe at rates that can exceed 100 mpy (mils per year) and is usually associated with copper levels on the pipe surface exceeding 5 mg/dm². In most cases, pipe failure by pitting occurred within 24–36 months after installation of new galvanized steel pipe. A mechanism for copper-induced pitting of galvanized pipe was discussed by Treweek et al. (1978).

Because the standard potential of copper is much higher than that of zinc (copper is more noble), copper ions will deposit on surfaces of galvanized pipe, acting as a cathode depolarizer. This raises the potential of the cathode surface and causes the zinc metal to oxidize into solution. This released zinc usually combines with oxygen in the water to form a zinc oxide coating on the pipe interior. Figure 18-16 shows that pitting rate increases in proportion to copper in the corrosion products on the pipe surface. Once a protective scale layer has been formed, however, deposition of copper on the metal

surface is minimized. This explains the observation that pitting corrosion is associated mainly with new pipe.

Effects of Potential Reversal. Under certain conditions, the electrochemical potential between zinc and iron can be reversed so that iron becomes the sacrificial metal in place of zinc (Hoxeng and Prutton, 1949; Hoxeng, 1950). Increased temperatures, bicarbonate concentration, and/or nitrate concentration may cause potential reversal, while increased chloride, sulfate, calcium, and silicate levels will mitigate against potential reversal. Figure 18-17a shows the effect of temperature on zinc–iron potential. At temperatures found in hot water systems, iron can be sacrificially corroded. Similarly, Figure 18-17b shows that at bicarbonate concentrations of 80 mg/L or more, iron can be sacrificially corroded.

Natural Scale Formation

Although $CaCO_3$ scale receives more attention when control of corrosion is discussed, evidence indicates that natural scales formed by the metals themselves are often important. These stable species are formed as oxidation products of corrosion of materials, which are thermodynamically favored. They can serve to protect the metal itself from further corrosion by inhibiting transport to the metal surface of species, such as inhibiting transport to the metal surface of species such as dissolved oxygen. Where iron is concerned, scales formed by both ferric and ferrous ion are important. The solubility of these scales in a typical water is shown in Figures 18-18a–f. Ferric hydroxide, ferrous hydroxide, and ferrous carbonate are formed from corroding iron. Both of the iron hydroxides from rather porous oxide scales which do little to reduce corrosion. The ferrous carbonate scale is a much denser, more protective scale.

The natural scales formed by certain other metals when exposed to natural waters are also of importance. These include scales formed by copper and zinc. As shown in Figure 18-18d, tenorite (CuO) is a thermodynamically stable scale formed by copper in high-pH water. This is a very dense, tough material which greatly reduces the rate of oxygen mass transfer to the cathode surface, thus reducing the rate of corrosion. The properties of tenorite, as well as copper's more noble potential, are the principal reasons for copper's excellent record of service as a

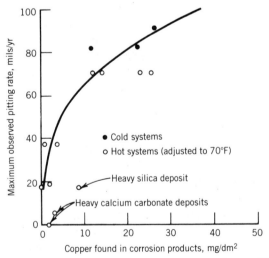

FIGURE 18-16. Copper deposition and pitting (Cruse, 1971). Reprinted with permission.

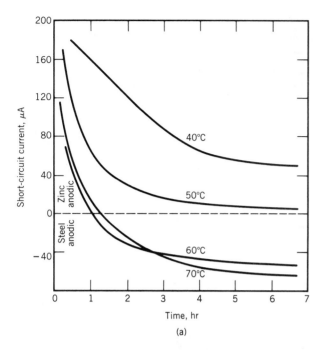

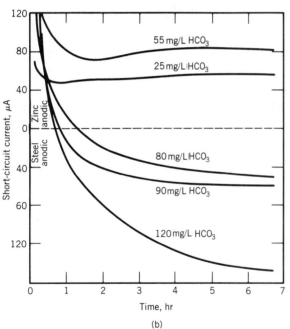

FIGURE 18-17. Factors influencing potential reversal between iron and steel (Hoxeng and Prutton, 1949). Reprinted with permission. Current flow between iron and zinc versus (a) temperature and (b) bicarbonate.

piping material. Figure 18-18d also shows, however, that the solubility of the tenorite scale dramatically increases with decreasing pH. As a result, low-pH waters are much more corrosive to copper than are alkaline waters. Neutral waters of high al-

kalinity also favor the formation of the basic copper carbonate malachite $[Cu(CO_3) \cdot Cu(OH)_2]$ rather than tenorite. In most waters, malachite forms as a metastable species which gradually transforms to the more stable tenorite. Though fairly insoluble in these waters, malachite is not as effective as tenorite in reducing corrosion.

Examination of Figure 18-18c shows that the chemistry of most natural waters favor formation of zinc carbonate or hydrozincite, $Zn_5(CO_3)_2(OH)_8$. Both of these materials are very effective at reducing the rate of oxygen mass transfer to the cathode surface. In fact, zinc itself is often used as a cathodic inhibitor; the good behavior of galvanized pipe also depends to a great degree on the protection of iron by the less noble zinc, which acts as a sacrificial anode.

The behavior of two other metals, lead and cadmium, are also of interest, not because they are widely used as water conduits, but because both are regulated by the drinking water standards.

Figure 18-18e shows the solubility diagram for lead in natural waters. It shows that in most natural waters, insoluble forms are stable: lead carbonate and lead hydroxide. Lead carbonate is a very dense scale and is effective in reducing oxygen transfer; hence corrosion of lead is reduced to very low levels after the scale is formed. Recent work (Schock, 1980) indicates that hydrocerussite $[Pb_3(OH)_2(CO_3)_2]$ is the predominant solid species formed at pH 7–11.

Figure 18-18f shows the solubility of cadmium, which occurs as a contaminant in the zinc coating of galvanized pipe. Cadmium can be stabilized by higher pH, which favors the insoluble carbonate form.

Characteristics of Scales

A number of researchers in the water field have tried to establish the significance of siderite in the formation of protective scales in water distribution pipes (Baylis, 1926; Larson and King, 1954). German investigators (Kolle and Rosch, 1980) have reported the existence of a "shell-like layer" in scales found in cast iron pipes. This layer seems to be important in the formation of a dense corrosion-resistant scale. Kolle and Rosch speculated that siderite is important as an intermediate in the formation of the shell-like layer but may not be part of the layer.

A model for scale formation has been proposed

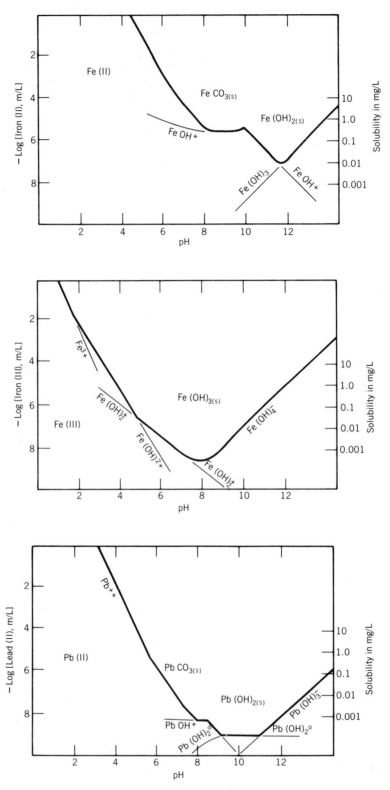

FIGURE 18-18. Solubility of common corrosion products. (approximate summary of stability of various oxidized species that might occur on surface of metallic water mains; $C_T = 2$ mm, 25°C. Cadmium is shown because it is an occasional contaminant of lead and zinc.

416

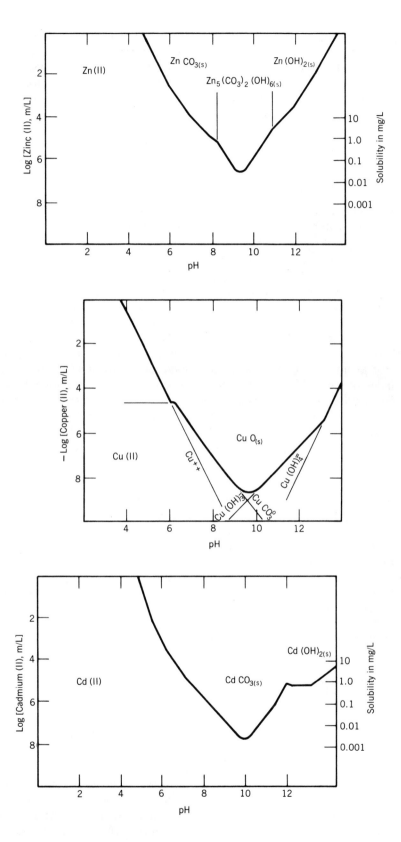

FIGURE 18-18. (*continued*).

by German researchers (Sontheimer et al., 1979) and is based on the concept that siderite formation is a key step in the formation of corrosion-resistant scales.

The formation of goethite (FeOOH) in the shell-like layer may involve the conversion of siderite while retaining its dense crystalline structure (Kolle and Rosch, 1980). The formation of calcite also appears to be necessary to form a good protective scale.

In order to test the siderite model, experiments were designed to determine the influence of various factors related to scale formation on the corrosion rate (Sontheimer et al., 1981). It was shown that the corrosion rate decreases as the proportion of Fe(II) in the scale increases, tending to support the model. Increasing calcium concentration also showed a protective relationship.

Dissolution of Concrete and Cement

Concrete is a composite material consisting of a Portland cement binder in which inert filler materials called aggregate are imbedded. Portland cement is primarily a mixture of calcium silicates and aluminates, including free lime. The composition of Type I Portland cement is shown in Table 18-5. Portland cement has the property of setting and hardening when mixed with water.

Aggregate used in concrete ranges from fine sand used in mortars to large stones 3–4 in. in size. Reactive aggregates may cause durability problems in water pipe and are to be avoided. Cement mortar is used as a protective liner in cast iron or steel water

TABLE 18-5. Basic Composition of Type 1 Portland Cement

Compound	Formula	Cement Composition (%)
Tricalcium silicate	$3CaO \cdot SiO_2$	42–67
Dicalcium silicate	$2CaO \cdot SiO_2$	8–31
Tricalcium aluminate	$3CaO \cdot Al_2O_3$	5–14
Tetracalcium alumino ferrite	$4CaO \cdot Al_2O_3 \cdot Fe_2O_3$	6–12
Calcium sulfate	$CaSO_4$	2.6–3.4
Free lime	CaO	0–1.5
Magnesium oxide	MgO	0.7–3.8
Volatiles	—	0.6–2.3

pipe. It is made by using a fine sand aggregate with a Portland cement binder.

Asbestos cement is used to make a low-cost water pipe. It is made by using asbestos fibers in place of sand or aggregate. This results in a relatively strong, dense pipe wall. Recently, there has been concern about the release of asbestos fibers from asbestos cement pipe subject to corrosion by low-pH, low-alkalinity waters.

The cement in reinforced concrete water pipe, cement mortar lining of steel and cast iron water pipe, and asbestos cement water pipe is subject to deterioration when subjected to prolonged contact with aggressive waters. At least two mechanisms are involved. The first is the dissolution of free lime and other compounds when in contact with low-pH, low-alkalinity waters. The second is chemical attack by aggressive ions such as sulfate and chloride.

Corrosion of cement pipe materials in domestic water systems is governed principally by solubility considerations. The following dissolution reactions have been proposed for the cement matrix of A/C pipe (Schock and Buelow, 1981) and should be applicable to the other cement-containing pipe materials. For free lime,

$$Ca(OH)_2(s) \rightarrow Ca^{2+} + 2OH^- \qquad (18-30)$$

for which $K_{so} = 10^{-5.20}$.

Possible dissolution reactions for the three major crystalline constituents of the cement matrix are:

For tricalcium silicate,

$$Ca_3SiO_5(s) + 5H_2O \rightarrow$$
$$3Ca^{2+} + H_4SiO_4 + 6OH^- \qquad (18-31)$$

For dicalcium silicate,

$$Ca_2SiO_4(s) + 4H_2O \rightarrow$$
$$2CA^{2+} + H_4SiO_4 + 4OH^- \qquad (18-32)$$

For tricalcium aluminate,

$$Ca_3Al_2O_6(s) + 6H_2O \rightarrow$$
$$3CA^{2+} + 2Al^{3+} + 12OH^- \qquad (18-33)$$

The net result is dissolution of calcium in the cement matrix with accompanying corrosion of the pipe material. The aggressiveness of a water toward dissolution of cement is related to the value of Langelier's index (LI), which measures the potential for precipitation or dissolution of calcium carbonate. Free lime in place in cement can be converted to calcium carbonate without weakening the concrete, when the LI of a water is positive. However, exposure to water with a negative LI can lead to loss of calcium and irreparable damage within a few months. The "aggressiveness index," a version of LI, has also been used to assess the potential for dissolution of concrete (AWWA, 1977).

Cement materials are also occasionally subject to chemical attack by water containing high concentrations of aggressive ions such as sulfates and chlorides. Normally this type of corrosion will not be significant with typical domestic water supplies. Chemical attack may be a problem with some high-TDS water supplies. A reaction of concrete with sulfate at high pH produces ettringite, a salt of calcium aluminate–calcium sulfate. Ettringite expands as crystals grow and can damage concrete when sulfate concentrations are extremely high.

Microbiological Aspects of Corrosion

Microorganisms can play a significant role in fostering corrosion of pipe materials. Bacteria have the ability, in certain situations, to form microzones of high acidity or high concentrations of corrosive species, to increase electrolytic concentration at surface sites to favor electron transfers, to mediate the oxidation of reduced chemical species, to disrupt the protective influence of surface films, and to mediate the removal of corrosion reaction products, enhancing the kinetics of the forward reaction. In particular, certain microorganisms can take advantage of local gradients in redox potential in order to obtain energy for their metabolic needs. Therefore, these bacteria can facilitate corrosion kinetics by accelerating the rate of redox reactions. Even anaerobic bacteria may thrive in otherwise aerobic water because of the formation of microsites on the surface of pipes, associated with pitting or other microscopic irregularities. When associated with pits, some bacteria can add significantly to the corrosion in the microzone.

The most significant bacteria involved in mediation of corrosion reactions are sulfate reducers, methane producers, nitrate reducers, sulfur bacteria, and iron bacteria. In particular, the sulfate-reducing and iron bacteria groups are nuisances in corrosive behavior.

Sulfate-reducing bacteria such as *Desulfovibrio desulfuricans* obtain their energy needs by reducing sulfate to sulfite, elemental sulfur, or other reduced forms of sulfur in anaerobic environments such as those that may occur within a corrosion pit. Bacteria in general can thrive at an interface between oxygenated and anaerobic areas because of the energetic difference associated with the oxygen gradient and the presence of oxidized species such as sulfate in an anaerobic region where sulfate is not thermodynamically stable. Sulfate reducers require ferrous iron and hydrogen gas as substrates. The uptake of these species, which are reaction products of the corrosion of iron metal, serves to lower the concentration of Fe^{2+} and H_2 in the microzone and thereby to enhance the kinetics of the forward direction of the corrosion reaction.

Similarly, hydrogen-oxidizing bacteria can assist the corrosion of iron by removing the hydrogen gas produced at the cathode. Iron bacteria such as *Gallionella* and *Sphaerotilus lepothrix* are aerobic organisms that mediate the oxidation of ferrous ions to ferric ions or to insoluble hydroxide forms; this serves to remove ferrous ions, which are the immediate reaction product of the oxidation of iron at the anode. Iron bacteria may also be involved in the formation of FeOOH (goethite), which can result in the mineralization of *Gallionella* colonies and the formation of encrustations (Kolle and Rosch, 1980). Nitrate-reducing bacteria and methane-producing bacteria may play a role in altering the pH of a surface site, affecting the driving force for a corrosion reaction.

Current understanding of the interactions of microbiological and electrochemical reactions does not permit quantitative prediction of the role of microorganisms in enhancing corrosion reactions. It appears likely that complex ecosystems involving a variety of genera can arise in microzones on the surface of pipes where corrosion reactions take place. Empirical evidence suggests a major role for microbiological reactions in some systems. For example, chlorination of well water supplies has reduced corrosion rates, suggesting that disinfection can inhibit the role of bacteria. Laboratory studies of the corrosion of cast iron shows that unsterilized water can lead to significantly more rapid corrosive attack than sterilized water. Similarly, aeration of otherwise anaerobic or microaerobic water can reduce the activity of anaerobic bacteria.

A practical consequence of the effect of microorganisms in promoting corrosion is the relationship between disinfection and corrosion. Water utilities can experience relatively severe corrosion in parts of the distribution system where water stagnates. The decline in chlorine residual and lack of scouring action in distribution dead ends results in increased growth of microorganisms on surfaces, especially where pitting has taken place or organic content is high. As a result, corrosive failures and main breaks tend to be more common in these locations. It is common practice for utilities to flush such dead ends routinely to minimize bacterial growth by scouring and by lowering the detention time of the chlorinated water.

Corrosion Indices

Corrosion indices used to date can be divided into three broad classes: (1) calcium carbonate saturation indices; (2) indices based on other solution properties expected to influence corrosion rates; and (3) empirical indices based on statistical correlations of corrosion rates with solution properties. The ideal corrosion index would allow prediction of the corrosivity of a water from more easily determined water quality parameters. Although a number of indices have been developed, none has demonstrated this ideal property, even under limited circumstances. Some of these indices give a general indication of the merits of water treatment alternatives, but none is a substitute for direct measurement of corrosion rates.

Calcium Carbonate Saturation. During the first half of the twentieth century, one of the most prominent developments in waterworks-related chemistry was the calcium carbonate saturation index developed by Professor Wilfred F. Langelier of the University of California at Berkeley (Langelier, 1936). Langelier's work was timely because it combined recently developed knowledge of chemical equilibria (Talbot and Blanchard, 1905) with the recent availability of practical pH measurement devices (Baylis, 1929) to provide a practical means for evaluating calcium carbonate saturation. Langelier named it the saturation index.

At the time, achieving calcium carbonate saturation was thought to be the principal means for controlling corrosion of iron distribution piping. If a solution were appropriately supersaturated with calcium carbonate, it would lay down a protective "eggshell" lining on the inside of the pipe, protecting it from the water. Earlier work had been conducted on this subject, notably by a German chemist named Tillmans (Tillmans and Hueblein, 1913; Tillmans, 1913).

Langelier proposed a "saturation pH" (pH_s) at which the total alkalinity and the calcium hardness would be at equilibrium with each other and with solid calcium carbonate. Without any simplifying assumptions, the pH_s may be found with the following relationships:

Definition of Alkalinity:

$$Alk = [HCO_3^-] + 2[CO_3^{2-}] + [OH^-] - [H^+] \quad (18\text{-}34)$$

Equilibrium for dissolution of calcium carbonate:

$$K'_{so} = [Ca^{2+}][CO_3^{2-}] \quad (18\text{-}35)$$

Equilibrium for dissociation of bicarbonate:

$$K'_2 = \frac{[H^+][CO_3^{2+}]}{[HCO_3^-]} \quad (18\text{-}36)$$

Equilibrium for dissociation of water:

$$K'_w = [H^+][OH^-] \quad (18\text{-}37)$$

Substituting:

$$Alk = \frac{[H^+]K'_{so}}{[Ca^{2+}]K'_2} + \frac{2K'_{so}}{[Ca^{2+}]} - \frac{K'_w}{[H^+]} + [H^+] \quad (18\text{-}38)$$

Rearranging and defining equilibrium $[H^+]$ as $[H_s^+]$, this equation in $[H_s^+]$ may be solved as a quadratic equation:

$$[H_s^+] = \frac{-b + \sqrt{b^2 - 4ac}}{2a} \quad (18\text{-}39)$$

where $a = K'_{so}$
$b = K'_2 2 K'_{so}[Ca^{2+}] - Alk \cdot K'_2$
$c = K'_w K'_2$

Where K'_{so} is the solubility constant for $CaCO_3$, K'_2 is the acidity constant for dissociation of bicarbonate, and K'_w is the dissociation constant of water. A similar derivation was recently published by Rossum & Merrill (1983).

Taking the logarithm:

$$pH_s = -\log[H_s^+] \quad (18\text{-}40)$$

TABLE 18-6. *Thermodynamic Data for the Carbonate System*

At ionic strength (I) approaching zero (Plummer and Busenberg, 1982):

$$\log K_1 = -356.3094 - 0.06091964T + 21834.37/T$$
$$+ 126.8339 \log T - 1684915/T^2$$

$$\log K_2 = -107.8871 - 0.03252849T + 5151.79/T$$
$$+ 38.92561 \log T - 563713.9/T^2$$

$$\log K_{so} \text{ (calcite)} = -171.9065 - 0.077993T$$
$$+ 2839.319/T + 71.595 \log T$$

For nonideal solutions with nonzero I, one may estimate activity coefficients

$$\gamma_B = \frac{\{B\}}{[B]}$$

by the Davies equation (Davies, 1962),

$$\log \gamma_B = -AZ_B^Z \left(\frac{\sqrt{I}}{1 + \sqrt{I}} - 0.3I \right)$$

where Z_B = charge on species B

A = Debye–Huckel constant, 0.5

I = ionic strength = $\frac{1}{2}\Sigma C_i Z_i^2$

C_i = Concentration of ith species

Table 18-6 gives necessary data for solving the equation. A simplified form of these equations has traditionally been used (APHA, 1985):

$$pH_s = pK_2 - pK_{so} - \log[Ca^{2+}] - \log[\text{Alk}] \quad (18\text{-}41)$$

Langelier then defined the $CaCO_3$ or LI saturation index (Langelier's index) as the difference between the measured pH and the pH_s; thus,

$$LI = pH - pH_s \quad (18\text{-}42)$$

The state of saturation with respect to calcium carbonate therefore depends on the value of the Langelier index:

LI < 0 solution undersaturated with $CaCO_3$ (will dissolve $CaCO_3$).

LI = 0 solution at equilibrium with $CaCO_3$.

LI > 0 solution supersaturated with $CaCO_3$ (will precipitate $CaCO_3$).

Figures 18-19a,b show early experimental data that illustrate the increased deposition of $CaCO_3$ and decreased weight loss from pipes as Langelier's index is made more positive.

It is a common practice for utilities to add lime or caustic soda to their treated water to achieve a saturation index that is near neutral or slightly positive (0 to +0.2), and there is some evidence that this practice will reduce red water complaints (DeMartini, 1938). On the other hand, the appropriateness of the saturation index for controlling corrosion or even the deposition of a $CaCO_3$ scale is subject to contradictory evidence (Stumm, 1956, 1960; Larson, 1970). This is illustrated by the data in Figure 18-20, which show a poor correlation between Langelier's index and the rate of formation of a scale. Although the saturation pH has a rational basis, the definition of the saturation index itself is empirical. Nevertheless, calcium carbonate chemistry is a subject of broad interest.

Since Langelier's original research, there have been a number of attempts to improve on the index (Larson and Buswell, 1942), to facilitate its determination (Hoover, 1938; Dye, 1944, 1952, 1958), or to demonstrate its use (DeMartini, 1938). Others have also proposed alternate indices associated with calcium carbonate saturation. Notable among these are the driving force index (McCauley, 1960), the momentary excess (Dye, 1952), the Ryznar stability index (Ryznar, 1944), and the aggressiveness index (AWWA, 1977). These indices represent concepts that are basically the same as Langelier's original principle.

Perhaps the calcium carbonate chemistry of a water is best characterized by three factors: (1) The free energy driving force of the precipitation reaction (ΔG); (2) the calcium carbonate precipitation capacity (CCPC); and (3) the calcium carbonate saturation buffer intensity (β_{Alk}^s).

The free energy (ΔG) is a quantitative measure of the energy available to drive the precipitation reaction. It can be shown that ΔG influences the rate of the precipitation reaction and that a certain ΔG threshold must be reached before crystal nucleation will ensue (Nielson, 1964). It can be shown that the ΔG for $CaCO_3$ precipitation can be defined as follows:

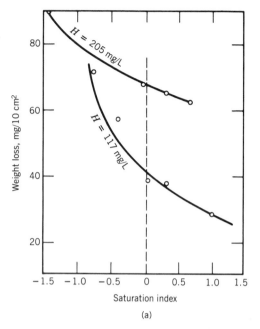
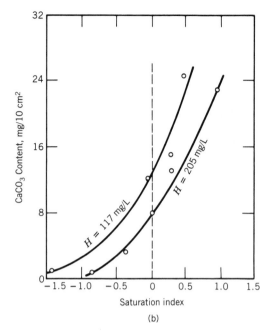

(a)

(b)

FIGURE 18-19. (a) CaCO₃ saturation versus corrosion. (b) CaCO₃ saturation versus CaCO₃ deposition. CaCO₃ saturation, corrosion, and CaCO₃ deposition for waters of moderate to high hardness and alkalinity; Langelier's saturation index seems associated with deposition CaCO₃ and reduction in rate of corrosion (Stumm, 1956). Reprinted with permission.

$$\Delta G = RT \ln \left\{ \frac{[Ca^{2+}][CO_3^{2+}]}{K'_{so}} \right\} \quad (18\text{-}43)$$

$$= RT \ln \left\{ \frac{[Ca^{2+}]C_t\alpha_2}{K'_{so}} \right\}$$

where

C_t = total inorganic carbon concentration

$$= \frac{\text{Alk} - \Delta}{\sigma} \quad (18\text{-}44)$$

σ = change coefficient (Ricci, 1952)

$$= \alpha_1 + 2\alpha_2 \quad (18\text{-}45)$$

Δ = electrical difference (Ricci, 1952)

$$= [OH^-] - [H^+]$$

$$= K'_w[H^+] - [H^+] \quad (18\text{-}46)$$

$$\alpha_1 = \frac{K'_1[H^+]}{F} \quad (18\text{-}47)$$

$$\alpha_2 = \frac{K'_1K'_2}{F} \quad (18\text{-}48)$$

$$F = [H^+]^2 + K'_1[H^+] + K'_1K'_2 \quad (18\text{-}49)$$

where K'_1 and K'_2 are the acidity constants for dissociation of H_2CO_3.

Again the necessary data for calculating ΔG are given in Table 18-6. The value of ΔG is a function of the solution alkalinity, calcium hardness, and pH, whereas Langelier's index is a function of the alkalinity and calcium hardness only. On the other hand, over the range of pH for which bicarbonate ion is the dominant species of the carbonate system (6.5–8.5), the ΔG of the reaction is related to Langelier's index as follows (Trussell et al., 1976):

$$\Delta G = -2.3RT(\text{LI}) \quad (18\text{-}50)$$

Calcium carbonate will spontaneously precipitate from solution at a ΔG of less than about −5 to −7 kcal/mole. When calcium saturation is used as a red water control technique, a ΔG of about −0.5 kcal/mole is probably sufficient. This technique should be used with caution for waters of low alkalinity and/or hardness because even though the water is supersaturated, its capacity to precipitate protective calcium carbonate is limited. This leads us to the need for a capacity index.

The calcium carbonate precipitation capacity (CCPC) is the amount of calcium carbonate that will precipitate or dissolve from the solution as it comes to equilibrium with solid CaCO₃. This parameter

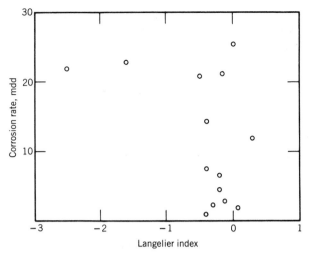

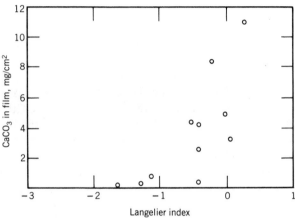

FIGURE 18-20. $CaCO_3$ saturation, corrosion, and $CaCO_3$ deposition for waters having a variety of qualities; Langelier's saturation index still correlates well with $CaCO_3$ deposition but not with corrosion. (Data from Stumm, 1960.)

was first proposed by Merrill and Sanks and termed the calcium carbonate precipitation potential (CCPP) (Merrill and Sanks, 1978). The magnitude of this capacity index can be derived as follows:

If the amount of $CaCO_3$ that will precipitate from a solution to achieve equilibrium is given by X moles/L, then the equilibrium condition is

$$[Ca^{2+}]_{eq}[CO_3^{2-}]_{eq} = K'_{so}$$
$$= ([Ca^{2+}] - X)(C_T - X)\alpha_2 \quad (18\text{-}51)$$

where K'_{so} is the solubility product for $CaCO_3$, C_T is the total inorganic carbon concentration, and α_2 is the fraction of inorganic carbon present as CO_3^{2-}, as previously defined. Alkalinity in the equilibrated system is given by

$$Alk - 2X = (C_T - X)\sigma + \Delta \quad (18\text{-}52)$$

where Δ is the difference $[OH^-] - [H^+]$, as previously defined. These two equations may be combined to eliminate the variable C_T, yielding a quadratic equation for X:

$$2X^2 + (\Delta - 2[Ca^{2+}] - Alk)X$$
$$+ ([Ca^{2+}]Alk - [Ca^{2+}]\Delta - \frac{K'_{so}\sigma}{\alpha_2} = 0 \quad (18\text{-}53)$$

The solution is given by

$$CCPC = \frac{-b + \sqrt{b^2 - 4ac}}{2a} \quad (18\text{-}54)$$

where $a = 2$
$$b = -2[Ca^{2+}] - Alk$$
$$c = Alk[Ca^{2+}] - \frac{K'_{so}}{2} - [Ca^{2+}]$$

The calcium carbonate buffer intensity (β^s_{Alk}) is a measure of the sensitivity of the calcium carbonate saturation of the solution to changes in alkalinity. This index is of interest because it measures the sensitivity of the calcium carbonate saturation of the solution to external changes. This buffer intensity can be estimated from the following equation:

$$\beta^s_{Alk} = \left[\frac{\partial(Alk)}{\partial S}\right]_{C_T}$$
$$= \frac{\partial(Alk)}{\partial[H^+]} \frac{\partial[H^+]}{\partial[CO_3^{2-}]} \frac{\partial[CO_3^{2-}]}{\partial S} \quad (18\text{-}55)$$

where S is the saturation with respect to calcium carbonate:

$$S = \frac{[Ca^{2+}][CO_3^{2-}]}{K'_{so}} \quad (18\text{-}56)$$

It can be shown that β^s_{Alk} can be calculated by:

$$\beta^s_{Alk} = \frac{K_{so}}{C_T[Ca^{2+}]K_1K_2} \cdot$$
$$\frac{[H^+]^2 + K'_1[H^+] + K'_1K'_2}{2[H^+] + K'_1} \cdot$$
$$\frac{C_TK_1([H^+]^2 + 4K'_2[H^+] + K'_1K'_2)}{F^2}$$
$$+ 1 + \frac{K'_w}{[H^+]^2} \quad (18\text{-}57)$$

where F is defined as for the free energy (18-49).

The significance of β^s_{Alk} is that higher values of this buffer intensity are associated with smaller changes in calcium carbonate saturation for a given exogenous change in alkalinity (as, for example, by local inhomogeneities in solution associated with electrochemical change sites). This means that local supersaturations of $CaCO_3$ would be expected at cathodic sites where hydrogen ion is being withdrawn from solution when β^s_{Alk} is larger. This reduces the protective effect of calcium carbonate scale formation; it is desirable to enhance scale formation by operating in a region where β^s_{Alk} is small, all else being equal. The β^s_{Alk} is smaller when C_T is larger or when $[Ca^{2+}]$ is larger. Figure 18-21 shows plots of ΔG, CCPP, and β^s_{Alk} as functions of pH for a water with $C_T = 2$ mM and $[Ca^{2+}] = 2$ mM. The figure shows how higher pH generally favors corrosion control by increasing ΔG and CCPP and lowering β^s_{Alk}. However, not all of these fundamental indices can be optimized under the same conditions. Also, there may be practical limitations to pH control.

Other Mechanism-Associated Indices. Two measures of solution properties have been based on an understanding of the mechanisms of the corrosion process. Larson's ratio (LR; Larson, 1970) is developed from the relatively corrosive behavior of certain ions, such as chloride and sulfate, and the protective properties of bicarbonate. It is alternatively defined as follows:

$$LR = \frac{[Cl^-] + [SO_4^{2-}]}{[HCO_3^-]} \quad \text{or}$$

$$LR = \frac{[Cl^-]}{[HCO_3^-]} \tag{18-58}$$

where $[Cl^-]$, $[SO_4^-]$, and $[HCO_3^-]$ are all expressed in milliequivalents per liter. Larson proposed the index including both sulfate and chloride in the numerator, but his original work was done with the ratio to chloride. Only chloride appears to have strong aggressive properties for all metals, so the chloride ratio form is probably more useful. Where ferrous metals are concerned, sulfate does behave as an aggressive ion, but even here it is likely that Larson's ratio overestimates the effect. Figure 18-22 shows 3-day rates of corrosion in neutral solutions of sodium chloride–bicarbonate versus Larson's ratio and short-term corrosivity of iron. Given the earlier discussion regarding the role of chloride in iron corrosion, this is not a surprising result.

After observing that drastic alterations in solution composition may occur in the vicinity of a corroding electrode, Stumm proposed that the buffer intensity, a measure of the solution's ability to resist changes in pH with the addition of acid or base, might be a good indicator of a solution's corrosiveness (Stumm, 1960). The buffer intensity is defined as follows:

$$\beta^{pH}_{Alk} = \frac{\partial_{Alk}}{\partial_{pH}} \tag{18-59}$$

or, for solutions buffered exclusively by the CO_2 system,

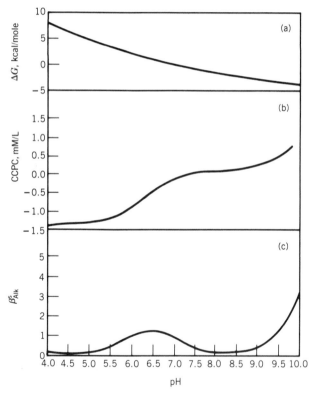

FIGURE 18-21. Parameters of $CaCO_3$ saturation versus pH. (a) Free energy driving force. (b) $CaCO_3$ precipitation capacity. (c) $CaCO_3$ saturation buffer intensity. Perhaps the three most fundamental descriptive factors are G, the free energy driving force of the precipitation reaction; the $CaCO_3$ capacity; the amount of $CaCO_3$ that would precipitate upon equilibration; and the saturation buffer intensity, the rate of change in $CaCO_3$ saturation with changes in alkalinity.

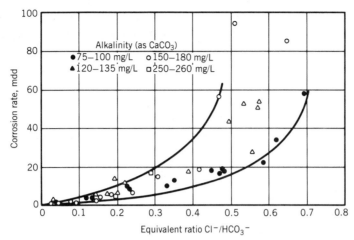

FIGURE 18-22. Effect of Larson's ratio on corrosion of mild steel. (Source: Larson, 1975.)

$$\beta_{Alk}^{pH} = 2300 \left\{ C_T K_1'[H^+] \right.$$

$$\left(\frac{[H^+]^2 + 4K_2'[H^+] + K_1'K_2'}{([H^+]^2 + K_1'[H^+] + K_1'K_2')^2} \right)$$

$$\left. + [H^+] + [OH^-] \right\} \quad (18\text{-}60)$$

where K_1' and K_2' are the acidity constants of the carbonate system and C_T is the total inorganic carbon concentration. Approximate values for β_{Alk}^{pH} can be estimated by graphical methods (Trussell and Thomas, 1971).

If the buffer intensity is large, it is more difficult for heterogeneity in pH to develop, which fosters the corrosion reaction. Therefore, all else being equal, the corrosion rate would be expected to decrease with increasing buffer intensity. Figure 18-23e illustrates this relationship as it is shown by data gathered from Stumm (Stumm, 1960). Stumm's data seem to indicate that buffer intensities greater than 0.5 meq/unit are desirable.

Empirical Indices. A number of empirical indices have been developed that attempt to relate solution properties and the corrosiveness of a solution. One of the earliest was the Riddick index, which includes a combination of a number of water quality parameters (Riddick, 1944). Including all the parameters, it takes the following form:

$$\text{Index} = \frac{75}{\text{Alk}} \left\{ CO_2 + 0.5(\text{hardness} - \text{Alk}) + [Cl^-] \right.$$

$$\left. + 2[NO_3^-] \left[\frac{10}{SiO_2} \right] \frac{DO + 2}{(DO)_{sat}} \right\} \quad (18\text{-}61)$$

All species in the equation are in milligrams per liter as shown except for hardness and alkalinity, which are in mg $CaCO_3$/L. Riddick proposed that waters having an index less than 5 are extremely noncorrosive, those between 6 and 25 are noncorrosive, 26–50 moderately corrosive, 51–75 corrosive, 76–100 very corrosive, and greater than 100 extremely corrosive. This index seems to have some meaning in soft waters like those often found in the northeastern part of the country where the index was developed, but it has been less effective elsewhere.

Pathak devised a nomograph to evaluate a rating system for copper pipe developed by Lucey (Pathak, 1971; Lucey, 1967). This index includes consideration of Na^+, Cl^-, SO_4^{2-}, NO_3^- ions, pH, and DO. The method has not been widely evaluated.

Israeli researchers developed an index to characterize the corrosiveness of the hard waters found in that region of the world. The index considers hardness, chloride, alkalinity, and sulfate:

$$\text{Index} = AH + B[(Cl^-) + (SO_4^{2-})]_e^{-1/AH}$$
$$+ C \quad (18\text{-}62)$$

where $A = 0.00035$
$B = 0.34$
$C = 19.0$
$H = \text{hardness}$

Waters having an index less than 200 are considered corrosive, waters having indices between 200 and 500 are moderately corrosive, and waters having indices greater than 500 are not very corrosive. This method has not been widely evaluated either.

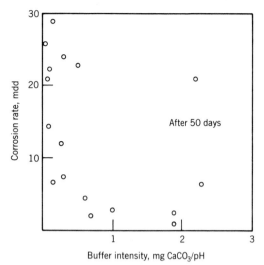

FIGURE 18-23. Relation between buffer intensity and corrosion of iron (data from Stumm, 1960). There appears to be a tendency for lower corrosion rates at higher buffer intensities. High buffer intensities will reduce the heterogeneity of pH's between anodic and cathodic site on the microscale.

Others have attempted to develop similar correlations (Singley, 1981), but none of any universal effectiveness has yet been developed. Although all the indices include factors that are certainly impor-

tant in determining corrosion rates of various metals, none can be expected to transport from one location or material to another effectively.

Summary

Corrosion indices based on calcium carbonate chemistry or on other mechanisms of corrosive attack are useful; they give us measures of solution parameters that are often important in the corrosion process. These include the capacity for $CaCO_3$ precipitation, the buffer intensity of the solution, and the ratio of aggressive to passivating ions. Empirical indices suggest other water quality factors important in the corrosion process, such as DO, SiO_2, and so on.

Table 18-7 summarizes the quality of several different water supplies, including several important corrosion indices and actual measured corrosion rates. It is clear that none of these indices give a definitive prescription for assessing the rate of corrosion. Although all these indices may be of some qualitative utility in comparing the corrosiveness of water supplies, there is little promise that any index will ever be developed that can be used to accurately predict the corrosiveness of a water supply

TABLE 18-7. Corrosion Rate in Certain Water Supplies

Utility	Source	Buffer Capacity (meq/pH unit)	Larson's Ratio $\frac{(Cl^-)}{(HCO_3)}$	LI	pH	Corrosion (mpy) Copper	Galvanized Iron	Black Iron
Contra Costa CWD	Bollman Effluent (JMM, 1980)	0.03	2.1	−0.0	8.4	0.18	1.6	8.+
Los Angeles DWP	Owens River (Streicher, 1956)	0.12	0.4	+0.3	8.3	1.1	2.5	7.7
MWD of southern California	Colorado River (Streicher, 1956)	0.13	2.1	+0.6	8.2	0.64	2.2	7.0
EBMUD	Mokelumne (Carns and Stinson, 1979)	0.08	0.3	−0.4	9.3	—	—	10.1
New York	Catskills (Gold and McCaul, 1975)	(0.08)	1.6	(−2.7)	(7.0)	0.02	0.4	12.4
Seattle	Tolt (Kennedy Engr., 1978)	0.04	2.8	−4.3	6.1	1.6	4.6	14.7
Seattle	Cedar (Kennedy Engr., 1978)	0.05	0.9	−2.2	7.3	0.5	4.2	11.7
Portland[a]	Bull Run (JMM, 1983)	0.10	0.3	−3.0	7.0	0.4	0.6	3.4
Portland	Bull Run (JMM, 1983)	0.10	0.3	−3.0	70	0.3	0.6	3.7
Army Corps[b]	Potomac River (JMM, 1983b)	0.11	—	−0.01	7.8	0.009	0.05	0.12
EBMUD[c]	Mokelumne River (JMM, 1983)	0.08	0.4	−2.7	7.1	0.3	1.3	2.5
EBMUD[d]	American River (JMM, 1983)	—	0.4	−2.5	7.3	0.2	1.4	4.8
EBMUD[d]	Sacramento River (JMM, 1983)	—	0.4	−1.4	7.4	0.5	2.7	8.5
EBMUD[d]	Clifton Court Forebay (JMM, 1983)	—	2.8	−1.2	7.4	0.3	2.5	7.3
EBMUD[d]	Indian Slough (JMM, 1983)	—	3.1	−1.0	7.5	0.2	3.3	6.2

[a] Supply with chloramination; other Portland study with free chlorine.
[b] Potomac Experimental Estuary Water Treatment Plant, treating 50/50 mix of secondary effluent and water from the Potomac River.
[c] Raw water supply prior to pH adjustment.
[d] Raw waters tested; not currently EBMUD supply.

quantitatively. There is no substitute for directly measuring corrosion rates.

TREATMENT ALTERNATIVES FOR CORROSION CONTROL

Because of economic losses associated with corrosion of water conduits, research has been devoted to developing effective corrosion control measures. General categories of such approaches to corrosion control include (1) control of environmental parameters affecting corrosion rate, (2) addition of chemical inhibitors, (3) electrochemical measures, and (4) considerations of system design. Fontana and Greene (1978) and Uhlig (1971) have presented complete discussions of these measures. This chapter is concerned primarily with internal corrosion of water distribution conduits, so the discussion will be limited to those control measures applicable to drinking water distribution. These include pH adjustment and addition of chemical inhibitors such as polyphosphates, zinc orthophosphate, and silicates.

pH Adjustment

Generally, pH adjustment is made in an attempt to achieve a positive value of Langelier's index. As explained previously, Langelier's index is the difference between actual pH of a water and pH_s, the theoretical pH at which the calcium hardness and the alkalinity would be in equilibrium. A positive value of the index indicates that the water will tend to deposit a calcium carbonate scale on the pipe surface, which will tend to protect the surface from corrosion. A negative value of the index indicates that calcium carbonate will tend to dissolve; also, the cement in a mortar lining or asbestos cement pipe will tend to dissolve in concrete conduits.

Adjustment of pH also has a significant effect on thermodynamic properties of the carbonate system, such as the free energy driving force, the calcium carbonate precipitation potential, the buffer intensity with respect to calcium carbonate saturation $(dAlk/dS)$, and the buffer intensity with respect to pH $(dAlk/dpH)$. As previously mentioned, these quantities may play a major role in the formation and behavior of the calcium carbonate protective scale.

In addition to affecting the carbonate system, pH is a key variable in the solubility of conduit materials such as lead, copper, and zinc. The thermodynamic stability of oxidized species such as metal oxides, hydroxides, and carbonates is strongly influenced by pH also, so pH may govern whether or not an insoluble natural protective scale can form or whether soluble species are favored as a metal corrodes. Therefore, pH adjustment can play a major role in stabilizing a pipe material such as lead: low pH favors dissolution into plumbic ions, whereas high pH favors formation of solid scales that deter further corrosion.

Water treatment practice to adjust pH has typically involved addition of lime [as CaO or $Ca(OH)_2$] or sodium hydroxide in order to achieve a positive value of the Langelier index. The addition of these alkaline materials also increases alkalinity; this has the side effect of decreasing the solubility of corrosion products such as lead carbonate and enhancing the formation of solid metal carbonate scales. Generally, pH adjustment may not be a satisfactory corrosion control measure for waters of very low alkalinity. In such cases, the buffer intensity is low; the additions of lime necessary to raise pH are small, but the resulting high-pH water does not have a large pH buffer intensity. Therefore, it is easier for corrosive microsites of low pH to form near the surface, as in pits, even though the overall thermodynamic tendency is for calcium carbonate scale to form.

In waters of high alkalinity, it becomes more difficult to adjust pH above 8.0 because of the more rapid precipitation of calcium carbonate in distribution lines. The interaction of pH change and alkalinity change has a complicated effect on buffer intensity, which may result in a lowered buffer intensity following treatment if this is not monitored carefully (Stumm, 1960).

Addition of Zinc Orthophosphate

Phosphate salts have been used as corrosion inhibitors in industrial water systems since the 1930s, with the application of dehydrated sodium phosphate. The use of zinc orthophosphate as a corrosion inhibitor in drinking water distribution systems is relatively recent. Zinc orthophosphate was first employed by the City of Long Beach (Murray, 1970), with dramatic results in reducing pitting and tuberculation by a fairly corrosive water. It is hypothesized that this inhibitor acts by forming a finely divided colloid in the water, which adsorbs in a thin film of insoluble zinc orthophosphate on the surface of conduits. At cathodic sites, this film is believed to act as a passivator, possibly inhibiting diffusion of oxygen, carbon dioxide, or cations. The zinc may play a role analogous to the passivating activity of the zinc layer on galvanized pipe. It may react with the hydroxide ions formed as cathodic reaction products to form zinc hydroxide at the site. The mechanism of cathodic protection is not fully understood.

A number of utilities have employed zinc orthophosphate successfully as a corrosion inhibitor (Kelly et al., 1973; Mullen, 1974), while others have been less successful (Kennedy Engineers, 1978). Mullen observed that addition of zinc orthophosphate at a dosage of 0.5 mg/L as Zn, plus pH adjustment to a range of 7.0 ± 0.2, resulted in a significantly lower corrosion rate than pH adjustment alone. He also found that, at lower temperatures, a corrosion rate of less than 2.0 mpy could be obtained with reduced zinc dosages and pH in the range of 6.8 ± 0.1. Recent research in Germany shows that orthophosphate without zinc is effective for controlling the corrosion of galvanized iron (Trussell and Wagner, 1985; Ryder and Wagner, 1985).

Other researchers (Buelow et al., 1980; Schock and Buelow, 1981) have observed that addition of zinc orthophosphate or zinc chloride, in combination with pH adjustment, significantly reduced deterioration of asbestos cement pipe exposed to soft waters of low alkalinity. The addition of zinc orthophosphate at dosages of 0.3–0.6 mg/L as Zn, with pH adjusted up to 8.2, resulted in extremely low rates of leaching of calcium from the pipe. A dense protective coating of hydrozineate $[Zn_5(OH)_6(CO_3)_2]$ may have played a major role in reducing loss of calcium from the concrete. Orthophosphate was believed not to have played a role in this protective mechanism; therefore, Schock and Buelow recommended that zinc chloride be used instead.

Addition of Polyphosphates

Polyphosphates include a variety of compounds such as crystalline forms of pyrophosphate $(P_2O_7^{4-})$, metaphosphate, and tripolyphosphate $(P_3O_{10}^{5-})$ salts and glassy forms of polyphosphate blends. The molecular structure of these compounds tends to be indeterminate because of the possibility of conversion of one form into another. A commonly used compound is dehydrated sodium phosphate, called sodium hexametaphosphate $(NaPO_3)_n$. Polyphosphates were introduced in the late 1930s for prevention of calcium carbonate scale formation by formation of a glassy scale on existing calcite surfaces and associated removal of carbonate crystal nuclei from contact (Rice and Hatch, 1939). It was then observed that polyphosphates could also act as corrosion inhibitors.

A certain amount of controversy has surrounded the use of polyphosphates for corrosion inhibition. It has been clearly established that polyphosphates can be effective in combining with and sequestering dissolved ferrous ions into stable dissolved complexes; this prevents the ferrous ions from further oxidizing to ferric forms and forming rust-colored insoluble hydroxides. Therefore polyphosphates can be useful in preventing the appearance of "red water" at the tap and reducing consumer complaints about apparent corrosion. Polyphosphate doses of 2–4 mg/L as P_2O_5 are appropriate for this purpose. Polyphosphates cannot, however, reduce the ferric oxide forms and therefore are less effective in controlling this problem when the ferrous ions have already been oxidized.

The conditions under which polyphosphates actually inhibit corrosion in drinking water appear to be limited. The mechanism of protection appears to be deposition of polyphosphate films composed of metaphosphate, ferric oxide, and other compounds at cathodic sites. The rate of cathodic deposition is strongly dependent on pH, peaking at pH 5.0, and decreasing significantly at higher pH (Lamb and Eliassen, 1954). Deposition of charged colloidal particles may also occur at anodic sites; such particles may be formed by reaction of polyphosphates with ferrous, zinc, and calcium ions.

Polyphosphates are ineffective for corrosion inhibition in stagnant water; the protection for corrosion increases with velocity (Larson, 1957). It appears that turbulence is required to maintain the polyphosphate film on conduit surfaces. As a result, the value of polyphosphates as corrosion inhibitors appears to be highest in controlled situations with flowing water, low pH, and high doses, as is the case in many recirculating industrial water systems. Polyphosphates may be ineffective in dead ends or service lines in a water distribution system.

Polyphosphates may also be effective only with certain types of water. In some natural waters, polyphosphates can greatly accelerate the rate of corrosion when added in small doses. This suggests that the practicing engineer should conduct extensive tests to determine the appropriate dose if polyphosphates are to be employed in a water distribution system.

Addition of Silicates

Sodium silicates have been studied as an alternative corrosion inhibitor. They were first employed in the 1920s to reduce corrosion of lead piping (Butler and Ison, 1966). As with polyphosphates, the molecular composition of silicates tends to be indeterminate, with a formula of $Na_2O \cdot nSiO_2$, with n a variable ratio; a silicate with n of 3.3 is suggested for alkaline waters. Typical dosages range from 8 to 12 mg/L as SiO_2 in alkaline waters.

The mechanism of silicates as corrosion inhibitors appears to proceed by formation of colloidal particles composed of silica (SiO_2) or polysilicic ions. The colloid slowly forms a film on surfaces; negatively charged silica micelles may be attracted to anodic sites. Metal ions can be incorporated into the film; ferrous iron can react with the silicate to form a ferrous silicate coating, while zinc silicate may form on galvanized pipe surfaces. Mixture of the metal corrosion products with the silica gel in the film appears to enhance the protective effect by affecting the nature of the film. Diffusion of oxygen to cathodic sites is inhibited by the film.

The degree of effectiveness of silicate as corrosion inhibitor depends on the characteristics of the water. Extensive tests by the Illinois State Water Survey (ISWS) (Lane et al., 1977) suggest that pH controls the silicate dosage required for effective control, with higher doses needed at a pH lower than 8.5. The concentration of calcium, magnesium, chloride, and other constituents affects the optimal silicate dosage. The presence of calcium may assist inhibition by the scale, while high magnesium concentrations may cause deposits of magnesium and decrease effectiveness. The ISWS concluded that silicates are the best means of inhibiting corrosion of galvanized steel and copper-based metals in domestic hot water systems, especially recirculating systems such as those used in commercial buildings. It is clear that for any particular water, detailed testing of the effect of silicate at different dosages is necessary. When added at too low a dosage, silicate may intensify the corrosion rate in some waters, while overdosage can affect the taste of the supply and cause discoloration of food.

METHODS OF CORROSION MEASUREMENT

Corrosion measurement techniques have been developed to determine corrosion rates in both the field and the laboratory because of the potentially high costs of corrosion damage. They are used to assess quantitatively the effectiveness of various corrosion control measures and the corrosion resistance of new pipe and lining materials. Until the early 1950s, corrosion measurements were almost exclusively made using weight loss methods, which are inconvenient and time consuming. Subsequently, electrochemical techniques were developed that can measure instantaneous corrosion rates and can be used to continuously monitor both laboratory corrosion studies and corrosion processes in operating water systems. Moreland and Hines (1978) published a good survey of corrosion monitoring methods. The principal methods of corrosion measurements currently in use in the water supply field will be described below.

Weight Loss Methods

Two weight loss corrosion rate measurement methods are in general use in the water field. The older is the coupon method and the newer is the machined nipple method developed by the ISWS. Both of these methods are described in ASTM Standard D2688 (ASTM, 1983).

Coupon Test. The coupon method employs a flat metal coupon. The coupon is located in the center of the pipe where the velocity is higher than at the pipe wall and where it may be subject to turbulence induced by the coupon holder, which may affect scale and oxide film formation or action of corrosion inhibitors. Coupon holders are normally installed in a bypass pipe loop to permit removal and checking of specimens without upsetting normal system operation. Multiple specimens are usually installed to allow determination of corrosion rate as a function of time. Coupons are $13 \times 102 \times 0.8$ mm ($0.5 \times 4.0 \times 0.032$ in.) for sheet metals and $13 \times 102 \times 3$ mm ($0.5 \times 4.0 \times 0.125$ in.) for cast metals.

Coupons are then installed in a test loop with flow velocities adjusted to match velocities in the system under consideration. Two time series of coupons are recommended. The first set of coupons should be removed at 4–7-day intervals to determine initial corrosion rates. The second set is used to determine a long-term steady-state corrosion rate. The first coupon of this set should be removed after 1 month with the remainder removed at 1–3-month intervals.

After removal, the coupons are subjected to a defined cleaning procedure before reweighing. Following ASTM D2688, corrosion rate for each coupon is calculated as follows (this equation provides a sufficiently accurate approximation of the actual corrosion rate for most corrosion studies):

$$P = \frac{H(W_1 - W_2)}{W_1 D} \times 1.825 \times 10^2 \quad (18\text{-}63)$$

where P = corrosion rate, mmpy
$\quad H$ = original thickness of the coupon, mm
$\quad W_1$ = original weight of the coupon, mg
$\quad W_2$ = final weight of the coupon, mg
$\quad D$ = exposure time, days

The relationship between corrosion rate in grams per square meter per day (g/m² · d) and penetration in millimeters per year (mmpy) is expressed as follows:

$$\text{mmpy} = \text{g/m}^2 \cdot \text{d} \times (365/d) \quad (18\text{-}64)$$

where d is the density of the metal, kg/m³ (for example, 7860 kg/m³ for mild steel). It should be recognized that the corrosion rate thus determined may deviate from corrosion rates in the actual system due to differences in flow regimes and differences in surface characteristics resulting from material composition, surface preparation, scaling, or biological growths.

Machined Nipple Test. The machined nipple test employs a short length of actual pipe material machined on the outer surface. This permits insertion in a PVC pipe sleeve connected into the pipe system under test with pipe unions. The specimen holder is designed so that a smooth flow line is maintained through the pipe specimen so as to simulate actual flow conditions. The inner surface of the pipe nipple is not altered or machined.

When monitoring corrosion in an existing system, inserts are made from the same material as used in the system being observed. A bypass pipe loop is constructed to permit removal and inspection of the pipe inserts without disturbing normal operations. A minimum upstream straight run of 9.4 mm (3 ft) shall be provided ahead of the ISWS tester to minimize nonuniform or turbulence. One insert should be exposed a minimum of 120 days before evaluation and the second should be exposed at least 12 months if possible. The corrosion rate is then computed according to ASTM D2668.

Rates of corrosion are expressed either as weight loss per unit of area per unit of time or the equiva-

lent rate of penetration. The accepted units are grams per square meter per day (24 hr) (g/m² · d) and millimeters penetration per year (mmpy). Calculation as g/m² · d may be made as follows:

g/m² · d = $117W/T$
  (for steel and galvanized specimens)

g/m² · d = $120W/T$ (for copper specimens)

where W = actual weight loss of insert, g
$\quad T$ = installation time, days

The relationship between corrosion rate in grams per square meter per day and penetration in millimeters per year is expressed in Eq. 18-64
The densities of the metals studied are:

Metal	Density, kg/m³
Steel	7860
Zinc (galvanized)	7150
Copper	8920

Electrical Resistance Method. The electrical resistance method of corrosion measurement is based on the principle that the rate of change of electrical resistance of a piece of metal is directly proportional to its rate of change of cross-sectional area,

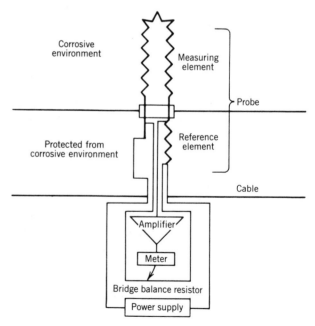

FIGURE 18-24.   Electrical resistance probe. (Courtesy of Magna Corp.)

which is in turn a function of the corrosion rate (Bovankovich, 1973). Electrical resistance also changes with temperature so some means of temperature compensation must be provided. This is done by providing a reference element connected in a Wheatstone bridge circuit in parallel with the measuring element. The reference element is of the same material as the measuring element but is protected from the corrosive environment. Figure 18-24 is a schematic diagram of an electrical resistance probe. The reference element is enclosed in the body of the probe to assure that it is maintained at the same temperature as the measuring element.

The electrical resistance method measures the integrated metal loss from which a corrosion rate can be calculated. The corrosion rate can be calculated from the slope of the plot by the following formula:

$$CR = (D_2 - D_1) \times \frac{PF \times 0.365}{T} \quad (18\text{-}65)$$

where CR = corrosion rate, mpy
D_2 = final dial reading, in.
D_1 = initial dial reading, in.
PF = probe factor
T = time, days

By letting $D_2 - D_1 = D$, the formula can be simplified to

$$CR = \frac{D \times PF \times 0.365}{T} \quad (18\text{-}66)$$

Note: 0.365 in Eqs. 18-65 and 18-66 is a conversion factor for changing inches to mils and days to years.

Linear Polarization Resistance Method. The linear polarization resistance method is based on the principle that at low corrosion potentials corrosion rate is essentially a linear function of polarization resistance. Commercially manufactured equipment based on this principle is available. It is normally calibrated to read corrosion rate directly as a function of corrosion current, usually at a fixed corrosion potential of 10 mv. The advantage of this method is that corrosion rates can be determined instantaneously.

Stern and Geary (1957) derived the basic equation on which the linear polarization method is based:

$$I_c = \frac{\Delta I}{2.3\,\Delta E} \frac{\beta_c \beta_a}{\beta_c + \beta_a} \quad (18\text{-}67)$$

This can be arranged to solve for the polarization resistance:

$$\frac{\Delta E}{\Delta I} = \frac{\beta_a \beta_c}{2.3 I_c (\beta_a + \beta_c)} \quad (18\text{-}68)$$

When corrosion is controlled by concentration polarization as when oxygen depolarization is controlling, β becomes very large and the equation simplifies to

$$\frac{\Delta E}{\Delta I} = \frac{\beta_a}{2.3 I_c}$$

where $\Delta E/\Delta I$ = polarization resistance
ΔE = polarization potential
ΔI = applied current
I_c = corrosion current
β_a = anodic Tafel slope
β_c = cathodic Tafel slope

Various researchers have presented data relating corrosion rate to linear polarization resistance (LPR) (Skold and Larson, 1957). Figure 18-25 is a log plot of corrosion rate versus LPR which indicates a linear relationship.

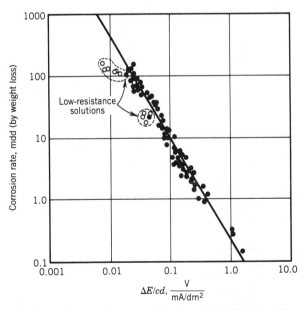

FIGURE 18-25. LPR versus weight loss. (After Skold and Larson, 1957.) Reprinted with permission.

$$\log \text{CR} = -K \log \left[\frac{\Delta E}{\Delta I} \right]$$

$$= K \log \left[\frac{\Delta I}{\Delta E} \right]$$

$$\text{CR} = K \left[\frac{\Delta I}{\Delta E} \right]$$

where $K = 2.3I_c$

Direct-reading corrosion meters using the LPR principle use this relationship. The value of E is held constant, usually 10 mv, and the instrument is calibrated to read corrosion rate directly as a function of current.

Stumm presented data relating corrosion current to LPR (Stumm, 1959). Stumm's measurements were made using the special flow cell wherein electrode potentials were measured potentiometrically using mercury–mercurion sulfate reference electrodes with the capillary tips in contact with the cast iron electrodes to eliminate infrared drop in the solution. This method eliminated the outliers shown in Figure 18-25 for low-resistance solutions.

REFERENCES

APHA, "Standard Methods for the Examination of Water and Wastewater," Section 203, APHA Wash. D.C., (1985).

ASTM A 120-82, "Standard Specification for Pipe, Steel, Black and Hot-Dipped Zinc-Coated (Galvanized) Welded and Seamless, for Ordinary Uses," in *Annual Book of ASTM Standards* (1983).

AWWA, "AWWA Standard for Asbestos-Cement Pressure Pipe, 4 in., through 24 in., for Water and Other Liquids," AWWA C400-77, rev. of C400-75, AWWA, Denver (1977).

Baylis, J., "Prevention of Corrosion and 'Red Water'," *JAWWA*, **15**, 598 (1926).

Baylis, J., "Tungsten Electrode for Determining Hydrogen-ion Concentration," U.S. 1,727,094 (Sep. 3, 1929).

Benson, J., and Klein, H., "Lead in Drinking Water. Investigation of a Corrosive Water Supply," *J. Env. Health*, **45**, 179–181 (1983).

Bethune, A., and Loud, N., *Standard Aqueous Electrode Potentials and Temperature Coefficients at 25C*, Clifford, A. Hampel, Skokie, IL (1964).

Bovankovich, J. C., "On-Line Corrosion Monitoring," *Mater. Prot. Perform.*, **7**, 20–23 (June 1973).

Buelow, R., Millette, J., McFarren, E., and Symms, J., "The Behavior of Asbestos-Cement Pipe Under Various Water Quality Conditions: A Progress Report," *JAWWA*, **72**, 91–102 (1980).

Butler, G., and Ison, H., *Corrosion and Its Prevention in Waters*, Reinhold, New York (1966).

Campbell, H., "Pitting Corrosion of Copper Pipes Caused by Films of Carbonaceous Material Produced During Manufacture," *J. Inst. Metals*, **77**, 345–356 (1950).

Campbell, H., "A Natural Inhibition of Pitting Corrosion of Copper in Tap Waters," *J. Applied Chem.*, **4**, 633–647 (1954).

Carns, K., and Stinson, K., "Corrosion Control and Water Quality Standards," presented at Operator's Forum at 79th Annual AWWA Conference, San Francisco, CA (1979).

Cohen, A., and Lyman, W., "Service Experience with Copper Plumbing Tube," *Mater. Prot. Perform.*, **11**, 48–53 (1972).

Cohen, M., "Dissolution of Iron," in G. Brubaker and P. Phipps, (eds.), *Corrosion Chemistry*, ACS Symposium Series, No. 89 (1979).

Cruse, H., "Dissolved-Copper Effect on Iron Pipe," *JAWWA*, **63**, 79–81 (1971).

Cruse, H., and Pomeroy, R., "Corrosion of Copper Pipes," *JAWWA*, **66**, 479–483 (1974).

DeMartini, F., "Practical Application of the Langelier Method," *JAWWA*, **30**, 85 (1938).

Donahue, F., "Corrosion Processes," in W. Weber (ed.), *Physiochemical Processes for Water Quality Control*, Wiley-Interscience, New York (1972).

Donahue, F., "Electrochemical Techniques in Corrosion Studies," in G. Brubaker and P. Phipps (eds.), *Corrosion Chemistry*, ACS Symposium Series, No. 89 (1979).

Dye, J., "The Calculation of Alkalinities and Free Carbon Dioxide in Water by the Use of Nomographs," *JAWWA*, **36**, 895 (1944).

Dye, J., "Calculation of Effects of Temperature on pH, Free Carbon Dioxide, and the Three Forms of Alkalinity," *JAWWA*, **44**, 356 (1952).

Dye, J., "Correlation of the Two Principle Methods of Calculating the Three Kinds of Alkalinity," *JAWWA*, **50**, 200 (1958).

Engell, H. J., and Stolica, N. D., *Z. Phys. Chem.*, **20**, 113 (1959).

Evans, U., *An Introduction to Metallic Corrosion*, Edmund Arnold, London (1948).

Fontana, M., and Greene, N., *Corrosion Engineering*, McGraw-Hill, New York (1978).

Foley, R., "Role of Chloride in Iron Corrosion," *Corrosion*, **26**, 58–70 (1970).

Geld, I., and McCaul, C., "Corrosion in Potable Water," *JAWWA*, **67**, 549 (1975).

Greene, N., and Fontana, M., "A Critical Analysis of Pitting Corrosion," *Corrosion*, **15**, 25t–31t (1959).

Greene, N., and Wilde, B., "Variable Corrosion Resistance of 18Cr–18Ni Stainless Steels" Influence of Environmental and Metallurgical Factors," *Corrosion*, **26**(12), 533–537 (1969).

Hatch, G., "Unusual Cases of Copper Corrosion," *JAWWA*, **53**, 1417–1428 (1961).

Hoar, T., *Discussion. Faraday Soc.*, **1**, 299 (1947).

Hoover, C., "Practical Application of the Langelier Method," *JAWWA*, **30**, 1802 (1938).

Hoxeng, R., "Electrochemical Behavior of Zinc and Steel in Aqueous Media—Part II," *Corrosion*, **6**, 308–312 (1950).

Hoxeng, R., and Prutton, C., "Electrochemical Behavior of Zinc and Steel in Aqueous Media," *Corrosion*, **5**, 330–338 (1949).

JMM, "Water Quality Study: Vol. 1," prepared for Contra Costa Water District, Concord, CA (Nov. 1980a).

JMM, "Water Quality Study: Vol. 2," prepared for East Bay Municipal Utilities District, Oakland, CA and Contra Costa Water District, Concord, CA (Oct. 1980b).

JMM, "Internal Corrosion Mitigation Study, Final Report," prepared for Portland Water Bureau, Portland, Oregon (1983a).

JMM, "The Operation Maintenance and Performance Evaluation of the Potomac Estuary Experimental Water Treatment Plant," prepared for the Baltimore Distict of the Army Corps of Engineers, Baltimore, MD (1983b).

Kelly, T., Kise, M., and Steketee, F., "Zinc/Phosphate Combinations Control Corrosion in Potable Water Distribution Systems," *Materials Protection Performance*, **7**, 28–30 (1973).

Kennedy Engineers, "Internal Corrosion Study: Phase 1 Report," prepared for Seattle Water Department, Seattle, WA (Oct. 1976).

Kennedy Engineers, "Internal Corrosion Study," for Seattle Water Department, Seattle, WA (1978).

Kenworthy, L., "The Problem of Copper and Galvanized Iron in the Same Water System," *J. Inst. Metals*, **69**, 67–90 (1943).

Kolle, W., and Rosch, H., "Untersuchungen an rohrnetzinkrustierungen unter mineralogischen Gesichtspunkten," *Vom Wasser*, **55**, 159 (1980).

Lamb, J., and Eliassen, R., "Mechanism of Corrosion Inhibition by Sodium Metaphosphate Glass," *JAWWA*, **46**, 445–460 (1954).

Lane, R., Larson, T., and Schilski, S., "The Effect of pH on the Silicate Treatment of Hot Water in Galvanized Plumbing," *JAWWA*, **69**, 457–461 (1977).

Langelier, W., "The Analytical Control of Anti-Corrosion Water Treatment," *JAWWA*, **28**, 1500 (1936).

Larson, T., "Report on Loss in Carrying Capacity of Water Mains," *JAWWA*, **47**, 1061–1072 (1955).

Larson, T. E., "Evaluation of the Use of Polyphosphates in the Water Industry," *JAWWA*, **49**, 1581–1586 (1957).

Larson, T., "Corrosion by Domestic Waters," Illinois State Water Survey Bulletin, 56, Urbana, Illinois (1970).

Larson, T., "Corrosion of Domestic Waters," Illinois State Water Survey, Urbana (1975).

Larson, T., and Buswell, A., "Calcium Carbonate Saturation Index and Alkalinity Interpretations," *JAWWA*, **34**, 1667 (1942).

Larson, T., and King, R., "Corrosion by Water at Low Flow Velocity," *JAWWA*, **46**, 1 (1954).

Larson, T., and Skold, R., "Laboratory Studies Relating Mineral Quality of Water to Corrosion of Steel and Cast Iron," *Corrosion*, **14**, 285t (1958a).

Larson, T., and Skold, R., "Current Research on Corrosion and Tuberculation of Cast Iron," *JAWWA*, **50**, 1429 (1958b).

Latimer, W., *Oxidation Potentials*, 2nd ed., Prentice-Hall, Englewood Cliffs, NJ (1952).

Levich, V., *Physiochemical Hydrodynamics*, Prentice-Hall, Englewood Cliffs, NJ (1962).

Lorenz, W., "Der Einfluss Von Halogenidionen Auf Die Anodische Auflosung Des Eisens," *Corrosion Sci.*, **5**, 121–131 (1965).

Lucey, V., "Mechanism of Pitting Corrosion of Copper in Supply Waters," *Brit. Corrosion J.*, **2**, 175–185 (1967).

Matsuda, S., and Uhlig, H., "Effect of pH, Sulfates and Chlorides on Behavior of Sodium Chromate and Nitrite as Passivators for Steel," *J. Electrochem. Soc.*, **111**, 156–161 (1964).

McCauley, R., "Controlled Deposition of Protective Calcite Coatings in Water Mains," *JAWWA*, **52**, 1386 (1960).

Merrill, D. T., and Sanks, R. L., "Corrosion Control by Deposition of $CaCO_3$ Films," AWWA Part 3, **70**, 12–18; Part 2, **69**, 634–640; Part 1, **69**, 597–599 (1978–1977).

Moreland, P., and Hines, J., "Corrosion Monitoring . . . Select the Right System," *Hydrocarbon Processing*, **57**(11), 251–255 (1978).

Mullen, E., "Potable Water Corrosion Control," *JAWWA*, **66**, 473–479 (1974).

Murray, B., "A Corrosion Inhibitor Process for Domestic Water," *JAWWA*, **62**, 659–662 (1970).

Nielsen, A., *Kinetics of Precipitation*, Macmillan, New York (1964).

Obrecht, M., and Quill, L., "How Temperature and Velocity of Potable Water Affect Corrosion of Copper and Copper Alloys: . . . In Heat Exchanger and Piping Systems . . . What is Corrosion? . . . Monitoring System Reveals Effects of Different Operating Conditions . . . Tests show Effects of Water Quality at Various Temperatures, Velocities . . . Different Softened Waters have Broad Corrosive Effects on Copper Tubing . . . Cupro-Nickel, Admiralty Tubes Resist Corrosion Better," in *Heating, Piping, and Air Conditioning*, Jan., 165–169; Mar., 109–116; Apr., 131–137; May, 105–113; Jul., 115–122; and Sep. 125–133 (1960).

Obrecht, M., and Quill, L., "How Temperature, Velocity of Potable Water Affect Corrosion of Copper and Its Alloys," *Heating, Plumbing, and Air Conditioning*, 129–134 (April 1961).

Pathak, B., in W. Ailor (ed.), "Testing in Fresh Waters," in *Handbook on Corrosion Testing and Evaluation*, Wiley, New York (1971).

Pocock, S., "Factors Influencing Household Water Lead: A British National Survey," *Arch. Env. Health*, **35**, 45–51 (1980).

Pourbaix, M., *Thermodynamics of Dilute Aqueous Solution, with Applications to Electrochemistry and Corrosion*, Edward Arnold, London (1949).

Plummer, L. N., and Busenberg, E., "The Solubilities of Calcite, Aragonite and Naterite in CO_2–H_2O Solutions Between 0 and 90°C, and an Evaluation of the Aqueous Model for the System $CaCO_3$–CO_2–H_2O," *Geochim. Cosmochim. Acta*, **46**, 1011–1040 (1982).

Rambow, C., and Holmgren, R., Jr., "Technical and Legal Aspects of Copper Tube Corrosion," *JAWWA*, **58**, 347–353 (1966).

Ricci, J. E., *Hydrogen Ion Concentration, New Concepts in a Systematic Treatment*, Princeton University Press, Princeton, N.J. (1952).

Rice, O., and Hatch, G., "Threshold Treatment of Municipal Water Supplies," *JAWWA*, **31**, 1171–1185 (1939).

Riddick, T. M., "The Mechanism of Corrosion of Water Pipes," *Water Works and Sewerage*, **88**, 291–298 (1944).

Ryder, R., and Wagner, I., "Corrosion Inhibitors," Proceedings of AWWA Research Foundation, U.S. German Workshop on Corrosion in Water Pipes, AWWARF, Denver, CO (1985).

Ryznar, T., "A New Index for Determining the Amount of Calcium Carbonate Scale Formed by a Water," *JAWWA*, **36**, 472 (1944).

Robinson, F., "Pitting Corrosion: Cause, Effect, Detection, and Prevention," *Corrosion Tech.*, vol 7 237–266 (1960).

Rossum, J., and Merrill, D., "An Evaluation of Calcium Carbonate Saturation Indexes," JAWWA, **75**, 95–100 (1983).

Schock, M., "Response of Lead Solubility to Dissolved Carbonate in Drinking Water," Proceedings of AWWA Research Foundation, U.S. German Workshop on Corrosion in Water Pipes, AWWARF, Denver, CO (1985).

Schock, M., and Buelow, R., "The Behavior of Asbestos-Cement Pipe Under Various Water Quality Conditions: Part 2, Theoretical Considerations," *JAWWA*, **73**, 636–651 (1981).

Sheiham, I., and Jackson, P., "The Scientific Basis for Control of Lead in Drinking Water by Water Treatment," *Inst. Water Eng. Sci.*, **35**, 491–515 (1981).

Sillen, L., "Graphical Presentation of Equilibrium Data," in I. Kolthoff and P. Elving (eds.), *Treatise on Analytical Chemistry*, Part I, Vol. 2, Wiley-Interscience, New York, Chap. 8 (1959).

Singley, J., "The Search for a Corrosion Index," *JAWWA*, **73**, 579–582 (1981).

Skold, R., and Larson, T., "Measurement of Instantaneous Corrosion Rate by Means of Polarization Data," *Corrosion*, **13**, 139–142 (1957).

Sontheimer, H., Kolle, W., and Rudek, R., "Aufgaven und Methoden der Wasser-chemie—dargestellt an der Entwidklung der Erkenntnisse zur bildung von Korrosionsschutzschichten auf Metallen," *Vom Wasser*, **52**, 1 (1979).

Sontheimer, H., Kolle, and Snoeyink, V., "The Siderite Model of the Formation of Corrosion-Resistant Scales," *JAWWA*, **73**, 572–579 (1981).

Stern, M., and Geary, A., "Electrochemical Polarization—A Theoretical Analysis of the Shape of Polarization Curves," *J. Electrochem. Soc.*, **104**(1), 56–63 (1957).

Streicher, L., "Effects of Water Quality on Various Metals," *JAWWA*, **48**, 219 (1956).

Stumm, W., "Calcium Carbonate Deposition at Iron Surfaces," *JAWWA*, **48**, 300–310 (1956).

Stumm, W., "Evaluation of Corrosion in Water by Polarization Data," *Ind. Engr. Chem.*, **61**, 1 (1959).

Stumm, W., "Investigations of Corrosive Behavior of Waters," *JASCE, San Div.*, **86**, 27 (1960).

Stumm, W., and Morgan, J., *Aquatic Chemistry—An Introduction Emphasizing Chemical Equilibria in Natural Waters*, 2nd ed., Wiley, New York (1980).

Szklarska-Smialowska, Z., "Review of Literature on Pitting Corrosion Published since 1960," *Corrosion*, **27**, 223–233 (1971).

Talbot, H., and Blanchard, A., *Electrolytic Dissociation Theory with Some of Its Applications*, MacMillan, New York (1905).

Tillmans, J., *Water Purification and Sewage Disposal*, Constable, London (1913).

Tillmans, J., and Heublein, O., "Investigation of the Carbon Dioxide which Attacks Calcium Carbonate in Natural Waters," *Gesundh. Ing.*, **35**, 669 (1913).

Treweek, G. P., Trussell, R. R., and Pomeroy, R. D., "Copper-Induced Corrosion of Galvanized Steel Pipe," presented at the AWWA Water Quality Technology Conference, Louisville (1978).

Trussell, R., Russell, L., and Thomas, J., "The Langelier Index," Proceedings of 5th Annual AWWA WQTC, Kansas City, Missouri (1976).

Trussell, R., and Thomas, J., "A Discussion of the Chemical Character of Water Mixtures," *JAWWA*, **63**, 49–51 (1971).

Trussell, R., and Wagner, I., "Corrosion of Galvanized Pipe," Proceedings of AWWA Research Foundation, U.S. German Workshop on Corrosion in Water Pipes, AWWARF, Denver, Co (1985).

Uhlig, H., *Corrosion and Corrosion Control*, 2nd ed. Wiley, New York (1971).

Uhlig, H., Triadis, D., and Stern, M., "Effect of Oxygen, Chlorides, and Calcium Ion on Corrosion Inhibition of Iron by Polyphosphates," *J. Electrochem. Soc.*, **102**, 59 (1955).

Wagner, C., and Traud, W., "Zeitschrift Fur Elektrochemie Und Angewandte Physikalische Chemie," *Z. Elektorchem.*, **44**, 391–454 (1938).

Whitman, W., Russell, R., and Altieri, V., "The Effect of Hydrogen Ion Concentration on Submerged Steel," *Ind. Engr. Chem.*, **16**, 1665 (1924).

— 19 —

Predesign

In the development of any major water treatment project, proper planning and predesign studies are essential for the timely completion of the project. Predesign studies include any preliminary investigations prior to start of detailed design required to define the specific elements of the project so that detailed design can proceed smoothly and efficiently. The objective of this chapter is to provide the design engineer with those guidelines and considerations necessary for preliminary investigations required to properly plan and implement a water treatment project.

The importance of predesign studies cannot be overemphasized. They are important because:

1. The preparation of predesign studies is a cost-effective investment since in relation to the total project costs (engineering plus capital costs) the preliminary investigations typically account for $\frac{1}{2}$–1 percent of the total project costs. Thorough evaluation of design concepts at this phase of the project can result in significant cost savings in the detailed design, construction, and operational phases of the project, since no major capital expenditures are made in predesign.

2. Predesign studies document the major engineering decisions for the project, including evaluation criteria, selection of treatment processes, and basis of design.

3. Predesign studies provide the owner an opportunity to review the engineering decisions before detailed design and construction begin. Preferences for specific type of equipment, operational procedures, architectural features, and so on, can be communicated to the design engineer at this time without expensive changes being incurred in the latter phase of the project through change order or major redesign effort.

4. Predesign studies provide a basis for the detailed design utilized by the design engineer and his project team. Design criteria for each design discipline is established during the preliminary phase. As soon as predesign studies are approved or accepted by the owner, the design team can start preparation of detailed plans and specifications.

5. Sufficient engineering evaluations prepared in the predesign studies are used to develop the scope of the project and thereby establish capital and operation and maintenance (O&M) cost estimates. The cost estimates are used by the owner to properly assess and plan the financial strategy required to ensure that adequate capital to construct the facilities is available and

that any adjustments in water rates can be made to offset O&M costs.

The discussions in this chapter apply to the planning, design and construction of new water treatment facilities. However, the same level of investigations would apply to projects that involve rehabilitation, modification, and expansion of existing facilities. In these cases, a thorough evaluation of existing plant operational procedures and records is preformed in lieu of specific investigations of source and plant siting. Both pilot and plant scale testing is beneficial in the development of treatment alternative schemes.

It should be noted that the processes normally encompassed by the term *water treatment* include turbidity removal, softening, removal of iron and manganese, THM treatment, color removal, treatment for taste and odor, and disinfection. However, the most widely employed process is for turbidity removal and is the process this chapter primarily addresses. Many of the topics discussed in this chapter have been presented in more detail in other chapters. However, a general overview of the steps required to prepare an adequate documentation of predesign activities is presented here.

WATER QUALITY OBJECTIVES

In any predesign study, the establishment of drinking water quality goals for treated water is a primary objective. The water quality goals are influenced by a combination of federal, state and local drinking water quality regulations, treatability of the raw water, and the basic objective of any municipal water treatment facility, that is, to provide a safe, aesthetically pleasing, economical water supply.

Drinking Water Regulations

The first step in establishing treated water quality objectives is the determination of the applicable water quality regulations. Water quality standards are promulgated on international, federal, and state levels (discussed in depth in Chapter 4). One consideration in the development of treated water quality objectives is the possibility of water quality standards being changed or modified in the future. Future changes may influence the reliability and flexibility of the proposed water treatment processes to meet the more restrictive standards.

Raw Water Characterization and Treatability

Adequate information regarding the raw water quality is a prerequisite for plant design. By combining the raw water quality data with the finished product requirements, the best type of treatment process can be identified. Information on the potential source water quality may or may not be readily available. Water quality data and information on the treatability of a particular source may be obtained from local water treatment plants, state or local health departments, United States Public Health Service, USGS, and State Water Resource Board. This existing information combined with additional studies on the raw water quality (both long- and short-term studies) provide the basis for the design. If this information is spotty or totally lacking, an extensive study for a minimum duration of 1 year may be necessary to cover the range of conditions to be expected. Periodic short-term studies are useful if they can properly characterize the range of conditions. If such studies are not practical or feasible, the alternative approach is to provide for a more conservative design.

To evaluate the physical, chemical, and bacteriological quality of the water source, specifically where purification is the prime concern, it is desirable to obtain as much long-term data on suspended matter and microbiological parameters as possible (i.e., turbidity, suspended solids, color, plankton, and coliform). If any special water quality problems exist, such as taste and odor, high color, radioactivity, and source contamination by industrial and domestic discharges, special historical studies should be conducted on the specific problem.

For a potential water source, general mineral quality, especially total dissolved solids, hardness, alkalinity, pH, manganese, iron, nitrate, fluoride, and arsenic, has an important influence on the type of treatment selected and on process performance. In the evaluation of general mineral quality, the deviation from electroneutrality should be calculated and checked for each mineral analysis. This deviation is a balance between the positive and negative charges in the solution. A mineral analysis and a calculation of the deviation is presented in Table 19-1. According to the principle of electroneutrality, these charges should be equal and the deviation should be zero. As a practical matter, a water analysis should be questioned whenever the deviation is greater than 0.02.

The method of data analysis is important in eval-

TABLE 19-1. *Mineral Analysis*

Cations	mg/L	meq/L	Anions	mg/L	meq/L
Ammonia–N	NA	0	Bicarbonate	265.3	4.35
Sodium	70.6	3.07	Carbonate	0.7	0.02
Potassium	6.5	0.17	Chloride	59	2.66
Calcium	41.7	2.09	Sulfate (DL = 0.6)	24	0.5
Magnesium	13.2	1.1	Nitrate–N (DL = 0.02)	0.5	0.04
			Fluoride (DL = 0.07)	0.3	0.02
			Hydroxide	0	0
Total cations		6.43	Total anions		6.59

$$\text{Deviation from electroneutrality} = \frac{\text{cations} - \text{anions}}{\text{cations} + \text{anions}} = 0.012$$

uating the quality of a water source as described in Chapter 2. Although arithmetic averages are useful in many instances, a statistical presentation is often more useful. The statistical presentation may be done by plotting data on an arithmetic or logarithmic probability scale. Generally, the two aspects of data analysis that are most important are (1) the method of describing the central tendency of the data (average, geometric mean, median) and (2) the degree of variability in the data (standard deviation, 90 percent value). For example, Figure 19-1, a log-normal probability distribution for turbidity, indicates that 95 percent of the time turbidity is 7.0 TU or less, and the geometric mean is about 2.7 TU (JMM/DMJM, 1977).

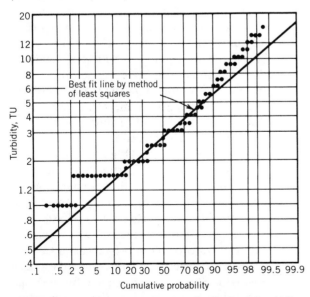

FIGURE 19-1. Log-normal probability distribution of turbidity. Data represents 589 weekly values recorded during the period 1963–1976.

In addition to establishing the historical raw water quality, projection of potential changes of quality in the future is necessary. This type of evaluation involves an extensive knowledge of the watershed area, as well as present and future resource and quality management programs. For example, if the watershed was to be significantly altered by major land development activities in the area, a special study should be undertaken to evaluate the impact of this development on raw water quality and additional treatment requirement. Additionally, if the raw water supply is withdrawn from a terminal reservoir, management alternatives of the reservoir require investigation. This investigation may include evaluation of aeration of the reservoir to eliminate or minimize destratification and selective intake of raw water from the reservoir.

The next step after the raw water has been characterized is to evaluate the treatability of the raw water. This can be based on past experience with this water source or, if limited information is available, by use of bench scale tests or pilot studies. The most popular bench scale test is the jar test, which evaluates the effectiveness of various coagulation, flocculation, and sedimentation treatments for a given raw water. Pilot studies are of value in the performance evaluation of coagulation–filtration (in-line filtration), coagulation–flocculation–filtration (direct filtration), or various other treatment alternatives. A more detailed discussion of bench and pilot testing is presented later.

Recommended Water Quality Goals

Treated water quality objectives are developed after consideration of the raw water quality, the quality required to assure an effluent that is not hazardous to health, the quality recommended by both local and recognized standards, and the quality attainable within the practical financial means of the water purveyor.

The procedure to develop water quality objectives involves the comparison of applicable water quality regulations with historical water quality data for the source of supply. For example, the data presented in Tables 19-2 and 19-3 were developed for the expansion of the Escondido-Vista Water Treatment Plant located in Escondido, California (JMM: Escondido Vista, 1981). The plant receives water from three different raw water supplies. Table 19-2 compares the water quality of the different sources with the California State Department of Health for Primary Drinking Water Regulations. California has primacy to enforce EPA Drinking Water Quality Regulations. In addition, historical treated water quality data are also presented in Table 19-2 to illustrate the ability of present facilities to meet existing water quality regulations. For existing facilities, these data are readily available and therefore included. Table 19-3 presents a comparison of water quality for the Secondary Drinking Water Regulations.

Based on information presented in Tables 19-2 and 19-3 and on the capability of existing and/or proposed treatment processes, treated water quality objectives were developed in Table 19-4 to assure that the treated water will be acceptable and easily disinfected.

REGULATORY REQUIREMENTS

To efficiently proceed with both the conceptual and detailed design phase, the design engineer should be thoroughly familiar with the various regulatory requirements. Furthermore, it is generally necessary to obtain approval of a regulatory agency, usually the state health department, before proceeding with design and construction of any water treatment facility. In fact, most states require the approval of plans and specifications for public water supply facilities before construction begins.

For this reason, meetings with the representatives of the necessary regulatory agencies in the early stages of the project, preferably at the start of the predesign studies and throughout the course of the project, are advisable. Both needless delays and the resulting loss of efficiency can be avoided if preliminary design concepts and criteria developed in the initial stages of the project are approved by the regulatory agencies. Subsequent meeting with the regulatory agencies can be helpful in refining the preliminary concepts and receiving further input into the applicable regulatory standards.

Another aspect of regulatory involvement in a water works project deals with the knowledge of local building and construction codes or laws. Safety standards, such as Occupational Safety and Health Administration (OSHA), are applicable in the design, construction, and operation of water treatment facilities. Other examples of regulations that require attention are (1) Uniform Building Code, (2) Uniform Plumbing Code, (3) Fire Protection Association Code, and (4) regionalized earthquake codes. If the water treatment plant involves a waste discharge (e.g., supernatant from sludge-drying beds), the appropriate water pollution control agencies should be contacted to determine the applicable discharge requirements and to obtain the necessary permits.

Design Criteria

The formulation of design criteria for a water treatment facility must generally correspond to state regulatory agencies' standards. Over the years, state review agencies have established minimum design requirements or standards based on good engineering practice, local conditions, and experience. The most widely known of these types of standards are the "Recommended Standards for Water Works" promulgated by the Great Lakes–Upper Mississippi River Board of State Sanitary Engineers, generally referred to as the "Ten-State Standards." This board includes representation from the states of Illinois, Indiana, Iowa, Michigan, Minnesota, Missouri, New York, Ohio, Pennsylvania, and Wisconsin. Many states have enforced identical or similar versions of these standards. Other states have established less formalized departmental guidelines that are used in the review of the design engineer's proposed design criteria and are more sensitive to site specific conditions on a case-by-case basis. Historically, state health officials have taken a reasonable attitude toward deviations from departmental guidelines or established standards when it has been shown that departures are necessitated by such factors as new developments in treatment processes equipment or technology, relative cost of treatment facilities required to meet these standards or guide-

TABLE 19-2. California State Department of Health Primary Drinking Water Regulations and Average Concentrations in Raw and Treated Water

Constituent	California Maximum Contaminant Level (mg/L)	Lake Wohlford[a]	Lake Dixon[a]	Lake Skinner[b]	Escondido-Vista Treated Water[c]
Inorganic					
Arsenic	0.05	0.004	0.004	0.003	<0.003
Barium	1.0	0.1	0.1	0.11	<0.1
Cadmium	0.01	<0.001	<0.001	<0.0001	<0.001
Chromium (total)	0.05	<0.01	<0.01	0.0002	<0.01
Fluorides	1.4–2.4[d]	0.25	0.30	0.27	0.18
Lead	0.05	<0.01	<0.01	0.0002	<0.01
Mercury	0.002	<0.0004	<0.0003	<0.0002	<0.0002
Nitrate (as N)	10	0.39	0.33	0.2	0.4
Selenium	0.01	<0.01	0.01	0.002	0.01
Silver	0.05	<0.01	<0.01	<0.005	<0.01
Organic					
Endrin	0.0002	ND[e]	ND	ND	—
Lindane	0.004	ND	ND	ND	—
Methoxychlor	0.1	ND	ND	ND	—
Toxaphene	0.005	ND	ND	ND	—
2,4-D	0.1	ND	ND	ND	—
2,4,5-TP Silvex	0.01	ND	ND	ND	—
TTHM[f]	0.1	—	—	—	0.069
Radioactivity (pCi/L)					
Gross alpha	15	<3	—	—	—
Tritium	20,000	—	—	—	—
Strontium-90	8	—	—	—	—
Others					
Turbidity	1 TU	14.7	1.4	2.5	0.15–0.25
Coliform bacteria	1/100 mL	8–130	5–79	—	<1/100 mL

[a] Lake Wohlford and Lake Dixon concentrations are from the following data: Inorganic chemicals—barium and silver values from one measurement taken in 1980, others are averages of three measurements taken in 1980; organic chemicals—one measurement taken in 1977; radioactivity—gross alpha measurement taken on May 15, 1980; turbidity—average of 3 years of data; coliform—range of values for 1980.

[b] Lake Skinner concentrations are from the 1979 Annual Report of the Metropolitan Water District of Southern California. No radioactivity or coliform data could be found.

[c] Escondido-Vista treated water concentrations are from the following data: Inorganic chemicals—barium and silver values from one measurement taken in 1980, others are averages of two measurements taken in 1980; organic chemicals—no data available, no contamination expected due to low levels in raw water; radioactivity—no data available; turbidity—average of 3 years of data; coliform—no bacteriological contamination of the treated water has ever been detected.

[d] Temperature dependent

[e] ND = not detected

[f] TTHM (total trihalomethane) data are from March 1981 quarterly samples by City of Escondido.

SOURCE: JMM Report for Escondido-Vista WTP Design Report (1981).

lines, localized water quality conditions, or other justified causes. Departures from the state-established standards must generally be supported by special engineering studies. For example, bench and/or pilot scale investigations can be conducted to verify or establish the feasibility of certain design parameters or concepts in question; or a detailed evaluation and analysis of raw water quality can be prepared to demonstrate the maximum (or worst) quality conditions so that the proper level of treatment can be established. Coordination with the state officials as to the scope of the engineering

TABLE 19-3. California State Department of Health Recommended Secondary Drinking Water Regulations and Average Concentrations in Raw and Treated Water

Constituent	California Maximum Contaminant Level	Lake Wohlford[a]	Lake Dixon[a]	Lake Skinner[b]	Escondido-Vista Treated Water[c]
Chloride	250 mg/L	32.5	85.3	89	38.2
Color	15 color units	53	5	—	3
Copper	1 mg/L	0.006	0.006	<0.005	0.01
Corrosivity	Noncorrosive	Noncorrosive	Noncorrosive	Noncorrosive	Relatively corrosive
Foaming agents	0.5 mg/L	0.03	0.03	—	0.02
Iron	0.3 mg/L	0.08	0.08	0.1	0.02
Manganese	0.05 mg/L	0.005	0.006	0.02	0.01
Odor	3 threshold odor numbers	8–12	8–12	—	<3
pH	6.5–8.5	7.9	7.6	8.2	7.4
Sodium[d]	20 mg/L	65.0	77.0	89.0	58.2
Sulfate	250 mg/L	114	155	199	97.6
TDS	500 mg/L	266	587	546	263
Zinc	5.0 mg/L	0.91	0.03	0.007	0.5

[a] Lake Wohlford and Lake Dixon concentrations are averages of three measurements taken in 1980. Odor data are from more frequent analyses conducted at the water treatment facility.

[b] Lake Skinner concentrations are from the 1979 Annual Report of the Metropolitan Water District of Southern California. No odor, color, or foaming agents data could be found.

[c] Escondido-Vista treated water concentrations are averages of two measurements taken in 1980. Odor data are from more frequent analyses conducted at the water treatment facility.

[d] EPA regulation issued August 27, 1980. Not technically a secondary regulation.

Source: JMM Report for Escondido-Vista WTP Design Report (1981).

study is essential to ensure that the necessary information is developed in the study.

Source Limitation

Most states have established classifications for various sources of water supply within their boundaries. The classifications are generally based on the sanitary conditions and the water quality of the supply. The two general classifications are (1) surface water supply and (2) groundwater supply. Surface water supplies can further be classified by the variation in water quality and the potential exposure to sanitary hazards. Most state agencies do not impose a limitation on the source of supply for domestic water use, but rather mandate the level or degree of treatment that is required for each particular source of supply. In California, for example, surface waters exposed to significant sewerage hazards or significant recreational use, shall receive, as a minimum, pretreatment, filtration, and disinfection. Qualifying what is a "significant sewerage hazard" or "significant recreational use" is done by the California State Health Department on a case-by-case

basis. Also, the degree of pretreatment required for a surface water supply is based on the California State Health Department's and the design engineer's previous experience with waters of similar quality and can be supplemented by bench and pilot studies.

SOURCE SELECTION

The procedure for selection of a raw water supply ranges from conducting extensive engineering evaluations including a water resource development investigation to simply applying for a service from a water wholesaler. In selecting the source of water, the design engineer must be aware of the quantity and quality of the source. In addition, the safety features of the source, along with the capital and operation and maintenance costs, are important elements to consider.

For surface water sources, such as streams, lakes, and reservoirs, the following items must be carefully studied for selection purposes:

TABLE 19-4. Treated Water Quality Objectives

Parameters	Objectives
Inorganic chemicals	
Arsenic, barium, cadmium, chromium, fluoride, lead, mercury, nitrate, selenium, silver	Maximum contaminant levels as specified by state regulations, listed in Table 19-2
Iron	0.3 mg/L
Manganese	0.05 mg/L
TDS	No substantial change from raw water
Hardness	No substantial change from raw water
Sodium	No substantial increase from raw water
Organic chemicals	
Pesticides and herbicides	Maximum contaminant levels as specified by state regulations, listed in Table 19-2
Total Trihalomethanes (TTHM)	0.1 mg/L
TOC	Less than raw water
Other organics (MBAS, etc.)	To comply with existing state recommended regulations and future regulations
Particulates	
Turbidity	0.2 TU
Particles (2.5–150 μm)	50 ml^{-1}
Microorganisms	
Total coliform bacteria	1/100 mL
Other microorganisms	Periodic monitoring
Radiological Parameters	
Radioactivity	Maximum contaminant levels as specified by state regulations, listed in Table 19-2
Other parameters	
Taste	None objectionable
Odor	1 threshold odor number
Color	5 color units
Corrosivity	Not corrosive

SOURCE: JMM Report for Escondido-Vista WTP Design Report (1981).

1. Quantity Aspects.

    a. Water quantity data over a sufficient period of time to permit statistical analysis of rainfall, runoff, and streamflow.

    b. Adequacy of safe yield to meet present and future demands.

    c. Degree of conservation measures by federal or state agencies including impoundments and future uses of tributary areas.

    d. A complete study of local water rights.

    e. Degree of land development in the watershed.

2. Quality Aspects.

    a. Water quality data over a significant period of time, preferably over the past 5 years to assess the physical, chemical, microbiological and radiological characteristics of the water. Seasonal and long-term variability in the quality of the source should be addressed.

    b. Risk assessment of potential supply contamination by accidental spillage of substances that may be toxic, harmful, or detrimental to domestic use.

    c. Degree of present and proposed future land development into the watershed.

    d. Degree of control and management of the watershed by the owner.

3. General

    a. Assessment of the reliability of the supply,

specifically with respect to adequacy of protection against flood, earthquake, sabotage, and other disruptions.

b. Degree of difficulty in construction of conveyance, pipelines, water treatment plant, and distribution network.

c. Environmental impacts.

d. Financial impact, both with respect to the cost of developing the source and the ability of the water agency to finance this development.

If groundwater is considered as a potential source of water supply, similar considerations as presented for surface water apply. The development of a groundwater source, however, usually requires additional studies on geological conditions, underground water tables, and drawdown characteristics of the water table due to pumpage and saltwater intrusion problems (in the coastal area). In addition, the question of water rights must be addressed, especially if groundwater basins are adjudicated.

In general, one of the most important factors to be considered in the selection of source is the quality of the raw water. Raw water should be of a high enough quality to meet drinking water standards with a minimal number of treatment processes and at the least treatment cost. Comparison of two types of sources from the quality standpoint reveal that the groundwater tends to be uniform in quality, free of turbidity, low in color, but has a greater potential for high concentrations of dissolved substances. Furthermore, groundwater is subject to organic contamination that may have a long-term adverse effect on water quality and may persist for years, if not decades. On the contrary, surface waters tend to vary in quality, with high fluctuations in turbidity and color. They also may contain taste- and odor-producing substances; however, the mineral content is usually less. Generalizations, as stated above, are no substitute for definitive evaluations required prior to plant designs.

TREATMENT PLANT SITING

Many factors require evaluation in the siting of a water treatment facility. Only after a thorough evaluation of all applicable factors have been made should the water agency proceed with site acquisition. The overall objective of siting facility is to minimize the overall cost to the water agency while at the same time providing a site that is safe from natural and man-made disasters. Site acquisitions should take into consideration both present and ultimate water supply and treatment requirements. Proper planning in the acquisition of a suitable site is therefore essential. If adequate land is not set aside initially, then the site selection and acquisition procedures for future plant expansions will become tedious, time consuming, and expensive. Moreover, additional engineering and legal evaluations will be required before land purchase. A detailed discussion of the principle factors in plant siting is presented in Chapter 20.

PLANT CAPACITY

The determination of the rated or nominal capacity of the water treatment plant is usually based on the maximum day water demand of the system. Factors required to determine the maximum day water demand include the forecast of the water demand in the region, the availability of initial capital investment, the design period, and the phasing of the ultimate capacity of the facility. For water systems with more than one source of supply or that cover a large geographical area, considerations should be given to strategically locating water treatment plants at more than one location to serve the various communities or pressure zones. Preparation and implementation of a water master plan is extremely useful in the determination of the plant capacity and potential plant locations.

The most widely used approach in determining water demands and therefore plant capacities is based on a per capita analysis. This approach includes evaluation of the present population, the population growth rate of the service area, and the projection of a water demand rate for the service area.

Population Estimates

Two types of population estimates are generally required for the operation and design of water supply and treatment systems: short-term estimates (1–10 yr) and long-term estimates (10–50 yr). The method used to project population for either of these two time periods is based on mathematical and graphical approaches and is at best a statistical guess. There are no exact solutions, even though seemingly sophisticated mathematical equations are used.

The most widely employed mathematical or graphical methods for extending past population data are:

1. Arithmetic progressive or uniform growth rates.
2. Constant percentage growth rate.
3. Declining rate of increase.
4. Graphical interpolation–extension.
5. Graphical comparison with growth rates of similar or larger cities.
6. A ratio method based on a comparison of local and national population figures for past census years.
7. The use of mathematical trends such as logistic curves.
8. Component method.

There is no method of predicting future regional growth patterns; nevertheless, population predictions are a prerequisite to determining the size of a complete water system and, in particular, water treatment plant capacity. Factors that influence the population and the rate of its growth include:

1. Birth–death rates.
2. Migration–immigration rates.
3. Annexation.
4. Urbanization.
5. Industrialization.
6. Commercial activity.

Sources of information used to develop population projections and growth rates include federal and local census bureaus, the planning commission, the bureau of vital statistics, local utility companies, and the chamber of commerce.

Table 19-5 presents a summary of four common methods used in predicting population. The first three methods are generally used to develop short-term population estimates. Graphically, the three segments of the growth rate curve, depicted as a typical S curve in Figure 19-2, are approximations of the arithmetic rate, the geometric rate, and the decreasing rate of increase. Figure 19-2 represents an idealized population growth pattern. For long-term population estimates, the logistic method is the most commonly used mathematical method; however, graphical methods are also used that extrapolate population data for a given community based on trends experienced by similar or larger communities. Several references (Fair et al., 1971; Clark et al., 1971; Medcalf and Eddy, 1979) present detailed discussion on formulation of the population estimation methods and examples of the various forecasting methods.

TABLE 19-5. Population Projection Methods

Method	Formula	Definition of Terms	Constant Formulation
Arithmetic	$P_n = P_2 + k_a(t_n - t_2)$	P = population t = time k_a = arithmetic growth rate	$k_a = \dfrac{P_2 - P_1}{t_2 - t_1}$
Geometric	$\log P_n = \log P_2 + k_g(t_n - t_2)$	k_g = geometric growth rate	$k_g = \dfrac{\ln P_2 - \ln P_1}{t_2 - t_1}$
Decreasing rate of increase	$P_2 - P_1 = (S - P_1)(1 - e^{-k_d(t_2 - t_1)})$	S = saturation population k_d = decreasing rate of increase constant	$k_d = \dfrac{-\ln[(S - P_2)/(S - P_1)]}{t_2 - t_1}$
Logistic (simplified)	$P = \dfrac{S}{1 + me^{bt}}$	m, b = constants P_0, P_1, P_2 = population at times t_0, t_1, t_2 n = interval between t_0, t_1, t_2	$S = \dfrac{2P_0 P_1 P_2 - P_1^2(P_0 + P_2)}{P_0 P_2 - P_1^2}$ $m = \dfrac{S - P_0}{P_0}$ $b = \dfrac{1}{n} \ln \dfrac{P_0(S - P_1)}{P_1(S - P_0)}$

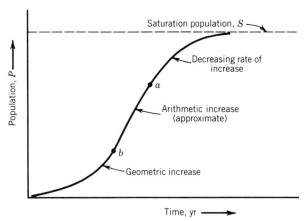

FIGURE 19-2.   Population growth curve.

TABLE 19-6.   Guide to Population Density

Area Type	Number of Persons Per Hectare	Number of Persons Per Acre
Residential, single-family units	12–86	5–35
Residential, multiple-family units	74–247	30–100
Apartments	247–2470	100–1000
Commercial areas	37–74	15–30
Industrial areas	12–34	5–15
Total, exclusive of parks, playgrounds and other large greenbelt areas	25–124	10–50

In most cases, a reasonable forecast period is 20 yr. Forecasting beyond 20 yr usually does not reflect accurate growth, especially for middle-to-small communities. More accurate population projections for small and middle-sized communities are obtained from local planning agencies. Since population forecasts, at best, are educated guesses, the probability of the forecasts being accurate is low. In general, the development of a range of population forecasts is suggested. In addition, engineering alternatives developed for these estimates should be adaptable to anticipated variations in the growth rate. As a minimum, the population at three growth rates should be estimated—low, medium, and high. In addition, utilizing more than one forecasting technique provides the design engineer with useful information as to the anticipated range of future population increases.

In addition to estimating the total population of a region, the distribution of the population is important. Population densities may be estimated from data collected on existing areas and from zoning master plans for undeveloped areas. Table 19-6 presents a guide that can be used to estimate population densities based on zoning if more reliable local data are not available.

Water Supply and Demand

The determination of the water demands for the area requires knowledge and experience in the interpretation of social, economic, and regional development trends as well as a sound judgment regarding the environmental impact to the region. In addition, a projection of the available quantity of water from the source supply is required to accurately predict the future water supply demand.

Review of past records and estimated future average water consumption rates across the United States indicate a wide range of values. The following factors are responsible for the nonuniformity:

1. *Climatic conditions.* Usually, warm dry regions have higher demands than cooler regions. In addition, the pattern of water usage is affected by the precipitation levels in the region.
2. *Economic conditions.* High-priced residential dwellings normally show water consumption rates significantly greater than medium- and low-priced units.
3. *Composition of region or community.* The type, magnitude, and blend of residential, commercial, and industrial development in the area affect local water use rates. Generally, industrial water requirements are exceedingly large.
4. *Water pressure.* Rates of water usage increase with increases in pressure.
5. *Cost of water.* A tendency toward conservation occurs when cost of water is high.
6. *Metering.* Communities that are metered usually show a lower and more stable water use pattern.
7. *Water quality.* Consumer perception of bad water quality can decrease the water usage rate.
8. *Sanitary sewer facilities.* The absence of community sewage facilities tends to decrease wa-

ter usage since the water user is impacted by the frequent cleaning of septic tank facilities.

9. *Water conservation.* Public awareness and implementation of water conservation programs by utilities tend to have an initial impact on the baseline water usage rates.

If no historical records of water demands are available, an estimate of the water demands is required. Table 19-7 shows the approximate per capita daily uses in the United States. Wide variations in the values presented in Table 19-7 are expected because of the factors discussed above. Detailed analysis and evaluation of water demand rates is therefore useful in developing total water demands. Tables 19-8 and 19-9 provide demand values for representative commercial and industrial components.

In addition to establishing average daily water demands, the fluctuation in the water consumption must be assessed. Water consumption changes sea-

TABLE 19-7. Normal Per Capita Water Consumption

Type of Consumption	Normal Range		Average	
	lpcd	gpcd	lpcd	gpcd
Domestic or residential	76–340	20–90	208	55
Commercial	38–492	10–130	76	20
Industrial	76–303	20–80	189	50
Public	19–76	5–20	38	10
Water unaccounted for	19–114	5–30	57	15
	227–946	60–250	568	150

SOURCE: G. M. Fair, J. C. Geyer, and D. A. Okun, *Elements of Water Supply and Wastewater Disposal*, 2nd ed., Wiley, New York (1971).

sonally, daily, and hourly. In general, fluctuations are greater for small communities than in large communities and when the period of time is short. Ex-

TABLE 19-8. Commercial Water Use

	Unit	Average Annual Demand (gpd)	Maximum Hourly Demand Rate (gpd)	Hour of Peak Occurrence	Ratio of Maximum Hourly to Average Annual	Average Annual Demand Per Unit	Ratio of Maximum Hourly Demand to R-40 Demand[a]
Miscellaneous residential							
Apartment building	22 units	3,430	11,700	5–6 PM	3.41	156 gpd/unit	2.2 : 1
Motel	166 units	11,400	21,600	7–8 AM	1.89	69 gpd/unit	4.0 : 1
Hotels							
Belvedere	275 rooms	112,000	156,000	9–10 AM	1.39	407 gpd/room	29 : 1
Emerson	410 rooms	126,000				307 gpd/room	
Office buildings							
Commercial Credit	490,000 ft^2	41,400	206,000	10–11 AM	4.89	0.084 gpd/ft^2	38 : 1
Internal Revenue	182,000 ft^2	14,900	74,700	11–12 AM	5.01	0.082 gpd/ft^2	14 : 1
State office building	389,000 ft^2	27,000	71,800	10–11 AM	2.58	0.070 gpd/ft^{2,b}	13 : 1
Shopping centers							
Towson Plaza	240,000 ft^2	35,500	89,900	2–3 PM	2.50	0.15 gpd/ft^2	17 : 1
Hillendale	145,000 ft^2	26,000				0.18 gpd/ft^2	
Miscellaneous commercial							
Laundries							
Laundromat	Ten 8-lb washers	1,840	12,600	11–12 AM	6.85	184 gpd/washer	2.3 : 1
Commercial	Equivalent to ten 8-lb washers	2,510	16,200	10–11 AM	6.45	251 gpd/washer equivalent	3.0 : 1
Washmobile	Capacity of 24 cars/hr	7,930	75,000	11–12 AM	9.46	330 gpd/car/ hr of capacity	14 : 1
Service station	1 lift	472	12,500	6–7 PM	26.5	472 gpd/lift	2.3 : 1

[a] Lot type R-40 (1 acre) peak hourly demand for single service is 5400 gpd.
[b] Exclusive of air conditioning.
SOURCE: Residential Water Use Research Project of The Johns Hopkins University and the Office of Technical Studies of the Architectural Standards Division of The Federal Housing Administration.

TABLE 19-9. Typical Rates of Water Use for Various Industries

Industry	Range of Flow (m³/MG)	Range of Flow (gal/U.S. short ton)
Cannery		
Green beans	50–70	1,200–17,000
Peaches and pears	15–20	3,600–4,800
Other fruits and vegetables	4–35	960–8,400
Chemical		
Ammonia	100–130	24,000–31,000
Carbon dioxide	60–90	14,000–22,000
Gasoline	7–30	1,700–7,200
Lactose	600–800	140,000–190,000
Sulfur	8–10	1,900–2,400
Food and beverage		
Beer	10–16	2,400–3,800
Bread	2–4	480–960
Meat packing	15–20[a]	3,600–4,800
Milk products	10–20	2,400–4,800
Whiskey	60–80	14,000–19,000
Pulp and paper		
Pulp	250–800	60,000–190,000
Paper	120–160	29,000–38,000
Textile		
Bleaching	200–300[b]	48,000–72,000
Dyeing	30–60[b]	7,200–14,000

[a] Live weight.
[b] Cotton.
SOURCE: Metcalf and Eddy (1979).

pressed as ratios to the average day demand, estimates of demand variations are as follows:

Ratio	Normal Range	Average
Maximum day–average day	(1.5–3.5) : 1	2.0 : 1
Maximum hour–average day	(2.0–7.0) : 1	4.5 : 1

Water treatment facilities are sized for the maximum day demand. Maximum hourly demand is handled by distribution system storage capacity.

Design Period and Phasing

A treatment plant is designed to serve the needs of the utility adequately for a number of years. Generally, a design period of 20–30 yr is used. Accurate records of past growth and water demands are especially valuable in forecasting the design period. Once the design capacity has been determined, a decision is required whether to provide a treatment plant for ultimate needs or to construct the facility in stages. Several factors should be considered in making this decision.

1. Life expectancy of the facility.
2. Regional growth pattern.
3. Trend of interest rates on a loan acquired to construct the facility.
4. Convenience of future expansion.
5. Development of future water treatment technology.
6. Performance of the treatment facility during early years.
7. Anticipated power costs and changes in future purchasing power.
8. Anticipated changes in drinking water quality requirements.
9. Handling of plant residual wastes (sludges).

If a decision is made to construct the plant in stages, a general rule of thumb is to provide from two to three phases. For example, if the ultimate capacity of a facility is 227 ML/d (60 mgd) the facility can be staged in two 113.5-ML/d (30-mgd) increments or three 75.7-ML/d (20-mgd) increments. The determination is based on the financial ability of the utility to pay for the initial phase, the initial water demand, and the rate of growth of the service area. Regardless of the phasing arrangement, as a general rule, steps to provide additional capacity should be taken at least 5 yr before the present capacity is reached to allow sufficient time for engineering investigations, design, financing, and construction.

Two general approaches are taken in the phasing of construction of water treatment facilities. The first approach is to provide initially those portions of the ultimate plant that are not conveniently or economically constructed in stages and to provide additional facilities in phases as the need develops. Structures that are constructed initially for ultimate capacity include intake structures, pump stations, control buildings, clearwells, chemical feed, and storage facilities. Facilities that can be added in the future include filters, sedimentation basins, and pretreatment units. One modification of this approach is to provide a modular concept whereby complete process trains are constructed in each phase. The second approach is to construct more than one treatment plant at a number of different locations to satisfy the quantity and quality needs of a region. Under certain geological, topographic, and regional development conditions, this approach often meets the utility's needs economically.

SELECTION OF TREATMENT ALTERNATIVES

The quality of the source water and finished water quality objectives form the basis for selecting treatment process alternatives. The following considerations all influence the selection of the treatment process scheme and facility designs:

1. Cost-effectiveness of the system both in terms of capital and O&M costs including off-site requirements (i.e., pipeline and storage facilities).
2. Overall system reliability.
3. Flexibility and simplicity of operation.
4. Ability to meet water quality objectives.
5. Adaptability of process to both seasonal and long-term changes in raw water quality.
6. Capability of process to be upgraded in cases where water quality and/or drinking water regulations are changed (e.g., if a direct filtration plant is designed, provisions for addition of future sedimentation basins should be included).
7. Capability of process to meet both hydraulic peaks and quality excursions (reserve capacity).
8. Availability of skilled operational and maintenance personnel.
9. Availability of major equipment items.
10. Postinstallation service and chemical delivery.
11. State and federal requirements.
12. Ease of construction of facilities.

In general, various types and combinations of treatment units are required to achieve the performance desired. The final selection of the most suitable treatment process scheme is achieved by comparative cost study of alternative schemes including an evaluation of the merits and liabilities of each alternative. Other chapters in this text discuss in detail candidate processes and process trains that are available to meet the desired water quality objec-

tives. Table 19-10 presents a general guideline for selection of candidate processes; however, site-specific analysis is required to develop the candidate processes that will produce a safe and aesthetically acceptable drinking water.

Experience acquired through treatment of the same or similar source waters provide an excellent guide in selecting the treatment process scheme. However, where experience is lacking or where there is desire to provide for a different degree of treatment, special studies are required. The special studies include bench-scale study in the laboratory, pilot plant testing, and plant-scale simulation testing. A detailed discussion of prototype studies is presented in the next section.

PROTOTYPE STUDIES

Although the evaluation of treatment processes utilized in water treatment plant design has been well established through decades of experience, prototype studies are still valuable in developing designs that can improve performance and/or reduce cost. This is especially true when an innovative process is considered. Principal considerations in determining the applicability of prototype studies for a given treatment application are:

1. Objective of prototype study.
2. Suggested scope of work to meet objectives, including regulatory agency requirements.
3. Scale of prototype equipment.
4. Cost justification for prototype study.

The discussion in this section is concerned with these considerations not only for turbidity (particulate) removal, but also for other water quality considerations, such as disinfection, organics, corrosion, and so on.

Objectives

The primary objective of any prototype study is the proper selection of the water treatment process that provides the necessary degree of removal to meet the water quality objectives at the lowest overall cost. The justification for meeting the prototype study objectives is based on the particular application that faces the design engineer and includes situations where (1) the technical feasibility of the treat-

TABLE 19-10. Process Selection Guideline

Water Quality Parameters	Process and Components	Applicability	Comments

Turbidity	In-line filtration
	  Coagulation
	  Filtration
		Low turbidity, low color	Greater operator attention required; shorter filter run than with direct filtration and conventional treatment; additional sludge-handling facilities may be required; pilot plant studies may be required; lower capital and O&M costs

	Direct filtration
	  Coagulation
	  Flocculation
	  Filtration
		Low to moderate turbidity, low to moderate color	Greater operator attention required; greater sludge-handling facilities may be required; pilot plant studies may be required; lower capital and O&M cost; better filter run time than in-line filtration but shorter than conventional treatment

	Conventional
	  Coagulation
	  Flocculation
	  Sedimentation
	  Filtration
		Moderate to high turbidity, moderate color	Detention time in sedimentation basins allows for adequate contact time for T&O and color removal chemicals; more operational flexibility and less operation attention required

	Microscreening	Removal of gross particulate matter (e.g., algae)	Process relies on straining mechanism; process could not meet water quality objective if used alone

Bacteria/virus	Chemical Disinfection
	  Chlorine
	  Chloramine
	  Chlorine dioxide
	  Ozone
	  Other chemicals
	    Bromine
	    Iodine
	    Potassium permangate
		Disinfection of surface and ground waters	THM potential with chlorine needs to be assessed, chloramine treatment not as potent as chlorine but does eliminate THM formation; cost of treatment: $Cl_2 <$ Chloramine $< ClO_2 <$ ozone; other chemicals such as bromine, iodine, etc., limited to small application

	Nonchemical disinfection
	  Ultraviolet
	  Ultrasonic
		Disinfection of surface and ground waters	Advantage of UV is that no residual is left, which is applicable to fish aquariums and hatchery disinfection; ultrasonic is expensive, some success when used with ozone in tertiary disinfection

Color	Coagulation	High color levels	Use of high coagulant dose and low pH (5–6) is cost-effective when high color levels exist

	Adsorption
	  GAC
	  PAC
	  Synthetic resins
		Moderate to low color levels	GAC bed life is in the order of 1–6 weeks; synthetic resins are expensive (capital and regeneration costs); PAC is used for handling short-term color problems (however, it is costly as a routine color control method)

	Oxidation
	  Chlorine
	  Ozone
	  Potassium permanganate
	  Chlorine dioxide
		Low, consistent color levels	Effectiveness: Ozone $> Cl_2 > ClO_2 > KMnO_4$; Cl_2 and $KMnO_4$ typically used for other purpose (disinfection and T&O control) but are effective for color control

Taste and odor	Source control
	  Copper sulfate
	  Reservoir
	  Destratification
		Used to prevent any T&O problems in plant	Most satisfactory way of controlling is to control the problem at the source; copper sulfate may require chelation at certain pH values

	Oxidation
	  Chlorine
	  Ozone
	  Potassium permanganate
	  Chlorine dioxide
		Low T&O levels	Chlorine may result in increased odor problems where odors are of industrial or algal origin; $KMnO_4$ widely used and very effective for odor control (however, overdosing may result in slightly pink color)

TABLE 19-10. *(Continued)*

Water Quality Parameters	Process and Components	Applicability	Comments

	Adsorption
	  GAC
	  PAC
		Low to moderate levels, GAC used for industrial odor sources	PAC, in slurry form, is usually added at coagulation process for moderate T&O levels and ahead of filters for low T&O levels; GAC commonly used for odors caused by industrial sources—bed life usually very long

Hardness	Lime–soda softening	Extremely hard waters	Most common method of removal of hardness
	Ion exchange	Removes not only hardness but also selective constituents	Very expensive, especially in large-scale facilities

Organics (THM)	Alternative distinfectants
	  Chloramine
	  Chlorine dioxide
	  Ozone
	Removal of precursors
	  Chlorine dioxide
	  Ozone
	  GAC
	  PAC
	  Coagulation
	Removal of THM
	  Ozone
	  GAC
	  PAC
		THMs	Chloramines are not as powerful a disinfectant as free chlorine; ozone offers no appreciable residual protection in distribution system; modification of chlorine addition points may reduce THM formation; PAC has been found to yield only partial removals at very high doses

ment (or removal) process is unknown, (2) the cost feasibility is uncertain, (3) a cheaper alternative may be effective, and (4) process refinements may be beneficial (Tate and Trussell, 1982).

Figure 19-3 illustrates simplified logic for determining when a prototype study is needed. If technical feasibility is not established, prototype scale testing is usually justified. Once technical feasibility has been resolved, the cost for prototype study should be compared with the potential savings.

There are a number of other cases where prototype studies may be needed. For example, a state approval agency may require a pilot study as a matter of policy for innovative treatment processes. Another example would be a situation wherein a process cannot be modeled sufficiently to lead to design criteria; an example is removal of organics from a multicomponent mixture that is not fully characterized and not amenable to modeling by current granular activated carbon (GAC) adsorption models. A third example might be a fleeting phenomenon which cannot be simulated, such as taste and odor episodes.

Technical Feasibility. The most basic objective for a prototype study is determining technical feasibil-

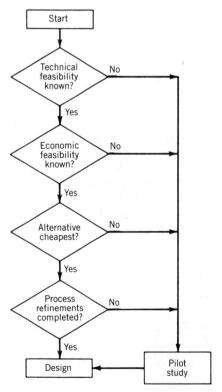

FIGURE 19-3. Logic for determining need of pilot study. (Source: Tate and Trussell, 1982.)

ity. If the process is found to be effective, the next issue is how effective it is. Determining what control variables affect performance and the approximate costs of the treatment process are also important.

Because of its inherent research and development features, a prototype study evaluating the technical feasibility of a process is likely to be one where the scope is not clearly defined at the beginning of the project. Staging is typical; for example, the first stage would include bench-scale tests, while later stages include pilot- and/or larger-scale tests. As a result, costs for such a study could be extensive.

The success of the proposed process is not guaranteed. However, there is a chance to advance fundamental understanding of treatment processes by undertaking such a study. Overall, there are greater risks and benefits associated with a technical feasibility prototype study than with the other types of pilot studies.

Examples of prototype study objectives addressing the technical feasibility issues are:

- Can ion exchange with activated alumina remove selenium?
- Will air stripping remove dibromochloropropane (DBCP)?
- Can PAC be used to remove color?

Economic Feasibility. For a prototype study investigating economic feasibility, there is usually an extensive evaluation of control variables. For example, filtration rate, empty bed contact time, and packing depth would be evaluated in studies of filtration for turbidity removal, GAC adsorption for organics control, and air stripping for organics control, respectively. In this type of study, there is also a more detailed evaluation of costs relative to a technical feasibility study and more sophisticated data analysis. Generally, only a single alternative would be evaluated in this case.

Examples of prototype study objectives addressing economic feasibility or practicality are:

- How much will it cost to use GAC for removal of carbon tetrachloride from this supply?
- How much ozone will be required to reduce color to acceptable levels?
- How much will it cost to remove nitrate by ion exchange?

Less Expensive Alternative. Many prototype plant studies are conducted to evaluate the feasibility of an alternative that is less expensive than another alternative already accepted as applicable to a given raw water quality and treated water quality goals. The central issue is the ability of the less expensive alternative to provide water quality equivalent to the known or more conventional alternative.

A prototype study of a less expensive alternative is characterized typically by a side-by-side comparison of two or more processes, such as direct filtration versus conventional treatment. There is a need for extensive historical water quality data to ensure that the less expensive and less conservative alternative will work under poor raw water quality conditions. Interaction between the utility, the consultant, and the approval agency is particularly important in this type of study. Finally, it is necessary to actually document the cost savings of the proposed process.

Examples of questions addressed by this type of prototype study are:

- Will direct filtration be as effective as conventional treatment with this water supply?
- Will chloramines meet disinfection requirements?
- Can ozone be used as a substitute for coagulation prior to filtration?

Process Refinement. This type of a prototype study is especially applicable to large utilities that have something to gain from minor changes in water quality or cost over a large system. In this case, a given process is known to be the best option for this water and the question of refinement is whether it can be made to be more reliable, less costly, or less energy intensive.

Process refinement prototype studies are necessarily site specific and may evaluate only one element of a treatment process train. Some of the work may be done at an existing full-scale plant, such as surface wash tests, coagulant selection, or the effect of chlorination on coagulation and filtration.

Examples include:

- Studying filtration rates.
- Developing better backwash or surface wash schemes for filtration.
- Adding a flash mixer to save on alum costs.
- Studying alternate filter media designs.

With a clear understanding of the primary objectives of the prototype studies, the design engineer must prepare a detailed experimental plan outlining the specific approach to be used, the parameters to be evaluated, and the schedule and manpower requirements to perform the testing. Adequate time needs to be alotted for equipment debugging and process optimization–stabilization.

Scale

Once the objective of the prototype study has been developed, the appropriate scale of the testing equipment needs to be determined. Table 19-11 summarizes the wide range of equipment sizes available to perform prototype testing. Full-scale prototype investigations can be at existing plants and are typically over 3.8 ML/d (1 mgd). On the opposite end of the size range are small-scale facilities, which are typically in the 0.06–0.63 L/sec 1–10 gpm) flow range. Bench tests, although often not considered as prototype plant tests, can produce considerable information for certain processes.

The selection of an appropriate prototype plant size depends principally on the process and the study objectives. Large-scale or full-scale tests evaluate scale-up reliability of equipment over the

TABLE 19-11.   Pilot Plant Scale Guidelines

Size	Approximate Flow	Examples of Pilot Tests
Full scale	>3.8 ML/d (>1 mgd)	Effect of launders on basin performance, effect of wastewater troughs on backwash performance
Large scale	0.38–3.8 ML/d (0.1–1 mgd)	Process reliability tests, ozone contactor configuration, backwash regime
Medium scale	0.63–3.15 L/sec (10–50 gpm)	Air stripping, backwash regime
Small scale	0.063–0.63 L/sec (1–10 gpm)	Filtration rate, empty bed contact time
Bench tests	Batch	Coagulant selection by jar test, ozone dose versus color, determination of disinfectant demand

SOURCE: Tate and Trussell (1982).

long term, and details of equipment design. Smaller-scale or bench tests are used to select alternatives or refine design details of a given process. Such tests are also typically used for side-by-side comparison of more than one process. Their inherent flexibility allows optimization of design criteria. Selecting an appropriate size of a prototype plant can best be illustrated through several examples based on the author's experiences.

Air Stripping.   The chemical engineering literature and information from packed tower manufacturers suggest that a minimum 10 : 1 ratio of column diameter to packing diameter is necessary to minimize wall effects. Since the smallest commercial packing diameter is 1 in., the prototype column size to evaluate air stripping is 1 ft in diameter. Typical water loading rates are 0.63–1.89 L/sec (10–30 gpm/ft^2), so pumps are sized to a maximum of 3.15 L/sec (50 gpm). The net result with the air stripper is medium-scale equipment.

Filtration.   Work done in England on filtration suggests that at least a 50 : 1 ratio of column diameter to media diameter will minimize wall effects. Since commonly used media diameters in water treatment are 0.5–1.5 mm, 3-in. filters are selected. To accommodate test surface loading rates of 48 m/hr (20 gpm/ft^2), the pumps for filtration are sized to 0.06 L/sec (1 gpm) maximum. The net result is small-scale equipment.

If the filter media deviates significantly from the design media size, then different-sized filters are necessary. For example, with 2 mm media tested in a study for the Los Angeles Department of Water and Power, 6-in. filter columns were designed. However, for a test with 1 mm GAC media for adsorption, the 3-in. filters are judged to be suitable.

One element of filtration that cannot be well modeled on a small scale is backwashing. With filters on the order of 3 in. diameter, backwashing commonly produces a piston flow of media in the initial stage of backwash rather than a fluidized bed. In addition, the backwash water overflow facilities in a prototype system of this size often do not accurately model the design used in a full-scale facility.

Corrosion Tests.   Home plumbing units typically have pipe that is 1 in. or less in diameter. In order to produce proper scale formation conditions, it is necessary to have a minimum velocity in the pipe of about 1 ft/sec. Therefore, corrosion test facilities have been designed for about 0.19 L/sec (3 gpm), with the result being small-scale equipment. Several tests have been successfully conducted with the Illinois State Water Survey coupon units, which use inserts of copper, iron, galvanized steel, or other materials. Weight loss measurements are used for evaluating corrosion rates in mils per year.

Ion Exchange.   Useful ion exchange breakthrough curves can be determined in any device that can contain the "exchange front." Typically, 1-in. columns and very low flow rates (10–50 mL/min) are satisfactory.

Test Parameters

The type of prototype testing and the stated objectives of the testing determine the test parameters to be evaluated. As indicated above, the scale of the prototype equipment limits the range of test parameters that can be evaluated. For example, small-scale filter plot units cannot properly model the backwash operation. In general, the development of test parameters for any unit process or treatment scheme includes an assessment of removal efficiencies and evaluation of operational parameters that influence removal efficiencies. A discussion of the test parameters utilized in both traditional and specialized prototype investigations are contained below.

Traditional Investigations.   Traditional prototype studies deal with unit processes utilized for turbidity (i.e., particulate) removal. Two basic types of prototype studies in this category are: (1) flocculation–coagulation–sedimentation process piloting and (2) filtration process piloting. Flocculation–coagulation–sedimentation process piloting usually involves the use of bench-scale test equipment (jar testing) and includes:

1. Selection of proper coagulant or a combination of coagulants.
2. A study on the effect of dilution of coagulant, especially alum (a maximum alum dilution before application should be 0.1 percent or the pH should be below 3.5).
3. Determination of optimum coagulant dosages for average and spike (i.e., peak) turbidity conditions.

4. Evaluation of chemical application sequence. This is especially important for polymer–alum and lime–alum applications (some cationic polymers lose their effectiveness when prechlorination is practiced; most anionic and nonionic polymers perform much better when they are introduced 1–5 min after alum addition).
5. Evaluation of mixing energy and mixing time on coagulation, flocculation, and subsequent filtration processes.
6. Effect of filter washwater recycling.
7. Effectiveness of compartmentalization on flocculation basins.
8. Improved flocculation for THM precursor removal.
9. Any special treatment for Mn, taste and odor, or other constituents.

In general, items 1–6 are easily evaluated using bench-scale test procedures (i.e., jar testing). Jar testing procedures are outlined in various publications (Camp, 1968; Hudson, 1981).

In filter piloting a more elaborate equipment setup is required, involving pilot-scale facilities. Pilot-scale operations should be designed to provide maximum flexibility for the intended study. Since the raw water quality may vary daily, the pilot testing program involving pretreatment should include comparison of parallel rather than sequential filter runs. In general, pilot-scale testing is divided into two categories: (1) optimization of chemical pretreatment and (2) evaluation of physical variables of filters.

The pretreatment evaluation on a pilot scale should incorporate data developed in the bench-scale evaluations. The level of pretreatment selected is based on the influent quality and its variability and on the ability of the selected pretreatment option to consistently meet water quality, operational, and economic objectives. The second category of pilot-scale testing evaluations would include investigation of the physical variables of filters such as:

1. Filter media size.
2. Bed depth.
3. Filter media type (mono-, dual, or multimedia).
4. Filtration rates.
5. Filter washing conditions (optional).

The interpretation of the results from bench- and pilot-scale testing is important since they are the basis for selection of the appropriate treatment scheme. The goals of the prototype studies therefore should be:

1. Selection of the treatment process (i.e., in-line filtration, direct filtration, or conventional treatment).
2. Selection of pretreatment that will provide the lowest practical level of colloidal destabilization.
3. Selection of the optimum economic combination of filter material, size, depth, and filtration rate.

In general, pilot studies do not cover a year-round period but rather are performed to meet established project schedules and water needs. Thus, the design engineer should use good engineering judgment in anticipation of seasonal quality variation, possible algae blooms, taste and odor potential, and degradation of raw water quality in the near future due to land development, watershed management changes, or other causes.

Specialized Investigations. These types of prototype studies involve investigations of removal or treatment of a specific water quality parameter, such as organics (THM), iron and manganese, selenium, corrosion, taste and odor, disinfection, and so on. The test parameters that are to be evaluated are dependent on the specific water quality parameter and treatment process under consideration. For example, if an air-stripping (packed tower) process is utilized for the investigation of THM (and other volatile organic) reduction, the following parameters require consideration:

1. Tower-packing depth.
2. Water-loading rate.
3. Air-loading rate.
4. Mass transfer constants for a particular volatile organic constituent.
5. Removal efficiencies (C/C_0).

These tests would be performed on a pilot-scale unit, since it is impossible to use a bench-scale tester for evaluation of these test parameters due to the physical limitation of this type of equipment.

Another example of a specific type of investigation is one of evaluation of disinfectant. The test parameters involved would include:

1. Type of disinfectant (Cl_2, O_3, ClO_2, etc.).
2. Dose.
3. Contact time.
4. Resultant removals (kills).

In the evaluation of adsorption process, such as GAC, the test parameters may include:

1. Empty bed contact time.
2. Carbon usage rate.
3. Type of carbon.
4. Selective organics and/or constituents to be removed.
5. Breakthrough concentrations.
6. Bed configuration, that is, single- or multiple-bed contactors.
7. Effect of other water quality and physical parameters (pH, temperature).
8. Regeneration methods.

Although there are theoretical formulas that can be used to design a GAC system, an empirical approach is suggested since several factors such as the type of organic solute and its concentration, the water quality in which it is dissolved, and the properties of the carbon affect the adsorption process.

In the consideration of corrosion testing, parameters should be used that predict the useful life of a given pipe material. The parameters should relate to the three general problems associated with corrosion, namely: (1) pipe material becomes so thin it cannot mechanically support the internal pressure; (2) localized corrosion (i.e., pitting) can cause a pipe to fail long before the entire pipe becomes too thin to support the internal pressure; and (3) the pipe becomes so congested with the mass of corrosion products that fluid flow is prevented or severely restricted. In the pilot testing for corrosion of pipe materials, test coupons are used. They are essentially a short section of pipe specially machined and painted to explore only the desired metal surfaces. The test parameters that are measured or observed are:

1. Coupon weight loss.
2. Physical dimension of the coupon.

3. Rate of weight loss per unit area.
4. Penetration rate.
5. Micrometer inspection.
6. Pitting factor (defined as the depth of the deepest pit divided by the average depth of all pits on the coupons).
7. Physical observations (color, striation patterns, and type of corrosion; i.e., modules, thin uniform film, or ridges).
8. Chemical composition of corrosion products.

In any specialized investigation, it is important to assess the variation of the raw water quality and its impact on removals (or treatment) efficiencies. Therefore, as in prototype testing for turbidity removal, the design engineer should evaluate the variation in raw water quality and use good engineering judgment in anticipation of variation in water quality. Included in these judgment processes could be the use of more conservative design criteria to meet treated water quality objectives.

Finally, the test parameter to be evaluated should be similar in magnitude to those found and available in full-scale facilities. In many cases, use of standard design parameters attainable by commercially available products and equipment is advisable.

Evaluation Criteria and Data Analysis

Prior to the start of prototype testing programs, criteria for the evaluation of data and acceptance of the test result needs to be established. In general, the evaluation parameters are similar to those used in full-scale plant operational evaluation and relate to both the finished water quality and the operational behavior of the process. For processes used for turbidity removal, they include:

1. *Removal efficiencies.* The traditional parameter is turbidity, although particle counts have been used to evaluate particulate removal efficiencies; finished water quality levels of 0.2–0.3 TU and 50–100 particles/mL (in 2.5–150-μm range) are reasonable objectives.

2. *Head loss accumulation.* For full-scale gravity filters, the total available head loss before filter backwash is in the range of 1.8–2.4 m (6–8 ft), excluding clean bed losses; the establishment of the terminal head loss for pilot-scale testing is similar and based on the design engineer's se-

lection of terminal head loss; for pressure filters, higher terminal head loss levels are possible.

3. *Economics of operation.* The selection of the "best" process operational parameter is a balance between attaining water quality objectives and minimizing the cost to achieve that quality level; the process performance needs to be evaluated so that the point of diminishing return is not reached. For example, higher chemical doses can result in better water quality; however, acceptable water quality levels are achieved with lower doses without experiencing the shorter filter runs normally associated with higher chemical doses.

4. *Production volume.* In addition to meeting designated goals for effluent quality, maximization of net water production per unit filter area is required. The principal factors that affect net water production are filtration rate, length of filter run, amount of water required for backwash, and amount of filter downtime associated with each filter cycle. As discussed in Chapter 8, the unit filter run volume (UFRV) or the volume of water filtered through each unit of filter area and the volume of backwash water required per unit of filter area are factors affecting net water production.

To develop a design having adequate production efficiencies, it is recommended that a UFRV goal of between 205,000 and 410,000 L/m^3 (5000 and 10,000 gal/ft^2) be set for any pilot studies. These UFRV correspond to filter run lengths of 16.7 and 33.3 hr, respectively, at a filtration rate of 12 m/hr (5.0 gpm/ft^2). With water backwash and surface wash, typical backwash volumes range from 240 to 480 m/hr (100–200 gal/ft^2). Air scour tends to reduce backwash volume; therefore, UFRV goals can be set somewhat lower.

Once the evaluation criteria is established, the experimental plan prepared and implemented and the test results generated, the method of data analysis is instrumental in evaluating trends in the behavior and sensitivity of select test parameters with respect to each other. For example, where the jar testing technique is used to evaluate coagulant doses for turbidity removal, use of isotopograms is extremely useful. From bench filter studies, optimization of coagulant doses was attained by preparing a matrix of alum and cationic polymer doses. Figure

19-4, a log–log isoturbidity topogram of alum and cationic polymer, reveals that increased alum doses improve effluent quality while increased cationic polymer doses tend to have a deleterious effect at low alum doses and minimal effect at higher alum doses (JMM, Casitas MWD, 1977).

Similar data analysis techniques are applicable to prototype testing for other water quality parameters and processes. For example, in the evaluation of breakthrough concentrations for GAC filters, analysis of the effluent organic concentration versus time at various empty bed contact times would establish breakthrough characteristics, carbon usage rates, and regeneration requirements. Figure 19-5 presents an example of a typical breakthrough curve (JMM, Contra Costa County Water District, 1980). Note that throughput volume as well as time

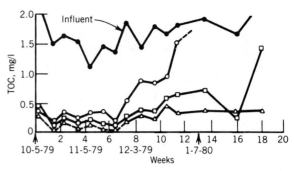

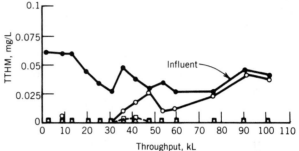

• Influent   o EBCT = 10 min   □ EBCT = 20 min   △ EBCT = 30 min

FIGURE 19-5. Breakthrough curves for TOC and THM on a carbon column at different empty bed contact times. (Source: JMM, November 1980.)

is used, thereby normalizing the data with respect to application rate.

Pilot Plant Design

The purpose of this section is to provide the design engineer with basic criteria and considerations for the design of pilot plant equipment. In general, the filtration and flocculation processes are very easily modeled on a pilot scale. The modeling of the sedimentation and aeration process does present some problems. In addition, the significance of good initial mix of coagulants is studied effectively on a pilot scale. The design of pilot units for each of the unit processes is discussed briefly below, as well as general considerations and instrumentation requirements for pilot plant design.

General Considerations.   Certain general considerations are required in the design of pilot plant facilities that make these facilities useful in process evaluation and selection. They include:

1. The equipment design is dictated by the anticipated use and objective of the pilot equipment. Permanently mounted trailer installations do

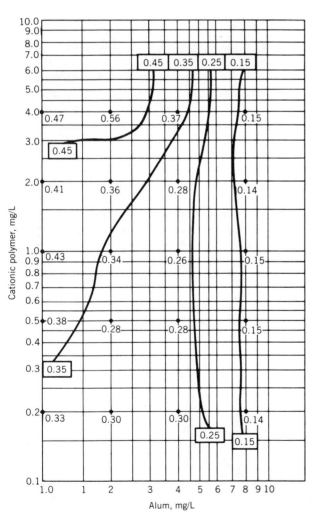

FIGURE 19-4. Alum–cationic polymer isoturbidity topogram (TU). (Source: JMM, Casitas MWD, June 1977.)

offer some advantages, however they tend to be somewhat bulky and inflexible. A modular approach provides the engineer with a reusable, flexible, and easily transportable configuration. In the module approach, each process module is self-contained, thus the experiment designer is free to vary the process train without concern for the interdependence of the various unit processes on each other (Lang, 1982).

2. At least two process trains are required to enable side-by-side comparison of various design parameters; otherwise, no control is available for data evaluation comparison.

3. Adequate bypass capabilities need to be provided around and within unit processes.

4. The effect of scale-down of facilities on system process performance should be considered, especially the "sidewall" effects in filtration systems.

5. Flexibility in the operation should be maximized. This is especially important with respect to multiple chemical addition points. In addition, provisions to add various types of chemicals are desirable.

6. All pumps and motors should be equipped with a variable-speed adjustment. For example, the speed of the flocculators should be adjustable to test effect of flocculation energy input for various treatment schemes.

7. Positive flow splitting is best achieved by use of weirs.

8. Accurate water and air flow meters are necessary. The use of rotometers in the proper flow range is recommended.

9. If positive displacement (peristaltic) pumps are used, a pulsation dampener is needed to eliminate the high and low surges created by the action of the pump.

10. Unlike full-scale facilities, the hydraulics of pilot plants should be designed to accommodate a wide range of flow conditions. For example, normal filtration rates in water treatment plants are on the order of 12–19 m/hr (5–8 gpm/ft^2); however, higher filtration rates (say, 29–36 m/hr or 12–15 gpm/ft^2) should be utilized in sizing the inlet and outlet piping of pilot equipment.

11. Provisions for the artificial injection of turbidity (or other water quality constituents) may be desirable to simulate periodic, extreme quality conditions.

12. When more than one process train is utilized, an influent structure may be required. The influent structure should include (1) a means of providing an even flow split to each process train, (2) a common location for the introduction of pretreatment chemicals and monitoring of influent water quality parameters, and (3) provisions for a strainer to remove gross debris such as leaves and twigs that could clog tubing.

Initial Mixing. The importance of initial mix to provide good rapid mix of coagulant chemicals has been demonstrated by various researchers (Stenquist, 1972; Letterman, 1973; Argaman, 1970). In full-scale plants, hydraulic flash mixing is extremely effective. In pilot plant facilities, an initial mixer as shown in Figure 19-6 or any commercially available

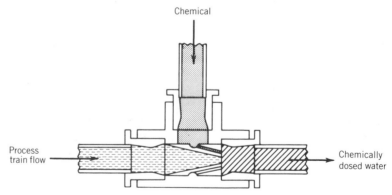

FIGURE 19-6. Cross-sectional view of turbulent initial mixer. (Source: JMM/DMJM, June 1977.)

(off-the-shelf) in-line static mixer has proven to be adequate.

Flocculation Basins. The following considerations are made in design of pilot flocculation basins:

1. It has been demonstrated that there is a scale-up problem between a pilot-scale model and the actual basins. The pilot basin generally gives a higher optimum G value than actual basins. For example, several pilot studies have indicated that the optimum G value is in the range of 50–70 sec^{-1}, but actual plant operation revealed that adequate flocculation was achieved with G values of 15–40 sec^{-1}.

2. If the sole objective of the flocculation basin is coagulant optimization for sedimentation, a bench-scale batch test (i.e., jar tests) provides the design engineer as equally good basic information as pilot-scale testing.

3. At least three compartments should be provided for each basin with the capability of bypassing each compartment. Compartmentalization is important because it affects performance and can reduce the energy requirement. In addition, by providing bypass capabilities, it is possible to vary the hydraulic detention time and therefore evaluate the effects of the mean velocity gradient G and Gt on process performance.

4. Each compartment should have a detention time of approximately 5 min and the capability to vary the G value, that is, the flocculation paddle speed in each compartment. This arrangement allows the pilot facility the ability to study various operational procedures such as tapered flocculation.

5. A clear material (Plexiglas) is preferred for the construction of the flocculation basin so that floc formation is easily viewed in the basin. Figure 19-7 shows a typical pilot flocculation basin.

6. The paddle area and configurations should be sized to provide the desired G value to be tested. The shape of the paddle is important. Generally, large paddles operating at slow speeds provide for formation of large settleable floc and are ideal when conventional treatment process trains are piloted. Small paddles at high speeds usually provide a dense floc, which is useful when piloting the direct filtration process. A calibration curve should be generated

FIGURE 19-7. Typical pilot flocculation basin. (Source: JMM/DMJM, June 1977.)

for each paddle size and configuration by the use of a torque meter, a stroboscope, and a tachometer and application of the following equation:

$$G = \sqrt{\frac{P}{V\mu}} \quad \text{sec}^{-1}$$

where P = power input (ft-lb/sec)
μ = absolute viscosity (lb-sec/ft^2)
V = volume of sample (ft^2)

Figure 19-8 shows a typical graph used for establishing the energy input (G value) in each flocculation compartment.

Sedimentation Basins. The following considerations are made in the designs of pilot sedimentation basins:

1. A hydraulic scale model study applying the similarity laws will simulate the flow characteristics of the tank, thereby allowing studies of inlet–outlet configuration and diffuser wall design (Kawamura, 1981).

2. Floc removal efficiency cannot be predicted well because of the scale-down and modeling problems of floc size and settling velocities. Therefore, the floc removal efficiencies of pilot sedimentation basins should be considered more as a *relative* removal value rather than an *absolute* removal value in the comparison of process train performances.

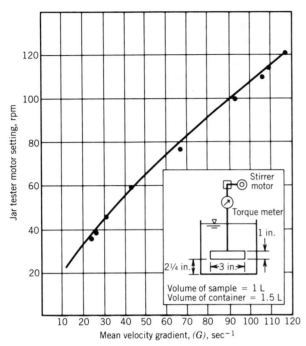

FIGURE 19-8. Velocity gradient calibration curve, jar test unit. (Source JMM, April 1980.)

3. The use of traditional design parameters (e.g., detention time or overflow rate) produces a basin that is disproportioned. For example, a circular clarifier design to treat a flow of 0.3 L/sec (5 gpm) with an overflow rate of 3785 L (1000 gal) per day per square foot and at a detention time of 2 hr would be 0.9 m (3 ft) in diameter and 3.4 m (11.2 ft) deep.

4. The flow distribution at the inlet of the basin should provide even hydraulic distribution of the flocculated water; therefore, diffuser walls are useful in accommodating this requirement.

5. The use of tube settlers warrant consideration since they have demonstrated good removal efficiencies.

6. Each pilot basin design should be tested for short circuiting prior to use. Tracer studies of the pilot sedimentation basin are recommended.

Filters. The following considerations are made in the design of pilot filters:

1. The general rule is to provide as many pilot filter units as possible. The number is sometimes affected by the budget. However, provid-

ing the maximum number of units reduces the time to make meaningful comparisons of test parameters and gives the design engineer a wider range of test variables to investigate. A minimum of three filters is recommended. However, additional pilot filters make test results easier to evaluate and compare since influent quality can vary from test to test.

2. It is recommended that the size of the column be a minimum of 75 mm (3 in.) in diameter to minimize sidewall effects for media sizes normally used in water treatment applications (up to 1.5 mm). From a quality standpoint, research has indicated that there is no difference in removal efficiencies between 75- and 200-mm (3- and 8-in.) diameter columns (JMM, Las Virgenes Municipal Water District and Triunfo County Sanitation District, 1979). If backwash studies are required, use of larger columns are preferred and in fact necessary to adequately model the backwash operation.

3. Figure 19-9 presents a general schematic of a pilot filter design. Basically, the filter hydraulics are controlled by pumping at a constant rate against a constant head with a peristaltic pump. The minimum water surface is kept above the media by a control weir in the effluent line. The level in the filter column rises with head loss buildup and corresponds to the head loss through the filter. Therefore, the available head loss is limited by the height of the column. In some experiments, higher head losses than can be attained from a gravity filter are required to study the various turbidity breakthrough phenomena. In these cases, pressure pilot filters are available. Figure 19-10 shows a pressure filter pilot module with associated instrumentation. The terminal head loss is no longer limited by the height of the column but rather by the influent pump characteristics. Note that the rotometer in this pilot filter design is positioned on the discharge line of the filter, thereby eliminating potential clogging of the system. Considerations should also be given for potential of shearing alum floc during the pump of filter influent. Experience has shown that use of polymers in combination with alum tends to reinforce the floc, thereby minimizing this concern. Gravity feed to the filters is another possibility.

4. The backwash system should be designed to provide both air and water backwash. An auxil-

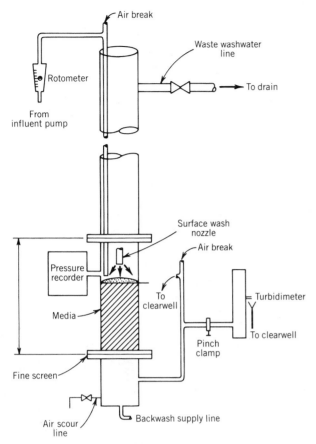

FIGURE 19-9. Typical design of a pilot filter. (Source: Sanks, 1980.) Reprinted with permission.

iary surface wash system can be included if only one media depth is studied since adjustment of the surface wash nozzle requires replumbing whenever the media depth is changed to provide the proper distance between surface wash nozzles and media surface.

5. Proper calculation of head loss of tubing is required.

6. The design of the waste backwash line should accommodate the *maximum* hydraulic rate anticipated.

7. Consideration for air locking of tubing should be made. Sizes of tubing should be a minimum of 12.5 and 25 mm ($\frac{1}{2}$–1 in.).

8. Proper design of inlet configuration is necessary so that the filter influent does not fall several feet, causing air binding in the filter, or the filter influent pump does not work against a varying head condition.

GAC Contactor. The design of a GAC contact column is similar to that of a filter column with the following exceptions:

1. The sizing of the GAC column is based on empty bed contact time at the selected flowrate.

2. Sample taps are provided at various depths to allow effluent sampling corresponding to selected contact times. A minimum of three taps are recommended.

3. A backwash system is not normally included in a GAC column if the column is specifically designed for adsorption studies. However, in actuality, the design of a GAC contactor and a filtration system is such that they are interchangeable and can be used for both filtration and adsorption studies.

4. The instrumentation on a GAC contactor is not as intricate as a filtration system. Since the influent to a GAC contactor is filtered effluent, the head loss buildup is insignificant. Also, continuous monitoring of organic compounds is impractical, very expensive, and near impossible. Manual sampling is the normal experimental practice, since organic removal levels can vary slowly over time and breakthrough is measured in terms of weeks instead of hours.

Ozone Contactor. The following considerations are made in the design of pilot ozone contactors:

1. The countercurrent flow arrangement is recommended. Usually a PVC or Plexiglas column is used.

2. The detention time should be in the range of 2–5 min. Adjustments in detention time are made by varying the flowrate into the contactor.

3. The contact column should be equipped with an efficient method of ozone distribution. A ceramic air stone and a submerged turbine running axially through the length of the column have proven satisfactory. Ozone transfer efficiency is about 85 percent with the air stone and increases to about 90 percent with the submerged turbine in operation.

4. Off-gases from the contactor should be properly vented.

Air-stripping Towers. In the design of air-stripping towers for use in removal of volatile organic com-

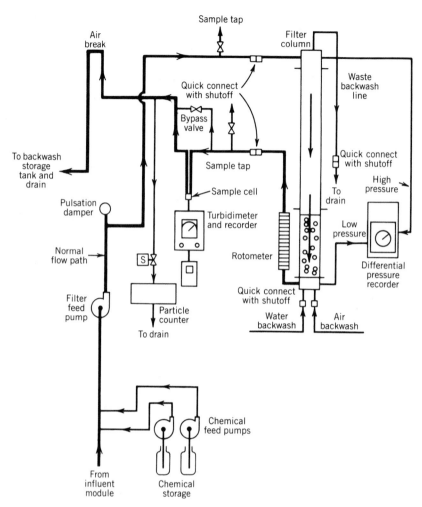

FIGURE 19-10.   Typical pilot filter—pressure type. (Source: JMM, February 1980.)

pounds, the following considerations should be made:

1. The most suitable type of air stripper is a counterflow packed tower design. Air is introduced into the bottom of the column and water on top. The water flowrate is measured by rotometers. The water is pumped at a constant rate to an overflow located above the highest possible position of the packing. A throttling valve controls the rate of flow bypassed to the stripping tower. The air flowrate is measured with an interchangeable orifice flow meter and throttled by a valve located on the blower outlet duct. Figure 19-11 shows a typical pilot air-stripping testing installation. Figure 19-12 shows a simplified cross section of an air-stripping tower.

2. The ratio of air to water should be adjusted over

a wide range so that the towers can be used for piloting removal efficiencies for a variety of organic compounds. Ratios of 30:1–150:1 are suggested.

3. The material of construction of the tower should be Plexiglas so that the distribution of air and water can be observed.

4. The tower should be built in sections (0.6-m or 2-ft sections are suggested) so that the height of the tower can be adjusted to investigate various packing depths. The packing depth and air–water ratio are functions of the required removal efficiency.

5. The diameter of the tower should be chosen to provide at least a 10:1 ratio between tower diameter and the diameter of packing media to minimize wall effects. Since commercially

FIGURE 19-11. Pilot air stripper. (Source: Lang, 1982.)

available packing media is in the order of 25 mm (1 in.) diameter, a packing tower diameter of 305 mm (12 in.) is suggested.

6. Flow distribution units are commercially available. In addition, wall wiper flow redistributor should be located at 2-ft intervals in the tower.

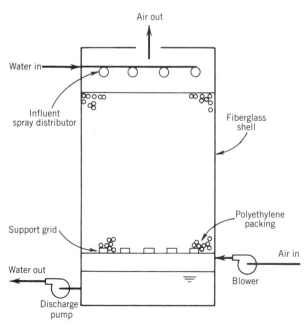

FIGURE 19-12. Air-stripping tower (forced draft). (Source: JMM, November 1980.)

These wall wipers intercept flow that has become entrained on the inner surface of the tower and redirect it back toward the center of the packing, thus minimizing wall effects.

Instrumentation and Control. The following instrumentation features can be incorporated into the pilot plant design:

1. Indication and continuous recording of (a) turbidity, both influent and effluent, and (b) head loss in the filter using a differential pressure meter or level-monitoring sensor. The turbidimeters should be adjustable over a wide range. Models are available which have ranges from 0.1 to 1000 TU. Therefore, they can be used for monitoring both influent and effluent turbidity levels. Flow-through turbidimeters with small sample cells provide fast response and easy calibration.

2. For pilot plants that are not manned 24 hr per day, overload protection should be provided. This can be a high-pressure limit switch on filter influent pumps or a contact probe at the base of influent pumps that would automatically shut the units off when leaks occurred.

3. As discussed in Chapter 2, particle counting is a useful parameter to monitor. If particle-counting equipment is available, particle counting can be automated to monitor effluent on a continuous basis. Otherwise, periodic sampling of effluent quality can be done manually.

4. For a packed tower system, thermometric probes should be included to measure ambient air temperature, air temperature entering and leaving the tower, and water temperature entering and leaving the tower.

5. Measurement of instantaneous flow is accomplished by the use of rotometers. Air flows are usually measured by orifice flow meters.

Case Studies

Department of Water and Power, City of Los Angeles. A comprehensive process evaluation project for the Los Angeles Department of Water and Power (LADWP) was conducted to determine the appropriate treatment of the Owens Valley water supply (JMM/DMJM, Vol. 4, 1977). The primary water quality treatment requirement was the removal of particulates. However, investigations were also conducted on various process alternatives for THM removal. The pilot plant was designed to include three process trains (in-line filtration, direct filtration, and conventional treatment) and nine filters that can be operated in parallel for comparative purposes. In addition, provisions for preozonation were included. A schematic of the pilot plant is presented in Figure 19-13. In addition to constant monitoring of influent and effluent turbidity, particle counts were also monitored.

The raw water turbidity for the Owens Valley supply during the evaluation period ranged from 1.5 to 6 TU. In addition, during latter phases of the study, higher-turbidity water was imported from Haiwee Reservoir to test the utilization of the selected process on higher raw water turbidity.

The following conclusions were made based on pilot evaluations of the Owens Valley water supply:

1. In the direct and in-line filtration process, chemical optimization studies indicated that a combination of cationic polymer and alum produced acceptable water quality (turbidity levels = 0.2 TU) and adequate water production levels (UFRVs greater than 205,000 L/m² or 5000

gal/ft²). Other chemical schemes were tested such as nonionic polymer, anionic polymer, and alum alone. However, neither quality nor operational criteria were met.

2. For direct filtration, the impact of flocculation mixing intensity was evaluated. Higher G values tended to increase the water production. However, the benefit of increased water production with higher G values, was not commensurate with substantial increases in costs to attain these values. Therefore, a G value of 70 sec⁻¹ was selected.

3. The levels of pretreatment were evaluated. Comparison of filtration results with and without flocculation indicated that the flocculation process did increase the water production 20–45 percent. Further addition of sedimentation increased the water production an additional 20–25 percent. However, results with direct filtration were within the criteria for water production (average UFRVs of 490,000 L/m², or 12,000 gal/ft²).

4. Various media types and depths were tested. A coarse media (1.5 mm effective size) at a depth of 2.4 m (8 ft) performed comparably to conventional dual media. However, a 50 percent

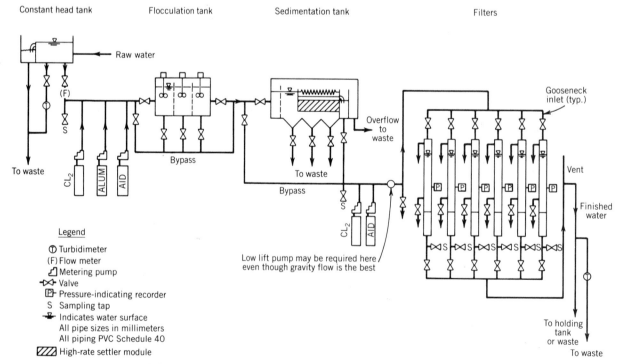

FIGURE 19-13. Pilot plant P&I diagram; a complete treatment process.

greater water production rate was attained with the coarse media.

5. The effect of filtration rate on water quality was also evaluated. Rates as high as 43 m/hr (18 gpm/ft^2) were tested. Results, presented in Table 19-12, indicated that there was little effect on performance with higher filtration rates.

Based on the pilot studies, the recommended process train included direct filtration with use of deep-bed, coarse-media filters at an initial filtration rate of 22 m/hr (9 gpm/ft^2). Subsequently, the city built a large-scale prototype filter and pretreatment unit to confirm the proposed scheme. The large prototype unit was designed to treat 9.5 L/sec, or 150 gpm, as opposed to 0.19 L/sec (3 gpm) on the pilot-scale unit. The large-scale prototype filter confirmed the feasibility of high-rate deep-bed monomedia filtration. Also, the use of preozonation did improve the filter run length time significantly and was included in the proposed process train. In addition, valuable data on supplemental air scour systems were also developed during the confirmation tests by LADWP.

City of Phoenix. As part of the predesign study prior to the 60-mgd expansion to the Val Vista Water Treatment Plant, pilot studies were conducted to evaluate optimum chemical types and dosages, media types, and filtration rates (JMM, Phoenix, 1980). The Val Vista Water Treatment Plant treats water from the south canal of the Salt River Project. The existing facility is a conventional treatment plant. Raw water turbidity ranged from 5 to 10 TU during the evaluation.

Both bench and pilot testing was conducted. Bench testing evaluated chemical coagulant types and dosages. Conclusions reached during the bench testing were that optimum alum dosages were in the range of 8–10 mg/L. In addition, cationic-type polymers used as flocculant aids produced the best settled water quality. Ferric chloride was determined to be less usable as an alternative coagulant. A cost comparison based on the jar test data indicates that the use of alum alone was the least expensive chemical scheme. However, due to the variability of the influent water quality, the plant expansion should provide for flexibility for alum and polymer addition.

Pilot filter studies were also conducted to evaluate filtration performance. The following conclusions were reached:

1. A nonionic polymer as a filter aid at a dose rate of 50 μg/L provided adequate effluent quality without excessive head loss.

2. Filtration rates as high as 19.2 m/hr (8 gpm/ft^2) were found to provide excellent water quality and filter production for lower influent turbidity values. However, at the higher influent turbidity values, the filtration rate of 19.2 m/hr (8 gpm/ft^2) did not meet effluent turbidity goals of 0.2 TU. A filtration rate of 14.4 m/hr (6 gpm/ft^2) provided both excellent effluent quality and water production.

3. A comparison of media types indicated that a dual media met both quality and head loss objectives. Two dual medias were evaluated: the existing dual media and a modified dual media. The existing dual media consisted of 50 cm (20 in.) of anthracite with an effective size of 0.8 mm and 10 in. of sand with an effective size of 0.45 mm. The modified dual media was 50 cm (20 in.) of anthracite with an effective size of 1.1 mm and 25 cm (10 in.) of sand with an effective size of 0.51 mm. The modified dual media proved superior to the existing dual media with respect to head loss buildup. During high solids loading conditions, the existing dual media was limited by head loss.

Based on the pilot studies, a design filtration rate of 14.4 m/hr (6 gpm/ft^2) using the modified dual media was recommended.

Aurora, Colorado. A pilot study was conducted to establish design criteria for a new high-rate direct filtration plant treating Rocky Mountain water from both the eastern and western slopes of the Continental Divide (JMM, Aurora, 1980). The objective of the pilot study was to optimize chemical dose and

TABLE 19-12. Effect of Filtration Rate of Filter Performance, LADWP Pilot Studies

Filtration Rate (gpm/ft^2)	Turbidity (TU)	Particle Count (mL^{-1})	UFRV L/m^3	UFRV gal/ft^2
4	0.14	23	630,000	15,500
8	0.15	17	665,000	16,300
12	0.15	22	675,000	16,600

flocculation performance, compare for candidate filter media over a wide range of filtration rates, compare direct filtration with both in-line and conventional treatment modes, and investigate the effect of chlorine addition on THM formation. Three treatability goals were used to evaluate performance of the various media and rate alternatives:

1. Effluent turbidity = 0.3 TU.
2. Filter effluent particle concentration = 50 particles/mL.
3. UFRV = 205,000 L/m^2 or 5000 gal/ft^2-run.

The raw water turbidity was generally below 10 TU with periodic levels above 15 TU.

The following general conclusions were reached based on pilot studies:

1. Optimum chemical dosage of 10 mg/L alum and 0.5 mg/L cationic polymer with a flocculation detention time of 15 min and mixing energy of 50 sec^{-1} produced the best water quality and filter production. However, at filtration rates equal to or greater than 19.2 m/hr, or 8 gpm/ft^2, the optimum chemical dosages were 6 mg/L alum and 1 mg/L cationic polymer.
2. A standard 750-mm (30-in.) dual media (500 mm, or 20 in., 1.1 mm anthracite and 250 mm, or 10 in., 0.51 mm sand) satisfactorily met all treatment goals and exhibited the best overall performance of the four candidate media tested. The four candidate media were 750 mm (30 in.) dual media, 1170 mm (46 in.) dual media, 750 mm (30 in.) mixed mdia, and 1170 mm (46 in.) anthracite monomedia.
3. Filtration rates of 19.2–24 m/hr (8–10 gpm/ft^2) produced excellent finished water quality.
4. Comparison of direct filtration to both in-line filtration and conventional treatment modes indicated that direct filtration met all treatability goals. Of the three processes, in-line filtration produced slightly higher turbidity and particle counts. Based on these tests, direct filtration was the selected process.
5. Investigations concerning the effects of temperature and point of chlorine addition on total THM formation indicated that all levels of THM measured were less than 50 percent of the EPA's maximum contaminate level (100 $\mu g/L$). Other variations in THM concentrations were observed with changes in temperature and point of addition.

Based on the result of the pilot studies, a direct filtration treatment process, consisting of chemical addition, rapid mixing, flocculation, and filtration was recommended and implemented for the proposed plant.

Contra Costa County Water District, California. As part of an extensive investigation to evaluate the various alternative processes required to meet and exceed present water quality goals, especially in organics, Contra Costa County Water District (CCCWD) conducted bench and pilot studies on various processes including (JMM, CCCWD, 1980):

- Chloramine.
- Ozone.
- Chemical dioxide.
- Aeration.
- Activated carbon.
- Biological activated carbon.

CCCWD's source of supply is the San Joaquin–Sacramento Delta. It owns and operates the 80-mgd Bollman Water Treatment Plant. CCCWD has consistently produced a high-quality water easily meeting state and federal regulations. However, recent regulation on trace organics contaminants has caused CCCWD to evaluate treatment alternatives. Therefore, the investigation of candidate processes to reduce and/or remove THMs is of particular interest. The maximum THM potentials of the CCCWD supply for raw and treated water is 400 and 200 mg/L, respectively. General conclusions reached by the prototype studies included:

1. Ozone did not reduce the coagulant demand.
2. Chlorine to ammonia (Cl_2–NH_3) ratios affected odor of the finished water.
3. Odor at high chlorine–ammonia ratios is often solely attributed to nitrogen trichloride (NCl_3). The nitrogen trichloride concentration is low when the Cl_2–NH_3 ratio is less than 3.
4. Chloramines do not affect ozone bactericidal efficiency.

In addition, extensive testing of GAC systems was also undertaken. Three carbon adsorption systems were evaluated with respect to THM and TOC removal: (1) virgin GAC; (2) regenerated GAC; and (3) biological activated carbon (BAC), which consisted of preozonating the influent to virgin GAC. Figure 19-14 shows the pilot ozone generator and GAC columns used for this study. Principal findings from the GAC and BAC pilot tests are as follows:

1. Effluent quality was the same for the three GAC systems. All three systems provided equivalent effluent quality with respect to THM and TOC. The effluent TOC concentration ranged from 0.25 to 0.35 mg/L, well below the EPA's proposed operational criteria of 0.5 mg/L. THM concentrations were generally less than $5\mu g/L$.

2. In-service times were the same for the three GAC systems. The service times, the time in which the system is in service prior to regeneration, were not substantially different for the three systems.

3. The optimal contact time was 20 min. The optimal empty bed contact time (EBCT), the detention time corresponding to the volume of carbon in the adsorption contactor, was approximately 20 min. Figure 19-15 shows the relationship between service time and bed depth for the regenerated carbon column.

4. Ozonation prior to GAC did not significantly increase the carbon adsorptive capacity. The column effluents were monitored for standard plate counts (SPC) related to biological activity within the carbon. The virgin GAC and BAC

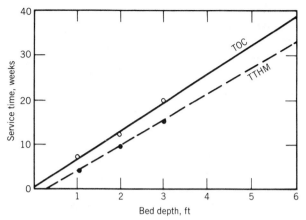

FIGURE 19-15. Bed depth–service time curve for regenerated carbon. (Source: JMM, November 1980.)

systems appeared to have similar degrees of biological growth while the regenerated GAC system showed very little growth; yet the TOC and THM removals were equal. The findings indicate that for the District's water, ozonation prior to activated carbon adsorption does not significantly enhance the capacity of the adsorption system.

PLANT LAYOUT

Once the selection of the treatment processes and plant site is made, the plant layout work is initiated. The initial step in preparing a plant layout is to define the specific process elements and support facilities required. The process design criteria is then developed based on past experience with the water supply, results of prototype testing, and good engineering judgment. Support facilities, such as the operations (or control) building, maintenance yard, and office space requirements, are usually based on the specific requirements of each water agency.

In the preparation of a plant layout, the following items warrant the design engineer's consideration:

1. The condition of the site, including slope of the land, soil conditions, cut-and-fill requirements, drainage, access to the site, and so on.

2. Anticipated traffic movement and operator flow.

3. Climatic considerations—for example, in mild climates the use of outdoor (uncovered) filters is acceptable; however, in very cold climates,

FIGURE 19-14. Pilot ozone generator and GAC columns. (Source: JMM, November 1980.)

filters must be covered to prevent problems with freezing.

4. Proximity of the operations and control building to major process units—various process units require more attention than others (in particular, chemical systems, filters, and flocculation basins require greater operator attention due to their higher degree of mechanical complexity). Therefore, locating these facilities next to the operations building helps ensure proper operator attention and centralization of plant control functions.

5. Integration of process elements to meet treated water quality objectives.

6. Difficulties in construction, especially from the viewpoint of the contractor, should be assessed and minimized.

7. The overall plant design should be as simple as possible. This includes a simple arrangement of treatment units and the minimal number of treatment units sufficient to provide adequate standby capability. For example, standby chlorination feed facilities are required since they are necessary to provide adequate reliability for disinfection. However, it would not be required to provide a standby rapid-mix system since failure of the rapid-mix facilities does not significantly affect overall plant operation and efficiencies.

8. Hydraulic balance in each treatment unit; the balancing of flows between the various treatment units without the need for throttling valves is very important; proper layout of the units will accomplish this goal.

9. Physical separation of certain structures is desirable to provide a certain degree of separation between massive concrete structures of different facilities such as the operations building, basins, and filters. The physical separation enables elimination of structural problems due to uneven subsidence of the structures, easy access for equipment installation and, in general, easier construction.

10. Separation of sanitary and sludge-handling facilities from the final treatment processes, that is, elimination of any potential hazards of cross-connections.

11. Easy discharge of the plant emergency overflow to a suitable watercourse and considerations to minimize damage inside and outside of the plant in case of a massive chemical spill.

12. Adequate space in the plant to deposit filter media, pipes, valves, and other materials.

13. Aesthetically pleasing structures and landscaping of the plant site.

14. Considerations for future expansion and process additions; included in this consideration are projections of future treatment capacity due to projected future water demands and potential addition of supplemental treatment requirements due to changes in regulations. In addition, emergence of new or improved technology in the water treatment field may impact future plant addition.

15. The length of chemical lines should be minimized to prevent clogging and operational problem; this requirement can be achieved by locating the chemical storage and feed equipment in close proximity to the point of application.

16. Simplified chemical delivery and handling facilities: Repeated manual rehandling should be avoided. Proper plant layout minimizes chemical handling requirements; in addition, turning radius and proper access should be considered in the routing of chemical delivery trucks.

17. Operator input: Discussions with plant operating staff is useful in determining operational preferences and requirements in the layout and design of the operation and maintenance facilities. For example, if a water agency has an existing main corporation yard, on-site maintenance facilities can be reduced to a minimum.

18. Orientation of the operations building: With the increased utilization of solar heating systems in residential, commercial, and industrial facilities, the proper orientation of the operations buildup will assure that maximum potential for use of solar heating is realized.

19. Isolation of equipment: Consideration should be given to isolate equipment that presents safety, noise, or heat problems. For example, noisy air compressors can be positioned or isolated at the plant so as not to disrupt other operational functions.

20. Provisions for bypass facilities: Plant layout should provide sufficient bypass both within and around treatment units so that operational flexibility is achieved. For example, where seasonal variations in water quality are experienced, it may be possible to operate a conven-

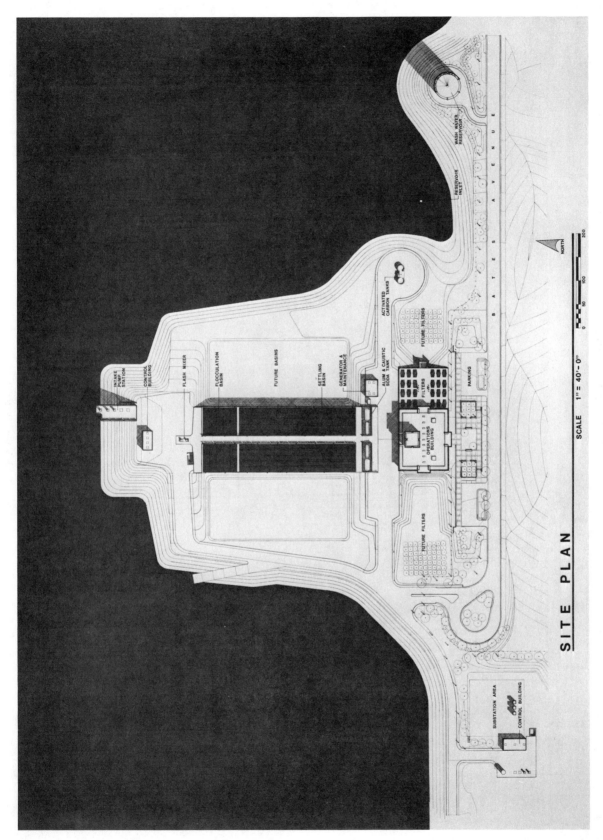

SITE PLAN

SCALE 1" = 40'-0"

FIGURE 19-16. Typical plant layout. Ralph B. Bollman Water Treatment Plant.

467

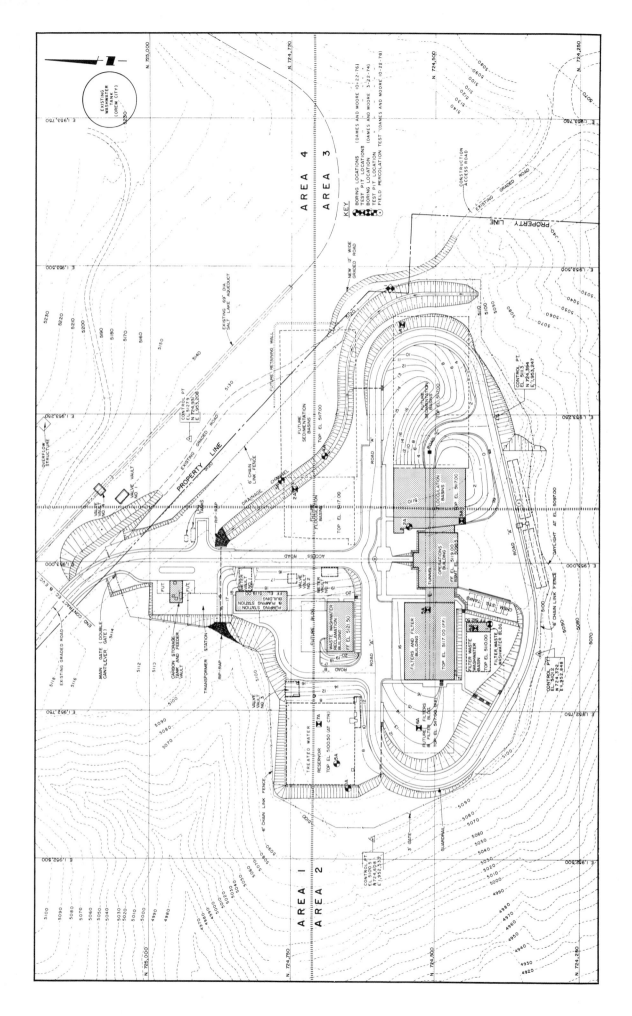

FIGURE 19-17. Typical plant layout, Utah Valley Water Purification Plant.

tional treatment facility in the direct, or in-line, filtration mode during periods of good water quality simply by diverting flow around the sedimentation and/or flocculation basins. In addition, provisions for bypassing the facilities allows for shutdown for maintenance purposes.

Most water treatment plants are publicly owned and are often visited by school children, public groups, and engineers from other cities, states, or countries. In addition, the siting of the treatment plant in a particular area may meet resistance from local citizens groups or other factions. The plant, therefore, should not only be functional and safe but should also blend into the surrounding environment and be visually pleasing when viewed from surrounding neighborhoods. Experience indicates that aesthetics of the plant may be adequately provided for if 2–3 percent of the total design budget is allotted to architectural design and landscaping.

Examples of two typical plant layouts are presented in Figures 19-16 and 19-17. Figure 19-16 depicts the layout for the Ralph B. Bollman Water Treatment Plant of the Contra Costa County Water District. This plant is a conventional treatment plant with an initial capacity of 300 ML/d (80 mgd) expandable to 900 ML/d (240 mgd). The layout provides for a centralized operations building and adequate traffic circulation to each major process element. Expansion of the facility is easily facilitated since the layout provides ample space for future units with minimal impact to the operation and accessibility of existing units. Figure 19-17 shows a plant layout of the Utah Valley Water Purification Plant, owned and operated by the Central Utah Water Conservancy District. This plant is a 190-ML/d (50-mgd) direct filtration facility with an ultimate capacity of 380 ML/d (100 mgd). Due to the severe winter weather experienced in Utah, the major treatment units are housed. The core of the plant is the operation building, which is centrally positioned. Adequate space is provided on the site for future installation of sedimentation basins, if needed, due to regulatory requirement or a drastic change in the water quality of present or future sources of supply.

GENERAL DESIGN CRITERIA

As part of the predesign work, the general design criteria for each design discipline needs to be estab-lished for use in the detailed design phase of the project. By providing this "up-front" effort, the basis for the detailed design is defined, thereby making the latter portions of the design effort operate more efficiently. Input from the owner can be requested at this time, which can include integrating the design criteria with the owner's standard method of approach (standard specifications and details). In addition, the owner may have specific knowledge of potential problem areas which can be communicated to the design engineer. Two specific preliminary investigations are required, which generate information used to develop site-specific requirements in the design criteria, namely, the soils report and the surveying investigations. Both investigations should be initiated as soon as the plant site has been selected.

Civil Design Criteria

The civil design of a proposed water treatment project involves site work based on the specific soil characteristics of the selected site. The soil conditions are investigated by a geotechnical engineer and the recommendations summarized in a design soils report. The soils report presents specific design criteria for the various site work components required.

Design criteria considerations for the various site work components are outlined below.

1. ROADS. Traffic loading should be established based on the type of traffic anticipated. Generally, design criteria for roadways is based on withstanding the maximum weight of vehicle traffic, which in the case of a treatment plant is chemical supply trucks. Other requirements to be defined include minimum width requirements for access and service roads and minimum turning radius. Access roads 20 ft wide and service roads 16 ft wide have proven adequate in most treatment plants. The minimum turning radius requirement is based on the maximum length of vehicle anticipated at the plant, which again is the chemical supply truck.

2. SIDEWALKS. Minimum width and design loads should be established, including reinforcement requirements, if needed.

3. PARKING. Provisions for adequate parking space should be included to handle both employees and visitors. Minimum requirements should be established based on staffing projections and anticipated volume of visitors.

4. LANDSCAPE AND AESTHETIC IMPROVE-
MENTS. Careful attention must be given to the
treatment plant landscaping. Criteria useful for
treatment facilities include the following:

- Utilize evergreen or native plantings to the
  greatest extent possible to reduce leaf fall and
  operator attention.
- Provide low-maintenance greenbelt areas wherever necessary or practical.
- Provide automatic sprinkler control systems for
  the irrigation system.
- Improve entry aesthetics by landscaping and
  plant title signs.
- Plant soil stability landscaping against any cut-
  and-fill embankments.
- Landscape with seasonal color variety using native trees.
- Avoid planting trees next to open treatment
  units to reduce leaf fall into units.

5. EARTHWORK. The civil design criteria should
include earthwork requirement (excavation, backfill, slopes, etc.). This information is site specific
and is usually defined in the soils report.

6. PLANT DRAINAGE AND FLOOD PROTECTION.
If the plant site is located adjacent to a major watercourse or in a flood plain, a hydrologic evaluation is
required to establish anticipated flood levels. Flood
levels for 50–100-yr storms are commonly used to
design flood protection systems. Plant drainage patterns should also be evaluated.

Structural Design Criteria

The development of the structural design criteria is
based on several codes and preliminary investigations. This includes, but is not limited to:

1. Local building codes.
2. Safety regulations, for example, OSHA requirements.
3. Soils report, including seismic potential.

The structural construction materials used for the
construction of the facility are based on both
strength requirements and the overall architectural
theme established for the facility. For each specific
type of material, appropriate design guides and criteria must be established. Materials typically used
are concrete for hydraulic and building structures,
structural and miscellaneous steel, aluminum, tim-
ber, and plastic or fiberglass.

In addition, specific structural requirements for
seismic intensity, freezing–permafrost levels, and
groundwater must be considered in developing the
structural design criteria.

Electrical Design Criteria

The development of the electrical design criteria is
based on the electrical load requirements for the
water treatment plant. In general, the major power
requirements at a water treatment plant are for
backwash pumps and treated water (high-service)
pumps. Local power companies need to be contacted at an early date to establish the type of service available at the site. Review of the power utility's energy pricing schedule and rates for peak and
off-peak usage may result in the use of design and/
or operational features that will reduce costly,
peak-period surcharges. In addition, phasing, voltage, and step-down requirements can be established
based on knowledge of the load requirements and
supply availability. It is preferable to have more
than one electric power service entering the plant.
If this is not practical, provisions for standby power
generation should be investigated. Whatever the
source of standby power, providing an automatic
transfer during power interruptions to all vital loads
necessary for normal plant operations is advisable.
In addition, in the development of the electrical design criteria, methods of power distribution should
be established. Generally, power distribution at a
plant site from structure to structure is underground. Inside structures, installation can either be
concealed or surface mounted. The number of applicable buildings and industry standards used include NEMA, IEEE, ANSI, and UL.

Instrumentation Design Criteria

Prior to any detailed design of the instrumentation
system, the basic control philosophy must be established. Discussions with the owner are necessary to
establish this philosophy and the corresponding design criteria. The information that needs to be established includes:

1. Period of time the plant will be manned.
2. Degree of automation required.

3. Degree of process control and monitoring required.

4. Type of signal to be transmitted.

5. Determination of local versus remote control.

6. Need for a graphic display panel.

7. Type, amount, and method of transmission of alarms.

8. Telemetry requirements.

9. Degree of computerization required.

To aid, establish, and organize instrumentation and control requirements, the tabulation form presented in Figure 19-18 is extremely useful. This form should be prepared by the design engineer and reviewed by the owner to verify the extent of instrumentation required. A more detailed discussion of instrumentation and control is presented in Chapter 23.

Mechanical Design Criteria

The establishment of the mechanical design criteria is interrelated with the electrical and the instrumentation design criteria. In the preliminary design phase, the objective of the mechanical design is to adequately define the major mechanical equipment so that adequate planning of the total project can be coordinated. Items that require attention include:

1. Definition of applicable codes and association guidelines and standards (e.g., ASTM).

2. Preliminary schedules of major pieces of mechanical equipment including items that may require prepurchase for fast track projects.

3. Establishment and coordination of equipment power requirement for the total project.

4. Coordination of the instrumentation and process control aspects of the project with the instrumentation and control engineer.

5. Early definition of the mechanical standards and guidelines site specific to the project, including general piping and valve schedules, material selection, and cathodic protection systems.

6. Standardization to existing equipment to reduce owner's parts inventory and maintenance instructions.

7. Assessment of any owner bias to equipment and method of operation.

To assist in the organized completion of the mechanical predesign effort, a mechanical equipment data sheet is useful. An example of such a data sheet is shown in Figure 19-19. The data sheet contains information that is used by other design disciplines (electrical, mechanical, structural) in the coordination of the detailed design.

Architectural Design Criteria

The architectural theme for the facilities should be developed to blend with the surrounding area and communities. It should be simple, economical, and in character with the environment. The importance for the architectural treatment of the facility cannot be overemphasized, since an aesthetically pleasing facility is a valuable public relations tool for the water agency. Additionally, the architectural theme will dictate the selection of the types of building materials and systems to be used. Minimizing maintenance requirements should be a goal in selecting the architectural theme.

Safety Considerations

Numerous safety standards influence the design and construction of the proposed treatment facility. OSHA and local codes should be referred to. Listed below are safety requirements that are most frequently encountered:

1. Aisles and passageways should be designed with a minimum width of 0.6 m (24 in.) and a minimum height of 2.0 m (6 ft, 8 in.).

2. Electrical installations should adhere to NEMA codes according to the voltage of the equipment.

3. Emergency eye and body flushing should be provided in areas where any hazardous and dangerous materials are handled.

4. At least two exits should be provided in areas designated by OSHA standards (e.g., chlorine feed rooms).

5. Ladders require safety devices, such as a cage or safety straps, if the length is over 6.0 m (20 feet).

6. Railings should be 106–114 cm (42–45 in.,) high to the top surface with midrail; at required exits, spacing rails should be 23 cm (9 in.), pits or depressions 76 cm (30 in.) above ground required guard railing.

REF. NO.	DESCRIPTION OF MEASUREMENT AND/OR CONTROL OPERATION (PURPOSE)	FIELD AND MCC			LOCAL CONTROL BOARD	MAIN CONTROL BOARD	COMMENTS
		MEAS'MT. RANGE	PRIMARY SENSOR	INSTRUMENT FUNCTION(S)	INSTRUMENT FUNCTION(S)	INSTRUMENT FUNCTION(S)	

JAMES M. MONTGOMERY
CONSULTING ENGINEERS, INC.

FIGURE 19-18. Instrumentation and control worksheet.

JOB NAME			JOB NO.		BY	DATE	

NAME OF EQUIPMENT							REVISIONS
TYPE OF EQUIPMENT							
LOCATION							
SERVICE							
SIZE	NO. OF UNITS		WEIGHT	IDENTIFICATION NO.			
SIZE OF CONNECTIONS							
MOTOR H.P.	ENCLOSURE		RPM	VOLTS	PHASES	AMPERES	
EQUIPMENT CAPACITY							
MATERIAL OR CONSTRUCTION							
EQUIPMENT CONTROL							
ACCESSORIES							

REMARKS	FURNISHED W/EQUIPMENT
	DISCONNECT SWITCH -
	STARTER--
	CONTROL -
	PRICE NET
	LIST
	$ APPROX

MANUFACTURER, MODEL NO., ETC.

EN-25 (9/73)

FIGURE 19-19. Mechanical equipment selection worksheet.

7. Stairs are required in areas where a 30–50 percent slope is found. Industrial stairs should have a useful width of 56 cm (22 in.). Exit stairs should have a width of 90 cm (36 in.) with a minimum rise–run ratio of 19–25 cm (7.5–10 in.).

8. Stop plates should be provided on all platforms 1.8 m (6 ft) or higher above working surfaces.

9. Adequate ventilation is to be provided in trenches, tunnels, pits, and other confined areas. Also, in storage areas of hazardous materials, the ventilation system shall be designed to minimize any harmful atmospheric concentration.

10. Warning signs are to be provided in areas where any dangerous or harmful materials are encountered and in areas where machinery stops and starts automatically.

Plant Utilities

Several utility systems are required in any water treatment plant. They include water, telephone, intercom, heating, ventilation and air conditioning (HVAC), sanitary wastes, chemical wastes, and electric power. In the predesign investigations, requirements for each of these systems need to be established. In addition, certain precautions must be exercised in the design of the plant utility system. For example, backflow prevention devices need to be provided on the portable water system to eliminate any potential cross-connections, such as in the injection water line for the chemical addition system. In-plant and outside telephone communication systems should be discussed with the owner to establish his basic communication requirements prior to start of detailed design. For sanitary wastes generated at the facility, either an on-site system or a tie-in to a local sanitary sewer should be provided. If the treatment plant site is located a considerable distance from existing sanitary sewer lines, judicious location of septic tank systems is necessary to prevent potential contamination of treated water.

PROCESS DESIGN CRITERIA

The development of detailed process design criteria is presented at length in Chapter 21. The objective of the predesign effort is to develop a treatment system that integrates the criteria for individual process components into an overall plant process design criteria that meets treated water quality requirements. The development of this criteria is based on experience with the specific water supply, past engineering experience, or, if this information is not available, through bench and pilot testing.

GENERAL DESIGN CONSIDERATIONS

In addition to the information presented above, the following general design considerations and guiding principles need to be evaluated during the predesign stages of the treatment plant project:

1. The potential for energy conservation and/or recovery should be evaluated. Specifically, use of solar heating in the operations building warrants consideration. Recent experience indicates that a payback period of 2–3 yr is possible. In addition, an evaluation of the total hydraulic head available both upstream and downstream of the treatment plant may reveal that there is potential for hydroelectric power generation.

2. The facilities should be designed for economic as well as durable construction. Design lives of up to 100 yr are not unusual.

3. Each treatment process should be balanced with the others to provide for effective and reliable production of high-quality water.

4. Strive for a structurally safe, durable, but economic design for construction.

5. Provide for extra hydraulic capacity in channels and pipes to accommodate future plant expansion or modification and peak hydraulic conditions. Beyond the initial design capacity, 50 percent excess hydraulic capacity is frequently desirable.

6. Provide for uniform distribution of plant flow by hydraulically balanced design and layout.

7. Simplicity in all aspects of plant process design and equipment is desirable.

8. Provide access to all process units, equipment, and valving, including adequate working area for equipment maintenance. Also, provide for centralized process operation and control.

9. Make provisions for an alternate source of raw water supply and power in order to secure an uninterrupted supply of good quality water.

10. Use reliable and simple process controls. Sophisticated systems should be avoided unless they are explicitly justified and within the technical limitations of operational staff.

11. Adopt well-established water treatment processes or techniques, but do not disregard recent developments in treatment alternatives and technology.

12. Provide only that degree of automation justified. Match the degree of automation and level of sophistication to the operation/maintenance personnel's capabilities and commitment to automation. Excess automation not only requires frequent maintenance and adjustment but also, in case of a failure of automatic controls and equipment, can result in a temporary shutdown of the treatment processes and water supply. In addition, if process automation is overcomplicated, operators become discouraged and do not maintain the equipment.

13. Minimize the use of imported equipment or materials.

14. Design for safety and operations and provide safety measures for plant emergency overflow and accidental massive chemical spills.

15. Consider access to the proposed site by various transportation means.

16. Assess available utilities and the ability to obtain technical service from the manufacturers of major equipment and chemical suppliers.

17. Locate bearings and other items that require maintenance and frequent lubrication above the water surface.

18. Provide sufficient flexibility in the operation of the facilities.

19. Provide drainage and flushing of each process unit.

20. Slope all floors to drains.

21. Provide the necessary access for removal of all major equipment.

PROJECT MANAGEMENT AND PLANNING

Project management and planning is one of the most important duties of the design engineer in accomplishing the project tasks efficiently. In recent years, program evaluation and review techniques (PERT) and critical path methods (CPM) have been developed and widely used in many areas. PERT is a statistical treatment of the uncertainties in the activities performance time and it defines the activities to meet specified schedules in various stages in the project. PERT also emphasizes the control phase of project management. The objective of CPM is to determine the duration of the project that minimizes both direct and indirect costs. Before these methods were developed, one of the most common methods to aid in project management was the use of bar charts. The bar chart method, however, is sometimes inadequate for the complex interrelationship associated with contemporary project management.

In predesign studies, the design engineer must consider the manpower, equipment, and facilities required to carry out the design program. The design engineer must devise a plan such that the cost and time required to complete the project are properly balanced and excessive demands on the available work force and facilities can be avoided. Both PERT and CPN, therefore, can prove to be very effective in providing for efficient project management and planning.

COST ESTIMATION

The importance of accurate cost estimation during the predesign stage is necessary to properly plan the financing of the project. Chapter 25 deals exclusively with the various cost-estimating techniques available to provide the design engineer with quick, preliminary construction and operation and maintenance cost estimates for the major items involved in a water treatment plant design. Development of cost estimates for specific water treatment projects, however, requires a good knowledge of the site-specific conditions.

FINANCIAL STUDIES

In addition to the engineering and technical aspects of any major water treatment project, the methods of financing the project need to be assessed to determine overall project feasibility. There are a variety of tools available to municipalities for the purpose of financing capital improvements. Financing options generally include pay-as-you-go funding and/ or several different methods of debt financing. Federal and state grants or loans may also be available from time to time for projects that meet specific criteria. Recently, privitization of water treatment facilities has gained some popularity as a viable financial mechanism.

Pay as You Go

Pay-as-you-go financing requires the accumulation of funding prior to the start of construction. Large projects are difficult to finance in this manner because of the time required to accumulate the funds as well as general public resistance to the accumulation of large reserves. Revenue sources for pay-as-you-go financing are generally obtained from a combination of user charges and fees levied on new development. The obvious advantage of this type of financing mechanism is the elimination of an interest cost component. However, when long-term borrowing rates are lower than the inflation rate of construction costs, there is little point in postponing necessary facilities by waiting to accumulate the funds. In such circumstances, debt financing with repayment secured from the same revenue sources represents the more logical solution. On the other hand, pay-as-you-go fund accumulation is particularly appropriate for the construction of replacement facilities.

Revenue Bonds

The issuance of revenue bonds is probably the most widely used method of financing large capital projects. Revenue bonds are often secured solely by a pledge of the revenues derived from the facility they are used to acquire, construct, or improve. However, revenue bonds issued by a nonprofit corporation and secured by a lease-purchase agreement do not require voter approval but are secured by a pledge of revenues derived from the financed facilities.

There is no legal limitation on the amount of authorized revenue bonds that may be issued, but from a practical standpoint, the size of the issue must be limited to an amount where annual interest and principal payments are well within the revenues available for bond service. The interest rate will depend on the degree of security provided and the status of the bond market.

One measure of revenue bond security is the "coverage" provided. Coverage is the ratio of net revenue to the annual bond service requirements. For revenue bonds to be salable, the issuer should pledge to maintain net revenue at least 1.25 times annual bond service. The marketability of the bonds will be enhanced if it can be shown that the actual coverage provided by the net revenues will exceed the pledged ratio.

A distinction must be drawn between the pledge to maintain excess revenues (coverage) for bond service and the use of such revenues for that purpose. Revenues pledged to the payment of bonds, but not needed to meet bond service, may be used for any lawful purpose. Frequently, these extra revenues are used for replacement and expansion of the existing facilities.

In addition, revenue bond buyers often demand further safeguards by the establishment of a reserve fund equal to the average or maximum annual bond service. This reserve is normally created from the proceeds of the bond sale. The reserve is maintained for the entire life of the bond issue to meet the annual principal and interest requirements in the event operating revenues are insufficient for bond service in any given year.

Federal and State Financial Assistance Programs

In the past, an extensive array of federal and state programs, including loans, grants, and other financial assistance for water and related resource development, have been available for specific purposes. Eligibility for financial assistance usually requires the municipality to demonstrate:

- Inability to obtain conventional financing.
- Economically depressed area.
- Substandard existing public service.
- Need for proposed facility.

The availability of federal or state funding is subject to many influences including current economic trends and legislative appropriations. Municipalities should always keep their options open. However, dependence on this type of funding in long-term financial planning is not recommended.

In summary, there is no single best approach to financing large capital improvement projects. The nature of the project, the beneficiaries, the municipal securities market, the municipality's debt rating, individual state legal constraints, and local policy considerations all influence financing decisions.

ENVIRONMENTAL ASSESSMENT

In conjunction with the evaluation of technical and economic factors affecting the project-planning and decision-making efforts, the impact of the project on the environment must be assessed. To this end, the National Environmental Policy Act (NEPA) of 1969 mandated that projects should be assessed for their overall impact to the community to ensure that balanced decision making occurred for the total public interest. Therefore, proper project planning should integrate considerations for the technical, economic, environmental, social, and other factors.

In the planning of a water treatment project, the regulations contained in the NEPA require that an environmental assessment of the project be made. Certain guidelines were established by this act. Additional guidelines were also established by state and local regulatory agencies. A thorough review of federal, state, and local regulations is required to adequately assess the environmental impact of the project. Therefore, a familiarity with the requirements of these regulations is essential.

The basic procedure for complying with the NEPA requirements includes initially the preparation of an environmental inventory. This includes a thorough description of the physical, biological, and cultural environment that is affected by the project. Once the environmental inventory has been established, the impact of the project is assessed. The assessment includes:

1. A prediction of the anticipated change in the environmental inventory.
2. Determination of a magnitude or scale of the particular change.
3. Advocations of the importance of the significant factors to the change.

Once the environmental assessment is made, the environmental impact statement is prepared, which summarizes the environmental inventory and the findings of the environmental assessment. The basic content of an environmental impact statement includes:

1. The environmental impact of the proposed project.

2. Any adverse environmental impact that cannot be avoided should the project be implemented.
3. Alternatives to the proposed project.
4. Relationship between local short-term uses of the human environment and the maintenance and enhancement of long-term productivity.
5. Any irreversible or irretrievable commitments of resources that would be involved in the proposed project should it be implemented.

Once the environmental assessment document is prepared, it is circulated for review and comment to federal, state, and local agencies and public and private interest groups. The final statement is a draft statement modified to include a discussion of problems and objections raised by the reviewers.

REFERENCES

Argaman, Y., and Kaufman, W., "Turbulence and Flocculation," *J. San. Eng. Div. ASCE*, **96**, 223 (1970).

ASCE, AWWA, and CSSE, *Water Treatment Plant Design*, AWWA, Inc., New York (1971).

Camp, T. R., "Floc Volume Concentration," *JAWWA*, **60**, 656–673 (1968).

Canter, L. W., *Environmental Impact Assessment*, 1st ed., McGraw-Hill, New York (1977).

Clark, J. W., Viessman, W., Jr., and Hammer, M. J., *Water Supply and Pollution Control*, 2nd ed., International Textbook Co., Scranton, Penn. (1971).

Fair, G. M., Geyer, J. C., and Okum, D. A., *Elements of Water Supply and Wastewater Disposal*, 2nd ed., Wiley, New York, 1971).

Hudson, H. E. *Water Clarification Processes Practical Design and Evaluation*, Van Nostrand Reinhold, New York (1981).

James M. Montgomery (JMM), Consulting Engineers, Inc., *Escondido Vista Joint Water Treatment Plant Expansion Design Report*, City of Escondido-Vista Irrigation District, California (April 1981).

JMM, *Pilot Filtration Studies at the Tapia Water Reclamation Facility*, Las Virgenes Municipal Water District and Triunfo County Sanitation District (September 1979).

JMM, *Report on Predesign Studies for the Proposed Quincy Water Treatment Plant; Vol. 1, Pilot Scale Treatment Study*, City of Aurora, Colorado (February 1980).

JMM, *Report on Studies With Pilot Plant System*, Casitas Municipal Water District (June 1977).

JMM, *Val Vista Water Treatment Plant Expansion Project Design Report*, City of Phoenix, Arizona (April 1980).

JMM, *Water Treatment Study, Vol. II*, Contra Costa County Water District (November 1980).

JMM/DMJM, Los Angeles Aqueduct Water Quality Improvement Program, *Water Quality Studies*, Vols. 2 and 4, Los Angeles Department of Water and Power (June 1977).

Kawamura, S., "Hydraulic Scale-Model Simulation of the Sedimentation Process," *J AWWA,* **73** (7), 372–378 (July 1981).

Lang, J., "Selection of Pilot Plant Equipment," Paper presented at Annual AWWA Conference, Miami, Florida (1982).

Letterman, R., Quon, J., and Gemmell, R., "Influence of Rapid Mix Parameters on Flocculation," *JAWWA,* **65,** 716 (1973).

Medcalf & Eddy, Inc., *Wastewater Engineering: Treatment Disposal Reuse,* 2nd Ed., McGraw-Hill, New York, (1979).

Sanks, R. L., *Water Treatment Plant Design for the Practicing Engineer,* Ann Arbor Science, Ann Arbor, MI (1980).

Stenquist, R., and Kaufman, W., *Initial Mixing in Coagulation Processes,* University of California, Berkeley, S.E.R.L. Report No. 72-2 (1972).

Tate, C. H., and Trussell, R. R., "Pilot Plant Justification," presented at 24th Annual Public Water Supply Engineers' Conference, April 27–29 (1982).

— 20 —

Plant Siting

Selection of a site for a water treatment plant invariably involves compromise and "trade-offs" of selection factors. There is never an "ideal" plant site. For this reason, an evaluation of all factors affecting the specific location of a water treatment plant is essential prior to undertaking aquisition proceedings on a specific piece of land or beginning design.

A number of the more important factors affecting site selection will be listed and some guidelines will be presented to assist in comparing alternative sites.

The principal factors to consider when looking for a plant site are:

1. *Planning and environmental constraints,* including present and future water demands, general environmental sensitivity, direction and rate of growth of the service area, quality of selected source water, zoning, and so on.

2. *Selected source,* that is, river, wells, lake, aqueduct, and so on, as it may affect selection of a specific site both from an elevation and geographic site selection standpoint. Quality of water from the selected source will also impact on the treatment process, which may in turn affect the site selection insofar as area requirements for chemicals, sludge disposal, and so on, are affected.

3. *Plant design factors,* including size, process selection, hydraulics, solids handling and disposal, and requirements for plant site storage.

4. *Site factors* relate to a specific site or group of sites being compared, including location with respect to service area, relation to system hydraulic grade line, required raw and finished water transmission piping, access, connections, and physical factors such as flooding, foundation conditions, cost of land, required excavation, and so on.

5. *Environmental factors* relate to a proposed specific site and include problems during construction and operations and cost of proposed mitigation measures, including architectural treatment, soundproofing, landscaping, and so on.

PLANNING AND ENVIRONMENTAL CONSTRAINTS

The need for a water treatment plant siting study normally develops as a result of external events unrelated directly to water supply. The more common events resulting in need for additional water treatment plant construction are:

479

- Population growth.
- Industrial water needs growth.
- Deterioration of source water quality.
- Increased quality standards.
- Deterioration or obsolescence of existing plants.

In each of these cases, a well-run water supply agency will have instituted an ongoing planning process well in advance of the actual need for the specific site study. This planning process normally will identify the direction and rate of growth, the population increase at specific planning milestones (10 and 20 yr are typical) and any of the other events listed above that have an effect on water and other utilities. Typically, this planning document or documents (commonly called the "general plan") will include the broad general framework within which the detailed planning for water, wastewater, drainage, power, roads, and other infrastructure must fit.

A common offshoot of the broad general plan is the water system master plan. This document is more detailed and is usually based on the growth factors listed above, plus consideration of hydraulics, fire flow, peaking, costs, and construction problems in general. Normally, the master plan identifies basic "skeleton" facilities but does not pick specific sites for any facilities.

The engineer with the responsibility for preparing a siting study should be keenly aware of these background documents because they will provide the basic background data for his work. Additionally, if the particular areas being considered for a plant have been zoned, the engineer must make sure that the zoning permits a plant or that the plant can be designed to be compatible with the zoning requirements.

In the absence of a formal planning document, the engineer responsible for a siting report must use his best professional judgment in picking a plant site after examining all constraints. This process should include discussions with representatives of the principal jurisdictions involved, including cities and counties, special districts, environmental groups, if appropriate, and the local and/or state health department. It is often possible to avoid misunderstandings or actual opposition to a plant-siting project if each of the principal groups who may be affected are brought into the process at an early date.

SOURCE FACTORS

Type and location of the raw water source have an important bearing on plant location and in some cases are the fundamental criteria.

Wells

Wells are normally the most flexible of sources. Often a well can be located right on the plant site provided that the aquifer is suitable hydrogeologically and has acceptable quality. This gives great latitude in the location of the plant. Other wells needed to satisfy total plant capacity can be located at a distance from the plant, as determined primarily by hydrogeologic factors, and then piped to the plant site. Selection of a site for the plant itself can be relatively independent of hydraulic considerations, since normally the plant occupies an elevation intermediate between the source (groundwater) and the distribution system hydraulic grade line. Double pumping is usually necessary in this case. An analysis of the comparative costs of pipelines, land, washwater and sludge disposal, and energy requirements for various candidate sites is usually sufficient to establish the most economical site.

Rivers

Rivers as a source often produce difficult problems in site selection. Selection of a plant site should be made simultaneously with selection of the point of diversion from the river. In many cases, the apparent most desirable point of diversion, be it from a river bank intake, an outlet structure in a dam, a submerged crib, or other diversion point, will be impossible or undesirable on closer inspection. For example, foundation conditions may be undesirable; the particular location may be exceptionally polluted, either from an upstream sewage plant or an eddy or from a poor-quality side stream; the intake structure may affect shipping lanes; the site may require cutting a levee; it may be subject to flooding; or there may be severe ice, sedimentation, or scouring problems. If any of these or similar problems occur, selection of the treatment plant site should not be completed until the intake problems are resolved. The best procedure is to analyze both intake and plant location at the same time.

Once the intake problem is resolved, there is a tendency to locate the plant as close to the river as possible. This is normally desirable to minimize length of raw water pipelines and length of discharge lines for backwash water and sludge if discharge to the river is acceptable. In addition, it is fairly common to have considerable level space adjacent to a river due to stream "meandering." This makes an attractive topographic site for a plant.

However, an economic evaluation of alternative sites, as shown in Figure 20-1, may indicate that construction further away from the river may be less costly. Some of the reasons to consider this are:

1. *Flooding.* Riverside structures may require special design considerations to cope with potential flood conditions.
2. *Groundwater.* A site near a river normally has high groundwater, which involves extra costs for design and construction. After construction, leakage problems from joints in deep structures may be a future annoyance.
3. *Poor foundation material.* A river bank area is frequently either swampy or fine-grained muddy subgrade. Design of foundations therefore may involve special types of footings, or pilings. Special precautions to prevent differential settlement of structures may also be necessary.
4. *Disposal of washwater and sludge.* Recent federal regulations requiring the elimination of discharge from water treatment plants into existing streams have also eliminated one of the principal reasons for locating plants near a river. Most plants now will be required to recycle or otherwise treat the waste streams from the plant internally. This means that on-site facilities for this purpose must be built on the plant site.

In the preliminary siting analysis, when rivers are utilized as sources, some effort should be made to investigate sites away from the river, at higher elevations, and with better foundation conditions. It may be found that a remote site located on the system grade line is most advantageous for the treatment plant.

Dams or Diversions

Dams or diversions as a treatment plant source result in their own set of siting challenges, primarily related to hydraulics. If the service area is located at a *lower* elevation than the minimum water level at the diversion, then a principal concern is the hydraulic profile. The plant site should be selected so that gravity flow through the plant can be achieved if at all possible. The design should be capable of peak future plant flow at minimum diversion level. In a preliminary siting evaluation, an allowance of about 15–20 ft should be provided for losses through a conventional water treatment plant, as shown in Figure 20-2, plus 20–40 ft (depending on the type) for the plant filtered water reservoir, if one is provided for system storage.

To avoid waste of energy, the plant site should be carefully selected so that plant structures can be built to provide a plant hydraulic grade line with neither excessive losses nor need for pumping. If the service area is *above* the dam or diversion source, pumping will obviously be required and the actual elevation of the plant is of lesser importance. However, in this case, topography permitting, con-

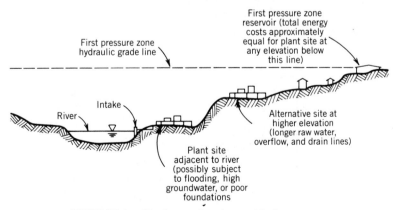

FIGURE 20-1. Site factors for plants with river sources.

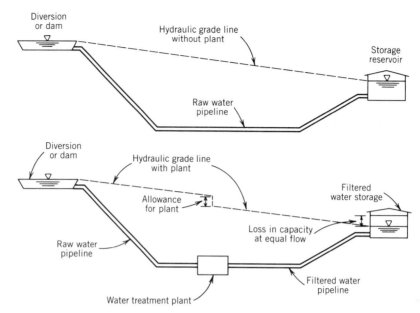

FIGURE 20-2.   Hydraulic considerations for supply from dam or diversion.

sideration should be given to locating the plant at an elevation from which one or more pressure zones in the service area can be served directly. If a filtered water reservoir is to be located at the plant site, double pumping can be minimized if service to one of the customer zones can be done directly from the filtered water reservoir.

Pressure Aqueducts

Pressure aqueducts as a source also require careful hydraulic analysis prior to selection of a plant site. If the plant in question is not the only recipient of

water from the aqueduct, the hydraulic grade line in the aqueduct may be beyond the control of the operating agency. In this case, it is important to locate the plant in such a way that maximum design plant flow can be attained regardless of the rate of flow in the main aqueduct. Aqueducts that flow under pressure from open control structure to open control structure have their highest hydraulic grade lines at highest flow. The plant must be sited low enough, as shown in Figure 20-3, so that it can take maximum flow when the flow through the aqueduct is at a minimum. If double pumping with its attendant control problems is to be avoided, the hydraulics

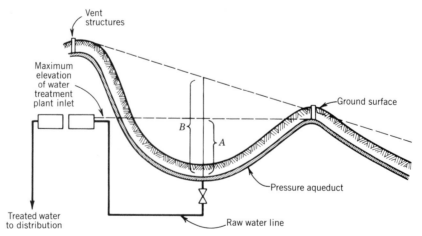

FIGURE 20-3.   Hydraulic considerations for plants supplied from pressure aqueducts: $A =$ available head at minimum aqueduct flow; $B =$ available head at maximum aqueduct flow.

should be carefully checked from the source to the distribution areas, including losses in the plant itself.

In siting a plant with a pressure aqueduct as supply, it is also important to establish any limitations in rates of withdrawal from the aqueduct. If there are limits on daily, weekly, or annual withdrawals, the site must take this into account through provision of appropriate storage or coordination with alternative sources.

One other consideration is possible generation of power from the flow of water into a plant from an aqueduct. In the past, it was seldom possible to justify hydrogeneration; however, recently, the dramatic increase in energy costs have made it feasible in some cases. It is not usually possible to utilize the generated power directly in the plant because of the obvious control problems and difficulty of matching load with production. However, many power companies are receptive to the purchase of this type of power directly into their grid. Present (1982) federal legislation in fact requires that such power be purchased at its replacement costs by the local power company.

PLANT DESIGN AND PROCESS FACTORS

One of the most common mistakes in selection of a particular piece of land for a water treatment plant is to ignore the many space needs of a plant. It is essential to carefully think through the construction, operation, and environmental needs, initially and in the future. Failure to include all these needs has resulted in plants with buildings jammed together, awkward to enlarge or modify.

Obviously, it is not always possible to acquire a site of adequate size in the optimum location. Compromises are common and/or understandable. However, these compromises should be intelligently approached, based on facts and analysis of actual detailed needs, prior to any commitment on site acquisition.

Prior to selecting a site for any size plant, preliminary determination must be made on the following items:

Capacity of plant, initial and future.
Presedimentation basins needed.
Dry or wet chemical storage, including truck access.
Sedimentation basins, initial and future.

Filtration rate.
Washwater storage or pumping.
Washwater reclamation or disposal.
Sludge disposal.
On-site raw and/or filtered water storage.
Low-lift and/or high-lift pumping stations.

Once these determinations have been made in a preliminary siting report, a rough sizing estimate can be made, including an additional allowance for access roads, setbacks, grading slopes, landscaping, and so on. Figure 20-4 shows typical site requirements for a conventional plant with a substantial amount of excavation. Figure 20-5 shows typical total area requirements for conventional water treatment plants, excluding raw or finished water reservoirs, sludge-drying beds, or sludge-holding ponds.

Problems with cramped sites occur most often with small plants, say under 5 mgd, where the peripheral land and appurtenances required are disproportionately large with respect to the actual treatment plant structures. In large plants, the same

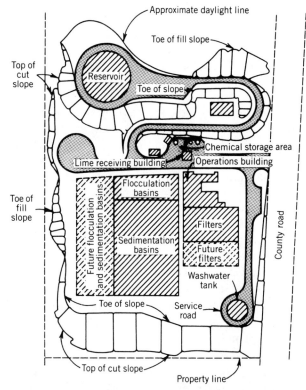

FIGURE 20-4. Land requirements; typical plant with significant cut slopes.

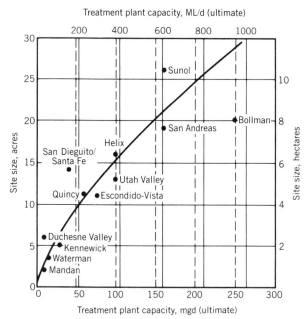

Treatment plant capacity, ML/d (ultimate)

FIGURE 20-5. Typical total area requirements without presedimentation, clearwell reservoirs, sludge-holding ponds, or drying beds.

or larger peripheral requirements occur, and no less care is required in assembling all the land needs into an efficient package. It cannot be emphasized too strongly that, wherever possible, provision of ample land is a fundamental part of the design process.

In the selection of a plant site, the treatment process itself is of primary importance. The physical facilities for making the selected process work (i.e., sedimentation basins, chemical feeding, filters, etc.) are susceptible to straightforward design and arrangement on any approved site. Process factors that may enter the site selection process are not always so obvious and may include such questions as these:

1. If the process includes lime, will local building or environmental restrictions permit a lime silo or tower or must this be underground? How will lime dust be controlled?

2. Are there any restrictions against transporting any of the plant chemicals through the streets to the plant?

3. If the plant process includes filters or gravity separation, will the method of disposal of the washwater or sludge be feasible from the proposed site?

4. Is the plant process likely to produce odors (H_2S removal, sludge disposal, ozone generation), and if so, will it be difficult or expensive to protect the neighboring properties from being affected?

5. Will any of the plant process units be excessively noisy (drive units, ozonators, engine generators, pumps, etc.), and if so, will protection of the plant to meet noise requirements be excessively costly?

Each of these items together with their solution should be considered as part of the preliminary siting report.

SITE FACTORS

Relation to Service Area

In preparing a preliminary siting report for the purpose of acquiring land, a basic consideration is the location of the plant with respect to the service area. Other things being equal, the ideal site would have the following qualities:

1. Close to the service area but far enough away so that raw and treated water lines are not needlessly duplicated.

2. At an elevation high enough so that most of the service area can be served by gravity.

3. Far enough away from the first customer (present and future) so that any postchlorination will have time to become effective.

4. Close to existing main transmission pipelines.

5. Sufficient space for adequate clearwell storage.

It is seldom possible for all of these factors to occur simultaneously. However, it is important to locate a plant with some knowledge of present and future service area limits, basic delivery system, service zone elevations and limits, and future land use, including future areas of peak demand.

Energy conservation is presently a fundamental fact of life, and careful site planning of a water treatment plant to minimize energy costs is required, not only inside the plant but also in delivery of water in the most direct manner from source to point of use. Plant location is only one element of overall energy planning which should (wherever possible) eliminate unnecessary double pumping, take advantage of gravity, and deliver water with minimum friction loss in a direct line from source to area of maximum consumption.

Hydraulic Factors

Additional hydraulic factors not included in the preceding general site discussion include (1) hydraulic grade lines, (2) raw water and finished water pipe sizing, and (3) head loss through the plant.

Hydraulic Grade Line.

A fundamental factor in selecting a plant site is the hydraulic grade line of the raw water supply as it relates to the service area. Since most water treatment plants (except for those with pressurized process units) have an open water surface on settling basins and filters, it follows that these units must be located at such an elevation that energy (i.e., head) is not wasted. If the raw water supply is from a canal, aqueduct, or river and the service area is below the point of diversion, it is worthwhile to carefully study the hydraulics of the entire system from source to service area. Based on that study and comparison with the ground topography, the optimum site for the plant should be selected.

It can take considerable ingenuity to weld all these factors into an acceptable site location; however, the justification in lowered energy costs more than compensates for the time and cost required.

Raw Water and Finished Water Pipe Sizing.

Another hydraulic factor in site selection involves the economic comparison between costs of raw water and filtered water pipelines. Storage is the key here. The farther *downstream* storage reservoirs can be located, the smaller the upstream pipelines can be. The worst case is that of not providing storage where the entire system, including the plant, must be capable of meeting "peak-hour" demand and the pipelines must be sized for this situation. Location of reservoirs at the plant are an improvement. Then the raw water lines and the plant normally would only be sized to meet "maximum day" demands. Similarly, if the reservoirs are located farther downstream, the pipe sizes from the plant to the reservoir need only meet maximum day demands, and those below would have to meet hourly peaks.

There are no firm rules in this analysis, only that all the alternatives should be examined, including the cost of the pipelines under various circumstances.

Head Loss through Plant.

In order to site the various plant units "on the ground" to see if the site is usable, it is necessary to block out the elevation of the water surface of the individual process units. This is a trial-and-error procedure. During design, a careful and precise computation of the plant grade line is necessary. However, during siting, it is possible to use some rules of thumb to see if the site is feasible.

Assuming that the plant inlet hydraulic grade line has been fixed, the water surface of the various units is roughly as follows:

Unit	Elevation of Hydraulic Grade Line	
	m	ft
Plant inlet	0	0
Sedimentation basins or clarifier	−0.3 to −0.6	−1 to −2
Filters	−0.6 to −1.2	−2 to −4
Filtered water reservoir (maximum elevation)	−4.5 to 6.1	−15 to −20

The range in hydraulic grade line at the filtered water reservoir outlet (exclusive of any high-service pumping) varies from a high elevation as indicated above to a low elevation matching the minimum operating level of the reservoir.

By utilizing the above rough figures and deducing from each one the expected depth of the structure, it is possible to establish probable foundation elevations of the units. From these elevations, it is possible to establish a plant site to minimize excavation, avoid rock or groundwater to the maximum extent possible, or identify any expected site problems.

This type of analysis can be done for all types of plants. It can also be done in reverse. That is, if the plant units on a site must comply with certain foundation requirements, this base can be used to establish the necessary incoming hydraulic grade line.

Land Availability

Negotiations for a particular piece of land should not begin until the general requirements for a site are established as outlined above. However, one of the considerations that should be investigated as soon as is practicable is the availability of land in the general area which otherwise seems best. It is usually difficult to find a piece of property at the right elevation and location and of the right size for sale at the time it is needed. In order to make a preliminary survey, contour maps and aerial photographs used together can quickly reveal any potential sites.

Each potential site (and it is important to have more than one) should be visually checked on the ground and evaluated in a preliminary way for technical feasibility prior to contacting the owner or agent. If it is technically feasible to use the site, a determination should be made of the purchase price for inclusion in the economics of the alternatives in the siting report. The appropriate price can be established through an independent appraisal, through a comparison with nearby property sold recently, or by accepting the asking price of the owner if the land is already for sale.

In most localities, public water agencies have the right to acquire property through the courts by "eminent domain" proceedings. Under these proceedings, if it can be demonstrated that it is in the public interest to locate a structure on a certain piece of property, the owner's land is "condemned." Then it is possible to acquire the property by payment of a court determined value.

Most agencies, however, avoid this unless there are no other alternatives. Condemnation frequently results in unfavorable publicity for the agency; in addition, in some locales, it is not permitted to commence construction or even enter the property until the court proceedings have been completed. Thus, if the owner contests the condemnation, lengthy delays can occur prior to construction.

Weather

Weather is a comparatively minor factor to consider during a siting study, although it has a great effect during detailed design. In areas subject to extremely cold winters, it is of course necessary to have the plant accessible at all times to staff and chemical delivery trucks. Access roads must be negotiable during storms or icing conditions and disposal of snow on the access roads must be planned. In some cases, at remote locations, staff housing may be provided for such eventualities. Similarly, in areas subject to heavy rains, drainage must be planned so that the plant is always accessible and the normal plant functions of filter backwash, sludge withdrawal, and chemical delivery can proceed without interruptions due to weather.

Access

In picking a water treatment plant site, it is essential to consider access to the plant.

During construction, it will be necessary to provide access for earth movers, concrete trucks, numbers of workman with their cars, heavy machinery and steel trucks, and many other units related to the construction. In many cases, it is also necessary to provide access for trucks carrying imported select backfill or subgrade material.

After the plant is in operation, access becomes even more critical. The plant is a vital utility, and year-around 24 hours per day accessibility is a must. The operating staff and supervisors will need to get to the plant at the change of each shift, 365 days a year. In addition, chemical trucks hauling chlorine and other plant chemicals must be able to enter the plant, discharge their loads, and leave without difficulty under all weather conditions. This requires limited grades and curves on access roads, particularly in areas subject to freezing, and a careful plan for routing trucks through and out of the plant after discharging the loads. Unless these conditions can be reasonably well assured during the site selection process, another site should be picked.

One point occasionally overlooked in the site selection process is the possibility of disruption of the access road itself during storms even though the plant site itself may be adequately protected. Adequate drainage facilities, culverts and proper elevation of the road are essential to prevent disruption of service. Snow removal must be considered and, if not done by another agency, planned by the water utility itself. In addition, thought should be given to the entire route operators have to travel to go from their homes to the plant.

Connections

A water treatment plant is tied into its environment through a variety of connections. In preparing a siting report, it is necessary to consider routing and alignment of each connection for a particular site because there is a cost implication to each. In some cases, lack of solution can render use of a site impossible. The most important connections are listed below:

- Raw water pipeline from source.
- Filtered water line to service area.

- Waste filter backwash line to point of disposal.
- Plant emergency overflow line.
- Plant drain from major units.
- Structure drains and underdrains.
- Sludge line to point of disposal.
- Site storm drainage lines.
- Power supply lines.
- Communication lines and telemetering.

Many of these connections require permits of one kind or another, which should be followed up during preliminary siting.

CONSTRUCTION DIFFICULTIES

Within the selected alternative sites, it is important to conduct studies to determine if there are any unusual or difficult site factors that will affect cost or suitability of the site. These can be done on a reconnaissance level for preliminary siting but must, of course, be done in detail prior to design. Among these factors are:

- Groundwater—high groundwater means high costs for dewatering, design of structures so they will not "float," and extra cost for possible unstable foundation material.
- Rock provides excellent foundation but costs several times as much as other suitable foundation material to excavate. It is also expensive if "overexcavation" and refill with select material is required. Pipe trench excavation is expensive.
- Soil Conditions should be analyzed to determine stability, expansiveness, bearing values, slope stability, and need for special foundations such as piling.
- Flooding should be analyzed under various conditions of flood frequency. Required mitigation measures should be identified. Routine removal of filter washwater, sludge, and storm drainage during flooding periods must be planned.
- Cost of land is always a major consideration, particularly if it may be necessary to include legal costs for condemnation.
- Volume of excavation, which is necessary to fit the plant into a given site, may vary widely and should be estimated early. This comparison should include a differential in cost for excavating rock, if any. For lowest cost, excavation

should be balanced with backfill. This may not be possible when considering the need to locate large hydraulic structures on undisturbed or well-compacted ground.

ENVIRONMENTAL FACTORS

Most water treatment plants will require that an environmental impact statement be prepared prior to commencement of design. Rules for preparation of such a statement vary from place to place and will not be addressed here. However, the environmental report will probably have to address at least the following factors which have commonly been required.

Noise during operation (pumps, engine generators, gear reduction drives, ozonators, chemical trucks).

Noise and dust during construction.

Erosion protection.

Danger from chemical spills (in-plant and during delivery).

Appearance (how will the plant look from all vantage points).

Odors (chemicals, sludge, etc.).

Site damage to sensitive species or archeological sites.

Growth inducement.

Sludge and waste washwater disposal effect on watercourses or bodies of water.

Night lighting impact on neighbors.

There may be some cost implications to the environmental analysis which will have to be reflected in the siting report for alternative sites. For example, appearance or noise control may not be critical if the plant is located in a heavy industrial area. On the other hand, a plant located in or near a residential area or area of public recreation will require the most stringent control on all of the factors listed above. In general terms, it is necessary to adjust the detailed design to allow for any environmental restrictions. Design techniques are available to deal successfully with almost any environmental restriction. However, the cost implications must be included in the evaluation of alternative sites.

CASE HISTORY 1

In the mid-1960s, the San Dieguito and Santa Fe Irrigation Districts in California joined forces to provide one water treatment facility to serve the two districts. The historical major source of supply for both districts had been San Dieguito Reservoir, a surface impoundment that also received water via an open channel flume from Lake Hodges, another surface water supply. A major pump station boosted water from San Dieguito Reservoir with a maximum water surface elevation of approximately 73 m (240 ft) to the upper pressure zones of the two districts.

The firm supply from these two surface water sources was inadequate to provide for the needs of the growing communities. An agreement with the San Diego County Water Authority was therefore executed to provide water from the authority's aqueduct. The aqueduct runs from north to south near the easterly extremities of the districts. The aqueduct's hydraulic gradient was approximately 900 ft, substantially above the highest pressure zone in either system. A pressure-reducing station at the aqueduct reduced the delivery pressure to approximately 157 m (515 ft). This source substituted for the San Dieguito Reservoir pump station to supply the higher-pressure zones directly when the local water supply was inadequate. Aqueduct water was also occasionally supplied directly to San Dieguito Reservoir to augment local supplies.

Historically, the main function of the district had been to supply irrigation water to a predominantly agricultural area. Development of the area and domestic water use had advanced to the point that by the mid-1960s both districts determined that treatment was essential.

The site selection for the proposed joint treatment plant proceeded through the following steps:

1. Initial and ultimate plant capacity, applicable processes, and preliminary unit layouts were determined. The initial plant capacity of 102 ML/d (27 mgd) with ultimate expansion to 151 ML/d (40 mgd) was determined based on projected community development. While water quality from the aqueduct and from the local supply were such that minimal treatment would be required through much of the year, at times complete treatment was necessary; therefore, the selected plant processes included rapid mixing, flocculation, sedimentation, and filtration. A 49-ML (13-mil-gal) reservoir was also included in the recommendation to act as a plant clearwell as well as to provide needed system storage for the 157-m (515-ft) pressure zone and emergency storage.

2. General site strategies considering the service area topography, existing system pressure zones, consumption demography, and hydraulic grade line conditions for both local surface water and aqueduct supplied were developed. A previous engineering study had selected a plant site to serve the San Dieguito Irrigation District adjacent to the aqueduct with a plant service elevation of approximately 167 m (515 ft). This elevation was determined to be proper considering the factors listed above although it required modification of the existing San Dieguito Reservoir Pump Station to enable this facility to deliver water to an elevation of approximately 165 m (540 ft) to allow for head losses through the plant and on-site clearwell reservoir. The plant influent elevation also required dissipation of head from the aqueduct, which has a maximum grade line elevation of approximately 287 m (940 ft) and a minimum of 268 m (880 ft) in this area. Based on the 165-m (540-ft) inlet elevation and the 157-m (515-ft) outlet elevation, a preliminary plant hydraulic profile was developed and corresponding process unit foundation elevations established.

3. General surveys of suitable sites utilizing USGS mapping and field altimeter surveys were conducted followed by more precise topographic surveys of selected promising sites. Based on this process, the previously recommended plant site was rejected. Apparently, head losses through the plant had not been considered in the previous analysis and foundation elevations throughout the plant had been based on a hydraulic elevation of 157 m (515 ft) throughout the plant resulting in the foundations of the pretreatment process units being substantially above any of the site topography. Several other potentially suitable sites were located and investigated; however, as a result of this screening process, it quickly became evident that a general site location immediately east of the previously proposed site was the only appropriate area for the plant.

4. In determining the precise plant site and unit process locations, other factors were considered, such as:

- Property ownership.
- Effects on future development of surrounding property.
- Plant access.

- Connections to the existing system.
- Soil conditions, including required rock excavation.

The general site selected slopes gradually east to west from an elevation of 170 m (560 ft) down to 149 m (490 ft). The foundation elevation of the major plant units were established as follows:

Unit	Foundation Elevation	
	M	F
Flocculation and sedimentation basins	160	525
Operations building	159	523
Filters	156	513
Clearwell reservoir	152	500

These elevations, as well as the required trenching for connecting conduits, were considered in relationship to general rock profiles underlying the site to minimize expensive rock excavation.

It was possible to locate the plant entirely on property under one ownership and to essentially limit the plant site to one corner of that property, allowing the owner to most effectively develop his remaining property.

Vehicular access to the plant was an important consideration due to the sloping nature of the site and the extent of cut-and-fill slopes.

By a thorough analysis in the plant siting phase of this project, a cost-effective site meeting the requirements of both districts resulted.

CASE HISTORY 2

The City of Kennewick, located immediately adjacent to the Columbia River in the State of Washington, has historically depended on a series of Ranney Collectors along the river as its primary water source. Like many cities of similar setting, Kennewick has progressively spread from the water's edge to higher ground. The capacity of the Ranney Collectors gradually declined over the years as the city grew. To augment its water supply, several potential sources were investigated, but in the late 1970s it was decided that a direct diversion of Columbia River water was the most economical and reliable solution to the Kennewick water supply

problem. A total of six possible sites were considered after an initial screening process.

Several unusual factors were encountered in the ensuing siting analysis, including:

- The Columbia Park site, on which previous distribution system development planning had been based, was located in a heavily used waterfront park controlled by the U.S. Army Corps of Engineers, which would not approve the use of such prime recreational property as a plant site.
- While the possibility of locating the treatment plant on high ground south of the city center had some real advantages, the only practical site (Olympia Street) was located too far from major transmission mains to be cost-effective.
- Although a total of four sites were originally studied in detail, potential acquisition problems with two of the sites, approval problems with the Columbia Park site involving the U.S. Army Corps and the high construction costs associated with the remaining site due to connections to the city's transmission mains, necessitated the study of two more potential sites (Columbia Drive and Union Pacific).
- The economics of land acquisition of several proposed sites was affected more by the anticipated delays in the process largely related to potential legal proceedings, and therefore escalation of project costs, than by the actual cost of property.

The actual site selection process proceeded as follows:

1. Site total area requirements were developed based on:
   a. Capacity requirements (28 ML/d or 7.5 mgd initial and 114 ML/d or 30 mgd ultimate) determined by projecting the water consumption needs of the city and subtracting projected available supply from the existing Ranney Collector system.
   b. Plant layouts considering applicable treatment processes recommended from a pilot plant water treatability study which initially included ozonation, flash mixing, flocculation, filtration, clearwell storage, waste washwater reclamation, and future provision of carbon adsorption and sedimentation processes plus space for plant expansion.

c. Construction and operation/maintenance access requirements.

2. Susceptibility to flooding of the site alternatives by the Columbia River was a critical factor analyzed in detail, and cost implications were evaluated.

3. Proximity to an acceptable intake location including influent pump station was considered for each alternative. This analysis included both water quality considerations and the difficulties of penetrating the levee bordering the river.

4. Environmental impacts of each site were evaluated, including the effects of

   a. traffic,
   b. aesthetics,
   c. resident location,
   d. growth effects,
   e. construction,
   f. water system reliability,
   g. interference with recreation, and
   h. property acquisition.

   An environmental impact statement was prepared in accordance with the State of Washington SEPA rules.

5. Potential construction problems were considered, including geotechnical considerations such as unstable foundation soils and groundwater levels.

6. Potential property costs and acquisition difficulties were analyzed.

7. Critical connections to existing major transmission lines were analyzed and costs evaluated.

Based on these criteria, the possible sites available were screened back to four potential sites and ultimately the Port site was recommended. Due to the potential acquisition problems, however, two other sites were then evaluated in the same general area as the Port site, and the Columbia Drive site was ultimately selected.

—21—

Facilities Design

BASIC DESIGN PRINCIPLES

Design Philosophy

Water treatment plant facility design is the combination and assemblage of all expertise relating to the water treatment field. In order to produce acceptable, workable facilities that can be built within the predetermined budget and on the selected site, it is necessary to utilize and blend the expertise of many separate disciplines. The successful project will utilize the skills of chemists, biologists, microbiologists, laboratory technicians, process engineers, civil engineers, mechanical engineers, electrical engineers, instrumentation specialists, structural engineers, architects, designers, management and administrative personnel, financial analysts, surveyors, construction specialists, operations and maintenance specialists, and many others.

To prepare for design, laboratory personnel and process engineers combine to establish a skeletal framework of the facility. First, it is necessary to gather raw water quality data, or to obtain data in the laboratory through analysis of raw water samples, to identify those constituents in the raw water that must be removed during treatment. Treated water quality goals are then established, and process engineers selectively determine and size those process elements (such as flocculation and sedimentation basins, filter facilities, disinfection facilities, etc.) that will most economically and effectively meet those water quality goals. At this stage of design, it is most efficacious to determine appropriate technology for the application studied and to introduce site-specific concepts that may provide for the best treatment possible. After the basic process scheme is determined, those personnel who are skilled in the preparation of detailed plans and specifications expand on the skeletal framework provided and emerge from the design process with a complete, interdisciplinary effort. Concurrent with this effort, management and administrative specialists examine the financial considerations necessary to build the facility and handle any legal responsibilities concerning, for example, right-of-way acquisitions.

Typically, a staff of construction specialists supervises plant construction, paying particular attention to safety, quality, and speed of construction. These construction specialists are necessary to ensure that the ideas and technology included in the detailed design are incorporated into the actual facility.

A review of the above yields the obvious conclusion that water treatment plant design requires a highly skilled, interdisciplinary team and is not simply the province of civil, mechanical, or electrical engineers. Successful design requires the input and expertise of personnel in fields as vastly different as microbiology and structural design.

Worldwide, there are differences in philosophy for determining the appropriate treatment scheme and facility layout. For example, in Europe, the responsibility for the layout of plant configurations and associated details is frequently assigned to equipment manufacturers or contractors who offer a proprietary "package." This is sometimes referred to as "turnkey" design and construction. In the United States, overall plant design (including all configurations and many details) is almost always provided by design engineers or consultants. Equipment manufacturers and contractors are responsible only for the design of certain specialized components. In either case, good coordination between the design engineer, the equipment manufacturers, and the contractor is critical.

For small municipal plants [less than 10 ML/d (2.5 mgd)] and for industrial and investor-owned plants, the U.S. approach is generally similar to the European philosophy. For larger municipal or publicly owned plants, however, the following procedure is normally adopted.

1. The designer evaluates alternative configurations early in the design process.
2. The designer establishes conceptual design criteria and prepares a detailed design report that includes the exact number, configuration, and dimensions of all facilities.
3. Design drawings and specifications are prepared complete with details such as reinforcing steel spacing, piping and valve layouts and dimensions, lists of all mechanical components, electrical wiring diagrams, and complete lists of all instruments.
4. Prospective contractors are invited to bid for construction of the facility. Alternative bids are not normally solicited.
5. Bids are reviewed and the construction contract is awarded.

Under the U.S. system, nearly all of the detailed engineering design is completed prior to the bidding phase. Engineering efforts required after receipt of bids are associated with review of manufacturers' equipment submittals and minor changes in design.

Innovations

Advancements and progress in plant efficiency, rates of operation, lower costs, and longer life can only be achieved by refinement of and deviation from previously established criteria and design. However, such deviations or innovations, even those that appear to be relatively minor, should be adopted only after considerable study. Whenever possible, innovations should be tested by pilot plants or full-scale tests on existing facilities. Designers should be sure that a previous change is functioning properly and is actually providing a measurable benefit before proceeding with the next innovation.

Process Criteria

Quality goals, testing, and process decisions should be translated into criteria tables prior to commencement of plant layouts. In general, process criteria are established in the following sequence:

1. Establish nominal, maximum, initial and ultimate hydraulic capacities.
2. Select unit processes and establish hydraulic design criteria for each process.
3. Select process chemicals and feed rates.
4. Calculate chemical storage requirements.

Hydraulic Design Criteria

Typical hydraulic design criteria are shown in Table 21-1. These criteria establish the number and basic dimensions of each unit process. Guidelines on factors to be considered in establishing the criteria are outlined in subsequent sections. In practice, it is usually necessary to modify the criteria as the design proceeds. After selection of the unit processes, establishment of hydraulic design criteria is the most important decision in the design process.

Chemical Design Criteria

Typical water treatment chemicals and their appropriate feed rates are listed in Table 21-2. Chemical storage requirements are set by space allowances, cost of transportation, effective storage lifespan, and on-site considerations such as convenience and ease of handling storage containers. Generally, storage must allow for a 15–30-day supply of each chemical at its established feed rate.

TABLE 21-1. Typical Hydraulic Design Criteria (Anstey Hill WTP Adelaide, South Australia)

Description	SI Units	SI Criteria	English Units	English Criteria
Plant capacity				
Maximum flowrate	m³/s	3.62	gpm	57,400
	ML/d	313	mgd	82.7
Flow metering				
Plant influent	Number	1	Number	1
Type	Magnetic flow (1050 mm diameter or 3.44 ft diameter)			
Range	m³/s	0.3–4.0	gpm	4800–63,000
	ML/d	25–350	mgd	7–92
Initial mixing				
Type	Pump injection (hydraulic nozzle diffusion)			
Pump capacity	L/s	60	gpm	950
Motor power	kW	5.5	hp	7.4
Energy input (G)	s⁻¹	1,000	s⁻¹	1,000
Flocculation basins	Number	2	Number	2
Compartments (each basin)	Number	3	Number	3
Compartment dimensions	mm	17,000 × 8600	ft	55.8 × 28.2
Average water depth	mm	4875	ft	16.0
Volume (each basin)	m³	2146	ft³	76,000
	ML	2.146	mg	0.57
Volume (total)	m³	4292	ft³	152,000
	ML	4.292	mg	1.13
Detention time at maximum flow	s	1180	min	20
Mixing energy, G_{max}	s⁻¹	90	s⁻¹	90
Vertical-shaft flocculators	Number	12 (total)	Number	12 (total)
Power (each)	kW	3.7	hp	5.0
Sedimentation basins	Number	2	Number	2
Inside dimensions (each)	mm	17,000 × 94,000	ft	56 × 308
Water depth	mm	5000	ft	16.4
Volume (total)	m³	15,980	ft³	564,000
	ML	15.98	mg	4.23
Detention time at maximum flow	s	4400	min	73
Basin loading at maximum flow	ML/m² · d	0.098	gal/ft² · min	1.67
Horizontal velocity at maximum flow	m/s	0.021	ft/min	4.1
Launder weir length	m	1250	ft	4100
Weir loading at maximum flow	ML/m² · d	0.25	gal/ft · min	14
Filters	Number	12	Number	12
Type	Gravity, dual-media 2 bays per filter			
Dimensions (each bay)	mm	3000 × 12,000	ft	9.8 × 39.4
Media area (each filter)	m²	72	ft²	775
Maximum filtration rate (all filters on line)	mm/s	4.2	gpm/ft²	6.2
Maximum filtration rate (2 filters off line)	mm/s	5.0	gpm/ft²	7.4
Maximum backwash rate	mm/s	15.0	gpm/ft²	22.1
Maximum static head	m	10.0	ft	32.8
Maximum surface wash rate	mm/s	2.1	gpm/ft²	3.1

TABLE 21-1. (*Continued*)

Description	SI Units	SI Criteria	English Units	English Criteria
Filter wash tank				
Diameter	mm	20,250	ft	66
Height (total)	mm	7000	ft	23
Available depth	mm	4000	ft	13.1
Available capacity	m^3	1250	ft^3	44,000
	ML	1.25	gal	330,000
Washwater pumps	Number	2	Number	2
Capacity (each)	L/s	165	gpm	2600
Power (each)	kW	30	hp	40
Waste washwater tank	Number	1	Number	1
Diameter	mm	20,000	ft	66
Depth	mm	6500	ft	21
Effective depth	mm	4000	ft	13
Capacity (effective)	m^3	1250	ft^3	44,000
	ML	1.25	gal	330,000
Recycling pumps	Number	2	Number	2
Capacity (each)	L/s	165	gpm	2600
Power (each)	kW	30	hp	40

Material Selection

The principal factors in material selection are economics and reliability. The life of the proposed facility, the availability of materials, and the evaluation of interest and depreciation costs versus operation and maintenance costs affect material selection. As an example, the following materials could be considered for the shell of a clarifier unit:

Material	Typical Applications
Steel (uncoated)	Temporary or short-life commercial
Steel (coated)	Small municipal or typical commercial
Reinforced concrete	Long-life commercial and major municipal
Brick or masonry	Where economical, such as developing countries
Aluminum	Portable or prefabricated for remote locations

Major municipal water treatment plants are usually designed and constructed for a plant life of 50 yr (or

more). The principal structures of such plants are generally constructed of reinforced concrete. Extensive use of stainless steel metal components, PVC piping (for diameters 200 mm or less), and low-maintenance exterior surfaces are also common. For commercial and industrial facilities, more use is generally made of coated carbon steel. Discussions of typical material specifications are given in the last section of this chapter, "Materials."

Simplicity in Design

Simple and basic designs are frequently the most economic and reliable solutions to water treatment problems. Designers have a tendency to include as much recent technology as possible into a design. There is merit in this progressive concept, but design is a process of compromises, and conscious efforts should be made to test innovations on a "one-at-a-time" basis. Efforts should also be made to minimize the number of units, systems, and components. For example, a filtration plant with a capacity of 400 ML/d (100 mgd) is best suited for 8–12 filters. The incorporation of more than 12 filters would be inappropriate. A large number of filters would increase both capital and operation and maintenance costs.

TABLE 21-2. Typical Chemical Feed Design Criteria

	Chemical Feed Rates (mg/L)		
	Minimum	Average	Maximum
Chlorine (preflocculation)	0	0.5	5.0
Chlorine (prefiltration)	1.5	2.0	5.0
Chlorine (postfiltration)	0	0.5	1.0
Aluminum sulfate (liquid 50% solution, 0.65 kg/L)	25	30	50
Sodium hydroxide (liquid 25% solution, 0.32 kg/L)	5	15	25
Lime (CaO)	10	15	30
Polymer (cationic)	0	0.5	2.0
Polymer (anionic)	0.01	0.02	0.05
Hydrofluorosilicic acid (22% $H_2SiF_6^-$, 0.27 kg/L)	0.5	0.7	1.0
Activated carbon (PAC)	0	2	10
Potassium permanganate	0.1	0.2	0.5

PROPRIETARY DESIGNS

Proprietary Versus Custom Design

Predesigned, standard, proprietary, or patented components are utilized in both the European and American design philosophies. In both approaches, it is impractical for the plant designer to design the many specialized components necessary in any water treatment facility.

A designer should never undertake the design of commercially available components (such as flow meters, chemical feeders, chlorinators, sludge-handling equipment, etc.) unless he is confident that he can improve on the design or provide a component that is more cost-effective. When costs are considered, the designer's engineering costs should be included. For this reason, proprietary equipment is generally utilized in small plants where design costs are proportionally higher.

On plants larger than 10 ML/d (2.3 mgd), the designer should carefully evaluate the use of proprietary facilities (such as flocculator–clarifier units) instead of custom-designed units. Economic factors such as capital costs, interest and depreciation, maintenance, reliability, and design costs should be considered. An economic evaluation can generally be made without obtaining formal tenders or bids for the facility. Evaluations are, however, frequently difficult because there may be conjecture on

reliability and efficiency factors for alternative designs. The experience of the designer then becomes an important factor in the decision of a proprietary versus a custom design. If design personnel are not experienced in the system under consideration, preference should be given to proprietary designs.

Types of Proprietary Facilities

Available proprietary facilities and components cover a range from complete plants to the smallest plant components such as filter bottom nozzles. The most commonly utilized proprietary facilities and equipment include:

Flocculator–clarifier units.
Mixing equipment.
Sludge removal equipment.
Filter underdrain systems.
Chlorinators.
Ozonators.
Tube settlers.
Fiberglass troughs and launders.
Chemical feeding systems.

Size of Proprietary Equipment

Water treatment plants, complete with flocculation, clarification, and filtration units, can be fabricated and transported as integral facilities ("skid mounted") in sizes up to 2 ML/d (0.5 mgd). Skid-mounted units have a high salvage value and are ideal for temporary facilities and for plants in remote locations. In addition, these proprietary units often have normal commercial and community applications. Skid-mounted units can be grouped to form large-capacity facilities. Plants in the range of 2–10 ML/d are generally proprietary packages. Plants larger than 10 ML/d are more likely to be broken into packages of several equipment suppliers or to be custom designed.

Cost of Proprietary Plants

In plant sizes of less than 10 ML/d, proprietary-designed plants are generally less expensive than custom-designed plants. For larger facilities, evaluations should be made to determine relative costs as outlined in Chapter 25. In general, custom-designed plants, which utilize concrete construction, may involve higher initial construction costs but may be judged economically more favorable because of lower depreciation and maintenance rates. Since competition generally results in lower costs, it is good practice to establish general criteria for facilities or components so that three or more suppliers can quote, tender, or bid each facility system (package) or component.

COMPONENT DESCRIPTION AND ORGANIZATION

General Components

Because of variations in plant capacity, raw water quality, site factors, and climate, no plant layout or design can be considered standard. There are, however, certain features and components of treatment plants which may be considered typical.

Municipal or community potable water plants are generally designed to provide a high degree of reliability. If the plant is the sole source of potable water, many critical components, structures, and systems may be duplicated or provided with parallel or standby facilities.

It is sometimes difficult to differentiate between the terms *structures, equipment,* and *components.* For the purpose of this text, components are considered to be parts of structures or systems. Components, generally, are specialized items of equipment not designed by the plant designer but selected and assembled to form a system or appurtenant part of a structure. Thus, for a clarifier (a structure), the launders, sludge removal units, valves, and gates are components.

Most components and equipment systems discussed in this text may be classified as mechanical. Other components and systems may be numerically more common, however, and represent a significant percentage of design and construction costs. The relative importance of each general discipline can be gauged from the relative amount of engineering and technical input from each specialized engineering classification, such as chemical, civil and hydraulics, structural, mechanical, electrical, and instrumentation engineering.

Organization

Only discussions of certain critical treatment systems are included in the remainder of this chapter. The components are discussed as parts of various systems, and emphasis is placed on selection rather than on design. Critical systems which are discussed further include:

- Flow metering.
- Chemical feeding.
- Mixing.
- Aeration.
- Flocculation.
- Sedimentation.
- Flotation.
- Filtration.
- Solids handling.
- Disinfection.
- Demineralization.

Design of biological process systems is not included in this text. Physical–chemical systems for wastewater recovery are, in most instances, very similar to the water treatment systems presented in subsequent sections.

FLOW MEASUREMENT

Introduction

The two principal applications for measuring flow are (1) process liquid flowrate measurement and (2) treatment chemical flow measurement. The four basic means of liquid flow measurement are (1) measuring differential pressure, (2) measuring direct discharge, (3) measuring positive volume displacement, and (4) flow velocity–area methods.

Differential pressure measuring devices include venturi tubes, Dall tubes, orifice plates for full flowing pipelines, and venturi flumes such as Parshall and Palmer-Bowlus flumes for open channel flow measuring. These devices are well suited for measuring process flows in water treatment plants and are widely used.

Direct-discharge measuring methods include magnetic flow meters, ultrasonic flow meters, propeller meters, and weirs. These methods are often applied for measuring process flows but have some limitations.

Positive volume displacement measuring methods are almost exclusively used as a form of chemical flow measurement. Various types of chemical metering pumps, such as plunger-type pumps, diaphragm pumps, and progressive cavity pumps, operate via positive displacement methods.

The flow velocity–area method is based on a measurement of the flow velocity. The measured velocity is multiplied by the cross-sectional area of the conveying system to calculate the flowrate. Flow velocity measurements can be achieved by various techniques, such as (1) current meters, (2) float measurements, (3) pitot tubes, and (4) tracers. These methods, however, are not frequently used in water treatment plants because of accuracy and redundancy problems.

The principal flow-measuring devices used extensively in water treatment plants are summarized in Table 21-3. Some important design considerations are also indicated.

Because of the extensive use of venturi-type meters, Parshall flumes, and propeller meters in water treatment plants, some useful information on these meter types are discussed in this section.

Venturi Meters

The operation of this flow-measuring device is based on Bernoulli's therorem. The meter itself consists of three basic parts: (1) inlet cone; (2) throat or constricted section; and (3) outlet cone. A piezometer ring is provided at the upstream end of the inlet cone and at the throat. The flowrate is based on the difference in pressure measured at these two points.

The ratio of the throat diameter to the inlet diameter is called the beta ratio and ranges from 0.39 to 0.73 depending on the manufacturer and model of the meter.

Venturi meters may be installed in any position, vertical, horizontal, or sloping, provided the line is full at all times. The length of uniform straight pipe immediately upstream of the meter depends on the beta ratio and falls in the range between 5 and 10 pipe diameters. Conditions downstream of the meter have little effect on flow measurements.

The practical minimum differential pressure for a reasonably accurate flow measurement is about 20 mm (0.79 in.). This limitation is due to the inaccuracy of secondary instruments, which detect and transmit the differential, below this level. In general, 2.0 m is the standard differential at the design flowrate. Typically, head loss due to a venturi meter is relatively small.

Venturi Flumes

Venturi flumes (usually Parshall or Palmer-Bowlus flumes) use the critical-depth principle to measure flowrates in open channels. Parshall flumes are suited for measuring large flowrates. Palmer-Bowlus flumes are usually small in size compared with Parshall flumes and are more commonly used to measure wastewater flows in sewers.

Advantages of the Parshall flume include: (1) no clogging due to suspended matter; (2) it can be made locally with concrete, metal, or wood; and (3) they have a wide flow measurement range. For these reasons, this type of flume is well suited for developing countries or remote areas. In addition, chemical flash mixing can be achieved at the hydraulic jump just downstream of the throat.

Like venturi meters, there is a need for a uniform straight channel upstream of the flume. The length of this channel should be a minimum of 3–5 times the channel width. Also, it is important to set the flume sufficiently high so that the downstream water level is low enough to maintain free discharge conditions to the diverging section of the flume.

Several equipment manufacturers such as BIF, Leopold, and Robertshaw provide plastic or fiberglass venturi flume liners or inserts of various sizes. These liners make construction of the flumes much easier and flow measurements more accurate.

If the flume is operating under free-flow conditions, flow measurements are usually made by de-

TABLE 21-3. Flow-measuring Devices and Their Applications

Method/Device	Proper Application	Design Considerations

Venturi-type meters
  Dall tube
  Herschel venturi
  Universal venturi
  Venturi inserts
	Raw water as well as finished water flow measuring; pipe should be flowing full at all times	Accuracy is within ±1% for a flow range of 1–10; provide a minimum of 5 pipe diameters of uniform straight pipe ahead of the meter; locate the venturi at a proper elevation so that no negative pressure occurs in the throat at maximum flowrate

Orifice Plates
  Concentric orifice
  Eccentric orifice
  Flow nozzle orifice
	Raw water as well as finished water; pipe should be flowing full at all times	Accuracy is within ±2% for a flow range of 1–4; Suitable for gas flow as well as liquid flow measurement; provide a minimum of 5 pipe diameters of uniform straight pipe ahead of the meter

Magnetic flow meters
  Fisher and Porter
  BIF
  Brooks
	Most any type of liquid including sludge; medium- to small-size pipe flow (<1.2 m diameter); pipe should be flowing full at all times	Accuracy is within a ±1% for flow velocity from 0.9 to 9 m/sec and within ±2% for flow rates from 0.3 to 0.9 m/sec; select the proper size in order to maintain the proper velocity; provide a minimum of 5 pipe diameters of uniform straight pipe ahead of the meter

Ultrasonic flow meters
  Sparling
  Badger
  Mapco
  Ocean Research Equipment
	Use for relatively clean water; apply only to liquid flow measurement	Accuracy of ±1% for a flow velocity range of 0.3–8 m/sec; there should be no air bubbles in the fluid to be measured

Propeller meters
  Sparling
  Rockwell
  Badger
	Use for relatively clean water; place in a readily accessible location; use for noncorrosive and not too viscous liquids	Accuracy of ±1% for a flow range of 1–10; provide a minimum of 5 pipe diameters of uniform straight pipe ahead of the meter; in general, minimum flow velocity should be 0.2 m/sec and maximum flow velocity should not exceed 1.8 m/sec

Venturi flumes
  Parshall flumes
  Palmer–Bowlus flumes
  Fisher Porter Co.
  BIF
  Leopold
	For relatively large flowrate measurements; use of clean as well as dirty water; use in an open channel only	Accuracy of ±5% for a flow range of 1–20; provide sufficient fall downstream to prevent "backwater" effects; freezing of the water may create problems

tecting one water level in the tapered accelerating section upstream of the throat. If operating conditions are not free flow, it is necessary to measure an additional water level at the throat; the proper flowrate must be computed from the two levels. In general, the maximum allowable flow velocity at the level-detecting location should be 0.5 m/sec for small flumes and no higher than 2 m/sec in large flumes.

Propeller-Type Meters

Propeller-type meters are of simple design and are relatively inexpensive in smaller sizes. For these reasons, propeller meters have been used exclusively as service meters for monitoring home water usage. The accuracy of propeller meters is within 2 percent of the true flow when properly installed and used within their rated range.

Basically, there are three types of propeller meters: (1) main-line or in-line type; (2) saddle type; and (3) intake meters, which are installed at pipe or conduit exits. Main-line meters are available up to 600 mm diameter. Saddle-type meters are available in sizes up to 1800 mm diameter. Intake meters are used for large flow-metering purposes and are available in sizes up to 2000 mm diameter. Installation of all three types of propeller meters is relatively easy. A uniform straight portion of pipe about 5 diameters in length or straightening vanes should be provided upstream of the meter.

CHEMICAL FEEDERS AND MIXERS

General

Chemicals such as coagulants, oxidants, and adsorbents play an essential role in water treatment. Thus, the design of chemical feed systems and mixers merit special attention.

Two basic types of chemical feeders are used in the water treatment field: liquid feeders (usually metering pumps) and dry chemical feeders. Liquid volumetric feeders are more widely used than dry feeders because of their compactness, accuracy, and the convenience of handling and storage of liquid chemicals. In some instances, gas feeders are used for oxidation, recarbonation, and disinfection purposes. Generally, chemical feeders must have a wide-range capacity to meet the variations in plant flowrate and chemical dosage.

Both mechanical mixers and hydraulic mixers are used in the water treatment field. Mechanical mixers range from variable low-speed flocculators to constant high-speed flash mixers. Hydraulic mixing devices include pump mixers, hydraulic jumps, and injectors.

Liquid Feeders

Several criteria should be considered in designing a liquid feeding system: (1) selection of materials for the meters, valves, and piping that are compatible with the metered chemical; (2) ease of operation (3) cost of the system; and (4) efficiency in continuous feeding. A continuous-feed system provides for better coagulation than does an on–off system. Metering pumps, because they are accurate and can be easily adjusted to plant flowrate, are generally preferred to orifices and weirs in liquid feeding systems.

The four major categories of metering pumps are (1) the diaphragm pump, (2) the plunger pump, (3) the combination diaphragm and plunger pump, and (4) the progressive cavity pump. The combination diaphragm and plunger pump is the most frequently used metering pump and is shown in Figure 21-1.

Diaphragm pumps are the most simple and inexpensive metering pumps, but have several disadvantages. Variation of both the suction and discharge pressure can affect the pump discharge rate significantly. Also, flow proportional feeding is poor. Finally, the small capacity of diaphragm pumps may be inadequate for flows greater than 100 L/hr (30 gph) per pump.

Plunger pumps provide good feeding accuracy, large capacity (approximately 1100 L/hr or 300 gph), and can easily match or pace the process flowrate. However, they are expensive and allow direct contact of corrosive or abrasive liquid to the plunger. In addition, the feeding solution will show a distinct pulsation pattern unless multiple pumps are run in parallel to smooth out the pattern or pulsation dampers are used.

The third type of pump is a combination of the diaphragm and plunger pumps. This pump avoids chemical contact with the plunger and thus minimizes corrosion problems. Like plunger pumps, the combination pump may create a distinct pulsation effect that may be minimized by a pulsation damper on the discharge line. Often, the pump contains a spring-loaded back-pressure valve, which ensures that the suction and discharge head variations do not affect the pumping rate.

The progressive-cavity pump is relatively expensive but is well suited to most chemicals except very corrosive or abrasive chemicals such as a slurry of activated carbon. Progressive-cavity pumps are sufficiently accurate for most metering functions and easily respond to changes in the process flowrate. However, if the pumps are operated without liquid, they may be damaged within a few minutes.

Chemical feed rate is typically adjusted to match plant flowrate and water quality. Techniques for changing the feed rate of metering pumps include (1) on–off controls activated by pulse duration signals and timers (not suited to certain chemical feeding applications, such as alum and other coagulants), (2) motor speed variation, and (3) stroke length variation by manual or automatic adjustment

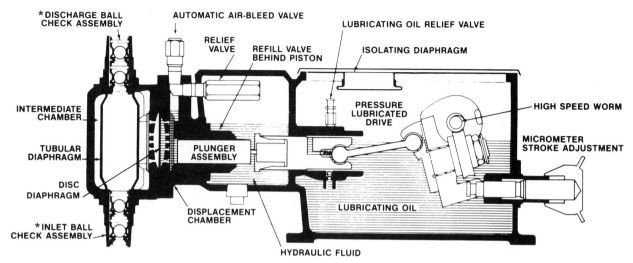

FIGURE 21.1. Tubular diaphragm–plunger combination pump. (Courtesy of Milton-Roy Company.)

through a pneumatic or electronic signal. In the usual open-loop proportional pacing system, the motor speed is controlled by a signal (4–20 mA) in proportion to flow, and stroke length of the feeder is set manually to determine the dosage rate.

A typical liquid feeding system with a metering pump is shown in Figure 21-2. If both the chemical dosage range and process flowrate range are fairly constant or in a narrow band (maximum 1–5 range each), a different feeding system where feed rate is

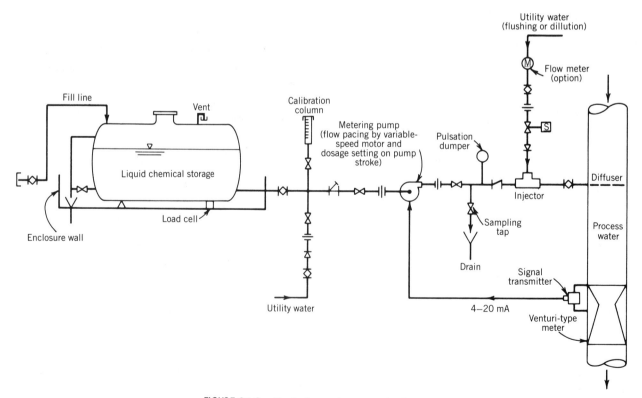

FIGURE 21-2. Typical metering pump feeding system.

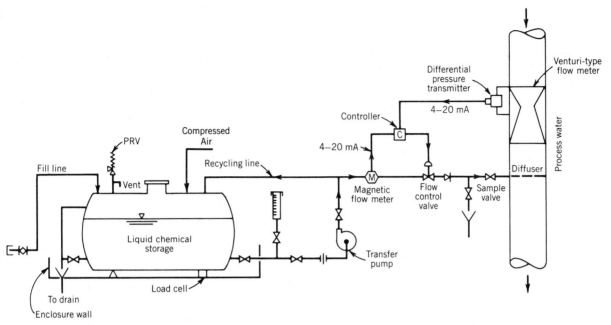

FIGURE 21-3. Flow-placing solution feed system (closed-loop control with magnetic flow meter).

controlled through a modulating valve, as shown in Figure 21-3, may be used. This system usually includes a magnetic flow meter for chemical measurement (normal range 1–30). The system is simple, inexpensive, and requires low maintenance, but the feeding accuracy generally drops off at the lower end of the feeding range. For a very aggressive chemical such as hydrofluosilicic acid, pumps can be replaced by pressurizing the chemical tanks with compressed air and feeding the chemical by pressure. Chloramination involves ammonia feeding together with chlorine application. A typical aqua ammonia feeding system is shown in Figure 21-4.

Special Considerations. To minimize nonuniform feeding patterns, the metered chemical can be diluted on the downstream side of a metering device or a pulsation damper can be placed after a pulsating metering pump or both. When pulsation pumps are used without a pulsation damper, the chemical feeding line should be sized for the pulsating flowrate and not the average hourly feeding rate. In this situation, the line typically requires twice the pipe diameter than if based on an average flowrate.

A pressure water line is also needed to flush out chemicals from the metering device in addition to the feeding line. Without the flushing system, dried chemicals can clog the components.

It is important that the designer consider both the miminum and maximum feed rate which a particular liquid feeder will be called on to deliver. A common design error is to size feeders for ultimate capacity, ignoring low flow conditions during the initial years of operation.

The important items to consider in liquid chemical storage system design are the corrosiveness, crystallization (or freezing) characteristics, stability, viscosity of the liquid chemical, chemical supplier situation, and weather conditions in the vicinity of the plant. After evaluating those conditions, the number and capacity of each storage tank and the proper materials for storage units can be determined.

Dry Chemical Feeders

Certain chemicals, such as lime, soda ash, and most nonionic and anionic polymers, are available only in dry form and, therefore, may require dry chemical feeders. In addition, dry chemical feed may be less expensive than liquid feed in remote areas due to reduced shipping weight. Two types of dry chemical feeders are typically employed: volumetric and gravimetric. Figure 21-5 illustrates a typical dry chemical feeding system. Volumetric dry feeders are relatively simple and inexpensive while gravimetric feeds are more complex and expensive. Because the bulk volume of a dry chemical is affected

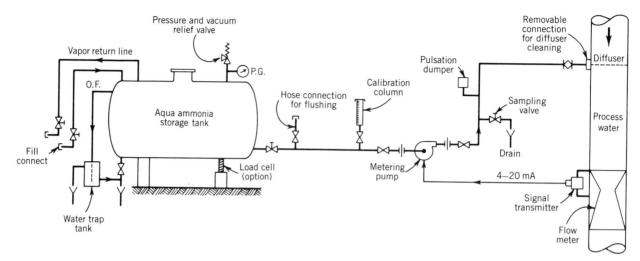

FIGURE 21-4.   Aqua ammonia feeding system.

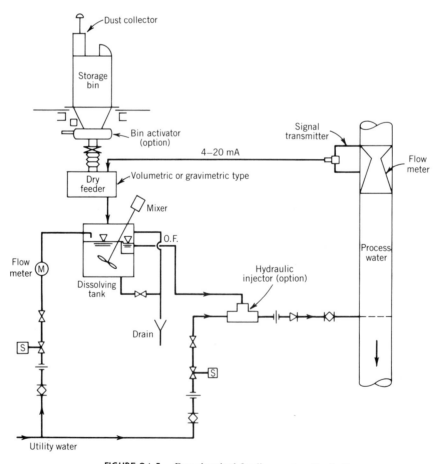

FIGURE 21-5.   Dry chemical feeding system (typical).

502

by the size and shape of the material, moisture content, and manufacturer, accurate control of chemical dose may be low for some applications.

Various types of volumetric feeders are available, the simplest being a vibrating bin unit with a screw-type conveyor. Dosing accuracy with this system may be poor, however. A rotating table design with a circular groove and adjustable knife height is more accurate. Most dry chemical feeder systems are proprietary in nature, and the design engineer should consider a number of alternative systems.

Gravimetric feeders are ideal dry feeders because, unlike volumetric feeders, dosage selection does not require calibration charts. In addition, in a plant using computer-assisted controls or a datalogging system, gravimetric feeders are easy to monitor. "Loss-in-weight" gravimetric feeders have been used extensively but are bulky and expensive. Recent models are mostly belt feeders, which can be easily paced to the process flow. Justification of high capital costs and the need for precision are two important criteria in selecting gravimetric dry feeders.

In addition to selecting the feeder itself, the engineer must also select appropriate dry chemical storage facilities, hoppers, and solution tanks. Most dry chemicals, especially lime, can bridge and flush (uncontrollable discharge) in the silo and hopper. Electric vibrators on the side walls of the equipment do not solve these problems completely. Proper dimensions and wall angles of the silo and hopper are critical design factors to prevent these problems.

The hopper sidewall angle should be at least 60° from the horizontal plane and positive silo outlet control system (such as a Vibra-Screw Bin Control System) may be most effective.

Dry Chemicals. The selection of the solution tank is dependent on the nature of the particular dry chemical. When inorganic chemicals are used, the solubility of the chemical at lowest water temperature at the maximum dosage rate should determine the size and water supply rate for a solution tank for a continuous dissolving system.

The most common inorganic chemical used in the water treatment field, other than alum, is lime. At large facilities, it is stored as quicklime (CaO), which is more compact than slaked lime $[Ca(OH)_2]$. Lime slakers must be specified with the feeders and chosen on the basis of the nature of the local quicklime. The amount of grit in the source limestone and the size and heat-generating characteristics of the lime are evaluated to select either a paste (Figure 21-6) or detention slaker.

Synthetic polymers, or polyelectrolytes, are used extensively in water treatment plants. The feeding system for dry polymers is a batch system that usually consists of a simple dry feeder, wetting device, aging tank, working solution tank, and a metering pump. For dry polymers, the wetting device and aging tank (30 min to 1 hr detention) are necessary to disperse the extremely viscous polymer uniformly. Without these systems, a high-molecular polymer produces "fish eyes," (lumps of powder that do not dissolve) preventing proper dis-

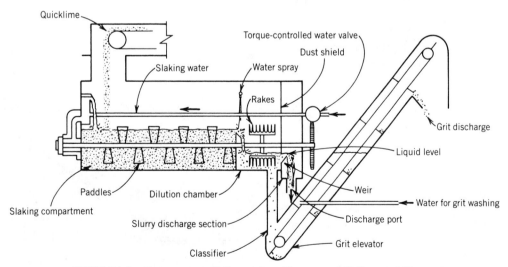

FIGURE 21-6. Paste or pugmill lime slaker. (Courtesy of Wallace and Tiernan.)

persion and use of the polymer. (Figure 21-12 shows a dry polymer feeding system.)

Gas Injection

Gas injection has many applications in water treatment, including (1) aeration for iron and manganese oxidation or taste and odor removal, (2) recarbonation by carbon dioxide gas after lime-softening process, (3) disinfection by ozonation or chlorination, (4) air-assisted filter backwashing, (5) compressed air mixing for a reservoir or tank, and (6) air flotation for scum removal or sludge thickening. For all applications, the most important design objective is uniform air distribution in the required area or volume of water. To ensure this uniform distribution in the injection system, sufficient head loss [more than 200 mm (8 in.) H_2O] should be created by each of the diffuser nozzles. The maximum gas flow velocity in the pipeline should be limited to 20 m/sec (60 fpm) for most applications.

If gas transfer is the main objective of injection, the size of gas bubbles, the volume of gas, and contact time are the important design criteria. [A minute bubble with a certain depth of water, for example, less than 5-mm ($\frac{1}{5}$-in.) individual bubbles with a 4-m (13-ft) water depth, would be practical design criteria.] When agitation or mixing is the purpose of air injection, coarse bubbles [5 mm ($\frac{1}{5}$ in.) or larger] are generally more effective.

Two physical properties important to gas injection system design are the compressibility of gas and its accumulation at the highest elevation in a piping system.

A gas injection system usually consists of a gas compressor or blower, piping, and a pipe manifold system with gas injection nozzles. A centrifugal or a positive-displacement blower may be used for a small facility (with an air requirement less than 0.5 m^3/min at 0.4 kg/cm^2), while a centrifugal blower is used for a large-capacity plant where gas injection or gas flowrate should be varied. For most applications, the tank depth should be 5 m (16 ft) or less.

Mixers

An analysis of mixing includes chemical diffusion as well as the effect of turbulence and eddies. To quantify the rate of chemical diffusion in a flowing body of liquid, the velocity gradient G is used as a standard variable in mixer design.

Addition of most chemicals requires only moderate mixing. A simple propeller mixer with an 80-rpm maximum speed should produce adequate dissolution of chemicals without causing molecular chain breakage in such processes as flocculation. In addition, moderate mixing avoids the formation of a large rotating motion of the water, which often directs flow to one side of the feeding channel. Motor horsepower and impeller size or type are determined by the size and shape of the tank and viscosity of the solution.

Flocculation is often provided by a slow paddle mixer, propeller, or turbine high-energy mixer. The paddle mixer, usually a long horizontal shaft with several submerged paddles as shown in Figure 21-7, is suitable for conventional flocculation, which requires a low-energy mixing (a G value of generally less than 50 sec^{-1}) to produce large flocs.

High-energy, vertical shaft mixers are used to form small, strong flocs prior to high-rate, multimedia filtraton systems. These mixers have few submerged parts.

A recently developed form of mixing system is a static in-line mixer (Figure 21-8). This efficient system consists of a series of baffles in a tube. This system is maintenance free but requires head loss (typically around 0.3–2.0 m or 1.0–6.6 ft). The only disadvantage of this system is the dependence of mixing intensity on flowrate, which may vary widely.

Initial (Flash) Mixing

The purpose of initial, or flash, mixing is to disperse the coagulant chemical quickly and evenly in the

FIGURE 21-7. Low-energy flocculator (Diemer Filtration Plant, Metropolitan Water District of Southern California).

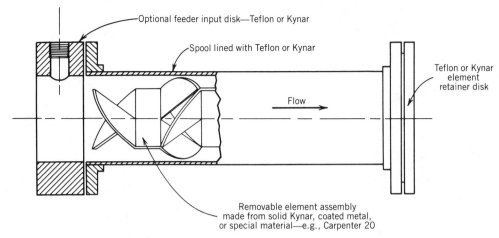

FIGURE 21-8. Motionless (static) mixer. (Courtesy of KOMAX.)

raw water. Alum as a coagulant, for example, must be dispersed very rapidly to coagulate the water properly. Initial mixing by mechanical mixers requires a G value up to 1000 sec^{-1}, or a Gt of 1000–2000, where t is the mixing time in seconds. Although the G value is the most commonly used design criteria for initial mixing, it is an arbitrary guide for design of such units.

Recent studies of this difficult design problem recommend diffuser grid systems, chemical jet systems, or in-line blender systems. The grid system requires extensive piping and small-diameter injection orifices that tend to clog, so practical use of this system is limited. The in-line blender system

and the countercurrent jet diffusion system, as illustrated in Figures 21-9 and 21-10, are both considered effective devices. Figure 21-11 presents a typical mechanical flash mixing system. Static, in-line mixers, such as that shown in Figure 21-8, are also used for flash mixing; chemicals are injected just upstream of the mixer through diffusers.

For effective flash mixing, it is usually helpful (particularly with inorganic salts such as alum) to dilute the coagulant to a concentration of 2.5–5 percent solution before injection into the raw water. Larger volumes are easier and quicker to disperse into a large body of raw water than smaller volumes. However, it should be cautioned that at con-

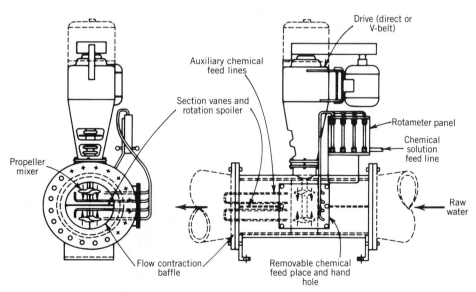

FIGURE 21-9. In-line blender flash mixer. (Courtesy of Walker Process.)

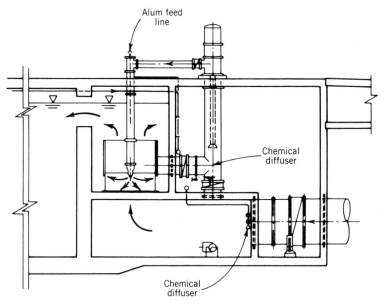

FIGURE 21-10. Pump injection–diffusion flash mixer.

centrations of less than 1 percent, alum may disso-ciate and alum floc or scale [Al(OH)₃] will form. This may clog feeding lines or orifices despite veloc-ities in excess of 10 m/sec. For this reason, alum or ferric salts generally should not be diluted below 2.5 percent prior to addition to raw water. This is par-ticularly critical under low-flow, low-alum-dosage conditions. In addition, the design should provide a confined, narrow path at the chemical injection point to improve chemical distribution. The flow velocity at the narrow path, however, should not exceed 3 m/sec (approximately 10 fps) unless excess static head is available at the plant inlet. When plant flow-rate exceeds 5 m³/sec (110 mgd), a multiple-jet sys-tem with multiple injection headers or a static mixer may be preferable. However, in case of multiple injection nozzles, the nozzles can become clogged if raw water is pumped through the nozzles. There-fore, the size of the orifices and the fine mesh screen in the pump suction line should be carefully se-lected.

The proper sequence of chemical injection proves important when more than one chemical is added to the raw water. Treatment of some raw waters requires prechlorination, pH adjustment, and polymer addition as a coagulant aid. The proper chemical sequence is best determined by testing the particular raw water and chemical combination in

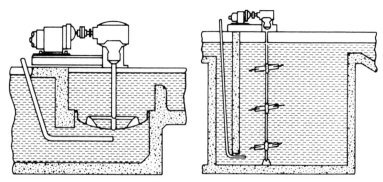

FIGURE 21-11. Mechanical flash mixing system. (Reprinted with permission from *Water Quality and Treatment*, American Water Works Association. Copy-right 1971, McGraw-Hill Book Company.)

the laboratory. For example, residual chlorine generally reduces the effectiveness of polymers and activated carbon. Anionic polymers are usually most efficient when they are fed to raw water after alum addition, once pin-point flocs have been formed. If a cationic polymer is used as a primary coagulant, the time of mixing becomes less important because polymers react directly with particulate matter. If THM formation due to chlorination is a problem, chlorine application points and dosage at each point should be carefully considered.

Coagulants and Flocculants

There are two basic types of coagulants and flocculants: inorganic salts and synthetic, high-molecular-weight organic compounds. Due to the negative surface charge of most naturally occurring particles, the most effective coagulants are compounds that have a positive charge of high valence. The most common inorganic salts of this type are aluminum and iron salts, such as aluminum sulfate (alum) and ferric chloride. High-molecular-weight, cationic polyelectrolytes are popular synthetic organic coagulants.

Flocculants are mostly organic in nature, but a few inorganic ones, such as aluminum hydroxide and activated silica, also exist. Among organic flocculants, synthetic high-molecular-weight anionic and nonionic polymers are the most popular and effective, but natural, high-molecular-weight organic substances such as chitosan, gum, seaweed extract (sodium alginate), and CMC (carboxyle methyle cellulose) have also proven to be quite effective.

Table 21-4 shows major coagulants used in water treatment processes. The EPA, as well as state health departments, regulate those chemicals that may be used in potable water treatment. Only approved additives should be used.

Applications

Coagulant is usually applied during flash mixing just upstream of the flocculation tanks. A second coagulant application point, between the sedimentation tanks and filters, is advantageous. There are typically two flocculant application points: (a) downstream of the coagulant application point where a pinpoint floc starts forming and (b) between the sedimentation tanks and filters as a filtration aid. In order to make an easy second chemical application

point for coagulant and flocculant, a common flow passage should run between the sedimentation tanks and filters. In the case of direct filtration, the second application point for coagulant and flocculant is a necessity. Flocculants should be properly diluted before they are added to process water. For alum, a stock solution of greater than 10 percent concentration not only minimizes storage space but also prevents deterioration. After metering, the stock solution can be diluted to 2.5–5 percent solutions by a simple hydraulic injector, but the pH of the alum solution should be maintained below 3.5, preferably below 3 before injection to the raw water.

The majority of nonionic and anionic polymers are generally available only in dry form. Because of their high viscosity, practical feed solution strength is generally less than 0.5 percent. This low-concentration polymer solution, however, does not sustain its effectiveness for a long period of time and should be used within one or 2 days. Longer storage not only reduces its effectiveness but also allows microorganisms to grow in the solution. Therefore, the capacity of the working solution tank should not allow for more than a 1-day detention time at the maximum-feed condition. A good washing system with high-pressure water should be provided for the tanks and the entire polymer feeding system including the metering pumps and the chemical feed pipeline. Every time polymer feed is stopped for more than 1 day, the whole system should be flushed by pressurized water.

Cationic polymers require the same precautions as anionic or nonionic polymers except their lower viscosity allows their solution concentrations to be higher. Normally, any metering pump can handle cationic polymer stock solutions without dilution. In addition, metering may be easier without dilution. After metering, dilution can be accomplished by a hydraulic injector. High-quality water should be used to prepare solutions for dilution water provided by the injector.

Doses

Coagulant or flocculant dosage depends on the characteristics of raw water colloids and suspended matter, and chemical composition of the water. Some physical factors such as water temperature, mixing conditions, and so on, also affect doses. Attempts have been made to establish a method to determine optimum coagulant dosage for any raw

TABLE 21-4. Chemicals Used in Coagulation Process

Chemical Name and Formula	Common or Trade Name	Shipping Containers	Suitable Handling Materials	Available Forms	Weight (lb/ft³)	Solubility (lb/gal)	Commercial Strength (%)	Characteristics
Aluminum sulfate, $Al_2(SO_4)_3 \cdot 14 H_2O$	Alum, filter alum, sulfate of alumina	100–200-lb bags, 300–400-lb bbls., bulk (carloads), tank truck, tank car	Dry-iron, steel, Solution-lead-lined rubber, silicon asphalt, 316 stainless steel	Ivory-colored powder, granule, lump liquid	38–15 60–63 62–67 10 (lb/g)	4.2 (60°F)	15–17 (Al_2O_3)	pH of 1% solution of 3.4
Liquid alum, $Al_2(SO_4) \cdot 49.6 H_2O$	Alum, liquor	4000-gal tank trucks, 10,000-gal tank cars	Same as above	Aqueous solution, approximately 50% solution	± 82.7	±5.5 lb dry alum/gal liquid alum	8 (Al_2O_3) 7.5–8.5% Al_2O_3	Freezing point is approximately 5°F (−15°C)
Ammonium aluminum sulfate, $Al_2(SO_4)_3(NH_4)_2 \cdot SO_4^- \cdot 24 H_2O$	Ammonia alum, crystal alum	Bags, bbls. bulk	Duriron lead, rubber silicon, iron stoneware	Lump, nut, pea, powdered	64–68 62 65 60	0.3 (32°F) 8.3 (212°F)	11 (Al_2O_3)	pH of 1% solution 3.5
Bentonite	Colloidal clay, volclay, wilkinite	100-lb bags bulk	Iron, steel	Powder, pellet, mixed sizes	60	Insoluble (colloidal solution used)		Very corrosive in liquid form
Ferric chloride, $FeCl_3$ (35–45 percent solution)	"Ferrichlor," chloride of iron	5–13-gal carboys, trucks, tank cars	Glass, rubber, stoneware, synthetic resins	Dark brown syrupy liquid		Complete	37–47 ($FeCl_3$), 20–21 (Fe)	
$FeCl_3 \cdot 6H_2O$	Crystal ferric chloride	300-lb bbls.	Same as above	Yellow-brown lump	60–64	6.8 (50°F)	59–61 ($FeCl_3$),	hygroscopic (store lumps and powder in tight container) no dry feed; optimum pH 4.0–11.0
$FeCl_3$	Anhydrous ferric chloride	500-lb casks; 100–300, 400-lb kegs	Same as above	Green-black powder	45–60	6.0 (32°F)	20–21 (Fe) 98 $FeCl_3$, 34 (Fe)	

Ferric sulfate, Fe2(SO4)3· 9 H2O	"Ferrifloc" ferrisul	100–175-lb bags, 400–425-lb drums	Ceramics, lead plastic rubber 18-8 stainless steel	Red-brown powder or granule	70; 72	Soluble in 2–4 parts cold water	90–94 (SO3), 25–26 (Fe); 55 (FeSO4), 20 (Fe)	Mildly hygroscopic coagulant at pH 3.5–11.0; Hygroscopic; cakes in storage; optimum pH 8.5–11.0
Ferrous sulfate, FeSo4·7H2O	Copperas, green vitriol	Bags, bbls. bulk	Asphalt, concrete lead, tin, wood	Green-crystal granule, lump	63–66			
Potassium aluminum sulfate, K2SO4Al2(SO4)3· 24 H2O	Potash alum	Bags, lead-lined bulk (carloads)	Lead, lead-lined rubber, stoneware	Lump granule, powder	62–67; 60–65; 60	0.5 (32°F); 1.0 (68°F); 1.4 (86°F)	10–11 (Al2O3)	Low, even solubility; pH of 1% solution, 3.5
Sodium aluminate, Na2OAl2O3	Soda alum	100–150-lb bags, 250–440-lb drums, solution	Iron, plastics, rubber, steel	Brown powder, liquid (27°Be)	50–60	3.0 (68°F); 3.3 (86°F)	70–80 (Na2), Al2O4 min.. 32 Na2, Al2O4	Hopper agitation required for dry feed
Sodium silicate, Na2OSiO2	Water glass	Drums, bulk (tank trucks, tank cars)	Cast iron, rubber, steel	Opaque, viscous liquid		Complete	38–42°Be	Variable ratio of Na2O to SiO2; pH of 1% solution, 12.3
Cationic polymer	Cat-Floc, Magnifloc-C, Pritloc-C, Nalcolyte, Percol	Drum, tank trucks	Type 18-8 or rubber lined steel tank fiberglass or plastic container	Liquid, viscous	65–75	Complete	Varies	Either acid or alkali in nature, dilution required before use
Anionic polymers	Cat-Floc, Magnifloc-C, Pritloc-C, Nalcolyte, Percol	50-lb bags, 35-gal drum for liquid form (not common)	Type 18-8 or rubber lined steel tank fiberglass or plastic container	Majority are dry form	44–46	0.1–0.5% solution recommended	Mostly full strength	Hygroscopic formation of "fish eyes," aging time required
Nonionic polymers	Cat-Floc, Magnifloc-C, Pritloc-C, Nalcolyte, Percol	50-lb bags, 35-gal drum for liquid form (not common)	Type 18-8 or rubber lined steel tank fiberglass or plastic container	Majority are dry form	44–46	0.1–0.5% solution recommended	Mostly full strength	Hygroscopic formation of "fish eyes," aging time required

water, but results show the traditional jar test to be the most practical method for dosage determination for ordinary flocculation and sedimentation processes. A pilot filter test provides the most reliable and practical method for dosage determination if direct filtration or in-line filtration is employed.

Feeding Considerations

Powdered activated carbon and potassium permanganate, often used for taste and odor removal, require some special attention in the feeding system. Carbon should be properly wetted before application. Potassium permanganate has a low solubility at water temperatures under 25°C (2.5–7.8 percent solubility by weight) and the dissolving tank should be sized properly.

Lime suspension transport pipes often clog with calcium carbonate deposits. Positioning lime feeders directly above the water process channel eliminates this problem. If transport of the lime suspension is necessary, a short open-channel feed line or rubber hose should be used for the feed line so that the scale can be removed by periodically hammering the hose. Regardless, avoiding a long feed line with excessive turbulence will minimize scale formation. If a long feed line is necessary, transport

water should be low in hardness (less than 20 mg/L as $CaCO_3$) and velocity in the line should be maintained in the 1–2 m/sec range.

Chemicals that have "hydroxide" molecules such as $Ca(OH)_2$, $NaOH$, and NH_4OH should not be fed with a hydraulic eductor system unless the hardness of utility water is less than 20 mg/L as $CaCO_3$. Otherwise, a softening reaction with the hardness of utility water takes place at the injector and the feed line. Consequently, severe scaling and clogging problems will occur in the feed line as well as injection system.

Many of the polymer feed considerations previously discussed are incorporated in Figure 21-12, which illustrates a typical polymer feeding system schematically.

AERATION EQUIPMENT

Introduction

Gas transfer or aeration processes are widely used in the water and wastewater treatment fields, providing services such as removal of undesirable dissolved gases or inorganic substances such as iron or manganese, solids removal, sludge concentration,

TABLE 21-5. Aeration Equipment Characteristics

Type	Oxygen Transfer Rate (kg O_2/kW·hr)	Hydraulic Head Required, m(ft)	Air Contact Time	Hydraulic Detention Time	Typical Applications
Spray	—	1.5–7.6 (5–25)	1–2 sec	—	CO_2 removal, taste and odor control, aesthetic value
Cascade	—	0.9–3 (3–10)	0.5–1.5 sec	—	CO_2 removal, taste and odor control, aesthetic value
Multiple-tray	—	1.5–3 (5–10)	0.5–1.5 sec	—	CO_2 removal, taste and odor control
Diffused air	0.5	—	10–30 min	10–30 min	Iron and manganese removal, CO_2 removal, taste and odor control, reservoir management
Hydraulic aspirator	1.5–3.5	3–6 (10–20)	<0.1 sec	—	Iron and manganese removal in pressure filter systems
Submerged turbine	1.5	—	2–6 hr	2–6 hr	Wastewater treatment
Mechanical aspirator	0.7	—	2–6 hr	2–6 hr	Wastewater treatment
Surface turbine	1.5–4.5	—	0.5–1 sec cyclically	2–6 hr	Wastewater treatment
Rotating brush	2.5	—	0.5–1 sec cyclically	1–7 days	Wastewater treatment

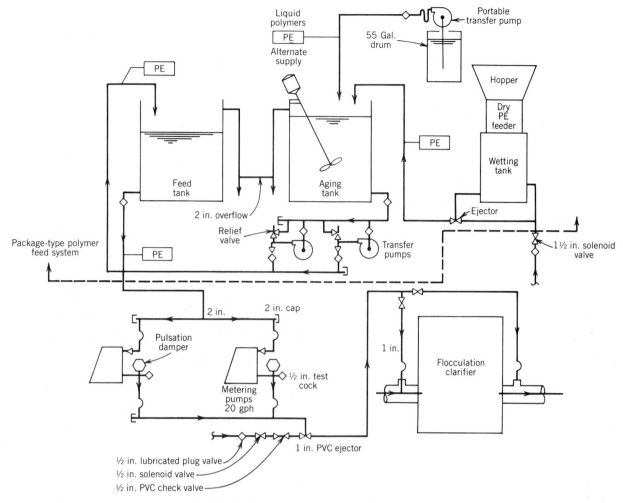

FIGURE 21-12. Typical polymer feeding system.

and mixing small reservoirs to eliminate problems associated with stratification.

Four basic types of aeration systems are discussed in this section: (1) droplet or thin-film-type aerators such as spray nozzles and multiple-tray aerators; (2) diffusion or bubble aerators using compressed air; (3) aspirator-type aerators; and (4) mechanical aerators. Characteristics and typical applications of these basic aeration systems are summarized in Table 21-5. Process design of aeration systems is developed in Chapter 11.

Droplet or Thin-Film-Type Aerators

This category of aerators is composed of spray and cascade or multiple-tray aerators. Tray-type aerators may either be of the natural draft type or the forced draft type. Spray aerators, in addition to

having aesthetic value, are usually quite efficient with respect to gas transfer. Their use, however, requires a large area (10–30 m²/100 L sec or about 50–150 ft²/mgd of flowrate) and favorable climatic conditions. Nozzles used in spray aeration systems vary in size, number, and spacing, according to head and area available for aeration. Nozzle diameters usually range from 2.5 to 3.8 cm (1–1½ in.) to minimize clogging.

Cascade and multiple-tray aerators operate on the basic principle of maximizing the exposure of the water to the atmosphere by forcing the water to flow over obstructions, usually coarse media such as coke or stone, 50–150 mm (2–6 in.) in diameter.

The cascade aerator shown in Figure 21-13 is commonly used in low-head, free-flowing wells. Cascade aerators are generally inefficient because of the thickness of the water flowing over the sides

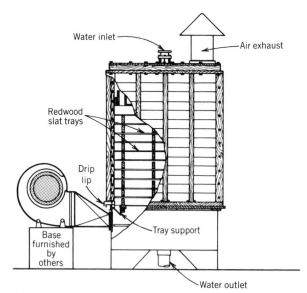

FIGURE 21-13. Forced-draft multiple-slat tower aerator. (Courtesy of General Filter Company.)

FIGURE 21-14. Natural draft coke tray aerator. (Courtesy of General Filter Company.)

and the small surface area of the cone. In addition, the air contact must rely on natural convection.

Wood slat towers are the most efficient of the multiple-tray-type aerators. The slat towers are mostly either forced or induced draft and are, therefore, enclosed in the wood, fiberglass, or metal shell. The slats are generally stacked 4–6 in. and centered vertically. The horizontal spaces between the slats are staggered so that the water trickling down one tray strikes the middle of the slat of the next tray. Commercially produced slat towers are designed for water-loading rates of roughly 0.014 m/sec (20 gpm/ft²) and air–water ratios of 25 : 50. The air blower is sized for a pressure drop of 0.33–0.5

Coke tray aerators, as shown in Figure 21-14, also are typically designed for natural draft. These provide somewhat more turbulence in air–water contact because of the large surface area of the coke. Coke tray towers are designed with 6 in. of coke on each bed. No more than two to four beds per tower give the unit an overall height of 4–6 ft. Water-loading rates in coke tray towers generally vary from 0.014 to 0.007 m/sec (10–20 gpm/ft²). The units are built with splash skirts to reduce the water loss and icing of the protective retaining screens. Influent water is generally distributed over the trays with a spray nozzle or special distribution trough.

Cascade and multiple-tray-type aerators encounter corrosion and algae and slime growth problems, particularly if the process water contains hydrogen

sulfide. Chlorination and copper sulfate treatment of the process water help control these problems. In general, cascade aerators are about half as efficient as multiple-tray aerators.

Diffusion-Type Aerators

Diffusion-type aeration systems (either diffused air or dispersed air systems) employ compressed air to force air into the liquid medium. In diffused air systems, the compressed air is generally introduced through porous membranes, porous plates or tubes, or wound fiber or metallic filaments such as the bubble column in Figure 21-15a. Dispersed air systems consist of mechanical mixers and a stationary air dispersion system, as shown in Figure 21-15b. The air dispersion system is usually provided with a sparger or orifice-type air dispersion apparatus. Diffused air systems generally require air filters to screen out particulates, but dispersed air systems usually do not require air filtration.

Diffusion-type aerators generally consist of rectangular tanks. With compressed air systems, the depth of these tanks is restricted to about 4.5 m (15 ft). Desired detention time usually ranges from 10 to 30 min. A spiral flow pattern is desirable in order to promote gas transfer while simultaneously minimizing flow short circuiting.

Air dispersion apparatus such as porous tubes or spargers are generally located a small distance above the tank bottom to reduce the pressure requirements of the air compressor. The amount of air required generally ranges from 10 to 100 L per 100 L of treated water (approximately 0.01–0.15 ft³/gal). Power requirements depend on the back pressure to

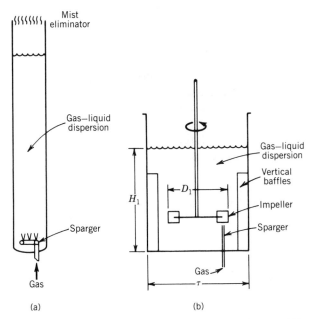

FIGURE 21-15. Common diffused aeration systems: (a) Bubble column; (b) agitated vessel.

the air compressors, but the average power requirement is about 20 kW for each cubic meter per second of flowrate (approximately 1 kW/mgd). Blowers used for aerating systems are typically either multiple-stage or single-stage centrifugal or rotary positive displacement. Rotary blowers are often used for small installations in the systems in which water depth varies significantly.

Turbine or impeller dispersion systems do not impose as much of a restriction on aeration tank geometry as do ordinary compressed air systems. The turbulent mixing in the vessel aids in the disintegration and increased service area of the gas; therefore, the gas transfer efficiencies are generally much better than simple diffused air systems. Oxygen transfer rates for compressed-air-type aeration systems are about 0.5 kg/kW·hr. Turbine-dispersion-type aeration systems have oxygen transfer rates of about 1.5 kg/kW·hr.

The advantages of diffusion-type aerators include (1) negligible head loss for the process water system, (2) less space requirements than for the droplet and thin-film-type aerators, and (3) better gas transfer efficiency. Diffused air systems may be extremely efficient for reservoir management. Many successful diffused air applications for reservoir destratification have been reported in the literatures (AWWA Committee Report, 1971; Symons,

Carswell, and Robeck, 1970; Laverty and Nielsen, 1970; Biederman and Fulton, 1971; Garton, 1978; Steichen et al., 1979).

Aspirator-Type Aerators

Air aspiration is commonly accomplished with either hydraulic aspirators or mechanical devices. A typical hydraulic aspirator is a type of hydraulic eductor or injector in which pressurized water flows through a throat similar to a venturi tube to create a low-pressure condition which in turn draws atmospheric air or gas into the water. A second type of aspirator is a mechanical aspirator that consists of a hollow-blade impeller that rotates at a speed sufficient to aspirate and discharge atmospheric air into the water.

Mechanical aspirators are capable of transferring about 0.7 kg O_2/kW·hr. Hydraulic aspirators are mostly small in size, but a single unit under favorable conditions and operating with atmospheric air produces twice the oxygen transfer rate of ordinary mechanical aspirators. If compressed air is supplied to the hydraulic aspirator, the transfer rate could be as high as 3.5 kg O_2/kW·hr.

Mechanical Aerators

Surface aerators and aeration pumps are the two basic types of mechanical aerators. Surface aerators may be of the turbine type or the brush type. Aeration pumps consist of a turbine mixer with a draft tube. A number of patented types of mechanical aerators can be purchased from manufacturers. Most are exclusively used for wastewater treatment. The turbine-type aerators have an oxygen transfer rate of 1.5–4.5 kg/kW·hr, and brush aerators generally have a rate of 2.5 kg/kW·hr.

Degasifiers and Deaerators

Henry's law states that the amount of gas dissolved in a liquid at a given temperature is proportional to its partial pressure in the gaseous (atmospheric) phase above the liquid (water) phase. This law governs not only the operation of the previously discussed aeration devices but the operation of degasifiers and deaerators as well. If volatile compounds are present in the water in greater concentrations than their equilibrium solubility with the atmosphere, these compounds can be removed (stripped) by aeration. The efficiency of the stripping process depends on the volatility of the constituents. Gases like carbon dioxide, which are highly volatile, are readily stripped. Higher-molecular-weight volatiles, such as the essential oils from algae, which give rise to taste and odor problems, are more difficult to remove. To optimize degasification, the most common approach is to use a packed tower arrangement with forced air ventilation. Air–liquid ratios for stripping of various constituents can be theoretically calculated as described in Chapter 11. Pilot-scale testing is also useful and frequently may be the only practical way of determining design parameters.

Deaeration, the removal of all dissolved gases in water (principally oxygen and carbon dioxide), takes advantage of the reduction in solubility of gases as the temperature of a liquid approaches its boiling point. Gases are nearly insoluble in boiling water. There are two basic approaches to deaerating water: (1) heat the water to boiling at atmospheric pressure or (2) reduce the pressure above the water until it boils at ambient temperature. The method used depends on the required water temperature following deaeration. Hot deaerated water is often used as boiler feed water. Vacuum deaeration is often used where pure but cool deaerated water is required as part of the process. The removal of the oxygen and carbon dioxide greatly reduces the corrosive properties of the water.

FLOCCULATION FACILITIES

Introduction

The objective of flocculation is to provide for an increase in the number of contacts between coagulated particles suspended in water by gentle and prolonged agitation. During agitation, particles collide, producing larger and more easily removed flocs. This process is discussed in detail in Chapter 6.

In designing flocculation facilities, several factors should be considered:

- Raw water turbidity and suspended particulates, including the particle size distribution and surface charge characteristics.
- Water temperature.
- Type of downstream treatment.
- Type of coagulant(s) used.
- Local conditions, such as the availability of repair personnel from equipment manufacturers.

Table 21-6 presents general design criteria for flocculation basins.

Several types of mixers have been designed for use in the flocculation process. The most common type of flocculator is the mechanical mixer. Reel or

---

TABLE 21-6. General Design Criteria for Flocculation Basins[a]

		Water Purification		Lime (and Soda Ash) Softening	
		River Water	Reservoir Water	River Water	Underground Water
Minimum mixing time, min	Conventional treatment	20	30	30	30
	Direct filtration	15	15	—	—
Range of energy input (G), sec^{-1}	Conventional treatment	10–50	10–75	10–50	10–50
	Direct filtration	20–75	20–100	—	—
Minimum number of stages	Conventional treatment	2–3	3	3	3
	Direct filtration	2	2	—	—

[a] (1) Tapered mixing is recommended. The range of energy input shown is typical of the range of energy input across the basin in a tapered energy flocculator. (2) For direct filtration, raw water should be low turbidity (less than 10 TU) most of the year. (3) The range of energy inputs shown are based on alum flocculation or alum addition with polymer as a coagulant aid. If ferric salts are chosen as a coagulant, the maximum G value should not exceed 50 sec^{-1}. If a cationic polymer is used as a coagulant, the required energy input may be 50 percent higher than shown above. (4) Tapered mixing and effective compartmentalization between stages are essential. (5) Frequently, there is a scale-up problem between pilot-scale flocculators and full-scale flocculators if G and Gt are used as the design criteria. Most pilot studies tend to give a significantly high G or Gt value as an optimum condition. (6) The design criteria shown are applicable to most types of flocculation units.

paddle flocculators are currently used for low- or medium-energy mixing, and propeller or turbine flocculators are used for medium- to high-energy mixing. In addition to mechanical mixers, baffled channel flocculation tanks can be efficient flocculation devices. This baffled channel concept is particularly attractive in developing countries where maintenance on mechanical equipment may be poor. Finally, several unusual flocculator designs are available, such as the "walking beam" flocculator, where paddles move up and down; the "flocsillator" process, where paddles swing gently; the "nu-treat" process, where large lattice work paddles make a nutating motion, combining rotation and oscillation; and a diffused air flocculator, which is similar to an aeration tank. A summary of various flocculator types is given in Table 21-7.

Basin Size and Shape

The size and shape of a flocculation basin is generally determined by the type of flocculator selected and the type of sedimentation process employed downstream. If mechanical flocculators are paired with rectangular, horizontal-flow sedimentation basins, the width and depth of the flocculation basins should match the width and depth of the sedimentation basins. Similar dimensions enhance constructability and reduce overall project costs.

If an upflow clarifier is selected (e.g., a sludge blanket clarifier) for the sedimentation tank, the flocculation tank is a cone-shaped frustum located in the center of the clarifier. The settling region is an inverted cone-shaped frustum located radially adjacent to the flocculation region.

In all cases, the size of the flocculation basin is determined by the required reaction or detention time. Typically, reaction times range from 15 to 45 min, depending on the characteristics of the raw water, the type of coagulant used, and the type of downstream treatment provided. For treatment of low-turbidity raw water in a cold region, the appropriate design flocculation time should be at least 30 min. A short mixing time (15 min) should only be adopted for direct filtration in warm regions or where the raw water can be easily flocculated year-round. A series of systematic jar or bench tests conducted in conjunction with pilot studies will aid in producing accurate design criteria. Although no theoretical relationship exists between basin area and water depth for optimal flocculation, the tank should be no deeper than 5 m (16.5 ft). Basins with depths in excess of 5 m often display unstable flow patterns and poor flocculation.

Baffled channel flocculators should be designed so that the cross section will provide the proper flow velocity. Adjustable baffles should be included to provide the desired degree of turbulence. Vertical-shaft floccuators should be situated to cover a square or circular mixing zone for maximum efficiency.

Baffled Channel Flocculators

Baffled channel flocculators provide plug-flow mixing conditions and are an effective flocculation system. Although baffled channel flocculators have some disadvantages, such as inflexible mixing and a large head loss across the basin, they are capable of producing good floc without any serious flow short circuiting if the plant flowrate is fairly constant.

A baffled channel flocculation tank should be designed for a channel flow velocity of 0.15–0.45 m/sec. Since tapered mixing is effective in forming large settleable flocs, the baffles should be properly arranged to reduce the mixing intensity and floc shearing force in the latter stages of the tank. Flow patterns for the baffled channel can be either "end around" or "over and under," but head loss is greater for the latter type. Since plant flowrate may vary in the range of 1:4 within a single day, achieving good mixing in the entire flow range may be difficult.

For baffled channel flocculation, the baffle arrangement can be computed based on an average G value of 20 or 30 sec^{-1} using the following formula:

$$G = \sqrt{\frac{g\rho h}{\mu t}}$$

where g is the gravity constant (9.8 m/sec^2), ρ is the mass density (1000 kg/m^3), h is the head loss (m), μ is the absolute viscosity (10^{-3} kg/m sec), and t is the retention time (sec).

The head loss in a baffled mixing channel from turbulence and friction on the sides of the channel can be calculated using the Chezy formula:

$$h = \frac{Lv^2}{C^2 R}$$

TABLE 21-7. Flocculator Design Guidelines[a]

Type of Flocculator	Basic Design Criteria	Advantages and Disadvantages
Vertical shaft	G value up to 100 sec^{-1}, maximum tip speed of 2 m/sec, approximately 5m × 5m to 10m × 10m basin surface area per unit, downward flow pattern preferable for propeller unit, stator baffles should be provided for turbine units	Easy maintenance and few breakdowns. Suitable for high-energy input. Suitable for direct filtration and conventional treatment. Many units required for a large plant. High capital cost for variable-speed reducers and support slabs.
Horizontal-shaft paddle	G value up to 50 sec^{-1}, maximum tip speed of 1 m/sec, number of paddles adjusted for tapered mixing, paddle area should not exceed 20 percent of tank section area	Generally produces a large-size floc. Simple mixing unit. Suitable for conventional treatment. Need for precise installation and maintenance. Difficult to increase energy input. Problems with leakage and shaft alignment.
Baffled channel	Tapered mixing by adjusting baffles, maximum flow velocity of approximately 0.75 m/sec, end-around baffle used when total head loss across tank is limited	Performs well if the plant flowrate is reasonably constant. Little maintenance. A lack of flexibility for mixing intensity. High head loss for the over-and-under baffle.
Diffused air and water jets	$G = 95–20$ sec^{-1} or $Gt = 10^5–10^6$, may be used for auxiliary mixing when plant is overloaded	Simple installation and less capital cost. Limited amount of operational data available. High local velocities for water jet flocculators. High operational cost for air diffuser flocculators.

[a] Mechanical flocculator units have more flexible mixing conditions. Proper compartmentalization and stator baffles improve the efficiency of mechanical flocculators. Thus, changing the unit or the blades may not always be necessary to achieve better flocculation.

where L is the length of the mixing channel, C is the Chezy coefficient, R is the hydraulic radius, and v is a mean flow velocity. Head loss around a baffle in a channel can also be computed assuming a 180° turn in the direction of water flow in a square pipe using the formula

$$h = 3.2 \left(\frac{v^2}{2g} \right)$$

Baffle Walls

The performance of mechanical flocculators may be improved through compartmentalization in order to minimize flow short circuiting. Baffle or diffuser walls are typically used for this purpose. Baffles can be specified in various shapes and arrangements, but the baffle opening must be sized correctly. Generally, a simple baffle wall with an opening rate of greater than 10 percent (of basin cross-sectional area) is very ineffective in preventing short circuit-

ing; a 2–5 percent ratio is optimal. When the opening falls below 2 percent, a high flow velocity is created through the opening. To prevent floc breakage, the flow velocity at the opening should not exceed 0.3 m/sec (1 fps) at the maximum plant flow, and head loss across the baffle should be about 8 mm (0.02 ft). The top of the baffle should be slightly submerged so scum does not accumulate behind the baffle. An opening should be allowed below the baffle to facilitate sludge removal after dewatering the tank. If the flocculation tank is designed as an integral part of the sedimentation basin, a diffuser wall should be provided at the end of the flocculation tank to assure uniform flow distribution into the sedimentation tank.

An example of a flocculator baffle wall is shown in Figure 21-16.

A different type of baffle can also be used as an antivortex and turbulence-producing device. A high-energy mixer in a confined tank or a high-speed turbine impeller may promote rotational

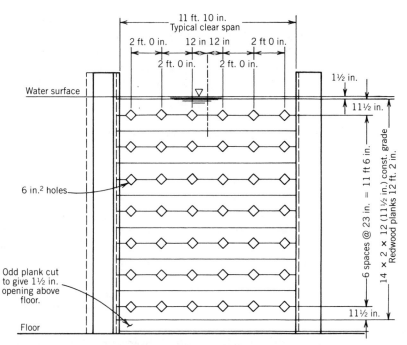

FIGURE 21-16. Baffle wall.

movement with possible vortexing about the impeller shaft. Since the rotational motion and vortex formation not only significantly reduce the mixing efficiency but also stress the mixer shaft bearings, the tank should be properly baffled. Vertical strips or baffles placed along the tank walls can break up rotational movement effectively. For a turbine mixing operation, the width of the baffles may be as small as one-tenth of the tank diameter; for propeller mixers, even smaller widths may be effective. In general, four baffles are sufficient. To reduce rotational motion in a flocculation tank, stationary strips called stator baffles are provided.

More detailed discussions of baffles and flow patterns can be found in chemical engineering handbooks and catalogs of mixer manufacturers. Figure 21-17 shows a guide for baffle location and size in mixing basins.

Mechanical Flocculators

Two basic types of mechanical flocculators are used in flocculation facilities: horizontal-shaft with paddles and vertical-shaft, high-energy flocculators. In recent years, vertical-shaft flocculators have been used with increasing frequency. The selection of one of the two types is mostly dependent on the type of filtration system used. Horizontal-shaft floc-

culators are typically employed if conventional rapid sand filtration is used. Conventional rapid sand filtration requires a high degree of solids removal by the sedimentation basin before filtration. The horizontal-shaft flocculators generally are suited to produce large and easily settleable floc

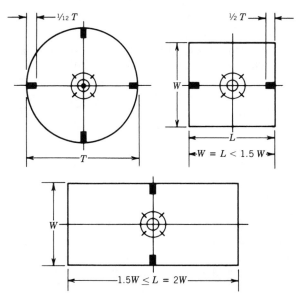

FIGURE 21-17. Schematic illustration of tank baffles. (Courtesy of Lightnin Mixing Equipment Co.)

with alum flocculation. However, they require more maintenance and expense mainly because bearings and packings are typically submerged. High-energy, vertical-shaft flocculators are the unit of choice for high-rate filtration systems. Since high-rate filters allow floc penetration into the filter bed, the desired type of floc for these filters is small in size but physically strong to resist high shear forces in the filter bed.

The basic design criteria for mechanical flocculators are the G value and t, the hydraulic detention time. Representative G values for horizontal-shaft paddle flocculator and vertical-shaft, high-energy flocculator designs are 30 and 80 sec^{-1}, respectively. For mechanical flocculators, the G value can be calculated with the following equations:

$$G = \sqrt{\frac{P}{V\mu}}$$

where P is the power input (J/sec or W), V is the tank volume (m^3), and μ is the absolute viscosity (approximately 0.001 kg/m sec at 20°C), or alternatively:

$$G = \sqrt{\frac{C_d A v^3}{2\nu V}}$$

where C_d is a drag coefficient of the paddle (approximately 1.8), A is the area of the paddle (m^2), v is the velocity of the paddles relative to that of the water, ν is the kinematic viscosity of water (approximately 10^{-6} m^2/sec), and V is the water volume in the tank (m^3).

For direct filtration and in-line or contact filtration, cationic polymers are often used as a coagulant, and high G values of 200 or 300 sec^{-1} have been recommended. The practical maximum G value, however, would be 75–100 sec^{-1} for effective mixing, optimum energy consumption, and overall plant efficiency.

When propeller or turbine vertical-shaft flocculators are specified, the maximum tip speed of the mixing blades should be kept below 2 m/sec (6 fps). When paddle flocculators are chosen, the paddle area should be kept below 20–25 percent of the tank section area to avoid the rotational movement of water. A paddle–tank area ratio of approximately 15 percent and a tip speed of 0.25–0.75 m/sec (0.8–2.5 fps) can serve as design guides.

Although it is not common practice in the environmental engineering field, displacement capacity (pumping capacity), Reynolds number, power number, and Froude number are used as design parameters for mixer selection in the chemical engineering field. The Froude number, however, is only used when a vortex formation is expected around the mixer axis.

Variable-speed drive units have been used extensively in mechanical flocculators. However, an infinitely variable speed is unnecessary; two- or three-speed drives are sufficient and require less frequent maintenance. All the drive units should be the same make and model for easy maintenance, even when tapered mixing is specified.

Corrosion of submerged metal portions of the flocculator assembly can be a serious maintenance problem. Specifying Type 18-8 stainless steel for all submerged portions of the flocculator assembly and a cathodic protection system for structural steel are common solutions to this problem.

Basin Inlet and Outlet Design

Another important consideration in flocculation basin design is uniform hydraulic loading to each tank. There are three basic types of tank inlet structures: a simple pipe connection to the tank, a weir inlet, and a submerged orifice inlet. (See Chapter 22.) The plant layout, especially symmetrization of tank layout to inlet line, and maintaining an appropriate flow velocity in the distribution pipe or channel will greatly minimize uneven flow distribution to each tank regardless of the inlet type selected.

As water leaves the flocculation tank after treatment, the formed flocs must not be broken before settling and should be distributed uniformly in the inlet of the settling tank. A simple solution to these problems is to build the flocculation tanks as an integral part of the settling tank and provide an effective diffuser wall between the two tanks. A permeable baffle with an opening ratio of about 5–7 percent and approximately 3–4-mm head loss across the baffle has proven effective.

EXAMPLE 21-1

Problem

A high-energy, vertical-shaft flocculator is designed to treat a daily maximum flow of 3.0 m^3/sec (259.2 ML/d) with two horizontal flow rectangular tanks. Each tank is 26 m long × 17 m wide × 5 m deep (85

× 56 × 16 ft). The flocculator system has three-stage flocculation (each stage to be compartmentalized by baffle walls) with tapered energy mixing. The energy input shall be 90–25 sec^{-1} as G from the first to last stage. Find (a) the number of flocculators required, (b) the required horsepower for the flocculators, and (c) the area of the openings on each baffle wall and head loss created by the baffle.

Solution

1. *Number of flocculators.* Each flocculator has the same approximate area of influence. Thus, two flocculators should be located in each of the three stages, making an influence area of 8.67 × 8.5 m (28.5 × 28 ft) for each unit. A total number of 6 flocculators is required for each tank, or 12 for the plant.

2. *Horsepower for the flocculators*
   a. First- and second-stage flocculators provide a maximum power input of 90 sec^{-1} G. Assume average water temperature to be 20°C.

$P = G^2 V \mu$
   $= (90 \text{ sec}^{-1})^2 \times (8.75 \times 8.5 \times 5 \text{ m})$
                                          $\times 10^{-3} \text{ kg/m sec}$
   $= 3013 \text{ J/sec or 3 kW (4 hp)}$

Assuming 80 percent overall efficiency,

$$\frac{3 \text{ kW}}{0.8} = 3.75 \text{ kW}$$

Use a 3.75-kW (5-hp) motor for each flocculator.
   b. Third-stage flocculators provide a maximum power input of 50 sec^{-1} G.

$P = (50 \text{ sec}^{-1})^2 \times (8.75 \times 8.5 \times 5 \text{ m})$
                                          $\times 10^{-3} \text{ kg/m sec}$
   $= 930 \text{ or } 0.94 \text{ kW (1.25 hp)}$
$$\frac{0.94 \text{ kW}}{0.8} = 1.2 \text{ kW}$$

Use a 1.5-kW (2-hp) motor for each flocculator. *Note:* For practical purposes, all flocculators may be equipped with a 3.75-kW motor. Each flocculator may also be equipped with a variable-speed reducer to cover the energy input specified.

3. *Baffle wall.* The baffle walls are between the first and second flocculators and second and third flocculators. To maintain 0.30 m/sec (1.0 ft/sec) flow velocity through the opening area, the required area is 3 m^3/sec ÷ 2 basins ÷ 0.30 m/sec = 5.0 m^2 for each baffle or 106 ft^3/sec ÷ 2 basins ÷ 1.0 ft/sec = 53 ft^2 for each baffle. Opening ratio is 5.0 m^2 ÷ (17 × 5 m) = 0.059, or 5.9 percent, which is acceptable.

Head loss through each baffle can be calculated as

$$h = \frac{1}{2g} \left(\frac{q}{ca} \right)^2 \quad (c \text{ assumed at } 0.7)$$
$$= \frac{1}{19.6} \left(\frac{1.5}{0.7 \times 5} \right)^2$$
$$= 0.051 \times 0.184$$
$$= 0.0094 m \quad \text{or } \sim 9.5 \text{ mm or } 0.37 \text{ in.}$$
$$\text{(acceptable)}$$

where c = coefficient of discharge
      a = area of orifice
      q = flow through the orifice
      h = head loss

To reduce head loss, a lower velocity could be provided through the baffle wall ports. However, do not decrease velocity so that the opening ratio exceeds 10 percent.   □

UPFLOW CLARIFIERS

Introduction

Upflow clarifiers are usually found in industrial and municipal applications where lime softening or softening–clarification is the major treatment process and uniform flowrates and constant water quality prevail. Upflow clarifiers are almost always proprietary units and can be grouped in three basic categories: (1) simple upflow clarifiers; (2) reactor-clarifiers; and (3) sludge blanket clarifiers. Although reactor-clarifiers with sludge recirculation and sludge blanket clarifiers are often classified more broadly as solids contact clarifiers, this section will discuss them separately.

Advantages of proprietary upflow clarifiers over horizontal-flow sedimentation basins are:

TABLE 21-8. Evaluation Summary and Typical Design Criteria for Clarifiers

Type of Clarifier[a]	Design Criteria	Advantages and Disadvantages	Application
Rectangular basin (horizontal flow)	Surface loading: 20–60 $m^3/m^2 \cdot d$ (0.3–1.0 gpm/ft^2) Water depth: 3–5 m (10–16 ft) Detention time: 1.5–3 hr Width–length ratio: 1 : 5	More tolerance to shock loads Predictable performance under most conditions Easy operation and low maintenance costs Easy adaptation to high-rate settler modules Subject to density flow creation in the basin Requires careful design of inlet and outlet structures Usually requires separate flocculation facilities	Most municipal and industrial water works; particularly suited to large-capacity plants
Upflow (radial)	Circular or square Surface loading: 30–45 $m^3/m^2 \cdot d$ (0.5–0.75 gpm/ft^2) Water depth: 3–5 m (10–16 ft) Settling time: 1–3 hr Weir loading: 170 m^3/m d (13,700 gpd/ft)	Economical compact geometry Easy sludge removal High clarification efficiency Problems of flow short circuiting Less tolerance to shock loads Need for more careful operation Limitation on practical size unit May require separate flocculation facilities	Small to mid-size municipal and industrial water treatment plants; best suited where rate of flow and raw water quality are constant
Reactor-clarifier	Flocculation time: ~20 min Settling time: 1–2 hr Surface loading: 50–75 $m^3/m^2 \cdot d$ (0.85–1.28 gpm/ft^2) Weir loading: (175–350 m^3/m (14,000–28,000 gpd/ft) Upflow velocity: >50 mm/min (2 in./min)	Flocculation and clarification incorporated in one unit Good flocculation and clarification efficiency due to a seeding effect Some ability to take shock loads Requires greater operator skill Less reliability than conventional clarifiers due to a dependency on one mixer Subject to upsets from thermal effects	Water softening; a plant that treats a steady quality and quantity of raw water
Sludge blanket	Flocculation time: ~20 min Settling time: 1–2 hr Surface loading: 50–75 $m^3/m^2 \cdot d$ (0.85–1.28 gpm/ft^2) Weir loading: 175–350 m^3m d (14,000–28,000 gpd/ft) Upflow velocity: >50/mm min (2 in./min) Slurry circulation rate: up to 3–5 times the raw water inflow rate	Good softening and turbidity removal Compact and economical design Adaptable to limited change in flowrate and raw water quality Sensitive to shock loads Sensitive to temperature change 2–3 Days required to build up the necessary sludge blanket Plant operation dependent on a single mixer Higher maintenance costs and a need for greater operator skill	Water softening; flocculation–sedimentation treatment of raw water that has constant quality and rate of flow; plant treating a raw water with low solids content

[a] Reactor-clarifiers and sludge blanket clarifiers are often considered as one category, solids-contact clarifiers.

1. The proprietary units have an economic and compact design.
2. Sludge removal is easier than in horizontal-flow basins.
3. High clarification efficiencies can be obtained due to slow inlet velocities, the seeding effect of previously formed sludge, and minor filtering action through the existing sludge zone.

Possible disadvantages of proprietary upflow clarifiers are:

1. The need for more careful operational control than with conventional horizontal-flow units.
2. Rapid loss of efficiency during hydraulic overloading and solids shock loading periods.
3. The tendency to exhibit operational problems when raw water turbidity and temperature continually fluctuate over a wide range.

A summary of design criteria and other data for the major clarifier types (including horizontal-flow sedimentation basins discussed in the next section) is given in Table 21-8.

Simple Upflow Clarifiers

The design of upflow clarifiers requires careful consideration of factors such as surface loading, uniform flow distribution into the settling zone, minimization of flow short circuiting from hydraulic and density currents, uniform withdrawal of clarified water, and sludge withdrawal without disturbing settling efficiency. For simple upflow clarifiers, the vertical-flow rise rate becomes an additional criterion; at any selected level, the flow rise must be less than the respective floc settling rate. Settling velocities of various flocs are presented in Table 21-9.

TABLE 21-9. Settling Velocity of Selected Flocs

Floc Type	Setting Velocity at 15°C	
	mm/min	ft/min
Small fragile alum floc	37–73	0.12–0.24
Medium-size alum floc	55–85	0.18–0.28
Large alum floc	67–92	0.22–0.30
Heavy lime floc (lime softening)	76–107	0.25–0.35

The most significant potential problem of center-feed circular clarifiers (as shown in Figure 21-18) is flow short circuiting. To minimize this problem, a peripheral-feed clarifier was developed that introduces inflow between the tank wall and an annular skirt. These two clarifier designs are shown in Figures 21-18 and 21-19, respectively. The peripheral-feed design allows the inflow to enter the settling zone near the tank bottom. The orifices in the annular inlet channel should be designed so that the head loss across each orifice inlet is approximately 10–15 mm. This provides uniform distribution and capacity for two to three times the loading permissible with center-feed clarifiers.

When high-rate settling modules are installed in a clarifier (see Figure 21-18) and the suspended solids are noncolloidal, the surface loading rate can be increased over those rates shown in Table 21-9.

Density currents can occur in clarifiers either from high influent solids loading or a temperature difference (as little as 0.3°C or 0.5°F) between the inflow and the ambient water. Experience shows that in circular center-feed clarifiers, placing the peripheral launder trough two-thirds to three-fourths of the radial distance from the center minimizes the

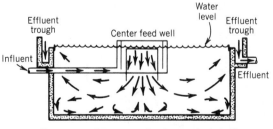

Flow pattern of the conventional center feed clarifier

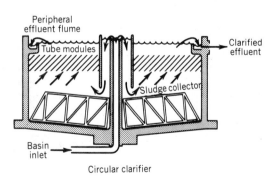

Circular clarifier

FIGURE 21-18. Center-feed clarifiers without (left, courtesy of Envirex) and with (right, courtesy of Neptune Microfloc) tube settler.

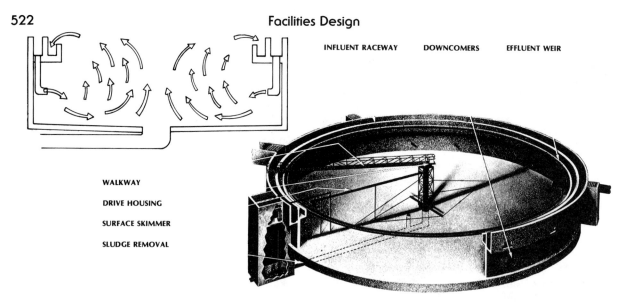

INFLUENT RACEWAY   DOWNCOMERS   EFFLUENT WEIR

WALKWAY

DRIVE HOUSING

SURFACE SKIMMER

SLUDGE REMOVAL

FIGURE 21-19. Peripheral feed–peripheral collect clarifier. (Courtesy of Ecodyne.)

density currents and produces better effluent. Recently developed peripheral-feed, peripheral-collection clarifiers have proven more efficient for high-solids influent water, which tends to cause distinct density currents.

In summary, peripheral-feed circular clarifiers are usually more effective than center-feed-type clarifiers and are now more prevalent. The best use of simple upflow clarifiers is clarification of waters with heavy, noncolloidal solids loading.

Reactor-Clarifiers

Reactor-clarifiers are center-feed clarifiers with flocculation zones built into the central compartment. Generally, these units contain a single motor-driven mixer, with optional controlled recirculation of the sludge slurry, followed by a settling zone in a separate outer annular compartment. One such unit is shown in Figure 21-20.

Due to the fact that a reactor-clarifier is a center-feed unit, it is subject to the particular hydraulic problems of this unit type. However, the flocculation compartment does make the design more economic and compact than that of simple upflow clarifiers. A minimum 20–30 min of contact time is usually provided in the flocculation zone, and for units without slurry recirculation, the surface area and detention time for the settling zone should be similar to that of simple upflow clarifiers.

Slurry recirculation units, frequently called solids contact clarifiers, are usually designed with a motor-driven impeller, which can recirculate up to 5 times the raw water inflow rate. The slurry concen-

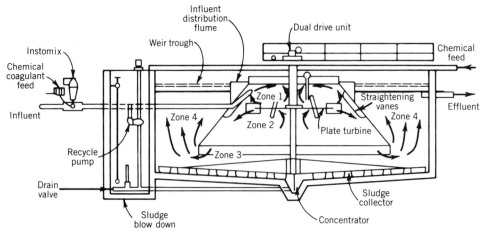

FIGURE 21-20. Sludge-recycling reactor-clarifier. (Courtesy of Walker Process.)

tration in the unit is controlled by an adjustable timer on a sludge blowoff line. When using alum, it is common practice to maintain the slurry concentration in the secondary mixing zone at 5–20 percent of the sludge volume found after 10 min of settling water taken from the primary flocculation compartment. The slurry concentration should be somewhat higher when used in a softening process.

Reactor-clarifier units without slurry recirculation are considered less efficient than the recirculation type and are sensitive to shock loads. Because these units do not have a large recirculation range (over 200 percent), they cannot provide sufficient buffer capacity as do the slurry-recirculation-type units. A severe hydraulic or water quality shock loading will cause floc carry-over or operational difficulty in maintaining a proper slurry concentration. However, treatment of industrial process water is a good application, since the initial capital investment is low and water quality and feed rates are usually uniform.

Sludge Blanket Clarifiers

Like slurry recirculation reactor-clarifiers, sludge blanket clarifiers are solids contact units. The difference between them, however, is the presence of a slurry pool or sludge blanket. The sludge blanket unit contains a central mixing zone for partial flocculation and a fluidized sludge blanket in the lower portion of the settling zone. Partially flocculated water flows through the sludge blanket where flocculation is completed and solids are retained. The sludge level is normally 2–3 m (6.5–10 ft) below the water surface, and clarified water is collected in launder troughs along the top of the unit. Sludge blanket clarifiers are made with or without slurry recirculation mechanisms.

Design criteria for sludge blanket clarifiers are essentially the same as for slurry recirculation reactor-clarifiers. However, the launder troughs should be sufficiently spaced to avoid sludge spillover. The spacing should be less than twice the distance between the water level and the top of the fluidized sludge blanket, approximately 4.5 m (15 ft). Usually, the troughs are arranged radially, as in many large unit designs, or in parallel.

Generally, sludge blanket clarifiers should be used only where the raw water characteristics and flowrates are relatively uniform. Given these parameters, the most effective applications are for lime softening and clarification of low-turbidity wa-

ter. These units may also be used for clarification of highly turbid water (exceeding 500 TU) if a sludge-scraping mechanism is provided.

SEDIMENTATION FACILITIES

Introduction

Sedimentation is one of the most basic water treatment processes. The following discussion is limited only to continuous-flow-type (as opposed to batch-type) sedimentation tanks. Plain sedimentation, such as presedimentation (e.g., grit chamber), and sedimentation following flocculation are the two basic applications for sedimentation facilities in water treatment. This section will discuss the design of both grit chambers and sedimentation tanks with emphasis on the latter.

Presedimentation Tanks (Grit Chambers)

In general, grit chambers should be located upstream of any raw water pumping facility (low-lift pumps) and as close as possible to the intake structure in order to avoid siltation problems in the plant intake pipeline. A horizontal-flow tank is commonly used for presedimentation. The shape of the chamber is usually rectangular with a contracted inlet in order to minimize turbulent flow. A divided single tank or two separate tanks are required at a minimum so that one tank or one side can be emptied for various purposes. Where sand carry-over is not a major problem, a single tank with a bypass pipeline would be satisfactory. Typical design criteria are shown in Table 21-10.

TABLE 21-10.   Typical Grit Chamber Design Criteria

Depth (without automated grit removal)	3.5–5 m (11.5–16 ft)
Depth (with grit removal)	3–4 m (10–13 ft)
Depth–length ratio	1 : 10
Width–length ratio	1 : 3–1 : 8
Surface loading	200–400 $m^3/m^2 \cdot d$ 3.3–6.6 gpm/ft^2
Horizontal mean flow velocity (at maximum daily flow)	0.05 m/sec (2 ft/sec)
Detention time	10–20 min
Minimum size of particle to be removed	0.1 mm

Assuming ideal design criteria, the required length of the tank can be estimated by the following equation:

$$L = K \left(\frac{h}{v_0}\right) v$$

where K is a safety factor (1.5–2), h is the effective water depth, v_0 is the settling velocity of the particle to be removed, and v is the mean water velocity at the maximum day flowrate. The settling velocities of various sizes of fine sand particles (specific gravity = 2.65) at water temperature 10°C (50°F) in still water are shown below:

Particle Diameter, mm	1	0.6	0.4	0.2	0.15	0.1	0.08	0.06
Settling Velocity, mm/sec	100	63	42	21	15	8	6	3.8

The bottom of the chamber should have a minimum of 1:100 longitudinal slope to facilitate basin draining and cleaning. A fine trash screen with approximately 20-mm openings is often provided at the front end of the grit chamber. Since the trash screen also acts as an effective diffuser wall, it should be installed close to the tank inlet. If a separate diffuser wall is specified, the total area of openings at the wall should be about 15 percent of the tank cross-sectional area. An example of a grit chamber facility for a 1.5-m³/sec (34.2 mgd) design flowrate is presented in Figure 21-21.

Sedimentation Tanks (Continuous-Flow Type)

General. Most sedimentation tanks designed for contemporary water treatment plants are horizontal-flow tanks with or without high-rate settler modules. These tanks can accept a shock load of either a hydraulic or water quality (e.g., turbidity) nature. Horizontal-flow tanks may be either rectangular or circular (plan view). Circular tanks may have either center feed with radial flow, peripheral feed with radial flow, or peripheral feed with spiral flow. Flow patterns for some of the various horizontal-flow sedimentation tanks are shown in Figure 21-22. Center-feed tanks often experience hydraulic short circuiting particularly when the peripheral collection channel is not equipped with radial weirs. To minimize short circuiting from density currents a peripheral-feed, peripheral-collect clarifier is commonly used (i.e., Kraus-Fall peripheral-feed clarifier) as discussed in the previous section. This type of clarifier is often used as a secondary clarifier for wastewater treatment where a high-solids influent always causes a distinct density flow. Similar density currents are also observed in water treatment clarifiers when the turbidity of the raw water ex-

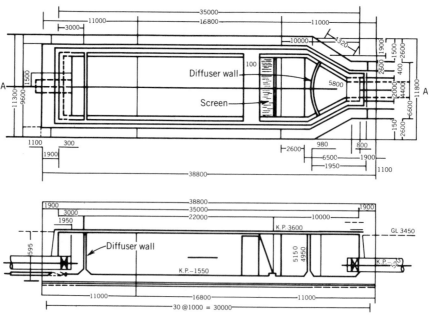

FIGURE 21-21. Typical grit chamber design.

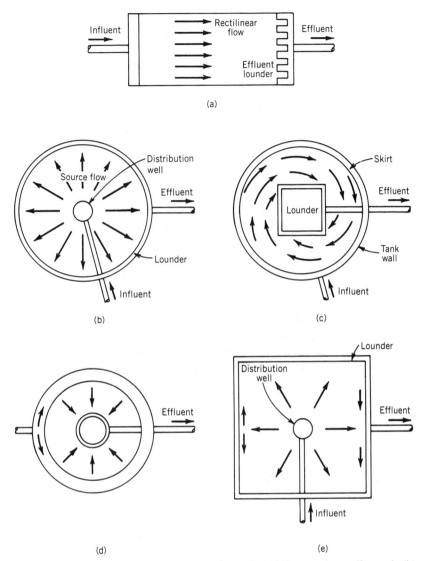

FIGURE 21-22. Flow patterns in sedimentation tanks: (a) Rectangular settling tank; (b) center feed, source flow; (c) peripheral feed, spiral flow; (d) peripheral feed, radial flow; (e) square, radial flow. (Reprinted from *Water Treatment Plant Design*. Copyright 1969, American Water Works Association, Inc.)

ceeds about 50 TU or the influent water temperature is more than 0.3°C colder than the water in the settling tank.

Configurations. Generally, sedimentation basins are rectangular with horizontal flow. One of the most cost-effective configurations for rectangular basins is to arrange the basins longitudinally side by side, utilizing common walls. The flocculation tanks can also be incorporated into the sedimentation basins by using the head-end portion of the basin as the flocculation tank. This arrangement minimizes

yard-piping and maintains a uniform flow distribution to each basin.

In areas with strong, predictable winds, sedimentation basins should be positioned so that the flow parallels the wind. In this way, the effects of wind currents on settling basin performance are minimized. Changes in water surface elevation are minimized when the wind blows across the length of the rectangular setting basin, as opposed to across the width. Wind-induced wave patterns on the surface of the water are much less severe, and the formation of circulation patterns which, in turn, affect settling rates, are greatly reduced. This design requires wave breakers (launders or baffles) at approximately 20–30-m (65–100-ft) intervals. Where sand storms are expected, a windbreak or cover over the basins may also be necessary. Trees may not be the best windbreak because leaves and small branches can drop into the basin and clog the sludge withdrawal system.

A minimum of two basins should be provided in order to be able to inspect, repair, periodically clean, and maintain one basin at a time while the other basin is in operation. An access road should be designed to and around the basins for operation and maintenance work and to facilitate any future additions. In the case of circular tanks, there will be some utilized land between the tanks and more yard piping is necessary than for a rectangular basin configuration.

Design Criteria. The basic design criteria for horizontal-flow sedimentation tanks are: (1) surface loading (overflow rate), (2) effective water depth, (3) mean horizontal flow velocity, (4) detention time, and (5) launder-trough weir loading rate. Table 21-11 provides some typical design criteria. Basin dimensions can easily be related to either the

Froude number or the Reynolds number. These numbers, however, have little value for actual basin design. The relationship between surface loading and the settling velocity of discrete particles was developed by Hazen (1904). Hazen's concept clearly indicates that the efficiency of an idealized, horizontal-flow settling tank is solely a function of the settling velocity of discrete particles and of the surface loading rate (the flowrate of the basin divided by the surface area) and is independent of the basin depth and detention time. This concept, however, is often inappropriate because the majority of settling basins treat flocculated suspended matter and do not have idealized flow pattern. Furthermore, flocculated or flocculent particles may increase in size in the basin and settle faster as they are carried through the basin. Therefore, the detention time is significant for most basin designs.

The basin can be made shallow with a large surface area, but there is a practical minimum basin depth necessary (2.5–3 m, 8–10 ft, minimum effective water depth) for mechanical sludge removal equipment. Also, other factors such as flow velocity, effect of wind and sun, and required basin area make this alternative less attractive. Effective water depth is particularly important for a basin with-

TABLE 21-11. Basic Design Criteria for Horizontal-Flow Sedimentation Tanks[a]

Description	Conventional-Type Sedimentation Basin	Sedimentation Basin for High-Rate Filters	
		Without High-Rate Settler	With High-Rate Settler
Surface loading			
For alum floc			
$m^3/m^2 \cdot d$	18–36	30–60	60–150[b]
gpm/ft^2	0.3–0.6	0.5–1.0	1–2.5[b]
For heavy flocs			
$m^3/m^2 \cdot d$	30–60	45–75	90–180[b]
gpm/ft^2	0.5–1.0	0.75–1.25	1.5–3[b]
Mean horizontal velocity			
m/min	0.15–0.9	0.3–1.2	0.05–0.13[b]
fpm	0.5–3	1–4	0.15–0.5[b]
Water depth, m	3–5	3–5	3–5
Detention time, min	120–240	60–120	6–25[c]
Weir loading			
$m^3/d \cdot m$	140–270	210–330	90–360[c]
gpm/ft	8–15	12–18	5–20[c]

[a] (1) High settler modules are commonly installed in the basin to cover up to 75 percent of basin surface area. (2) Cold weather regions generally use a conservative design criteria. (3) Weir loading requirement is somewhat questionable since some recent studies have shown that the long launders often do not yield better tank performance.
[b] These figures are for the surface loading of the settler module not the basin.
[c] These figures vary depending on the kind of settler selected.

out mechanical sludge removal facilities since the basin must provide adequate volume for sludge deposit.

With an efficient flocculation process, about 70 percent of floc will settle within the first one-third of the basin length. Estimated sludge height for well-flocculated water under normal conditions and without mechanical sludge removal mechanism would be 2–3 m (6.5–10 ft) at the influent end of the basin, but only 0.3 m in the last half of the basin.

An optimal weir loading rate is difficult to specify. Even 100 percent lower weir loading may not result in a significant improvement of sedimentation efficiency. This is particularly true for primary settling tanks used in wastewater treatment.

Inlet Structure. The inlet to a sedimentation tank should be designed to distribute the flocculated water uniformly over the entire cross section of the tank. It should be noted also that uniform or equal distribution of flow to each individual flocculation–sedimentation basin is also essential. The flow pattern in the tank is strongly controlled by inertial currents, density flows, and wind direction. An appropriately designed inlet and outlet can improve the tank performance significantly. A well-designed inlet permits water from the flocculation tank to enter directly into the sedimentation tank without channels or pipelines. The flow velocity in a pipe or flume can be either too slow or too fast depending on the daily and seasonal plant flow variations and may cause floc settling or breakage to occur in the pipe or flume. The permissible flow velocity to maintain floc suspension generally ranges from 0.15 to 0.60 m/sec (0.5–2 fps). The diffuser wall, discussed in an earlier section ("Baffle Walls") is one of the most effective and practical flow distribution methods used at the tank inlet. An extensive hydraulic scale model study of the diffuser wall is reported by Kawamura (*AWWA Journal*, 1981). The openings should be small holes [100–200 mm diameter circular (4–8 in.) or equivalent] or identical size evenly distributed on the wall. When sedimentation tanks are fed from a common channel, the basin inlet structure may consist of either weirs or submerged ports, with a permeable baffle about 2 m (7 ft) downstream. The hydraulics for these inlet facilities are discussed in Chapter 22.

Outlet Structures. Water leaving the sedimentation basin should be collected uniformly across the width of the basin. Experiments show that density flow, and not a high approach velocity due to short weir length, is primarily responsible for sludge being carried over the effluent weir. However, some studies have shown that the effectiveness of long launders is questionable. Long skimming troughs are commonly used as outlet structures for sedimentation basins. They have at least three advantages for rectangular sedimentation tanks: (1) a gradual reduction of flow velocity toward the end of the tank; (2) minimization of wave action from wind; and (3) skimming of clearer portion of water located in the middle of the tank when a distinct density flow occurs in the basin. A high-rate settler such as a tube settler module can be easily installed under the long launders, significantly increasing the tank loading rate without adding additional basin volume.

The water level in the tank is controlled by the launders. V-notch weirs are commonly attached to the launders. Submerged orifices or weirs have sometimes been used on the outlet structure of the sedimentation for a rapid sand filtration system, rather than using free-discharge weirs. Freely discharging weirs have a tendency to break fragile alum floc, resulting in turbidity breakthrough in rapid sand filters. For high-rate filter designs (dual, multimedia and coarse monomedia), there is little concern over floc breakage since the high-rate filters require a small, strong floc, and the settled water is typically reflocculated by means of polymer addition (filtration aid) before filtration.

In the past, installation of a permeable baffle at the tank outlet was a popular design. The effect of the effluent baffle, however, is often not beneficial and in fact, may have an adverse effect on basin performance (see Figure 21-23). Upflow at the discharge end of the basin may result in floc carryover; use of a baffle at the effluent end of tank is seldom practiced today.

Settling-Zone. Setting zone design depends on five basic design parameters: (1) settling characteristics of the suspended matter (usually floc); (2) surface loading; (3) tank depth; (4) width–length ratio; and (5) Reynolds and Froude numbers of the tank. The width–length ratio of the tank should range from 1:3 to 1:5. In general, long, narrow, and shallow basins are preferred because of minimal short circuiting and instability. Reynolds and Froude numbers are not good design criteria from a practical viewpoint. However, the following values are the recommended requirements:

$$\text{Re} = \frac{V_0 R}{\nu} < 2000 \quad \text{and} \quad \text{Fr} = \frac{V_0^2}{gR} > 10^{-5}$$

where Re = Reynolds number
    V_0 = displacement velocity in the tank, m/sec
    R = hydraulic radius, m
    ν = kinematic viscosity, m²/sec
    Fr = Froude number
    g = gravity constant, 9.8 m/sec²

Sludge Handling. The bottom of a sedimentation tank is either sloped toward a sludge hopper or may have no slope, depending on the method of sludge removal. For manual sludge removal systems utilizing pressurized water for flushing, the bottom slope should be at least 1:300 to ensure gravity movement of sludge. If mechanical sludge scraper equipment is used, the bottom slope should be 1:600 although a flatter slope is permissible. When a mechanical sludge remover using suction is specified, the entire tank bottom is usually level. If local labor is inexpensive or if funds for investment are limited, sedimentation tanks may be designed without mechanical sludge removal. However, provisions should be made for possible future installation of the units. Small plants, where only a small amount of sludge accumulation is anticipated, may be designed without mechanical sludge collection. In general, sludge removal is much easier for solids contact clarifiers than for rectangular horizontal-flow tanks.

When mechanical scraper units are provided, the velocity of the scraper should be kept below 0.3 m/min (1 fpm) to prevent resuspending the settled sludge. For suction sludge removal units, the velocity can be 1 m/min (3 fpm) since the principal concern is not the resuspension of settled sludge but the disruption of the settling process.

Manufacturers produce several types of mechanical collectors for rectangular and circular sedimentation tanks. The major types of mechanical collectors for rectangular tanks, ranked from most to least expensive, are (1) traveling bridge with sludge-scraping squeegees and a mechanical cross collector at the influent end of the tank, (2) traveling bridge with sludge suction headers and pumps, (3) chain-and-flight (plastic material) collectors, and (4) sludge suction headers supported by floats and pulled by wires. The operation and maintenance cost is usually highest for the chain-and-flight collectors if the length of basin exceeds 60 m (200 ft). However, it is one of the most suitable types of sludge removal mechanisms when used with high-rate settler modules, although traveling bridge collectors may also be used if channels for the sludge header passage are included. It should also be noted that a tank equipped with high-rate settler modules must provide continuous sludge removal because of the high sludge accumulation rate produced by the high-rate settler modules.

Traveling bridges can span up to 30 m (100 ft), and in many cases, the capital investment for sludge removal equipment can be significantly reduced by using one bridge to span two or three tanks, since tank width is often less than 15 m (50 ft). Both the drain and sludge drawoff pipelines should have a minimum diameter of 150 mm (6 in.) to prevent clogging problems.

Figures 21-24 through 21-26 illustrate the various sludge collection mechanisms previously described.

High-Rate Settlers

There are two types of practical high-rate settler modules: parallel-plate settlers and tube settlers. Both have a detention time of less than 20 min but

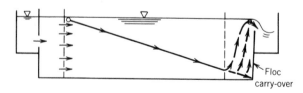

FIGURE 21-23. Drawback of effluent permeable baffle.

FIGURE 21-24. Chain-and-flight-type sludge collector. (Courtesy of Tsukishima KK.)

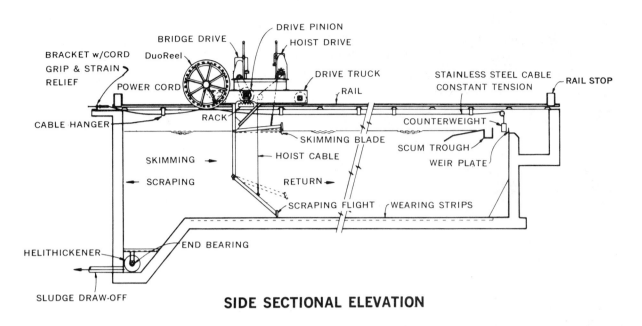

SIDE SECTIONAL ELEVATION

FIGURE 21-25. Traveling bridge sludge collector. (Courtesy of Walker Process.)

have a settling efficiency comparable to that of a rectangular settling tank with a minimum 2-hr detention time. In parallel-plate settlers, water enters the settler module horizontally, parallel to the plates, and is turned upward. In tube settlers, water almost always flows upward at an angle of 60° to the horizontal plane. Usually settled floc or sludge does not adhere to the plates or tubes, but rather slides down to the tank bottom for subsequent mechanical sludge removal. However, some flocs, especially biological floc, may adhere to the modules. If the plates are designed with a scouring device, water jets or compressed air may be included. For certain applications, downflow settlers can be used to enhance the self-cleaning action. Figures 21-27 and 21-28 demonstrate the tube settler modules described.

The design of high-rate settlers includes consideration of (1) the settling characteristics of suspended matter, (2) surface loading, (3) flow velocity in the settler module, (4) detention time in the settler module, (5) the Reynolds and Froude numbers, (6) selection of the sludge collector unit, and (7) design of settler supports and launders.

The surface loading for high-rate settlers is primarily controlled by the settling characteristics of the suspended particles to be removed and the por-

FIGURE 21-26. Cable-operated underwater bogies sludge collector. (Courtesy of Tsukishima KK.)

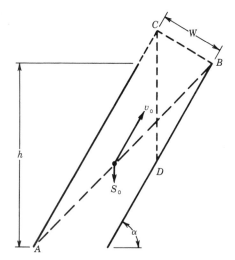

FIGURE 21-28. Settling action analysis in lamella-type sloped plate settler.

tion of the total tank surface area to be covered by the settler modules. In cold regions where alum floc is to be removed, the maximum surface loading should be limited to 150 m³/m² · d (2.6 gpm/f²). Pilot testing may help establish design criteria, but actual design criteria should be more conservative to allow for poor inlet conditions, density flow, inappropriate coagulant dosage, or other negative factors. Flow velocity at the inlet of the settler module should be less than 0.6 m/min (2 fpm). In a horizontal-flow tank, the front one-quarter length of the basin is generally free from settler modules to allow for better inlet flow conditions. The flow velocity allowed in the settler is closely related to the surface loading. For any high-rate settler, an upflow velocity of less than 0.1 m/min is acceptable for most alum or lime floc.

The detention time in the settler module varies depending on the type of settler modules used. For plate settlers, the width of the plate spacing is directly proportional to detention time. Tube settlers are designed for 3–6-min detention times, while tilted plate settlers are designed for 15–25 min.

While Reynolds and Froude numbers have only limited use in ordinary settling tank design, they are good design guides for high-rate settlers. The Reynolds number for the settler design should be kept below 200 and the Froude number should be approximately 10^{-5}.

No definite relationship exists between the weir loading rate and the type of settler modules used. However, an extensive launder system not only helps to collect clarified water uniformly from the area covered by the settler modules but the modules can also be hung from the launders, thereby eliminating an elaborate support system. Launders are usually spaced 3–4 m on center.

EXAMPLE 21-2

Problem

The settling system for a continuous horizontal flow is designed to treat a flow rate of 1 m³/s (23 mgd) and remove discrete particles (silt) larger than 0.02 mm in diameter using a high-rate settler module system (tube settlers). The settling velocity of the 0.02-mm-size silt was measured in a laboratory as 0.62 mm/s (1.46 ipm) in a quiescent condition at 10°C.

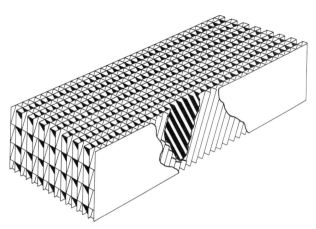

FIGURE 21-27. Tube settler module (60°) (Courtesy of Microfloc, Inc.)

Determine the size of basins and the basin area that needs to be covered with the tube settler module. The settler module consists of a series of 50-mm (2-in.) square honeycombs tilted at a 60° angle and has a vertical height of 0.6 m.

Solution

1. *The area required for the settler modules.* The design surface loading for the settlers can be established from the settling velocity of the particle to be removed:

$$S_0 = \frac{Q}{A} \frac{W}{h \cos \alpha + W \cos^2 \alpha}$$

Since $\alpha = 60°$

$$S_0 = \frac{Q}{A} \frac{W}{0.5h + 0.25W}$$

For these tube settlers:

$$S_0 = \frac{Q}{A} \frac{0.05}{(0.5 \times 0.6) + (0.25 \times 0.05)} = \frac{0.16}{A}$$

The settling velocity of the silt measured in a laboratory for S_0 is not used because it was measured under ideal conditions (quiescent and carefully controlled conditions). Actual conditions in the basin are not as good as ideal settling conditions because of many factors such as fluctuations in temperature and flow rate, wind effects and density flow from heavy solid concentration, and so forth. Also, basin water temperature can be lower than the 10°C at which the settling velocity was measured. Therefore, a safety factor of 0.7 is applied to determine the S_0 for the design:

$$S_0 = (0.00062 \text{ m/sec} \times 0.7) = 0.00043 \text{ m/sec}$$

$$S_0 = 0.00043 = \frac{0.16}{A}$$

$$A = 368.7 \text{ m}^2 \quad [\text{say } 370 \text{ m}^2 (3980 \text{ ft}^2)]$$

Surface loading for the basin area covered by the tube settler modules is:

$$\frac{Q}{A} = \frac{(1 \times 60 \times 60 \times 24) \text{m}^3/\text{d}}{370 \text{ m}^2}$$

$$= 233.5 \text{ m}^3/\text{m}^2 \cdot \text{d} (3.4 \text{ gpm/sf})$$

Flow velocity in the settler module is

$$v_0 = \frac{Q}{A \sin \alpha} = \frac{60 \text{ m}^3/\text{min}}{370 \times 0.866}$$

$$= 0.187 \text{ m/min} \quad (\text{or } 0.0031 \text{ m/sec})$$

2. *Sizing the basins.* Two identically sized basins are required for the settling system. A basin width of 6 m (approximately 20 ft) is selected, because a standard width of a chain and flight sludge collection system is 6 m. The water depth of the basin is 3.7 m (12 ft) since the sludge collector system must be installed underneath the skimming troughs (approximately 0.5 m depth) and the tube settlers (approximately 0.6 m depth). Additionally, a space of approximately 0.5 m is needed between the bottom of the launder troughs and the top of the tube settlers. Thus, approximately 2 m of depth is left for the sludge collection system.

As a rule of thumb, the front portion of the basin (one-quarter of the basin length) is left as an open space in order to improve the inlet conditions to the settler modules.

Length of the basin(s) covered by the tube settlers is as follows:

$$L = 370 \text{ m}^2 \div (2 \times 6 \text{ m}) = 31 \text{ m} (100 \text{ ft})$$

A total length of the basin(s) is, therefore:

$$\frac{3}{4} L = 31 \text{ m thus } L = 41.3 \text{ m} (135.5 \text{ ft})$$

A mean velocity is given as:

$$60 \text{ m}^3/\text{min} \div (2 \times 3.7 \text{ m} \times 12 \text{ m}) = 0.68$$
$$\text{m/min} (2.2 \text{ fpm})$$

3. *Reynolds number and Froude numbers.* Hydraulic radius is given as:

$$R = \frac{A}{P} = \frac{0.05^2}{4 \times 0.05} = 0.0125$$

Thus,

$$\text{Re} = \frac{v_0 R}{\nu} = \frac{0.0031 \times 0.0125}{0.00000131} = 30$$

$$\text{Re} = 30 < 2000 \quad (\text{Good})$$

$$\text{Fr} = \frac{v_0^2}{g \cdot R} = \frac{0.0031^2}{9.8 \times 0.125} = 8.8 \times 10^{-5}$$

$$\text{Fr} = 8.8 \times 10^{-5} > 10^{-5} \quad (\text{Good})$$

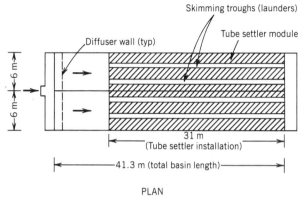

PLAN

LONGITUDINAL SECTION

Design Criteria (summary)

- Design flow rate — 1 m³/sec (23 mgd)
- Minimum particle size to remove — 0.02-mm diameter (silt)
- Total number of basin — 2
- Size of each basin — 12 m(W) × 41.3 m(L)
- Water depth — 3.7 m
- Basin detention time — 30.6 min
- Surface loading of basins — 174 m³/m² · d (3 gpm/sf)
- Area covered by tube settlers — 370 m² total (6 m × 31 m area for each basin)
- Detention time in the settlers — 3.7 min
- Surface loading for the basin area covered by the settlers — 233.5 m³/m² · d (3.4 gpm/sf)
- Basin mean velocity — 0.68 m/min

FLOTATION

Introduction

Flotation, in general, is an effective unit process used to separate solid or liquid particles from a liquid phase and has been used in the mineral industry for many years. In the environmental engineering field in the United States, the process has been applied primarily to wastewater treatment works mainly for sludge thickening or grease removal. In potable water treatment, sedimentation is the most commonly used prefiltration process. However, since flotation is effective in removing hydrocarbons, fibers, and algae, attention has been directed toward flotation particularly in the United Kingdom and Packham and Zabel have been active on this subject.

Types of Systems

There are four basic types of flotation techniques: (1) dispersed air–froth flotation; (2) dispersed air–foam flotation; (3) dissolved air flotation; and (4) electrolytic flotation. Figure 21-29 shows these processes schematically.

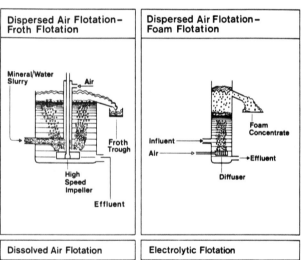

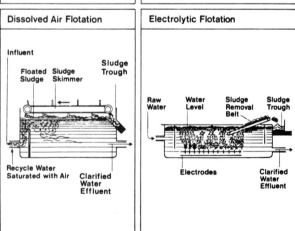

FIGURE 21-29. Four basic flotation techniques. (Reprinted with permission from Zabel, T., "Flotation", International Water Supply Association Twelfth Congress, 1978.)

Dispersed air flotation, termed froth flotation, requires violent agitation, usually by means of mechanical mixers, for aeration and dispersion of the processed liquid. Another type of dispersed air flotation, commonly called foam flotation, supplies air bubbles by diffusing air through porous media or sparagers into the flotation cell.

In the dissolved air flotation process, the air bubbles are generated by a reduction of the pressure of a liquid stream saturated with air. The air bubbles by this process are fine bubbles and their size is usually less than 0.1 mm. No mechanical agitation of the process water is provided.

In recent years the most widely used dissolved air flotation system is pressure flotation. In this process, air dissolved in water under pressure is released in the form of small bubbles by discharge to atmospheric pressure.

Another form of the dissolved air flotation process is the microflotation process, which has been developed in recent years. In this process, the entire volume of water is subjected to increasing pressure. This is achieved by passing the water down and up through columns about 10 m deep. In the downward pass, the water is aerated and the air dissolves in the water due to increased hydrostatic pressure. As water rises in the return pass, the dissolved air is released in the form of minute air bubbles.

Electrolytic flotation, or electroflotation, uses electrolysis to generate small bubbles of oxygen and hydrogen similar in size to those produced in dissolved air flotation. A major problem of this process is the fouling of the electrodes.

Flotation System Design

Because most foam separation systems use dissolved air flotation (DAF), this section will be limited to the design of those systems. Design data specific to the water, waste, or sludge to be treated should be generated from pilot plant or bench-scale systems operated under conditions that are as similar as possible to those anticipated in the field. Only in this way can the various process parameters such as coagulant or polymer dosages, air requirements, and loadings be optimized. Once this information has been finalized, detailed design can proceed.

The required surface area of a DAF unit is based on the selected solids loading. This area is then checked against the maximum allowable hydraulic loading. If the hydraulic loading is exceeded, considerations should be given to altering or adding flocculation facilities, selection of a different or new polymer or coagulant, or varying the operating conditions of the unit.

DAF units are designed in rectangular and circular configurations and are commonly purchased as complete packages from equipment manufacturers.

Proper baffling is necessary to prevent short circuiting in a DAF unit. Typically, the energy of the influent flow stream is dissipated by directing that flow toward a fixed plate or at the tank bottom. The pressurized flow is introduced at the same location and near the bottom of the suit. An over-and-under weir is generally used to prevent carry-over of solids into the subnatant. A schematic of typical baffling in rectangular DAF units is shown in Figure 21-30.

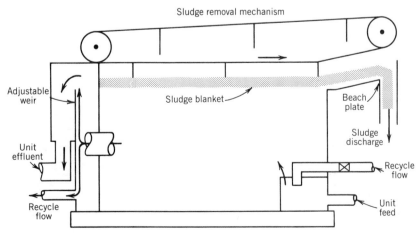

FIGURE 21-30. Typical baffling arrangement.

DAF mechanical equipment typically includes a pressurizing pump, an air supply, a retention tank, chemical feed equipment, float and settled sludge removal equipment, and suitable operator controls. A surface skimmer is provided to scrape the floated solids over a dewatering bench and into a hopper from where it can be pumped to subsequent solids-handling operations. Penetration of the blades into the float blanket is controlled by an adjustable weir on the unit effluent and by adjustments of the wipers themselves. The system is typically driven by variable or intermittently operated constant-speed drives with a torque-limiting device for overload protection. A typical circular DAF system is shown in Figure 21-31.

In addition to float removal, means must be provided to remove settleable solids that are not amenable to flotation or that may settle during periods of unit upset. Circular units use the same drive as that of the float skimmers and mount raker arms and blades at the bottom of the unit to convey the sludge into a central hopper. Rectangular units utilize chain-and-flight systems to sweep settled materials to a central hopper. In both units, the settled solids are pumped or hydrostatically displaced to the skimmings hopper. Air is furnished to the unit by a compressor design to operate at the desired pressure of the system. A pressurizing pump is also necessary to deliver water to the pressurizing vessel. This unit is almost universally a centrifugal pump. The retention tank in which the air is dissolved in the water should be designed to safely withstand the working pressure of the system. The controls for a DAF unit are dependent on the system, the degree of automation required, and the equipment manufacturer's design. They include, at a minimum, a pressure controller for the pressure vessel and flow meters for the feed and float streams.

GRANULAR FILTERS

Introduction

Granular filters include slow sand filters, rapid sand filters, and high-rate filters [which may utilize coarse monomedia (other than sand), dual media, or multimedia]. The three basic types of granular filters require different types and degrees of pretreatment. A common design error is to provide for a type and degree of pretreatment that does not match the type of filter selected. One example is the combination of conventional flocculation and sedimentation basins with high-rate filters. While this combination will almost certainly produce finished water that meets the National Primary and Secondary Drinking Water Regulations or equivalent drinking water standards, the cost involved for plant construction and overall plant operation may be much higher than is necessary to provide for high-quality water. If the raw water quality is good throughout the year, direct or in-line filtration with

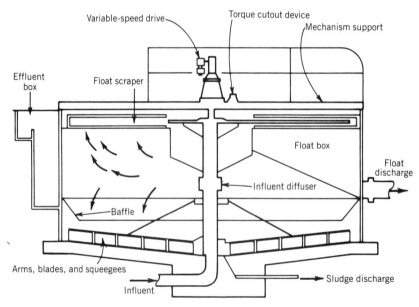

FIGURE 21-31.   Dissolved air flotation system.

high-rate filters may be preferable. Thus, the designer should review all pertinent data, and the pretreatment scheme should be considered prior to selection of a filter type.

Filter design involves determination of the following items:

1. Type, size, and number of filters.
2. Filtration rate and terminal head loss.
3. Filter flow control scheme.
4. Media depth, size, and material.
5. Filter washing and related systems.

These items are discussed in the following sections.

Selection of Type, Size, and Number of Units

The water treatment plant designer may select any one of a number of types of granular filters for use in a plant. While slow sand filters are still employed (particularly in Europe), high-rate filters almost exclusively dominate the field in the United States. For small plants [>0.5 m³/sec (10 mgd)], however, rapid sand automatic backwash filters are frequently used. This filter type is a proprietary product marketed by a number of manufacturers. In small plants, it offers a number of advantages. These include (1) lower capital costs, (2) low head loss across the filter, (3) a good filtered water quality, and (4) simple operation and maintenance. For large plants [>0.5 m³/sec (10 mgd)], the gravity, downflow, sand, and coal dual-media filter in a reinforced concrete structure is the most common type of filter selected. The design engineer has a number of choices for backwash systems for this filter system. The most common are (1) water backwash with water jet surface wash (either fixed grid or rotating arm) and (2) air and water backwash (generally referred to as air-scour wash). In small installations or industrial or wastewater treatment applications, the use of steel tanks or pressure filters may offer advantages. Steel shell filters, such as proprietary gravity or pressure filters, are usually less costly for small plants (>1 mgd), because of the reduced amount of waste backwash water per filter and the ability to handle higher solids loading.

Large plants (generally greater than 19–38 ML/d or 5–10 mgd) utilize a minimum of four filters, while smaller plants may utilize as few as two. Maximum filter size is generally determined by the economic sizing of the filter backwash facility and possible difficulties in providing uniform distribution of backwash water over the entire filter bed. The practical maximum size of one high-rate gravity filter is about 100 m² (1100 ft²). However, proprietary, automatic backwash filters, extremely large plants (over 350 ML/d), and slow sand filters may have bed areas much larger than this.

One note of caution is that when a plant is designed with a small number of filters (less than four), the filtration rate in the remaining filters will increase substantially when one filter is out of service for backwashing. Thus, the designer must take into consideration the filtration rate with at least one filter out for backwashing. In a plant with only two filters, it is obvious that the filtration rate may double if one filter is out of service. Thus, only if the design filtration rate is low enough is it possible to utilize as few as two filters with no adverse effects.

Upflow filters in which the flow passes upward through the media bed may also be used. These have an obvious advantage because coarse-to-fine filtration can be achieved with a single medium such as sand. Thus, there is no need for more costly multimedia such as anthracite coal and garnet. One disadvantage of this filter type is expansion of the filter bed as it becomes clogged, allowing previously removed solids to escape in the effluent. An extra deep bed (>1.8 m) with coarse sand (1.5–2 mm) or a restraining bar system at the top of the bed have been used to overcome this problem. Upflow filters have found wide application in process water, wastewater, cooling water treatment and other applications which are free from human health considerations.

Filtration Rate and Terminal Head Loss

The effect of filtration rate on effluent quality can vary widely depending on the specific application. In filtering floc resulting from alum and polymer coagulation or biological floc at reasonably low influent solids concentration, the effect of filtration rate on effluent quality is generally not significant for rates up to 6.94 mm/sec (10 gpm/ft²) and reasonable filter run lengths may be achieved. With weak chemical floc such as alum floc or with high concentrations of poorly flocculated biological flocs, filter effluent quality tends to degrade at filtration rates above 3.47 mm/sec (5 gpm/ft²). Higher rates tend to increase solids penetration (if media size is properly selected) and head loss buildup per unit volume filtered may actually decrease at higher rates due to lower solids removal rates.

For gravity filters, the available head loss is generally less than 3 m. Certain other types of filters (such as proprietary valveless or monovalve package filters) do not provide more than 1.8 m of available head loss, and automatic backwash filters typically do not provide more than 0.4 m of available head loss. If terminal head loss is much above 3 m, the design may utilize pressure filters, although this is probably not practical in large plants.

Head loss development in filter beds is determined by the amount and location of retained solids in the bed. The average solids capture (mass per unit plan area of filter media) during a filter run may range from 555 to 5550 $g/m^2/m$ of head loss (0.035–0.35 $lb/ft^2/ft$) depending on the many factors involved.

The filtration rate and terminal head loss for a particular filter type and filter media design should be selected by analyzing the economic trade-offs between filter size, operating head requirements, and filter run lengths, all within the limits dictated by effluent quality requirements.

Filter Flow Control Scheme

The five basic types of filter rate control schemes are (1) constant-rate filtration, (2) declining rate filtration, (3) constant-level filtration, (4) equal-loading filtration, and (5) constant-pressure filtration. In recent years, the most popular control schemes for gravity filters are declining rate, constant level, and equal loading; while declining rate and constant pressure are used for pressure filters. Declining rate, constant-rate, constant-level, and equal-loading with rising level filters are shown schematically in Figure 21-32.

Equal-loading filters use an influent flow-splitting device to distribute the plant flow among all operating filters equally. Constant-level filters have a flow control (or modulating) valve in each filter effluent pipe to maintain the water level above the filter bed at nearly constant level. Equal-loading with rising level filters do not have any flow control valves but provide a filter effluent control weir for the common filter effluent line. In addition, filter shells for this type of filters are made deep enough so that the water level can provide the operating head needed when the filter bed becomes clogged. A deep water depth above the filter bed in the later part of the filtration cycle is also desirable as "air-binding" problems are suppressed. Constant-rate filter control is usually achieved with a venturi-type flow

tube and a flow-modulating valve (commonly called flow controller) for each filter. The equal-loading filters described above are also constant-rate filters if total plant flow remains constant. Declining rate filtration systems do not include flowrate controllers, but they may include a hydraulic restriction such as an orifice plate in each filter effluent line to limit the initial filtration rate to about twice that of the average filtration rate. All filters (in the declining rate mode) are served by a common influent header or channel so that the cleaner filters can assume more of the total flow than filters that are clogged from extended service.

A filter flow control scheme that has been developed in recent years is equal-loading self-backwash filtration (see Figure 21-32). This type of filter control scheme will probably become more popular because of its simplicity, low energy requirements, and cost-effectiveness. It eliminates the need for a backwash header, pumps, and storage tank and backwash is initiated typically by turbidity breakthrough, time or head loss build-up. This is described in Chapter 23.

Media Depth, Size, and Material

As described in Chapter 8, current filtration theory emphasizes the transport and attachment steps in describing the filtration process. Suspended particles in the filter influent may be attached to the surface of filter media due to diffusion, interception, gravity settling, impingement, and so on, which are all related to the surface area of the filter media. Therefore, filter efficiency is a function of certain physical characteristics of the filter bed, which include bed porosity and the ratio of bed depth to media grain diameter. The selection of the appropriate media depth and size is thus of crucial concern to the design engineer. Currently, there are two basic ways of selecting the proper depth and size of filter media: (1) through use of pilot plant studies and (2) through use of existing data from filtration facilities treating similar waters. There have been many pilot studies in recent years evaluating the most effective filter-media size and depth for various waters especially examining dual- or multimedia filters. These studies may provide additional data for selecting media.

There are many possible combinations of filter media size, d, and depth, L. The relationship between L and effective size d_e (10 percent finer by sieve analysis) of many high-rate filters reported in

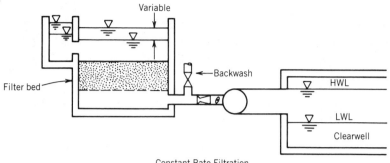

Constant Rate Filtration
(Flow meter & flow modulation valve)

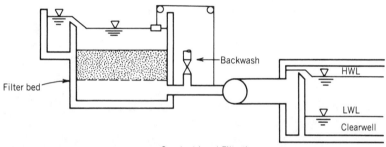

Constant Level Filtration
(Influent control, level sensor & modulating valve)

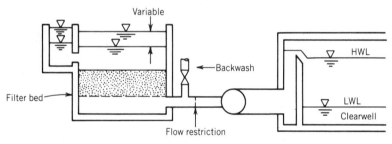

Declining Rate Filtration
(No influent control, no modulating valve, an orifice plate)

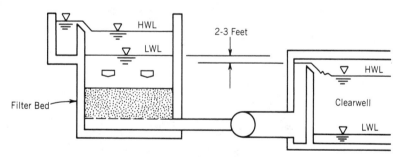

Rising Level, Self-Backwash Filters
(Influent control, no modulating valve, no backwash piping)

FIGURE 21-32. Four basic filter control systems.

537

the literature are shown in Figure 21-33. In this figure, the average effective size for dual- or multimedia was computed as a weighted average. The data in the figure reveal that a clear relationship exists between L and d with some exceptions. The major reasons for the scatter in Figure 21-33 are: (1) porosity of each filter is not the same; (2) variation in the objectives of purification and raw water characteristics; and (3) pretreatment methods are not the same for each filter. The relationship between L and d_e shown in Figure 21-33 can be expressed as $L/d_e =$ constant.

The surface area of filter media per unit of filter plan area, a (m^2), can be expressed by the following equation:

$$\Sigma\, a = \frac{6(1-f)}{\phi}\frac{L}{d}$$

where d = average diameter of filter media grain
ϕ = measure of sphericity (ranging from 1.0 to 0.7 and 0.8 for ordinary filter media)
f = porosity of bed

Since, for ordinary high-rate filters, $f = 0.45$, $\phi = 0.8$, and $L/d = 950$ (all approximate), the total surface area of filter media per square meter of filter plan area is on the order of 2800 m^2 (30,000 ft^2). Recent filtration theory (described in Chapter 8) has determined that the efficiency of collection of the filter may be described by the following equation:

$$\text{Efficiency} = \frac{C_L}{C_0} = e^{-\eta L/d}$$

where η is a constant for the bed, C_0 is the influent particle concentration, and C_L is the effluent concentration for the packed bed. This equation indicates that as the media used becomes coarser, the required depth is increased and, conversely, as the media become finer, the depth required is reduced.

The specific deposit, defined as the volume of material deposited per unit of filter depth during the filter run, can be expressed by the equation

$$\sigma = \frac{v}{L}(C_0 - C_L)\,\Delta t$$

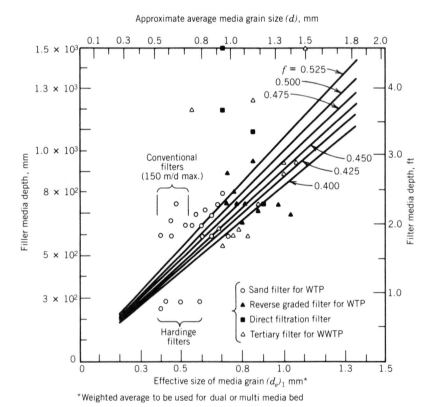

FIGURE 21-33. Relationship between depth of media and size of media. *Weighted average to be used for dual- or multimedia bed (f values based on $l/d = 950$.)

where σ = specific deposit
v = filtration rate
Δt = length of time filter has been operating

Experimental data have shown that for a given filter and water quality, turbidity breakthrough usually occurs at the same level of specific deposit. This level of deposit is usually referred to as the effective deposit, σ_e, and can be defined as follows:

$$\sigma_e = \frac{v}{L}(C_0 - C_L)\,\Delta t_b$$

where Δt_b is the filter run time to turbidity breakthrough.

This relationship allows comparison of the ability of various filter media to remove particulates. It follows that higher filtration rates or higher concentrations of particulates in the influent will reduce the length of time to turbidity breakthrough, and that deeper beds will give longer runs.

For good quality raw water, C_0 is in the vicinity of 6×10^4 nL (10^{-9} L) of particle volume per liter, C_L is around 1×10^2, and a value of η of about 5×10^{-3} may be used.

As far as dual- or multi-media filters are concerned, the combination of media grain sizes and specific gravity is a very important factor. If mismatched media combinations are used, only part of the filter media would be adequately cleaned at a particular backwash rate. Figure 21-34 indicates the proper combinations of common filter materials for a particular backwash rate and also shows the appropriate backwash rate at a water temperature of 20°C.

EXAMPLE 21-3

Problem

An existing set of conventional rapid sand filters consisting of 650 mm depth of 0.4 mm effective-size sand are to be converted to dual-media high-rate filters to improve filter efficiency without changing the total media depth that is governed by the physical limitation of filter cell dimensions and wash trough height. Maximum backwash rate is also limited to 0.6 m/min (24 in./min).

Determine the size and depth of both sand and

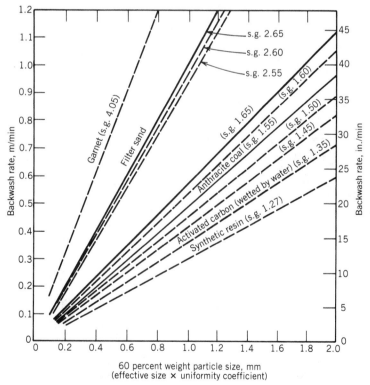

Figure 21-34. Appropriate filter backwash rate at a water temperature of 20°C (68°F).

coal layers assuming that the pretreatment process is also improved by the addition of a new flash mixer and high-energy flocculators.

Solution

Set the design conditions as follows

1. New filter bed is to be reverse-graded dual-media bed with sand and coal grains.
2. Provide 200 mm of sand layer with 0.43 mm effective size (AWWA Standard classification—see B100.2.1) and 1.3 uniformity coefficient in order to ensure a good quality filtrate and a reasonable length of filter run. Coal layer, therefore, has 450 mm depth. Larger-size sand media cannot be used because of the limited maximum backwash rate. A specific gravity of 2.65 is used for the sand media.
3. Assume that overall porosity of the total filter media is 0.475. By use of Figure 21-33, d_e = 0.68 mm (weighted average d_e for both sand and coal grains) from the point of intersection between L = 650 mm and f = 0.475. Effective size of coal grains, therefore, may be computed through weighted average computation as follows:

$$\frac{450d_e + 200 \times 0.43}{650} = 0.68$$

Therefore, d_e = 0.79 mm, say, 0.8 mm for the coal media. To determine the uniformity coefficient of anthracite (specific gravity = 1.60) from Figure 21-34, draw a horizontal line from 0.6 m/min wash rate, drop a vertical line from the intersection of the horizontal and the anthracite line of s.g. 1.60, and read the 60 percent weight particle size on the horizontal axis. With an effective size of 0.8 mm, a uniformity coefficient of 1.4 should be used.

EXAMPLE 21-4

Problem

Assume that an average particulate volume in a raw water is 60,000 nL/L and filtered water should contain an average of 100 nL of particulates per liter or less. Also, assume that a pilot filtration study concluded that media constant, η, is 0.005, and 1.2 mm

effective-size coal is the most efficient filter media size from the viewpoint of filter efficiency and head loss development during filtration. Determine the filter bed depth for 1.2 mm effective-size anthracite coal filter bed.

Solution

The equation describing filter collection efficiency can be rewritten as follows:

$$\frac{L}{d} = \frac{1}{\eta} \ln \frac{C_0}{C_L}$$

Since the values of η, C_0, and C_L are known, L/d can be obtained easily.

$$L/d = \frac{1}{0.005} \ln \frac{60,000}{100}$$
$$= 1280$$

Thus, L = 1.2 × 1280 = 1536 mm or 60 in. media depth.

Wash Troughs

There are two distinct types of filter backwash systems: fluidized bed backwash (which is the most common system in the United States), and partial fluidized backwash with air scour (which is commonly used in Europe). Partial fluidized bed backwash with air scour is typically employed in filters with no backwash troughs. A single overflow weir or side channel weirs provide for discharge of backwash waste.

Spacing and elevation of troughs for fluidized bed backwash systems should be carefully determined so that (1) dislodged suspended matter will be washed away efficiently and (2) filter media will not be carried out. As a rule of thumb, the height and spacing of troughs can be determined based on the filter configuration as shown in Figure 21-35 on the following criteria:

Height of trough: $(0.75L + P) < H_0 < (L + P)$

Spacing of trough: $1.5H_0 < S < 2H_0$

Wash troughs are generally of two basic types: those with a shallow but wide cross section and a slight V-shaped bottom and those with a U-shaped cross section. The wide section troughs result in a

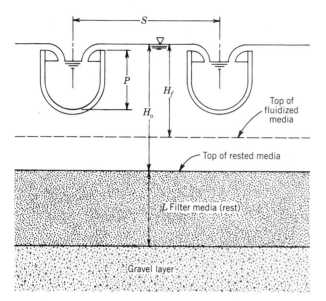

FIGURE 21-35. Height and spacing of wash troughs.

higher upflow velocity when flow gets above the trough bottom elevation and the U-shaped troughs allow for thinner walls because of a higher moment of inertia and greater structural integrity. The bottom of the wash trough should not be flat because

froth and suspended matter is often trapped under the trough bottom and may never be washed out.

In either case, the troughs should be large enough to carry the maximum expected wash rate with 5–10 cm free-fall into the trough at the upper end. They should also provide a free-fall to the main collection outlet gullet at the lower end. The bottom of the trough may be either horizontal or sloping.

The required cross-sectional area of the wash trough for a given design flow can be quickly estimated from Figure 21-36. For troughs that have level inverts and rectangular cross section, required trough height can be computed by the following formula:

$$\text{Minimum trough height} = \left(\frac{Q}{1.4B}\right)^{2/3} + \text{free board}$$

where Q is the total flow rate of discharge (m³/sec), B is the inside width of the trough (m), and, freeboard should be a minimum of 50 mm (2 in.).

Filter Underdrains

Filter underdrainage systems differ primarily due to the different filter-washing systems and filter types

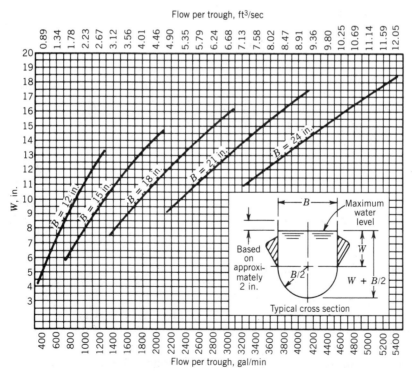

FIGURE 21-36. Wash-through sizing diagram. (Courtesy of Leopold Co.)

employed. For instance, small size filters, including steel plate fabricated filters, are usually equipped with a fixed-pipe grid-type underdrainage system because of simplicity and cost-effectiveness. Filters for most municipalities are usually large in size and a variety of underdrainage systems are used including false bottoms with strainers, underdrain blocks, precast concrete underdrains, teepee-type underdrains, and porous plates. For large filters, the uniform distribution of flow becomes the most important factor in selecting the type of underdrain. Individual designer preference based on experience with past installations is frequently a deciding factor. The majority of underdrains installed are of the proprietary, underdrain block-type due, perhaps, to the availability of a manufacturer's warranty for performance of the underdrain system. Pipe grid systems are seldom used in large filters despite the fact that they are cost-effective and easy to install. Their lack of use is perhaps attributable to the perception that fixed-grid pipe underdrainage systems are outdated and outmoded. Figure 21–37 shows examples of underdrains commonly used in water treatment facilities.

Selection of proper underdrain is also affected by the size and type of filter media as well as the type of filter wash system adopted. For air-scour washing filters, false bottoms with long leg strainers have been the most popular and accepted type of underdrainage system. However, debris remaining in pipes and conduits can clog the strainers from inside during backwash and filter underdrain can be loosened and lifted unless rather large slits are used in the strainers. A thin gravel layer may also be necessary to prevent media from clogging the strainer openings. Orifice sizing, spacing, and underdrain gravel layers depend on backwash rates, type of underdrain system used, media size, and hydraulic conditions.

Uniform backwash flow distribution, durability, and cost are the three most important factors in selecting filter underdrains. In order to achieve an even distribution of flow, (1) the orifice openings should be small enough to introduce a controlling loss of head and (2) the flow velocity in the pipe or channel in the underdrainage system should be reasonably low and uniform throughout the entire filter area.

Loss of head in the underdrainage system during backwash ranges from 0.1 to 3 m (0.3–10 ft) depending on the type of underdrain and backwash rates. For a false-bottom-type underdrain system, the re-

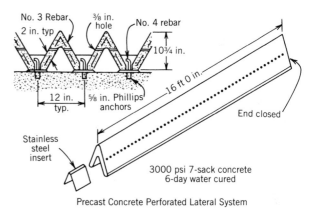

Precast Concrete Perforated Lateral System

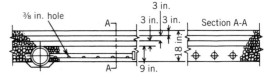

Manifold and Perforated Lateral System

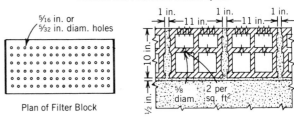

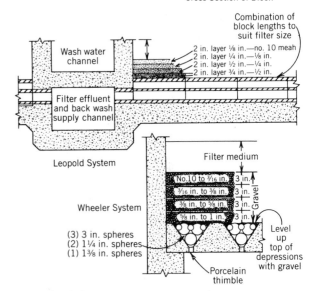

FIGURE 21-37.   Filter underdrain systems.

quired head loss could be small (0.1 m for some systems) because the unbalance of pressure underneath is self-compensating in the plenum if the inlet is properly designed. Perforated pipe grid systems and other perforated lateral systems without a plenum underneath usually require smaller orifices to create a controlling head loss of at least 0.6 m (2 ft).

The ratio of orifice area to bed area varies from 0.2 to 1.5 percent depending on the type of filter underdrain, backwash rate applied, and type of filters. Pipe grids with strainer type systems usually have a small orifice opening ratio (less than 0.5 percent), but false-bottom-type systems generally have a higher opening ratio (over 1 percent) without sacrificing a uniform distribution of flow. A few modern high-rate filters (especially for nondomestic use) apply up to a 900-m/d (15 gpm/ft^2) filtration rate. The underdrainage system for these filters should be a low-head-loss type or the excessive initial head loss in the filtration cycle will significantly reduce the available head for gravity filtration.

Filter Washing and Related Systems

Selection of the filter-washing system depends on the characteristics of the raw water, the type of filter selected, the size and material of the filter media, and the type of auxiliary media scouring selected for filter washing.

Most physical factors of filter design are related to the filter-washing process. The major factors include (1) the size and specific gravity of the media, (2) the type and arrangement of the filter underdrainage system, (3) the washwater trough design, (4) the size and elevation of the washwater tank (if necessary), (5) the wash rate control system, and (6) the type and capacity of auxiliary scour. Although numerous papers have discussed many variations of the filter-washing process, there are three basic types of washing systems: (1) complete fluidization of filter media by a high backwash rate with or without surface washing; (2) air-scouring wash with partial fluidization of the media by backwash; and (3) combinations of 1 and 2.

Although most water treatment plants provide more than 10 m (33 ft) of static head for backwash by elevated tanks, most filters only require 1.2–2 m (3.9–6.6 ft) static head at the filter bottom. More than 80 percent of the available head is dissipated by delivery piping, the throttling valve, and flow controller in order to ensure a relatively constant backwash rate.

In the case of low-head backwash systems, such as valveless or monovalve filters which utilize filtered water from other filters or from their own storage, a slight change in the available static head will affect the backwash rate quite significantly. For instance, a drop of 1 m (3 ft) water level in an elevated backwash tank results in about a 15 percent reduction of backwash rate for a conventional-type filtration plant without a flow controller, but the same condition will create almost a 75 percent reduction of the backwash rate for most low-head-type backwash filters.

Certain filters such as multicell-type pressure filters and automatic backwash gravity filters require

TABLE 21-12. Typical Design Criteria for Various Backwash Systems

	Fixed-Jet-Type Surface Wash	Rotary-Jet-Type Surface Wash	Air Scour Type
Pressure at discharge point			
kPa	100–200	420–680	28–50
kg/cm^2	1–2	4.2–6.8	0.28–0.5
psi	15–30	60–100	4–7
Flowrate			
Water			
m^3/m$^2 \cdot$ min	0.12–0.17	0.03–0.06	0
gpm/ft^2	2.9–4.1	0.7–1.5	0
Air			
m^3/m$^2 \cdot$ min	0	0	0.5–1.3
cfm/ft^2	0	0	1.5–3.5
Duration of washing, min	4–8	4–8	8–15
Backwash rate			
m^3/m$^2 \cdot$ min	0.55–1.0	0.55–1.0	0.25–0.70
gpm/ft^2	13.5–22.5	13.5–22.5	6–17

rather small amounts of water for one backwash cycle; the capital savings for the backwash system and the wash waste recovery system is significant.

The choice between high-rate backwash with a hydraulic surface wash system or an air scour system has been the subject of much discussion. Both systems have advantages, and the first system has a long history of successful use with rapid sand filters and high-rate filters in the United States. However, air scour washing has been gaining popularity mainly due to its effectiveness for cleaning tertiary filters for wastewater treatment.

Design criteria for water-jet-type surface wash systems and air scour washing systems are shown in Table 21-12.

PRECOAT (DIATOMACEOUS EARTH) FILTERS

Introduction

Precoat, or diatomaceous earth (DE), filters can be exceptionally efficient in removing particles from a fluid. These filters are typically classified as special proprietary components. Design of precoat filters should be undertaken only by experienced personnel.

Application

Precoat, or DE filters can remove particles as small as 0.1 μm from fluids without the need for particle destabilization (or coagulation). Precoat filters are compact, lightweight, and normally less expensive than conventional sand and dual-media gravity filters, but they are not widely used for large water treatment plants. The term *precoat* refers to the feeding of the filtering media into a stream of water so as to coat a fine cloth or screen, the septum, with the media, as described in Chapter 8. Initially, the media is added to coat the septum with a layer approximately 3 mm in depth. Additional media may be added during the filter run (body feed) and the total media thickness may ultimately reach 10 mm or more. Normally, high removal efficiencies can be obtained without pretreatment or destabilization. In the most common application of DE filters, swimming pools, addition of metal ion coagulants could also cause a buildup of dissolved solids due to recirculation.

Inadequate precoating or body feed, as described in Chapter 8, can allow turbidity to pass the filter.

Media can also slough off the septum from improper or discontinuous flow conditions. Septa of precoat filters are usually designed with vertical surfaces. Media can slough off the vertical surface from discontinuous flow during formation of the coating or from other factors. When the septa are arranged horizontally, a disruption of the usual combination of upflow and downflow can cause media to fall off the underside. Precoat filters are considerably smaller than granular media filters, requiring only 5–50 percent of the space required for conventional gravity filters. Normally, even in nonfreezing climates, precoat filters are enclosed in a permanent building together with media storage and slurry makeup tanks.

Design Criteria

Head loss across the precoat should be determined, based on experience or pilot tests. The maximum head loss across vacuum DE filters is limited to about 6 m because of practical limitations. If head loss of more than 6 m is permitted, negative pressures will develop in the effluent chamber. On pressure DE filters, however, head losses of up to 75 m may be used. Design filtration rates (loading rates) are usually from 1 to 3 gpm/ft² (60–180 m/d), based on total septum area. Backwash rates are generally 7.5–15 gpm/ft² (450–900 m/d). Pilot plant tests should be conducted to confirm design criteria, media type and finished water standards. The initial coating of media (precoat) is applied to the septum as a slurry of 0.3–0.6 percent media, carried in a flow of recirculated water. Body feed rates may be in the range of 1–3 mg/L/NTU. Filter runs range from 12 to 150 hr in length, depending on turbidity, media, and the type of filter. (Pressure filters usually permit longer filter runs.) Backwash water requirements are roughly proportional to filter run times and usually less than 1 percent of the water treated is required for backwash. Additional sluicing and supplemental jets may be required to remove all spent media from the vicinity of the septums during backwashing.

Chemical Feeding

In addition to feeding media into the precoat filter, feeding of various pretreatment chemicals may also be required. Chlorine, for example, may be used before filtration but is usually not essential. The equipment manufacturer's limitations for pH and chlorine should be followed to protect the septum and other internal materials and coatings. Coagulants may improve removal efficiencies but can also cause rapid head loss development. Metal ion coagulants and/or polymers should only be incorporated after demonstrating their feasibility by pilot or full-scale testing.

Solids Disposal

The volume of wastes produced by precoat filtration is normally several times greater than equivalent direct-filtration, granular-media filters. Based on an average raw water turbidity of 5 NTU, it is estimated that generated solids would reach 15 kg/ML (15 mg/L). Spent DE media is dewatered on sand beds or decanted and dried in lagoons. In dry climates, it dewaters readily to 50 percent solids and is easily handled as a wet cake. Additional drying may cause dust problems. For ultimate disposal, it can be used as landfill or spread and harrowed in as a soil conditioner.

SLOW SAND FILTERS

Application

Slow sand filters may be used in either the in-line mode or preceded by clarification. Normally, a good quality water with turbidities not exceeding 20 NTU may be applied directly to slow sand filters. The principal advantage of slow sand filters is that they operate efficiently without predestabilization. They are reliable and can be operated with relatively unskilled personnel. Table 21-13 summarizes the advantages and disadvantages of slow sand filter usage. Table 21-14 summarizes the design criteria of slow sand filters.

SLUDGE-HANDLING METHODS

Various types of sludge (solids) handling equipment and methods have been employed to dewater and dispose of lime and alum or ferric water treatment plant sludges, as discussed in Chapter 13. Experience has indicated that a dewatered solids concentration of 35 percent or higher is typically required for successful sludge disposal in a landfill. Solids-handling equipment includes mechanical devices such as gravity thickeners, vacuum filters, centrifuges, and filter presses. Water plant sludges can

TABLE 21-13. Advantages and Disadvantages of Slow Sand Filters

Advantages
 Efficient performance with no chemical destabilization
 Little operational sophistication necessary
 Turbidity reductions of 90 percent
 Coliform reductions of 85–99 percent
 Reduction of organics
 Ability to use a broad range of filter media
Disadvantages
 Greater site requirements than rapid sand filters and higher costs
 Poor algae removals
 Lower color reductions
 Treated water turbidities of less than 1.0 NTU may not be attainable in certain applications and color removal may be low or ineffective. If coagulation and settling precedes slow sand filters, the problems relative to algae, color, and turbidity may not be a problem, but the complexity of operation is increased and rapid sand filters would be a more economical solution.

also be dewatered by nonmechanical methods such as sludge lagoons and drying beds, which are most attractive where ample land space is available and favorable climatic conditions exist.

Gravity Thickening Systems

Gravity thickening is typically accomplished in a circular tank designed and operated similar to a solids-contact clarifier or sedimentation tank. Sludge is introduced into the tank and allowed to settle and compact. Gentle agitation of the sludge prior to settling is often provided to open channels for water to escape and promote densification of the solids. The thickened sludge is collected and withdrawn at the bottom of the tank. Figure 21-38 shows a typical solids-contact unit often used for gravity thickening. Supernatant is normally returned to the main water treatment process flow. A properly designed and operated gravity thickening system can produce a 2–6 percent alum sludge. Softening sludges can thicken to even greater concentrations depending on the calcium carbonate and magnesium hydroxide content in the sludge. Lime sludges that are predominantly calcium carbonate can be thickened to 30 percent solids and higher. At minimum, the following should be evaluated for every

TABLE 21-14. Design Criteria for Slow Sand Filters

Criteria	Description
Loading rate	Design rates (filter loadings) are normally in the range of 3–15 m/d (3.2–16 U.S. mgd/ acre). The lower rates are applicable to turbid >20 NTU water and/or fine filter media.
Media size	Filter media can be less uniform than media for rapid sand filters. Because the media is not hydraulically backwashed, the finer fraction does not end up on the top of the bed. The maximum uniformity coefficient should not exceed 2.5 in order to minimize head loss through the media. The effective size is normally in the range of 0.25–0.35 mm, but coarse media (up to 0.4 mm effective size) may be used for low-turbidity waters at high rates.
Depth of media	Depth of filter media is normally about 1.0 m (39 in.), supported on 0.5 m (10 in.) of gravel
Submergence and outlet control	A slow sand filter is operated continuously under submerged media conditions, generally about 1.2–1.5 m (4–5 ft). To avoid air binding, the maximum head loss across the filter should not exceed the submergence. Buildup of head loss to maximum is slow, varying from 2 or 3 days to 6 months. Automatic control valves on the outlet structures are not required and manually adjusted weirs, outlets, and/or valves are satisfactory.
Inlet conditions	Inlet structures may be located at one end or side of the filter. Careful consideration should be given to preventing scouring or erosion of media during filling of the filter. Gradual filling of the filter may be necessary until the depth above the media is sufficient to allow maximum inlet rates and no disturbance of the surface.
Underdrainage	Underdrains have traditionally been open joint clay or concrete pipe but plastic piping (PVC) with orifice or slot openings are satisfactory. The diameter of underdrain pipes should be based on a maximum velocity of 0.6 m/sec (2 ft/sec). Underdrains may be spaced up to 2 m on center. Thorough hydraulic analysis of underdrains and central collection gallery should be made for large filters more than 4000 m² (1 acre), or high rates more than 10 m/d (10 mgd/acre).
Hydraulic control and head loss measurement	The filtered water outlet is a hydraulic conrol structure designed to maintain submergence of the media under all conditions. The difference in level of water in the collection gallery and the level above the media represents the head loss through the media. It can be monitored by instrumentation, but because the loss of head buildup is so slow and relatively predictable, most plants do not measure the data. Control of the effluent valve and/or level control weir or device can be adjusted manually.
Cleaning of media	Cleaning of filter media is performed by scraping 50–75 mm (2–3 in.) of media off the top. The dirty media is conveyed to a stockpile or basin where it is cleaned by jetting and/or mechanical washing. The process of scraping off media may be continued until the effective sand depth is 400–500 mm (16–20 in.), at which time the washed sand is spread back into the filter, leveled, and the filter returned to service.

gravity thickener: minimum surface area requirements, hydraulic loading, drive torque requirements, floor slope, and total tank depth. Typical design criteria for gravity thickeners is shown in Table 21-15.

Vacuum Filters

Vacuum filtration of sludge involves the application of a pressure differential across a sludge cake to yield an increase in solids concentration. The process uses a rotary drum suspended in a container of thickened sludge, as depicted in Figure 21-39. Sections of the drum are alternatively subjected to suction and pressure during each revolution or cycle. The surface of the drum is covered with a filter medium through which the water is drawn as the drum passes through the sludge. Solids are deposited as a cake on the medium surface and the cake is further dewatered as it rises out of the liquid sludge

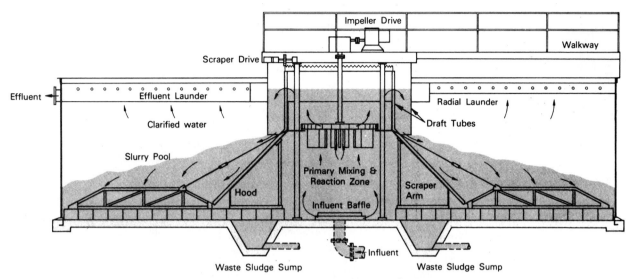

FIGURE 21-38. Solids-contact clarifier typically used for sludge thickening. (Courtesy of Infilco Degremont Inc.)

and air is drawn into the cake. The sludge cake that develops on the filter medium is then mechanically scraped off before the cycle is repeated. The vacuum filtration process is divided into three basic phases: cake formation, cake dewatering, and cake discharge. The combination of these three steps is called the filter cycle time, an important variable in establishing the dewatered solids yield from a vacuum filter. Another important consideration of vacuum filter design is the selection of a proper filter medium. Efficient sludge solids capture must occur without resulting or binding of the filter medium. The filter media may be either a stainless steel mesh, a synthetic or natural fiber fabric, or a coil-spring-type media. The more common fabric media materials are nylon, polypropylene, and polyethylene.

Dewatering of alum sludges by vacuum filtration

has been shown to produce a solids cake of up to 20 percent if a DE precoat is employed. Vacuum filtration of lime softening sludges has been more successful. Final lime sludge cakes having solids concentrations in the range of 45–65 percent have been reported. Softening sludges that are predominantly lime dewater so well by vacuum filtration that sludge-conditioning chemicals often may not be required. For a specific sludge, filter leaf tests are used to help determine the appropriate conditioner and dose, filter media, and cycle time. Typical design and operation criteria for vacuum filters is presented in Table 21-16.

Centrifuges

Centrifuges are used both to thicken and to dewater sludges. The centrifuge is basically a sedimentation

TABLE 21-15. Gravity Thickening Design Criteria

Mixing energy	10–50 sec^{-1}
Mixing time	10–20 min
Surface loading rate^a	
Design	0.5–1.0 gpm/ft^2
Maximum	1.0–1.5 gpm/ft^2
Chemical feeds^b	
Cationic polymer	1–4 mg/L
Anionic/nonionic polymer	1–4 mg/L
Tank depth	3–4 m

^{a} Intermittent operation allowable.
^{b} Not required simultaneously.

TABLE 21-16. Typical Design and Operation Criteria for a Vacuum Filter

Parameter	Alum Sludge	Lime Sludge
Feed solids, %	1–6	5–30
Flowrate, gal/ft^2/hr	2–5	2–5
Dry solids yield, lb/ft^2/hr	0.75–1.25	4–20
Cake concentration, %	15–25^a	45–65
Drum speed, rpm	0.2–0.5	0.2–0.5
Operating vacuum, in. Hg	10–25	10–25

^{a} With polymer and precoat.

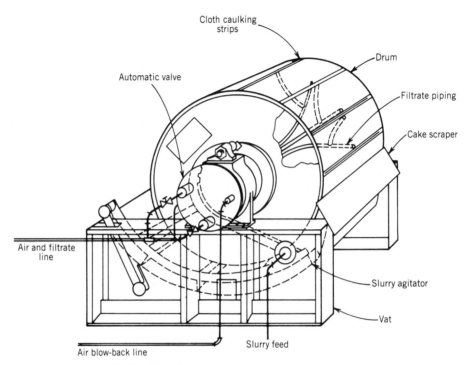

FIGURE 21-39. Cutaway view of a rotary drum vacuum filter.

device in which the solids–liquid separation is improved by rotating the liquid at high speeds to increase the gravitational forces subjected on the sludge. A common centrifuge design for sludges is the continuously discharging solid bowl machine shown in Figure 21-40. The two principal elements

of this centrifuge are the rotating bowl, which is the settling vessel, and the conveyor discharge of the settled solids. As the bowl rotates, centrifugal force causes the slurry to form an annular pool, the depth of which is determined by the adjustment of the effluent weirs. A portion of the bowl is of reduced

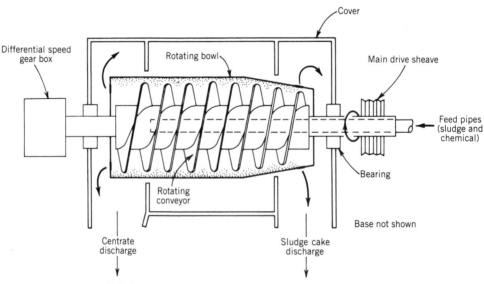

FIGURE 21-40. Solid bowl conveyor discharge centrifuge.

diameter so that it is not submerged in the pool and thus forms a drainage deck for dewatering the solids as they are conveyed across it. Feed enters through a stationary supply pipe and passes through the conveyor hub into the bowl itself. As the solids settle out in the bowl, they are picked up by the conveyor scroll and continuously carried along to the solids outlets. Clear supernatant at the same time continuously overflows the effluent weirs.

Application of centrifuges in the water treatment field has normally been in dewatering lime softening sludges. Operational data from four plants using centrifugation of lime sludges showed that the cake solids concentrations were in the range of 55–65 percent suspended solids by weight. However, solids concentrations of 35–50 percent are more typical. In centrifugation of predominantly alum sludges, solids concentrations of 12–15 percent have been obtained. Polymer doses of approximately 3 lb/ton of feed solids are typical. Feed solids concentration for alum sludge centrifugation is in the range of 1–6 percent and 10–25 percent for lime sludge.

Filter Presses

Filter press dewatering is achieved by forcing the water from the sludge under high pressure. Although the filter press produces high solids concentration and low chemical consumption, its disadvantages include high labor costs and limitations on filter cloth life.

A filter press consists of a number of plates or trays supported in a common frame. During sludge dewatering, these frames are pressed together either electromechanically or hydraulically between a fixed and moving end. A filter cloth is mounted on the face of each plate. Sludge is pumped into the press until the cavities or chambers between the trays are completely filled. Pressure is then applied forcing the liquid through the filter cloth and plate outlet. The plates are then separated and the sludge removed. A cutaway view of a filter press is illustrated in Figure 21-41.

Pilot studies and operating results of a full-scale filter press installation using alum sludge in Atlanta, Georgia, have produced cakes of 40–50 percent solids using a 10 percent lime dosage and precoating. Filter press installations in Europe also report that with proper pretreatment using lime or polymer, cakes as dry as 30–50 percent can be produced from alum sludges. Attempts to filter alum sludges

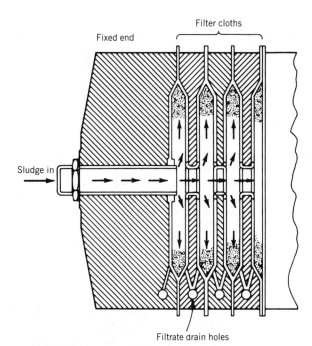

FIGURE 21-41. Cutaway view of a filter press.

without any conditioning agents have proved impractical. Lime sludges have been reported to readily dewater to above 50 percent solids without sludge conditioning.

Sludge Lagoons and Drying Beds

As described previously, a nonmechanical means of handling water treatment plant sludges consists of dewatering in sludge lagoons or drying beds. Lagoons are commonly of earthen construction and are equipped with inlet control devices and overflow structures. Wastes with settleable solids are discharged into the lagoons from which the solids are separated by gravity sedimentation. Sludge lagoons can be classified by their mode of operation: permanent lagoons and dewatering lagoons.

Permanent lagoons, which have 2.5–4 m (8–13 ft) side water depths, act as a final disposal site for settled water solids. They are decanted periodically to allow a thin sludge layer to be dewatered and dried. This cycle is repeated until the lagoon is filled with solids. At that point, the lagoon is covered to reclaim the land and new lagoons are required.

Dewatering lagoons, which have 1–2-meter side water depths, are used to store and concentrate sludge discharges and produce a 0.3–1.0-m (1–3-ft) layer of sludge. Sludge and filter washwater is periodically discharged into the lagoon, thereby causing the decant of supernatant from the lagoon. When the lagoon is full of solids (usually at a 4–12 percent concentration), the sludge layer is allowed to drain and air dry for removal and ultimate disposal. The lagoons are used repeatedly as a dewatering device. Dewatering lagoons are generally sized to settle and store the sludge solids for 4 to 6 months to a year. The volume requirement to store the alum sludge for the desired period can be determined using an estimated sludge storage of 4 percent solids. With proper climatic conditions, solids concentrations of 40 percent can easily be obtained in sludge lagoons with alum sludges. Lime sludges have been dewatered to 50 percent solids concentration by the sludge lagoons handling method.

Drying beds are shallow structures with a sand and gravel underdrainage system from which filtrate is collected. These beds require large areas of land to spread thin sludge layers. Sludge is normally applied in 15–30-cm (6–12-in.) layers, and dewatering on the drying beds is achieved by gravity drainage through the underdrain and evaporation from the sludge surface. After the sludge layer is dewatered, it is removed for final disposal. The dewatered solids concentration obtained from the sand drying bed is similar to that obtained from the lagoon dewatering method, but the drying bed process can achieve a cake more rapidly. Drying beds usually are designed with 15–23 cm (6–9 in.) of 0.3–1.2-mm-size sand over an underdrain system composed of 30 cm (12 in.) of gravel and 15–20 cm (6–8 in.) drain tiles to handle the filtrate. The specific characteristics of the sludge will determine the bed surface requirements and the need for sludge conditioning.

Sludge Piping and Equipment Layout

There are many items to be considered in the design of a sludge-handling and disposal system. The following considerations do not include all areas of concern but address many of the unforeseen problems experienced in the design and operation of sludge-handling facilities. All sludge-carrying pipelines should follow good sewage design practices in order to avoid plugging. Slopes on all gravity lines and pump suction lines should not be less than 2 percent. Careful attention must be given to cleanouts, and provisions must be made for water, and if possible, air pressure flushing. Side branches, dead ends, and reverse-flow conditions can cause serious maintenance problems. Where possible, sludge

TABLE 21-17. Dewatering Solids Concentrations Obtainable by the Various Dewatering Methods for Water Treatment Sludges

Dewatering Method	Typical Dewatered Solids Concentrations (%)	
	Alum Sludge	Lime Sludge
Gravity thickeners	2–6	30
Vacuum filters	20	45–65
Centrifuges	12–16	35–50
Filter presses	30–50	50
Sludge-drying lagoons	20–40[a]	30–50[a]

[a] Dependent on length of drying time.

should be piped by a single-purpose pipeline. On lines with more than 60 m between clean-outs, provisions should be made for positive pipe-cleaning methods.

Summary

A summary of dewatering capabilities of water plant sludges by the aforementioned solids-handling equipment is provided in Table 21-17. It is clear that lime softening sludges dewater more readily than alum sludges. Final solids concentrations obtained by the various dewatering methods will vary depending on the type of sludge and sludge conditioning employed.

ADSORPTION

Introduction

The design of adsorption systems for control of organics or taste and odor problems is primarily concerned with activated carbon as the adsorbent of interest. Both powdered activated carbon (PAC) and granular activated carbon (GAC) are used in water treatment. PAC is commonly used for intermittent control of taste and odor problems, and there has been much recent interest in PAC for removal of influent contaminants, which may be related to accidental chemical spills. GAC use in water treatment, however, is not common in the United States; previous experience with GAC is limited to control of persistent taste and odor problems. In these cases, gravity sand filter media is usually replaced or supplemented with activated carbon. The GAC process is more complex and requires considerably more capital and operational costs than PAC. The process of adsorption is discussed in Chapter 9.

Powdered Activated Carbon Feed Systems

Currently, a rational design procedure is not well defined for PAC feed systems; designs are usually based on experience and generally accepted design criteria. Since use of the system is intermittent, little work has been done to optimize the process.

PAC may be applied in the water treatment process in any location prior to filtration. However, a significant amount of PAC breakthrough can occur if PAC is fed prior to filtration. Following adsorption, most of the PAC is removed through sedimentation or filtration. In most water treatment plants, dosage is set by trial and error, but jar tests may help optimize the dosage. In general, taste- and odor-causing compounds are adsorbable; experience has shown that 2–5 mg/L PAC is often sufficient. A typical upper design range is 10 mg/L, but in stubborn cases, some systems have been designed with capacities up to 20–100 mg/L. The designer's judgment must be used with respect to variability, intensity, and anticipated duration of system use when developing design criteria.

PAC may be fed as either a dry chemical or a slurry. It is most convenient to store PAC and feed it as a slurry. Generally, PAC is very fine (90 percent passing 300 mesh, AWWA Standard D-600-66) and can be delivered in bags or bulk (drums, steel containers) or covered hopper railroad cars. It may also be delivered by hopper cars or trucks, which utilize compressed air to fluidize the carbon. When delivered in this manner and stored as a slurry, problems with carbon dust are minimized at the plant.

Slurry tanks are usually constructed of concrete and equipped with a mixer to keep the carbon in suspension. PAC is typically stored and fed as a slurry at a concentration of approximately lb/gal. Volumetric metering or proportioning pumps are recommended. Alternatively, a dry feeder unit with a wetting device may be used. Because PAC slurries are highly corrosive, all materials should be PVC, stainless steel (i.e., type 304 SS), rubber coated, or plastic lined. All PAC slurry feed lines should be of ample size to provide enough carrier water for suspension of the PAC and should also be provided with fittings to flush the line immediately on shutdown. Care should be taken to avoid contact of dry PAC with oxidizing agents, such as chlorine, potassium permanganate, or gasoline. Activated carbon dust is also an electrical conductor. Other safety considerations include provision of protected coverings and respirators for operators. In addition, activated carbon adsorbs oxygen, and adequate ventilation should be provided for storage areas (at least 1 room change per minute).

GAC Systems

The adsorption process analysis and conceptual design considerations for granular carbon systems were presented in Chapter 9. This section is concerned with contactor and support facility design. Much of this material has been drawn from experience with GAC in advanced waste treatment and industrial waste treatment and experience in European water treatment plants. Although it is felt that this information is applicable to advanced water treatment processes, caution should be taken by the designer since experience with GAC for control of organics is limited in the United States.

GAC Process Location. For most water treatment applications the generally accepted practice is to apply the highest-quality water to the GAC system. This implies that, as a minimum, the raw water has been through a pretreatment step before GAC. One alternative scheme is to replace the filtration step with GAC for turbidity and suspended solids removal. This alternative is particularly attractive for retrofit of existing plants. However, elimination of the filtration process prior to GAC may result in poor efficiencies with respect to both turbidity and suspended solids as well as organics removal in the adsorption process. In addition, it should be recognized that filters insulate the GAC process from pretreatment upsets. Optimization of organics removal in pretreatment and filtration is also desirable to lower the organic loading on GAC. Replacement of filter media with GAC may also not be feasible due to the limited empty-bed contact time available. In any case, this should be evaluated on a site-specific basis. In all plants, GAC effluent will require final disinfection due to inevitable microbiological growth within the GAC bed. Disinfection in the pretreatment or filtration processes is advisable to reduce the microbiological load to the bed. Any residual disinfectant will be lost within the first few layers of the carbon bed.

Contactor Configurations. Three type of beds could be used for the application of GAC in water treatment:

1. Upflow expanded bed.
2. Gravity, or pressurized, downflow fixed bed.
3. Moving, or pulsed, bed.

Each of the three types of beds can be oriented in parallel or in series with other units, or a combination thereof. Examples of several possible configurations are shown in Figure 21-42 and are described below.

Downflow Fixed Beds in Series. In this reactor configuration, the flow is downward through adsorbers connected in series. The first bed in the series receives the highest adsorbate loading and thus

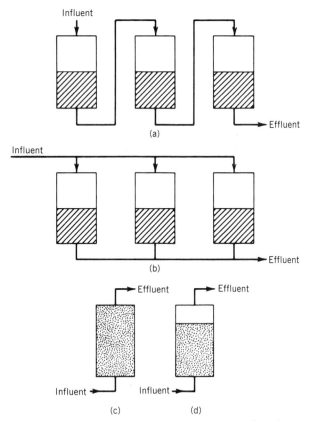

FIGURE 21-42. Contactor: configurations: (a) Downflow, in series; (b) downflow, in parallel; (c) moving bed; (d) upflow expanded bed.

is the first to be exhausted. The final bed in the series acts as a polishing step and receives a very light loading. When the first bed is exhausted, it is taken out of service and the exhausted adsorbent is replaced with fresh adsorbent. The bed is then put back in service as the last in the series, so that the cleanest water sees the freshest bed. The adsorbent thus moves countercurrent to the flow. The efficiency of the adsorbent is maximized by keeping the water with the lowest concentration of contaminants in contact with the cleanest adsorbent. This design is most suitable for systems that require small amounts of adsorbent, that is, have long column runs before exhaustion, but require countercurrent efficiency in order to achieve high degrees of removals. If backwashing is provided, a limited amount of suspended matter can be handled directly on the lead bed without affecting performance.

Moving Beds. Moving beds are a refinement of the downflow fixed-bed in-series concept. Water flows upward through the bed and is withdrawn from the top while adsorbent flows downward and is withdrawn periodically or continuously from the bottom of the bed. Fresh adsorbent in an amount equal to that withdrawn enters at the top of the bed. This system is most useful for applications that require large amounts of adsorbent, that is, have short column runs before exhaustion and countercurrent efficiency. Suspended matter should be removed prior to application to moving beds to prevent fouling and clogging of the bed. This configuration has also shown promise where synthetic resins are used as the adsorbent.

Downflow Fixed Beds in Parallel. In this reactor configuration, the flow is downward through adsorbers connected in parallel. The start-up of the beds can be staged so that exhaustion of the beds occurs sequentially. The effluent from fresh and partially exhausted beds can then be blended to extend the life of the beds. This arrangement is simpler to operate than beds in series and is used in large-volume plants where effluent blending will meet treatment requirements. No complicated system of piping and valving is necessary for parallel beds as compared to beds in series for switching the position of beds as they become exhausted and regenerated. If backwashing is provided, a limited amount of suspended matter can be handled without adversely affecting system performnce.

Upflow Expanded Beds. Upflow beds may operate in series or in parallel. Upflow operation expands the beds by about 10 percent of their packed volume and allows the bed to pass suspended solids without developing an excessive pressure drop. While this characteristic is important in wastewater applications, in water treatment, pretreated water generally has a low turbidity and suspended solids concentration. Release of carbon fines in the column effluent has been reported to be an operational problem. For these reasons, the upflow expanded bed configuration does not show promise for use in water treatment.

Column Design. Once conceptual design criteria have been selected based on the process analysis methodology presented in Chapter 9, the detailed design of contactors and support facilities can begin. The first consideration is contactor design flow. In general, one should consider overall plant capacity, economics, and anticipated plant expansions. In addition, plant recycle flows from backwash water, scrubber flows, or other water demands should be included. Any impact from potential changes in treated water quality must be addressed. Because most GAC contactor installations lend themselves to a modular approach, it is advisable to expand facilities with new units in a staged manner.

Both gravity and pressure contactors have been used for GAC. Use of pressure contactors usually permits more flexibility as the system can be operated with higher head loss, but gravity contactors, however, have an advantage of common wall construction. The number of contactors must be adequate to allow for one or more to be out of service for repair and/or backwashing. Pressure vessels are limited to 12 ft diameter and 60 ft length for shipment. For small GAC facilities, more than one contactor should be planned, even where one is sufficient, to take advantage of parallel operation for blending of effluent concentrations.

Flow distribution in downflow GAC contactors is generally not a critical problem. However, the designer should take care to prevent short circuiting or channeling through the bed. Inlet and outlet screens are recommended to prevent GAC media loss. Typical hydraulic design criteria are a surface loading rate of 4.8–24 m/hr (2–10 gpm/ft^2) and a bed depth ranging from 3 to 9 m (10–30 ft). During hydraulic design of the contactors, careful consideration should be given to the head loss characteristics of the selected GAC. Most GAC suppliers provide guidelines for head loss development in the bed or, alternatively, pilot tests may be required to obtain head loss data.

Backwashing Systems. Backwashing of downflow GAC contactors is required to reduce bed head loss. An increase in head loss is caused by deposit of suspended solids or turbidity and/or microbiological growth within the GAC bed. For downflow columns, a 10–50 percent bed expansion is recommended. Backwashing criteria are typically 30–50 m/hr (12–20 gpm/ft^2) for 10–15 min. Some installations where microbiological growth is anticipated to be a significant concern may consider air scour use. In general, waste backwash water should be recycled to the head of the plant.

Carbon Storage. The carbon storage facilities required depend on whether regeneration will be conducted on or off-site. At least two vessels for carbon storage are required, one for spent GAC and one for fresh GAC. In-plant carbon transport between contactors and storage vessels is best accomplished with a hydraulic system by slurrying the GAC at a concentration of approximately 120 g/L (1 lb/gal). The capability to fluidize the bed within each contactor to slurry the carbon must be provided. Either slurry pumps or eductor systems with separate pumping can be used for carbon transport. Construction materials should be stainless steel (type 304 or 410) and rubber lined. Provision for GAC dewatering to a concentration of 40–50 percent moisture before regeneration must be provided.

Makeup GAC delivery to the plant is usually by hopper railroad car or truck with pneumatic delivery to the carbon storage bin. If the carbon is moved with air, it must be washed to remove fines. Because the carbon slurry is quite abrasive, care must be taken to use long-radius elbows in piping. In general, the use of steel is sufficient if adequate drains, clean-outs, and flushing are provided. It may be less expensive to periodically replace carbon steel piping than to use expensive corrosion-resistant materials. Valves that do not require a positive seat such as rotary-type valves (ball or plug) are recommended. Knife gate valves are particularly successful in this application. Globe or gate valves should be avoided.

ION EXCHANGE FACILITIES

Configurations

An ion exchange unit consists of a closed vessel containing a bed of resin and the appurtenances required to direct product water or regenerant through it. The resin bed can be fixed in place or in motion, moving either continuously or intermittently. Anionic and cationic resins can be utilized in separate columns or comingled in one bed. Product water can flow through the bed cocurrently or countercurrently to the direction of regenerant flow. The primary configurations for use in water treatment are single-resin, fixed-bed, cocurrent and countercurrent flow, mixed-resin bed, and continuous countercurrent bed ion exchangers. These configurations are discussed in more detail in Chapter 10.

Cocurrent regeneration is most common as it is convenient to introduce and remove the regenerant and product water through the same distribution manifolds. However, countercurrent regeneration allows more complete utilization of the resin's exchange capacity before exhaustion.

Regeneration of mixed-resin ion exchange beds is more complicated than conventional beds because the anionic and cationic resin beads are equally distributed through the bed and must be separated during regeneration.

In continuous countercurrent ion exchange systems, regeneration is carried out in a separate vessel. There are two principal types of these exchangers, fluidized bed and pulsed flow systems. In fluidized bed exchangers, the resin is moved continuously either hydraulically or mechanically. In pulsed bed systems, resin movement occurs in 10–20-sec pulses during which time the flow of product water through the system is halted. Pulses usually occur every 5–10 min. During each pulse, a small increment of resin is moved throughout the system.

In general, the size of vessels used for ion exchange beds are within the bounds feasible for transport by rail and truck. This makes it advantageous to specify ion exchange systems as prefabricated units that require only a minimum of on-site fabrication. Thus, multiple-bed ion exchange plants are generally laid linearly in plan both to reduce overall width for shipment and to take advantage of straight runs for the pipe and valve manifolds connecting the vessels. Very-large-capacity plants may consist of interconnecting rows of linear ion exchange modules.

Equipment Sizing

Sizing of ion exchange reaction vessels is a function of the flow required and the volume necessary to treat it. Maximum and minimum flowrates (liters per second or gallons per minute per square foot) for the particular resins utilized and minimum bed depth will also affect the size of the vessel. Generalized design data for typical resins is presented in Table 21-18. The maximum flowrate for the resin and the required flow determine the bed surface area. Bed depth is determined by the resin volume required to treat the flow or minimum bed depth, whichever is greater. For designs in which the resin bed is backwashed and fluidized during regeneration in the ion exchange vessel, it is necessary to allow free board for this expansion. An allowance of 50–100 percent of the bed depth is customary.

Should conditions dictate that a continuous flow of product water is necessary, some accommodation must be made to ensure this supply during the regeneration period. This can be accomplished with a single-column system by oversizing the system capacity and providing a storage tank for product delivery during regeneration. With multiple-column systems, storage can be dispensed with, provided the on-line exchangers have sufficient capacity to deliver the required flow.

Column Details

Except for continuous-counterflow ion exchangers, ion exchange vessels are almost identical to vertical pressure filters. Even distribution of water entering and leaving the ion exchange column is critical and, for columns more than a few inches in diameter, requires a manifold with flow control orifices at the top and bottom of the column.

The upper manifold usually takes the form of a pipe grid. It is necessary that the orifices in the influent flow distribution manifold be designed so the particles trapped in the resin bed, which are subsequently removed during backwash, can readily pass through them.

A common design for the underdrain manifold is a pipe grid similar to the distribution manifold. To conserve resin, it is common to fill the lower dished

TABLE 21-18. Design Data for Fixed-Bed Ion Exchanger[a]

Type of Resin	Maximum and Minimum Flow (gpm/ft²)	Minimum Bed Depth		Maximum Operating Temp.		Usable Capacity (meq/mL)	Regenerant	
		cm	in	°C	°F		Per m³ Resin	Per ft³ of Resin
Weak-acid cation	8 max., 1–2 min.	60–75	24–30	120	250	0.5–2.7		110% theoretical (HCl or H_2SO)
Strong-acid cation	7–12 max., 1–2 min.	60–75	24–30	120	250	0.8–1.5	80–240	5–15 lb NaCl
						0.5–0.9	32–190	2–12 lb 66°Be H_2S
						0.7–1.4	80–480	5–30 lb 20°Be HCl
Weak- and intermediate-base anions	4–7 max., 1–2 min.	75–90	30–36	38	100	0.8–1.1	32–64	2–4 lb NaOH
Strong-base anions	5–7 max., 1–2 min.	75–90	30–36	38–50	100–120	0.4–0.8	64–128	4–8 lb NaOH
						0.5–1.0	80–240	5–15 lb NaCl
Mixed cation and strong-base anion (chemically equivalent mixture)	8–12 max.	90–120	36–48	38	100	0.3–0.5	Same as cation and anion (based on mixture) individually	

[a] Represent the usual ranges of design for water treatment applications. For chemical process applications, allowable flowrates are generally somewhat lower than the maximums shown and bed depths are usually somewhat greater.

SOURCE: Perry and Chilton, *Chemical Engineer Handbook,* 5th ed., McGraw-Hill, New York (1963).

head of the ion exchange column with concrete so that the pipe grid can rest on a flat floor. The grid is wrapped with a mesh material to retain the resin or the grid itself can be constructed of well screen.

An alternative design for the underdrain is to construct a false floor in the bottom of the vessel. The floor can be constructed of porous plates or it can be a solid membrane pierced by a regular grid of screen nozzles, which serve to distribute the flow and contain the resin.

Regeneration Facilities

Regeneration facilities for an ion exchange system consist of those appurtenances required for contacting the resin with the regenerant. In ion exchangers other than the continuous-counterflow systems, regeneration occurs inside the ion exchange vessel. The facilities required for regeneration include regenerant storage, pumps to transfer regenerant from storage to the ion exchange vessel, a resin backwash system, and facilities for the disposal of spent regenerant.

The most commonly used regenerants are acid, caustic, and brine. Acid and caustic are most commonly stored as liquids while brine is usually stored

as the solid salt and dissolved as needed. Liquid chemical storage tanks are generally of steel construction; mild steel is completely satisfactory for caustic storage while either an appropriate coating or stainless steel is necessary for acid storage. Tanks should have fill connections that are designed to minimize spillage and must be provided with adequate vent and drain connections. Concrete tanks are convenient for storing salt. Piping connections are provided so that the tank can be flooded to dissolve the salt and the resultant brine pumped directly from the tank for regeneration. Such tanks are frequently constructed below grade so that trucks or rail cars delivering salts can discharge directly into them without additional handling. If it is necessary to locate salt storage tanks above ground, some means should be provided to elevate the salt into them.

Pumps for transferring regenerants are generally end suction centrifugal pumps designed for chemical service. They should be constructed of corrosion-resistant material suitable for the liquid being pumped. Either packing or mechanical seals can be specified for these pumps; however, a flushed doubled-seal arrangement or lantern should be used to prevent chemical leakage at the pump shafts. Backwashing can be done in several ways. With multiple-vessel ion exchange systems, it is sometimes feasible to utilize product water from the on-line exchangers to backwash the one being regenerated. In this case, backwash pumps are not necessary. If it is necessary to store water for backwash, this is usually done in an elevated storage tank or a clearwell. Relatively low capacity pumps can fill an elevated tank for gravity backwash, but with a clearwell, it is necessary to size the backwash pumps for the full flow.

Depending on the nature of the spent regenerants and the site conditions, it may be possible to discharge spent regenerant directly to a sewer or receiving water. Otherwise, it may be necessary to store spent regenerant for off-site disposal or on-site treatment to meet discharge conditions. The waste may also be discharged to evaporation ponds.

DISINFECTION

Introduction

The destruction of waterborne pathogens, that is, disinfection (as opposed to sterilization), is generally accomplished through the addition of chemical reagents. Typical chemical reagents added to water include various forms of chlorine (e.g., sodium hypochlorite–NaOCl, calcium hypochlorite–Ca(OCl)$_2 \cdot 4H_2O$, and chlorine gas), chloramines, ozone, and chlorine dioxide. Design of feed facilities for each of these chemicals is discussed herein.

Chlorine Feed Systems

Chlorine feed systems may be divided into four basic types:

1. Direct solution or dry chemical feed systems.
2. Gas-to-solution systems.
3. Direct gas injection.
4. On-site generation and feeding.

Hypochlorites and chlorine may be generated electrolytically from brine or salt using specially designed cells and an ordinary power source. Several equipment manufacturers can supply packaged units that include a feeding system. This alternative is particularly attractive for the places where the local codes do not permit transportation of hazardous chemicals such as chlorine in the plant site area

or where the chemical supply is not reliable because of special site locations or restrictions. The capacity of the units range from 10 kg/d to 24,000 kg/d (22 lb/d to 53,000 lb/d).

Direct Solution or Dry Chemical Feed Systems. Sodium hypochlorite (NaOCl, 12–15 percent available chlorine) is the most commonly utilized solution for disinfection. It is used extensively for swimming pools and small water treatment plants (to 5 ML/d capacity), is safer to handle than chlorine gas, and is available commercially in many quantities, from 4-L plastic bottles to 8000-L tank trucks. Calcium hypochlorite [Ca(OCl)$_2 \cdot 4H_2O$], a white granular or tablet material with 70 percent available chlorine, can be fed directly to water as powder or it can be converted to a solution and pumped, educted, or gravity fed to the application point. Calcium hypochlorite increases pH and hardness, which in small plants can simplify chemical feeding and eliminate the need for pH adjustment with lime or caustic soda. Clogging of solution piping with calcium carbonate may be a problem.

Gas-to-Solution Systems. The chlorine gas-to-solution system is the most commonly used disinfection method in the United States. The basic system may include one or more chlorine gas cylinders, evaporators, chlorinators, and injectors, plus a piping system.

The injector is the most essential component of the system. The injector creates a vacuum that permits chlorine gas to be metered and handled below atmospheric pressure and ensures the safety and reliability of the entire chlorine feed system. The injector also converts the gaseous chlorine to a concentrated chlorine solution of 100–3500 mg/L.

Direct Gas Injection. Chlorine gas can be injected directly into the main flow, thus eliminating the solution phase of the system. This type of system may be significantly more efficient for bacteria or virus kills, especially when other soluble constituents in the water (such as ammonia) may be competing for the chlorine. However, direct injection is not a common procedure.

Chlorine Gas Storage

Chlorine gas in a pressurized liquid form is stored in steel cylinders. The pressure in storage cylinders varies with the ambient temperature as shown in

TABLE 21-19. Vapor Pressure of Liquid Chlorine

°C	°F	kPa	psi
−20	−4	74	10
0	32	273	39
20	68	582	83
40	104	1045	149
60	140	1705	243

Table 21-19. Cylinder capacities range from 45-kg (100-lb) cylinders to 16-ton (35,000-lb) tanks. Liquid chlorine containers should never be completely filled; space must be provided for the liquid to expand. Because of the variations in the liquid chlorine pressure due to temperature, quantities in cylinders can only be gauged reliably by weight.

Most plants that have maximum chlorine feed rates of 10–1000 kg (22–2200 lb) per day use 900-kg (2000-lb) cylinders. For plants using more than 1000 kg (2200 lb) per day, consideration should be given to on-site bulk storage of chlorine in stationary tanks, tank trailers or rail tanks. The minimum quantity of chlorine stored at a plant is often established by state or local regulatory agencies. Typically, a 15–30-day supply based on average feed rates and maximum plant flow is provided. Minimally, two cylinders or tanks should be provided. All chlorine storage facilities should be covered. Where the ambient temperature falls below 0°C, storage areas for cylinders should be completely enclosed and heated. Enclosed rooms must be provided with a floor-level exhaust fan, an outward-opening door with panic hardware, adequate lighting, windows adjacent to or in doors, and safety equipment. Other equipment should not be installed or stored in the chlorine storage room. In small plants of less than 200 kg/day (400 lb/day) chlorine usage, chlorinators may be installed in the chlorine storage room if regulations permit. The large 900-kg (2000-lb) chlorine cylinders should be stored on a storage rack and prevented from rolling. Cylinders should be provided with retainers or chained down to prevent movement during earthquakes. Smaller 45- and 68-kg (100–150-lb) cylinders should be provided with safety chains to prevent them from falling from a vertical position (see Figure 21-43). Lifting equipment for unloading 900-

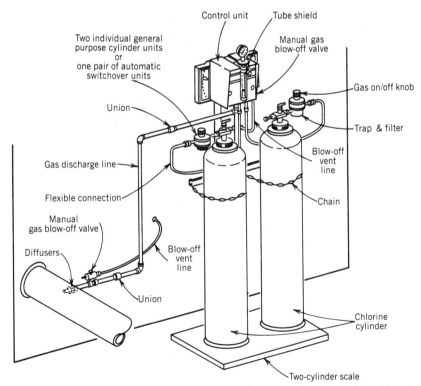

FIGURE 21-43. Typical detail of a direct-feed chlorinator (maximum capacity = 136 kg/day) with two-cylinder scale and rigid discharge piping. (Courtesy of Wallace & Tiernan.)

kg (2000-lb) cylinders (and transferring cylinders from inventory to use) is an essential feature of all chlorine storage facilities. Manual or electric, 3600-kg (4-ton) hoists may be used. A cross section of a typical chlorine storage room including hoist equipment is shown in Figure 21-44.

Chlorine cylinders that are in use are typically mounted on scales with two or three cylinders on each scale (or load platform). Similarly, stationary tanks and tank trailers should be mounted on load cells to indicate and record the weight of chlorine in storage. Scales are proprietary components and may be purchased for local or remote weight indication. Low-weight alarms or contacts are frequently incorporated in scales or weight indicators. In seismic areas, cylinders should be chained to scale platforms to minimize damage during earthquakes. Chlorine gas or liquid is transferred from cylinders to a manifold by means of flexible 15 mm (0.5 in.) diameter copper tubing. Chlorine gas or liquid is transferred to chlorine evaporators or directly to chlorinators by rigid seamless steel or wrought-iron piping. Valves should be provided on the cylinder where the flexible copper tube connects to the manifold and as may be required to maintain two separate manifold systems. Piping for liquid chlorine or chlorine gas should be Schedule 80 seamless steel pipe, normally 20 to 40 mm diameter ($\frac{3}{4}$ to $1\frac{1}{2}$ in.). Joints may be welded, screwed, or flanged. Valves should be metal-seat-tested to 4000 kPa (600 psi).

Auxiliary Chlorination Equipment

Maximum chlorine withdrawal rate from a 900-kg (2000-lb) cylinder is temperature dependent, but, in general, if it exceeds 150 kg (330 lb) per day, chlorine

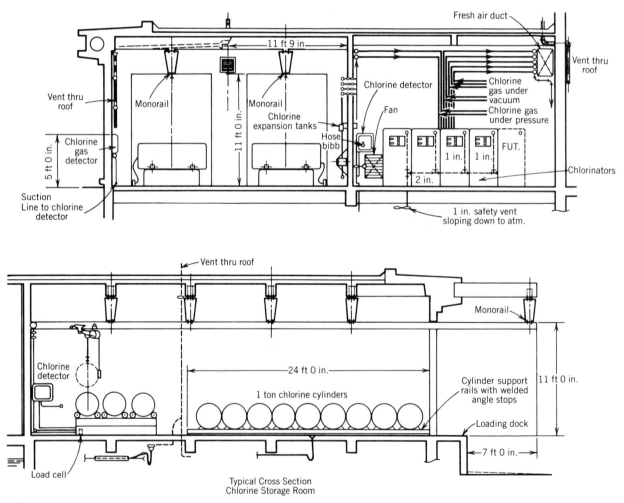

FIGURE 21-44. Typical cross sections of chlorine storage room and chlorinator room (top) and chlorine storage room (bottom).

evaporators are required to furnish latent heat of evaporation to the liquid chlorine unless two cylinders are connected in parallel. Chlorine expansion chambers should be provided on all liquid chlorine pipelines. The chambers allow the chlorine to expand should the chlorine become isolated in the piping system and temperatures increase. Pressure reducing valves (PRVs) are required upstream of chlorinators on the inlet piping. It is also advisable to provide a chlorine strainer upstream of the PRVs. Evaporators, expansion chambers, PRVs, and strainers are proprietary components.

Chlorinators

Chlorinators are manual or automatic metering devices for chlorine gas and are proprietary components. In small plants of 1–25 kg/d (2–55 lb/d) chlorine usage, chlorinators may be mounted on or adjacent to the chlorine cylinders. In larger plants, the chlorinators are generally installed in a room separate from the chlorine cylinders. Chlorine injectors, located downstream of the chlorinator, create a suction in the chlorine supply pipeline which draws chlorine gas from the chlorinator. Metering is usually effected by controlled pressure and/or variable orifices. The feed rate may be manually or automatically regulated. A visual flow indicator (rotometer) or a more sophisticated chlorine gas metering device is normally an integral part of the chlorinator. Chlorinators are available with maximum feed rates ranging from 50 to 3600 kg/d (100–8000 lb/d) and with operational ranges of from 0.05 to 1.0 of maximum capacity.

Injectors

Injectors are proprietary components that utilize a pressurized water supply and the venturi principle to create a suction in the chlorine supply pipeline. Chlorine gas is drawn into the throat or low-pressure area of the injector and mixed with the carrying water. Chlorine dissolves readily in water to a maximum of 3500 mg/L. Injectors may be located adjacent to the chlorinators with the solution transported to the point of diffusion or adjacent to the point of diffusion with the chlorine gas conveyed from the chlorinator to the injector. The quantity and pressure of water required to operate the injector depends on the design of the injector as well as the required chlorine feed rate and back pressure of the solution line. Back pressure on chlorine solution

lines can be reduced by designing solution lines of adequate diameter. The normal velocity in solution lines is approximately 1–2 m/sec (3.3–6.6 ft/sec). After injection, chlorine solution may be proportioned to several feed locations by throttling valves and flow indicators (rotometers) on two or more solution lines.

Chlorine gas can be injected directly into the main plant flow by injectors submerged in a basin or conduit. Direct gas injection may be more efficient than chlorine solution addition, especially where ammonia or other chlorine demand components need to be satisfied.

Diffusers

Chlorine should be dispersed rapidly into the main plant flow. In clear filtered water with low chlorine demands, dispersion may not be necessary if the flow is turbulent and undivided for a short distance downstream of the diffuser. In waters with chlorine demands greater than 1 mg/L or with coliform counts in excess of 22/100 mL, diffusion and supplemental mixing should be provided. In small pipelines with less than a 500 mm diameter, chlorine solution may be injected at the side of the pipe, as shown in Figure 21-43. The inlet should extend 0.2 diameters clear of the pipe wall to minimize corrosion of pipe walls or linings. In larger-diameter pipes, diffusers, normally perforated PVC pipes, are designed to disperse the chlorine across the total cross section of the flow. Perforations in diffusers should be designed to produce about 2 ft of head loss at each orifice giving 0.13–0.26 L/sec (2–4 gpm) flow rate for each orifice.

Supplemental flash, or hydraulic, mixing should be provided for applications where high chlorine demands may interfere with disinfection, such as chlorination of secondary wastewaters and polluted raw water flows to water treatment plants. Typical energy requirements (G) for supplemental chlorine mixing are on the order of 400 sec^{-1}.

Chlorine Contact Basins

In order to provide sufficient contact time for disinfection (or to meet the requirements of regulatory agencies), contact or detention times of 10–120 min may be required before distribution or discharge of the water depending upon the level of residual chlorine. Long conduits minimize short circuiting and circular, square, or wide and deep rectangular basins are most prone to short circuiting. If the risk of positive coliform counts in the treated water plant effluent is low, a clear well or treated water reservoir, with flowthrough piping, is generally satisfactory as a contact basin. Balancing tanks, or reservoirs with a single inlet/outlet do not lend themselves to providing chlorine contact time. Chlorine contact basins should be covered to prevent sunlight from interfering with disinfection. In most water treatment plants, chlorine is added upstream of the filters and a positive chlorine residual is maintained through the filter media and effluent piping. In such cases, the actual flow time from the underside of the fine filter media to the clear well may be considered as contact time.

Where providing a sufficient chlorine contact time is critical, prevention of short circuiting should be incorporated into the design of the chlorine contact structure. The normal approach is to provide a long serpentine plug-flow basin with a length-to-width ratio of 200 or greater. An equally effective structure is a compartmentalized basin with at least four compartments and baffles designed to distribute the flow across the total cross section of the basin. Compartmentalized basin baffles should develop 3–5 mm ($\frac{1}{8}$–$\frac{1}{4}$ in.) of head loss across each baffle under maximum-flow conditions.

Sampling and Monitoring

Whenever chlorine is added for disinfection, effluent chlorine residual should be continually monitored. Chlorine residual analyzers are proprietary equipment that can monitor free or combined (chloramine) residuals. Sampling pumps, required to provide a continuous sample stream to the analyzers, are normally designed for flows of 0.15 L/sec (2.4 gpm). When analyzers are used for pacing chlorine feed rates (closed-loop residual control), two or three alternative sampling points downstream of the diffuser should be provided in order to minimize a time lag problem.

Most chlorine analyzers transmit a signal output to an indicator–recorder or a data logger. Low- and high-level alarm contacts are normally connected to the analyzer output signal. The alarm contacts may provide audible or visual alarms or may be sequenced to shut down the plant after a prolonged low chlorine residual condition. Low chlorine residual is generally caused by power outages, low pressure in the injector water supply, low chlorine gas pressure, or an insufficient chlorine inventory on-

line. It may be more reliable to monitor the causes of low chlorine residual than the effects. On unstaffed plants, therefore, the previously mentioned causes should be monitored in addition to the chlorine residual. Furthermore, the plant should be designed to shut down before the failure is reflected at the chlorine residual analyzer.

Chlorine Safety Equipment

Wherever chlorine is stored or piped, the following safety equipment should be installed:

Shower and eye wash unit adjacent to storage or handling equipment.

Gas masks and instructions outside access doors.

Chlorine gas detectors in chlorine storage and chlorinator rooms.

Floor-level fans with outside overriding ''on'' switches.

In addition to these safety items, plant personnel should be completely trained in the handling, operation, and emergency procedures of the chlorination system.

Chloramine Application

When chlorine and ammonia (NH_3) are both present in water, they react to form products collectively known as chloramines (see Chapter 12 for reaction chemistry). The application of ammonia in the process stream in relation to chlorine varies, and there appears to be no general rule of thumb based on a survey of various utilities' experience. The raw water quality, particularly its THM formation potential characteristics and microbiological quality, and health department policies influence the point of application decision.

The actual location of ammonia application points in the existing treatment process are in two spots, the flash mix basin and/or after filtration. The points of chlorine addition in the treatment process vary quite a bit, but the two typical locations are at the flash mix basin and/or after filtration. Since ammonia will normally remain in its original state until reaction with chlorine, an excess of ammonia can be carried through the treatment plant and will react at any point of chlorine addition. Thus, it is not usually necessary to provide more than two application points for ammonia.

In cases where chloramines have been used to maintain a longer-lasting residual in larger distribution systems, ammonia is added to the filtered water effluent with chlorine added upstream of this point, usually in the flash mix basin. In this way, free chlorine is maintained in the flocculation–sedimentation basins and in the filters to help reduce growths and reduce the chlorine demand in the water, which allows for a longer-lasting chloramine residual.

Many utilities carry a combined chlorine residual through their entire treatment process. Ammonia is added either upstream or downstream of chlorine in these cases, but the time before forming a combined residual is very short, 1 or 2 sec at most.

Chlorine–Ammonia Weight Ratio

A survey of more than 50 U.S. utilities using chloramines revealed that most utilities applied a chlorine–ammonia nitrogen weight ratio in the range of 3:1–4:1. The utilities added excess ammonia in relation to chlorine to ensure that only monochloramine was being formed.

Total Chlorine Residual Monitoring

Approximately one-third of the utilities surveyed used continuous total chlorine analyzers to monitor the residuals leaving their treatment plants. These analyzers measured the total chlorine residual and did not differentiate between species. In treatment plants without continuous monitors, samples from various points in the treatment process are analyzed in the laboratory using the DPD, or amperometric, titration methods.

The actual concentration of chloramines leaving treatment plants varies from site to site depending on the specific conditions of each distribution system. Generally, the range of concentrations leaving treatment plants varies from 0.5 to 2.0 mg/L.

Forms of Ammonia

Ammonia is a colorless gas with a very pungent irritating odor. It is liquified for convenience in storage and handling. Liquid ammonia boils at $-28°F$ at atmospheric pressure but is normally stored in pressure vessels at ambient temperatures. Most of the ammonia used in the United States is produced synthetically by the Haber-Bosch process by reacting nitrogen and hydrogen at high temperatures and pressures in the presence of a catalyst. Nitrogen is obtained by liquifying air, and hydrogen is obtained by decomposition of natural gas. Because natural gas is a coproduct of oil production, many oil refineries use this natural gas to produce ammonia. Consequently, some of the major oil companies are major ammonia producers.

Ammonia is commercially available in four forms: anhydrous ammonia, which is commonly stored and transported as a liquid in pressure vessels; aqua ammonia, most commonly a 20–30 percent solution of ammonia in deionized or softened water; ammonium sulfate and ammonium chloride, both of which come in granular form.

Anhydrous ammonia is the least expensive of the four ammonia forms to purchase, but it requires pressurized storage tanks which are relatively expensive. Aqua ammonia is more expensive than anhydrous ammonia, but the storage tanks are somewhat less expensive. Because aqua can be stored in nonpressurized or low-pressure tanks, it is less hazardous to handle than anhydrous ammonia. In addition, aqua ammonia feeding equipment is less expensive than anhydrous ammonia feeders.

The salts of ammonia are the most expensive forms. Ammonium sulfate, which is cheaper than ammonium chloride, is the typically preferred salt. Ammonium sulfate and ammonium chloride both involve feeding a dry material through feeders that are difficult to maintain. The salts are extremely hydroscopic and tend to cake up, causing problems with feeding equipment. Because of these problems and the higher chemical costs, neither of these are investigated further.

Ammonia Storage and Shipment

As previously mentioned, anhydrous ammonia is stored and shipped in rather expensive pressure vessels. In contrast, aqua ammonia can be stored and shipped in nonpressurized or low-pressure vessels. For bulk storage and long-distance shipping, the lower shipping and handling costs of the more concentrated anhydrous ammonia more than compensate for the expensive pressure vessels required.

Anhydrous Ammonia Containers. Anhydrous ammonia shipping containers comply with ICC regulations, which require a minimum design working pressure of 265 psig with safety valves set to release at 225 psig. Valves and fittings are normally rated at 300 psig.

For bulk shipments, anhydrous ammonia is normally shipped in 80-ton rail tank cars, 25-ton rail tank cars, and 20-ton tank trailers. Smaller shipping containers include 50-, 100-, and 150-lb cylinders and a 1730-lb (625-gal) portable tank, a popular size for agricultural applications. Another popular size is a 1000-gal (4600-lb) tank, a common vendor-supplied item.

Permanent (stationary) storage tanks for anhydrous ammonia can be custom fabricated to any size desired. They will typically have the same pressure restrictions as the shipping containers and are usually made from carbon steel. No special restrictions apply to the materials other than no copper, bronze, or brass. Tanks should be sheltered from the sun to prevent excessive pressure buildup and/or venting to the atmosphere of ammonia vapor.

Aqua Ammonia Containers.

Aqua ammonia is not commonly shipped long distances, hence the largest transport vessel commonly used is a 7500-gal (approximately 25-ton) tank trailer. There seems to be less standardization in on-site aqua ammonia storage tanks since only nonpressurized or low-pressure tanks are required. Again, there are no restrictions on the type of materials for storage vessels other than no copper, bronze, or brass.

Depending on the concentration of aqua ammonia, excessive temperatures (>38°C) will cause ammonia vapor to come out of solution and exert vapor pressure. This should be considered in design, and a slightly pressurized storage tank with relief vent and water trap may be necessary to keep vapors from escaping to the atmosphere if a high concentration of aqua ammonia is stored. Nonpressurized tanks with vents to a water adsorption system are also used for aqua storage.

Ammonia Handling.

Ammonia is a very hazardous material to handle. Exposure to high concentrations of ammonia vapor can be fatal. These effects of ammonia are due to its great solubility, which results in severe corrosion of mucous membranes that come in contact with ammonia vapor. Personnel who work regularly with ammonia should be provided with safety equipment, including the following:

1. Safety shower and eye wash fountain.
2. Industrial gas masks.
3. One-piece rubber or neoprene suits.
4. Rescue harnesses.
5. High-pressure water system with fog nozzles for controlling ammonia leaks.

Ammonia Application

The method of applying ammonia to the process stream will depend on the type of ammonia stored, the flowrate of ammonia required, and the pressure of the water to be treated at the point of ammonia application. Details of these methods are outlined below based on the type of ammonia stored on-site.

Anhydrous Ammonia.

Ammonia can be fed directly as a gas into the flow of water or it can be diluted with carrier water through an eductor (injector) and applied as a solution into the process stream. Two commercially available methods of anhydrous ammonia application, direct-feed and solution-type ammoniators, are described below.

Direct-Feed Ammoniators. A direct-feed ammoniator is a self-contained unit consisting of an ammonia pressure-regulating valve, pressure gages, a vent with a relief valve, a rotameter, a control valve, and a back-pressure valve mounted inside a modular cabinet. An actuator and feed control system would also be included if automatic controls were installed. The direct-feed ammoniator meters gaseous ammonia into the process stream under positive pressure. The high pressure from the storage tank is regulated by the pressure-reducing valve to about 40 psi. At this controlled pressure, the gas flows through a rotameter, which indicates the flow directly in pounds per 24 hr. The gas then passes through the control valve used to vary the flow of ammonia. This valve can be adjusted manually or automatically with a signal from a flow meter. Finally, the gas passes through a back-pressure valve, which maintains a constant back pressure on the system. This pressure is limited to a range of 15–18 psi and limits the pressure of a line to which anhydrous ammonia will be added to about 15 psi. This valve will dampen pressure variations at the discharge to keep feed rates constant as well as prevent the backflow of water into the metering apparatus. The gas is then fed directly into the water with some form of diffuser. Figure 21-45 shows a typical flow diagram for the use of anhydrous ammonia with a direct-feed ammoniator. This type of ammonia feed is the most commonly used in the

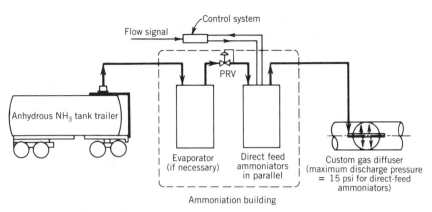

FIGURE 21-45.   Ammoniation system: Direct gas feed; anhydrous storage.

United States. (An evaporator is shown in the figure, even though it is necessary only for high withdrawal rates.)

Solution-Type Ammoniators.   Solution-type ammoniators are typically recommended when higher feed rates or greater discharge pressures prohibit the use of direct-feed ammoniators. V-notch ammoniators are available in up to 4000-lb/day capacities (the direct gas feed ammoniators have a maximum capacity of 1000 lb/day) and can operate up to 150 psi discharge pressure.

With this method, anhydrous ammonia is withdrawn under pressure from the storage vessel, me-

tered under a vacuum by the solution feed ammoniator, dissolved in soft water with an eductor (injector), and applied as a solution through a conventional chlorine solution-type diffuser. Figure 21-46 shows a flow diagram with a solution feed ammoniator. An evaporator is shown, but its use is only necessary for high withdrawal rates.

The addition of anhydrous ammonia to hard water will cause precipitation of calcium and magnesium hydroxides and carbonates. This water-softening reaction is caused by an increase in pH from high ammonia concentrations. These precipitates will build up in the solution pipelines and reduce their carrying capacity and will clog injectors. To

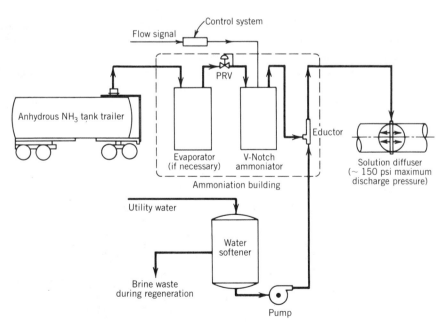

FIGURE 21-46.   Ammoniation system: V-notch solution feed; anhydrous storage.

avoid the scaling and precipitation problems, the injector water usually must be softened, unless the hardness of the water is already less than 30 mg/L $CaCO_3$. This is typically done by sodium cycle ion exchange with the resin regenerated with salt.

Aqua Ammonia. Aqua ammonia can be fed to the process stream very simply by a metering pump (positive displacement diaphragm or progressive cavity type). The ammonia is already in solution when it is delivered and the liquid must be transferred to the water to be treated. This operation would be similar to feeding liquid alum, liquid polymer, or caustic soda.

Figure 21-47 shows a typical arrangement for feeding aqua ammonia. The storage vessel would be a permanent, on-site tank that should have enough storage capacity for at least 10 days at maximum day usage. The storage vessel should have liquid-level monitoring with or without a transmitter to determine the inventory of aqueous ammonia in storage. The storage tank should be slightly pressurized (up to 25 psi) to prevent the escape of ammonia vapors into the atmosphere if higher concentrations of aqua ammonia (25 percent or greater) are used. The metering pump should be fairly close to the storage and, if convenient, be sheltered from the sun and elements.

Because of the relatively low flows of ammonia in relation to the process flow, care must be taken to ensure that the ammonia will be properly diffused prior to chlorine application, if preammoniation is predicted. This may require that the ammonia injection point be far upstream from the chlorine application point or that an adequate form of mixing is present a short distance downstream of the chlori-nation point. The point of introduction of ammonia into the process stream will be subject to scaling and precipitation due to localized high-pH conditions at the diffuser; therefore, design of the ammonia diffuser should incorporate methods for easy cleaning.

All piping materials and fittings used to carry anhydrous or aqueous ammonia should be made of Schedule 80 seamless steel or black iron. Diffusers should be 316 stainless steel. Some aluminum alloys are suitable for appurtenances in ammonia systems. By no means should copper, brass, bronze, or galvanized fittings, pipes, or valves be used for any contact with ammonia. Moist ammonia will corrode copper, silver, zinc, and their alloys. PVC piping can be used, but some experience with other utilities has shown that plastic ball valves have a tendency to freeze or jam when used with aqueous ammonia. Some chemical reaction might be taking place with the plastic and moving parts affected the most. Iron or steel valves with stainless steel trim or all stainless valves are preferable.

Ozonation

Ozone is attractive for water purification for many reasons. It is a strong oxidant that reacts rapidly with most microorganisms and organic substances in water, does not impart tastes and odors, and does not aid in the formation of THMs.

The disadvantages of ozone use include (1) high capital and energy costs of an ozone generation system (relative to a chlorination system); (2) diffusion of the ozone gas into the main plant flow is difficult because open-channel flow systems are not well suited to ready application of any gas; and (3) ozone

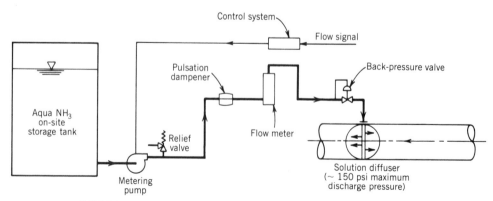

FIGURE 21-47. Ammoniation system: Aqua ammonia feed and storage.

does not produce a stable residual with which to maintain the bacteriological integrity of the water distribution system over a few-day period.

Most full-scale ozonation plants utilize air as the source of oxygen because pure oxygen is generally uneconomical on a full-scale or continuous-use basis. When ozone is produced, the resulting gas mixture is generally 0.5–1.5 percent by volume for O_3 produced from air and 1–2 percent by volume for O_3 produced from oxygen. Gas injection is accomplished through spargers, porous piping or plates, or venturi-type eductors or injectors. The depth of contact basins is typically 3–5 m. Ozone transfer efficiencies of 90–95 percent can be achieved. Ozonators, like chlorinators, are proprietary components and are available in capacities of up to 500 kg/d (1100 lb/d).

Unlike chlorine, residuals of ozone are not directly proportional to the original ozone dose. Ozone has a self-destruction or exponential decay characteristic, which limits residuals and application rates. In low ozone-demanding waters, development of residuals is limited because rates of self-destruction are directly proportional to a positive exponential of the residual ozone concentration. Thus, practical concentrations of residual ozone in waters are limited to approximately 1.5 mg/L. Existing European water treatment plants are designed to produce residuals in the range of 0–1.0 mg/L. Short-term or low-rate ozonation applications may not produce any measurable ozone residuals.

Ozone Contact Time

Total ozone contact time can be divided into three phases.

Phase 1: Demand phase (ozone transfer without residual).

Phase 2: Residual maintenance phase (measurable residual).

Phase 3: Decay phase (possible ozone and oxygen stripping).

In pilot plant and laboratory work, all contact may take place in a single column or basin. In full-scale plants, however, the phases should be physically separated to prevent hydraulic short circuiting from reducing the designated contact times.

The time required for phase 1, typically from 1 to 4 minutes, is dependent on the application rate and the type of water to be treated. Phase 2 contact time is critical for microbiological inactivation; maximum inactivation occurs when the product of ozone residual (in mg/L) and contact time (in min) exceeds 5. With residuals of 0.5–1.0 mg/L, phase 2 contact time should be 5–10 minutes.

Ozone Contact Basin Design

Based on published data, European practice, and air pollution restrictions, the parameters presented in Table 21-20 are guidelines for design of ozone contact basins. Figure 21-48 shows a typical section through an ozone contact basin.

TABLE 21-20. Design Guidelines for Ozone Contact Basins

Feature	Criteria
Depth	3–5 m (10–16 ft)
Compartmentalization	Not less than two stages in series plus degasification stage
Detention time	5 min/stage
Degasification	Excess ozone or oxygen to be removed by exhaust fan in final stage (detention time 30–120 sec)
Basin covers	All contact basins to be covered and provided with exhaust piping to ozone gas strippers
Ozone gas scrubbers	All residual ozone in off gases to be removed by heat, activated carbon columns, recirculation chemicals, or submerged disposal in a lagoon or basin
Installed application rates	1.0–4.0 mg/L for potable water plants, depending on water quality; piping to be arranged for possible delivery of all ozone to one stage
Application piping	Cleaning system to be included for spargers, orifices or porous tubing, or plates. Stainless steel pipings.
Air volume	50 L/g ozone
Air rate	0.6 L/sec per mg/L ozone per ML/d (4.8 cfm per mg/L ozone per mgd)
Energy requirements for ozonators and compressors	30 kW · hr/kg ozone (13.6 kW · hr/lb ozone)

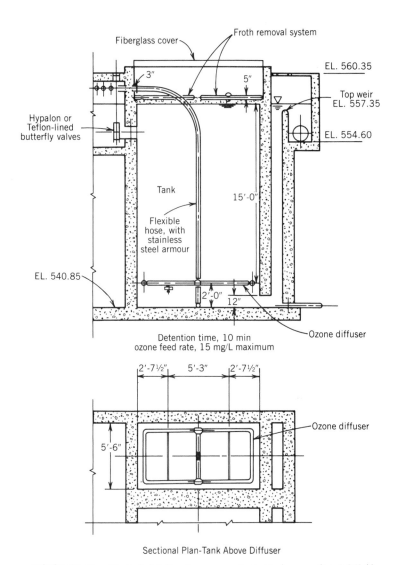

FIGURE 21-48. Typical details of ozone contact: Basin capacity, 3 ML/d.

Ozone Generators and Controls

Several types of ozone generators are available, the most common being plate or tube units. Applied rates of ozone feeding should be established by laboratory tests and full-scale trials. Feed rates of ozone generators may be adjusted by voltage controls, frequency controls, or both. Operating the generators under maximum voltage and turning off spare units during low demand can improve the efficiency. On major water treatment plants, at least two ozone generators should be provided. Most plants have four units with one provided as a standby. Automatic controls usually include manual rate setting, pacing of total plant flows by electronic signals, and signal splitting to apportion feed rates to the number of ozonators on line. In addition, control systems should include monitors for ozonator safety contacts and alarms. Noise and venting of ozone-generating equipment may also pose design problems.

Chlorine Dioxide

Chlorine dioxide has been used for many years in European water treatment plants and has recently received much attention for potable water applications in the United States. Investigators have discovered several properties of the chemical that make it useful for water treatment:

1. Chlorine dioxide does not react with ammonia or related nitrogenous compounds.
2. Chlorine dioxide does not react with organic compounds in water to produce THMs.
3. Chlorine dioxide does not react with water as does chlorine, making it easily separable from water by mild aeration, leaving no residuals.

Chlorine dioxide is typically generated on-site. When large quantities are needed, chlorine dioxide is generated from sodium chlorate ($NaClO_3$) while for small-production uses, including water and wastewater processes, sodium chlorite ($NaClO_2$) is a more cost-effective source. The most common method of generating chlorine dioxide for water treatment is the chlorine–chlorite process. A sodium chlorite solution and a chlorine solution are placed in a container filled with porcelain Raschig rings for 1 min to produce chlorine dioxide.

Smaller installations that use hypochlorite systems can also generate chlorine dioxide with the chlorine–chlorite process by acidifying the two solutions with sulfuric acid. The sulfuric acid, however, must remain separated from the solid sodium chlorite to prevent explosions. In France, the CIFEC method uses an enrichment loop (or recirculation process) in the chlorine system and a longer reaction time to generate chlorine dioxide.

A chlorine dioxide feeding system is quite similar to a chlorine feeding system. The feeding system may be paced manually or automatically to the plant flow rate. It is important, however, to maintain the chlorine solution concentration from the chlorine feeder at 500 mg/L or higher to complete the reaction between the chlorine and the sodium chlorite. Figure 21-49 illustrates a typical flow-proportional system for generating chlorine dioxide. An automatic control for residual chlorine dioxide can be made similar to the automatic chlorine residual control with a compound loop, but has not been used extensively.

Other Disinfectants

A number of alternative disinfectants have been studied or used in actual treatment applications. These include bromine, iodine, potassium permanganate, ultraviolet irradiation, and others. None of these methods have seen widespread practice and their related technology will not be discussed here.

IRON AND MANGANESE REMOVAL

Introduction

A discussion of the principles and chemistry of iron and manganese removal processes is given in Chapter 15. The two processes are similar and are often

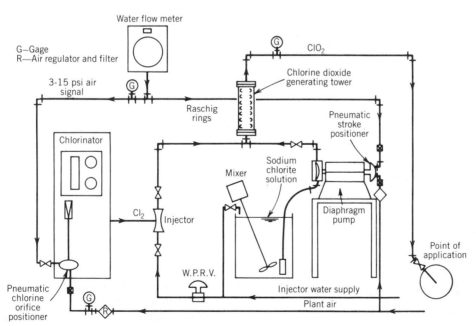

FIGURE 21-49. System for chlorine dioxide generation and feed control (White, 1978). Reprinted with permission.

considered as one process. However, there are several differences in the two processes which are discussed herein.

The purpose of iron and manganese treatment is to oxidize divalent iron (Fe^{2+}) and manganese (Mn^{2+}) from the soluble ferrous and manganous forms, or alternatively, to prevent the oxidation of the ferrous and/or manganous forms by sequestering. The first alternative (oxidation and removal of insoluble oxides) is the most commonly used and cost-effective process.

The presence of iron or manganese in drinking water is not a health problem. However, problems are associated with aesthetics (appearance and taste), practical considerations, such as staining, and industrial uses. Threshold values for industrial and commercial uses of iron and manganese are listed in Table 21-21. Iron- and manganese-bearing waters, with no dissolved oxygen, may appear clear when drawn from a well or stratified reservoir, and their clarity may be maintained in a distribution system. However, when the water is exposed to air in a glass, bathtub, washer, swimming pool, or in an industrial process, it will become turbid due to precipitation of the iron and manganese resulting from oxygen transfer. Iron-bearing waters will have an oxide red or brown color; manganese-bearing waters will appear black or gray. In surface waters, it may be preferable to treat the source by aeration and/or destratification in the impoundment. Similarly, groundwater sources should be evaluated to determine if iron- or manganese-bearing waters can be excluded from the well by minimizing flows from certain formations.

Pilot Testing

Although the principles for removal of iron and manganese are well established, experience shows that reaction times and removal efficiencies vary with different waters. Optimum design of facilities should be preceded by bench tests, pilot plant tests, or full-scale testing, if feasible.

The factors which may need to be determined by pilot tests include:

- Oxidation chemical selection.
- Aeration versus chemical oxidation.
- Sequestering feasibility.
- Reaction and contact time requirements.
- Media selection.

TABLE 21-21. Maximum Iron and Manganese Concentrations for Industrial and Commercial Uses

Industrial Use	Maximum Concentration or Threshold Range (mg/L)		
	Mn	Fe + Mn	Fe
Air conditioning	0.5	0.5	—
Baking	0.2	0.2	0.2
Brewing	0.1	0.1	0.1–1.0
Canning	0.2	0.2	—
Carbonated beverages	0.2	0.1–0.2	0.1–0.2
Cooling water	0.2–0.5	0.2–0.5	0.5
Confectionary	0.2	0.2	0.2
Dyeing	0.0	0.0	—
Electroplating	—	—	trace
Food processing, general	0.2	0.2	0.2
Ice	0.2	0.2	—
Laundering	—	—	0.2–1.0
Milk industry	0.03–0.1	—	—
Oil well flooding	—	—	0.1
Photographic processing	0.0	0.0	0.1
Pulp and paper			
Ground wood	0.5	1.0	0.3
Kraft pulp	0.1	0.2	—
Soda pulp	0.05	0.1	0.1
Kraft pulp, unbleached	0.5	—	—
Kraft pulp, bleached	0.1	—	—
Fine paper pulp	0.05	0.1	0.1
High-grade paper pulp	0.05	0.1	—
Plastics (clear)	0.02	0.02	—
Rayon pulp	0.03	0.05	—
Rayon manufacturing	0.0–0.02	0.0	0.05
Sugar manufacturing	—	—	0.1
Tanning	0.2	0.2	0.1–2.0
Textiles, general	0.1–0.25	0.1–0.25	0.1–1.0
Textile dying	0.25	0.25	—
Textile bandages	0.2	0.2	—
Wool scouring	1.0	1.0	—

- Filtration rates.
- Filter backwash rates and frequency.

Where highly colored or highly polluted waters are to be treated, determinations of both reaction and contact time requirements are important. On major plants, evaluation of air stripping versus air diffusion may provide valuable criteria.

Oxidation Processes

Preoxidation of ferrous or manganous materials is a commonly practiced pretreatment process. Oxidation may be achieved by aeration, chlorination, ozonation, or potassium permanganate oxidation.

Aeration. Aeration of iron- and manganese-bearing water has two purposes: (1) to transfer oxygen to the water for oxidation of iron and manganese and (2) to remove any volatile organics that may be present and that reduce the efficiency of subsequent processes due to their oxidant demand. It should also be noted that oxidation of manganese-bearing water by aeration is generally not effective below pH 9.5.

Air-diffusion-type aerators for iron and manganese oxidation operate typically at air–water volume ratios of 0.75:1.0. Normally, air is applied at depths of 3–5 m. Average oxygen transfer efficiency is in the range of 5–10 percent. Similar oxygen transfer efficiencies can be effected by mechanical (surface-type) aerators.

After aeration, oxidation of divalent iron and manganese may require from 5 to 60 min contact time to permit formation of a filterable floc. Contact time requirements may be based on similar local experience or on laboratory and/or pilot testing. In some cases, coagulation and clarification may be required prior to filtration.

Chlorination. Chlorination is widely used for oxidation of divalent iron and manganese. Chlorination is generally a more rapid oxidation process than aeration, especially under conditions of organic interference with the oxidation process. The insoluble material may be highly dispersed, and long contact times (30–60 min is normal) and/or coagulation and clarification may be required. In addition, the formation of THMs in highly colored waters may be a problem. Chlorine feed rates and contact time requirements can be determined by simple jar tests. However, for effective oxidation of iron and manganese, the effluent should contain free chlorine at approximately 4.0 mg/L.

Ozonation. Ozone may be utilized for iron and manganese oxidation but is rarely practiced. Ozone may not be effective for oxidation in the presence of humic or fulvic materials. When ozone is applied to water, excess air or oxygen is also applied in sufficient quantities to supersaturate the dissolved oxygen content of the water. The excess transferred oxygen is of concern due to its effect on accelerating corrosion rates and outgassing via effervescence. Care must also be taken to control overdosing due to ozone's ability to oxidize Mn^{2+} to Mn^{7+} or permanganate. The formation of permanganate and the resulting pink water are drawbacks to the use of ozone for iron and manganese removal.

Potassium Permanganate. As an oxidant, potassium permanganate ($KMnO_4$) is normally more expensive than chlorine and ozone, but for iron and manganese removal, it has been reported to be as efficient and may require considerably less equipment and capital investment. Permanganate gives an easily detected pink color to water with overdoses in the 0.05-mg/L range. Due to this fact, the dose range is critical in avoiding consumer complaints.

The normal process for utilizing potassium permanganate involves adding it as a solution ahead of a filter. Actual feed rates may need to be adjusted based on bench-scale and pilot plant studies. After addition of the potassium permanganate, the oxidized water is delivered to a specially prepared filter. The reaction time or required contact time after oxidant addition is typically 5 min at 20°C or 10 min at 1°C. The filter media may be natural green sand, but silica sand and/or anthracite may also be used. Before silica or anthracite becomes efficient, they need to be coated with manganese oxide. Under normal conditions, the coating can be applied by controlled operation for several days with optimum potassium permanganate feed rates. Partial or marginal treatment may occur during the coating process. Once the coating is effected, satisfactory removals are usually maintained. The process is more efficient at pH values above 7.5.

The filtration process generally used for iron and manganese removal is pressure filtration. Filtration rates normally vary from 240 to 480 m/d (4–8 gpm/ft²). Backwash rates typically range from 480 to 1200 m/d (8–20 gpm/ft²) depending on media size, temperature, and supplemental scour. Greensand media requires periodic regeneration with potassium permanganate solution. Media effective size and depth are similar to those used in normal filtration applications. However, greensand media usually have a very small effective size, less than 0.3 mm.

Sequestering Processes

Sequestering of soluble iron and manganese is the opposite of oxidation. In chemistry, the term *sequester* means to "bind up" or "complex" so as to prevent normal chemical reactions (in this case oxidation). The chemical used for sequestering in water treatment is sodium hexametaphosphate $(NaPO_3)_6$ (SHMP), commonly known as polyphosphate, glassy phosphate, or liquid glass. This chemical is available in crystal, granular, or liquid forms and is highly soluble. Chemical addition should occur before the water has a chance to come in contact with air or chlorine to ensure that the iron and manganese are still in the soluble state. However, sequestering agents are aggressive compounds with respect to metals, and they may dissolve precipitated iron and manganese or corrode metallic pipe materials.

SHMP is high in phosphates (66 percent P_2O_5) and releases the phosphate for biological stimulation. Therefore, it should be used with caution where treated water is stored in open reservoirs; otherwise, algal blooms and slimes may result. Similarly, the use of polyphosphates may adversely affect receiving waters and cause biological stimulation. It should also be noted that sequestering does not remove the iron and manganese but merely holds it in an aesthetically acceptable condition that will degrade with time and appear on analytical tests as possibly being present above the water quality standard.

Feed rates can be determined by testing and are typically less than 2 mg/L. Sequestering may be considered for waters if the iron content is in the range of 0.3–1.0 mg/L and/or if the manganese content is between 0.05 and 0.1 mg/L. Consideration must be made for the additive demand of these compounds. The chemical feed equipment for SHMP normally includes a dissolving tank and a proportional metering pump. Standby facilities should be provided. The solution may be fed directly into a pipe, or water flushing may be utilized to transport the solution to the feed point.

Lime Treatment

Lime treatment is effective in removing both iron and manganese. Almost total removals are effected if the water is preaerated if the pH exceeds 9.5 during the process and if sufficient alkalinity (>20 mg/L as $CaCO_3$) exists. Softening is normally less economical than other listed processes used for Fe and Mn removal because of high capital costs. High Fe and Mn removals preclude recalcination of lime sludge due to the impurity of the sludge.

Other Processes

When iron is associated with organics, or if there is significant color of organic nature present, large dosages of coagulants may be required to remove iron or manganese after oxidation. Activated silica, a coagulant chemical that may be effective in such conditions, may be formed from sodium silicate (Na_2SiO_3) and chlorine. Sodium silicate is diluted to a 1.5 percent concentration before activation with chlorine. At concentrations above 1.5 percent, jelling may occur. The chemical is difficult to handle, activate, and feed, and maintenance and operation costs are high.

Another process, which has been demonstrated on a pilot plant basis for iron and organics removal, is the proprietary "Sirofloc" process. This process involves feed rates of 1 percent (10,000 mg/L) of fine 1–2-μm magnetite (a naturally occurring iron oxide, Fe_3O_4, with a specific gravity of 4.9–5.2). Although the magnetite is 72 percent iron, it is stable, inert, and maintains magnetic properties.

Before addition to water, magnetite is activated with caustic soda. After addition and mixing, colloidal materials attach to the surface of the magnetite particles, and the magnetic floc rapidly settle because of the density of magnetite. Settling times are typically on the order of 15 min. The settled magnetite is collected and washed clean of iron and other materials on magnetic plates before reactivation and reuse.

Iron may also be removed in association with ion exchange processes. As an example, the City of Santa Monica, California, removes approximately 1.0 mg/L of iron from a 7.5-mgd groundwater flow by a process of aeration and filtration through 30 in. of ion exchange media. Softening is provided during the filtering stage. The principal design parameters for this system include:

Aeration: 10 min contact at 0.1 cfm/gpm (0.75 L/sec air per L/sec).

Contact: 60 min open unbaffled basin.

Filtration: 4 gpm/ft² (235 m/d).

Media: 0.4 mm (effective size) polystyrene resin.

The resin requires periodic acid treatment due to fouling of the surface by oxidation. Effective resin life is approximately 12 yr.

MEMBRANE PROCESSES

Applications

Membrane processes are used in the field of potable and industrial process water treatment primarily for the removal of dissolved solids or demineralization. In this process, a feedwater containing appreciable concentrations of dissolved solids is transformed into two streams, product and brine. The product stream, generally 75–85 percent of the feed, contains a relatively low concentration of dissolved solids while the reject or brine possesses a high dissolved salt concentration. Either the entire process flow or a portion of the process flow may be treated, followed by blending, to produce the desired dissolved solids concentration.

Demineralization can be divided into two specific applications, treatment of brackish water (1000–10,000 mg/L feed) and treatment of seawater (over 35,000 mg/L feed). The distinction is made since each represents a different application of membrane equipment. Membrane processes also are used to produce ultra-high-purity water for industries such as the electronic industry, which requires TDS concentrations of less than 0.05 mg/L. A third application is specific ion removal where the ionic composition of a water may be acceptable with the exception of one or two constituents. An example is the removal of chloride in process waters of certain chemical industries. A number of additional benefits may be realized by treatment with membrane processes, including removal of organics, color, nutrients, and residual suspended solids. Membrane processes also have wide applications in the chemical industries as a separation process.

Methods

Reverse osmosis (RO) and electrodialysis (ED) are two true semipermeable membrane processes currently available for the removal of dissolved electrolytes from water. Ultrafiltration is sometimes considered as a third membrane process used for separation of "impurities" from water. However, in ultrafiltration the predominate mechanism for separation is selective sieving through pores, as opposed to RO and ED where the transport process is diffusivity. Due to the difference in transport mechanisms involved, ultrafiltration is limited to the removal of compounds with molecular weights greater than 500. In light of these differences, this discussion will be limited to diffusion-controlled membrane processes.

ED is a process in which electrolytes are transported through permselective (polarity-selective) membranes as a result of the application of electric energy and the consequent attraction of ionic species to charged electrodes. In this process, the dissolved solids pass through the membrane leaving the low-TDS water on the "feed side" of the membrane. In RO, a pressure in excess of the bulk solution osmotic pressure is applied to the feed side of the membrane. As a result, water passes through the membrane and electrolytes are rejected. ED and RO processes are described in more detail in Chapter 10.

The widespread acceptance and use of RO has helped lower the costs associated with the process. As a consequence, the use of ED is now fairly uncommon in the field of water treatment in this country.

Design of an RO treatment facility should only be attempted if the designer is thoroughly familiar with available equipment and has received considerable input from the manufacturers.

Reverse-Osmosis Systems

Membrane Types. The three major types of membranes used are the cellulose acetate class (CA membranes), the polyamide membranes (PA membranes), and thin-film composite membranes. The characteristics of these membranes are discussed in Chapter 10.

Membrane Configuration. Modern reverse-osmosis membranes are constructed in modular form. Two types of module configurations are commonly used: spiral-wound and hollow fine fiber. In each case the modules are mounted in containment pressure vessels into which the pressurized feedwater is pumped. Other configurations, although not as common, include the flat sheets and tubular construction.

Spiral-Wound Modules. The basic construction of a spiral-wound module is shown in Figure 21-50. Membranes sheets are manufactured in an envelope

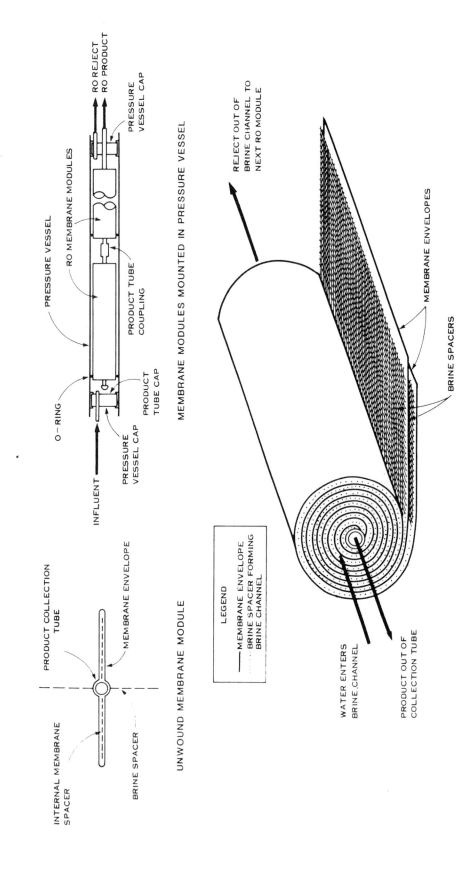

RO REJECT
RO PRODUCT

PRESSURE VESSEL CAP

RO MEMBRANE MODULES

PRESSURE VESSEL

PRODUCT TUBE COUPLING

O — RING

PRODUCT TUBE CAP

PRESSURE VESSEL CAP

INFLUENT

MEMBRANE MODULES MOUNTED IN PRESSURE VESSEL

PRODUCT COLLECTION TUBE

MEMBRANE ENVELOPE

INTERNAL MEMBRANE SPACER

BRINE SPACER

UNWOUND MEMBRANE MODULE

LEGEND

—— MEMBRANE ENVELOPE
········· BRINE SPACER FORMING BRINE CHANNEL

REJECT OUT OF BRINE CHANNEL TO NEXT RO MODULE

MEMBRANE ENVELOPES

BRINE SPACERS

WATER ENTERS BRINE CHANNEL

PRODUCT OUT OF COLLECTION TUBE

PARTIALLY WOUND MEMBRANE MODULE

FIGURE 21-50. Spiral-wound module construction.

572

configuration and an internal membrane spacer is provided to ensure that the internal surfaces of the membrane do not touch each other. The use of plastic mesh for the spacer provides a flow path inside the envelope. The open ends of the envelopes attach to a tube known as a product collection tube. The tube is perforated in the area where the envelopes are attached. Mesh brine spacers are placed between the envelopes. By rolling the membrane envelopes around the product collection tube, the brine spacer forms a spirally shaped channel (brine channel) whereby RO feedwater comes in contact with the membrane surface as it passes through the module. Spirally wound membrane units typically are approximately 100 cm (40 in.) in length. The cylinders are available in diameters ranging up to 30 cm (10 in.). Typically, numerous spiral-wound modules are coupled together at the product tube and are placed in a pressure vessel.

In operation, the pressurized RO influent enters one side of the pressure vessel and encounters the first membrane module. As the water flows through the module under pressure, a portion of the water passes into the membrane envelope and into the product collection tube, leaving a majority of the salts behind. The concentrated reject flows to the next module in series, and the process is repeated. When the water emerges from the last membrane in the series, the highly concentrated reject exits the pressure vessel. Spiral-wound modules have a product recovery (defined as product flow divided by feed flow) of approximately 5–15 percent for each element and a pressure drop of approximately 70 Pa (10 psi) per element plus 200–275 Pa (30–40 psi) pressure drop in the product piping.

Hollow Fine Fiber Modules. The construction of a typical hollow fine fiber module is shown in Figure 21-51. Unlike spiral-wound systems, the membranes are processed into hair-thin hollow fibers. In a typical module, there are usually over 1 million fibers. The fibers are folded and suspended lengthwise in the module. Due to the folding, the open ends of the fibers are all at one end of the module. The feed enters the module and permeates the interstitial spaces between the fibers. A portion of the feed passes into the hollow fiber, leaving a majority of the salts behind. Once inside the hollow fiber, the desalinated water flows to a common collection header. The concentrated brine exits from the opposite end of the membrane. Hollow fine fiber modules typically have a product recovery rate of

50–60 percent of the feed and a pressure drop in excess of 1400 Pa (200 psi).

Basic System Layout. The basic components of an RO system include a high-pressure pump, feed control valve, bank of RO modules, and brine control valve, as shown in Figure 21-52. Required pressures vary between 1700 and 6000 Pa (250–850 psi) depending on the feed quality, flux, recovery temperature, and membrane assembly. The greater the number of modules in parallel, the greater the membrane surface area and thus the product output. The brine control valve, through throttling, controls the product recovery. Limitations on recovery are governed by the concentration of the brine, the flux of the membrane, the available pressure, and the desired flowrates in the various portions of the membrane. Scaling, due to precipitation, will occur if solubility products in the brine are exceeded or if turbulence (to disrupt concentration boundary layers) is not maintained in the brine channel.

The system shown in Figure 21-52 is termed a one-pass system. Normally, the product recovery of a one-pass brackish system is approximately 45–55 percent. For a seawater system, the product recovery is generally 20–35 percent.

To increase the product recovery, two- and three-pass systems are utilized. In multiple-pass systems, the pressurized brine from the first pass is fed to a subsequent stage or stages where additional recovery occurs. In multiple-pass systems, brackish water can be processed to provide recoveries of 85–90 percent.

Pretreatment. Agents such as precipitates, colloids, microorganisms, and particulates may damage the membrane or effect the membrane process. Thus, pretreatment is frequently needed in RO applications (see Chapter 10). Figure 21-53 shows a typical pretreatment train for a feedwater low in suspended matter (turbidity less than 1.0) and low in sparingly soluble solids. A low-pressure pump transfers the feed through the various pretreatment processes and provides an acceptable NPSH at the high-pressure pump suction.

pH adjustment is very important in prolonging the life of the membrane. Membrane hydrolysis is typically least in the pH range of 4–5. However, it is generally accepted practice to adjust the pH to a range of 5.0–6.0 to minimize the costs associated with acid addition. This pH range yields a near-optimum rate of membrane hydrolysis. In addition

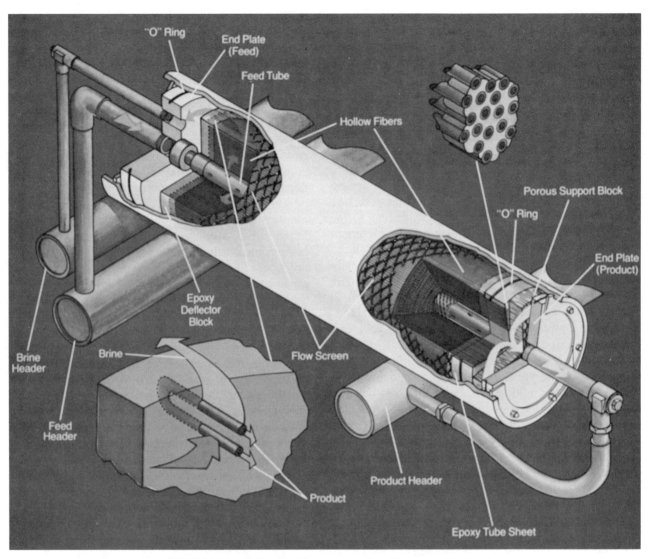

FIGURE 21-51.   Hollow fine fiber construction. (Courtesy of DuPont.)

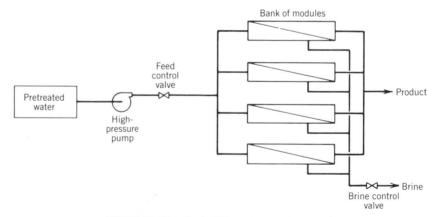

FIGURE 21-52.   Basic RO treatment components.

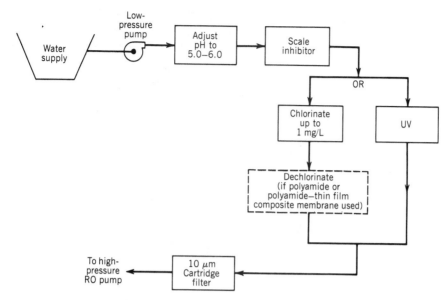

FIGURE 21-53. Typical RO pretreatment train.

to controlling the rate of hydrolysis, pH adjustments also lessen the potential for metal precipitation and calcium carbonate precipitation.

To further limit the precipitation of sparingly soluble solids, the addition of a scale inhibitor such as SHMP is typically part of the pretreatment process. SHMP dosages are usually in the range of 2–5 mg/L. Disinfection is also typically included in the pretreatment train to limit biological activity in the membranes. As an additional safeguard against fouling due to particulates, a cartridge filter (influent turbidity less than 1.0 NTUs) is usually employed. Nominal filter sizes used range from 5 to 20 μm with typical loading rates of from 4.8 to 9.6 m/hr (2–4 gpm/ft^2). Maintenance and flushing of membranes are design considerations that also must be addressed. However, specifics will not be covered here as they are highly varied.

Materials of Construction. The selection of materials is an important factor in RO design. Consideration must be given to both high pressure and corrosivity. In high-pressure systems [up to 7000 Pa (1000 psi)], stainless steel is typically used. In low-pressure systems [>700 Pa (100 psi)], PVC is frequently used due to its corrosion-resistant characteristics.

Although stainless steel has been used in the past for pressure vessel design, modern systems utilize reinforced fiberglass. This material must be shielded from the sun to avoid deterioration. In the design of seawater plants in particular, care must be taken to avoid situations where stagnant seawater will remain in contact with stainless steel as serious corrosion can occur.

RO Design Process. Due to the variance in equipment, methods, and manufacturers, the particulars of the optimum system design are not always readily apparent. For this reason, procurement of RO systems is usually accomplished by means of a ''functional specification.'' By this method, an engineer is retained to study the system requirements, design the pretreatment processes, design the RO system support facilities, and refine the basic requirements of the RO system. The functional specifications outline the operating requirements of the system, physical constraints of the system, and warranty agreements between the manufacturer and the owner. Bid proposals are returned by the interested manufacturers which outline the particulars of the system being supplied, estimates of system product quality as a function of time, system capital costs, and system operating costs as a function of time. The proposals are typically reviewed by the engineer to determine the optimum life cycle cost system.

Electrodialysis

Commercial ED units have primarily been used for the treatment of brackish water. The use of ED on seawater has been extremely limited due to the high energy cost as compared to other processes (e.g., RO). This process is also discussed in Chapter 10.

System Components.

As described in Chapter 10 and illustrated in Figure 10-21, the heart of the ED system is the membrane stack. The membrane stack consists of an alternating series of cation-permeable and anion-permeable membranes. Electrodes are located on either side of the membrane stack. When current is applied, electrolytes migrate toward the electrode of opposite charge. Through the use of polarity-selective membranes, two streams are formed, one dilute and the other concentrated. In practice, spacers are placed between the membranes to form flow channels.

Two types of membranes are currently available. In one unit, termed homogeneous, the ion-selective transfer sites are located throughout the membrane. A heterogeneous membrane has powdered anion or cation exchange resins uniformly distributed throughout the membrane film. The film is then treated to produce microcracks in the plastic. These crevices permit transport of the ion, while the ion exchange resins provide the selectivity. The electrodes used in ED are generally constructed of niobium or titanium with a platinum coating. Degradation of the anode can occur due to oxidation, while scale typically forms at the cathode. To equalize the electrode wear, the electrodes reverse change several times per hour (known as ED reversal process). By this process, scale formed during the cathode operation is removed by acid generated during its anode cycle. However, with time, both electrodes must be replaced due to oxidation.

Other essential ED equipment includes a DC power source and pumped circulation system. The DC power supply is typical provided by a rectifier, which converts alternating current to direct. The circulating system serves to process the feedwater through the units (see Figure 10-22). Pumps and pipes are typically PVC. Pump discharge pressure heads are typically 350–500 Pa (50–75 psi).

Pretreatment.

Many of the pretreatment processes of ED are similar to RO, although less rigorous. In ED, the likelihood of precipitating sparingly soluble salts is reduced by pH adjustment and addition of a threshold agent. Cartridge filters are utilized as a precaution against plugging due to partic-

TABLE 21-22. Electrodialysis Pretreatment Requirements

Parameter	Limitation
Chlorine	Some combined residual required to limit microbial activity but no free chlorine should be present
Iron	Less than 0.3 mg/L
Manganese	Less than 0.1 mg/L
Hydrogen sulfide	Less than 0.3 mg/L
Turbidity	Less than 2.0 NTU

ulate matter. Biological growth in the membranes should be limited to avoid membrane fouling and high differential pressure in the feed channels (biological degradation not being a problem). Table 21-22 illustrates various pretreatment requirements.

Limitations and Operational Problems.

Prior to selection of ED as a process in a treatment plant, the various limitations and operational problems should be thoroughly understood. One such limitation of ED is the TDS of the feed stream. As previously mentioned, ED is typically used in brackish water applications only. Scale, formed in the membrane stacks due to polarization, the supersaturation of the brine stream, or other factors, can be another potential problem. Scale fouls the membrane surfaces, blocks passages in the stack (changing flow patterns), and creates areas of resistance. These areas of high resistance, called "hot spots," occur when the feedwater flow is stopped or slowed in its passage through the product cell. The slowly moving water then becomes highly desalted due to the longer period of exposure to the electromotive force. The highly desalted water has a low conductivity and offers a high resistance to current flow. Hot spots can consume excess power and reduce the efficiency of the stack.

Leaks in ED units can also be a problem. As previously mentioned, spacers are placed between membranes. The membranes and spacers are stacked like cards of a deck, and in typical units, this can amount to 1200–1800 membranes and spacers. Therefore, manufactural tolerances are a critical concern. Units typically have no serious problems if manufactured properly. However, leakage can occur after a unit has been disassembled for cleaning and reassembled.

MATERIALS

Introduction

Materials and standards vary considerably depending on application (i.e., commercial or municipal), size, and location. In general, large community or publicly owned facilities are designed for a life of 50 yr or more, smaller plants (less than 5 ML/d) may be designed for 25 yr of life, and commercial or industrial plants may be designed for as low as 10 yr of use. In developing countries, where capital funds are less available, it may be cost-effective to use less expensive materials that may require extensive manual maintenance. In addition, many countries emphasize the use of local materials. Classes of construction, applications, and types of materials used for various types of treatment plants are presented in Table 21-23.

U.S., Japanese, and Canadian plants extensively use Type 18-8 stainless steel for submerged components such as ladders, shafts, bolts, nuts, and other metal components. In Europe, submerged components may be constructed of more traditional materials, such as bronze alloys. European plants tend to be of more elaborate design than U.S. plants. Extensive tile finishes in filters, for example, are incorporated in Europe where bare concrete surfaces are often used in the United States. U.S. plants tend to be more industrial in appearance, more spread out, and simply designed, while European plants tend to be more institutional in appearance and more enclosed.

Steel, concrete, piping, large-diameter valves, and major electrical components generally comply with national standards and codes. Many types of equipment such as chlorinators, chemical feeders, mixers, analyzers, instruments, and motor operators for valves are often interchangeable in Europe, Japan, and North and South America.

Standards and Codes

In the United States and Canada, many types of equipment, piping, materials, and coatings are covered by standards of the American Water Works Association. The other major associations and organizations that establish standards for materials and equipment in North America include the American Concrete Institute (ACI), the American Standards Association (ASA), the American Society of Civil Engineers (ASCE), the American Society of Mechanical Engineers (ASME), the American Society for Testing Materials (ASTM), the National Electrical Manufacturers Association (NEMA), the Uniform Building Code (UBC), and Underwriters Laboratories (UL). Whenever possible, materials and equipment should conform to these standards.

TABLE 21-23. Applications of Materials

Class	Application	Basin and Filter Walls	Typical Coatings	Submerged Components	Control Equipment
A	Major publicly owned facilities in western countries (includes power associated treatment)	Structural concrete	Minimum	Type 18-8 series stainless steel and PVC	Automated monitoring and control
B	Minor or temporary facilities in western countries	Welded steel or concrete	Low-maintenance, 10-yr life coatings	Coated steel and PVC	Automated monitoring and control
C	Commercial or industrial facilities	Welded steel	Commercial coatings	Coated steel and PVC	Minimum automation, thorough alarm monitoring
D	Publicly owned and locally financed facilities in developing countries	Concrete, steel or masonry	Minimum to basic coatings	Coated steel	Manual control, basic alarms

Piping Materials

Selection of the most appropriate piping materials, fittings, and valves for plant flow conduits, chemical solution lines, and sampling lines minimizes maintenance and improves plant reliability. Typical piping materials vary for each application and chemical compatibility should be of primary concern. PVC piping, corrosion free for almost all water and chemical solution applications, is used extensively in water treatment plants.

Corrosion

Corrosion and subsequent maintenance problems can be significantly reduced by proper material selection, coatings, and the use of cathodic protection. Knowledge of the treatment process and awareness of zones of low pH and locations where free chlorine or other oxidants may be applied is important. Low pH can be a major cause of corrosion in water treatment plants. When metal ion coagulants are added to the main plant flow, a low water pH will etch concrete and severely corrode the uncoated ferrous components. Adjustment of pH during the process may minimize long-term corrosion problems. For example, adding lime or caustic soda before filtration reduces corrosion and etching of concrete components in filter underdrains. Adjustments, however, may hinder treatment, and dual pH correction (pre- and postfiltration) is often required. Usually, etching does not occur where the water has a positive Langelier index. Where etching of concrete surfaces is a potential problem, epoxy coatings can protect the surfaces.

When bare or coated steel components connect to concrete-imbedded steel, the steel coatings can rapidly deteriorate. This corrosion is due to the creation of an electrical potential between the imbedded steel and the imperfections of the coated steel. In such instances, imposed cathodic potentials may be required. Steel imbedded in concrete is usually well protected, but partially coated submerged steel creates problems. Complete isolation or insulation from reinforcing steel of partially imbedded ferrous components cannot be assured. Partially exposed components should, if possible, be Type 18-8 stainless steel. Insulation from other submerged steel or ferrous components should be provided with visible insulators.

Stainless steel is often used as a noncorrosive metal for many applications with excellent results, but it should be used with caution. For instance, type 316 stainless steel will corrode in an environment with no free oxygen such as an anaerobic sludge or activated carbon tank. Also, two stainless steel parts rubbing together can cause a serious abrasion problem.

Bimetal systems will corrode under a wide range of pH and Langelier index conditions and should not be used. All copper piping should be insulated from ferrous pipes and systems, and bronze and copper alloys should be limited to less than 2 percent zinc.

Architectural and Structural Materials

Architectural materials cover a broad range of considerations. Continuity of materials and finishes between structures and buildings may be aesthetically and economically logical. Low-maintenance, unpainted exterior surfaces are common on water treatment plants. Anodized aluminum for windows, vents, and handrailing is common.

Reinforced concrete is the most commonly used structural material for major plants. Water-retaining reinforced concrete walls sometimes show signs of moisture penetration. Discoloration of exterior walls is probable unless the interior (wetted) surfaces are completely sealed and coated. Small hairline cracks in these walls generally seal off from leaching and subsequent redeposition of free lime, but cracks that do not heal will require repair from the interior. Water-retaining concrete walls and slabs should be provided with construction joints and "water stops" (flexible PVC preformed strips) cast into the concrete every 6–81 m. Where the process water and soil contains a significant amount of sulfate, a special type of cement such as type 2 or type 5 should be used.

Structural steel can be used for steel tanks, sludge rakes, traveling bridges, bulkheads, and equipment supports.

REFERENCES

Akazawa, H., "A Study on Horizontal Flow Sedimentation Basins," *JJWWA*, **395**, 7 (Aug. 1967).

AWWA, ASCE, CSSE, American Water Works Association Inc., New York (1969).

AWWA Committee Report, "Water Treatment Plant Sludges—An Update of the State of the Art, Parts 1 and 2," *JAWWA* **70** (Sept. and Oct. 1978).

AWWA, Inc., *Water Quality and Treatment,* McGraw-Hill, New York (1971).

AWWA Research Foundation Report, "Disposal of Wastes from Water Treatment Plant, Parts 1 to 4," *JAWWA* **61, 62** (Oct. 1969 to Jan. 1970).

Baylis, J. R., "Nature and Effect of Filter Backwashing," *JAWWA,* **51,** 1433 (Nov. 1959).

Biederman, W. J., and Fulton, E. E., "Destratification Using Air," *JAWWA,* **63,** 462 (July 1971).

Bond, A. W., "The Behavior of Suspensions," *ASCE Proc.,* **SA3,** 57–85 (May 1960).

Bond, A. W., "Upflow Solids Contact Basin," *ASCE Proc.,* **SA6,** 73–91 (Nov. 1961).

Brown & Associates, *Unit Operation,* John Wiley & Sons, New York, (1950).

Calise, V. J., "Design of Water Clarifiers and Softeners," *Water Sew. Works,* **102**(7), 238–244 (June 1955).

Camp, T. R., "A Study of the Rational Design of Settling Tanks," *Sew. Works J.,* **8,** 742 (1936).

Camp, T. R., "Sedimentation and Design of Settling Tanks," *ASCE Trans.,* **111,** 895–958 (1946).

Camp, T. R., "Studies of Sedimentation Basin Design," *Sew. Ind. Wastes,* **25,** 1 (1953).

Committee Report, "State of the Art of Water Filtration," *JAWWA (U.S.A.),* **64,** 662–665 (Oct. 1972).

Culp, G. I., and Culp, R. L., *New Concepts in Water Purification,* Van Nostrand Reinhold Co., New York/Melbourne (1974).

Culp, K. Y., Hsiung, K., and Conley, W. R., "Tube Clarification Process, Operating Experiences," *ASCE, J. San. Eng. Div.,* **95,** 829–847 (1969).

DuHamel, N. Y., "What is High Rate Filtration," *W. W. Eng.,* 47 (Jan. 1970).

EPA Design Manual, "Suspended Solids Removal by Granular Media Filtration," Draft (Jan. 1975).

Fishcherstrom, C. H. H., "Sedimentation in Rectangular Basins," *ASCE Proc.,* **47,** 768 (1955).

Foust, A. S., et al., *Principles of Unit Operations,* John Wiley & Sons, New York (1960).

Fujita, K., "Hydraulics of Filtration Theory," *JJWWA,* **445,** 2 (Aug. 1973).

Fujita, K., "Study on Filter Media Size and Depth," *JJWWA,* **485,** (Feb. 1975).

Garton, J. E., "Improve Water Quality through Lake Destratification" *W. W. Eng.,* 42 (May 1978).

Graber, S. D., "Outlet Weir Loading for Settling Tanks," *WPCF,* **46,** 2355 (Oct. 1974).

Hamann, C. L., and McKinney, R. E. "Upflow Filtration Process," *JAWWA (U.S.A.),* **60,** 1023–1039 (Sept. 1968).

Hansen, S. P., and Culp, G. L., "Applying Shallow Depth Sedimentation Theory," *JAWWA* **59,** 1134–1148 (Sept. 1967).

Harris, R. H., "Monitoring and Control of Filtration in Water Works," *AWWA Willing Water,* 4 (Aug. 1970).

Hawkins, F. C., Judkins, J. F., and Morgan, J. M., "Water-Treatment Sludge Filtration Studies," *JAWWA* **66,** 653 (Nov. 1974).

Hazen, A., "On Sedimentation," *ASCE, Trans.* **63,** (1904).

Hernandez, J., and Wright, J., "Tube Settler Design," *Ind. W. Eng.,* **7**(9), 25–27 (Sept. 1970).

Hirsch, A. A., "Functional Effluent Baffling," *W & WW,* **113,** R-141-2, R145-6, R149-50 (Nov. 1966).

Hubbs, S. A., and Paroni, J. L. Optimization of sludge dewaterability in sludge-disposal lagoons. *JAWWA,* **66,** 658 (Nov. 1974).

Hubert, L. N., "Alum Sludge Disposal—Problems and Successes," *JAWWA,* **69,** 335 (June 1977).

Hudson, H. E., Jr., "Density Conditions in Sedimentation," *JAWWA,* **64,** 382–386 (June 1972).

Hudson, H. E., Jr., *Water Clarification Processes,* Van Nostrand Reinhold, New York (1981).

Huisman, L., "Sedimentation and Flotation" Delft University of Technology, Delft, The Netherlands (1973).

Hyde, R. A., Water Clarification by Flotation—(4) Design and Experimental Studies on a Dissolved Air Flotation Pilot Plant Treating 8.2 m^3/h of River Thames Water," WRC Technical Report TR13, Medmenham, The Centre (November 1975).

Ives, K. J., "Significance of Filtration Theory," *J. Inst. Water Eng. (London)* **25**(1), 13–20 (Feb. 1971).

J. M. Montgomery, Consulting Engineers, Inc., "Report on Design and Operation of R. D. Bollman Water Treatment Plant" (Feb. 1971).

Kalinski, A. A., "Settling Characteristics of Suspensions in Water Treatment," *JAWWA,* **40,** 113–120 (Feb. 1948).

Kavanaugh, M., "Mechanisms and Kinetic Parameters in Granular Media Filtration," Doctoral Thesis, University of California, Berkeley (1974).

Kawamura, S., "Studies on Horizontal Flow Settling Tank By Hydraulic Models," *JJWWA, Proc. 3rd Ann. Conf.,* 8 (1952).

Kawamura, S., "A Study on a Grit Chamber by a Hydraulic Model," *JJWWA,* **222,** 11 (Apr. 1953).

Kawamura, S., "Evaluation of Actual Settling Basin Performance—Before and After Basin Modifications," Hanshin MWD Annual Report (1954).

Kawamura, S., "A Pilot Study on Double Filtration for Rapid Sand Filtration," *JJWWA, Proc. 5th Ann. Conf.* (Oct. 1954).

Kawamura, S., "A Pilot Study on Filter Bed and Underdrainage System," *JJWWA, Proc. 14th Ann. Conf.* (May 1963).

Kawamura, S., "Coagulation Considerations," *JAWWA,* **65,** 417 (June 1973).

Kawamura, S., "Design and Operation of High Rate Filters—Part 1," *JAWWA,* **67,** 535–544 (Oct. 1975).

Kawamura, S., "Considerations on Improving Flocculation," *JAWWA,* **68,** 328 (June 1976).

Kawamura, S., "Hydraulic Scale-Model Simulation of the Sedimentation Processes," *JAWWA,* **73,** 372–379 (July 1981).

Krasanskas, J. W., "Review of Sludge Disposal Practices," *JAWWA,* **61,** 225 (May 1969).

Laverty, G. L., and Nielsen, H. L., "Quality Improvements by Reservoir Aeration," *JAWWA,* **62,** 711 (Nov. 1970).

McGauhey, P. H., "Theory of Sedimentation," *JAWWA,* **48,** 437–448 (1956).

Miyakita, T., "A Study of Flow Patterns in Horizontal Flow Settling Tank by Hydraulic Model," Osaka Suido Jigyo Kenkyu (1949).

Mueller, H. M., Jr., "High Rate Sedimentation & Filtration/The Equipment," *Water Wastes Eng.* **7**(6), 65–67 (June 1970).

Murphy, K. L., "Tracer Studies in Circular Sedimentation Basins," Proceedings of the 18th Industrial Waste Conference, Purdue University, pp. 374–390 (1963).

Nakagawa, Y., "A Study on Sedimentation Basin," *JJWWA,* **257**(22) and **258**(9) (March and Apr. 1956).

O'Melia, C. R., and Crapps, D. K., "Some Chemical Aspects of Rapid Sand Filtration," *JAWWA,* **56,** 1326–1344 (Oct. 1964).

O'Melia, C. R., and Stumm, W., "Theory of Water Filtration," *JAWWA,* **59,** 1393–1412 (Nov. 1967).

Packham, R. F., and Richards, W. N., "Water Clarification by Flotation—(2) A Laboratory Study of the Feasibility of Flotation," WRA Technical Paper TP88, Medmenham, the Centre (March 1976).

Packham, R. F., and Richards, W. N., "Water Clarification by Flotation—(1) A Survey of the Literature," WRA Technical Paper TP87, Medmenham, the Association (November 1972).

Packham, R. F., and Richards, W. N., "Water Clarification by Flotation—(3) Treatment of Thames Water in a Pilot-scale Flotation Plant," WRC Technical Report TR2, Medmenham, The Centre (February 1975).

R. Rhodes Trussell, "Establishing Treatment Requirements and Design Criteria," 5th Environmental Engineers Conference, Montana (1976).

Rich, L. G., *Unit Operations of Sanitary Engineering,* John Wiley & Sons, New York (1961).

Sanks, R. L., *Water Treatment Plant Design,* Ann Arbor Science Publishers, Ann Arbor, Michigan (1979).

Sludge Disposal Committee Report, "Lime Sludge Treatment and Disposal," *JAWWA,* **73,** 608 (Nov. 1981).

Sparham, V. R., "Improved Settling Tank Efficiency by Upward Flow Clarification," *WPCF,* **42,** 801–811 (May 1970).

Steichen, J. M., Garton, J. E., and Rice, C. E., "The Effect of Lake Destratification on Water Quality," *JAWWA,* **71,** 219 (Apr. 1979).

Symons, J. M., Carswell, J. K., and Robeck, G. G., "Mixing of Water Supply Reservoirs for Quality Control," *JAWWA,* **62,** 322 (May 1970).

Tambo, K., "General Characteristics of Sludge Blanket Type Clarifiers," *JJWWA,* **386,** 38 (Nov. 1966).

Tambo, K., and Shoji, M., "Settling Characteristics of Sedimentation Tank with Tilted Plate Settler," *JJWWA,* **459,** 6 (Dec. 1972).

Tchobanoglous, G., and Eliassen, R., "Filtration of Treated Sewage Effluent," *JSED, ASCE,* **96,** 243–265 (Apr. 1970).

Tesarik, L., "Flow in Sludge—Blanket Clarifiers," *ASCE,* **SA6**(93), 105–120 (Dec. 1967).

Thirumurthi, D., "A Break-Through in the Tracer Studies of Sedimentation Tanks," *WPCF,* **41,** R-405–R-418 (Nov. 1969).

Tolman, S. L., "Sedimentation Basin Design and Operation," *Public Works,* **94**(6), 119–122 (June 1963).

Viraraghavan, T., "Tube Settlers for Improved Sedimentation," *Pollut. Eng.,* **39** (Jan. 1973).

Weber, W. J., *Physiochemical Processes,* Wiley-Interscience, New York (1972).

Westerhoff G. P., and Daly, M. P., *"Water Treatment Plant Waste Disposal, Parts 1 and 2, JAWWA,* **66,** 319, 379 (May and June 1974).

White, G. C., *Disinfection of Wastewater and Water for Reuse,* Van Nostrand Reinhold, New York (1978).

Yamaguchi, M., "Recent Development Related to Preclarification," *Int. Water Supply Assoc., 10th Ann. Conf. Proc.* (Aug. 1974).

Yao, K. M., "Theoretical Study of High Rate Sedimentation," *JWPCF,* **42,** 218–228 (Feb. 1970).

Yao, K. M., "Hydraulic Control For Flow Distribution," *ASCE, Proc. Sanit. Eng. Div.,* **98,** 275–285 (Apr. 1972).

Yao, K. M., "Design of High Rate Settlers," *ASCE San. Eng. Div.,* **99,** 621–637 (Oct. 1973).

Yao, K. M., Habibian, M. T., and O'Melia, C. R., "Water and Wastewater Filtration," *Envir. Sci. Tech.,* **5,** 1105–1112 (Nov. 1971).

Young, E. F., "Water Treatment Plant Sludge Disposal Practice in the United Kingdom, *JAWWA* **60,** 717 (June 1968).

Zabel, T. F., and Rees, A. J., "Flotation," International Water Supply Congress, Kyoto (1978).

Zabel, T., "Flotation," International Water Supply Association Twelfth Congress, Kyoto, pp. F1–16 (Oct. 1978).

Special Plant
Hydraulic Topics

Since the majority of treatment processes are functionally dependent on hydraulic factors, hydraulic analysis is an integral part of process design. Water entering and leaving each process unit must be carefully controlled to provide optimum hydraulic efficiency for treatment. In a broad sense, plant hydraulics is concerned with the flow characteristics in all treatment units, connecting conduits, and plant emergency overflow controls. There should be no dispute that in a process train, any low hydraulic treatment efficiency or control will ultimately affect the overall efficiency of the total process. It is therefore the objective of this chapter to address those hydraulic problem areas such as rapid mixing, flow distribution, currents and short circuiting, special hydraulic head losses, and longitudinal dispersion.

RAPID MIXING

Rapid mixing (or initial mixing), which provides the complete homogenization of coagulation chemicals with the plant influent stream to be treated, is often required upstream of the flocculation process. This type of mixing may also be required for disinfection processes. The initial stage of the coagulation process in water treatment consists primarily of two mechanisms: (1) adsorption of the soluble hydrolysis species on the colloid and (2) destabilization or sweep coagulation where the colloid is entrapped within the precipitating metal hydroxide. Detailed coagulation kinetics were discussed in Chapter 6. Unlike the flocculation process, which requires gentle mixing for a relatively long detention time to promote particle agglomeration resulting from interparticle collisions, rapid mixing must be completed within a few seconds (Sank, 1978; O'Melis, 1972; Hahn and Stumm, 1968; Letterman et al., 1973).

Presently, the standard design guidelines are based on the study reported by Camp and Stein (1943) and adopted by the water industry (ASCE, AWWA, and CSSE, 1969). The equation used to calculate the mechanical power needed to facilitate this mixing is

$$G = \left(\frac{P}{\mu v}\right)^{1/2} \tag{22-1}$$

where G = velocity gradient, sec^{-1}
P = power input, W (ft-lb/sec)

μ = dynamic viscosity, N-sec/m^2
    (lb-sec/ft^2)
v = volume of mixing chamber, m^3 (ft^3)

It is interesting to notice that Eq. 22-1 is originally derived for the flocculation design. As opposed to flocculation, rapid mixing requires that the coagulant be dispersed uniformly into the relatively high velocity receiving stream in a very short period of time. Therefore, care must be taken not to apply Eq. 22-1 indiscriminately in the rapid mixing process in a flow stream where hydrodynamic effects of the receiving stream cannot be ignored. The inadequacy of characterizing the rapid mixing operation by the average mixing velocity gradient G is also discussed by Letterman et al. (1973) and Vrale and Jorden (1971).

Through the years many devices have been employed to provide the rapid mixing needed for chemical dispersion. These include rapid mixing chambers equipped with rotary mixing devices such as propellers or turbines, the hydraulic jump, the Parshall flume, and in-line blenders. In recent years, the application of in-line blender-type mixers is gaining more attention. A number of in-line rapid mixers were studied by Vrale and Jorden (1971) in the laboratory. Stenquist and Kaufman (1972) approached the rapid mixing problem by using a biplanar, square-mesh of grid bars in a 5-cm(2-in.)-diameter pipe turbulent flow, tubular reactor. Chemicals were introduced into the receiving stream through injection orifices on the grid pointing downstream. It was concluded by Stenquist and Kaufman that an estimated injection orifice density of 0.15 cm^{-2} (1.0 in^{-2}) or more would be needed to produce a significant improvement in flocculation performance. Installation of jet injection blending in a number of water treatment plants was reported by Kawamura (1976). In the designs mentioned, G values of 700–1000 sec^{-1} with mixing times of approximately 1.0 sec were used. A more detailed summary and physical arrangements of the above-discussed devices were given by Amirtharajah (Sank, 1971).

Rapid mixing by jet injection was discussed in detail by Chao and Stone (1979). Their approach is to inject the solute through multi-orifice jets perpendicular to the receiving stream. The jets can be introduced into the receiving pipe flow in several ways. Two simple and practical methods are injection through a tube placed in the center of the pipe or injection through the pipe wall around the pipe periphery (Figure 22-1).

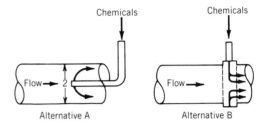

FIGURE 22-1. Alternative concepts of multijet injection into a pipe (Chao and Stone, 1979).

The basic advantage of alternative B is that it minimizes obstruction of the receiving flow. However, unless each application is carefully analyzed, uniform distribution of jet discharge around the pipe circumference may be difficult to achieve. On the other hand, in alternative A, an equal discharge through each orifice of a specific size around the supply tube can be assured. Another factor that should be considered is the turbulent nature of the pipe flow before it is disturbed by jet injection. For turbulent pipe flow, the turbulent velocity intensities in the longitudinal, radial, and circumferential directions generally increase from the center of the pipe toward the wall. It is thus assumed that a jet introduced at the center of the pipe and terminating in the wall region instead of being introduced from the wall, as illustrated in alternative B, would receive more mixing and would be transported back into the center of the pipe more effectively.

The basic data used by Chao and Stone in their analysis was derived from the experimental study reported by Pratte and Baines (1967). A jet issuing from an orifice perpendicular to the receiving flow was shown to be deflected by the receiving flow into three zones, as illustrated in Figure 22-2. Briefly, in the potential core zone, the mean velocity is greater than elsewhere in the flow field. The turbulent shear region increases from zero at the jet outlet to the point at which the entire jet cross section is covered, resulting in entrainment of surrounding fluid from the cross flow. Immediately downstream from the jet potential core, the wake is subjected to the induced radial spiral motion and grows in size to provide additional jet entrainment. Beyond the potential core, the cross flow and jet flow mix rapidly. This rapid entrainment of cross flow into the jet flow causes an increase in effective jet diameter, with a corresponding decrease in jet radial velocity. At this point, the higher cross-flow momentum forces the jet flow to bend downstream. This region is defined as the zone of maximum deflection. In the

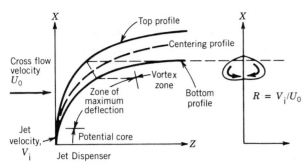

FIGURE 22-2. Definition of jet and axes (Pratte and Baines, 1967).

third region, which follows the zone of maximum deflection, the twin turbulent vortices are carried along at the cross-flow velocity with increasing size but decreasing angular velocity. Similar jet behavior was observed and reported by Chan et al (1976).

Given the experimental results reported by Pratte and Baines and Chan et al., it is suggested that the principal chemical injection jets be designed with a trajectory such that, as the jet approaches the pipe wall (Figure 22-1, alternative A), the zone of maximum deflection is nearly developed, as illustrated in Figure 22-2. From their experimental data, empirical equations were derived for the jet trajectory. The top, centerline, and bottom profiles, respectively, were given as

$$\left(\frac{X}{DR}\right)_t = 2.63 \left(\frac{Z}{DR}\right)^{0.28} \quad (22\text{-}2)$$

$$\left(\frac{X}{DR}\right)_c = 2.05 \left(\frac{Z}{DR}\right)^{0.28} \quad (22\text{-}3)$$

$$\left(\frac{X}{DR}\right)_b = 1.35 \left(\frac{Z}{DR}\right)^{0.28} \quad (22\text{-}4)$$

where X = radius of the pipe
D = jet diameter
$R = V_j/U_0$
V_j = jet velocity
U_0 = cross-flow velocity
Z = distance downstream from orifice

A plot of the centerline penetration along the curvilinear axis, as reported, indicates the profile to be two straight lines with a joining curve (Figure 22-3). These two straight lines appear to be the vertical centerline in the potential core and the centerline in the vortex zone. It was postulated that the end of the zone of maximum deflection takes place at

$\xi/DR \simeq 5$. The corresponding value of X/DR is approximately 2.80. Substituting this value in Eq. 22-3 gives $Z/DR \simeq 3.04$. For the present discussion, it is then assumed that at the point where $X/DR = 2.80$ and $Z/DR = 3.0$, the deflected jet will reach the end of the maximum deflection and approach an approximately parallel direction to the pipe flow. If the pipe diameter conveying the receiving stream to be treated is known, a combination of individual jet diameters and velocities can be selected to cover a portion of the pipe cross-sectional area and thus form a jet grid intercepting the incoming stream. As the deflected jets extend close to the pipe wall, the slight inclination of the jet axis to the longitudinal direction of the pipe will tend to deflect the jet back toward the pipe centerline and provide further mixing with the receiving stream. The total length in which rapid mixing takes place is assumed to be approximately three times the length required to reach the end of the maximum deflection zone; that is, $Z/DR \simeq 10$. Using this information, time of mixing and gross velocity gradient G can also be estimated.

Design Application

The following example illustrates the application of the principles previously discussed to the design of the rapid jet mixing system illustrated in Figure 22-4. The water treatment plant is assumed to have a design flow of 420 l/sec (10.0 mgd) with influent conveyed to the plant through a 600-mm (24-in.) pipe. Rapid mixing is required to mix alum with the raw influent upstream of the flocculation process.

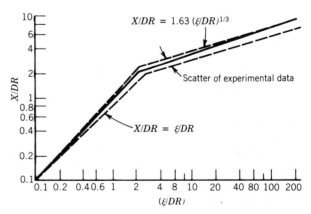

FIGURE 22-3. Centerline penetration versus distance along ξ axis (Pratte and Baines, 1967).

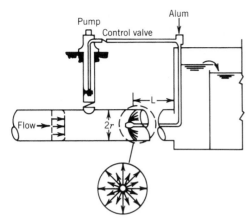

FIGURE 22-4. Jet injection rapid mixing system. (Reprinted from *JAWWA*, Vol. 71, No. 10 (October 1979, by permission. Copyright 1979, The American Water Works Association.)

STEP A: Influent pipe diameter = 0.6 m (2.0 ft); influent line velocity = 1.5 m/sec (4.93 fps).

1. Assume that the top profile of the deflected jet reaches the pipe wall at the end of the maximum deflection zone. Then, $Z/DR = 3.0$ and $(X/DR)_t = 3.58$.
2. Let X = radius of pipe, $R = V_j/U_o = 5$, and $V_j = 7.50$ m/sec (24.6 fps); then $D = 1.68$ cm (0.66 in.).
3. Assume eight jets spaced at 45°.

STEP B: Check the approximate area of pipe covered by the eight jets at the end of the maximum deflection zone. From Eqs. 22-2 and 22-4, the radial spacing covered by each individual jet at the end of the maximum deflection zone can be estimated. This gives $\Delta X/DR = 1.28 \times (Z/DR)^{0.28}$. Therefore, $\Delta X \simeq 0.15$ m (0.50 ft) (A in Figure 22-5).

STEP C: To cover zone B, as indicated in Figure 22-5, it is suggested that a second row of eight jets 22.5° out of phase with the first row be used and that the second row of jets be sized to reach the end of the maximum deflection at $r = x = 0.15$ m. Since the jet velocity–line velocity ratio remains constant, the jet diameter, D, of the second row will be 0.84 cm (0.33 in.). Correspondingly, this will cover an annular mixing area (B_1 in Figure 22-5). The total mixing area covered by A and B_1 is approximately 0.264 m² of a total pipe cross-sectional area of 0.282 m². Therefore, roughly 94 percent of the flow area is effectively covered by the jets as it passes the injec-

tion point. Based on this analysis, it is postulated that an effective rapid mixing is taking place between the alum and the receiving water.

STEP D: Check for G value,

$$G = \left(\frac{P}{\mu v}\right)^{1/2}$$

where $P = Q \times \Delta H \times \gamma_w$, with $Q = 8 \times 0.61 \times (0.00022 + 0.00055) \times 7.50 = 0.0101$ cm/sec, with discharge coefficient = 0.61; $H = V_j^2/2g = 2.81$ m: $\gamma_w = 1000$ kg/m³; $\mu = 1.15 \times 10^{-4}$ kg-sec/m² V = area of pipe $\times 10Z = 0.282 \times 10 \times 0.084 = 0.237$ m³; and $G = 1020$ sec⁻¹.

STEP E: Time of duration of mixing:

$$T \simeq \frac{10 \times 0.084}{(0.420 + 0.101)/0.282} = 0.55 \text{ sec}$$

STEP F: The minimum distance that the jets should be placed inside the effluent pipe outlet is $L = Z = 10(DR) = 0.84$ m. Therefore, use $L = 1.0$ m.

Based on the discussions and the example presented in the previous section, it is felt that the jet type mixer proposed provides a number of advantages over mechanical mixers and wake-type arrangements for rapid mixing. The system is simple, inexpensive, and easy to operate. The liquid jets form an axisymmetric jet screen and generate their own wake, which causes entrainment and spreading, and provides efficient rapid mixing with the receiving stream without creating any significant head loss.

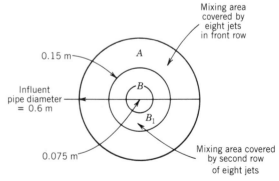

FIGURE 22-5. Mixing area covered by jets (Chao and Stone, 1979).

FLOW DISTRIBUTION

In the design of water treatment plants, flow distribution channels are used extensively to distribute incoming flows to parallel process units such as flocculation and sedimentation basins and filters. Depending on the physical arrangement, various types of devices can be employed to achieve the desired flow distribution. Channels have been most commonly used to distribute the flows with free discharge weirs or submerged orifices as control (see Figure 22-6). Pipe manifolds have also been used to distribute the incoming flow as illustrated in Figure 22-7. Basically, the flow in the distribution channels or conduits are a steady, spatially nonuniform flow. A control is therefore used to counterbalance the variation of the hydraulic energy such that the incoming flow can be evenly distributed to the process units. If the distribution channel or conduit and associated flow control devices are not properly designed, poor process performance can result.

General hydraulic mechanics of flow distribution control was studied by Camp and Graber (1968). Broad functional hydraulic interpretations were given for a number of flow phenomena. By assuming a constant-discharge coefficient for flow through all the distributing orifices, Camp and Graber derived a very simple distribution formula (Camp and Graber, 1968; Davis and Sorensen, 1969):

$$\frac{h_0}{h_0 - h_f} = \frac{1}{m^2} \qquad (22\text{-}5)$$

where h_0 = Head loss through orifice
h_f = frictional loss between orifices
m = ratio of discharges at the two orifices considered

For example, Eq. 22-5 indicates that if the range between the downstream and upstream orifice discharges is designed to be 5 percent or less, then $h_0 = 10\ h_f$ with $m = 0.95$. Eq. 22-6 has been used as a general rule of thumb.

However, studies conducted on discharge distribution along a diffuser pipe revealed that the discharge coefficient through an orifice in a distribution conduit is a function of the ratio of the local velocity head and the total head (Vigander et al., 1970; Koh and Brooks, 1975). This subject was further reviewed by Chao and Trussel (1980). It is suggested that, for a sharp-edged orifice, the discharge coefficient be expressed as

$$C_D = 0.611 - 0.291\left(\frac{V^2}{2g}\right) \qquad (22\text{-}6)$$

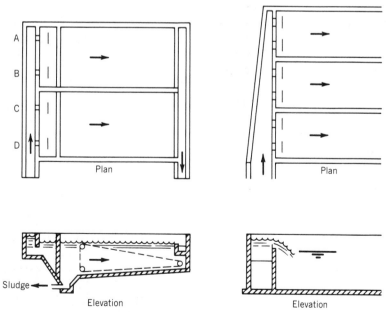

FIGURE 22-6. Flow distribution controlled by submerged orifices or overflow weirs. (Reprinted with permission from Fair, G., et al., *Water Purification and Wastewater Treatment and Disposal*, Copyright 1968 John Wiley & Sons, Inc.)

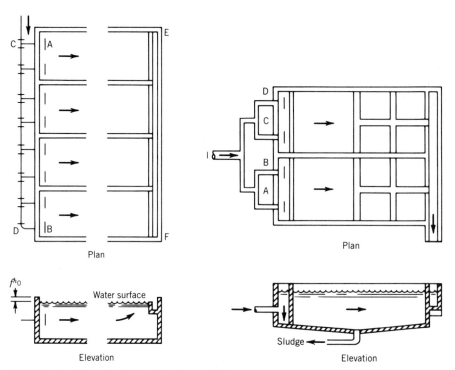

FIGURE 22-7. Flow distribution by pipe manifold arrangement. (Sources: Davis and Sorenson, 1969 and Fair, G., et al., *Water Purification and Wastewater Treatment and Disposal*, copyright 1968, John Wiley & Sons, Inc.   Reprinted with permission)

in which V is the local average velocity in the distribution conduit where the orifice is located and g is the gravitational acceleration.

As illustrated in Figure 22-6b, overflow weirs have been used widely to distribute flows. In a broad sense, this type of flow resembles the flow over a side weir spillway. Based on the study by Subranmanya and Awasthy (1972), discharge distribution along a distribution channel controlled by side weirs was examined by Chao and Trussel (1980) in detail. Similar to the coefficient of orifice discharge, the weir discharge coefficient varies with the local channel Froude number (Subranmanya and Awasty, 1972). For a rectangular side weir, this coefficient can be expressed as

$$C_w = 0.611 \times \left(1 - \frac{3F_u^2}{F_u^2 + 2}\right)^{1/2} \quad \text{with } F_u < 1.0$$

$$(22\text{-}7)$$

in which F_u is the Froude number of the channel flow immediately upstream of the weir and equals V_u/gh_u where V_u and h_u are the local average velocity and depth of flow in the distribution channel upstream of the weir. Figure 22-8 shows the experi-

mental data reported by Subranmanya and Awarthy. Based on their study, Chao and Trussell suggested that the two most viable alternatives to provide a uniform discharge distribution over the side weirs is either by employing a very wide feed channel with some kind of mechanical means installed in the channel to prevent solids deposition, if required, or by employing a tapered channel. To illustrate the application of the discussed method, the example used by Chao and Trussell (1980) is reproduced:

EXAMPLE 22-1

Problem

A water treatment plant is designed to handle a flow of 56,800 CMD (15 mgd). The flow enters from one end of the influent channel, which distributes it to the sedimentation basins through six side weirs. The channel is 14.63 m (48 ft) long and 1.219 m (4 ft) wide. The side weirs are 1.219 m (4 ft) wide and placed 1.219 m (4 ft) apart as shown in Figure 22-9. Estimate the discharge distribution through all weirs.

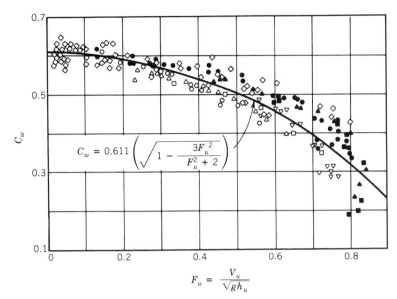

$$C_w = 0.611 \left(\sqrt{1 - \frac{3F_u^2}{F_u^2 + 2}} \right)$$

$$F_u = \frac{V_u}{\sqrt{gh_u}}$$

FIGURE 22-8. Variation of C_w with F_u. (Source: Chao and Trussell, 1980.)

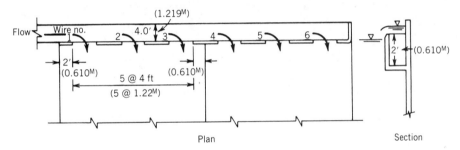

FIGURE 22-9. Example distribution with side weirs.

Solution

The procedure to calculate the discharge over each weir is presented as a flow diagram in Figure 22-10 and can be easily computerized. Results of the calculations are given in Table 22-1.

Table 22-1 clearly shows the uneven distribution of the discharge over the side weirs along the distribution channel. Discharge over weir 6, the last weir, is approximately 30 percent higher than weir 1, the first weir, for the design flow of 56,800 CMD (15 mgd). Overall, the second basin, which takes flow from weirs 6, 5, and 4, will receive approximately 17 percent more flow than the first basin. As expected, the discharge per weir decreases continuously against the flow direction in the distribution channel. This phenomena is caused by the changes of flow depth in the channel and the magnitude of discharge coefficients.

TABLE 22-1. Flow Distribution: Rectangular Channel[a]

Weir Number	y_w, ft (m)	C_w	q_w, ft³/sec (m³/sec)	$Q_w = \Sigma q_w$, ft³/sec (m³/sec)
6	2.476 (0.755)	0.610	4.29 (0.121)	4.29 (0.121)
5	2.468 (0.752)	0.607	4.21 (0.119)	8.50 (0.240)
4	2.455 (0.748)	0.603	4.04 (0.114)	12.54 (0.354)
3	2.439 (0.743)	0.597	3.81 (0.109)	16.35 (0.463)
2	2.420 (0.738)	0.590	3.56 (0.101)	19.91 (0.564)
1	2.398 (0.731)	0.588	3.30 (0.093)	23.21 (0.657)

[a] All weirs set at same elevation.

SOURCE: Chao and Trussell, 1980. Reprinted with permission.

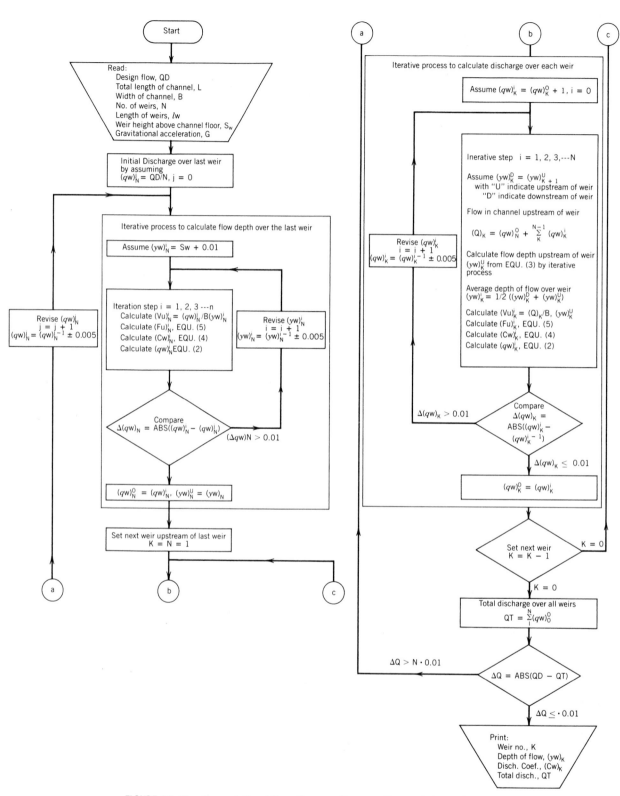

FIGURE 22-10. Computational flow diagram. (Source: Chao and Trussell, 1980.)

Many alternatives can be employed to modify the channel and/or the weir geometric arrangement to accomplish the desired uniform discharge distribution. These alternatives include (1) no modification of the channel geometry but adjusting the weir elevations; (2) widening the channel and/or deepening the channel to make the channel Froude number small enough so that the discharge coefficients for all weirs will nearly remain constant with a nearly level water depth throughout the channel; (3) tapering the channel such that the Froude number will be kept nearly constant, which in turn will give a constant-discharge coefficient; and (4) minor adjustment of weir elevations with tapered channel modification. A discussion of these alternatives is presented below:

ALTERNATIVE 1. Adjust weir elevation without any modifications of the channel: Weir 6 used as control with weir crest set at 0.61 m (2 ft) above the channel floor, as shown in Figure 22-9. In order to increase the discharge over the weirs upstream of weir 6, it is obvious that all weirs except weir 6 have to be lowered to compensate for the decreasing water depth in the distribution channel and the change of discharge coefficient. Results of this calculation are presented in Table 22-2.

From Table 22-2, it is noted that the adjustment of weir elevation is substantial. Between weirs 1 and 6, 3.0 cm (1.2 in.) is required to offset the change of discharge coefficients and flow depth in

the channel. Due to the fact that design flow generally is considered to be the maximum flow, flows lower than design flow will result in more discharge over the upstream weirs, which is certainly undesirable.

ALTERNATIVE 2. Widen and/or deepen feed channel: By substantially widening and/or deepening the feed channel, one would be able to minimize the effect of the channel Froude number with low channel velocities. If this alternative is used, consideration should be given to any suspended solids that are carried by the incoming flow. If the flow is solids laden, settling could be a problem and some mechanical means, such as aeration, must be provided to keep the solids in suspension.

ALTERNATIVE 3. Tapered feed channel: The changes of flow depth and discharge coefficient are caused by the change in the kinetic energy of the flow in the channel. If the channel can be modified to balance this change, it is possible that a constant discharge will prevail over the weirs.

Theoretically one can set all weirs at the same elevation and taper the channel to compensate for the change of channel flow. This approach requires that F_u or V_u and y_u be kept constant. The only way to accomplish this is to reduce the channel width in accordance with the decrease of flow at upstream sections of each weir. In practice, the channel width is tapered at a uniform rate from the channel entrance to the channel downstream end. With constant y, the relation between the width of the channel and the flow is linear. In the present example, if the downstream end width of the channel is 0.610 m (2 ft), the upstream end width should be $0.610 \times 6 = 3.66$ m ($6 \times 2 = 12$ ft). This is too wide to be practical. Once again, the flow velocity throughout the channel is low and causes the same settling problems discussed in the previous alternative.

To illustrate this alternative, it is assumed that the widths at the upstream and downstream end of the channel are 6 and 2 ft, respectively. The 0.610-m (2 ft) width is assumed to be the minimum channel width to provide maintenance space. Again, we start by assuming that all weirs are set at the same elevation, that is, 0.610 m (2 ft) above the channel bottom (see Figure 22-11). The flow distribution was calculated and is presented in Table 22-3. Basically, the procedure of this calculation is identical to the procedure of Figure 22-10 except that the downstream flow depth at a weir in question is adjusted from the depth of the previous weir calculation to

TABLE 22-2. Flow Distribution (Alternative 1)

Weir Number	y_w, ft (m)	S_w, ft (m)	C_w	q_w, ft³/sec (m³/sec)	$Q_w = \Sigma q_w$, ft³/sec (m³/sec)
6	2.445 (0.745)	2.00 (0.610)	0.610	3.875 (0.110)	3.875 (0.110)
5	2.441 (0.744)	1.99 (0.606)	0.607	3.875 (0.110)	7.750 (0.220)
4	2.431 (0/741)	1.98 (0.603)	0.603	3.875 (0.110)	11.625 (0.330)
3	2.416 (0.736)	1.96 (0.597)	0.596	3.875 (0.110)	15.500 (0.440)
2	2.396 (0.730)	0.94 (0.591)	0.586	3.875 (0.110)	19.375 (0.550)
1	2.368 (0.722)	1.90 (0.579)	0.574	3.875 (0.110)	23.250 (0.660)

SOURCE: Chao and Trussell, 1980. Reprinted with permission.

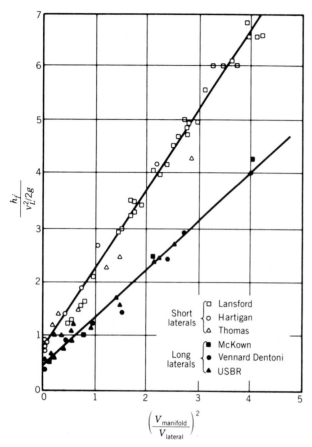

FIGURE 22-11. Port entry head loss coefficient as function of ratio of conduit velocity to lateral velocity (Hudson et al., 1979). Reprinted with permission.

compensate for the change of channel width. For a constant channel width, it is assumed that there is no change of flow depth between consecutive weirs.

As shown in Table 22-3, the discharge distribution over the weirs is greatly improved as compared to the flow distribution in the rectangular channel with the constant width shown in Table 22-1. Discharge over the last weir, weir 6, is now approximately 12 percent higher than the first weir, weir 1. Overall, the second basin, which takes flow from weirs 6, 5, and 4, will receive about 7 percent more flow than the first basin.

ALTERNATIVE 4. Tapered channel with minor adjustment of weir crest elevations: Similar to the modification described in alternative 1, each of the weir crest elevations can be adjusted such that a uniform discharge over the weirs can be attained. The results from these adjustments are presented in Table 22-4. It is shown that weir 1 must be approximately 1.2 cm (0.5 in.) below weir 6 to offset the

TABLE 22-3. Flow Distribution (Alternative 3)

Weir Number	y_w, ft (m)	C_w	q_w, ft³/sec (m³/sec)	$Q_w = \Sigma q_w$, ft³/sec (m³/sec)
6	2.461 (0.750)	0.609	4.09 (0.116)	4.09 (0.116)
5	2.456 (0.748)	0.605	4.00 (0.113)	8.09 (0.229)
4	2.444 (0.746)	0.603	3.90 (0.110)	11.99 (0.339)
3	2.444 (0.744)	0.600	3.82 (0.108)	15.81 (0.447)
2	2.437 (0.742)	0.598	3.74 (0.106)	19.55 (0.553)
1	2.431 (0.741)	0.596	3.66 (0.104)	23.21 (0.657)

SOURCE: Chao and Trussell, 1980. Reprinted with permission.

change of discharge coefficients and flow depth in the channel. This adjustment is relatively small as compared with the similar adjustment presented in alternative 1. With this minor adjustment, flow is then uniformly distributed through all six weirs.

For the given flow of 56,800 CMD (15 mgd), it appears that this alternative offers the best solution in flow distribution. However, caution must be exercised. If the 56,800-CDM (15-mgd) flow is the maximum flow that will be reached in the distant future, and the present flow is much lower, even this minor adjustment of weir elevations would result in uneven flow distribution due to the lower kinetic energy prevailing in the channel at the

TABLE 22-4. Flow Distribution (Alternative 4)

Weir Number	y_w, ft (m)	S_w, ft (m)	C_w	q_w, ft³/sec (m³/sec)	$Q_w = \Sigma q_w$, ft³/sec (m³/sec)
6	2.445 (0.745)	2.00 (0.610)	0.609	3.875 (0.110)	3.875 (0.110)
5	2.432 (0.741)	1.98 (0.604)	0.605	3.875 (0.110)	7.850 (0.220)
4	2.428 (0.740)	1.98 (0.604)	0.602	3.875 (0.110)	11.625 (0.330)
3	2.425 (0.739)	1.97 (0.600)	0.600	3.875 (0.110)	15.500 (0.440)
2	2.419 (0.737)	1.96 (0.597)	0.597	3.875 (0.110)	19.375 (0.550)
1	2.414 (0.736)	1.96 (0.597)	0.595	3.875 (0.110)	23.25 (0.660)

SOURCE: Chao and Trussell, 1980. Reprinted with permission.

present time. Therefore, the design engineer must take this factor into consideration.

Based on the hydraulic alternative of the example, it appears that alternative 2 wide and/or deep channel, and alternative 3, tapered channel, provide a better solution to the problem of distributing flow over influent channel side weirs. Selection of the proper alternative at this point must also consider the operation andmaintenance problems, cost, and other associated factors, in addition to the hydraulic considerations discussed. ☐

In addition to the distribution channel with overflow weirs, flow can also be distributed by means of pipe manifolds in water and wastewater treatment plants. Similar to the problems discussed in the previous sections, unequal distribution of flow through the outlet laterals can take place with improper design. Pipe manifold hydraulics were studied by a number of investigators (Hudson et al., 1979; McNown, 1954; Miller, 1986).

In the study of dividing flow controlled by manifold with square-edged laterals (as illustrated in Figure 22-7a) by Hudson et al. (1979), it was found that the head loss at the lateral entrance can be expressed as

$$\frac{h_f}{V_L^2/2g} = \phi \left(\frac{V_m}{V_L}\right)^2 + \theta \qquad (22\text{-}8)$$

where h_f = loss entering lateral
V_L = average velocity in lateral
V_m = average velocity in manifold
ϕ = slope of line
θ = intercept of Figure 22-11.

Figure 22-11 reproduces the experimental data collected by Hudson et al and shows that the functional relationship as expressed in Eq. 22-8 also depends on the length of lateral in question. In the two distinct groups, the short lateral lengths are taken as those less than three pipe diameters, while long lateral lengths are substantially greater than three pipe diameters. Values of ϕ and θ are (McNown, 1954)

Lateral Length	ϕ	θ
Long	0.9	0.4
Short	1.67	0.7

An iterative technique that has to be employed to estimate the flows in each lateral was discussed by

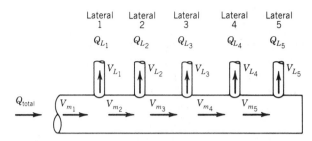

FIGURE 22-12. Example of dividing flow situation (Hudson et al., 1979).

Hudson et al. (1979). To achieve an even flow distribution, the design criteria requires the same manifold–lateral velocity ratio at every junction point.

In the example given by Hudson et al. (see Figure 22-12), the calculation shows that for a flow of 14.16 L/sec entering the 10 cm diameter manifold with five 5 cm diameter laterals connected to the manifold at 90° with square-edged entrance, the discharge distribution is approximately:

Lateral	Discharge (L/sec)
1	1.93
2	2.41
3	2.89
4	3.17
5	3.60
	14.00

The example indicates that lateral 1 discharges about 14 percent while lateral 5 discharges about 25 percent of the total flow. Techniques to correct this uneven distribution of discharge were discussed by Hudson et al. (1979). These include use of much-larger-diameter manifold and smaller laterals, a successive reduction of lateral diameters along the manifold, or a tapered manifold to maintain a constant velocity in the manifold. This alternative is consistent with the solutions discussed in the channel flow distribution. Therefore, problems involving sedimentation and excessive head losses should be evaluated in the final design.

CURRENTS AND SHORT CIRCUITING IN SEDIMENTATION BASINS

Settling of suspended particles in sedimentation basins has been studied extensively. Theory and design criteria were presented in Chapters 7 and 21, respectively. Unfortunately, sedimentation basins seldom perform in accordance with ideal settling theory. One of the most important factors that often causes the poor performance of the sedimentation process is the uncontrolled currents taking place in the basin. These currents can be generated by a number of causes, which include density currents resulting from density differences between the influent water and the water already in the basin, convective currents caused by temperature differences in the basin, eddy currents resulting from the motion of sludge cleaning equipment, wind-induced surface currents, and direct short circuiting from high-velocity jets resulting from poor design at the inlet and outlet. These currents can distort the idealized flow pattern in the basin and reduce the hydraulic efficiency of the settling process.

Density Currents

When feedwater is entering the sedimentation basin, it can form a surface or a bottom density current, as illustrated in Figure 22-13, depending on the relative densities of the feedwater and water in the basin. Under these flow conditions, flow-through velocity will depart from the idealized average basin velocity, which equals the total incoming flow divided by the total cross-sectional area in the basin. Studies were made to estimate the steady propagation velocity of these density currents (Harleman, 1961; Kao, 1977). A general equation was derived by Kao (1977) for calculating this velocity:

$$U = \left[2 \frac{\Delta\rho}{\rho_0} gh + (\beta - \beta_d)gh^2 \right]^{1/2} \quad (22\text{-}9)$$

where $\Delta\rho$ = density difference in ambient fluid and incoming flow = $\rho_d - \rho_0$
ρ_0 = density of incoming flow
ρ_d = ambient fluid density in the basin
g = acceleration of gravity
h = depth of density current
$\beta = -\dfrac{1}{\rho_0}\dfrac{d\rho}{dz}$
$\beta_d = -\dfrac{1}{\rho_0}\dfrac{d\rho_d}{dz}$
z = vertical coordinate

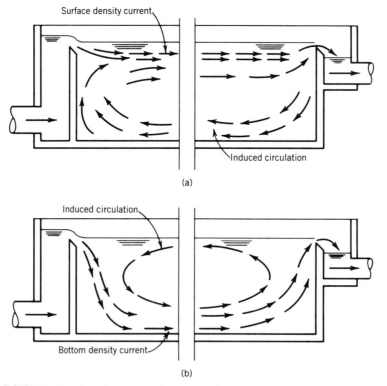

(a)

(b)

FIGURE 22-13. Density currents in a conventional sedimentation basin: (a) Surface; (b) bottom.

If both the incoming flow and ambient fluid is not stratified, $\beta = \beta_d = 0$, then Eq. 22-9 becomes

$$U = \left(2\,\frac{\Delta\rho}{\rho_0}\,gh\right)^{1/2}$$

Short circuiting caused by density current has been observed in many water treatment plants (Harleman, 1961; Kao, 1977; Camp, 1946). Methods used to minimize the effects of density currents were reported by a number of investigators (Camp, 1946; Fitch and Lutz, 1960; Harleman, 1961; Hudson, 1972; Kao, 1977; Sank, 1978).

It is commonly recognized that special attention be given in the design of inlet and outlet arrangements, which will provide some degree of control in minimizing the density currents effects. At the inlet, the following techniques have been practiced: (1) distribute the feed flow uniformly through the basin cross section in the plan perpendicular to the flow by employing diffuser walls; (2) devices that

will break up the feed stream and dissipate the energy by turbulence. Improvements can be made in the basin to control the density currents. These include tube settlers, redistribution baffles, or intermediate diffuser walls. Launders extending into sedimentation basins have been commonly used to control the effluent flow distribution. This arrangement is certainly a more effective control for bottom density currents than surface density currents.

Model studies were conducted by Kawamura (1952, 1953, 1981). In his study, a 1 : 25 scale model was used with dynamic similitude modeled following Froude's law. Temperatures in the tank and the feed flow were kept constant with the feed flow about 0.3°C lower than the water in the tank. A 0.1 percent fluorescent dye solution was used to trace the movement of the density current as the feed flow travels across the tank. Fronts of the density current were observed and marked on the glass wall of the tank at 1-min intervals. Without diffuser walls in the tank, the density current flows along the bot-

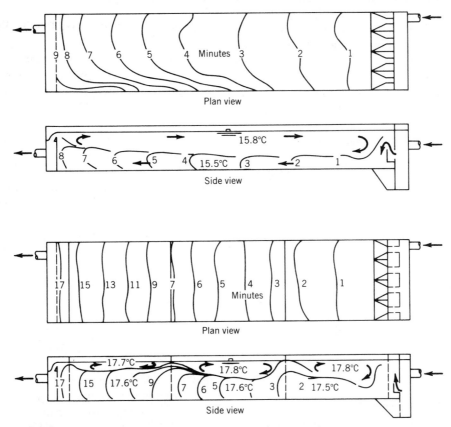

FIGURE 22-14. Density current model study. (a) Existing tank under density flow conditions. (b) Modified tank under density flow conditions. $Q = 134$ mL/sec. (Source: Kawamura, 1981.) Reprinted with permission.

tom of the tank at a relatively shallow depth and takes about 9 min to reach the outlet, as shown in Figure 22-14a. Under these conditions, the top two-thirds of the tank depth was not effectively used. To improve the hydraulic efficiency, diffuser walls with approximately a 7 percent net opening were added as shown in Figure 22-14b. Under these conditions, head losses are created at the diffuser walls and thus force the retardation and mixing of the density current with the ambient water. Figure 22-14b clearly demonstrates the improvement in the flow distribution in the tank, which should increase the efficiency of hydraulic performance of the settling tank.

Wind-Generated Currents

When wind is blowing over a body of water, energy from the wind imparted to the water drags the water at the surface in the direction of the wind with the water near the bottom moving in an opposite direction, thus forming a circulation (see Figure 22-15). Strong currents induced by wind can cause short circuiting and resuspension of already settled solids in sedimentation basins. This has been observed in treatment plants in locations where strong winds occur frequently. Wind-induced currents were studied by Baines and Knapp (1965) and Hidy and Plate (1966) in laboratory wind-water tunnels. Liu and Perez (1971) studied this problem by numerically solving the governing equations of motion with certain assumptions of the kinematic eddy viscosity of water. In the analysis, a rectangular basin with a length twice its width is considered. The basin has a flat bottom with a water depth 1/100 of its length. The Karman–Prandtl relation for turbulent flow over smooth surfaces was used in their analysis to simulate the shear stress acting on the water surface. The predicted drift with a sustained steady wind was reported by Liu and Perez and is reproduced in Figure 22-16. The drift velocity as shown in Figure 22-16 is by no means small enough to be neglected.

Therefore, when long and shallow rectangular settling basins are used, their orientation with the

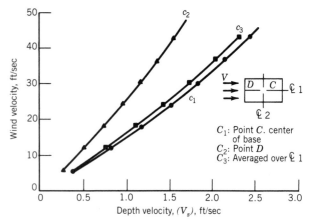

FIGURE 22-16. Variation of drift velocity with wind speed (Liu and Perez, 1971). Reprinted with permission.

local prevailing wind direction should be considered. It is suggested that the longitudinal axis of the basin be placed perpendicular to the prevailing wind direction and thus minimize the effects of the wind-induced currents.

SPECIAL HYDRAULIC HEAD LOSSES

For logical or convenient reasons, different performance parameters have been employed for various hydraulic components. This variation is particularly applied to meters, valves, pipe bends, and other hydraulic components that cause local hydraulic losses. For instance, equivalent pipe lengths have been used in expressing head losses for valves and bends. A special flow coefficient (C_v), defined as the flow of water through a valve in terms of gallons per minutes at 1.0 lb/in.[2], has been used to describe the throttling characteristics of a valve. To minimize the confusion, a universal dimensionless head loss coefficient has been used by many investigators to express local hydraulic head losses. As illustrated in Figure 22-17, the head loss coefficient, K, through a hydraulic component is defined by

$$K = \frac{\Delta H}{V^2/2g} \qquad (22\text{-}10)$$

in which ΔH is the head loss through the component expressed in feet (or meters) of water, V is the average flow velocity in fps (or mps) in the conduit upstream of the component, and g is the acceleration of gravity. Since K is dimensionless, it can be used in any unit system.

FIGURE 22-15.   Schematic of wind-driven draft current.

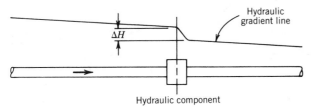

FIGURE 22-17. Sketch of local head loss through a hydraulic component.

In most hydraulic textbooks and technical publications, head losses caused by hydraulic components are referred to as minor losses. This definition is relative in nature. For example, the friction loss in a long pipeline is much higher and more important than the head losses created by valves and bends that make up part of the entire system. However, the opposite will apply in a water treatment plant. Flow through a water treatment plant is regulated at a number of control points, and those losses, in most instances, account for the majority of the total hydraulic losses through the plant. Therefore, attention should be exercised in the analyses of these hydraulic components. It is assumed that the reader is familiar with the basic hydraulic theorems and their applications. Hydraulic devices such as sharp-crested weirs and orifices, which have been extensively treated in textbooks (Chow, 1959; Henderson, 1966; Streeter, 1975) and handbooks (Chow, 1964; Davis and Sorensen, 1969; King and Brater, 1963), will not be discussed here. Instead, this section shall be devoted to head losses caused by special hydraulics devices and arrangements commonly encountered in water treatment plant designs but not discussed in standard textbooks.

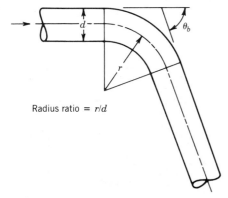

FIGURE 22-18. Illustration of circular cross-sectional bend.

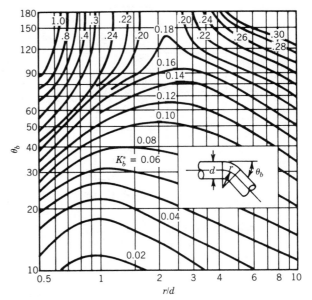

FIGURE 22-19. Loss coefficients, K_b^*, for circular cross-section bends (Re = 10^6) (Miller, 1978). Reprinted with permission.

Bend Losses

Bend losses for closed conduits with circular cross sections have been studied by many investigators. The bend loss coefficient K_b is expressed as

$$K_b = \frac{\Delta H_b}{V^2/2g} \qquad (22\text{-}11)$$

in which ΔH_b is the bend head loss and V is the average velocity in the pipe. As shown in Figure 22-18, K_b is a function of the bend geometry, r/d, θ_b, and the flow Reynolds number, Re ($=vd/\nu$, where ν is the kinematic viscosity of the fluid). Test data were collected by Miller (1978a) and reproduced in Figures 22-19 and 22-20. Figure 22-19 gives the ba-

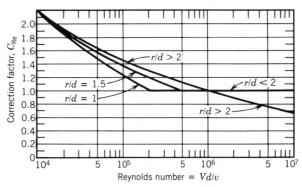

FIGURE 22-20. Reynolds number correction factors (Miller, 1978). Reprinted with permission.

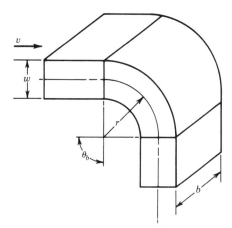

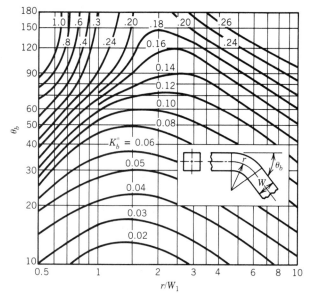

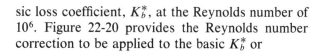

FIGURE 22-21. Illustration of bend geometry
with rectangular cross-section (Miller, 1978).
Reprinted with permission.

FIGURE 22-23. Loss coefficient K_b^* for square cross-section
bends (Re = 10^6). (Miller, 1978). Reprinted with permission.

sic loss coefficient, K_b^*, at the Reynolds number of
10^6. Figure 22-20 provides the Reynolds number
correction to be applied to the basic K_b^* or

$$K_b = C_{Re} K_b^* \qquad (22\text{-}12)$$

For closed rectangular cross-sectional conduits
with circular arc bends, as shown in Figure 22-21,
the parametric factors that affect the head loss are
the radius ratio r/w, the aspect ratio b/w, the bend
deflection angle θ_b, and the flow Reynolds number

Re. Under these conditions, Re = UD_r/ν, with D_r
(hydraulic diameter) = $2bw/b + w$. As with the circular cross-sectional case, the basic loss coefficient
K_b^* is first developed (Miller, 1978) at a Reynolds
number of 10^6 for three aspect ratios, as shown in
Figures 22-22 through 22-24. Figure 22-20 provides
the Reynolds number correction factors.

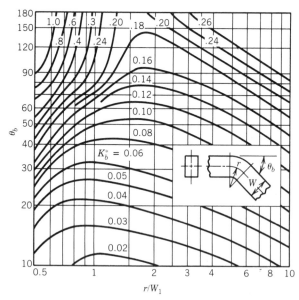

FIGURE 22-22. Loss coefficient K_b^* for aspect ratio 0.5 rectangular bends (Re = 10^6) (Miller, 1978). Reprinted with permission.

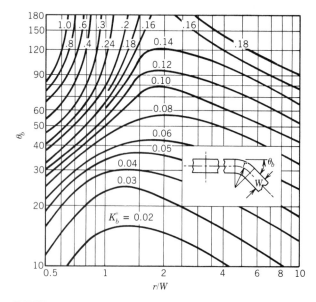

FIGURE 22-24. Loss coefficient K_b^* for aspect ratio 2 rectangular bends (Re = 10^6) (Miller, 1978). Reprinted with permission.

Rectangular Open-Channel Bend Losses. There is very limited information available in calculating bend head losses for rectangular open-channel flows. Superelevation and related head loss studies were conducted by Soliman and Tinney (1968) and Shukry (1950). A resistance coefficient similar to Eq. 22-10 was reported (Shukry, 1950). This coefficient was found to vary with the angle of bend deflection θ, ratio of the depth of flow y to channel width b, ratio of the radius of the bend γ_c to channel width b, and the flow Reynolds number Re, where Re $= UR/\nu$ with U equal to mean velocity, R equal to the hydraulic radius, and ν is the kinematic viscosity. Shukry plotted the loss coefficient (which is equivalent to K_b in Eq. 22-10) in parametric form (reproduced in Figure 22-25).

Dividing and Combining Flows. Flow phenomena in branching conduits and manifolds were investigated in detail by the ASCE Task Committee on Branching conduits (Rhone, 1973). The terminology used for the various configurations is given in Figure 22-26. The most practical information was presented by Miller (1978). Based on the defini-

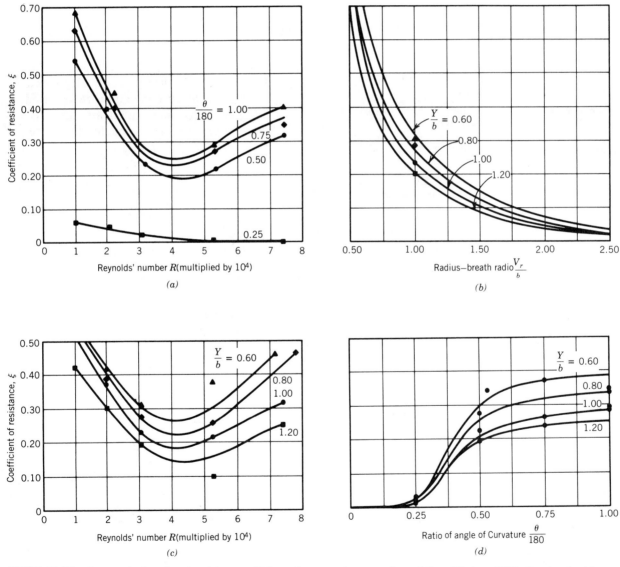

FIGURE 22-25. Parametric function: bend loss coefficient of rectangular open-channel flow (Shukry, 1950). Reprinted with permission.

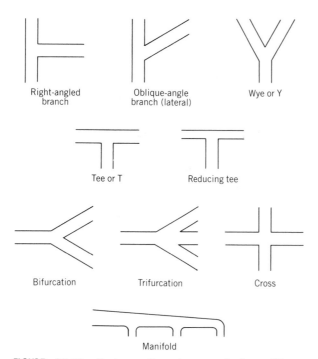

FIGURE 22-26. Basic configurations terminology (Rhone, 1973). Reprinted with permission.

tions provided by Miller (see Figure 22-27), the loss coefficients are

Combining flow:

$$K_{13} = \frac{(U_1^2/2g) + h_1 - (U_3^2/2g) + h_3}{U_3^2/2g} \quad (22\text{-}13)$$

$$K_{23} = \frac{(U_2^2/2g) + h_2 - (U_3^2/2g) + h_3}{U_3^2/2g} \quad (22\text{-}14)$$

Dividing flow:

$$K_{31} = \frac{(U_3^2/2g) + h_3 - (U_1^2/2g) + h_3}{U_3^2/2g} \quad (22\text{-}15)$$

$$K_{32} = \frac{(U_3^2/2g) + h_3 - (U_2^2/2g) + h_2}{U_3^2/2g} \quad (22\text{-}16)$$

A number of the loss coefficient plots are reproduced in Figure 22-28 through 22-36. Head losses at trifurcations in closed conduits were reported by Rao et al. (1967). As shown in Figure 22-32, the loss coefficient from the main pipe to the branching

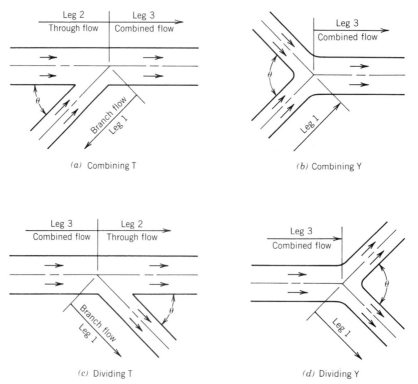

FIGURE 22-27. Geometric parameters for combining and dividing junctions (Miller, 1978). Reprinted with permission.

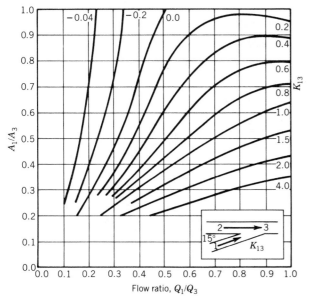

FIGURE 22-28. Combining flow, branch angle 15°, loss coefficient K_{13} (Miller, 1978). Reprinted with permission.

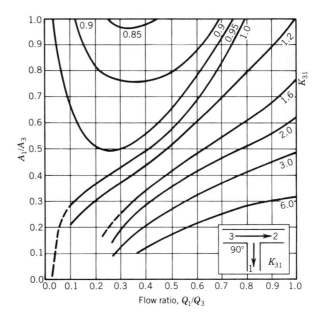

FIGURE 22-30. Dividing flow, branch angle 90°, loss coefficient K_{31} (Miller, 1978). Reprinted with permission.

pipes is estimated to be

$$\frac{h_L'/V_m^2}{2g} = K_V' = 0.96 \sin^2 \theta_b + 0.90 \left(\cos \theta_b - \frac{V_b}{V_m}\right)^2$$

where V_m = average velocity in supply main
       h_L' = head loss between main and branch pipe

θ_b = angle between main and branch pipe in flow direction
V_b = average velocity in branch pipe

Dividing and combining flows in open channels were studied by Taylor (1944) and Law and Reynolds (1966). Due to the complication of flow regimes and downstream controls, no generalized form was available for direct applications. To obtain an accurate estimate of flows in this type of branching flows, a model study was recommended by these investigators.

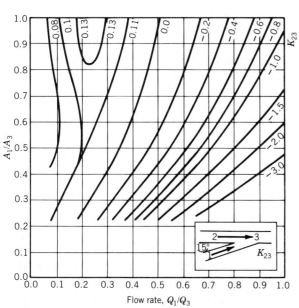

FIGURE 22-29. Combining flow, branch angle 15°, loss coefficient K_{23} (Miller, 1978). Reprinted with permission.

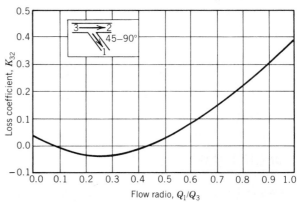

FIGURE 22-31. Dividing flow, branch angles of 45°–90°, loss coefficient K_{32} (Miller, 1978). Reprinted with permission.

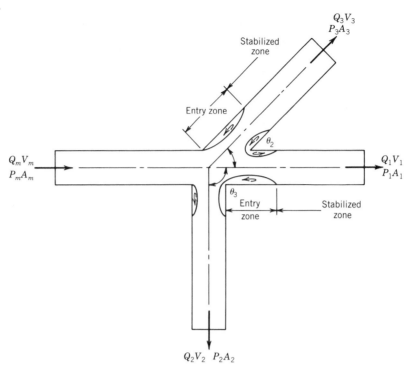

FIGURE 22-32.   Typical trifurcation (Rao et al., May 1967).

Valves.   Several types of valves are extensively used in water treatment plants to control flows. The geometry of a particular type of valve is not necessarily similar among different manufacturers. This may also apply to the same type valve of different size from the same manufacturer. To provide proper control, valves are often operated in partially open positions. Under these conditions, a complete set of valve characteristics is required. If a valve is expected to dissipate a relatively high head continuously, care must be exercised to ensure that it will not be operated in the cavitation range. Loss coefficients for a number of valves are reproduced from Miller (1978) (Figures 22-33 through 22-36). The values presented are typical for some commonly used valves. Accurate values have to be obtained from the manufacturer of the valve actually used.

LONGITUDINAL DISPERSION

In water treatment design, there are unit processes in which relatively slow reactions are required between the chemicals and the water. Chlorination is

one of those applications. Unlike the rapid mixing and flocculation processes, in which mixing is generally promoted by mechanical means, chlorination is usually carried out by introducing chlorine solution into the water stream through single- or multiple-port arrangements. Self-mixing then takes place in the receiving streams. Longitudinal dispersion is the spreading or mixing of the chlorine or other chemicals in the receiving stream along its course. Depending on the physical geometry of the reactor and the flow characteristics, mixing can vary significantly. Furthermore, rather than move downstream as a slug, the injected chemicals may travel at different speeds along the length of the flow course, some parts moving faster and some slower than the mean velocity. The uniformity of the mixing and the residence time that each water parcel receives in the reactor is of importance to the treatment process.

The first important study of dispersion in turbulent shear flow was published by Taylor in 1954. Taylor asserted that although the primary mechanism for dispersion in shear flow is the variation in convective velocity within the cross section, the process could be described by a one-dimensional

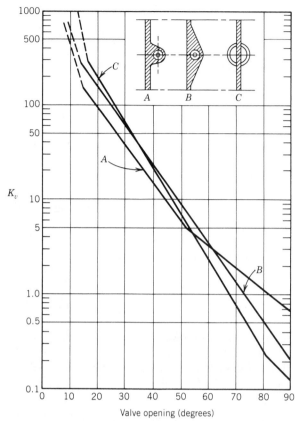

FIGURE 22-33. Loss coefficients for butterfly valves (Miller, 1978). Reprinted with permission.

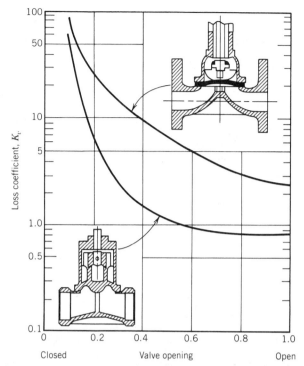

FIGURE 22-34. Diaphragm valve loss coefficients (Miller, 1978). Reprinted with permission.

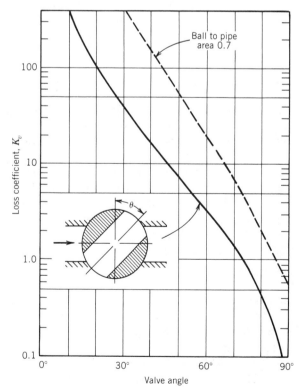

FIGURE 22-35. Ball valve coefficients (Miller, 1978). Reprinted with permission.

Fickian diffusion equation:

$$\frac{\partial \bar{C}}{\partial t} + \bar{U}\frac{\partial \bar{C}}{\partial x} = D_L\frac{\partial^2 \bar{C}}{\partial x^2} \qquad (22\text{-}17)$$

where $\bar{C}$ = cross-sectional mean concentration
$\bar{U}$ = cross-sectional mean velocity
t = time
x = distance in direction of flow
D_L = dispersion coefficient

Equation 22-17 states that the rate of change of concentration is determined by convection and dispersion. Dispersion coefficients for long straight pipes and very wide two-dimensional channels obtained by Taylor and Elder were discussed in Chapter 12. Since in water treatment and supply facilities pipes are commonly used in chlorination or disinfection processes, more discussion is presented in the following sections on longitudinal dispersion in long pipes.

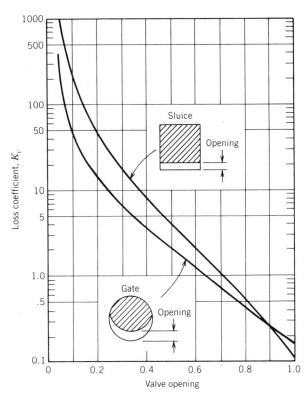

FIGURE 22-36. Gate and sluice valve loss coefficients (seat area = pipe area) (Miller, 1978). Reprinted with permission.

$$C_v = \left[\frac{1}{A} \int_A \left(\frac{C}{\bar{C}} - 1 \right)^2 dA \right]^{1/2} \qquad (22\text{-}18)$$

where A = pipe cross-sectional area
C = tracer concentration at a point in the cross section
$\bar{C}$ = cross-sectional average concentration

For a given flow system, C_v decreases with the increase of mixing distance. To study the effectiveness of mixing, Holley et al. introduced tracers into the flow either as a simple source at the wall or by injecting it into the pipe as a jet. Empirical equations and graphs were developed relating the degree of mixing and distance of travel. It was reported in their studies that more effective mixing would be accomplished by injecting the tracer into the pipe. The optimum injection is to have the tracer jet penetrate to the center of the pipe. For a simple source and a Darcy–Weisbach friction factor, f, of 0.02, the mixing distance required to produce a C_v of 0.01 is approximately 150 pipe diameters. If f is reduced to 0.01, then a corresponding mixing length of approximately 210 pipe diameters is required. However, if the tracer is introduced by jet injection into the center of the pipe, the mixing distance required to produce the same C_v of 0.01 is only about 50 pipe diameters (with $f = 0.02$). This reduction of mixing length is apparently the result of the combination of initial mixing due to the injection and ambient mixing due to the pipe flow. An important finding reported in their studies is that no significant reduction of mixing length is observed by multijet injection. Readers are advised to study the original publications (Ger and Holley, 1976; Holley, 1969; Holley and Ger, 1978; Fitzgerald and Holley, 1981) regarding multijet applications.

In Chapter 12, a dispersion number, $d = D_L/UL$, is introduced in which D_L is the dispersion coefficient, U is the average velocity in the pipe, and L is a characteristic length. Physically, this dispersion number can be interpreted as the ratio of dispersion effect to convection effect. In an identical flow system where D_L and U remain constant, then a decrease of the dispersion number, which corresponds to an increase of uniformity of concentration distribution, is in direct proportion to the distance traveled. Therefore, under certain conditions where sufficient pipe length is available and the reaction is relatively slow, the ambient turbulence in a pipe flow can ultimately provide complete mixing of the added miscible substances.

A series of studies on pipe mixing have been conducted by Holley and his colleages (Ger and Holley, 1976; Holley, 1969; Holley and Ger, 1978; Fitzgerald and Holley, 1981). In analyzing the degree of uniformity of the concentration distribution at each cross section, a single parameter was used. This is the commonly used coefficient of variation, C_v, of the concentration distribution, defined as

REFERENCES

ASCE, AWWA, and CSSE, *Water Treatment Plant Design*, American Water Works Association, Inc., New York (1969).

Baines, W. D., and Knapp, D. J., "Wind Driven Water Currents," *J. Hydr. Div., ASCE*, **91**, 205–221 (March 1965).

Camp, T. R., "Sedimentation and the Design of Settling Tanks," *ASCE Trans.* **3** (1946).

Camp, R. T., and Graber "Dispersion Conduits," *J. San. Eng. Div., ASCE*, **94**, 31–39 (February 1968).

Camp, T. R., and Stein, P. C., "Velocity Gradients and Internal Work in Fluid Motion," *J. Boston Soc. Civil Eng.*, **30**, 219–237 (1943).

Chan, D. T. L., Lin, J. T., and Kennedy, J. F., "Entrainment and Drag Forces of Deflected Jets," *J. Hydr. Div., ASCE* **102**, 559–568 (May 1976).

Chao, J. L., and Stone, B. G., "Initial Mixing by Jet Injection Blending," *JAWWA*, **71**, 570–573 (October 1979).

Chao, J. L., and Trussell, R. R., "Hydraulic Design of Flow Distribution Channels," *J. Environ. Eng. Div., ASCE*, **106** (2), 321 (April 1980).

Chow, V. T., *Open-Channel Hydraulics*, McGraw-Hill, New York (1959).

Chow, V. T. (ed.), *Handbook of Applied Hydrology*, McGraw-Hill, New York (1964).

Davis, C. V. (ed.), *Handbook of Applied Hydraulics*, McGraw-Hill, New York (1969), Chap. 38.

Davis, C. V., and Sorensen, K. E., *Handbook of Applied Hydraulics*, 3rd ed., McGraw-Hill, New York (1969).

Fitch, E. B., and Lutz, W. A., "Feedwell for Density Stabilization." *JWPCF*, **32**, 147–156 (February 1960).

Fitzgerald, S. D., and Holley, E. R., "Jet Injections for Optimum Mixing in Pipe Flow," *J. Hydraul. Division, ASCE*, **107**, 1179–1195 (October 1981).

Ger, A. M., and Holley, E. R., "Comparison of Single-Point Injections in Pipe Flow," *J. Hydraul. Div., ASCE*, **102**, 731–746 (June 1976).

Hahn, H. H., and Stumm, W., "Kinetics of Coagulation with Hydrolyzed Al(111)," J. Coll. Interface Sci., **28**, 134–144 (1968).

Harleman, D. F., "Stratified Flow," in V. Streeter (ed.), *Handbook of Fluid Mechanics*, McGraw-Hill, New York (1961).

Henderson, F. M., *Open Channel Flow*, The Macmillan Company, New York (1966).

Hidy, G. M., and Plate, E. J., "Wind Action on Water Standing in a Laboratory Channel," *J. Fluid Mechanics*, **26**, (4), 651–688 (1966).

Holley, E. R., "Unified View of Diffusion and Dispersion," *J. Hydraul. Div., ASCE*, **99**, 621–631 (March 1969).

Holley, E. R., and Ger, A. M., "Circumferential Diffusion in Pipe Mixing," *J. Hydraul. Div., ASCE*, **104**, 471–485 (April 1978).

Hudson, H. E., "Density Considerations in Sedimentation," *JAWWA*, **64**, 382–386 (June 1972).

Hudson, H. E., Uhler, R. B., and Bailey, R. W., "Dividing Flow Manifolds with Square-Edged Laterals," *J. Environ. Eng. Div., ASCE*, **105** (4), 745 (August 1979).

Kao T. W., "Density Currents and their Applications," *J. Hydrol. Div., ASCE*, **103**, 543–555 (May 1977).

Kawamura, S., "Studies on Horizontal Flow of the Settling by Hydraulic Models," Proceedings of the 3rd Annual Congress, Japan Water Works Association (1952).

Kawamura S., "A Study on a Grit Chamber by a Hydraulic Model," *J. Japan WWA*, **222**, 11 (April 1953).

Kawamura, S., "Considerations on Improving Flocculation," *JAWWA*, **68**(6), 328 (June 1976).

Kawamura, S., "Hydraulic Scale-Model Simulation of the Sedimentation Process," *JAWWA*, **73**(7), 372 (July 1981).

King, H. W., and Brater, E. F., *Handbook of Hydraulics*, McGraw-Hill, New York (1963).

Koh, R. C. Y., and Brooks, N. H., "Fluid Mechanics of Wastewater Disposal in the Ocean," *Annual Review of Fluid Mechanics*, **7**, 187–211 (1975).

Law, S. W., and Reynolds, A. J., "Dividing Flow in Open Channel," *J. Hydraul. Div.*, ASCE **92**(HY2), 207–232 (March 1966).

Letterman, R. D., Auon, J. E., and Gemmell, R. S., "Influence of Rapid-Mix Parameters on Flocculation," *JAWWA*, **65**, 716–722 (1973).

Liu, H., and Perez, H. J., "Wind-Induced Circulation in Shallow Water," *J. Hydraul. Div.*, ASCE **97**(HY7), 923, (July 1971).

McNown, J. S., "Mechanics of Manifold Flow," *Trans. ASCE*, **119**, Paper No. 2714 (1954).

Miller, D. S., *Internal Flow Systems*, British Hydrodynamics Research Association, Cranfield, England (1978).

O'Melis, C. R., "Coagulation and Flocculation," in W. J. Weber (ed.), *Physiochemical Processes for Water Quality Control* Wiley-Interscience, New York (1972).

Pratte, B. D., and Baines, W. D., "Profiles of the Round Turbulent Jet in a Cross Flow," *J. Hydr. Div., ASCE* **93**(HY6), 53–64 (Nov. 1967).

Rao, N. S. L., et al., "Pressure Losses at Trifurcations in Closed Conduits," *J. Hydr. Div., ASCE* **93**(HY3), 51 (May 1967).

Rhone, T. J., "General Considerations of Flow in Branching Conduits," Proceedings of the 21st Annual Hydraulic Division Specialty Conference, ASCE (Aug. 1973).

Sank, R. L. (ed.), *Water Treatment Plant Design*, Chapter 8, Ann Arbor Science Publishers, Ann Harbor, Mich. (1978).

Shukry, A., "Flow Around Bends in an Open Flume," *Trans. ASCE*, **115**, 751–788 (1950).

Soliman, M. M., and Tinney, E. R., "Flow Around 180° Bends in Open Rectangular Channels," *J. Hyd. Div., ASCE*, **94**, 893–908 (July 1968).

Stenquist, R. J., and Kaufman, W. J., "Initial Mixing in Coagulation Processes," SERL Rept. 72-2, University of California, Berkeley, CA (February 1972).

Streeter, V. L., *Fluid Mechanics*, McGraw-Hill, New York, (1975).

Subranmanya, K., and Awasthy, S. C., "Spatially Varied Flow over SideWeirs," *J. Hydraul. Div., ASCE*, **98**, 1–12 (January 1972).

Taylor, E. H., "Flow Characteristics at Rectangular Open Channel Junctions," *Trans. ASCE*, **109**, 893–912 (1944).

Taylor, G. I., "The Dispersion of Matter in Turbulent Flow Through a Pipe," *Proceedings of the Royal Society of London*, Vol. 233, Royal Society, London (1954).

Vigander, S., Elder, B. A., and Brooks, N. H., "Internal Hydraulics of Thermal Discharge Diffuser," *J. Hydraulic Division*, **96**, 509–527 (February 1970).

Vrale, L., and Jorden, R. M., "Rapid Mixing in Water Treatment," *JAWWA*, **63**, 52–58 (1971).

— 23 —

Process Control

OVERVIEW

Basic Considerations

The understanding and knowledge of a process and the interdependence of the subsystems and components is essential to process control. The designer of the process control for a facility such as a water treatment plant has to take an approach of system design rather than equipment design. Process control requires the interdependence of many components such as the capability and characteristics of the control equipment, the environment in which the equipment will operate, and the human factors involved in equipment operation.

This chapter discusses various process control strategies and instrument systems, types of physical and chemical measuring elements, computers, and final control elements that are integral parts of the process control system for a facility.

Overview

The main purposes of process control are to produce a better product at lower cost; to be alerted to any pending process or equipment malfunction; and to provide data acquisition for study and analysis of plant behavior to improve plant operation, aid in understanding of plant behavior, and provide input for the design of plant expansion and/or design of future plants.

Factors that influence the design of a specific plant are:

- Size of plant.
- Difficulty of treatment.
- Number of control points.
- Interdependence of process elements and controls, including process time delays.
- Number and capability of operating personnel available.
- Availability of qualified maintenance personnel, either owner furnished or contracted from outside companies.

Once these factors have been identified, the selection of appropriate methods of control, data acquisition, and data display follow.

Methods of Control

Controls can be manual, semiautomatic, automatic, and supervisory. The least complex, most reliable, and easiest method of control to perform the required tasks should be applied. Manual control is the simplest method of control. Therefore, it should be used whenever optimum control can be achieved by this method. Examples for application of manual controls in a water treatment plant are for flocculators, mixers, and certain valves (either direct, by turning a handle, or indirect, by manually manipulating an electric, hydraulic, or pneumatic valve operator). On continuous operating units, such as flocculators, consideration has to be given to whether a unit should restart automatically after a power outage or be restarted manually. Automatic restart after power failure is preferred for operations convenience unless the safety of personnel or process upset could be affected by automatic restart or unless severe power surges are expected to be encountered (multiunit starting). Manual control should be included for semiautomatic, automatic, or supervisory control as a local override for backup, testing, and safety (lockout). An example of semiautomatic control is manual initiation of filter backwash sequence or other batch processes. Automatic controls involve the use of sensors, switches, timers, instruments, controllers, and control logic devices (relays, programmable controllers) to automatically control the process and/or equipment. Examples of this are fully automatic filter operation, plant rate of flow control, chlorination, and sludge handling. Supervisory control refers basically to remote controls, either in-plant or remote from the plant. It can be in the form of manual, semiautomatic, and automatic control and includes the use of feedback information as the basis for a control decision.

Role of Process Monitoring and Control in Total Plant Design

Process monitoring and control play an important part in plant design but an even more important part in plant operations. However, it cannot, as previously mentioned, substitute for correct plant process design. The amount of calculations required and performed are small compared to the amount spent for the hydraulics, structural, and chemical design effort of a typical treatment plant. The design time for the process control and monitoring part of a water treatment plant is in the range of from 2 to 5 percent of the total plant design time. Construction cost is also in the range of 2–5 percent of total cost. This excludes cost of flow meters (primary devices), chlorination equipment, and chemical feeders, which are usually not supplied under process control equipment.

Other Considerations. Standard, field-proven instrumentation and control equipment, suitable for the intended use, should be selected. In addition, the environment in which the equipment has to operate has to be considered, such as process media, ambient temperature, humidity, dust, and vibration. A lab instrument will very likely not perform for a long time when installed inside a damp pipe gallery, an underground manhole, or next to a vibrating engine.

Human engineering refers to the part of engineering where the criteria and decision making during design is governed with the plant operating personnel in mind. Next to the plant's laboratory, no other equipment is nearly as much encountered and operated by personnel on a daily basis as the process control equipment. Therefore, human engineering has two main objectives: (1) to ease the plant operators' tasks and (2) to ease maintenance personnel's tasks. When there is a choice in selecting a process or equipment, the operator's preference should be taken into consideration.

CONTROL STRATEGIES

Before any control is applied to a process, it is important to evaluate and understand the process and to know the requirements of the accuracy to which the process is to be controlled. A frequent problem encountered in the process control of water treatment plants is too fast a reaction of the controls. This is either caused by process noise pickup such as rapid pressure fluctuation or due to the large process lag times caused by the low velocities and large volumes in the basins and filters of a water treatment plant. Fast response times are only required for control applications such as pressure surge suppression on pumping stations where a large amount of kinetic energy is present.

Treatment process automatic controls can be of two basic types. The two basic types are discrete control and continuous control. Discrete control correlates equipment status (on/off) and status changes with a preset value, or program of events. Discrete control may be represented by an automatic start/stop sequence such as high pumping activity based on clearwell level. Or, discrete control may be represented by a logic predefined program of events for a more complex process such as filter backwashing. The operation may be initiated manually by an operator using a pushbutton or automatically by an internal-process-generated event such as filter high lead loss.

Programmable controllers are commonly used to perform discrete control tasks. The PC replaces multiple components (relays and timers) in hard-wired logic networks. To program a PC requires knowledge and experience with the design of multiple component ladder networks. Being programmable means that the PC can accept process-generated data to alter the state of internal circuitry to modify its function.

Continuous control is called continuous because it requires an analog measurement for its input and manipulates a final control element as its output. Continuous control may take the form of feedback, feedforward, floating, proportional (P), proportional plus integral (PI), or proportional plus integral plus derivative (PID). P, I, and D are tuning constants that can be applied to a particular control application for the controller to maintain the desired setpoint. The particular process and measurement lag times and response times determine the tuning constants required.

In continuous process control, there are two types of process variables: (1) the uncontrolled variable (sometimes called the wild variable) and (2) the controlled variable. For example, the flow through the water treatment plant is held constant at a "set point," regardless of the level in the lake, by the use of a controller for the plant influent valve. In this case, the water level is the uncontrolled variable, the plant flow the controlled variable.

In another case, the chemical feed flow should be changed automatically to keep a constant ratio between plant flow and chemical flow (constant 1 milligram per liter dosage). In this control loop, the plant flow is the uncontrolled variable and the chemical flow becomes the controlled variable.

Feedback Control

The classical fundamental function in a control loop is the feedback function. Feedback in a control loop is accomplished by "feeding back" the controlled variable to the controller for the purpose of detecting the difference between the set point (or desired value) and the actual value. This difference is called deviation.

The great value of feedback in the control loop is, as the name implies, the constant checking and rechecking of the actual system output against the desired output value. Therefore it is self-correcting,

taking automatic care of system disturbances and (limited) unknowns within the process loop. The limitation or drawback of feedback control is the fact that the corrections can only be made after a deviation has occurred, and that feedback elements such as meters, transmitters, and transducers are required.

Feedforward Control

Feedforward control is a control strategy whereby the process is controlled without the use of feedback and therefore control action is initiated before a deviation in the process occurs. Although the principle of feedforward control is simple, the word has only come into usage within the last 20 years. An example of feedforward control is utilized for the chemical feeding process in most water treatment plants. The chemical feeder output is controlled in proportion with the plant flow and by a ratio station (either a separate unit or part of the chemical feeder). The ratio station is set manually for dosage level control. The system is used for (a) its simplicity, (b) the fact that more accurate control is not required, and (c) the fact that the time lags in the system are not critical. Feedforward usually is applied to either very simple control problems or to complex but predictable control loops since there is no automatic feedback for corrections.

Two-Position Control

Two-position control is the simplest form of automatic feedback control, although it is usually not referred to as feedback control due to its simplicity. It is also called on–off control, even though two-position control does not necessarily have to be limited to on–off control. This type of control is utilized where close control is not required and where the system capacity is large compared to the energy or volume into and out of the system. An example is pumping on–off control based on reservoir level.

A control band of practically any width can be established with on–off control; the limitations are the capacity of the pump, the allowable rate of on–off cycling, and the rate of system demand. When a pumping station with several pumps is involved, the principle can still be on–off control. However, by setting the control points at different levels for each pump, this on–off control application can approach true proportional control.

Floating Control

Floating control is a variation of on–off control with two additional elements introduced: a second (negative) "on" position and a dead band. This control is applicable where the control action is reversible such as heating–off–cooling for an air-conditioning system or opening–off–closing of a valve. Both of these applications are three-position control, as all floating controls are.

When applied to valve control, the sizing and type of valve, and especially the operating time (time required to move the valve from fully open to fully closed or vice versa) of the valve, have to be evaluated much more closely than for straight on–off control. For the floating control mode, the valve is usually selected for throttling application and the operating time is specified to be of sufficient length to move the valve within its throttling range. Floating control with sampling time is often used to extend the traveling time of the valve effectively and to provide an adjustable range. Compared to on–off control, floating control allows a tighter control (narrower range) and, in most cases, a lesser degree of cycling.

Floating Proportional Control

The floating proportional controller is the on–off control equivalent to the conventional analog controller with proportional and automatic reset feature. The advantages of floating proportional control over straight analog control are the inherent simplicity of on–off control and the fact that there is no change in the control action of the final control element during the time when the control signal is lost. This limits sudden upsets in the process caused by the controls. A limitation is that there is always a dead band present and, therefore, a true set point does not exist. The set point marked on this type of controller is, in effect, the center point of the dead band. In practical applications, these limitations often prove helpful to stabilize the system by ignoring "noise." When the adjustable dead band is set at a very low value, a "set point" exists for all practical purposes.

Proportional Control

Continuous analog control is the classical mode of control and is the most widely used control strategy. It differs from the two-position (on–off) con-

trol and floating control in that it applies a continuous analog signal (current, voltage, air pressure, etc.) to the control element. Therefore, it provides inherently smooth transition from one operating level to the next. Response to a deviation from the set point caused by a change in the process is practically instantaneous without an inherent dead band requirement. The family of these control strategies includes proportional, reset and derivative control. Proportional control is the simplest of the three and it is always used in connection with reset and/or derivative control.

Proportional control is defined as a control mode where the output of the controller is directly proportional to the deviation from the set point. The degree of proportionality is called the proportional band. Proportional band is defined as the percentage of full-scale change in input required to change the output from 0 to 100 percent. Proportional control can best be understood by comparing a controller with a transmitter. A transmitter is a controller with a fixed proportional band of 100 percent and the set point at 50 percent. Its output signal would be the input to a valve operator. With a 100 percent proportional band, the transmitter could also serve as a level indicator. Proportional control is extensively used for filter level control in a filtration plant. In this application, a level transmitter acting as controller is the most secure choice. A process is controlled more precisely when the proportional band is small since a small change in the process variable creates a large change in control action (valve travel). This has led to the expression of "gain" instead of proportional band. Gain is defined as the inverse of proportional band but is expressed as per unit rather than percent. A 50 percent proportional band is therefore equivalent to a gain of 2. The process control will be unstable if the proportional band is set too narrow.

Since proportional control has a band around the set point, it follows that there is only one point of operation where there is no deviation from the set point. This point is at the 50 percent output signal level. For any other output signal level, there is an associated deviation. This deviation is called offset, and it is an important feature of proportional control.

Proportional Plus Reset Control

Proportional control alone cannot eliminate deviation from the set point over the control range of the proportional band. To eliminate this inherent limitation, reset control (also referred to as integral) is added. The reset action of a controller will constantly change the controller output so long as there is a deviation from the set point. The rate of output change is dependent on the amount and duration of the deviation (error signal). Reset rate is expressed either in repeats per minute or in the inverse function, namely, time. A reset rate of two repeats per minute means that every 30 sec the controller repeats the control action required to bring the process back to the set point. The reset control would repeat the action as long as there is an offset. Therefore, if the process is shut down or the final control element does not respond (for example, due to power outage), the reset will drive the output signal off scale. This is called controller "windup." Controllers usually have antiwindup circuits. Reset action should only be added to a control loop when the process needs a wide proportional band for stable operation and when the offset of proportional control cannot be tolerated.

Proportional Plus Reset Plus Derivative Control

Derivative control action is the response of the controller to the rate of change of the input. Derivative action should be applied to slow responding processes since it enhances the speed of response. Derivative action should therefore not be used on fast responding processes (such as flow loops) or where noise problems (e.g., turbulence) exist. Derivative action is sometimes used in the industry on discontinuous systems where shutdown would cause reset windup. In water filtration plants, derivative action is seldom applied.

Accuracies

Accuracy is the degree of conformity of an indicated value to the actual value of the parameter. In the instrumentation and controls field, accuracy includes the influence of dead band, hysteresis, temperature and other ambient conditions, linearity, and any other error. Accuracy should never be considered as equal or similar to repeatability. An instrument can have good repeatability but unacceptable accuracy.

Accuracies are always expressed in percent. Accuracy statements become more nebulous when the measurement is taken indirectly from a differential

pressure producing flow-metering device, as is the case with a venturi tube or an orifice plate. One of the three following statements is usually used to define or to specify the accuracy of a component or system:

1. Percent of range (of maximum process variable).
2. Percent of actual value.
3. Percent of range of (maximum) measured variable (differential pressure in a flow meter).

Figure 23-1 should be referred to for a comparison of these accuracy statements.

System components have to be compatible in accuracy with each other to a certain extent and also suitable for the media or process variable. An example is the measurement of the level of a large reservoir. A measuring element is used that has the ability to detect a level change of 6 inches. The output of this element is the input to a digital indicator

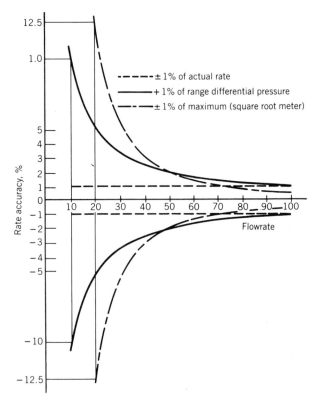

FIGURE 23-1.   Graphical comparison of three numerically equal accuracy statements.

having four digits, with the least significant digit representing $\frac{1}{10}$ inch or 0.1 inch. The last two digits of this indicator will be unstable. The same condition would result if the measuring element is ideal, but the level of the reservoir is unstable due to 6-inch-high ripples.

Also, the designer should be aware that the accuracies of equipment published by manufacturers cannot necessarily be counted on in the field. These published accuracies are derived in the laboratory or on the test strand. It is impractical to expect performance of equipment in the field which is better than 30 or 50 percent of the published guaranteed accuracies.

A measurement or control loop is comprised of several elements, each element having its own accuracy. To arrive at a system accuracy statement, it is first necessary to determine that all component statements are of the same type. A common method to combine accuracies of a series of connected components is to calculate the mean-root-square of the sum of the accuracies.

$$\text{System accuracy} = \pm \sqrt{(\text{acc}_1)^2 + (\text{acc}_2)^2 + (\text{acc}_3)^2 \cdots}$$

This assumes that not all components have the worst accuracy in the same direction, which is not always the case.

INSTRUMENT SYSTEMS

The development and use of instrumentation in the industry dates back to the 1920s and 1930s, when indicators and recorders for the basic process variables of temperature, level, pressure, and flow were developed. The type of instruments available today reflect the advancement of engineering and manufacturing of instruments starting with mechanical devices, then hydraulic, pneumatic, and finally electrical and electronic equipment.

Mechanical

Mechanical instrumentation systems consist of indicators, controllers, and actuators operated directly by the media via floats, pulleys, and levers. Mechanical instrumentation systems are rarely used today. However, some mechanical instrumentation systems can still be found operating today in older water treatment plants.

Hydraulic

With hydraulic instrumentation, the signal carrier can be any fluid. Items to be considered in the selection of the fluid include availability, ambient temperature, corrosion, scaling, viscosity, and toxicity. Oil, meeting most of the requirements for a good signal carrier, is extensively used in the industry. For waterworks applications, water is utilized as much as possible for obvious reasons of availability and media compatibility should leaks occur. Corrosion, scaling, and possible freezing are the main drawbacks when water is used. Fluid, being noncompressible for all practical purposes, provides positive control response with essentially no lag time. It can also serve the dual function of control element.

The basic simplicity of hydraulic systems, and the fact that many times no outside power source is required, will keep this approach to control and instrumentation viable in the future for self-contained equipment. Typical applications are valve controls based on liquid level, pressure or flow, or a combination thereof, and variable-speed drives for pressure or flow control.

Pneumatic

For years, pneumatic instrumentation has been one of the two types which has had almost universal acceptance. For many large users, pneumatic instrumentation is still the standard, even though electronic instrumentation has rapidly gained acceptance in most industries and is replacing pneumatic instrumentation. Pneumatic control system features are summarized in Table 23-1.

Electric

Electric instrumentation was the forerunner of electronic and the counterpart of pneumatic instrumentation. The two types of signals most widely used for these systems are electrical pulse-rate and electrical pulse-duration. In both cases, the signal is created by the opening and closing of a switch and is therefore digital in nature. In the case of the pulse-rate signal, the analog value of the process is expressed is the number of switch actions per time interval, whereas the analog value for the pulse-duration signal is proportional to the time length of

TABLE 23-1. Pneumatic and Electronic Control System Features

Feature	Pneumatic	Electronic
Transmission distance	Limited to a few hundred feet	Practically unlimited
Type of signal	20–100 Pa (3–15 psi) practically universal	Varies; 4–20 mA most common
Control valve capability	Controlled output operates control valve directly	Most control valves require an electronic/pneumatic or electric converter
Compatibility with digital computer or data logger	Pneumatic to electronic converters required for all inputs	Easily arranged with minimum added equipment
Reliability	Superior, if energized with clean dry air	Excellent under usual environmental conditions (less than 95 percent humidity)
Reaction to very low (freezing) temperatures	Inferior, unless air is completely dry	Superior
Reaction to electrical interference	None	No reaction with system if properly installed
Operations in hazardous locations	Completely safe	Intrinsically safe equipment available Equipment must be removed for most maintenance
Reaction to sudden failure of energy supply	Superior—capacity of system provides some safety margin; backup fairly inexpensive	Inferior—electrical failure may disrupt and backup fairly expensive
Dynamic response	Slower but adequate for almost all situations	Excellent—valves frequently become the limiting factor
Operations in corrosive atmosphere	Superior—air supply becomes a purge for most instruments	Inferior unless consideration is given
Measurement of all process variables	Many pneumatic-type measurements have to be converted to electrical	Excellent
Performance of overall control systems	Excellent if transmission distances are reasonable	Excellent—no restriction on transmission distances

each switch closure. Typically, these signals are 0–30 pulses/min, 0–20 pulses/sec for pulse-rate signals, and 0–12 sec for pulse-duration signals.

Electronic

The development of the electronics industry has brought about the development of on-line analytical instruments and the application of computers and microprocessors to plant operations and signal manipulation. Once equipment with temperature limits of up to 80 and even 105°C are readily available, a major limitation for electronic instrument application and source of drift or failure will be eliminated. Electronic control system features are summarized in Table 23-1.

Analog. Today most process signals for electronic instrumentation are direct current, either milliamps or voltage (4–20 mA, 1–5 Volts, are the industry standards). A current signal is immune to the effect of losses (voltage drop) in the line when applied within the limits of the power supply. This signal is also resistant to outside electrical noise pickup, and allows two-wire connection between transmitter and receiver. Since a current loop is a series circuit, all components in the loop have to be connected in series. This provides assurance that all components in the loop are subject to the same signal level. The disadvantage of the series loop is that, if the circuit is interrupted at any point along the loop, all components lose the signal. Therefore, parallel voltage loops are often utilized when several components are within a common enclosure, such as the main control board of a plant. In a parallel-loop configuration, the disconnecting of one unit does not affect others.

Digital. Equipment has been developed over the last 15 years for telemetry systems to utilize digital signals instead of analog. Digital signals are being used more and more for in-plant applications. Microprocessor-based scanners and multiplexers are used to allow cost-effective multiplexing and signal manipulation and to produce compatible digital signals. Digital signals, as used in the waterworks industry, consist of a train of on (1) and off (0) logic pulses forming binary digital signals.

The advantages of the digital data transmission is its consistent accuracy and security, its simple binary nature, and its application for microelectronics. The disadvantages are complex message structure and the inability to accept "graceful" degradation.

PROCESS MEASURING ELEMENTS

No process control can begin without proper measuring or detection devices. The different types of measuring elements are discussed below.

On-line

On-line elements are installed such that they are connected to the process media. This allows continuous measurement of the process variable without interaction of operating personnel. For process control, on-line devices are preferable over off-line. Some measuring elements are by nature on-line devices, such as for pressure, flow, level, and temperature. Others, used for the measurement of density, turbidity, pH, conductivity, chlorine residual, fluoride residual, DO, and TOC, have been developed for on-line applications.

Off-line

Off-line measuring elements are the typical laboratory instruments. They are used to either measure process constituents, for which on-line equipment is not available, or to serve as calibration devices for on-line measuring elements. Off-line equipment is used to detect and/or measure water for taste, odor, hardness, alkalinity, THM, and EPA priority pollutants. The results of the measurement of these constituents are used to change process control such as changing the source of water to the treatment plant or changing dosage control of chemical feeders.

Point Measurement

Point measurement is a noncontinuous type of measurement used mainly on two-position control and alarm applications. It is simple and almost always more accurate than variable measurement. The most frequently used true point measurement units are electrode probes with induction or electronic relays, mechanically actuated limit switches, proximity switches, certain capacitance probes, and some sonic devices. Pressure, flow, and sonic switches are not true point measuring devices since the point detection device is a continuous variable measurement element.

Variable Measurement

Variable measurement is a continuous type of measurement and is the basis for any indication/recording of process variables and for most controllers. It is available for limited applications in mechanical and hydraulic units and practically unlimited applications in pneumatic, electric, and electronic units.

PHYSICAL MEASUREMENTS

Level

The devices first used for measuring water level, the staff gage, dip stick, and sight gage are still the most reliable and accurate. Unfortunately, they provide only a local indication, and therefore different means of measuring level have been developed for the transmission of a pneumatic or electric signal for remote indication. The most commonly used methods of level measurement are described below.

Float Type. The method commonly used in the past to measure water level in a tank or reservoir is the float-operated level indicator and/or transmitter. It uses a float connected by cable to a counterweight device to detect water level. As the float rises or falls with water level, the float cable winds or unwinds on a drum that is mechanically connected to the transmitter. The mechanical motion is converted to either a pneumatic signal (3–15 psi), to an electric signal (4–20 mA), or to a pulse-duration or pulse-frequency signal for level indication or telemetering. The advantage of the float-actuated level measuring device is its simplicity. Disadvantages are:

1. It must be mounted over the water level to be measured and, generally, a stilling well must be provided.
2. In cold climates, where ice may form on the surface of the water, a float is not practical.
3. Mechanical maintenance must be provided to ensure relatively friction-free motion of the float and cable assembly.

Pressure-actuated Level Transmitter Type. A pressure transducer connected to measure the water pressure at the base of a tank will read water level directly if calibrated in centimeters (or meters) of water. If the pressure transducer is connected below the tank, zero suppression must be used to measure level directly. Where the level in a pressurized tank is to be measured, a differential pressure transducer must be used with the low-pressure tap connected to the top of the tank.

Bubbler Unit Type. Since many reservoirs or basins are buried underground or located such that a direct pressure connection near the bottom is difficult or impossible to manage, an alternate method of measuring head, called a bubbler system, is often used. A small tube is inserted from the top of the tank or reservoir to within 2 or 3 in. of the bottom. Gas, usually air, at a low flowrate is forced through the tube, producing bubbles and generating back pressure equal to the hydrostatic head. This pressure can then be measured by a conveniently located pressure transducer. The bubbler system has the disadvantage that a source of pressurized air (either from a compressed air tank or from a tank and compressor unit) must be supplied and more mechanical components are required. It has the advantages of simple installation and of not introducing the process liquid into the pressure-measuring instrument, which may be located anywhere near the tank or reservoir. In addition, bubbler systems can be used to measure the level in the elevated tanks whose height demands zero suppression beyond the capability of pressure transducers.

Pressure Bulb Type. The pressure bulb, which consists of a sealed cavity covered by a neoprene diaphragm, can be installed in the liquid at or near the bottom of the tank or reservoir. The pressure, proportional to the height of water above the bulb, is sensed by the bulb and transmitted to the pressure-measuring element through an air-filled pressure-tight tube. Because small tubing 3 or 1.5 mm inside diameter ($\frac{1}{8}$ or $\frac{1}{16}$ in.) must be used, response time is slow, particularly where long tubing runs up to 100 m (300 ft) must be used.

Sonic Type. Sonic-type level devices measure the height of the water by transmitting a sonic pulse that is reflected from the water surface. The time between the transmission of the pulse and its reception is a measure of the distance of the transponder (the device transmitting and receiving sound pulses) from the liquid surface. Since the transponder is mounted above the water surface, typically over a Parshall flume or a weir, this distance is inversely related to water level. A 4–20-mA output can be calibrated directly in water level. Since the speed of sound in air increases with temperature (0.18 percent per °C), temperature correction is required to obtain accurate measurement. With temperature correction, an accuracy of 1.0 percent of full scale is claimed.

Pressure

Pressure is defined as a force per unit area. There are basically two methods of measuring pressure. The first method, used in the manometer, balances the gravitational force or weight of a liquid head against the pressure to be measured. The second method measures the distortion in a metal caused by the strain induced by the pressure to be measured. Most commercial pressure gages or transmitters use one or the other of these two basic principles.

Pressure, as measured, can either be absolute (measured with reference to the pressure of absolute vacuum, which is zero), gage (measured with reference to the ambient atmospheric pressure), or differential (where the difference between two pressures is measured). All pressure transducers can be considered as differential pressure transducers, since, in the case of gage pressure, the second pressure is atmospheric pressure, and in the case of absolute pressure, the second pressure is a sealed, evacuated chamber.

When very low ambient pressure is to be measured, methods other than those mentioned above must be used. However, pressure measurements in the close to absolute vacuum range are not required in the water industry, and therefore no details will be given on these devices.

Manometers are usually made from a U-shaped

glass tube. Pressure can be read directly from gradations etched on the side of the tube. To obtain a signal capable of being transmitted to an indicator or recorder, mercury (for its high specific gravity) and a float are used to detect the change in level. Although popular in the past, the high cost of mercury and the improvements made in other methods of pressure measurement have led to manometers being replaced by other types of pressure sensors in most cases.

Almost all pressure and differential pressure transducers used in the water industry rely on the distortion in a metal exposed to pressure to measure that pressure. The sensing elements include standard bourdon tubes (C shaped), twisted, helical, and spiral bourdon tubes, pressure capsules, diaphragms, and bellows. In each case, the pressure introduced into the bourdon tube, bellows, or to one side of the diaphragm causes a small motion proportional to the pressure introduced. In standard, non-transmitting pressure gages, this motion is translated into the motion of a pointer on a scale. In a pressure transmitter, this motion is converted into a pneumatic or electric signal for transmission to a recorder or controller. Over the last few years, strain gages and wire resonance-type units have been used more for pressure transmitters. Strain gage pressure transmitters are available with stated accuracies of 0.25 percent of span and zero shift of less than 0.8 percent of upper range limit. Other units have stated accuracies of 0.5–1 percent of range. Ranges from 0 to 3 in. of water and 0–10,000 psi can be supplied.

In selecting a pressure transducer for a specific application, several factors must be considered, such as process liquid (e.g., corrosivity), range, span, and zero suppression–elevation. To protect the pressure element from corrosive process fluid, temperature, and/or pressure surges, a sealed system is used. This consists of a diaphragm unit and a sealing fluid. The sealing fluid fills the area between the diaphragm and the pressure-sensing element.

Flow

Of the three primary measurements, level, pressure, and flow, commonly required in the water industry, flow is usually the most difficult and expensive to measure. Many different ways of measuring flow are available and each has its advantages and limitations. A summary of the different types of flow-measuring devices is given below, together with some of the advantages and drawbacks of each method.

Differential Pressure or Loss of Head Meters.   Probably the most widely used method of flow measurement both for liquids and for gasses is a differential-pressure-producing element. Many types are available, such as orifice plate, venturi tube, and modified venturi tube (insert tube). The differential pressure output signal, which is the decrease in pressure resulting from increased velocity due to an area restriction, is proportional to the square of the velocity or flowrate. The basic formula for the loss of head type of flow meter is

$$Q = K\sqrt{h/\rho}$$

where Q is the flowrate, K is a constant typical for the meter, h is the head loss, and ρ is the density of the flowing liquid or gas. Since fresh water is an incompressible fluid of constant known density, the formula reduces to $Q = k\sqrt{h}$. Where venturis, orifices, or insert tubes are used, the range of flow must be restricted to 3:1 or, at most, 4:1 due to this nonlinearity. At one-third full-scale flow, assuming a 1 percent basic accuracy in the differential pressure element, the actual flowrate can be measured to an accuracy of ±5 percent.

Where head loss is not an important factor, a simple orifice plate in the line will provide the differential pressure drop. Orifice plates are universally used in gas flow measurement, but the high and unrecoverable energy loss (40–80 percent) limits their usefulness in water systems. In water systems, the more efficient venturi-type meters (5–20 percent energy loss) are commonly used. This saving in energy is made at a considerable increase in cost, size, and inflexibility. Flow tubes are used when savings in cost of pumping equipment and energy justify the higher capital investment.

One of the major advantages of the head loss type of meter is that the primary element, the venturi tube, has no moving parts and requires practically no maintenance. Once calibrated, its flow constant will not change, and it is not easily damaged by foreign objects in the pipe.

Positive-Displacement Meters.   Positive-displacement (PD) meters operate by trapping a known volume of fluid and passing this known volume from inlet to outlet. The number of known volumes passed in a given time is proportional to the flowrate. The most common use of displacement meters in the water industry is the ordinary household water meter, which operates on the rotating disk principle. Other uses of PD meters include the metering of dosages of chemicals in water treatment (piston type).

A PD meter is extremely accurate provided it is properly maintained. The rangeability of PD meters is very large, especially with high-viscosity liquids where leakage around the meter or slippage is very low. Most PD meters are used in measuring liquids such as fuel oils, which have a relatively high commercial value compared to water, and where extremely accurate metering is important.

Disadvantages of these meters are: the close tolerances required by the moving parts, making regular maintenance and service essential; the high cost, particularly in large diameters; limited throughput for a given size; and damage caused by overspeeding, that is, flowrates in excess of the design rate.

Magnetic Flow Meters.   Magnetic flow meters operate on the principle that the electromotive force introduced in a conductor moving through a magnetic field is proportional to the velocity of the conductor. In the case of magnetic flow meters, the fluid in the pipe acts as the conductor. A magnetic field is produced across the tube in which the liquid is flowing by exciting coils mounted around the pipe. The electromotive force induced by the flowing liquid, which is proportional to the velocity, is picked up by two small electrodes mounted on each side of the pipe.

Because the electrodes are mounted flush with the inside of the pipe, there is no obstruction of flow, and thus no head loss. Also because of the obstructionless nature of the meter, slurries, pulps, and liquids containing solids can be handled effectively.

One of the disadvantages of magnetic flow meters is their high cost, but since the cost does not increase as much with size as with venturi tubes, the larger sizes may be cost-effective. Of course, magnetic flow meters cannot be used with gasses or nonconductive liquids (conductivity should be greater than 5 μmho/cm for best operation, but magnetic flow meters can be used where conductivity is as low as 0.1 μmho/cm). Calibration is required and drift may be a problem. Where fouling of electrodes is a problem, ultrasonic electrode cleaning can be provided. Accuracy of magnetic flow meters is good, typically as high as ±0.5 percent of

measured flowrate for the range of 1.2–7.6 m/sec (4–25 fps) to ±2 percent in the 0.3–0.6-m/sec (1–2-fps) range. Rangeability is on the order of 10:1.

Propeller and Turbine Meters.   Propeller or turbine meters consist basically of a bladed rotor mounted in a pipe and driven by the flowing liquid. The rate of rotation of the propeller or turbine is coupled mechanically or magnetically to a pickup which produces a pulse or pulses for each rotation. This pulse train can be counted electrically to indicate total flow and converted to a 4–20-mA signal to indicate flowrate.

Propeller and turbine meters have good accuracy (±0.35 percent to ±2 percent of measured flowrate is typical) and good rangeability (approximately 10:1) and can be used to measure flow in the range of from 1 to 10 m/sec (3–30 fps). Some head loss results but is not high (typically 0.5–1.5 velocity heads).

The disadvantages of these meters are that they are subject to wear of moving parts and permanent damage caused by overspeeding. Particles in the flow stream can also damage the rotor blades or bearings. Screens commonly are used to prevent this problem from occurring. The meters are also sensitive to turbulent flow of swirl in the flow stream. With straightening vanes, integral to the meter or installed separately, required lengths of unobstructed piping upstream and downstream can be reduced. Turbine meters are relatively expensive, particularly in small sizes.

Ultrasonic Flow Meters.   Ultrasonic flow meters are based on the principle that sound travels at a constant speed in a liquid. When a sonic pulse is directed downstream in a flowing liquid, the speed of liquid flow is added to the constant speed of propagation of sound. When the sonic pulse is directed upstream, the speed of liquid flow is subtracted from the speed of sound. The ultrasonic flow meter effectively subtracts the speed of transmission upstream from the speed of transmission downstream to obtain a result proportional to the velocity of the flowing liquid. The associated amplifier and converter converts the flow signal, usually a frequency or time difference, to a standard 4–20-mA signal proportional to flowrate.

Ultrasonic flow meters are obstructionless and can be provided with accuracies of ±1 percent of measured flowrate over a range of 10–100 percent of full-scale flow. Rangeability is higher than either magnetic or turbine flow meters [flows from 0.8 to 9 m/sec (0.25–30 fps) can be measured]. The ability to strap the transducers to the outside of the pipe makes installation simple and inexpensive; however, reliability suffers greatly. Although expensive for small pipe sizes, the ultrasonic flow meter can be utilized for pipe sizes as large as 6 m (20 ft) in diameter, and since the cost is not dependent on pipe size, this type of meter is economical for large pipe sizes.

The disadvantage of this meter is sensitivity to changes in the fluid composition (such as the percentage of particulates) and other variables, such as air, which can distort or prohibit the propagation of sound waves. The fluid flow profile must also be considered. Although the output curves are linear with flow within specified limits, there are two distinct slopes for turbulent and laminar flow conditions. Consequently, unnecessary errors can be introduced by improper location of the transducers in close proximity to valves, pipe bends, or where fluids pass through a laminar to turbulent transition zone.

Vortex Meters. Vortex meters are based on a natural phenomenon known as vortex shedding. When a flowing liquid flows past an unstreamlined or bluff body, vortices are shed from alternate sides of the bluff body. The frequency at which these vortices are shed is proportional to the flow velocity of the liquid. Small probes mounted on either side of the bluff body measure the pressure changes produced by the vortex shedding phenomenon, and these signals are amplified and converted to a standard 4–20-mA signal proportional to flowrate.

The stated accuracy of a vortex meter is within ±1 percent of flowrate. Rangeability varies but is typically 12 : 1. The head loss is typically two velocity heads. Cost is moderate but it is not usable with viscous, dirty, abrasive, or corrosive fluids.

Variable-Area Meters (Rotameters). Rotameters utilize the same principle as differential pressure meters, that is, the relationship between kinetic energy and potential (or pressure) energy. In the variable-area meter (rotameter), the differential pressure is constant while the restriction area varies with flow. The typical rotameter consists of a glass tube tapered in inside diameter with a float inside the tube. Fluid flows through the tube from bottom to top carrying the float upward. As the float rises, the annular area between float and tube increases

until the lifting force produced by the differential pressure between upper and lower float faces balances the effective weight of the float. The equilibrium position reached by the float gives a direct linear indication of the flowrate.

In its simplest form, the glass tube of the rotameter is inscribed with a scale so that flow may be read directly. Where high pressures are involved, the tube can be made of stainless steel or other alloy, and the float position indicated by an extension with an indicating scale. Where an electronic signal is required, a converter can be magnetically linked to the float (or the float can drive the core of a linear differential transformer) and a standard 4–20-mA DC output is produced.

Rotameters have a typical accuracy of ±2 percent of maximum flow and a rangeability of 10 : 1. The main advantage of a rotameter is that it can be used for very low flowrates, as low as 1 mL/min full scale for water (approximately 60 mL/min for air).

Disadvantages are that rotameters must be mounted vertically, are sensitive to changes in fluid density, and are limited in maximum flowrate measurable, which, for the largest sizes, is 20 L/min (5 gpm) of water. For low flows, especially where indication only is required, they are relatively inexpensive. In water industry instrumentation, rotameters are consistently used in measuring the rate of air flow in bubbler systems and sometimes in the measurement of chemical solution flow.

Flumes and Weirs. For very large flows, the most practical method of measuring flow is with a flume (typically a Parshall flume) or weir. In either case, the rate of flow is a function of the water level at a specific point in the flume or near the weir. The water level can be measured by a float or bubbler system installed in a stilling well beside the weir or flume, by a specially designed float riding on the surface of the water, or by a sonic-type level meter mounted above the weir or flume.

Standard types of weirs include the V-notch weir ($22\frac{1}{2}°$, $45°$, $60°$, and $90°$), the rectangular weir, and the Cipolletti weir. The relationship between height and flow is nonlinear, and special characterizing cams or electronic function modules are usually used to produce an output that is linear-proportional to flow. Because the head loss in weirs is approximately four times as great as the head loss in a Parshall flume, Parshall or Palmer-Bowles flumes are more commonly used, even though they are more expensive to construct. In addition, the

Parshall flume is self-scouring so sludge and silt cannot build up to destroy their accuracy, as with weirs.

Parshall flumes have been constructed with throat widths of from 25 mm to 12 m (1 in. to 40 ft) for measuring flows up to 5.5 mL/d (1.5 mgd). Tables have been prepared for discharge rate versus head for Parshall flumes with throat widths from 25 mm to 3 m (1 in. to 10 ft). However, under normal field conditions, the correlation between level and flow for weirs and flumes has a basic error of at least 3–5 percent. The rangeability of a Parshall flume is probably on the order of 20:1.

Miscellaneous Flow-measuring Techniques. Many other methods of measuring flow have been developed, but most have little or no application in the water industry. Some of these devices include thermal flow meters and the target flow meter.

The most commonly used point flow device is the pitometer. With this instrument, the flow at a point is measured by the differential pressure created across two small orifices, one of which faces into the flowing stream and one of which faces in the opposite direction. Pitometers are used to obtain flow profiles across the diameter of pipes and may also be used to obtain spot checks of approximate flowrates in, for example, distribution systems, where accuracy of total flow is not important and where the small size of the pitot tube makes insertion in to the pipe easy. A variation of the pitot tube known as the Annubar attempts to solve the problem of the pitot tube by measuring the point flow at four points on the diameter of the pipe in which the liquid is flowing. The differential pressure at these four points across the flow profile is averaged by the Annubar to produce a differential pressure proportional to the flowrate. The accuracy is approximately ±2 percent. Since the Annubar is a head-producing device (the differential pressure output is proportional to the square of the flowrate), the rangeability is limited to approximately 4:1. Head loss is approximately 8 percent of differential. With flow tubes, venturis, and so on, care must be taken in choosing the location point to avoid turbulent flow conditions. Two major advantages of the Annubar flow-measuring device are its low cost and its ease of installation, even in lines under pressure.

Filter Head Loss

The successful design and operation of the water filtration plant is often evaluated on the basis of the length of filter runs. Since the length of the filter run is determined by the maximum allowable head loss across the filter, it can easily be seen that filter head loss measurement is an important function in the instrumentation and control of the treatment plant. As the name implies, the filter head loss measurement is, in principle, straightforward and is accomplished either by a level or a differential pressure meter. Principally, the same transducer can be used for filter head loss as can be used for liquid-level or differential pressure flow-measuring devices.

In filter designs where there is no modulating filter effluent valve, which is the case in low-head self-backwashing filters, head loss is reflected in a water level above the filter media. Therefore, head loss measurement will not be discussed further here. The simplicity of controls for self-backwashing filters has contributed to the increased popularity of this type of filter.

Following is a detailed description of the differential pressure type of head loss measurement as is utilized in the more traditional filter design with sand or dual-media filter beds. With this method, the level above the filter is held relatively constant and head loss is offset by the modulating action of the filter effluent valve, thereby keeping a constant overall head loss between the top of the filter and the downstream side of the filter effluent valve. Even though the measuring device utilized is straightforward, the application is somewhat unique and the designer or instrumentation technician often has a problem visualizing the application and important aspects of this type of measurement. The differential pressure type (DPT) of head loss measurement has additional limitations when one side of the measurement, namely the measurement of the filter level above the filter, is also utilized for the control of the filter effluent valve. The rangeability of the instruments used determines the limitation of the correct application. The pressure below the filter is measured by direct connection to the filter effluent piping or the filter drain piping. Care must be taken to ensure that the location of the tap into the pipe is not at the point where the water flow influences or prevents the correct measurement of static pressure.

Figures 23-2 and 23-3 show two of several arrangements of filter head loss and filter level control measurement. The 300-mm (12-in.) level control band shown is typical; the 5000-mm (200-in.) dimension shown between the filter level and the location of instruments is only applicable to filters with deep

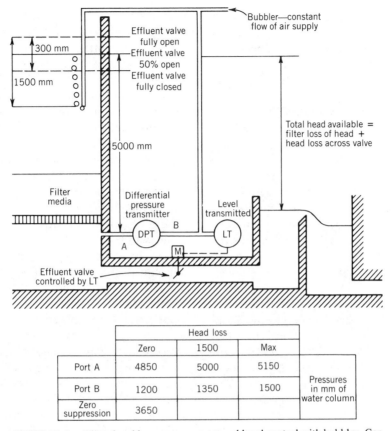

	Head loss			
	Zero	1500	Max	
Port A	4850	5000	5150	
Port B	1200	1350	1500	Pressures in mm of water column
Zero suppression	3650			

FIGURE 23-2. Filter head loss measurement and level control with bubbler. Conditions shown when effluent valve is half open and filter loss is half of maximum.

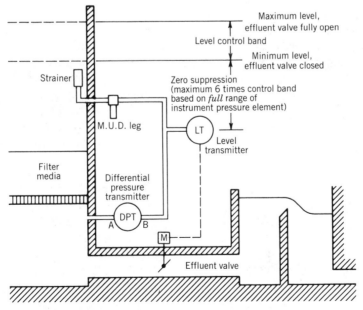

FIGURE 23-3. Filter head loss measurement and level control with sediment trap and strainer.

filter beds, but it was chosen to illustrate the limitations of the instrumentation more vividly. When the bubbler arrangement is used, the submergence of the bubbler tube [shown in Figure 23-2 as 1500 mm (60 in.)] has to be balanced between the limitations of the zero suppression capability of the head loss instrument and the zero suppression capability of the filter level instrument. For the arrangement shown in Figure 23-2, the zero suppression limitation only applies to the level measurement. Most instruments have a zero suppression limitation of 5 : 1. This means that for an instrument with a span of 6000 mm (240 in.), the maximum suppression could be 5000 mm (200 in.). It has to be remembered that this limitation is based on the range of the pressure capsule installed in the instrument. For example, for a control span (control band) of 300 mm (12 in.), a pressure capsule of 1500 mm (60 in.) rating or less would have to be used to stay within the 5 : 1 range of zero suppression for level control. The table in Figure 23-2 shows the pressures to which ports A and B of the differential pressure transmitter are subjected to at various stages of filter operation. The easiest way to accomplish the correct zero suppression adjustment for the example shown is to close the effluent valve, thereby establishing zero head loss and adjust the zero suppression until the transmitter output indicates zero differential (4 mA on a 4–20-mA instrument). No zero suppression is required when the "wet" connection is used to the differential pressure transmitter as shown in Figure 23-3. The disadvantage of the wet connection is the possibility of dirt plugging one side of the differential pressure transmitter. Obviously, no zero suppression is required when a float-operated level transmitter is utilized.

Turbidity

The measurement and monitoring of the turbidity of the water is another major "yard stick" to measure the effectiveness of operation of a water treatment plant. Although the word *turbid* is clearly defined, (turbid: having sediment or foreign particles stirred up or suspended; muddy; cloudy; turbid water), the units of turbidity measurements have been changed many times over the years. Since the 1960s, the official name for turbidity units has been changed from Jackson turbidity unit (JTU) to Formazin turbidity unit (FTU) to simply turbidity unit (TU), to nephelometric turbidity unit (NTU). Turbidity units have changed over the years because the measurement of turbidity using different criteria results in different turbidity values depending on the type of suspended matter (particles) in the media. The sizes, indices, absorption coefficients, and viewing angles of particles influence the reading of the particular design of a turbidity meter.

The principle of operation of turbidity meters is to detect and measure the scattered portion of a light beam, or the portion of a light beam that has not been absorbed by the turbid sample through which the light beam has passed. Most of the turbidity meters installed to date in the water industry are of the innate scatter type where the photocell measures the "perpendicular" portion of the scattered light, which is at a 90° angle from the light source beam. However, other types of meters have come on the market measuring the "forward" or "reverse" portion of the scattered light. Most of the turbidity meters manufactured today include automatic means to compensate for the variation in the light source intensity (fluctuations are due to supply voltage and aging of lamp). Figure 23-4 shows three of the most common types of turbidity meters. The innate-scatter-type meter is designed for low turbidities of up to 30 NTU. It has a fairly linear direct response, zero signal to zero turbidity, and dissolved color does not register as turbidity but can cause negative errors. This type of meter should not be used where high turbidity levels can be expected, since the meter goes "blind" at high turbidity levels and therefore will register low turbidity values. The surface scatter type is used for a great range of turbidities. It does not go "blind" at high turbidities, but it has to be installed completely vibration-free, otherwise false (high) readings will result. The response is nonlinear. The absorption type is suitable for high turbidities. It has maximum output at zero turbidity (reverse response), response is nonlinear, and colors will register as turbidity.

Successful operation of most turbidity meters requires an air-bubble-free sample. Where there is a possibility of the presence of air bubbles in the sensing line, an air bubble trap should be installed ahead of the meter. The sample required for the various types ranges from 0.5 to 1.5 L/min (0.13–0.4 gpm) for the low range type to 3–6 L/min (0.3–0.6 gpm) for the absorption type.

Temperature

Temperature has a great effect on some processes and equipment operation. Therefore, it is measured

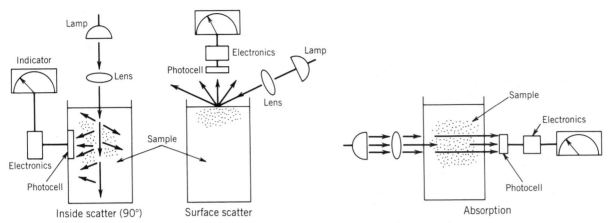

FIGURE 23-4. Types of turbidity meters.

to monitor and/or control and/or equipment conditions. The most common methods of temperature measurement include thermocouples, filled (bulb) systems, resistance elements, bimetallic units, and thermistors. Pyrometers, thermagraphics, acoustical thermometry, and quartz crystal thermometers are other methods of temperature measurement, but are not used for process control in water treatment plants.

The thermocouple operates on the principle that an electromotive force is generated when heat is applied to one of the junctions of two dissimilar metals. The electromotive force is proportional to the temperature difference of the two junctions. The thermocouple is usually inserted into a thermowell, which in turn is mounted within the process fluid. This provides protection for the couple and convenience of removal without disturbing the process. Thermocouples are available for temperatures from −150 to 1300°C and with an accuracy of ±1 percent of full span.

The filled system or thermal bulb operates on the principle that the absolute pressure of a gas is proportional to the absolute temperature. A temperature-sensitive bulb is immersed in the medium to be measured and the bulb is connected to a pressure-actuated device via capillary tubing. Compensation is often needed to correct for ambient variations along the capillary and/or the inside of the pressure-actuated device (case compensation). The filled systems are available as liquid, vapor, or gas type with temperature ranges of −90–300°C, −240–250°C, and −220–800°C, with an accuracy of ±1 percent of full span.

The resistance temperature detector (RTD) is based on the fact that most electrical conductors change their resistance with a change in temperature. Resistance materials include platinum, nickel, copper, nickel–iron, and tungsten. These thermometers are normally operated as one leg of a Wheatstone bridge. Similarly, RTDs are for applications that require the average temperature of an area rather than a point and where high accuracy, stability, and good sensitivity are required, such as for bearing and winding temperature measurements on large electric motors and generators. Typical resistance values of these units range from 10 to 500 Ω, with some units available up to 1200 Ω. Accuracies are within ±0.5°C or ±0.5 percent measured temperature.

Bimetallic thermometers are used extensively for local temperature indication and on–off controls such as thermostats. They operate on the principle that metals expand or contract with temperature changes and that the coefficient of change is not the same for all metals. Two dissimilar metals are bonded together, causing the bimetal unit to move under the influence of temperature changes. Accuracies are from ±0.5 to ±2 percent of temperature.

Thermistors are semiconductors made from carbon, germanium, silicon, and mixtures of certain metallic oxides that exhibit high temperature coefficients. In addition to temperature devices, thermistors are used for voltage regulation. Large changes in resistance allow long lead wires and their small sizes (2.5 mm typically) are often advantages. Their temperature–resistance characteristic is nonlinear, but they are good for narrow measurement spans. Accuracies of ±0.15°C can be achieved.

Particle Size

Particle size analysis has been performed by water quality personnel using a light microscope. Over the last several years, other methods have been developed to produce commercially available on-line instruments for particle counting and size analysis. At present, three principal methods are used. They are electrical zone sensing, light scattering, and light interruption. Electrical zone sensing uses a measuring cell with test tube, electrodes with electrolyte, and an orifice. The change of resistance between the electrodes due to the volume of electrolyte being displaced by the particulate is measured. The light-scattering method uses either a laser beam or a beam of white light. The light interruption method uses the measurement of the magnitude of the interruption of the light by the particle. This magnitude is directly proportional to the cross-sectional area of the particle.

Although on-line particle size analyzers have been installed in water treatment plants, the analyzers are still in the development stage. Requirements for sample preparation and measurement response time are two important aspects. Particle counting and size measuring, once fully developed, will be a more meaningful method for filter breakthrough.

CHEMICAL MEASUREMENTS

pH

pH is defined as the negative logarithm of the hydrogen ion concentration or activity, and it covers the acidity and alkalinity of solutions over a range of 0–14. pH instruments require three electrodes: a pH-sensitive electrode, a reference electrode, and a temperature-compensating electrode. The pH-sensitive electrode develops an electrical potential based on the hydrogen ion activity. The reference electrode is required to measure the ambient electrical potential present. To compensate for this, a temperature-compensating electrode is installed to offset the influence of temperature. These three electrodes are usually mounted in an electrode chamber for installation in a sample line or submerged in a tank, pond, or channel.

Modern pH-metering systems are often provided with signal preamplifiers installed within the electrode assembly to minimize electrical noise pickup in the high-impedance electrode circuits. Also, ref-

erence electrodes are now available for most applications, which eliminates the need for electrolyte needed in conventional reference electrodes of the flowing junction type. The accuracy of pH analyzers are in the range of 0.1 pH ($\pm$0.05).

Ion-selective Electrodes

Ion-selective electrodes are the result of further development of the ion electrodes for pH and DO measurement. These new electrodes make possible measurements of ion activity and can be classified into solid-state, liquid ion exchange, and gas-responsive electrode systems. A list of available electrodes for water treatment plant application is given in Table 23-2.

Conductivity

The increasing importance placed on water quality and the greater utilization of water treatment processes such as demineralization, desalinization, and water softening has increased the need for conductivity meters to monitor and control the purity of the process stream.

Electrical conductivity is a measure of the ability of a material to allow electric current to flow. It is defined as the reciprocal of resistivity. The present international unit for conductance is siemens (S), replacing the designation mho. Conductivity is specific conductance, (in siemens per centimeter). Typical conductivities of various water samples are: pure water, 4×10^{-8}; distilled water, 10^{-6}; raw water, 10^{-4}; and seawater, 0.33 S/cm.

Microsiemens per centimeter is the commonly used dimension for the industrial and waterworks field. Measurement of the conductivity is usually accomplished with alternating current (60–100 Hz) utilizing an AC Wheatstone bridge arrangement or

TABLE 23-2.   Ion-Selective Electrodes

Electrode	Concentration Range (Molarity)	Temperature Ranges (°C)
Bromide	5×10^{-6}–1	−5–100
Calcium	10^{-5}–1	−5–50
Copper (cupric)	10^{-7}–1	−5–100
Fluoride	10^{-6}–1	−5–100
Potassium	10^{-7}–1	−5–40
Water hardness	10^{-6}–1	−5–40

other similar circuits for direct reading meters, in conjunction with a conductivity cell. The conductivity cell consists essentially of metal plates or electrodes positioned in an insulated body structured to allow the measurement of the fluid sample. Electrodeless sensors are available for the measurement of the conductivity of solutions where abrasion, clogging, or corrosion of the cell can create a service and maintenance problem. The electrodeless measurement utilizes two toroid coils, with the solution to be measured acting as the inductive coupling for the two coils. The strength of the electrical coupling is proportional to the conductivity of the solution.

Conductivity measuring requires corrections for temperature variation, since the conductivity of a solution changes with temperature. The temperature coefficient varies nonlinearly with the temperature and is different for different solutions. Most on-line process conductivity meters include automatic temperature compensation. The type of solution and operating temperature range have to be specified for the proper selection of the type of conductivity cell because conductivity cells are designed with dimensions and for electrical frequency of the power supply to bring the measured resistance and impedances of the cell lead wires and analyzer within an optimum range. Mechanically, the unit has limitations for pressure, temperature, flow velocity, and the presence of solids. Chemical resistance considerations have to include the electrical potentials which will be imposed on the electrode material in addition to the chemistry of the solution to be measured.

Accuracies for these analytical instruments are in the range of ±1 percent of full scale and stability of ±1 percent per month.

Chlorine Residual

Of all the chemicals used in water treatment, chlorine is practically a standard ingredient, utilized for oxidation but primarily used for disinfection. Most chemical controls are feed-forward without direct feedback. The residual chlorine analyzer allows closed-loop feedback control for chlorine, in addition to its role as indicator–recorder of the adequacy or excess of chlorination. On-line residual chlorine analyzers operate either on the amperometric or the polarographic sensing principle.

The amperometric unit receives the process fluid through a sample line. The flow sample is filtered, a thermistor determines the temperature, a reagent is added by a feed assembly to maintain constant pH, and the buffered sample flows through the gold/copper electrode cell where a continuous electric current is produced which is linear–proportional to the chlorine content contained in the sample. Either free residual (potassium bromide addition to buffer) or total residual (potassium iodide and sodium hydrozide addition to buffer) chlorine concentration can be measured. Interference with the measurement can be caused when manganese, iron, nitrite, chromate, and other substances are in the sample. Also, a filter or strainer is recommended when settleable solids are present in the sample.

The passive polarographic-type sensor probe houses the gold/silver electrodes, the electrolyte, and the thermistor, which are protected from the process flow by a gas-permeable membrane. When chlorine permeates the membrane, it causes an electric current linear–proportional to the chlorine concentration. This current is fed into a remote (up to 300 m or 1000 ft) analyzer–transmitter. The sensor can be installed directly in the process stream (up to 100 m or 300 ft) submergence, or equivalent pressure, but a velocity of 0.3 m/sec (1.0 ft/sec) across the membrane is required, otherwise a cleaner–agitator attachment must be supplied. Sensors are available for either free chlorine or total residual chlorine measurement. A wet chemistry rather than a stand-alone sensor should be utilized if (1) total free chlorine is to be measured and pH of sample is between 5 and 7 and is fluctuating more than 0.2 pH units or pH is 7 or higher and (2) total residual chlorine is to be measured in seawater or in wastewater with ammonia content less than four times the chlorine concentrations. The wet chemistry unit consists of an enclosure housing a sensor, a reagent container (buffer acetate for total free chlorine or buffered potassium iodide for total residual chlorine), pumps for sample and reagent mixing, and sensor face cleaning.

The chlorine residual analyzers are available for operating ranges of from 0–0.5 to 0–20 mg/L, ambient and sample temperature limits of 0–50°C, and stated accuracy of 2–3 percent of full-scale measurement.

Dissolved Oxygen

Dissolved oxygen is an important parameter for the evaluation of the quality of surface waters. Sufficient DO in the water is necessary for the direct oxidation of pollutants, for the life of microorganisms which biologically stabilize organic pollution, and for the survival of aquatic life. The location of the DO probe in the process stream and the type of process (river water, aerated sludge, pure oxygen treated sludge) is important for meaningful measurement of DO and for equipment selection. Several electrode systems are popular, such as gold–silver, gold–copper, platinum–lead, silver–zinc, silver–iron, and silver–lead, operating as passive or active electrodes forming a polarographic or galvanic cell.

In the selection of DO systems, special attention has to be placed on service requirements, that is, stability (drift) of measurement, and requirements for agitation devices. Almost each manufacturer has a patent (or patent pending) for special aspects of its analyzer, and claims of performance features often conflict with tests made by users during online operation of this equipment.

Published performance features vary widely. For instance, measuring ranges can vary from 0–5 to 18–119; response times from 90 percent in 20 sec at constant temperature to 90 percent in 7 min; accuracy from 0.1 to 5 percent of full scale; operating temperature limits from -29 to 70°C; temperature compensation from -30 to 70°C; pressure ranges (submergence) from 7.5 to 200 m (656 ft); stability from ± 1 percent/24 hr to ± 0.25 percent/month; minimum velocity of process from 0.1 to 0.5 m/sec (0.4–20 in./sec); and maintenance from once a week to practically never.

Total Organic Carbon

Organic compounds in water or wastewater supply can result in objectionable taste, color, or odor and can cause toxicity. The TOC analyzer was developed to supplement and/or replace biochemical oxygen demand (BOD) and chemical oxygen demand (COD) instruments. The TOC analyzer of today is a continuous-flow on-line monitor. Some analyzers are based on removal of inorganic carbon from the sample stream, oxidizing the carbon to CO_2, which is then measured by titration, manometrically, or in an infrared spectrophotometer. A limitation of these units is the requirement of sample filtering, which can result in serious errors. Some newly developed analyzers permit the measurement of unfiltered samples by using a process whereby a nephelometric detector is utilized instead of an infrared spectrophotometer. These units can measure total

carbon (TC) by direct combustion of the entire sample, inorganic carbon (IC), and organic carbon (OC) by performing a sequence of sparging and separate analyses. The analyzers are available to measure ranges of 0–10 up to 0–5000 mg/L carbon.

Density

The principal use of density measurement in water treatment plants is for the control of sludge removal from settling basins. Several principles have been utilized to develop density measurement instruments for liquids. These principles include buoyancy (displacement-type units), mass weight (U-tube and balance beam type), vibration amplitude (U-tube caused to vibrate by a pulsating force and measurement of the amplitude), gamma ray absorption (radioactive gamma ray source with ionization detector or geiger counter measurement), light refraction, scatter or absorbance (measurement of refractive index, amount of scatter, or absorbance units of the media), and ultrasonic (ultrasonic energy is attenuated as a function of suspended solids when transmitted through the liquid stream).

The most common density meter utilized in the water–wastewater industry has been the nuclear radiation (gamma ray absorption) type. The advantages are high accuracy, of up to ± 1 percent of span, and great range, of from 0.01 to 0.25 SGU (specific gravity units), depending on application and proven design. The disadvantages are handling radioactive materials, federal certification requirements, warm-up time from a cold start of several hours, and buildup of solids on pipe walls, which causes drift. Within the last few years, on-line meters have been developed based on the light scatter and transmittance, or light transmittance principle, as well as meters based on the ultrasonic principle. These meters measure suspended solids rather than density. The light transmittance–scatter meter is similar in construction to a turbidity meter, with special considerations to prevent fouling of the light source and photocells. The ultrasonic type is constructed similar to the ultrasonic flow meter, but with a different arrangement of the transducer and measurement of energy transmission loss rather than time of travel, as is done by a flow meter. Advantages of the light and sonic systems are ease of installation, lower cost than radioactive units, and fast response after startup. Disadvantages are fouling and abrasion of photocells, and air entrainment, causing misreading or failure of measurement on a sonic meter. Published ranges are 10^{-5}–0.1, 2 + 10^{-3}–3, 0–4, 1–10, and 0–15 percent suspended solids. Accuracies are ±5 percent of range. Density and suspended solids meters have typical operating temperature and pressure ranges of 0–50°C and 0–25 m (0–80 ft) to 150 m (500 ft) of head, respectively.

COMPUTERS

In the late 1960s, several water utilities installed digital computers to supervise the operation of a water distribution system. At first, these systems were used to collect data, prepare reports, and alert the system operator when measured variables, such as line pressure or level, exceeded preset limits. Today the digital computer has been accepted by the water industry as a logical alternative in the design of a Supervisory Control and Data Acquisition (SCADA) system for monitoring and control of a water distribution system or a water treatment plant.

A computer system can be divided into two major parts, the hardware and the software. The hardware consists of the computer itself, with its core or semiconductor memory, disk memory, and peripherals such as printers, plotters, cathode-ray tube displays, magnetic tape memory storage units, and other associated input and output equipment. The software consists of the programs, sets of instructions telling the computer what to do, and the associated data on which the instructions will operate.

Computer Hardware

The basic hardware of the computer consists of the central processing unit (CPU), which executes the instructions, and the computer memory where data and instructions are stored. The memory of a computer system usually consists of three types, namely, working memory, disk memory, and magnetic tape. Working memory consists of high-speed core or semiconductor memory where data and instructions are stored or brought for execution. Less frequently used programs are stored in disk memory where they wait to be moved into working memory for execution and where the data base of the system is stored and periodically updated. Large amounts of historical operating data is normally stored on magnetic tape for future analysis. A

typical SCADA system might have a 256,000-byte (one byte is eight bits of information) working memory, 10 or 20 million bytes of disk memory, and a magnetic tape memory storage unit.

In selecting a computer for an on-line SCADA system, several other characteristics besides memory size and speed are important:

1. The computer should have an adequate repertoire of instructions to enable it to be easily programmed and to do its tasks efficiently.
2. Word length should be 16 bits or 32 bits. Shorter word lengths lead to inefficient programming.
3. A hardware floating point processor should be provided. The cost of this feature has reached a point where it is not economical to do without it.
4. Hardware stack-handling ability should be provided to permit reentrant subroutines and the automatic (hardware-controlled) nesting of subroutines and interrupts.
5. A multilevel priority interrupt system with vectored interrupts should be provided. An on-line system, unlike a business or engineering computer system, always requires a large number of interrupts that must be serviced rapidly and efficiently.
6. Automatic shutdown (without loss of memory) must be provided together with automatic start-up when power returns.
7. An additional essential feature is commonly known as a watchdog timer. This device is reset routinely by the computer program as long as it is operating normally, much as a night watchman in a building punches time clocks as he makes his rounds. If the program should halt or go into a loop, then the watchdog timer would not be reset and would eventually (in several seconds) time out and automatically reload and restart the program.
8. A real-time clock and calendar.
9. Error checking and correcting codes and procedures should be used when writing to or reading from memory. If an error is detected, the data or program should be reread until the reading is error free or, if this proves impossible, shifting to a backup computer or disk will take place.

The on-line control computer must be extremely reliable, but even more important, if it should fail, it should attempt to recover automatically without operator intervention. If this should fail, it should alert the operator to switch over to backup equipment.

Peripherals.　Peripherals consist of the equipment, such as cathode-ray tube (CRT) displays, printers, disk drives and magnetic tape readers, that are connected to a computer system to improve its utility and to allow for better communication between man and machine. The peripherals most commonly used in SCADA computer systems for the water industry include printers, CRT displays, and programmer's equipment.

The CRT display with its associated keyboard is the most important interface between the operator and a SCADA system. Using the keyboard of the CRT display unit, the operator can enter data, change parameters, such as alarm limits, request displays of operating data, and command pumps at remote stations to turn on and off.

The typical CRT for a water treatment plant will include symbols for valves and pumps as well as the facility for drawing in piping, reservoirs, tanks, and so on. With this feature, the operator can call up a graphic display of a process area complete with live data indicating present conditions of equipment and process values. If a light pen is provided, the operator can simply point it to the symbol of a pump to command the pump to start or stop and watch the result of his action as the pump changes color and the flow and pressure readings change accordingly.

The CRT display, together with its keyboard and light pen, is the operator's window to the treatment plant or water distribution system, presenting him with operating, historical, or predicted data in the form most easily understood, whether graphic diagram or map, trend display or bar graph, or a simple tabular display of data.

For preparing reports and operations or alarm logs, the SCADA system requires one or more printers, in addition to the programmer's equipment. Normally two printers are provided—an alarm and operations logger and a report printer. If printing in color or printing of graphs or symbols is required, color copiers can be provided. With these color copiers, the operator can request a printout of anything appearing on the screen of the CRT and will receive a full color copy of the display.

In order to prepare new programs for the computer system or to troubleshoot problems in existing programs, the programmer requires equipment to edit, compile, assemble, debug, and load programs into the computer. These devices are usually employed for the preparation of programs. For writing and editing programs, the CRT display with keyboard is used. For storing the source program (the one written by the programmer) and for preparing the object program (the one the computer will actually execute), a dual diskette or "floppy disk" system is used. For listing programs, the programmer requires a medium-speed printer.

Computer Software

The software for a computer system consists of all those instructions stored in computer memory that tell the computer what to do. The software for a SCADA system is the hardest part of the system to specify, the most important factor determining whether the system lives up to expectations or not, and the part that will probably cause the most problems during startup, operation, and system expansion.

The software of an on-line computer system can be roughly divided into three parts. The first package of software includes editors, assemblers, compilers, debugging programs, loaders, and maintenance programs. This package is normally supplied by the computer manufacturer. These programs are not used during normal operation of the computer system but are used as required to modify or add to the existing system program.

The second group of programs is called the operating system. This is the master control program for the system and is responsible for scheduling programs to be run on a time or priority basis, transferring programs from core, or semiconductor, memory to disk memory or vice versa, handling interrupts according to priority, and, in general, supervising the entire operation of the system. The operating system may be supplied either by the computer manufacturer or by the SCADA system supplier.

The third group of programs is the applications programs, which are designed to perform to the functional specifications of the user and may be either standard programs of the system supplier modified to suit the application or programs specially written for the particular application. Useful application programs are discussed in the next section. The applications programs will include scaling and alarm checking of data; accumulating running time of pumps and motors; computing flows, reservoir tank volumes, or dosage rates; preparing displays for the CRT screen; and preparing reports for the report printer. Useful application programs are discussed in the next section.

The flexibility of the operating system will be a major function in deciding whether the system can grow with the water system or treatment plant expansion and change with operational requirements. Without this capability, the software will seem like hardware and will be as difficult to modify, thereby losing one of the major advantages of a computer-based SCADA system.

Applications Programs. One of the prime functions of a digital acquisition system is to acquire up-to-date and reliable digital measurements of analog variables (e.g., pressure, level, flow, etc.) and to store these data in memory for further processing. The actual process of measuring the analog variable, whether it be a 4–20 mA signal from a standard instrument transmitter, a pulse duration or pulse frequency signal, or an encoded digital measurement, varies considerably. However, once the analog variable is stored in memory as raw digital data, the treatment of the data is similar, no matter what the original analog source.

It is important that the operator should have the ability to inhibit the scanning of data from any input point and of reactivating any input without interfering with other system functions. When a transducer is being calibrated or is out of service, the operator should be able to remove this input from the norn.al scan and enter an estimated value manually.

Whenever possible, the measured analog input should be checked for validity. Validity checks may be done by rejecting values that are outside reasonable limits or that change suddenly, unless the change is confirmed in a subsequent scan. In digital systems, error-detecting codes may be incorporated that allow the hardware or software to reject spurious signals.

Random changes in a measured variable, such as pressure pulsations, can make the received data unreliable or unusable. Digital filtering with an operator-adjustable time constant will smooth this pulsation data.

Where the measured value is an arbitrary nonlinear function of the desired variable (e.g., the millivolt output of a thermocouple), the calibration curve can be stored in computer memory as a table of values and the computer programmed to obtain the temperature by interpolation. Square root extraction is often required in flow measurement.

To be of use to the operator, the raw data collected by the SCADA system must be reduced to engineering terms. The computer must multiply the raw data by a scaling factor and add a zero suppression factor to reduce the raw data to properly scaled engineering units.

Checking for alarms is one of the most important functions of a SCADA system. Whenever an input variable exceeds an alarm limit, the system will alert the operator by displaying an alarm message on the CRT screen, printing an alarm report on the alarm printer, and sounding an audible alarm, if required. When a variable previously in alarm returns to normal, a similar report can be made. In order to avoid excessive alarms when an input is near an alarm limit, the alarm program should be provided with an operator-adjustable dead band.

Another of the most valuable features of a computer-based SCADA system is the ability to alert the operator when the rate of change of a variable exceeds a preset limit. Both the maximum amount the variable may change and the time over which this change is to be considered are operator adjustable. When a trend alarm occurs, the operator will be alerted and the alarm printer will print a report, as for the high and low alarms and alerts.

Whenever flowrates are being monitored by a SCADA system, total flow is generally required. By summing up the flowrate periodically, for instance, once each minute, the computer can integrate this data to get the total flow over an hour, day, month, or other period.

One useful function of SCADA systems is maintaining running averages of input variables. Usually, 5-min, hourly, daily, and monthly averages are maintained. Another commonly used program computes the maximum and minimum of each variable during each hour, day, and month. It is advisable to be able to select an adjustable period over which the variable is to be averaged before checking for maximum or minimum.

In addition to the applications programs listed above, there will be special applications programs tailored to the requirements of the water distribution system or water treatment plant being monitored. The SCADA system will be expected to produce all the daily and monthly reports on system operation, such as the amount of water produced and chlorine and chemical usage. Water quality reports and water quality summaries can be produced from manual entry of laboratory and field test data and statistical analysis of this data. Another useful feature of the computer-based SCADA system is the ability to produce maintenance reports based on the number of hours of operation of pumps and other equipment (or on the number of "starts") and alert the operator, usually by daily or weekly reports, when preset maintenance milestones have been reached. The on-demand automatic production of these daily or monthly reports, in a neat and standardized manner, is an important function of a computer-based system.

By constantly monitoring equipment energy demand, the computer can determine the effect on demand charges of turning equipment on or off. This information is especially valuable where time-of-use energy rates are in effect, since turning on a pump too soon may result in higher energy costs. Turning on a pump too late may result in unsafe operating conditions.

Where suction pressure, discharge pressure, flowrate, and power usage are measured for each pump or each group of pumps, operating curves for each pump or group of pumps can be developed, efficiencies calculated, and incipient pump problems detected at an early stage.

MICROPROCESSORS

What a microprocessor is, what it consists of, and the differences between a microprocessor and a microcomputer, must be briefly explained. A microprocessor incorporates the functions of the central processor of a digital computer on a single chip (or in a few cases, two chips). The microprocessor chip incorporates the arithmetic–logic unit, the instruction decoding and execution, the memory address calculations, and the interrupt-handling procedure. In order to produce a working microprocessor system, memory chips, input–output chips, peripheral handlers, and sometimes even clocks, as well as power supplies and peripheral equipment, must be added.

Memory for microprocessors is also produced on chips and consists of two basically different types: RAM (random-access memory) and ROM (read-only memory). In addition to memory chips, input–output chips are also required. These normally provide 16 lines of programmed input and output, which can be used (with buffering) to drive relays or lights or to accept signals from sensing contacts or push buttons. The tendency is to make these input–output chips programmable to a certain extent and even to include ROM or RAM on the same chip.

The microcomputer differs from the micropro-

cessor in that ROM and RAM, as well as limited input–output capability, is manufactured on the same chip as the microprocessor. While of much more limited capability than the microprocessor system described above, it is the most economical solution for high-volume applications that are not very complex in their requirements.

One of the most obvious uses of microprocessors in water treatment is in filter control, where the entire backwash sequence can be programmed and controlled by microprocessors. Any relay network used for timing and sequential control, such as for pumping station control and sludge removal, where the relay network to be replaced is complicated and has a limited number of output contacts, may be replaced to advantage with a microprocessor. Microprocessor equipment designed for this use can be programmed by an engineer or technician by simply entering symbols for relay contacts, timers, and coils in a ladder diagram form. The ability to add to or change the relay network in the field without doing any rewiring is a major advantage of these microprocessor-based units. The low power requirements of microprocessors makes supplying battery backup during power failure a simple task, and their small size makes packaging the system for a hostile environment much easier.

A second area where microprocessors are making an impact is in instrumentation equipment, such as recorders and controllers. Microprocessor-based strip chart recorders are much more flexible than the older analog types and can print charts on blank paper while recording. Because of its speed and capacity, a single microprocessor can act as multiple controllers with each controller individually configured and tuned to suit the process being controlled.

In the field of computers and solid-state devices, and especially in the field of large-scale integrated solid-state devices, any evaluation of a particular component or system such as a microcomputer or a microprocessor system will be valid for a few months or a year at most. Development is proceeding so rapidly that last year's new product may be replaced this year by a unit with twice the capability and speed at half the price.

FINAL CONTROL ELEMENTS

The final control elements in a water treatment plant are the direct interface of the control action with the process. This section limits the discussion to valve and pump controls, since these require the process designer's input, whereas final control is inherent in other equipment, such as chlorinators and chemical feeders.

Valve Controls

Successful process control with a valve is contingent on proper sizing and selection. The principal types of valve operators are hydraulic, pneumatic, and electric.

Hydraulic Operators:

Advantages. Direct acting, predictable (precise) motion and position due to noncompressible media, operating (water) pressure often available at the location, simple adjustment of response time.

Disadvantages. Corrosion (especially of needle valves), limited speed of action, energy storage more difficult in many cases, consideration of discharge fluid necessary (e.g., oil).

Pneumatic Operators:

Advantages. Direct acting, clean operation, some can be readily interfaced with standard pneumatic instruments, exhaust of air is no problem, great flexibility in speed of operation, simple energy storage for operation (pressure tank), no corrosion problem from operating media.

Disadvantages. Compressible operating media (air) can cause operator to be unstable and/or provisions have to be made to limit shock when applied for high-speed operation.

Electric Operators:

Advantages. Predictable and positive operation (speed and position), especially with induction motor drive, simple interface with other electrical equipment, clean operation.

Disadvantages. Will not operate during power outage [emergency generator, battery, or uninterruptible power supply (UPS) unit is required] or will change position (single-acting solenoid type). Motor-driven operators need reduction gears, making the operator more complex. Variable operating speed and modulation operation requirements add complexity.

Any valve is an impedance device and, as such, is inefficient for throttle control. Where valve control creates large energy losses, consideration should be given to energy recovery systems (such as turbine–generator units) or variable-speed pumping should be applied for water flow control.

VARIABLE-SPEED PUMP CONTROL

Varying the speed of a pump is often the best way to control plant flow. The speed can be varied with slip devices (energy inefficient but reliable, field proven, and cost-effective) or nonslip devices (more energy efficient and, therefore, increasingly more cost-effective). However, before variable-speed drives are considered, multispeed application should be investigated. The two principal multispeed devices are electric motors (2-speed or 3-speed) and gear drives. Either of these devices is superior to any electrical, magnetic, mechanical, or hydraulic variable-speed system in efficiency, lower first cost, simplicity, and ease of maintenance. Often, variable-speed units are applied when the requirement is for constant-speed operation with multispeed features. One example is the flocculator drive in a water treatment plant. In this case, the flocculator is operated at a constant speed with only seasonal switching to a different speed. Operation of a variable-speed unit on a constant-speed basis can lead to failures, especially on mechanical belt or pulley drives. This can result in a groove in the sleeve and rapid deterioration of the pulley when the unit is changed to another speed. Multispeed applications are also often superior to variable speed for pumping stations, fans, and blower drives.

Slip Drives

Drives based on the principle of speed regulation due to slippage are inherently not energy efficient unless slip (heat) losses can be utilized for heating. Drives with slip loss speed regulation are:

1. Electric-wound rotor motors.
2. Electric induction motors with voltage regulation.
3. Magnetic clutches.
4. Hydraulic clutches.

Electric-Wound Rotor Motors. An electric-wound rotor motor is an AC induction motor having a special rotor with slip rings and brushes. This motor is suitable for variable-speed applications with a wider speed range and higher efficiency than can be achieved utilizing variable-voltage regulation.

Electric Induction Motors with Voltage Regulation.

This type of variable-speed drive is the simplest and most rugged overall. It is based on the fact that changing the supply voltage to the motor changes the output torque by approximately the square of the voltage and, in turn, the slip (speed). However, this drive application is limited by the small speed range and the relatively low efficiency even at full speed, since a high-slip squirrel cage induction motor has to be used.

Magnetic Clutches.

The magnetic clutch is a unit connected mechanically between the constant-speed motor and the load unit (pump, fan, blower, etc.). Speed change is accomplished by changing the strength of the magnetic action, which changes the available torque. The full-load speed, and therefore full-load efficiency, is lower than some other slip devices (e.g., wound rotor motors) since two slips are present during full-speed operation. This drive, like other slip drives, is challenged today by the nonslip variable-speed drives due to change in the cost-effectiveness of solid-state power switching devices. This is especially true on applications where large speed ranges are required and the equipment has to be operated at lower speeds for long durations.

Hydraulic Clutches.

The hydraulic clutch is similar in operation to the magnetic clutch, except that the friction between the two rotating members is mechanical or hydraulic with the associated wear. Its advantage is the independence of outside electric power and the electric power's technically more complex control. Its disadvantage is the more limited control strategy, such as ramp control, which is readily available on electric controls.

Nonslip Drives

Whenever slip losses cannot be tolerated, other means of variable-speed control have to be employed. In the case of electric motor drives, the first method is changing the motor speed itself. The common types of motor speed changing units based on the nonslip principle are:

1. Wound rotor motors with energy recovery.
2. Induction motors with frequency regulation.
3. DC motors.

Wound Rotor Motors with Energy Recovery.

Wound rotor motors with energy recovery can be listed in this category since slip losses are electrically recovered, as previously discussed.

Induction Motor with Frequency Regulation.

The induction motor is essentially a constant-speed device whose base speed is determined by the frequency of the supply voltage and the number of pole pairs of the motor. Changing frequency, together with the voltage, on an induction motor is an almost perfect speed-changing principle. The advantages are operating flexibility, wide speed range, and higher efficiency at lower speeds. The disadvantages are sensitivity to high ambient temperature, complex equipment, and the high harmonics content of electric power, thereby running at higher temperatures.

DC Motors.

The DC motor is very well suited for variable-speed operation. Its drawbacks are the mechanically more complex construction with associated higher price and maintenance (commutators) and the requirements of DC power supply. Today, the DC power supply requirement is accomplished by solid-state power rectifier units, such as thyristors (SCRs). Large DC motor drives have been and are extensively used in other industries such as steel mills, but it is the unique construction, operation, and maintenance features, among others, that has kept, and might keep, the DC motor drive application limited in the waterworks field.

Other Nonslip Variable-Speed Devices.

Variable pulley drives are one of the mechanical variable-speed devices available. Most of these devices depend to a great extent on reliable, proven mechanical design and construction features, making these drives more susceptible to changing reliability of operation based on manufacturer and model. Caution should be exercised when these units are operated at the same speed for long durations due to wear of the equipment (for example, a groove in the pulley). Engine drives, as mentioned before, are excellent for variable-speed operation. It is the complexity, initial price, large space requirement, high maintenance, and high noise level that has kept, and is keeping, engine drives limited to special applications.

GRAPHIC PRESENTATION

Nomenclature and Symbols

The field of instrumentation is comparatively young, and there had not been an organized approach to nomenclature and symbols until 1949, when the Instrument Society of America issued the Standard of Instrumentation Symbols and Identification. This was revised and reissued in 1968 and again in 1973. As stated in the standard, "to be of real value, this document should not be static, but should be subject to periodic review." This general objective is even more apparent in the instrumentation field, which covers a wide variety of applications, ranging from the aerospace industry to residential housing. It is primarily because of this that no standard nomenclature or identification letters have yet been developed to cover all the requirements in the water treatment field (see Table 23-3). Standard identifications are still missing for turbidity, pH, and chlorine, to name the most common. Table 23-4 shows symbols and nomenclature commonly encountered and used in water treatment plant applications.

Process Flow Diagrams

Process flow diagrams are an essential tool in developing an organized presentation of the process. Depending on the nature of a project, the process diagram can be laid out in the form of a plant flow layout showing the physical arrangement of the plant or the hydraulic relationship in the form of hydraulic profiles, as shown in Figure 23-5.

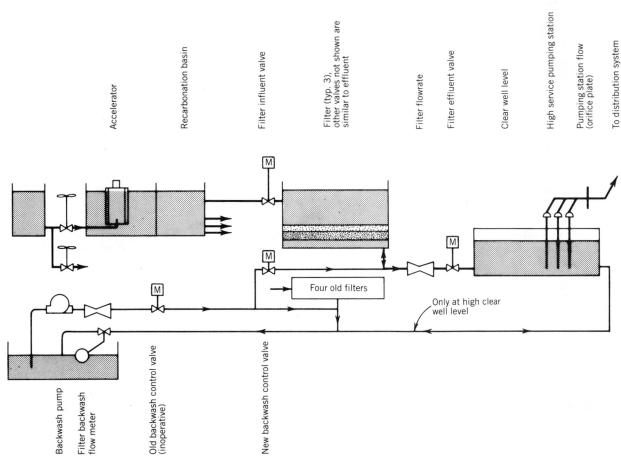

FIGURE 23-5. Process flow diagram.

TABLE 23-3. Meanings of Identification letters[a]

First Letter		Succeeding Letters		
Measured or Initiating Variable	Modifier	Readout or Passive Function	Output Function	Modifier
A Analysis		Alarm		
B Burner flame		User's choice	User's choice	User's choice
C Conductivity (electrical)			Control	
D Density (mass) or specific gravity	Differential			
E Voltage (EMF)		Primary element		
F Flowrate	Ratio (fraction)			
G Gaging (dimensional)		Glass		
H Hand (manually) initiated				High
I Current (electric)		Indicate		
J Power	Scan			
K Time or time schedule			Control station	
L Level		Light (pilot)		Low
M Moisture or humidity				Middle or intermediate
N User's choice		User's choice	User's choice	User's choice
O User's choice		Orifice (restriction)		
P Pressure or vacuum		Point (test connection)		
Q Quantity or event	Integrate or totalize			
R Radioactivity		Record or print		
S Speed or frequency	Safety		Switch	
T Temperature			Transmit	
U Multivariable		Multifunction	Multifunction	Multifunction
V Viscosity			Valve, damper, or louver	
W Weight or force		Well		
X Unclassified		Unclassified	Unclassified	Unclassified
Y User's choice			Relay or compute	
Z Position			Drive, actuate or unclassified final control element	

[a] Applies only to the functional identification of instruments.

SOURCE: Instrument Society of America.

Piping and Instrumentation Diagrams

The piping and instrumentation (P&I) diagram is an extension of the process flow diagram showing the instrumentation requirements necessary to control, supervise, and record the behavior of the process. Its advantage is to give an overall view of the requirements and the interrelationships of the processes. The drawbacks are that the presentation becomes too cumbersome and complex and leads to repetition showing, for example, flocculators, filters, and so on, many times over. Also, the P&I diagrams on large processes cannot be presented on one drawing, thereby losing some of this overall view. An example of a P&I diagram is presented in Figure 23-6.

Instrumentation Loop Diagrams

Instrumentation loop diagrams are extensively used in the instrumentation field (see Figure 23-7). The loop diagram differs from the P&I diagram in that it takes the essentials for a specific measurement or control application and presents it in the form of a separate diagram, removed from the overall process. Such a loop diagram is comparable to the "functional diagrams" used in the electrical control field, whereas the P&I diagram is comparable to a wiring diagram. The advantages of the loop diagram are its simplicity, the ease of presentation and, therefore, understanding. Loop diagrams are usually well suited for conventional water treatment plants. Loops for such subsystems as filter level

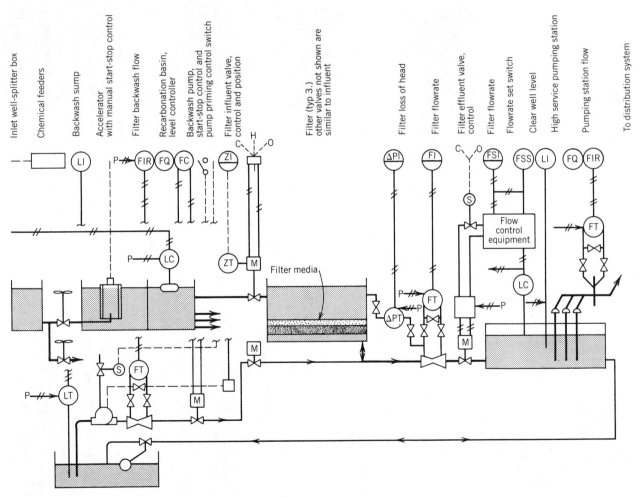

FIGURE 23-6. P&I diagram.

TABLE 23-4. Nomenclature and Symbols

Symbols			Nomenclature		Nomenclature
◯	Instrument or other component to be mounted in field (for locations, see plans)	ACL	Residual chlorine analysis element	MCB	Main control board
⊖	Instrument or other component to be mounted inside main control board or other panel as marked	ApH	pH analysis element	PI	Pressure indicator
⊘	Instrument or other component to be mounted on MCB front panel or other panel as marked	ATU	Turbidity analysis element	pHIR	pH indicator–recorder
◯◯	Single Instrument or other component having multiple functions	CLIT	Residual chlorine analyzer–indicating transmitter	pHIT	pH-indicating amplifier–transmitter
(dashed circle)	Instrument or other component to be furnished and installed by others in the future	CLIR	Residual chlorine indicator–recorder	pHN	pH signal isolator
		CLH	Residual chlorine signal isolator	pHSL(H)	pH switch low (high)
(propeller meter symbol)	Propeller meter	CLSH	Residual chlorine switch high	PRV	Pressure-regulating valve
(venturi meter symbol)	Venturi meter	CLSL	Residual chlorine switch low	PS	Pressure switch
(valve symbol)	Valve	DPI	Loss of head indicator	SP	Set point
		DPIR	Loss of head indicator–recorder	TuI	Turbidity indicator
(line symbol)	Pneumatic pressure signal	DPN	Loss of head signal isolator	TuIR	Turbidity indicator–recorder
(line with arrows)	Pulse frequency electric signal	DPSH	Loss of head switch high	TuN	Turbidity signal isolator
		DPT	Loss of head sensing transmitter	TuSH	High-turbidity switch
		FSL(H)	Flow switch low (high)	TuT	Turbidity transmitter
		FCV	Flow control valve	WE	Weight-sensing element
		FE	Flow-sensing element	WT	Cylinder weight (load cell) sensing transmitter
		FI	Flowrate indicator	WI	Weight indicator
		FIC	Flow-indicating controller	WSL	Weight switch low
				WX	Weight signal transducer (type as shown)

Symbol	Description	Code	Description	Code	Description
	Electronic instrument signal, except as noted	FIR	Flowrate indicator–recorder	ZI	Valve position indicator
	Hydraulic line	FN	Flowrate signal isolator	ZT	Valve position transmitter
	Process piping	FQI	Flow-integrating totalizer		
	Sonic signal	FT	Flowrate transmitter	TE	Temperature-sensing element
	Power supply; 120V, 60H: or 24VDC, as required	FX	Flow signal transducer (type as shown)	TI	Temperature indicator
	Related device performs linearizing or square root function	FY	Flow signal linearizer of summator (type as shown)	TIR	Temperature indicator–recorder
	Related device performs summating function	FIK	Flowrate setter	TT	Temperature transmitter
	Equipment designation used on mechanical drawings	LI	Level indicator		
	Relay contact	LIC	Level-indicating controller		
	Motor-operated device	LIR	Level indicator–recorder		
	Indicated components are not specified as part of the instrument package	LN	Level signal isolator		
	Future equipment	LIT	Level-indicating transmitter (type as shown)		
	Solenoid valve	LSL(H)	Level switch low (high)		
		LT	Level-sensing transmitter (type as shown)		
		LX	Level signal transducer (as required)		
		LY	Level signal linearizer		

633

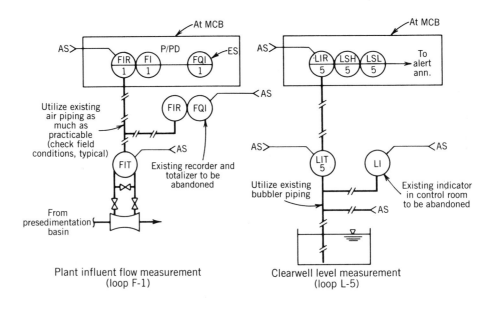

Plant influent flow measurement
(loop F-1)

Clearwell level measurement
(loop L-5)

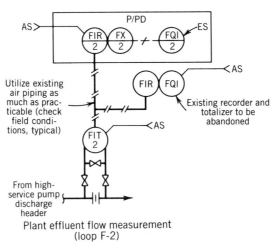

Plant effluent flow measurement
(loop F-2)

FIGURE 23-7. Instrumentation loop diagrams.

control, plant influent flow control, turbidity measurement, and so on, fall into this category. On very complex and large systems, it is appropriate to utilize P&I diagrams and instrument loop diagrams.

REFERENCES

Andrews, W. G., and Williams, H. B., *Applied Instrumentation in the Process Industries*, Gulf Publishing Co., Houston, Texas, Vol. 1 (1979), Vol. 2, 2nd ed. (1980), Vol. 3, 2nd ed. (1982), Vol. 4 (1979).

Barrett, D. F., and Brown, A. E., "Why Use Ultrasonic Flowmeters?" *Instruments and Control Systems,* **50**(5), 55–56 (May 1977).

Lomas, D. J. "Selecting the Right Flowmeter," *Instrumentation Technology,* **34** (5), 55–62 (May 1977); **34** (6), 71–77 (June 1977).

Norton, H. N., *Handbook of Transducers for Electronic Measuring Systems*, Prentice-Hall, Englewood Cliffs, New Jersey (1969).

Spink, L. K., *Principles and Practice of Flow Meter Engineering*, 9th rev. ed., Foxboro Co., Foxboro, Massachusetts (1967).

—24—

Operation and Maintenance

The design of a water treatment plant establishes the reliability, flexibility, operational labor requirement, and degree of automation and control of the facility. It is essential, therefore, that these four factors be carefully considered at the beginning and throughout the design period in order to provide optimum operating and maintenance conditions.

The relative weight to be given each factor will, of necessity, be influenced by local conditions that prevail at the plant site. A basic weighting factor is usually the initial capital cost and operational budget allocated to the facility by its owner. Reliability is the major consideration in the design of any water supply facility since the dependable output of acceptable potable water quality is the primary role of the facility. Flexibility should be provided to assure satisfactory operation under whatever conditions may occur and to allow innovation and testing of future technological advances. The operational labor requirement will be determined by the amount and sophistication of the mechanical equipment and the level of automation and remote control. Availability of qualified operating and maintenance personnel, and their salaries, will determine the stress to be placed on labor-saving controls and automation.

PRINCIPLES OF OPERATION

Reliability

Every water treatment facility must be so designed that, when properly operated, it can produce continuously the design rate of flow and meet the established water quality standards.

All equipment must function satisfactorily throughout the entire potential range in flowrates, from the lowest to the highest. Frequently, equipment designed to operate satisfactorily within the nominal or average flow conditions will not perform accurately at extreme high or low flowrates. Facilities must be provided for the operator to determine what quantity of water is being treated so that the chemical feeders may be properly adjusted without relying on frequent laboratory jar tests. In plants where future expansions are planned, equipment can be developed in modules to provide flow metering within acceptable accuracy limits.

Provision must be made to alert the operators to changes in raw water quality, whether gradual or rapid. This may be accomplished by: (1) providing sufficient laboratory facilities for the operator to make periodic tests of the critical constituents in the source water; (2) continuous recorders with an alarm to alert the operator when unusual raw water conditions occur; and (3) periodic detailed analysis by a competent laboratory of trace metals, pesticides, organics, and so on.

Similar provisions must be made to alert the operator when the quality of the finished water quality does not meet prescribed standards, either as a result of a change in the raw water quality, a change in the rate of flow without corresponding adjustment of the chemical feed, or from a malfunction of some of the chemical feeding equipment.

Flexibility

It is essential that the water treatment facility be designed to operate continuously with one or more pieces of equipment out of service for maintenance or repair. This generally requires that two or more units be provided for all major equipment. In the case of essential pumps, motors, chemical feeders, and so on, a spare unit to serve as standby should be provided.

The treatment of water is not an exact science. Improvements in the treatment process are continually being made. It is, therefore, very desirable that treatment facilities be so designed that they can be arranged to test potentially improved methods of treatment. This usually requires inclusion in the original design of sufficient gates, valves, and connections to allow bypassing or isolating critical operating equipment during a test period. Provisions for this flexibility would also likely facilitate modifications to the existing facilities for maintenance or process alteration.

With the possibility that new chemicals may be developed in the future for treating water, or that points of application of a given chemical may change, facilities should be designed to enable the operator to provide controlled chemical treatment from all chemical feeders to all presently considered potential points of application.

Labor Intensity

The design of a water treatment plant will be influenced by the local labor conditions. If local qualified labor is plentiful and relatively cheap, labor-saving devices would be kept to a minimum. If labor costs are high, a balance between the cost of the labor-saving features and the saving in labor cost must be made.

The type of equipment considered for inclusion in the plant must be determined by the ability and availability of local technicians capable of operating and maintaining the equipment. This is particularly true with respect to complicated automation and control equipment.

The method of plant operation and staffing will be influenced by the ability to obtain qualified operating and maintenance personnel as well as the timeliness of major repairs made locally. In a small remote installation, it is essential that the staff include technicians capable of maintaining and repairing most of the mechanical, electric, and control equipment in the plant. In a large plant, which is part of a major water system, it may be possible to rely on technicians from outside the plant staff for specialized or major repair work.

Automation and Control

The degree of automation and control included in a water treatment plant depends primarily on the decision regarding full- or part-time plant staffing. It is essential that a plant manned only a portion of the time include all necessary equipment and controls so that the prescribed quality of water can be produced continuously in the quantities required to meet the system demands.

If a plant is to have an operating staff in attendance at all times, the cost of installing and operating the automatic equipment versus the cost of providing manpower to manually control and maintain the equipment to provide the same functions must be determined. Depending on fluctuations in quality of the raw water supply and in the flow throughout the day, automation is often found to be practical for control of chemical feed rates, even when a full-time operator is in attendance.

A basic factor in the decision regarding the level of automation to provide a plant must be control reliability, which is dependent on the availability of local technicians for maintaining the equipment. Malfunctioning automatic equipment can cause many operating problems due to false reliance.

Some automatic equipment is difficult to operate manually unless specifically designed for ready conversion to manual control. Sufficient controls should be provided to assure reliable operation of all equipment in the plant. In other words, "fail-safe" operation should be provided.

PRINCIPLES OF MAINTENANCE

Maintenance is an essential key to efficient plant operation. Maintenance, as it pertains to water treatment plant operation, may be defined as the art of keeping plant equipment, structures, and other related facilities in a condition suitable to perform the service for which they were intended. To keep the plant operating efficiently as intended by the designer requires two general types of maintenance at a plant: preventive maintenance and corrective maintenance.

To provide a satisfactory maintenance program to bring the breakdown maintenance to a reasonable level, the overall maintenance policy should be as follows:

1. Responsibility for maintenance must be clearly defined and vested in competent personnel.
2. Management must state its maintenance objectives and emphasize their importance by properly allocating sufficient labor and budget.
3. Proper tools, spare parts, test instruments, and maintenance shop facilities must be provided.
4. Preventive action must be planned and scheduled.
5. An adequate system of written or computerized records and reports must be available to permit control of the program.

Preventive Maintenance

Preventive maintenance is the planned or scheduled maintenance of the "fix it before it breaks" type. Successful preventive maintenance programs obviate unscheduled equipment failures or emergencies and, in turn, reduce overall operation and maintenance costs.

Every maintenance program should start with good housekeeping. The plant and all equipment should be maintained in a neat, clean, and orderly condition. Visual appearances reflect the policies of supervision. It is essential that operational personnel take pride in the appearance of the plant and the quality of its product.

The single most important item of a preventive maintenance program is lubrication of the equipment with the lubricant to be used designated by the manufacturer. To ensure that equipment is lubricated properly, a schedule must be developed from the manufacturer's service recommendations and then followed precisely. The lubrication schedule is an integral part of the routine preventive maintenance program.

Preventive maintenance also means keeping equipment clean, in state of good order and proper operation, and free of excessive vibration, overloads, and improper heating. It also includes a periodic check for wear and the replacement or repair of parts before breakdowns occur. Such inspections may require the complete disassembling and reassembling of the equipment on a routine, scheduled basis. A maintenance schedule of this type must be developed for each piece of mechanical and electrical equipment in the plant.

It is important to make a clear and definite assignment of responsibility for carrying out maintenance work. When maintenance is everybody's business, equipment breakdown becomes nobody's fault. Assignment of the responsibility for maintenance of a particular piece of equipment should preferably be made to one person who has the capacity and is allowed the time for the job. The person assigned needs to have knowledge of such maintenance factors as what constitutes excessive vibration, when ball bearings should be replaced, what is a loose fit, and similar points. This knowledge is acquired only by experience and education.

Except in very small plants with only one operator, it is preferable to assign the maintenance responsibility to someone other than the operator. An individual with both operating and maintenance responsibilities is likely to neglect one or the other. However, the overall responsibility for operating and preventive maintenance should be handled by persons permanently assigned to the plant. This may not be practical for major corrective maintenance in some plants.

One responsibility of the preventive maintenance program must be to maintain an adequate inventory of spare parts so that repairs can be made without delay. The types of spares to provide and the number of each to maintain on hand should be determined from actual maintenance experience with the equipment, where such historical repair frequency is available, together with consideration of the time required to get delivery after ordering the spare parts. For a new plant, an estimate can be made of the spares that may be required, and the inventory can then be adjusted later in accordance with actual operating experience.

Corrective Maintenance

Corrective maintenance is the repair of equipment after failure and usually is of an emergency nature. If service is to be continuous, a water plant should not let its major pieces of machinery run until they fail and only then make needed repairs. Proper record keeping will provide insight into the wear of equipment, such as decreasing pump efficiency or noisy bearings. Since equipment has the uncanny habit of breaking down just when it is needed most or when it is most difficult to repair, plant management should be on the lookout for any decline in efficiency.

The responsibility for making major repairs will differ at each plant. Frequently, major repairs are handled by personnel outside the regular operating staff of the plant. This system allows operating personnel full attention to operating the facility at an overload condition while maintenance progresses on the malfunctioning equipment. In a large water system, there may be a general maintenance department assigned to handle major repairs. Or such repairs could be handled by outside contractors, for either a large or small water system.

The method of handling a specific repair job may depend on the urgency to get the equipment back into operation because of its process importance to the treatment train. Whenever corrective maintenance is required, an analysis should be made to determine whether the broken down item should be repaired or whether it should be replaced by a new unit.

The degree of reliance on preventive maintenance versus corrective maintenance may depend on the relative cost of maintaining an expensive spare parts inventory compared to the cost of periodic replacement of the specific piece of equipment.

All maintenance work, whether preventive or corrective, must be performed in full coordination and cooperation with the operating staff of the plant to ensure no interruption in the supply of treated water to the system.

Maintenance Record System

A minimum of records and forms are required to administer and carry out an effective maintenance program, yet such records are very important to successful maintenance. The preventive maintenance operations, even in a small plant, are too numerous to rely on an individual's memory for performance. The maintenance program must be planned and scheduled and then tabulated and calendared on individual record forms or on a computer.

It is vital that an effective, comprehensive maintenance system be established to standardize the above-mentioned points. The goal of the maintenance system should be to provide accurate, convenient records on each piece of equipment in order to streamline preventive and corrective maintenance. An example of a proven card maintenance system consists of six key record forms which direct activity: (1) a numerical file of inventory cards; (2) a numerical file of preventive maintenance cards for each piece of equipment; (3) maintenance data sheets; (4) a file of spare parts cards for various items and equipment; (5) daily maintenance sheets or maintenance log cards; and (6) maintenance work order and report. A computerized system utilizing the same general six categories may be utilized instead of the card system.

Equipment Inventory Cards

As a first step in setting up the maintenance record system, each piece of equipment is assigned an inventory number. The treatment plant is divided into sections, and each section is assigned a block of 1000 numbers; then each item within the section requiring maintenance is allotted an individual number within the block. Sufficient open numbers in each block remain to provide for any additional equipment that may be required in the future. In any case, the assigned numbers will serve to identify each item of equipment in all plant records described below and are used on the spare parts cards. Each inventory card (see Figure 24-1) has two numbers on it: (1) an equipment inventory number (four-digit number) and (2) and equipment schedule number (i.e., ME-35, T-12, M-16, P-63, etc.). The equipment inventory number as shown on the inventory card should be stenciled on the equipment. The schedule numbers refer to each piece of equipment on the drawings. If further expansion takes place beyond the capacity of the system, a fifth digit may be added to the inventory numbers to expand the system by a factor of 10.

All the name and place data for each piece of equipment (driver and driven unit) is recorded on each inventory card. Any other pertinent data, such

MAINTENANCE MANUAL NO. 1010		SCHEDULE NO. P-75	
MANUFACTURER Worthington			
SERVICE REPRESENTATIVE Worthington Pump Sales, 280 Harbor Way, S. San Francisco, CA. (415) 871-6455 94080			
DRIVER		**DRIVEN UNIT**	
MAKE Fairbanks-Morse		MAKE Worthington	
MODEL	TYPE Vert.	MODEL 14 MN	TYPE MixFlo-MN
ID NO. 483 925-40		SERIAL NO. 1743-14	
CODE	DESIGN	SIZE 14"	RPM 1150
FRAME. 445 T	S.F. 1.15	GPM 7600	TDH 51
HP 125	RPM 1200	IMP. DIA. 15"	MATERIAL CG I
HZ 60	PHASE 3	SEAL NO. Crane type I MATERIAL Tung. Carb.	
VOLTS 480	AMPS 156	SLEEVE 3"ID/4-1/2"OD MATERIAL SS	
INSULATION CLASS B			
SHAFT END BRG. SKF 7783		SHAFT END BRG. JHM 522649	
OPPOSITE END BRG. SKF 7683		OPPOSITE END BRG. 98400	
EQUIP. NO.	DESCRIPTION	LOCATION	
1010	Inlet Pump No. 1	Inlet Pump Station	

FIGURE 24-1. Equipment inventory card.

as oil capacities, bearing numbers, unique modifications to a particular piece of equipment, and so on, also are recorded on the inventory card.

Preventive Maintenance Cards

There could be one or more preventive maintenance cards (Figure 24-2), depending on the amount of maintenance to be performed for each piece of equipment in the plant that requires preventive maintenance. Listed on each card is all pertinent information with respect to preventive maintenance, including instructions for each task, frequency, materials used, and a reference number, which indicates a page in the corresponding maintenance manual. All additions, deletions, or corrections to the maintenance schedule should be entered promptly on the appropriate cards.

The equipment manufacturers' manuals are normally bound under separate cover and contain the maintenance information provided by the various manufacturers for their equipment. The manual will contain drawings, exploded views, charts, tables,

and so on, pertinent to a particular piece of equipment. The sections are numbered to correspond to the equipment inventory numbers for each piece of equipment. The reference numbers on the preventive maintenance cards refer to the page numbers in the corresponding equipment manufacturers' literature.

If required information is not contained in the published literature, the manufacturer or his local representative must be contacted. Work not specifically covered in the manual should not be performed without first consulting the manufacturer or his representative.

Maintenance Data Sheets

Maintenance data sheets (Figure 24-3) are included as a reference for the operators. They contain all maintenance data that appears on the preventive maintenance cards. When changes are made in the maintenance procedures, the manuals should be changed to correspond. A penciled-in change is sufficient for the maintenance data sheets until the

PREVENTIVE MAINTENANCE CARD				
		MAINTENANCE MANUAL NO. 1010		
ITEM	WORK TO BE DONE	REF.	FREQ.	MATERIALS
1	Check mechanical seal for proper water flow		D	
2	Hose down pump and area		W	
3	Grease spicer shaft "U" joints	P. 37	M	UNOBA EP-2
4	Grease upper and lower pump bearings	P. 11	M	UNOBA EP-2
5	Check pump for proper operation through all speeds		M	
6	Grease motor bearings	P. 14	S/A	UNOBA EP-2
7	Check amp draw on motor		A	

EQUIP. NO.	DESCRIPTION	LOCATION
1010	Inlet Pump No. 1	Inlet Pump Station

FIGURE 24-2.   Preventive maintenance card.

manuals can be corrected. However, the master cards and operators cards should be changed permanently and not penciled-in. The operators should ignore any changes that do not appear on the master cards as permanent changes or that are made by other operators. Any unauthorized changes in the maintenance procedures are to be reported immediately to the shift supervisor for correction.

Too much emphasis cannot be placed on following manufacturers' recommended maintenance procedures and methods as well as using recommended materials.

Spare Parts Cards

A spare parts card (Figure 24-4) should be maintained with each equipment inventory card for each piece of equipment. Separate inventories are not normally maintained for duplicate pieces of equipment; however, the cards will show the same quantity for each item.

The spare parts cards contain a list of spare parts on hand for each piece of equipment, along with a part number. For convenience, and to help save time when reordering parts, the manufacturers'

and/or suppliers' part numbers will be used. When a part has been used and a replacement must be ordered, this card should serve as the control for ordering and restocking.

Daily Maintenance Sheets

Daily maintenance sheets (Figure 24-5) are used as a vehicle for each shift to record maintenance in the plant's permanent records. At the end of each shift, the daily maintenance sheets are turned in and filled out with all maintenance *completed* notated for that shift. It is very important that the sheets be filled out carefully and completely.

Maintenance Log Cards

Maintenance log cards (Figure 24-6) may be used along with the daily maintenance sheets or by themselves to record maintenance completed for that day. The log card has a space for each day of each month covering one year. As maintenance is completed each day, it is recorded by task number in the appropriate square and initialed by the operator doing the maintenance. This information may also be

DESCRIPTION __Inlet Pump No. 1_____   EQUIPMENT NO. __1010__

_____   MAINTENANCE MANUAL NO. __1010__

DIRECTIONS	REFERENCE	FREQUENCY	MATERIALS
1. Check mechanical seal for proper water flow		D	
2. Hose down pump and area		W	
3. Grease spicer shaft "U" joints	P. 37	M	UNOBA EP-2
4. Grease upper and lower pump bearings	P. 11	M	UNOBA EP-2
5. Check pump for proper operation through all speeds		M	
6. Grease motor bearings	P. 14	S/A	UNOBA EP-2
7. Check amp draw on motor		A	

FIGURE 24-3.   Maintenance data sheet.

transferred from the daily maintenance sheets, if they are used, along with the log cards. When the log cards are filled (the end of December), those cards are removed and placed in permanent storage and new cards for the new year are put into the maintenance files.

The cards described above are kept in two sets of files: (1) the master file and (2) the operator's maintenance books.

1. The master file contains:
   a.   Equipment inventory cards

PART NO.	DESCRIPTION	QUANT.	PART NO.	DESCRIPTION	QUANT.
TP-1	Mechanical seal crane # Type-1	1			
14 MN-1	Wearing ring	1			
14 MN-9	Sleeve-SS	1			
(NAME OF EQUIPMENT)					
EQUIP. NO. 1010	SPARE PARTS IN STOCK				

FIGURE 24-4. Spare parts card.

   b. Master copies of the preventive maintenance cards
   c. Spare parts cards

The Master File is kept in the operations building
and is to be kept up to date by the maintenance
staff.

2. The operator's maintenance books contain:
   a. One copy of the preventive maintenance
cards
   b. Maintenance log cards

Those books that correspond to the general section will be kept in a building in that section. Any
changes made in the maintenance procedures
should be recorded on both cards (the one in the
operator's book and the one in the master file) and
on the data sheets as described above.

The operators will check daily the maintenance
book in their assigned area and perform the maintenance required. When the maintenance is completed, the operator records it on the daily maintenance sheet, which is turned in to his supervisor at
the end of his shift. Daily maintenance is performed
as described in the general routine maintenance
outline.

Importance of Records

It must be emphasized finally that any maintenance
record system, whether card or computer, if it is to
fulfill its function properly, must be kept up to date
faithfully and consistently. Service requirements
can be expected to change as the equipment ages,
flowrates change, and modifications are incorporated in the treatment process. Operations experience will dictate needed additions, deletions, and
corrections in the records. If changes are not entered properly and accurately as need is ascer-
tained, the entire system will soon become obsolete
and will lose much of its value. If a card maintenance record system is used, it is necessary to
check at regular intervals to see if any maintenance
cards have been removed from the books and
lost or misplaced. Similar precautions must be
taken to ensure that computer records are not destroyed.

DATE _4-30-75_   SHIFT _Day_   OPERATOR _J. Smith_

EQUIP. NO.	EQUIP. DESCRIPTION	MAINTENANCE COMPLETED	COMMENTS
1010	Infl. Pump No.1	1, 2, 3, 4, 5	Excessive Seal Leakage
1030	Infl. Pump.No.3	1, 2, 3, 4, 5	
1060 1065	Inlet Sump Pumps No. 1 & 2	1, 3, 4	No. 2 Pump Motor brg. noisy
1080 1085	Pump rm. exhaust Fans No. 1 & 2	1, 2	

FIGURE 24-5.   Daily maintenance sheet.

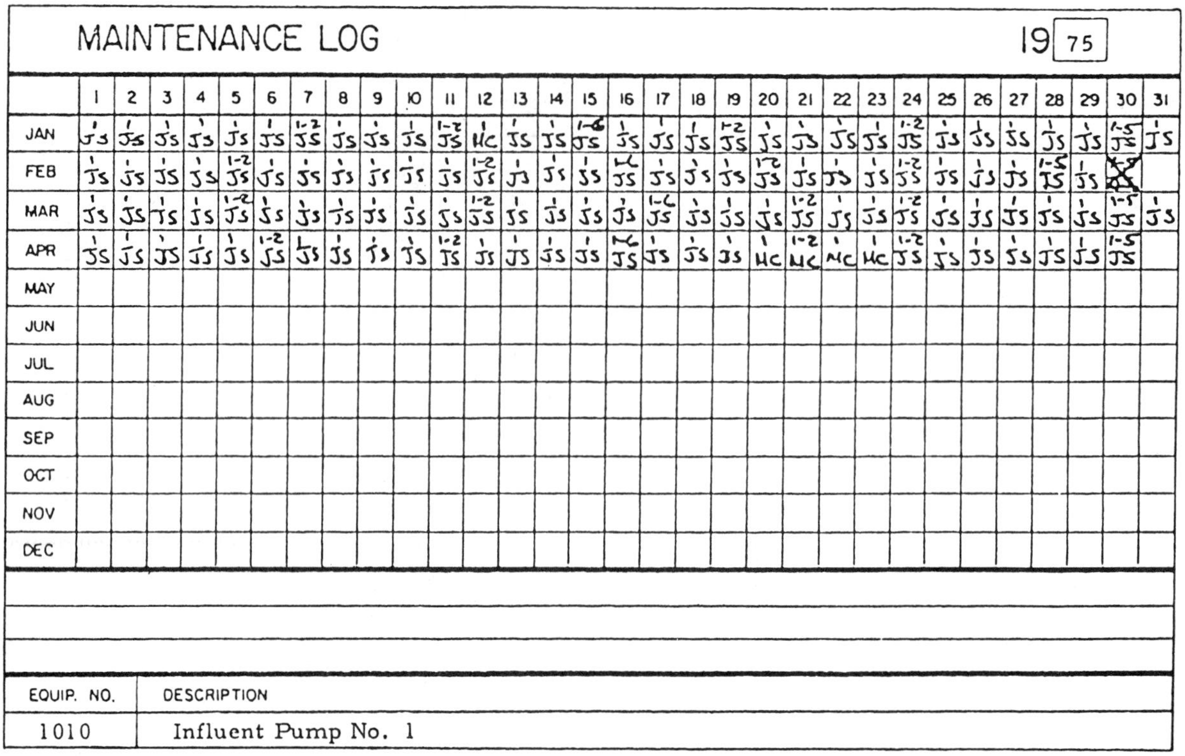

MAINTENANCE LOG   19 | 75

	1	2	3	4	5	6	7	8	9	10	11	12	13	14	15	16	17	18	19	20	21	22	23	24	25	26	27	28	29	30	31
JAN	JS	JS	JS	JS	JS	JS	JS	JS	JS	JS	JS	HC	JS	JS	JS	JS	JS	JS	JS	JS	JS	JS	JS	JS	JS	JS	JS	JS	JS	JS	JS
FEB	JS	JS	JS	JS	JS	JS	JS	JS	JS	JS	JS	JS	JS	JS	JS	JS	JS	JS	JS	JS	JS	JS	JS	JS	JS	JS	JS	JS	JS		
MAR	JS	JS	JS	JS	JS	JS	JS	JS	JS	JS	JS	JS	JS	JS	JS	JS	JS	JS	JS	JS	JS	JS	JS	JS	JS	JS	JS	JS	JS	JS	JS
APR	JS	JS	JS	JS	JS	JS	JS	JS	JS	JS	JS	JS	JS	JS	JS	JS	JS	JS	JS	HC	HC	HC	HC	JS	JS	JS	JS	JS	JS	JS	
MAY																															
JUN																															
JUL																															
AUG																															
SEP																															
OCT																															
NOV																															
DEC																															

EQUIP. NO.	DESCRIPTION
1010	Influent Pump No. 1

FIGURE 24-6. Maintenance log.

PLANT START-UP

The successful operation of a water treatment plant is dependent not only on the basic design of the plant but more importantly on the capability of the individuals operating the plant. The quality of water produced from the best-designed plant may be unacceptable if the operators in charge of the plant do not operate the plant as designed. It is therefore essential that a comprehensive start-up program be established during the initial operation to train operators, ''debug'' equipment, and modify operational procedures to meet the requirements of both the treatment train and the equipment.

Operator Training

Anyone operating a water treatment plant requires special training. Such training is often obtained by working up through the ranks in an existing plant or by training courses and seminars. Regardless of the experience of an operator, the start-up of a new plant requires special instruction in the design details of the new facility.

Whenever possible, the supervisory personnel who will be operating the new plant should be assigned full time during the final construction phase of the plant. If some of the operating personnel could work as inspectors during construction, the advantage would be even greater. It is essential that the operator be present at the plant during equipment testing in order to learn how to operate the specialized equipment. It is common practice for the design engineer to conduct a formal operational instruction program that includes both classroom instruction and instruction on the equipment's practical application.

Construction Responsibilities

The design, construction, and initial operation of a water treatment facility requires the successful coordination of the design engineer, construction supervision team, operators, and the governing body for which the facility is built. Areas of responsibilities can vary among facilities, but a detailed understanding of each party's responsibility should be determined prior to design.

Responsibility of Design Engineer

The design engineer has the responsibility to provide a plant that will produce the quality and quantity of water desired. It should also be his responsibility to see that the actual construction agrees with the engineer's design by making periodic visits to the plant during construction. In addition, the owner of the plant would be well advised to include in the design engineer's contract the requirement that the engineer provide the full-time services of a qualified plant operator to supervise the start-up and operation for 3–6 months and to provide special training to the staff. Experience has shown that it generally takes from 6 months to a year to get the "bugs" out of a new plant and have it operating smoothly.

Construction Supervision

The resident engineer and inspectors on a plant during construction are responsible for seeing that the detailed designs and specifications of materials, equipment, and construction are followed; to witness the operation of the equipment after it is installed; and to provide coordination and liaison between the design engineer and the construction activities. The resident engineer may either be an employee of the owner or of the design engineer. Experience has shown that the construction activities are handled more efficiently if the resident engineer is from the design engineer's staff, with an intimate knowledge of the engineer's design methods and specialties of the design staff, in order that questions arising during construction can be answered promptly.

Special precautions must be taken to eliminate the possibility of improper operation of any of the equipment by the contractor that would cause damage to the treatment facilities—for example, the introduction of backwash water to a filter at a rate sufficient to disrupt the filter media. These problems can be eliminated by specifying in the construction contract that the equipment shall only be operated by, or in the presence of, the resident engineer or a member of the operating staff.

It is the prime responsibility of the construction contractor to complete the entire facilities according to the design engineer's plans and specifications within the time allotted. The contractor should also be responsible for the satisfactory operation of all facilities during the guarantee period, which is usually one year, provided, of course, that any malfunction is not the result of negligence or improper operation.

Operating Staff

The plant operating staff is responsible for operating the treatment facilities after they are completed by the contractor. The transition between construction and operation can be made significantly smoother if the senior members of the operating staff are present at the end of the construction period while equipment is being tested. This permits the operators to become familiar with equipment and to learn some of the problems that may occur while placing it in operation. It also permits the operators to make sure that the equipment will operate as specified for the treatment process.

PLANT OPERATION

The ability of a plant to produce the quantity and quality of water for which it is designed is the joint responsibility of the design engineer and the plant operating staff. A plant may be designed that can hydraulically and operationally produce the desired quantity and quality of water, but if it is not operated properly, the results may be completely unsatisfactory.

The staff required and the necessary ability of the various staff members will depend on the type of treatment facility being operated. The need to provide close supervision 24 hr a day, seven days a week, will depend on how the quality of the raw water fluctuates and the method of treatment used. A plant designed primarily for turbidity or color reduction from a consistent raw water quality source, such as that originating from a large reservoir, will require much less supervision than a softening or iron and manganese removal plant treating water from a source that varies greatly in quality from hour to hour or day to day.

Treatment Process

A so-called conventional treatment plant—rapid mixing, flocculation, sedimentation, filtration, and disinfection—has the capability of absorbing shocks from changes in rate of flow or quality of raw water more readily than a solids-contact type of plant or a direct filtration plant. Therefore, much closer control and more qualified supervisors are needed for the latter two plant types. The solids-contact units are much more difficult to keep in balance than a conventional plant when malfunction occurs in the chemical feeding system, when there is significant change in flow through the plant, or when the raw water quality varies. Direct filtration plants with short-period contact or flocculation basins and no sedimentation basins must maintain close control of the chemical feed rates, mixing, and flocculation in order to maintain uniform quality of finished water.

Water Recycle

In any plant using filters, a determination must be made whether to recycle the filter washwater. This amounts to about 2–4 percent of the total plant flow. In an area where water supplies are limited or where it is difficult to find a satisfactory location to discharge washwater, recycling washwater may be a necessity and/or economically beneficial.

Where conventional treatment is provided and raw water quality is such that a significant dosage of coagulant is always required, it is possible to return the filter washwater directly back to the plant intake without any special treatment. Under these conditions, the best practice is to discharge the washwater to a holding basin or reclamation clarifier where it can be pumped back at a reasonably uniform rate so as not to cause rapid, radical change in flow or abnormal change in required coagulant dosage. With this method, the heavy sediment in the washwater will settle out in the basin and can be discharged periodically to the same location as the sludge from the sedimentation basins.

Where direct filtration is practiced, or where the raw water is of such low turbidity that little coagulant is required for normal treatment, return of the filter washwater without treatment to the plant intake can cause more major operating problems. With an influx of diatoms, for example, the concentration of the objectionable constituent can build up to a point where they will plug the filters, creating short filter runs and possible turbidity breakthroughs. If the raw water supply is such that these conditions could develop, a separate small washwater reclamation plant should be constructed to coagulate and settle the suspended matter from the wa-

ter before it is returned to the plant intake. In some cases, when the washwater reclamation treatment process is efficient, the water is returned directly to the filter influent; however, return to the plant intake generally provides better operation.

Safety

The safety of personnel and equipment should be of particular concern to the operators of every water facility. Industries have effectively prevented accidents and have shown that people can work safely when they know how to do their jobs properly and have safe places in which to work. Therefore, safety programs, possibly better designated as effective accident prevention programs, should be actively carried out at all plants, not just once but on a continuing basis. These programs have as their purpose the elimination of as many accidents as possible, both on and off the job. There are many references and texts that can serve as guides for such programs.

Both people and equipment are involved in accidents. Accident prevention must provide protection against human failure or actions and against equipment failure and involvement. Both labor and management must be convinced that most personal injuries can be prevented if they collectively assume the responsibility for taking the necessary steps, which involve the following:

1. Initiating and maintaining interest in safety.
2. Keeping adequate injury records.
3. Locating and correcting hazards.
4. Making equipment, plant arrangements, and working methods safe.
5. Controlling work habits.

Of particular concern in a water treatment plant is the safe handling of the operating chemicals, most of which are dangerous if not handled properly. These include chlorine, ammonia, fluorides, alum, soda ash, lime, ferric chloride, carbon dioxide, and others. A special period of instruction on safe chemical handling should be given every person employed at the plant, from the common laborer to the plant supervisor, with a refresher lecture at least annually.

Other subjects requiring special consideration in all safety studies include:

1. Guard rails
2. Fan ventilation
3. Ladders
4. OSHA rules
5. Emergency showers
6. Leak detectors
7. Fire prevention

OPERATION OF UNIT PROCESSES

Much valuable information to assist in optimizing the pretreatment processes—rapid mixing, flocculation, and sedimentation—can be obtained by bench tests in the laboratory and by pilot plant tests as described in Chapter 19. The plant treatment train can be simulated by jar testing in the laboratory. This enables the operator to adjust chemical additions, points of application, and energy input to raw water quality changes or to investigate new products and technology. It is essential that adequate laboratory facilities and equipment be provided at every water treatment plant regardless of size. The equipment should be sufficient to enable the necessary operating control tests to be made and the equipment should be used regularly.

Chemical Feeding and Application

Chemicals to be used and general treatment processes are determined by the process design, quality of raw water, and required quality of the finished water. Based on laboratory jar tests—or preferably pilot plant tests—the quantities, types, and points of application of various chemicals can be closely estimated. Starting with indicated jar test or pilot plant results, plant feeders should first be set at indicated optimum chemical dosages. Chemical efficiency in full-plant operation for a properly designed facility is generally higher than laboratory or pilot plant tests indicate, thereby permitting a lower dosage of chemicals to obtain optimum results.

After a sufficient period of full-plant operation to permit equilibrium to be established, observation and tests should be made to correlate the jar or pilot plant test results with the full-plant-scale results, and thereby select the chemical dosages and plant operating procedures to obtain the desired finished water quality.

The laboratory jar testing procedures are particularly essential when quality of the raw water fluctuates. A common fault in plant operation is a tendency to economize by reducing the quantity of chemicals applied below that required for efficient operation. This results in lower quality of finished water than would be obtained otherwise. There is also a tendency for an operator to increase chemical feed if required finished water quality is not obtained. In low-dosage operation, the best technique may be to drop the feed rate.

Adjustment, calibration, and regular checking of chemical feeders to ensure correct chemical dosage to the water is essential. Installation of dry feeders should be such that a timed sample can be obtained and then weighed to determine feed rate. Likewise, where the chemicals are fed as a liquid, provision must be made for collecting discharge from the feeder or control device to ascertain feed rate. The calibration settings and curves provided with various feeders cannot be relied on to be accurate for a prolonged period of time. Encrustation and packing may affect dry feeders, while slippage and wear can change liquid feed rates. In dry chemical storage bins, bridging can occur, which will shut off or reduce chemical flow to the feeders.

Problems are common when lime is fed for pH correction or softening. Lime will react with the hardness-forming constituents in water added to slake the lime or make a lime slurry. The reaction produces a calcium carbonate precipitate that may adhere to the sides of the carrying pipe or trough and gradually restrict flow. The pipe or trough must be inspected regularly and cleaned when necessary. The rate of encrustation can be reduced if it is possible to provide soft water for the slaking or slurry process or by returning some of the calcium carbonate sludge to the lime slakers. Studies should be made to determine if these methods will be cost-effective. It has also been observed that some of the plastic materials are much less likely to be encrusted with the calcium carbonate scale than are metal or asbestos cement lines. The rate of encrustation appears to be directly proportional to the velocity of flow; the higher the velocity, the more rapid and more dense the encrustation. For example, propellors for mechanical rapid mixers in a lime softening plant encrust very rapidly.

Special precautions must be taken to ensure the proper operation of fluoride feeding equipment. A prolonged application of a substantial overdose of fluoride to the water could be a serious concern. Most state health authorities require a recording fluoride analyzer to be installed to check the concentration wherever fluoride is applied. This analyzer must be checked and calibrated regularly to ensure accuracy.

To provide adequate disinfection to ensure that pathogen-free water is delivered at all times requires close control of plant chlorination facilities. If chlorine is provided as a hypochlorite, its feeding is handled the same as any other liquid chemical.

The handling of chlorine gas requires special precautions for it is very toxic and very corrosive. A chlorine leak must be stopped immediately for the chlorine will rapidly corrode most metals with which it comes in contact, thereby endangering any mechanical or electrical equipment in the area. A leak in a ferrous metal line carrying chlorine liquid or gas will increase at an alarming rate in the presence of atmospheric moisture. One of the best indicators of operation efficiency is the lack of evidence of a chlorine odor in the vicinity of the chlorine storage and feeding facilities. The source of a chlorine leak can be determined easily and rapidly by means of a bottle or swab of liquid ammonia. When the chlorine and ammonia fumes combine, a white cloud is produced.

Initial Mixing

Recent research has shown that the most efficient use of chemicals is obtained when the chemical is instantaneously dispersed into the entire body of water being treated. This is particularly true of coagulants, coagulant aids, and chemicals used for pH correction.

Various types of rapid mixing methods and devices are described in Chapter 21. Each device will have its operation peculiarities, which must be considered in order to obtain optimum utilization of the chemicals. Rapid dispersion of the chemicals is generally obtained by some type of diffuser in combination with hydraulic turbulence. In operation, it is essential to determine that diffuser ports are kept open and that diffuser discharge is at the point of most efficient mixing. This is generally beneath the surface of the water.

Flocculation

Flocculation provided hydraulically, using fixed baffles in a basin, does not permit any operator ad-

justment of the energy input to correspond with changes in the quality of the raw water, physical or chemical, varying dosages of chemicals, or rates of flow through the plant.

In flocculation basins where mechanical flocculation is provided by either horizontal- or vertical-shaft equipment, equipment should be made so that rotation speed can be adjusted to provide varying energy input to the water. Although an infinite speed adjustment is desirable, a two-, three- or four-step adjustment is satisfactory and may require less maintenance than the wide-range variable-speed equipment.

Using laboratory jar tests or pilot plant tests as a guide, the flocculators should be adjusted to provide a decreasing energy input to the water as it passes through the basin. With the plant in full operation, tests should be made to determine the optimum setting for each set of flocculators, observing the results visually and by laboratory examination. Tests should be made to establish the best operating speeds for the various flocculators. These may vary as (1) raw water quality changes with a corresponding change in chemical dosage, (2) temperature of water varies, or (3) rate of flow changes substantially. By recording results obtained under different operating conditions over an extended period, future adjustments can be made as conditions change without having to do additional testings.

Solids-Contact Units

Flocculation and sedimentation and sometimes rapid mixing processes are included in solids-contact units. These units provide for flocculation in one zone and clarification in another zone as the water rises through a blanket of the flocculant precipitate. There is generally some adjustment possible in the flocculating zone by varying the speed of rotation of a flocculating paddle, the rate of circulation by pumps, or timing of the pulsating operation. These adjustments must be made by the operator, following instructions provided by the equipment manufacturer or as determined from actual operation, to provide the best results.

No adjustment is possible in the settling zone of the solids-contact units except as the level of the sludge blanket is controlled. This level is very sensitive and difficult to control.

Close operator supervision is required with the operator being particularly knowledgeable about the effect of each adjustment on the operation.

Some solids-contact units are particularly difficult to control if they are operated only a portion of each day. It may require a significant period of time for the sludge blanket to again be fluidized and reach equilibrium as the unit is first started.

Sedimentation

A horizontal-flow-type sedimentation basin, if properly designed, will require little operator supervision other than to ensure regular removal of the sludge. Tests should be made at varying flowrates to determine if good laminar flow is being obtained. If not, baffles should be installed, or adjusted, so as to produce the desired flow pattern. Once properly adjusted, further changes should not be necessary.

In sedimentation basins containing sludge removal equipment, whether scraper type or vacuum type, sludge should not be allowed to accumulate. It should be removed on a regular schedule to prevent overloading the equipment and to ensure a clean basin that will minimize carry-over of floc to the filters.

The schedule for sludge removal from sedimentation basins that are cleaned manually will depend on the design of the basin, the type and volume of sludge produced, and the quality of settled water going to the filters. The latter is one of the most critical items.

Basins should be designed so as not to require cleaning more frequently than two or three times a year. A long, narrow basin will usually permit a greater volume of sludge to be stored before cleaning is required.

Aluminum hydroxide used for plain sedimentation will produce a greater volume of sludge than will lime used for softening. Basins where alum is used will, therefore, require more frequent cleaning. The cleaning cycle may be controlled by the quality of sludge produced. If sludge contains a high percentage of putrifiable organic matter, water passing through a basin may pick up tastes and odors if the sludge is not removed at frequent intervals. In this case, it may be necessary to clean the basins more often than would be necessary strictly from the volume standpoint, in order to minimize the possibility of tastes and odors developing in the settled water.

The clarity of the settled water to be maintained will depend on the type and efficiency of filters in the plant. Plant design and operation should be balanced between pretreatment and filters to achieve the desired finished water quality most efficiently and at the lowest cost.

Flotation

Flotation practices in water treatment are presently used on a limited basis. Flotation has been used as a solids-removal-dependent process for removal of low-density particles, such as algae. Since water with a large concentration of algal blooms would not readily settle out, utilization of a form of flotation is a natural consideration. Of the four types of flotation systems, dissolved air flotation (DAF) is the most commonly used in water treatment.

When beginning operation of a DAF unit, it is necessary to force all water from the retention tank with air by running the compressor for a short time. Recycled flows can then be started and balance obtained in the retention tank to the design pressure specifications (275–620 k Pa or 40–90 psi). Once the unit is operating in this fashion, the retention tank will be one-third to one-half full of water with the remaining volume taken up with air. (Any increase of water volume will create water surging within the basin of the DAF unit, thereby upsetting the float.) After balancing the retention tank, the water to be treated is introduced into the DAF unit.

Operation and maintenance of a DAF treatment system is relatively simple once the unit is operating within the design parameters. It is of utmost importance, however, for the operator to operate at the design pressure of the DAF unit in the specific plant. The retention tank and auxiliary pumping components do not always function correctly at pressurization zones outside its design range.

Optimum operation of the DAF unit will require the plant staff to determine the proper cycling rate of the upper scrapers. The scrapers are operated in a slow deliberate manner, generally allowing one sweep of the complete surface area every 30 min. Many installations include an off–on timer, which allows the scrapers to run 10–15 min every half hour. This type of operation is beneficial in developing the desired float level. Proper use of either system, or combination of both systems, is dependent on solids loading rate, operator experience, and solids removal efficiency. Solids removal should be 95–98 percent, while average float will vary from 10 to 20 cm (4–8 in.) in accordance with the total solids concentration. When taking the graduated cylinder tests, the operator should note how long the float remains at the top. A continuous roll of sludge could indicate a problem in chemical feed or a prolonged detention of solids within the reactor.

Removal efficiency can be obtained through total suspended solids laboratory analysis and by observing 1000-mL graduated cylinder tests, which dictate the rise level of the sludge.

Granular Filters

The design of the various types of granular filters is covered in detail in Chapter 21. These include slow sand; conventional rapid sand; deep-bed sand; dual-media sand and coal; mixed-media sand, coal, and garnet; and granular carbon.

Operation and maintenance problems of these filters are more or less the same. They need to be filled slowly with upflow water until the media is covered in order to remove the air entrapped between the granules and to prevent the surface from being scoured by the entering water. This upflow filling of the unit is required every time the water in the filter is allowed to drop below the top of the media. This procedure eliminates air binding, which can restrict the flow of water through the filter.

Slow Sand Filter.
A slow sand filter, either new or recently scraped, should be placed in operation as follows: after first having been slowly filled upflow as described above, the source water is allowed to enter and filtration proceed at low rates until a film has collected on the surface of the sand. This requires several days. The best procedure is to increase the rate of filtration gradually during a period of four to seven days before the maximum rate is reached. This period of "conditioning" the filter will be longer for a unit containing new sand than for one from which the sand has been scraped. The water from a unit being conditioned should be wasted until the quality of the effluent indicates that the sand has developed the necessary film.

In general, the effectiveness of filtration is proportional to the depth and fineness of the filter sand and inversely proportional to the rate of filtration. With a sand depth of 0.9 m (3 ft), the rate of filtration should not exceed about 3.7 m/day (4 mgd/ acre), otherwise bacteria may penetrate the filter and be present in the effluent. Lower rates should be used with filters of lesser depth.

When the accumulation of material in the sand media is such that the maximum permissible loss of head is reached, filtration should be stopped to permit the filter to be "cleaned." The filter should be drained to a point sufficiently far below the surface to permit walking on the sand. The amount of sand to be removed from the surface of the filter depends on the depth the majority of the suspended solids penetrate. Ordinarily, from 0.6 to 2.5 cm ($\frac{1}{4}$–1 in.) of the surface sand is scraped by the use of flat shovels and placed in the hoppers of water ejectors, which force the mixture of sand and water to the sand-washing equipment. The usual procedure is to continue to intermittently scrape the filters as required until the remaining depth of sand is about 0.8 m (2.5 ft), after which about 15 cm (6 in.) of washed sand is replaced in the filter to restore the normal depth.

Rapid Sand Filters.
The media in a new rapid sand filter must be backwashed to remove the excess fines which, if left in the unit, will restrict the flow through the unit. In the commercial screening of the media, some fines always remain. These fines are removed after the media is installed in the filter unit by backwashing at the maximum design rate for 10–15 min, draining the water to a sufficient distance below the top of the media so it can be walked upon, and then skimming the fines manually from the media with flat shovels. The amount that must be removed will depend on the quantity of fines. The depth of scraping required can be determined by observation and usually amount to 1.25–2.5 cm ($\frac{1}{2}$–1 in.) of the surface. The backwash and skimming operation should be repeated two or three times or until no significant amount of fines are observed at the end of the backwash. In dual- or mixed-media filters, the backwashing and skimming operation is required after the installation of each separate layer of media. In a deep-bed, coarse-media filter, the skimming operation may not be satisfactory but should be attempted by backwashing with water alone at the highest rate possible after stopping the application of air and water.

In regular operation, when the media in a rapid sand filter becomes clogged with suspended matter so there is insufficient head to maintain the designed rate of flow, the media does not have to be removed for cleaning as in a slow sand filter but is cleaned by backwashing. This process cleans the deposited material from the media and discharges it to waste, leaving the filter clean.

The backwash operation is accomplished either by water alone or by a combination of air and water. Either method, when properly designed and operated, will provide efficient backwash results on conventional sand, dual-media, and multi-media filters.

On deep-bed, coarse-media sand filters, and on filters used in tertiary wastewater treatment facilities, air and water backwash is required for satisfactory washing.

When using water alone for backwashing, the efficiency of washing and thus the ability to reduce mud ball formation, is improved by using a surface wash system to assist in cleaning the upper portion of the filter bed. Best results with the surface wash are obtained when the surface wash jets are directed about 64° from vertical at 2–3 gpm/ft² for 1–3 min before the main backwash is started. The rate and duration will depend on the tenacity of the material adhering to the filter media. The surface wash should continue during backwashing and then be shut down before the backwash water is stopped in order to permit the media to be hydraulically graded.

Even though a surface wash system is provided, it is essential that the plant operator examine the filter media regularly by digging out small sections of sand or coal to determine if mud balls are being formed. If evidence of their formation is found, a higher and possibly longer backwash coupled with rodding or jetting during backwash should be used to break up the balls. If they should still persist, it may be necessary to break them up by hydraulic probes or to replace the media.

Special precautions must always be taken when backwashing a filter to prevent the disruption of the media, particularly when the sand, coal, or finer media are supported on a coarse gravel layer. A rapid application of the backwash water can cause a globular surge that will completely overturn the media in certain spots. This disruption is more likely to occur when there is entrapped air present in the filter bed. Therefore, when air entrainment occurs, before the backwash is started the filter influent and effluent valves should be closed and the filter allowed to stand until the major portion of the air is released. Then the backwash water should be started slowly and built up to the desired rate.

Mixing the media in spots or boils changes the hydraulic flow characteristics of the filter bed, both on the downflow filtering cycle and the upflow backwash cycle. During filtration, the disrupted areas may permit turbidity breakthrough and thus cause a lower-quality filtered water to be produced. This is true because, during the backwash cycle, the less resistant area of the boils will receive a disproportionate amount of washwater, depriving the remainder of the filter of an adequate backwash. This condition creates nonuniform head loss across the surface area in the filtration mode.

The backwash rate with water alone should be established sufficiently high to develop at least 20 percent expansion of the filter media. The rate required to accomplish this will vary with the size and specific gravity of the filter media and water temperature. Because of the change in water temperature from summer to winter, it may be necessary to adjust the backwash rate seasonally.

The air and water backwash systems used may vary. Air and water may be introduced simultaneously and used together throughout the backwash, or air may be introduced alone initially followed by a water backwash. In the latter case, air is shut off before the end of the wash, allowing the water to hydraulically grade the media before the unit is placed back in operation. Air alone sequencing has the effect of pushing the suspended solids to the bottom of the bed, thus forcing the following water wash to lift the solids a greater distance to disposal. It is therefore recommended that air alone, if used, be utilized for only a short period before starting the water backwash.

For satisfactory operation of a filter plant, it is essential that the head loss through each unit be known at all times. This is usually obtained from head loss gages located on each filter. In filters with dual- or multimedia, provision of measuring the head loss at the interface of the various media is desirable in one or all filters in order to obtain more precise control of the filtering operation. Head loss equipment should be regularly maintained. On declining rate filters, the head loss of the entire bank of filters is indicated by the level of water in the filters, but the head loss through any one filter is not known unless a separate head loss device is installed on each filter. Thus, it is difficult from an operating standpoint to determine when a filter should be backwashed. Backwashing on a time basis may not be satisfactory if the raw water quality is subject to frequent change. On self-backwash filters with uniform filter inflow, head loss is indicated by the water level in each filter unit.

The length of filter run, that is, the length of time between backwashing, will depend on the quality of water entering the filter and the size and type of media in the unit. With relatively clear water going to the filter, runs well in excess of 100 hr may be possible before the head loss reaches the maximum permitted. If filters are open and exposed to sunlight, long periods between backwashing may permit undesirable algae growths to develop in the filter media. Under these conditions, the filter should be backwashed on a time cycle rather than on a head loss basis. Filter runs in excess of about 24 hr are considered good.

Development of bacteria and algae growths in a filter bed can be greatly retarded and controlled by maintaining a chlorine residual in the water entering the filter. Under most operating conditions, the chlorine residual desired in the filtered water can be obtained by controlling the residual in the filter influent. This ensures a reasonably bacteria-clean filter bed at all times. Some control of algae on the media can be obtained from slug chlorine doses in the filter influent or the filter backwash supply.

Operation of proprietary filters, such as the valveless, self-backwash, automatic backwash, and pressure types require the same general operating controls described above but with minor modifications pertinent to the specific proprietary design.

Ion Exchange

Conventional ion exchange units operate about the same as downflow granular filters, except some may be operated upflow and others may be so-called continuous units. Normally, water entering an ion exchange unit is clear, either having been filtered first or having come from wells. There are a few installations where ion exchange units have been used satisfactorily, both as filters and as softeners, with conventional filter backwashing being used to remove the accumulated suspended matter deposited from the water followed by regeneration of the ion exchanger. Backwashing of an ion exchange unit not used as a filter is done primarily to loosen the bed to increase regeneration efficiency.

The ion exchange media is not discarded when its exchange capacity is exhausted but is renewed by bringing a regenerant into contact with the exchanger. The regenerant used will be either cationic or anionic, depending on the ion exchange process used. This is described in more detail in Chapter 10.

The methods of applying the regenerant (its concentration, contact time, rinse time, etc.) will vary with each installation. The recommendation of the manufacturer of the specific ion exchange equipment being used should be followed to the letter in order to maintain the efficiency of the exchanger. Many regenerants are corrosive and require special construction materials and special precautions in handling. Maintenance problems in an ion exchange

plant are much more severe than in a conventional filter plant.

Disposal of the waste regenerant and rinse water may present special problems in order to comply with local, state, and federal regulations regarding quality of waste discharges.

Adsorption

Granular carbon filters to remove organics from water by adsorption are designed and operated similar to granular sand filters. As with ion exchange units, water entering a granular carbon filter is generally clear in order not to clog the pores of the carbon and reduce its adsorptive capacity. There have been a few instances where a carbon filter is used for both filtration and adsorption, but the capacity of carbon to adsorb organics could be reduced by fouling the materials with fine suspended matter from the water.

Carbon in a filter has limited capacity to adsorb organic compounds. When laboratory tests of water being processed show that undesirable organics are no longer removed, the carbon must be reactivated. This requires its removal from the filter unit and processing by a heat steam or chemical process to remove the accumulated organics and reopen the pores of the carbon so it will again adsorb organic constituents from the water.

In small plants, a study should be made to determine the relative cost of removing carbon from the filter unit, transporting it to and from a place where it is reactivated, and replacing it in the unit, compared to the cost of removing it and replacing it with new carbon. In large plants, such a study could indicate the economy of providing a carbon reactivation plant at the site. With the probable great increase in use of granular carbon filters required for compliance with the more stringent water quality regulations of federal and state agencies, new and improved methods for handling and reactivating carbon will be developed.

Solids-Handling Systems

The solids to be handled in a water treatment plant consist primarily of the chemicals used in, and the sludge produced from, the treatment process. These are discussed in detail in Chapter 13.

Chemical Handling. From an operating standpoint, the main concerns in handling chemicals used in the treatment process are that (1) an adequate supply always be available, (2) the chemicals required for efficient treatment be continuously and accurately controlled, and (3) the chemicals be handled in a safe manner at all times.

The first requires that a running inventory be maintained to show the quantity of each chemical currently on hand. Then, based on estimates of the quantity of each chemical that will be required for treatment in the future and the maximum length of time it will take to obtain delivery of the chemical to the plant, an order should be placed for replenishment of the supply.

The second requires regular checking of the chemical feeders to ascertain that the desired dosage of chemical is actually being applied to the water. If not, adjustments must be made to correct the feed rate. To ensure the continuous, accurate operation of the chemical feeders requires that the preventive maintenance program be pursued vigorously.

The third requires that every member of the operating staff be thoroughly trained in the safe handling of all chemicals. The staff should be given regular refresher lectures. In addition, periodic inspections by the plant supervisors should be made to ascertain that the chemicals are being safely handled as directed.

Sludge Handling. From an operating standpoint, the handling of sludge produced in the sedimentation basins is primarily to (1) provide sufficient space in the sedimentation basins for proper settling of the flocculant material to maintain the desired turbidity of the water leaving the basin and (2) prevent the accumulation of putrefiable sludge that could impart tastes and odors to the water passing through the basin.

The maintenance of any sludge removal equipment (i.e., scrapers or pumps) is not significantly different from the maintenance of any other mechanical equipment, except that sludge may contain material that would cause rapid abrasion of the metal parts. Air binding and corrosion-erosion are frequent problem with sludge pumps.

Where no sludge removal equipment is furnished, the accumulated sludge in a basin can be rapidly and efficiently removed provided that (1) an adequately large drain line is provided, (2) the floor and drain troughs in the basin have sufficient slope to permit rapid runoff of water and sludge, and (3) a sluicing system of large-capacity, high-pressure fire hoses or water monitors is provided around the entire basin. Methods of sludge handling are discussed in Chapter 13.

Operation Records and Reports

Records. Keeping accurate records of performance is an important part of good water treatment plant operation. Making complete, clear, and concise records of what has happened and what has been accomplished will be valuable in meeting future operation situations and interpreting results of treatment.

Records are necessary to provide a check on things done or to be done. Equipment in water treatment plants requires periodic service, some daily, some weekly, and other monthly or yearly. The records, if properly kept, show when service was last performed and when time for service approaches. From these records, a schedule can be developed so that nothing will be overlooked. Such a record system was previously discussed in this chapter.

Details of day-to-day experience provide a running account of plant operation and thus have an important historical value. Functions of various records cover a wide spectrum of activities and values. A summary of the main functions would include the following:

1. Satisfaction of legal requirements.
2. Aid the operator in solving treatment problems.
3. Provide an alert for changing source water quality.
4. Proof that the final product is acceptable to the consumer.
5. Proof that the final product meets plant performance standards and complies with the applicable regulatory agency drinking water standards.
6. Aid in answering complaints.
7. Anticipate routine maintenance.
8. Provide cost analysis data.
9. Provide future engineering design data.
10. Determine equipment, plant, and unit process performance.
11. Provide the basis for monthly or annual reports.

The detail to be included in record keeping depends entirely on intended use of the data. Size of the installation, type of treatment, and type of installations auxiliary to the treatment plant will control the amount of necessary record keeping.

Some of the basic items of information that should be considered to record are as follows:

Surface Source	Ground Source
Air temperature	Source water temperature
Source water temperature	Source water quality
Rainfall data	Well logs
Runoff data	Static levels
Height of reservoir versus storage	Pump test data
Capacity	Pump performance curves
Raw water quality	Drawdown levels
Raw water quantity used	Observation well data
Quantity of raw water released for other users	Quantity pumped
	Pumping levels
	Recovery rates

Treatment

Amount of water treated	Amount of washwater used
Chemical dosage; kg/hr, kg/day	Length of time washwater was applied
Total amounts of chemical used	Rate of wash
Quantity of water filtered	Cycle times for filters
Final head loss before filter was washed	Number of filters in service
Numbers of hours filters in service	Daily or more frequent results of chemical, physical, and bacteriological laboratory tests

Record systems can be either simple or complex. They should be realistic and apply to operating problems involved at each particular treatment facility. A careful plan should be made to determine what data are essential and useful, and then forms should be prepared on which the information may be quickly entered.

Once made, the records should be carefully preserved and filed where they can be located. The question that always arises is how long records should be kept. Obviously, they should be kept as long as they may be useful with due consideration given to the historical value of some types of data. Data that might be used in the future as a basis of design for plant expansion or new facilities should be kept indefinitely. Likewise, results of laboratory analyses will always be pertinent and should be kept indefinitely. Some other types of data may not be useful beyond a year. A good plan is to decide at the time a record form is set up how long the record should be kept. Record maintenance has recently been simplified and reduced since the inception of microfilm, microfiche technology, and computerization.

Reports. The maintenance of complete and accurate records, while essential in the daily evaluation of water treatment plant performance, does not provide the supervising and governing bodies with the complete picture of the status of the water treatment plant operations. A monthly, quarterly, or annual report summarizing and evaluating data included in the various records will acquaint those interested, including the public, with the plant's operations as well as accomplishments, problems, and plans for the future.

The report should be accurate, complete, concise, and easily understandable by the layman. It should not only include data on effectiveness of water treatment but also information on cost of labor, chemicals and power, and any other items that may have a bearing on the treatment plant operation. Charts, tables, and photographs add to the presentation and are generally better comprehended than long, wordy descriptions. The report should conclude with a summary of items presented in the body of the report, together with recommendations for correcting any deficiencies observed during the reporting period and a description of future plans.

REFERENCES

American Water Works Association, *Water Utility Management,* AWWA Manual No. M5, Denver, CO (1980).

American Water Works Association, *Tailgate Safety Lectures,* AWWA No. M16, Denver, CO (1978).

American Water Works Association, *Basic Water Treatment Operators's Manual,* AWWA No. M18, Denver, CO (1971).

American Water Works Association, *Water Quality and Treatment,* 3rd. ed., Denver, CO (1971).

American Water Works Association, *Water Chlorination Principles and Practice,* AWWA No. M20, Denver, CO (1973).

American Water Works Association, *Safety Practice for Water Utilities, AWWA No. M3,* Denver, CO (1983).

Cleasby, J. L., Arboleda, J., Burns, D. E., Prendville, P. W., and Savage, E. S., "Backwashing of Granular Filters," *JAWWA,* **69,** 115–126 (February 1977).

Cox, C. C., *Operation and Control of Water Treatment, Process,* World Health Organization, Geneva (1964).

Department of Air Force, *Maintenance and Operation of Water Plants and Systems,* Air Force Manual AFM 85-13 (1959).

Hoover, C. P., and Riehl, M. L., *Water Supply and Treatment,* National Lime Association, Washington, DC (1976).

New York State Department of Public Health, *Manual of Instruction for Water Treatment Plant Operators,* New York.

Texas Water Works Short School, Austin, Texas *Manual for Water Works Operators* (1975).

Wagner, E. G., and Banoix, J. M., *Water Supply for Rural Areas and Small Communities,* World Health Organization, Geneva (1959).

Water Treatment Handbook, Degremont, Paris, France (1979).

— 25 —

Construction and Operating Cost Estimating

Cost estimations are best performed by individuals with supernatural powers of foresight and insight and who possess an unvarying record of good fortune. Others must rely on chance, experienced judgment, and hard work.

This chapter presents practical approaches to cost estimating for both preliminary and detailed engineering estimates of construction, operation, and maintenance costs. The subject of cost estimating is broad and complex. In a book of this scope, it is not possible to include enough cost data to permit actual project estimates to be prepared. However, sufficient references are included to provide an adequate source of information on water treatment and associated costs for practically any need.

COST ESTIMATING

Cost information is typically required at several stages of a water treatment project, as indicated in Figure 25-1. At the inception of a project, approximate values are necessary for general discussions of feasibility. Feasibility estimates are usually a few degrees better than wild guesses only because the person responsible for them usually has some familiarity with the project in general. Rarely does the estimator have available such basic information as the required capacity or the approximate time period of construction.

The preparation of engineering estimates takes place at later phases of the project. Early in the planning stages, preliminary estimates are developed for major project components. These preliminary estimates are usually included in a feasibility or planning study prepared for the owner by the engineer and are based on carefully prepared analyses of project needs and projections. Generally, a typical or traditional treatment process is identified, sized, and used as the basis for the discussion of costs. These estimates are frequently used for arranging project funding and for securing engineering design services.

Two types of estimates are usually prepared in the design phase of a project. The first are cost estimates used in comparing and evaluating process alternatives. These estimates require sufficient detail and accuracy of total construction costs and operation and maintenance costs to provide a basis for sound decisions concerning alternatives. The esti-

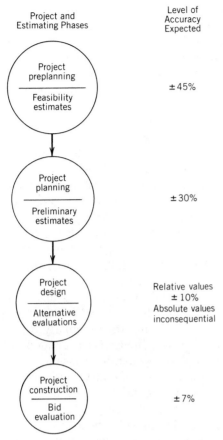

Project and
Estimating Phases

Level of
Accuracy
Expected

Project
preplanning

Feasibility
estimates

± 45%

Project
planning

Preliminary
estimates

± 30%

Project
design

Alternative
evaluations

Relative values
± 10%
Absolute values
inconsequential

Project
construction

Bid
evaluation

± 7%

FIGURE 25-1. Cost estimating levels and accuracy.

mates are commonly made using a component cost approach based on data developed for one place and time and extrapolated to another place and time. The second type of estimate made in conjunction with a project design is the detailed estimate of construction costs. This detailed estimate is used to compare and evaluate bids received for the construction of the project.

Each type of cost estimate mentioned above requires a different degree of accuracy and hence requires different techniques and different amounts of labor to prepare. Specific techniques used to accomplish these extrapolations are discussed in detail in the following sections. The emphasis is on the mechanics of preparing estimates and describing sources of cost information rather than on the presentation of extensive cost data and statistics. The techniques of estimating described here should be applicable in spite of economic trends and upheavals or other unforeseeable events. The basic concepts and methods are unaffected by such variables. Because new cost data are published regularly to

reflect new process modifications, equipment, and materials applications in water treatment, the estimator is encouraged to seek the most current information available.

Cost estimating can be one of the more challenging aspects of a project. It requires a dedication to detail and a willingness to search out and critically analyze existing data. The estimator should also have an appreciation of the uncertainties associated with financial planning, especially the economics of the construction industry. The result of these efforts will be more reliable, accurate cost information from which decisions can be made with confidence.

CONSTRUCTION COSTS

The capital costs of water treatment plants vary widely due to variations in plant capacity, treatment process, design criteria, site and weather conditions, land costs, operations building characteristics, and year and location of construction. Other factors influencing costs that may not be as evident include competition among bidders and suppliers, permit costs, and quality of plans and specifications with regard to description of existing site conditions and stability of the local and nationwide economic conditions.

Many of these cost variables cannot be specifically accounted for in either feasibility or preliminary estimating procedures. For this reason, at the outset, the estimator should make a general survey of factors influencing costs and their order of magnitude so that a balanced degree of accuracy can be achieved in estimating each of the various components of the project. It makes little sense to perform a material take-off down to counting nuts and bolts for estimating one phase, and for another, base the estimate on a percentage of overall project cost.

Preliminary construction cost estimates are based on historic data from other treatment plants constructed at various locations and times. The basic approach is to adjust data from other plants to account for different circumstances applying to the project of interest. The accumulation of cost data from existing water plants is normally accomplished by disaggregating the overall project costs into appropriate categories to separate out as many components as possible that could vary from one project to another. Costs incurred in different years and at different locations are adjusted to a common base using an appropriate cost index. Component costs are then expressed as a function of a significant variable such as plant design capacity, or better, maximum hydraulic capacity. A plot of component costs versus treatment capacity will frequently reveal a somewhat ordered pattern that can then be used to obtain cost estimates for that component in a treatment plant whose capacity falls within the range of applicability of the data.

Frequently, the estimator has neither the time nor the necessary data available to develop cost curves appropriate for a specific project. In these circumstances, cost information published by federal agencies, by private organizations, or in professional journals such as those listed in the bibliography at the end of this chapter may be used to advantage. At the predesign stage of a project, it is necessary to tailor the estimating procedures employed according to the availability of cost data and cost curves. Particular care should be exercised, when using cost data processed by others, to be familiar with its derivation and applicability so that it can be used intelligently. Large project components can easily be completely overlooked or possibly included more than once through careless application of published data.

Indexes

The most widely used mechanism to adjust cost estimates from one geographic location and time period to another place and time period is the cost index. As used in cost estimating, an index is simply a calculated numerical value that is a function of an established quantity of material and labor. The index number changes with geographic location and time as the prices of the index factors vary, thus providing a single numerical value to indicate trends with time and the relative value of the index factors from place to place. The popularity of the index among cost estimators is due in part to the ease of applications and the availability of a large number of indexes to fit a wide range of estimating needs. Also, the indexes have proven to be quite reliable when used in appropriate circumstances.

In spite of the fact that specific indexes are frequently used in inappropriate circumstances and indexes are used to indicate relationships that were never intended, the results are often satisfactory. Nonetheless, the estimator should always bear in mind the degree of extension applied in each circumstance. For example, indexes generally are not intended to measure differences in costs of construction between different cities because they do not account for variations in productivity, construction practices, or the competitive atmosphere. Yet, indexes are commonly used for this purpose, based on the assumption that such differences are insignificant.

A number of the figures presented in this chapter are based on an *Engineering News-Record* construction cost index (CCI) of 1000. This is done as a matter of convenience, and the estimator should appreciate the assumptions that were used in the adjustment of the specific cost data. Foremost is the fact that the CCI was assumed to be representative of each component even though the labor, equipment, and materials associated with the components were markedly different from the labor and materials used to derive the index.

The most widely used indexes in the construction industry are the CCI and building cost index (BCI), which are published regularly by *Engineering News-Record*. The CCI was created in 1921. It was designed as a general-purpose index to reflect the variation in costs of construction materials and labor. The items included in this index are the average cost of 200 hr of common labor, 1.128 tons of Portland cement, and 1088 board feet of 2×4 lumber and the mill price of 2500 lb of structural steel shapes. The increase in the 20 U.S. cities average CCI between 1973 and 1979 is shown in Figure 25-2.

The BCI, which was introduced in 1938, simply substitutes an average cost of skilled labor for the 200 hr of common labor used in the CCI. The cost of 68.38 hr of an average of carpenter, bricklayer, and structural iron worker wages is used. The CCI and BCI for the United States are updated on a weekly basis, while the city-by-city index values for 22 cities in the United States and Canada are published monthly. Over the years, as the cost of labor generally increased more rapidly than the cost of material, these two indexes have shown an increasing percent attributable to labor costs.

One index developed specifically for water utility construction is also shown in Figure 25-2. The Handy-Whitman index of public utility construction costs added a separate publication of water utility construction costs in 1957. The water utility index is subdivided into 23 general sections, such as pumping plant structures and equipment, large water treatment plant equipment, steel transmission mains, and installed filter gallery piping. Also, each

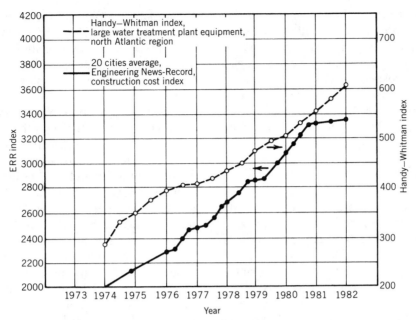

FIGURE 25-2. Construction cost indexes.

index number is determined for six geographic locations of the coterminous United States.

In response to the needs of the massive sewerage construction grants program established by PL 92-500, the 1972 Amendments to the Federal Water Pollution Control Act, the U.S. EPA developed three cost indexes that include values starting in late 1973. The small-city conventional treatment (SCCT) index was based on a cost breakdown for a 0.22 m³/sec (5 mgd) conventional activated sludge sewage treatment facility; the large-city advanced treatment (LCAT) index on a model 2.19 m³/sec (50 mgd) activated sludge sewage treatment plant utilizing lime clarification and multimedia gravity filtration for advanced treatment; and the complete urban sewer system (CUSS), based on a sewage collection system serving a population of 1700 over a 70-hectare (170-acre) area. These EPA indexes are published semiannually in *Engineering News-Record* for 25 cities across the United States. They are based on a detailed, material take-off and labor breakdown to which are applied current wage and price information.

In addition to the six indexes briefly described above, many other cost-indexing systems are available. Two other widely used indexes in the construction industry are (1) the Dodge BCI, which is compiled twice each year from a survey of costs of specific classes of labor and materials in 183 cities in the United States, and (2) the quarterly R. S. Means

CCI, which is derived from the current costs of 80 materials, 24 classes of labor wages, and 9 types of construction equipment. These indexes are tailored to reflect changes in costs for buildings and are published regularly in *Engineering News-Record*.

The U.S. Department of Labor's Bureau of Labor Statistics monitors changes in the nation's unemployment rate, wages, and fringe benefits of various classes of labor, as well as costs of various materials and produced goods. The Federal Highway Administration maintains a composite bid price for roads financed in part with federal funds. Other indexes of both general and special interest are compiled and published by various construction companies, public bodies, appraisal companies, engineering firms, and periodical publishers. In addition to the CCIs and BCIs, over 20 indexes are regularly reported in *Engineering News-Record*.

In spite of the number and variety of indexes available to the cost estimator, seldom is it possible to find a single index that can be used for anything other than general preliminary estimates for water treatment projects. Indexes rarely reflect the degree of disaggregation necessary or the desired balance between components or the time period or the specific location concern. For these reasons, estimators must have a working familiarity with many different indexes so that the best available one can be applied to each individual situation. The best index would be the one derived from a composite of materials and/or labor that most closely matches the materials and labor expected to be used in the project of interest. Frequently, several different indexes will be used for various elements of a complex project. Similarly, unit costs of specific items such as a process chemical or a major piece of mechanical equipment can be projected as individual items rather than incorporated as part of a larger component.

Total Cost Approaches

The subdivision of total project costs into smaller components and the presentation of the disaggregated data as a function of various project parameters is a logical approach to isolate factors responsible for the variance in the aggregated data. A simple tabulation of total project cost for a long list of water treatment projects would be of little value to the estimator. Inclusion of some measure of plant capacity would improve the table somewhat. Grouping the plants according to principal treatment processes and indicating the year of construction would begin to reveal some semblance of order, but not much. Figure 25-3 shows the typical variation that would occur in a graph of construction cost per unit of capacity, adjusted for construction date, as a function of the plant size. These data are actual costs incurred in the construction of 24 water treat-

ment plants in the southwestern United States between 1949 and 1977. The costs of land, right-of-way, reservoirs, pumping stations, and engineering are not included. It is evident that a cost estimate based on this type of display would be of limited value.

Further subdivision of projects to isolate and focus on the costs of those portions of the water treatment plants that are directly involved in the treatment of water yields relationships that are useful in making initial overall project estimates. Construction cost curves such as Figure 25-4 can be used as guides for estimates of a very general nature such as might be useful in the earliest planning stages of a proposed project. The estimator would be expected to exercise considerable judgment to adjust a value as appropriate to fit a particular situation.

The data used to develop the curve shown in Figure 25-4 are from three sources. The Dodge Guide to Public Works and Heavy Construction Costs data were adjusted using the CCI. These cost data were for total cost to the owner under average conditions for water supply treatment facilities employing clarification, filtration, and chlorination. The second set of data is from projects designed by James M. Montgomery, Consulting Engineers, Inc. (JMM), between 1949 and 1979. The construction costs were incurred at 19 water treatment projects in the southwestern United States. The remaining set is from a paper by Louis Koenig (1967). Both JMM and Koenig's data were adjusted to exclude the highly variable costs of land and right-of-way, raw and finished water reservoirs, and intake and treated water pumping stations. The treatment plants were all conventional, employing flocculation, sedimentation, and filtration. Data were adjusted using the CCI.

The data from JMM and Koenig were further adjusted to eliminate some of the variance due to the difference in design criteria. Costs of sedimentation tanks are directly related to their size, and their size is determined by design criteria, which allow for a considerable degree of discretion. Capital costs of the plants were adjusted to reflect the cost of a sedimentation basin that would provide for 90 min detention at the design flow, using a unit cost for sedimentation tank volume. In addition, since nominal plant capacity is directly proportional to filter loading rate and the nominal loading rate varies widely within an allowable range, the nominal plant capacities were adjusted to reflect a filter surface loading rate of 14.4 m/hr (6 gpm/ft²).

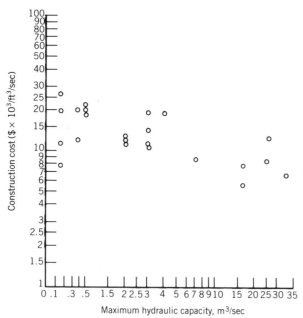

FIGURE 25-3. Water treatment plant unit construction cost (adjusted to ENR CCI of 1000).

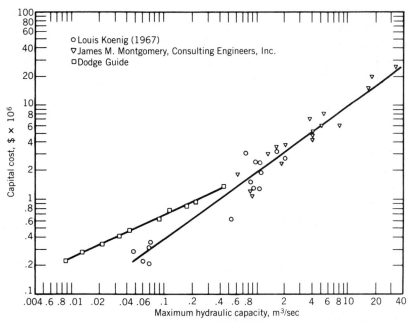

FIGURE 25-4. Conventional water treatment plant cost (adjusted to ENR CCI of 1000).

Component Cost Approach

Separation of a water treatment project into its major component parts enables the estimator to accumulate cost data for the components actually involved in a specific project. Some degree of variability remains in the basic data due to such things as the imperfect methods of adjusting historic data to other places and times or due to difference in construction methods and efficiencies. However, it is usually possible to obtain reliable average volume for construction costs for various processes that would be incurred under average conditions. Table 25-1 presents a breakdown of percentages of total capital cost that might be expected in major components of a water treatment plant. Several items normally included in a project were not included because of their highly variable costs. These items were general site and civil work such as excavation, paving, and yard piping, low and high service pumping facilities, and the finished water storage reservoir.

It is essential for the estimator to review the derivation of any unit of cost estimates he might use to avoid mistakes in application of data. A number of the components discussed below may or may not be included in the basic unit process cost data obtained from any particular reference.

Water Treatment Processes. Data for construction costs of specific water treatment components are normally obtained from one of two sources: either a series of typical detailed designs is made and detailed cost estimates are prepared for each design or historic data for actual construction of project components are adjusted to obtain comparable values. Each method in general will produce values satisfactory for most preliminary estimating needs.

A series of typical construction cost curves for eight representative water treatment unit operations

TABLE 25-1. Percent of Total Cost Attributed to Components

Water Treatment Plant Components	Percent of Total Construction Cost
Flocculation and sedimentation basins	20–40
Filters and appurtenances, backwash water storage and pumping, washwater reclamation	20–40
Operations and administration building	10–30
Electrical and telemetry	10–20
Miscellaneous chemical tanks, small structures	10–20

SOURCE: James M. Montgomery, Consulting Engineers, Inc.

is presented in Figures 25-5 through 25-7. These curves were published by the U.S. EPA in 1978, and costs are representative of January 1978 costs. They were developed using both historic cost data and detailed engineering estimates and include costs of excavation and site work for the processes, manufactured equipment, concrete and steel, labor, piping, electrical, and instrumentation and housing. Included in these curves are the costs of subcontractor overhead and profit and 15 percent allowance for contingencies.

Operations Buildings, Shops, Garages.

The costs of administrative and other treatment plant buildings are generally obtained by determining space requirements and applying unit building costs. Administration building costs can cover a wide range depending on the intended use by the owner. It is not uncommon among water utilities to find operations buildings that are impressive or even majestic, an expensive manifestation of community pride. While structures of that stature are rarely built today, water plants are popular destinations for school field trips and civic organizations, and the administration or operations building should be a model of efficiency and cleanliness. Unit costs for various classes of buildings are readily obtained

from such periodicals as the Means Building Construction Cost Data.

Land.

The quantity of land required for a water treatment plant is generally a function of the plant's capacity and the process involved. Table 25-2 presents approximate minimum land requirements for the treatment process units, operations building, chemical storage, and parking. No allowance is made for raw and treated water pumping stations or storage or for additional areas for maintenance yards, maintenance sheds, and general landscaping. Some consideration must also be given to the shape and slope of land since arrangement of treatment process units can influence other project costs. The data in Table 25-2 were obtained from measurements of land areas occupied by structures of actual operating water treatment plants.

Site Work.

While most cost curves for treatment components include costs of site preparation for the individual units, it is sometimes necessary to add a separate estimate for an assortment of items constructed at the site as part of the overall project. This group would consist of such things as general site clearing, grading, intercomponent piping, and

TABLE 25-2. Approximate Minimum Water Treatment Plant Land Requirements

| | Maximum Hydraulic Capacity | | | | | | | | | | | | | |
Process	ML/d 45	mgd 11	ML/d 90	mgd 22	ML/d 180	mgd 45	ML/d 360	mgd 90	ML/d 900	mgd 230	ML/d 1700	mgd 460	ML/d 2000	mgd 570
Conventional treatment[a]														
Hectares	0.6	—	0.8	—	1.2	—	1.9	—	3.2	—	5.0	—	6.5	—
Acres	—	1.5	—	2.0	—	3.0	—	4.7	—	7.9	—	12.4	—	16.0
Direct filtration[b]														
Hectares	0.4	—	0.6	—	0.8	—	1.2	—	2.0	—	3.2	—	4.0	—
Acres	—	1.0	—	1.5	—	2.0	—	3.0	—	5.0	—	7.9	—	10.0
Conventional treatment with sludge-drying beds														
Hectares	0.9	—	1.9	—	3.7	—	7.7	—						
Acres	—	2.2	—	4.7	—	9.1	—	19.0						
Direct filtration with sludge-drying beds														
Hectares	0.8	—	1.5	—	3.0	—	6.5	—						
Acres	—	2.0	—	3.7	—	7.4	—	16.0						

[a] Coagulation, flocculation, sedimentation, filtration, disinfection.
[b] Chemical conditioning, filtration, disinfection.
SOURCE: James M. Montgomery, Consulting Engineers, Inc.

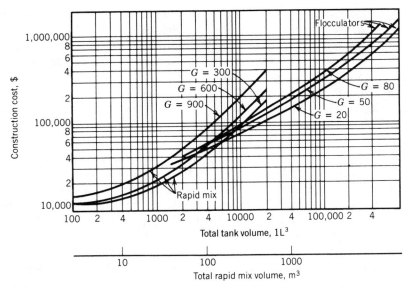

FIGURE 25-5. Construction cost of rapid mix and horizontal paddle flocculator. (Source: EPA, January 1978.)

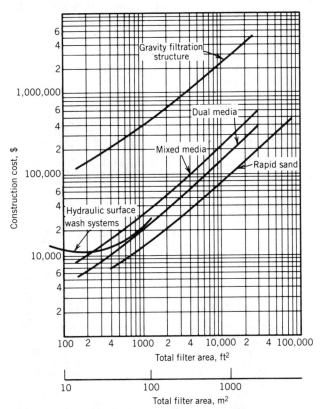

FIGURE 25-6. Construction cost of filtration media, gravity filtration structure, and hydraulic surface wash systems. (Source: EPA, January 1978.)

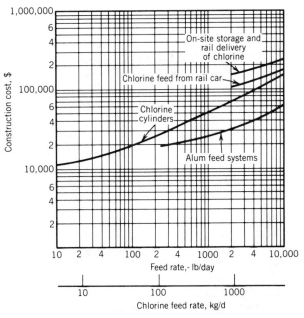

FIGURE 25-7. Construction cost of chlorine storage and feed systems and alum feed systems. (Source: EPA, January 1978.)

associated appurtenances, electrical wiring, lighting, parking, sidewalks, fencing, and site landscaping. Under average circumstances, this group will add about 10 percent to the total capital cost of a water treatment plant. Also, it should be noted that costs of site work can be significantly greater than this for unusual conditions requiring blasting, large amounts of cut or fill, dewatering, and preserving archaeological discoveries.

Engineering. One of the general methods commonly used to determine the compensation of consulting engineers is to base the compensation on percentage of construction cost. The American Society of Civil Engineers (ASCE) has published the curve shown in Figure 25-8, which relates the median compensation for basic engineering services for projects of above-average complexity. The curves represent median compensation for certain preliminary engineering, project design, and construction phase activities. Special services frequently required by the owner such as land surveys, soils investigations, and construction supervision are not included in this figure. The construction of new water treatment plants is considered by ASCE to be of above-average complexity. It should be em-

phasized that the curve shown in Figure 25-8 is only a general guide and is subject to discussion and negotiation in each individual circumstance.

To obtain total cost to the client for all needed engineering services, estimates for services such as surveying, soils investigations, and construction supervision must be added to the total for basic engineering services.

Interest During Construction. As the construction of a project progresses, the owner generally makes incremental payments to the contractor based on work accomplished. The money to make these payments may be obtained by the owner from a variety of sources, most of which require the payment of interest. This interest is a real cost to the owner and should be included in overall project cost estimates. Where funds are borrowed from an institution or obtained from the sale of bonds, the associated interest rate should be used. The interest cost is also a function of the length of the construction period, the pattern of payment to the contractor, and the overall project cost. The U.S. EPA (1978) published the cost curve shown in Figure 25-9, which can be used to estimate the cost of interest during construction as a function of other cost totals.

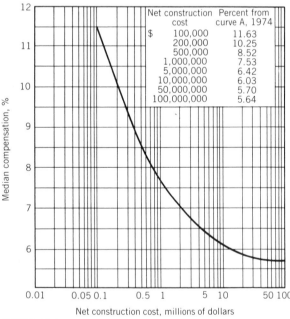

FIGURE 25-8. Median compensation for basic engineering services expressed as percentage of construction cost for projects of above-average complexity. (Source: American Society of Civil Engineers, 1981.)

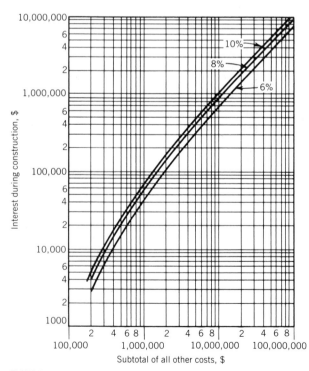

FIGURE 25-9. Interest during construction for projects greater than $200,000.

General Contractor and Overhead and Profit. One important cost to the owner that is occasionally overlooked by estimators (but never by contractors) is the general contractor's overhead and profit. This item is sometimes included in the water treatment process unit cost curves, but not always. General contractor overhead would include such items as cost for maintaining office space, secretarial and administrative personnel costs, and costs associated with preparing bids, bid bonds, insurance, permits, and idle equipment. Table 25-3, published by the U.S. EPA (1978), shows approximate values for general contractor overhead and fee as a percentage of total construction cost. This table can be used to obtain preliminary estimates.

Administrative, Legal, and Fiscal. While the costs of administrative, legal, and fiscal services on a project may not be directly related to the overall capital cost of a project, in the absence of any specific data for a particular job, the overall cost can be used as a basis for making a reasonable estimate. Legal and fiscal services typically include contract preparation, land acquisition, and when bonds are involved, consultants and bond attorney fees. Administrative expenses to the owner are incurred in such activities as arranging for professional services and participating in numerous consultations and project status reviews. Figure 25-10, published by the U.S. EPA (1978), can be used as a guide to determine the cost for legal, fiscal, and administrative services on a water treatment plant project.

Detailed Estimates

Final cost estimates for water treatment plant projects are prepared by the engineer after the detailed project design has been completed. These estimates

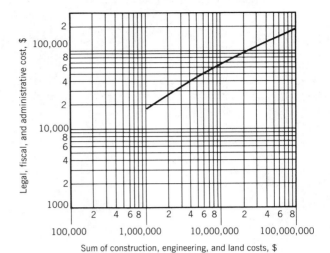

FIGURE 25-10. Legal, fiscal, and administrative costs projects greater than $1,000,000. (Source: EPA, 1978.)

normally are used as an aid in evaluating bids received for project construction. Detailed estimates are prepared by the engineer in essentially the same manner as a contractor would make an estimate before submitting a bid.

From the project design plans, a material take-off is prepared. This itemized list of project material and mechanical equipment and a summary of the necessary excavation and site work form the backbone of the estimate. Labor time and grade estimates are made based on these quantities. Current cost estimates for specific materials and equipment are obtained from local suppliers. Current labor and wage and fringe benefit rates are obtained for general classes of labor and for specialized skills. The prices are then applied to each item in the list of job components and summed for a total of direct project costs.

The judgment and experience of the estimator is then applied to the project to evaluate overall uncertain, subjective aspects. General economic instability, labor unrest, seasonal weather conditions, special contract conditions, time to complete the project, liquidated damages, short-term inflation, and other contingencies must be factored to obtain a final dollar value. Also, estimates for contractor overhead and profit must be included, as well as a lump-sum estimate for the cost to the contractor of moving equipment to the site, setting up, and finally leaving the site at the end of a job.

Figure 25-11 presents a typical detailed cost estimate summary sheet prepared for a water treatment plant in southern California. Note that the costs of

TABLE 25-3.   General Contractor Overhead and Fee Percentage

Total Construction Cost (in millions)	Overhead and Fee Percent of Total Construction Cost
$ 1–2.5	12
2.5–10	10
10–25	9
25–100	8.5

Source: EPA (August 1978).

JAMES M. MONTGOMERY, CONSULTING ENGINEERS, INC.
555 E. WALNUT STREET, PASADENA, CALIFORNIA 91101

COST ESTIMATE

PROJECT: SO. CALIFORNIA WWTP   DATE 26 OCT 78   ENR: 3430

JOB NO.: 1983.0600   CLIENT: SO. CALIF. MWD   TYPE OF ESTIMATE: DETAILED   SHEET 1 OF 38   EST. BY: TJ

DESCRIPTION:

REF. ITEM		QUANTITY	UNIT	M/H	MATERIAL UNIT	MATERIAL TOTAL	LABOR/EQUIPMENT UNIT	LABOR/EQUIPMENT TOTAL	TOTAL COST UNIT	TOTAL COST TOTAL
	SUMMARY									
1	SITE WORK									1874625
2	RAPID SAND FILTERS									3249950
3	FLOCCULATION & SEDIMENTATION									3720050
4	OPERATIONS BUILDING									391600
5	TELEMETRY BUILDING									14275
6	INFLUENT STRUCTURE									10500
7	INFLUENT PUMPING STATION									1008000
8	RESERVOIR PUMPING STATION									641825
9	RESERVOIR									2122225
10	WASH WATER TANK									47000
11	YARD PIPING									4467850
12	ELECTRICAL & TELEMETRY									5843550
13										
14	SUBTOTAL									24729950
15										
16	MOVE ON & MOVE OFF									100000
17	OVERHEAD, PROFIT,									
18	CONTINGENCIES @ 20%									4946000
19	ESCALATE COSTS 2 MOS. (2.5%)									618375
20										
21	TOTAL DETAILED ESTIMATE									31394325
22										
23										
24										
25										
26										
27										
28										
29										
30										

PAGE TOTAL

AC-2 (7/79)

666

FIGURE 25-11. Detailed cost estimate summary for typical project.

JMM JAMES M. MONTGOMERY, CONSULTING ENGINEERS, INC.
555 E. WALNUT STREET, PASADENA, CALIFORNIA 91101

COST ESTIMATE

PROJECT	SO. CALIFORNIA WWTP		DATE 26 OCT 78		ENR: 3420
JOB NO.: 1983.6100	CLIENT: SO. CALIF. MWD		TYPE OF ESTIMATE: DETAILED		SHEET 14 OF 28 EST. BY: TJ

DESCRIPTION: INFLUENT PUMPING STATION SHT.# M-8,9

REF. ITEM	DESCRIPTION	QUANTITY	UNIT	M/H	MATERIAL UNIT	MATERIAL TOTAL	LABOR/EQUIPMENT UNIT	LABOR/EQUIPMENT TOTAL	TOTAL COST UNIT	TOTAL COST TOTAL
1	EXCAVATION STRUCTURAL	19	CY						5	95
2	BACKFILL STRUCTURAL	16	CY						3	48
3	18"Ø MLCP PIPE 21' SPOOL	1	EA			810		150	960	960
4	18" —"— 4' SPOOL	7	EA			470		125	595	4165
5	18"Ø 90 deg EL	6	EA			900		120	1020	6120
6	18"Ø 90 deg EL w/1 MJ	2	EA			1290		150	1440	2880
7	18" CHECK VALVE V-32	1	EA			5800		300	6100	6100
8	14" OUTLET	2	EA						1020	2040
9	12" PLUG VALVE (1)/MOTOR OP. V-45	1	EA			5722		500	6222	6222
10	16" B. FLY VALVE Q-350	2	EA			1450		200	1650	3300
11	PIPE SUPPORTS	14	EA						125	1750
12	16" CAST IRON PIPE 1' SPOOL	4	EA			248		80	328	1312
13	16"x12" CAST IRON ECC RED FLG	2	EA			925		130	1055	2110
14	16" VICTAULIC COUPLING	2	EA			111		25	136	272
15	PRESSURE GAUGE	4	EA						125	500
16	BOOSTER PUMP 200HP	2	EA			27000		3000	3000	60000
17	3" AIR VAC VALVE	2	EA						550	1100
18	VALVE BOX AND COVER	2	EA						75	150
19	CONCRETE PUMP PADS	6	CY						125	750
20	4" CAST IRON SOIL PIPE 61 LF	1	EA						425	425
21	GRAVEL POCKET	1	LS						75	75
22	6"Ø MLCP PIPE 10' w/1 MJ	2	EA			95		40	135	270
23	18" VICTAULIC COUPLING	1	EA			130		20	150	150

PAGE TOTAL: $ 100744

AC-2 (7/79)

FIGURE 25-12. Detailed cost estimate itemized worksheet for typical project component.

667

each of the first 12 items listed were developed from a complete material take-off from the final project plans, including labor estimates.

An example of the detailed estimate prepared for the influent pumping station, item 7 in the cost summary, is shown in Figure 25-12. In this example, the material and labor estimates were obtained from an estimating manual that combined an average labor cost with time required for installation to obtain a unit installation cost for most of the items. Cost estimates and labor requirements for the pumps were obtained from a manufacturer's representative.

OPERATION AND MAINTENANCE COSTS

The costs associated with operating and maintaining water treatment plants generally vary almost as widely as construction cost, and for similar reasons. The principal cost components of operation and maintenance (O&M) activities are labor, materials, chemicals, repairs, and energy for both processes and enclosures.

Accurate estimates of the magnitude of O&M costs are needed during preliminary stages of project work to aid in evaluating various alternatives. It also is helpful to the owner to have reliable early estimates of operating expenses to that arrangements can be made to meet these ongoing expenses.

As with construction cost estimating, the task of developing O&M cost data can be accomplished during the preliminary phases using functions developed from the reported experiences of actual plants. The degree of disaggregation of the total plant expenses into smaller systems or processes depends on the particular data source.

When necessary, detailed O&M estimates can be prepared after the final design has been completed and specific unit processes, equipment, and plant layout have been established. At this stage, more definite labor requirements can be determined from pumping estimates and housing requirements, chemical usage can be derived from estimates of dosage rates, and material requirements and repair frequencies can be estimated.

Component O&M Costs

The U.S. EPA (1978) has published a series of O&M cost functions for a number of commonly employed water treatment unit processes. The

functions were developed from a combination of actual plant operating data and from assumptions based on experience. The published curves generally include energy requirements, labor requirements, maintenance material costs, and total O&M costs. They are presented as a function of the most commonly used design parameter for each particular component. Chemical costs were specifically excluded.

Examples of the cost curves for total O&M costs shown in Figures 25-13 and 25-14 are used to estimate O&M cost in the preliminary cost estimating example. The total cost curves include labor requirements at a rate of $10.00 hr^{-1}, electric energy requirements at a rate of $0.03 kW \cdot hr^{-1}$, and maintenance material costs (excluding process chemicals) as needed for each process. Energy requirements include both enclosure energy at an average of 1100 kW $\cdot$ hr/m^2/yr (100 kW $\cdot$ hr/ft^2/yr) and process energy. Similar cost curves for water treatment

TABLE 25-4. Suggested Water Treatment Plant Staff Requirements

Small plant [under 37.85 ML/d (10 mgd)]
  Treating surface water, 24 hr/d
    1   plant superintendent and chief operator
    4   plant operators (one per shift, three shifts per day, four 42-hr work weeks)
    1   maintenance mechanic
    3   general utility help (duties would include chemical handling, grounds care, operators' vacation and holiday relief, mechanics helper and basin cleaning)

Larger plant [7500–30,000 ML/d (20–50 mgd)]
  treating surface water, 24 hr/d
    1   plant superintendent[a]
    1   assistant plant superintendent[a]
    3   chemists and bacteriologists[a]
    4   chemical building operators
    4   high-service electric pumping station operators
    1   stenographer
    4   low-service pumping station operators
    4   filter plant operators
    2   basin men
    5   maintenance mechanics
    10   general utility helpers
    1   supplies and store keeper

[a] Employees that should be highly qualified (certified where required) in their duties.
SOURCE: American Water Works Association (1971).

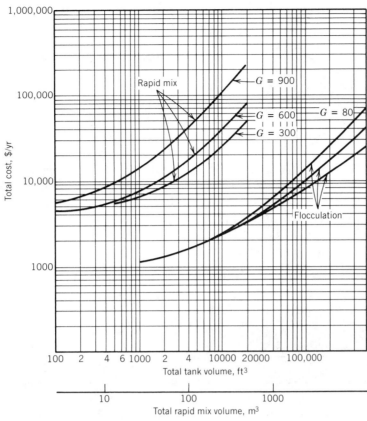

FIGURE 25-13. Total operation and maintenance costs for rapid mix and flocculation. (Source: EPA, 1978.)

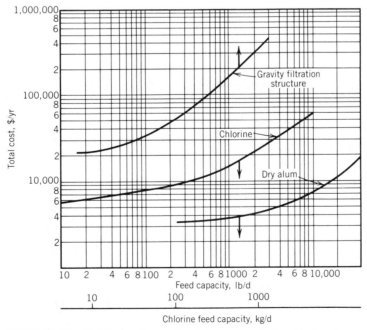

FIGURE 25-14. Total operation and maintenance costs for chlorine storage and feed systems utilizing cylinder storage for dry alum feed systems and gravity filtration structure.

669

plants of smaller capacity have also been published by the U.S. EPA (1977).

Factor O&M Costs

Another approach to determining O&M costs for a preliminary cost estimate can be termed the *factor approach*, in which the major costs associated with water treatment plant operations are lumped into a few major categories.

In a comprehensive study of the cost of water treatment published by Koenig in 1967, it was shown that the cost of labor, energy (excluding heating), and chemicals for typical plants accounted for between 91 and 95 percent of the total of all O&M costs. Even at the preliminary stages of a water treatment plant study or design, it is possible to develop reliable estimates for labor, pumping energy, and chemical requirements. Unit costs for

these items can also be obtained to produce the needed O&M cost data.

Personnel Requirements. While there are no national standards for either the number or the qualifications of staff needed to operate and maintain water treatment plants, the American Water Works Association (1971) has suggested minimum staffs that might be needed to operate filtration plants. These suggested personnel levels, shown in Table 25-4, can be adjusted as necessary to fit particular circumstances. Current water and salary information can usually be obtained from existing water utilities.

An alternative approach to estimating staff requirements would be to judiciously apply information developed for other, similar plant activities such as that published by the U.S. EPA (1973) for municipal wastewater treatment plants.

TABLE 25-5. Building Energy Requirements[a]

	Electric Energy (kW · hr/ft²/yr)									
	Lighting		Ventilation		Heating		Cooling		Total	
City	Per m²/yr	Per ft²/yr	Per m²/yr	Per ft²/yr	Per m²/yr	Per ft²/yr	Per m²/yr	Per ft²/yr	Per m²/yr	Per ft²/yr
Seattle	188	17.5	24	2.2	640	59.4	2.2	0.2	853	79.3
Salt Lake City	188	17.5	24	2.2	1550	144.0	8.6	0.8	1770	164.5
Omaha	188	17.5	24	2.2	1680	156.3	9.7	0.9	1914	177.9
Minneapolis	188	17.5	24	2.2	2150	199.4	7.5	0.7	2365	219.8
Chicago	188	17.5	24	2.2	1575	146.4	8.6	0.8	1795	166.9
New York	188	17.5	24	2.2	970	90.3	7.5	0.7	1191	110.7
Boston	188	17.5	24	2.2	1120	104.4	4.3	0.4	1340	124.5
San Francisco	188	17.5	24	2.2	435	40.5	5.4	0.5	653	60.7
Denver	188	17.5	24	2.2	1610	149.5	17.2	1.6	1840	170.8
St. Louis	188	17.5	24	2.2	1255	116.6	25.8	2.4	1492	138.7
Las Vegas	188	17.5	24	2.2	770	71.6	17.2	1.6	1000	92.9
Nashville	188	17.5	24	2.2	760	70.6	21.5	2.0	993	92.3
Washington, D.C.	188	17.5	24	2.2	845	78.3	17.2	1.6	1072	99.6
Los Angeles	188	17.5	24	2.2	300	27.7	5.4	0.5	515	47.9
Phoenix	188	17.5	24	2.2	255	23.7	25.8	2.4	492	45.8
Albuquerque	188	17.5	24	2.2	870	80.6	12.9	1.2	1092	101.5
Dallas	188	17.5	24	2.2	470	43.8	60.3	5.6	745	69.1
Tampa	188	17.5	24	2.2	99	9.2	34.4	3.2	345	32.1
Atlanta	188	17.5	24	2.2	590	54.9	16.1	1.5	820	76.1
Miami	188	17.5	24	2.2	24	2.2	34.4	3.2	280	25.8
Averages	188	17.5	24	2.2	875	81.3	17.2	1.6	1100	102.6

SOURCE: U.S. Environmental Protection Agency (August 1978).

[a] Building module used was 6 × 12 × 4 m (20 × 40 × 14 ft), with a winter inside design temperature of 20°C (68°F), a summer inside design temperature of 24°C (75°F), and a ventilation rate of 6 changes per hour.

Energy Requirements. The two principal consumers of energy in a typical water treatment plant are the pumps and building air conditioning. Pumping costs are readily calculated directly from knowledge or estimation of average water flows, head differentials from the raw water source to the plant influent, pump efficiencies, and unit energy cost. Each of these factors is normally available to the estimator at the preliminary design stage. To the treatment plant pumping cost can be added the high service pumping cost if it is to be considered as part of the treatment plant operation.

Heating and cooling expenses for the operations building are naturally dependent on the local climate. Table 25-5, published by the U.S. EPA (1978), gives electric energy requirements for an 800-ft^2 building module located in various cities in the United States. It can be used in conjunction with an estimate of total building area and unit energy cost to obtain a total for enclosed space and annual costs.

Chemical Requirements. Only rarely is a water treatment project undertaken that will treat a raw water not used by any other water utility. Since the

TABLE 25-6. Component Cost Estimates

Component	Unit Design Value	Construction Costs[a]	Total O&M Costs[b] ($/yr)
Alum feed system	94 kg/hr (214 lb/hr)	$ 44,000	$ 5,000
Rapid mix tank	10.4 m^3 (368 ft^3)	78,000	32,000
Flocculation tank	625 m^3 (22,070 ft^3)	140,000	30,000
Rapid sand filters	171 m^2 (1840 ft^3)	620,000	38,000
Filter media	171 m^2 (1840 ft^2)	19,000	Negligible
Surface wash system	4 × 43 m^2	4 × 14,000	Negligible
Chlorination	150 kg/d (342 lb/d)	33,000	5,000
Total		$990,000[c]	$114,000

[a] January 1978.
[b] Assuming 80 percent capacity factor.
[c] Note that no allowance has yet been made for the fact that the units are to be added to an existing plant, nor for interconnecting with existing units and equipment.

TABLE 25-7. Capital Cost Tabulation

Item	Cost
Total construction cost	$ 990,000
Special site preparations	0
Alterations to existing sedimentation tanks	0
Contingencies at 15 percent	150,000
Contractor overhead and profit at 12 percent (Table 25-3)	140,000
Land	0
Subtotal	$1,280,000
Engineering at 13 percent[a]	170,000
Legal, fiscal, administrative (Figure 25-11)	20,000
Subtotal	$1,470,000
Interest during construction (Figure 25-10)	100,000
Total capital cost	$1,570,000
Estimate: $1,570,000 × $\frac{3675}{2675}$ =	$2,160,000

[a] Includes 7.5 percent for basic services, 55 percent for special services.

quality of the raw water, its variability, and the types of treatment unit processes employed are the primary determining factors in the types and quantities of chemicals used in the treatment plant, the experiences of water plants using the same or a similar raw water source can be a valuable source of data for estimating purposes. The cost of chemicals used in the treatment plant follows directly from expected concentrations, the volume treated, and chemical costs. Expected chemical costs can be readily obtained from prospective suppliers.

TABLE 25-8. Annual Costs[a]

Item	Annual Cost	Annual Unit Cost $/ML	Annual Unit Cost $/mgd
O&M costs (Table 25-5 value × 1.4)	$160,000	$18.00	$ 4.75
Alum cost (657,000 kg/yr)	98,000	11.00	2.90
Chlorine costs (44,000 kg/yr)	9,000	1.00	0.26
Amortized capital at 8 percent, 20 yr	220,000	25.00	6.60
Total annual cost	$487,000	$55.00	$14.51

[a] Assuming 80 percent capacity factor.

TABLE 25-9. Cost Comparison Among GAC Systems

System Type	Design Flow		Peak Capacity		Empty Bed Contact Time (min)	Costs (cents/1000 gal)[a]											
						Reactivation Frequency											
						0.1 mo		0.5 mo		1 mo		2 mo		4 mo		6 mo	
	ML/d	mgd	ML/d	mgd		Per m³	Per 1000 gal	Per m³	Per 1000 gal	Per m³	Per 1000 gal	Per m³	Per 1000 gal	Per m³	Per 1000 gal	Per m³	Per 1000 gal
Sand replacement	26	7	38	10	13.4	—	—	—	—	3.5	13.2[b]	—	—	—	—	—	—
	265	70	380	100	14.3	16.5	62.4	4.2	15.8	2.5	9.3	1.5	5.8	1.0	3.9	0.8	3.1
	265	70	380	100	28.6	31.3	118.6	7.6	28.8	4.4	16.7	2.7	10.2	1.8	6.7	1.4	5.4
	265	70	380	100	42.9	45.6	172.6	10.9	41.4	6.3	23.7	3.8	14.4	2.5	9.4	2.0	7.6
	265	70	380	70	10	11.9	45.1	3.1	11.7	1.8	7.0	1.2	4.4	0.8	2.9	0.6	2.4
	265	70	380	70	20	22.5	85.1	5.6	21.1	3.2	12.3	2.0	7.6	1.3	5.0	1.1	4.1
	265	70	380	70	30	32.8	124.1	8.0	30.1	4.6	17.4	2.8	10.6	1.8	7.0	1.5	5.6
Postfilter absorber	26	7	38	7	18	—	—	—	—	—	—	4.4	16.7[c]	—	—	—	—
	265	70	380	70	10	8.6	32.7	3.1	11.8	2.3	8.7	1.8	6.9	1.6	5.9	1.5	5.5
	265	70	380	70	20	15.3	57.8	4.9	18.6	3.4	13.0	2.6	9.9	2.1	8.1	2.0	7.5
	265	70	380	70	30	21.7	82.0	6.6	24.9	4.5	16.9	3.3	12.5	2.7	10.1	2.4	9.2

[a] June 1977.
[b] 1.2 mo frequency.
[c] 2.4 mo frequency.

EXAMPLE 25-1

Problem

A preliminary capital cost estimate and total annual cost estimate are to be prepared for the following project: An existing water treatment plant is to be expanded by the addition of a new process train using dry alum as coagulant, a rapid mix basin, flocculation tanks with horizontal paddles, rapid sand filters with surface wash mechanisms, and disinfection using 1-ton cylinders of chlorine gas. The existing sedimentation basins are to be modified by others to accept the expanded flow; the existing clear wells and backwash pumping system are to be used. The following design materials are assumed for the preliminary estimate:

Plant capacity: 30 ML/d (7.93 mgd).

Alum feed capacity: 75 mg/L.

Rapid mix tank detention time: 30 sec.

Rapid mix tank velocity gradient: 900 sec^{-1}.

Flocculation tank detention time: 30 min.

Flocculation tank velocity gradient: 50 sec^{-1}.

Filtration rate: 7.2 m/hr (3 gpm/ft^2).

Chlorine feed capacity: 5 mg/L.

January 1978 CCI: 2675.

Construction period CCI: 3675.

O&M Cost escalation to date of initial plant operation: 1.4.

Interest rate: 8 percent.

Allowance for contingencies: 15 percent.

Facility expected life: 20 yr.

Solution

Use Figures 25-5 through 25-7 and 25-13 and 25-14 to obtain estimates for construction costs and O&M cost as shown in Table 25-6. Project these construction costs to the estimated construction period using the ENR 20 cities CCI given in Table 25-7. Project the O&M costs to the plant start-up date using the 1.4 multiplier, which was derived from an assessment of recent water inflation rates, electric utility price projections, and a projection of basic chemical prices as shown in Table 25-8. Use chemical cost estimates of $15.00/100 kg for alum and $20.00/100 kg chlorine.   □

SPECIALIZED TREATMENT PROCESSES

Occasionally the estimator is faced with the task of estimating the cost of a water treatment system that employs specialized unit operations or processes such as ion exchange, reverse osmosis, electrodialysis, or carbon adsorption. Because these processes are not as common as those used in conventional treatment, data are more difficult to obtain, and extrapolation across time and to different locations is further complicated by differences in design parameters and applications.

With the possible exception of carbon adsorption, the specialized treatment processes frequently employ packaged assemblies and proprietary equipment contained or housed in conventional structures. Approximate estimates for the equipment can be obtained from manufacturers, and costs of auxiliary features such as connecting piping, pumping facilities, and containment structures can be readily estimated.

It is difficult to prepare preliminary estimates for GAC adsorption systems because of the number of alternatives involved in the design and the dependence of the actual design on the results of pilot testing programs, which are rarely available in the preliminary stages of a project. Symons (1978) discussed the influence on cost of some of the more significant variables in GAC systems, including the use of postfilter adsorbers as opposed to sand replacement systems, the contact time, reactivation frequency, carbon cost, reactivation costs, and the rate of inflation. Results of this study, shown in Table 25-9, show significant decreases in water cost for decreasing contact times and increasing reactivation frequency.

REFERENCES

American Association of Cost Engineers, 308 Monongahela Building, Morgantown, West Virginia 26505.

American Society of Civil Engineers, Consulting Engineering, *A Guide For the Engagement of Engineering Services*, ASCE Manual No. 45 (1981).

American Society of Professional Estimators, 5201 North 7th Street, Suite 200, Phoenix, Arizona 85014-2865.

American Water Works Association, *Water Quality and Treatment*, 3rd ed., McGraw-Hill, New York (1971).

Anonymous, "EPA Switches to New Cost Indexes," *J. Water Pollut. Control Fed.*, **48**(2), 233–235 (1976).

Dickson, R. D., "Estimating Water System Costs," in R. L. Sanks (ed.), *Water Treatment Plant Design,* Ann Arbor Science, Ann Arbor, MI (1978).

1984 Dodge Guide to Public Works and Heavy Construction Costs, Annual Edition No. 16, Leonard A. McMahon (Author), Percival E. Pereina (Editor), P.O. Box 28 Princeton, New Jersey 08540 (Copyright 1984). McGraw-Hill Information Systems Co.

Engineering News-Record, McGraw-Hill, Inc., New York.

Gumerman, R. C., Culp, R. L., and Hansen, S. P., "Estimating Costs for Water Treatment as a Function of Size and Treatment Efficiency," U.S. Environmental Protection Agency, Municipal Environmental Research Laboratory, Cincinnati, OH, EPA-600/2-78-182 (August 1978).

Guttman, D. L., and Clark, R. M., "Computer Cost Models for Potable Water Treatment Plants," U.S. Environmental Protection Agency, Municipal Environmental Research Laboratory, Cincinnati, OH, EPA-600/2-78-181 (September 1978).

Hinomoto, H., "Unit and Total Cost Functions for Water Treatment Based on Koenig's Data," *Water Resources Research,* **7**(5), 1064–1069 (1971).

Koenig, L., "The Cost of Water Treatment by Coagulation, Sedimentation, and Rapid Sand Filtration," *JAWWA,* **59**(3), 290–336 (1967).

Larson and Associates, "Desalting Seawater and Brackish Water: Cost Update, 1979," Department of Energy, Oak Ridge National Laboratory, Oak Ridge, TN, ORNL/TM-6912 (August 1979).

Logsdon, G. S., Clark, R. M., and Tate, C. H., "Costs of Direct Filtration to Meet Drinking Water Regulations," paper presented to American Water Works Association Annual Conference, Atlantic City, NJ (June 1978).

Orlab, G. T., and Lindorf, M. R., "Cost of Water Treatment in California," *JAWWA,* **50**(1), 45–55 (1958).

R. S. Means Co., Inc., "Building Construction Cost Data," Duxbury, MA (1984).

Symons, J. M. (ed.), *Interim Treatment Guide for Controlling Organic Contaminants in Drinking Water Using Granular Activated Carbon,* U.S. Environmental Protection Agency, Municipal Environmental Research Laboratory, Cincinnati, OH (January 1978).

The Trade Book, P.O. Box 2366, Inglewood, CA (1984).

U.S. Department of the Interior, *Manual for Calculation of Conventional Water Treatment Costs,* Office of Saline Water Research and Development, Report 917, INT-OSW-RDPR-74-907, NTIS PB 226 791 (March 1972).

U.S. Environmental Protection Agency, *Estimating Staffing for Municipal Wastewater Treatment Facilities,* Office of Water Program Operations, Washington, DC (March 1973).

U.S. Environmental Protection Agency, *State of the Art of Small Water Treatment Systems,* Office of Water Supply, Washington, DC (August 1977).

Whitman, Requardt and Associates, *Handy-Whitman Index of Water Utility Construction Costs,* (July 1, 1984).

Index

675